NMR Spectroscopy for Probing Functional Dynamics at Biological Interfaces

New Developments in NMR

Editor-in-chief:
William S. Price, *Western Sydney University, Australia*

Series editors:
Bruce Balcom, *University of New Brunswick, Canada*
István Furó, *Industrial NMR Centre at KTH, Sweden*
Masatsune Kainosho, *Tokyo Metropolitan University, Japan*
Maili Liu, *Chinese Academy of Sciences, Wuhan, China*

Titles in the series:

1: Contemporary Computer-Assisted Approaches to Molecular Structure Elucidation
2: New Applications of NMR in Drug Discovery and Development
3: Advances in Biological Solid-State NMR
4: Hyperpolarized Xenon-129 Magnetic Resonance: Concepts, Production, Techniques and Applications
5: Mobile NMR and MRI: Developments and Applications
6: Gas Phase NMR
7: Magnetic Resonance Technology: Hardware and System Component Design
8: Biophysics and Biochemistry of Cartilage by NMR and MRI
9: Diffusion NMR of Confined Systems: Fluid Transport in Porous Solids and Heterogeneous Materials
10: NMR in Glycoscience and Glycotechnology
11: Fast NMR Data Acquisition: Beyond the Fourier Transform
12: Cross-relaxation and Cross-correlation Parameters in NMR: Molecular Approaches
13: Contrast Agents for MRI: Experimental Methods
14: NMR-based Metabolomics
15: Modern Methods in Solid-state NMR: A Practitioner's Guide
16: Paramagnetism in Experimental Biomolecular NMR
17: Optimizing NMR Methods for Structure Elucidation: Characterizing Natural Products and Other Organic Compounds
18: Field-cycling NMR Relaxometry: Instrumentation, Model Theories and Applications
19: Hybrid MR-PET Imaging of the Brain: Systems, Methods and Applications
20: NMR Methods for Characterization of Synthetic and Natural Polymers
21: In-cell NMR Spectroscopy: From Molecular Sciences to Cell Biology
22: Long-lived Nuclear Spin Order: Theory and Applications

23: NMR and MRI of Gels
24: Advanced Diffusion Encoding Methods in MRI
25: NMR and MRI of Electrochemical Energy Storage Materials and Devices
26: NMR Spectroscopy for Probing Functional Dynamics at Biological Interfaces

How to obtain future titles on publication:
A standing order plan is available for this series. A standing order will bring delivery of each new volume immediately on publication.

For further information please contact:
Book Sales Department, Royal Society of Chemistry, Thomas Graham House, Science Park, Milton Road, Cambridge, CB4 0WF, UK
Telephone: +44 (0)1223 420066, Fax: +44 (0)1223 420247
Email: booksales@rsc.org
Visit our website at www.rsc.org/books

NMR Spectroscopy for Probing Functional Dynamics at Biological Interfaces

Edited by

Anirban Bhunia
Bose Institute, India
Email: bhunia@jcbose.ac.in

Hanudatta S. Atreya
Indian Institute of Science, India
Email: hsatreya@iisc.ac.in

and

Neeraj Sinha
Centre of Biomedical Research Lucknow, India
Email: neeraj.sinha@cbmr.res.in

ROYAL SOCIETY OF CHEMISTRY

New Developments in NMR No. 26

Print ISBN: 978-1-83916-209-1
PDF ISBN: 978-1-83916-570-2
EPUB ISBN: 978-1-83916-571-9
Print ISSN: 2044-253X
Electronic ISSN: 2044-2548

A catalogue record for this book is available from the British Library

The Royal Society of Chemistry is a charity, registered in England and Wales, Number 207890, and a company incorporated in England by Royal Charter (Registered No. RC000524), registered office: Burlington House, Piccadilly, London W1J 0BA, UK, Telephone: +44 (0) 20 7437 8656.

For further information see our website at www.rsc.org

Printed in the United Kingdom by CPI Group (UK) Ltd, Croydon, CR0 4YY, UK

Preface

Nuclear Magnetic Resonance (NMR) has immense potential in providing atomic-level information on the mechanism of action of diverse biological systems such as globular proteins, membrane proteins, lipids, nucleic acids, biofibrils and biomaterials. While some applications of the high-resolution NMR spectroscopic technique correlate the structure with function, others are concerned with the dynamics of biomolecules interacting with other bio- or non-biomolecular systems such as nanomaterials. Recent advances in this field have led to the advent of novel approaches and techniques which go beyond the conventional NMR experiments used in structural biology. Thus, the integration of diverse techniques is necessary to bridge the gap between structural and functional studies so as to provide a comprehensive understanding of dynamic NMR. The initial planning of this book was carried out with Prof. Anirban Bhunia and Prof. H. S. Atreya. However, the untimely demise of Prof. Atreya in 2020, while we were in the middle of this compilation process has been an inexpressible loss to us. We reorganized and started with Prof. Neeraj Sinha, who incorporated several important applications of solid-state NMR in the book.

The book has been compiled in such a way that the initial section deals with the basics of dynamics and relaxations of NMR spectroscopy. The next section deals with various applications in the solution state NMR methodologies including protein–ligand interactions, saturation transfer difference (STD) NMR, dynamics of protein–nanoparticle interactions, and large protein complexes. The final part of the book contains applications of solid-state NMR concerned with protein–mineral interfaces, G Protein Coupled Receptor (GPCRs), protein fibrils, and biomaterials such as bone, cartilage, *etc.* Chapter 1 discusses the basics of triple-resonance experiments for assignments and structure elucidation of biomolecules. Additionally, the authors also discuss the characterization of sparsely

New Developments in NMR No. 26
NMR Spectroscopy for Probing Functional Dynamics at Biological Interfaces
Edited by Anirban Bhunia, Hanudatta S. Atreya and Neeraj Sinha

Published by the Royal Society of Chemistry, www.rsc.org

populated protein conformations using recent NMR methodologies. In Chapter 2, Harindranath Kadavath and Roland Riek discuss the principles of exact nuclear Overhauser enhancement (eNOE) and its recent application in multistate structural ensembles. L. Jaremko and co-workers (Chapter 3) have provided various NMR spin relaxation mechanisms and NMR methodologies used to probe these relaxation rates. This chapter nicely illustrates the investigation of molecular motions over practically all timescale ranges.

In Chapter 4, Linda Nicholson and co-workers present the recent advances of NMR lineshape analysis in probing protein dynamics; and determining the kinetics and thermodynamics of multistate reaction schemes with examples. Ipsita Chakraborty and Rangeet Bhattacharyya (Chapter 5) present an interesting case study that uses a simple, real-time NMR method of measuring the relaxation of solvent to quantitatively monitor the dynamics of the aggregation of smart polymers. Cristinal Airoldi and co-workers (Chapter 6) present an overview of the current status of the applications of NMR spectroscopy to monitor ligand–receptor interactions using STD NMR, WaterLOGSY, trNOESY, DOSY-based, and relaxation filters. In addition, they have commented on the weaknesses of these methods and described on-cell NMR methods in molecular recognition using HR-MAS. Chapter 7 by Angulo and Monaco explains several important STD NMR techniques that monitor interactions and presents the latest developments in this field.

Bhunia and co-workers in Chapter 8 focus on Dark State Exchange Saturation Transfer (DEST) NMR as a tool to map slow chemical exchange at functional protein interfaces. Adeptly, the DEST experiment is introduced as an extension to the original STD technique and the details of the NMR pulse sequence and quantitative data analysis using Bloch–McConnell equations are explained. Finally, its application to several IDPs is discussed. Chapter 9 summarizes the information on different chemical biology tools and techniques used for characterizing molecular interactions among nanoparticles and proteins. Surajit Bhattacharjya in Chapter 10 describes NMR structural and dynamical studies of leucocyte-specific β2 integrin cytosolic tails with cognate proteins. The transient nature of the tail–tail and tail–protein interactions are refractory to other methods for determining its atomic-resolution structure. The NMR based 3-D structures of α and β cytosolic tails and tail/tail complexes; in addition, the complexes of β2 tail with integrin activator and repressor proteins are discussed in this chapter. Chapter 11, by Hiller and Burmann, explains the dynamics of big complexes such as chaperones and their client proteins. The authors elegantly describe the various NMR approaches including CSPs, NOEs, hydrogen–deuterium exchange, relaxation, PRE, and also other biophysical experiments to study the lifetimes of the complexes and understand the kinetics of the interactions.

Rui Huang in Chapter 12 discusses how NMR spectroscopy, specifically ^{13}C-methyl-TROSY, can be used to probe co-operativity in large proteins and protein–protein complexes. This chapter includes the general principle of the methyl-TROSY technique, the challenges in the sample preparation of the NMR-active methyl group labels on the specific amino acids, along with different biosynthetic labelling approaches, e.g., *E. coli vs.* cell free. Chapter 13 by Gooley and co-workers dwells on the various applications of NMR methods to study GPCR structure, dynamics, and ligand interactions. The authors provide an overview of GPCR structure, followed by the structural rearrangements associated with receptor activation. Subsequently, various expression systems in vogue, labelling strategies, as well as membrane-mimicking systems used for encapsulating GPCRs to express and purify receptors in manners suitable for NMR studies are described. Prototypical adrenergic and adenosine receptors, and more recent opioid receptors are discussed also. Chapter 14, by Hansen and co-workers, comprehensively surveys the study of intrinsically disordered proteins (IDPs) using CPMG-based NMR techniques in conjunction with molecular dynamics (MD) simulation. Koh in Chapter 15 discusses the recent developments in NMR strategies for higher-ordered structure analysis of biopharmaceuticals, improvement of protein druggability, and understanding of the cell permeabilization activity of medium-size molecules.

The methodology of solid-state NMR spectroscopy is increasing rapidly in terms of ultra-high-speed magic angle spinning. This has opened up new frontier in terms of applications. The final section of the book covers these important developments. In Chapter 16, Nishiyama describes methods to locate hydrogen atoms in solids using fast MAS solid-state NMR and their application for atomic-resolution structural studies. One of the advantages that solid-state NMR spectroscopy provides is that we can structurally study biological systems in their native environment. Chapter 17, by F. M. Marrassi and co-authors, describes various methods to mimic native protein environments. These systems include membrane proteins and biominerals such as bone and cartilage. Chapter 18, by S. J. Opella and co-authors, describes solid-state NMR methods for membrane proteins in their native environments. The system chosen is membrane receptor CXCR1 and its interactions with agonist IL8. Chapter 19, by Markus Weingrath and co-authors, describes the application of probe antibiotics in membranes. The remainder of the chapters deal with diverse applications of solid-state NMR to the systems such as antimicrobial peptides (in Chapter 20), different native biomaterials such as bone, cartilage, teeth (in Chapter 21), and the lipid phase of stratum corneum (in Chapter 22). Overall, we have tried to cover recent methodology developments as well as diverse interface biological applications.

Firstly, we would like to thank the Royal Society of Chemistry for giving us the opportunity to edit this book. Special thanks are due to Liv Towers and Janet Freshwater for their timely support and understanding. Thanks are also expressed to all the authors and reviewers for their excellent contributions and making this compilation a success. We thank Dr Janarthanan Krishnamoorthy and Miss Dipita Bhattacharyya for their tireless support at various stages of editing this book. Finally, we extend our thanks to many other people, who contributed to this book in bringing it to its current form, despite this being our first experience in editing a book.

Anirban Bhunia
Bose Institute, Kolkata, India
Neeraj Sinha
Centre of Bio-Medical Research, Lucknow, India

Contents

New Developments in NMR No. 26
NMR Spectroscopy for Probing Functional Dynamics at Biological Interfaces
Edited by Anirban Bhunia, Hanudatta S. Atreya and Neeraj Sinha

Published by the Royal Society of Chemistry, www.rsc.org

CHAPTER 1

Theory and Applications of NMR Spectroscopy in Biomolecular Structures and Dynamics of Proteins

KOUSIK CHANDRA,[a,†] ABDUL HAMID EMWAS,[b] SAMAH AL-HARTHI,[a] ZEYAD AL-TALLA,[c] DINA HAJJAR,[d] ARWA ABDULAZIZ MAKKI,[d] GHADA KHOUQEER[e] AND MARIUSZ JAREMKO*[a]

[a] King Abdullah University of Science and Technology (KAUST), Biological and Environmental Sciences & Engineering Division (BESE), Thuwal 23955-6900, Saudi Arabia; [b] King Abdullah University of Science and Technology. Core Labs, Thuwal 23955-6900, Kingdom of Saudi Arabia; [c] King Abdullah University of Science and Technology (KAUST), Ali I. Al-Naimi Petroleum Engineering Research Center (ANPERC), Thuwal 23955-6900, Saudi Arabia; [d] University of Jeddah, College of Science, Department of Biochemistry, Jeddah, Saudi Arabia; [e] Imam Mohammad Ibn Saud Islamic University (IMSIU), Riyadh, Kingdom of Saudi Arabia
*Email: Mariusz.jaremko@kaust.edu.sa

1.1 Introduction

Nuclear magnetic resonance (NMR) is a potent analytical technique with superior advantages, including its nondestructive nature, high reproducibility,

[†]These authors contributed equally.

New Developments in NMR No. 26
NMR Spectroscopy for Probing Functional Dynamics at Biological Interfaces
Edited by Anirban Bhunia, Hanudatta S. Atreya and Neeraj Sinha

Published by the Royal Society of Chemistry, www.rsc.org

and inherently quantitative methods.[1–3] The main strengths of NMR spectroscopy are that researchers can study a molecule at its atomic level, and several NMR measurements can be done on the same sample.[4–7] Thus, NMR can be used to identify and quantify different molecules within sample mixtures.[8–14] NMR can be utilized for molecular identification and elucidating different structures of the same molecule and hence is widely used in stereochemistry to identify different isomers of a molecule.[15–18] Moreover, coordination information on each atom and electron density and the memetic environments of neighbor's atoms can be extracted from the NMR spectrum.

The applications of NMR spectroscopy are no longer limited to the liquid state; it is now routinely used for solid states also.[19–25] Thus, NMR spectroscopy is broadly employed in a wide range of applications, including organic[26–32] and inorganic chemistry,[33–39] biology, physics,[40–46] and medical applications.[47–49] A simple one-dimensional (1D) proton NMR spectrum can be detected in a few minutes with minimal sample preparation, offering a high-throughput method appropriate for detecting the progress of organic reactions and for studies that include a large number of samples such as metabolomics and drug-discovery studies.[50]

One of the essential advantages of NMR spectroscopy is that samples can be detected under mild conditions without a vacuum or high temperatures. Researchers can measure NMR spectra separately by detecting selectively only one type of atom using 1D NMR experiments or the correlation between two types of atoms using 2D NMR experiments. Furthermore, the bond coordination of three or more different types of atoms can be detected using 3D and higher dimensional NMR methods.[51–53]

Indeed, one can carry out NMR measurements for biological molecules under physiological conditions such as body temperature, pH, and pressure. This was done first by Richard Ernst, who, based on an idea by Jeener, demonstrated two-dimensional (2D) NMR spectroscopy.[54,55] This opened up a new domain with many new experimental possibilities to obtain chemical shifts of coupled homonuclear and heteronuclear spins and the interactions between pairs of nuclear spins. This also forms the basis of the transfer of magnetization, which enables the basic building blocks for polarization transfer from one spin to another spin. The second breakthrough came from Anil Kumar, Ernst, and Wüthrich with the discovery of nuclear Overhauser effect experiments whereby spatial connections from adjacent spins to one spin are observed through the interplay of cross-relaxation and dipolar interactions between them.[56] This experiment and some other related 2D NMR experiments developed by the same group formed the basis for solving protein structures.

In 1985, the first protein structure of proteinase inhibitor IIA from seminal bull plasma was reported by the group of Wüthrich.[57] In the 1980s and 1990s, there was an explosive growth of high-resolution NMR in terms of developing new pulse sequences; instrument hardwire improvements in terms of higher magnetic fields, cryogenically cooled probes, *etc.*, and also better sample preparation, which involved the use of isotopically labeled proteins (^{1}H, ^{13}C, and ^{15}N). Slowly NMR became an increasingly powerful technique for protein structure determination. Currently, about 10% of structures being deposited in the Protein Data Bank are solved using NMR. With an increase in dimensionality, measurement time increases rapidly, and there is constant effort to meet both the requirement of optimum sensitivity and desired resolution.

Over the past two decades, numerous advances have been made to achieve three- and four-dimensional (3D and 4D) spectroscopy in a reasonable amount of time.[58–61] There is an area of research for FAST NMR methodologies that have found profound application in the last two decades.[62,63] Perhaps the most exciting frontier is the application of NMR to investigate protein dynamics, especially for large molecular machines, which will undoubtedly lead to new fundamental insights in biology. NMR is the method of choice for dynamic molecular investigation; herein we focus on the theory and applications of different NMR approaches for biological molecules, focusing on protein dynamics. Since historical dynamics came first, we first discuss the basic theory and experimental details of protein dynamics. The details of structure elucidation follow this. Subsequently, we cover the recent development of the study of NMR dynamics, which includes structural information of sparsely populated states that are functionally most relevant.

1.2 Methodologies Involving Protein Structure Determination: A Comparative Study

As noted earlier, one of the most important advantages of NMR spectroscopy is the ability to elucidate the atomic-level structure of important biomolecules of interest, such as proteins.[64] Moreover, NMR is the only method that can be used for structural elucidation of biomolecules together with their complexes in solution under physiological conditions (with strictly defined temperature and pH, and ambient pressure). Thus, the determination of protein structure in its native environmental conditions is essential for a wide range of applications such as understanding protein functions, protein–protein and protein–drug interactions, and drug development. Initially, NMR experiments, such as 2D ^{1}H–^{1}H NOESY, were successfully applied to distinguish between the secondary structures of proteins.[65–68] Later, in 1982, the first-ever sequence-specific assignment for a small protein, basic pancreatic trypsin inhibitor, was published,[68] and this led to the determination of the first protein structure in solution in 1985.[57] Figure 1.1 shows how one can use NOESY data to distinguish an α-helix from other secondary structures.

The presence of secondary motifs within the protein structure is usually confirmed by the experimental data provided by NMR studies of peptides and proteins. The backbone vicinal coupling constants between amide protons NH and H^{α} protons $[^{3}JH_{N}H_{\alpha}]$ in the random coil are approximately equal to 7 Hz, which is close to the 8.5 Hz typical of β-sheets and far from the 4.2 Hz typical of an α-helix. In the case of the NOE (nuclear Overhauser effect), the random coil motifs exhibit strong sequential $H_{\alpha}(i)$ – $HN(i+1)$ effects typical of β-sheets, while $NH(i)$ – $NH(i+1)$ are fragile or not observed at all.[69–71]

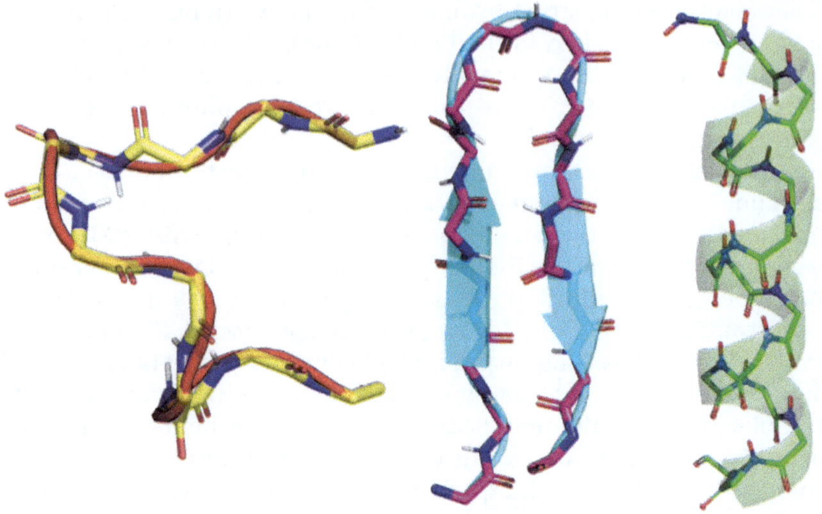

Figure 1.1 The secondary structures of protein highlighting the connectivity of interstrand NOEs.

Table 1.1 The distribution of protein and nucleic acid structures solved by currently available techniques. Data were taken from the Protein Data Bank (www.pdb.org), and the last update was May 17, 2021.

		Molecular type				
		Proteins (only)	Protein/ NA	Nucleic acids (only)	Other	Total
Exp. method	X-Ray	138 515	7275	2172	149	156 318
	NMR	11 693	270	1358	31	13 389
	EM	5090	1738	58	3	7655
	Multiple methods	168	3	8	0	185
	Neutrons	69	0	2	0	71
	Other	32	0	1	0	37
	Total	155 567	9286	3599	183	177 655

Nowadays, the structures of different biomolecules solved by either NMR or other techniques are deposited in the Protein Data Bank (PDB) (website: www.pdb.org). There are over 300 000 structures of different biomolecules (of which 155 218 and 3594 are structures of proteins and nucleic acids, respectively) deposited in the PDB (April 29, 2021). Statistics of the PDB are presented in Table 1.1.

The relatively large disparity between the number of structures solved by NMR and X-ray techniques has several different causes, besides the fact that

NMR is a newer technique used in structural studies for important biomolecules. These causes include the following:

- The size limitation of objects suitable for NMR studies.[69] The average protein studied using NMR is *ca.* 100–150 amino acids long and requires double labeling. In proteins with over 180 residues, triple labeling with ^{2}H, ^{13}C, and ^{15}N isotopes is very often required. This is because the amount of information in spectra is too much and the relaxation behavior of larger systems that broaden the signal makes it difficult to detect and study.
- Relatively higher costs of protein production and measurements with the spectrometer.

The structure determination by NMR requires backbone and side-chain assignment as a prerequisite step. To achieve this, proteins need to be isotopically enriched. Isotope labeling requires cells expressing the protein in minimal media composed of ^{13}C-labeled glucose and ^{15}N-labeled ammonia salt. There can be an additional requirement of deuterium labeling for higher molecular size complexes or selective methyl labeling to resolve the highly overlapped methyl region. These requirements increase the sample production cost. Spectrometer hours are expensive because of relatively high cryo liquid consumption. The resonance assignment process typically involves recording multiple triple resonance experiments and other 2D and 3D experiments. The conventional method of recording experiments leads to higher time consumption. The analysis procedure is relatively tedious and time-consuming. The sample stability in solution is an issue as the sample should be stable for at least several weeks to complete the minimal set of multidimensional NMR experiments for assignment.

A tremendous amount of research has been carried out to overcome these difficulties, and these developments have led to different areas of research altogether. These developments are in different areas, such as ease of sample preparation, better labeling strategies, economically viable unlabeling strategies, improved hardwire settings, various fast NMR methodologies, *etc.* It is most common that researchers prefer to determine the high-resolution structure of their protein by X-ray, then assign only the protein backbone atoms with experiments based on NH detection, and study protein interactions and dynamics by solution-state NMR techniques.

In comparison to X-ray, NMR offers several possibilities that are not accessible to X-ray methodology. The most important advantages of NMR are:

- studying the protein structure in a native-like water environment;
- ability to test the impact of pH, temperature, ionic strength, and other factors on protein structure, stability, and interactions with ligands;
- studying protein dynamics in aqueous solutions (backbone by relaxation of ^{15}N atoms and side chains by ^{13}C-methyl relaxation);
- observation of chemical kinetics and estimation of thermodynamic parameters, binding constants, *etc.*;

- uncertainties of one type of experiment can often be resolved by different types of experiments done on the same sample, while in the case of X-ray, one crystal provides one consistent dataset that a different type of experiments cannot validate.

There are several routine steps in high-resolution protein structure elucidation based on multidimensional nuclear magnetic resonance techniques. First, when the target protein is known, and its molecular weight is under 20–25 kDa, and the expression protocol in *E. coli* is established, the following well-established scheme is usually applied:

1. Expression of the nonlabeled protein in *E. coli* in minimal M9 media (there are also other less common expression systems available, such as *Pichia pastoris* yeasts);
2. Defining the solution conditions, like buffer, pH, temperature, and salt concentration;
3. Expression of the double-labeled ^{15}N, ^{13}C protein in *E. coli* in minimal M9 media;
4. 3D experiments for backbone assignments and side-chain assignment;
5. Experiments delivering constraints for structure elucidation;
6. Determination of the structural constraints and structure calculation;
7. Structure validation and refinement.

The standard 3D experiments for double-labeled proteins for side chains and sequence-specific backbone assignments are based on HN-detected experiments. The protein solution usually contains 90%/10% H_2O/D_2O, which still allows for observation of the NH backbone and side-chain amides, locking the ^2H signal, but efficient suppression of the water peak at $\sim$4.7 ppm is required.[69] The experiments connecting the selected atoms of residues and between the residues are based on magnetization transfer on magnetically active nuclei.

For the sequence-specific backbone assignments, the simplest experiments used for double ^{15}N- and ^{13}C-labeled protein of MW under 20–25 kDa[69–72] are:

- 3D HNCA;
- 3D HN(CO)CA;
- 3D H(CA)NH;
- 3D HNCO;
- 3D HN(CA)CO;
- 3D CBCA(CO)NH;
- 3D CBCANH;
- 3D HNCACB.

The above-mentioned experiments are proton-detected ^{15}N-edited experiments where 2D [^{15}N, ^1H] HSQC serves as the fingerprint of the associated protein molecules. These experiments rely on one- and two-bond *J*

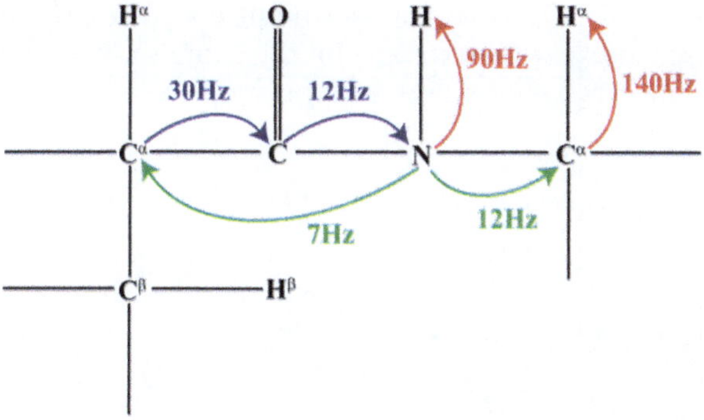

Figure 1.2 The approximate values of coupling constants between various backbone nuclei.

couplings, which are used to affect multiple magnetization transfer steps. The couplings are shown in Figure 1.2.

These couplings are largely independent of the backbone conformation of the proteins. However, there are changes in the values in the case of moving from proteins to nucleic acids. The nomenclature of the triple resonance experiments describes the associated magnetization transfer during the experiment, along with the frequency labeling. The nuclei, which are used only for transfer without frequency labeling, are kept in brackets. As an example, in the HNCO experiment, the magnetization starts from the $H^{N}_{(i)}$ proton and gets transferred through the directly attached $N_{(i)}$ atom to the $CO_{(i-1)}$ carbon atom and returns the same way to the $H^{N}_{(i)}$ nucleus. In HN(CA)CO, the magnetization passes through C^{α} atoms after nitrogen; however, the frequencies are not labeled. The detailed magnetization transfer pathways are presented in Scheme 1.1. All the experiments starting with HN are "out and back" experiments, while the last three experiments starting with $H^{\alpha/\beta}$ are "out and stay" experiments.

The magnetization transfer steps involve INEPT and will discuss the HSQC pulse sequence first in-depth, as it is the simplest of all, to understand INEPT and reverse-INEPT. Subsequently, we discuss in detail one triple resonance experiment.

The basic pulse sequence of HSQC is shown in Figure 1.3, which is composed of an INEPT sequence to transfer the magnetization from I to S spin. This is followed by the chemical shift evolution of S spin, during which I is decoupled by applying a 180° pulse in the middle. The last portion (c to d) represents the reverse INEPT block, which transfers the polarization back from S to I. Up to position a, I magnetization undergoes evolution under coupling, and an antiphase I magnetization (I_xS_Z) is generated. Simultaneous application of two 90° pulses makes it I_zS_x, which is antiphase S magnetization. Thus, I magnetization is transferred to S. Subsequently, this antiphase term undergoes chemical shift evolution (Ω_S) for the t_1 period while the coupling is refocused by application of 180° I pulse at the midpoint of the evolution period t_1. Hence at position c, the coherence terms of

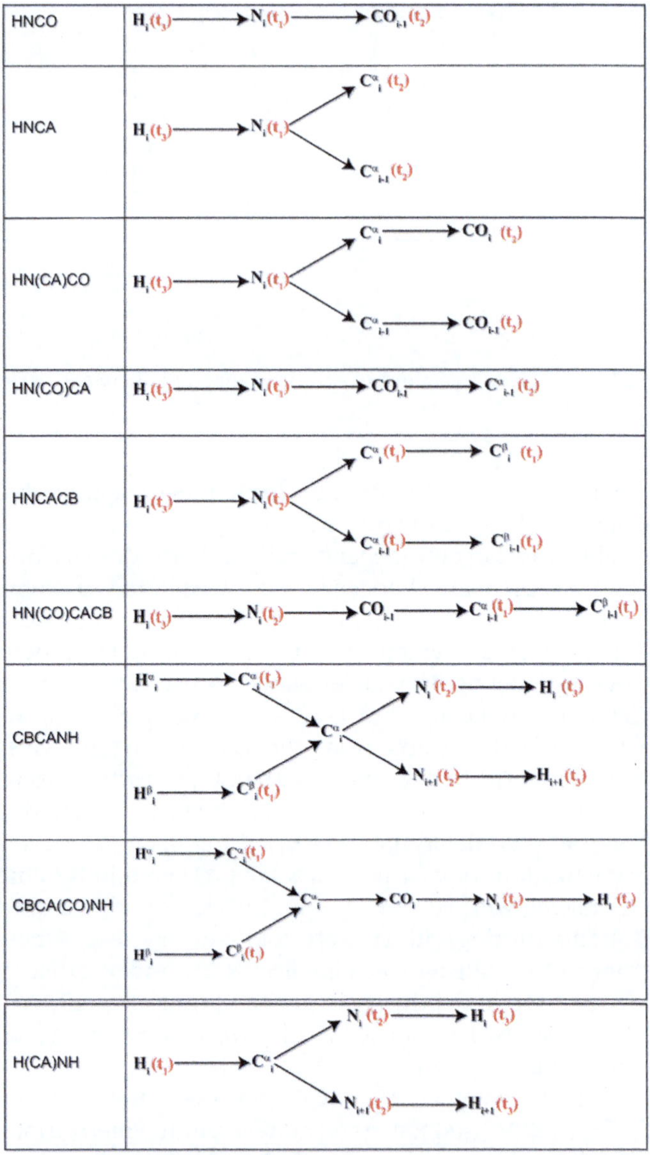

Scheme 1.1 The detailed magnetization transfer pathways for different experiments. (t_3) represents the detection dimension, while (t_1) and (t_2) represent the other two indirect dimensions in respective cases.

interest are $I_zS_x\cos\,\Omega_St_1 + I_zS_y\,\sin\,\Omega_St_1$. After this, the back INEPT starts, and two 90° pulses are applied. This results in the generation of I_xS_x cos $\Omega St_1 + I_xS_z\,\sin\,\Omega St_1$. Among these two terms, the double quantum coherence (I_xS_x) is not observable. The second term encoded by antiphase I magnetization (I_xS_z) with sine modulation of S spin chemical shift becomes the

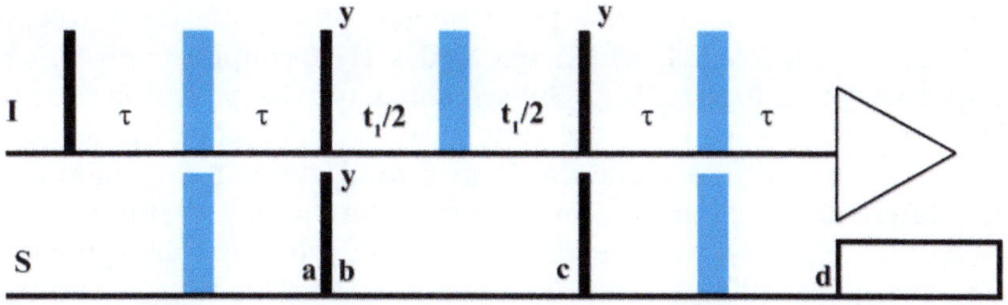

Figure 1.3 The pulse sequence of HSQC. I and S represent any two heteronuclear spins. The τ is $(1/4J)$, where J is the associated coupling constant. The thick blue and thin black bars represent the 180° and 90° pulses, respectively.

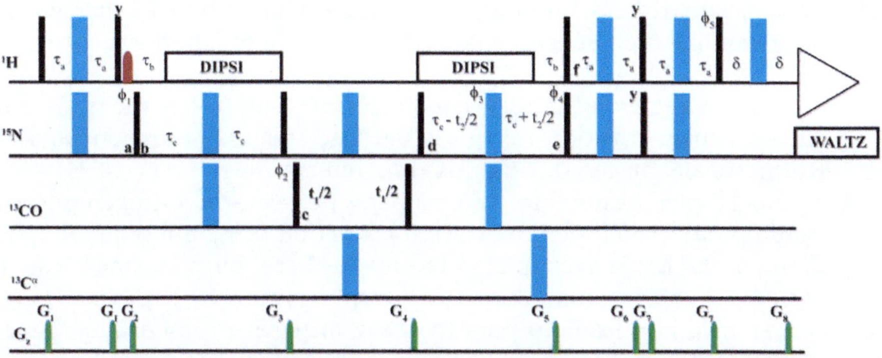

Figure 1.4 The pulse sequence of an HNCO experiment. Thin black bars and wide blue bars represent 90° and 180° pulses, respectively. The red-colored, shaped proton pulse is a water-selective 90° pulse. The DIPSI decoupling scheme is applied on the proton channel with lower power (5 kHz) compared to the hard pulse. The carbon pulses are applied as semiselective pulses. The transmitter offsets for the carbonyl and Cα carbon channels are set to 175 and 55 ppm, respectively. The coupling between the Cα and the nitrogen during chemical shift evolution of the nitrogen is removed by application of a single 180° pulse on Cα. The delays τ_a, τ_b, and τ_c are set to 2.3 ms $[1/(4J_{NH})]$, 5.5 ms $[1/(2J_{NH})]$, and 12.4 ms $[1/(4J_{NC})]$, respectively. The delay δ is required for the application and recovery of gradient G8 and is set to 0.5 ms. The phase cycle implemented is $\phi_1 = x, -x$; $\phi_2 = 4(x), 4(-x)$; $\phi_3 = 2(x), 2(-x)$; $\phi_4 = x$; $\phi_5 = x$, $\phi_{rec} = 2(x, -x), 2(-x, x)$. Gradients are usually sine-shaped with the following strengths and duration: G_1 (5 G cm^{-1}, 0.5 ms), G_2 (15 G cm^{-1}, 2 ms), G_3 (20 G cm^{-1}, 0.75 ms), G_4 (5 G cm^{-1}, 0.2 ms), G_5 (30 G cm^{-1}, 1.25 ms), G_6 (5 G cm^{-1}, 0.3 ms), G_7 (10 G cm^{-1}, 0.2 ms), G_8 (27.8 G cm^{-1}, 0.125 ms).

main interest. The antiphase term is evolved under J coupling for a period of $(1/2J)$ in order to generate in-phase I_y magnetization before detection at position d. During detection, the S spin is decoupled. Thus, in the ^{1}Hσ^{13}C/^{15}N HSQC experiments, the magnetization gets transferred from the more sensitive nucleus (I:^{1}H) to the less sensitive nucleus (S:^{13}C/^{15}N).

A detailed analysis using product operators is as follows (see Figure 1.4). Up to position b, ^{1}H magnetization goes through INEPT and gets transferred to the nitrogen. Thus, the first pulse on the proton creates a transverse magnetization ($H_z \rightarrow -H_y$), which undergoes H–N coupling and generates antiphase magnetization ($2H_xN_z$). Subsequently, two 90° pulses on the proton and nitrogen channel generate the antiphase nitrogen magnetization ($2H_zN_y$).

In the next part of the sequence (from b to c), the magnetization evolves under different Hamiltonians containing evolution due to three different couplings: J_{HN}, J_{NC}, $J_{NC\alpha}$, and the nitrogen chemical shift. In this segment, only the J_{HN} and J_{NC} couplings evolve. Coupling evolution due to $J_{NC\alpha}$, and the nitrogen chemical shift evolution are refocused during this segment. The net result in this segment is that the antiphase proton-nitrogen magnetization is refocused to in-phase nitrogen magnetization ($2H_zN_y \rightarrow N_x$), and simultaneously, antiphase nitrogen-carbonyl carbon magnetization is generated ($N_x \rightarrow 2N_yC_z$). Since the Hamiltonians in NMR commute with each other, we can separately examine their effects, which is independent of how they are treated.

- Chemical Shift Evolution: The chemical shift evolution of the transverse nitrogen magnetization refocuses because the 180° nitrogen pulse is symmetrically placed during this time interval.
- Proton–Nitrogen Coupling: The coupling between the nitrogen and the amide proton is allowed to evolve for a period of τ_b. Subsequent application of the DIPSI decoupling prevents further evolution. Since τ_b is set to $1/(2J_{NH})$, the evolution due to this coupling ($2H_zN_y \rightarrow 2H_zN_y\cos(\pi J_{NH}\tau_b) - N_x\sin(\pi J_{NH}\tau_b)$) leads to pure in-phase nitrogen magnetization generation ($2H_zN_y \rightarrow N_x$).
- Nitrogen–Carbonyl Coupling: The coupling between the nitrogen and the carbonyl carbon occurs for the entire period as 180° pulses were applied to both spins. This leads to the generation of two terms from pure in-phase nitrogen magnetization;

$$N_x \rightarrow N_x\cos(\pi J_{NC}\ 2\tau_c) + 2N_yC_z\sin(\pi J_{NC}\ 2\tau_c).$$

Since $\tau_c = 1/(4J_{NC})$, only the antiphase term survives, which is $2N_yC_z$.

- Nitrogen–C$^\alpha$ Coupling: Since a 180° pulse is applied only to the nitrogen during this period, this coupling is refocused over the $2\tau_c$ period.

  In the following period (c to d), the first antiphase carbonyl magnetization is created by simultaneous application of a 90° pulse on the nitrogen and carbonyl channels ($2N_yC_z \rightarrow -2N_zC_y$). After this, the chemical shift for the carbonyl is recorded, and the following terms are generated.

$$-2N_zC_y \rightarrow -2N_zC_y\cos(\omega_C t_1) + 2N_zC_x\sin(\omega_C t_1)$$

  Out of these, only the cosine term is detectable. Thus, during the chemical shift evolution, the coupling with nitrogen and C$^\alpha$ is suppressed by applying the 180° pulses in the respective channel. Subsequently, the magnetization is transferred back to nitrogen by simultaneous application of 90° pulses at point d ($2N_zC_y\cos(\omega_C\ t_1) \rightarrow 2N_yC_z\cos(\omega_C\ t_1)$).

Subsequently, in the following period (d to e), antiphase nitrogen magnetization with respect to carbonyl is transferred to in-phase nitrogen magnetization with chemical shift labeling. Simultaneously, antiphase nitrogen magnetization is generated concerning the proton. The following terms are generated:

$$2N_yC_z \rightarrow -N_x \rightarrow -2N_yH_z$$

During this time, coupling with nitrogen and C^{α} is suppressed by applying the 180° pulse in the C^{α} channel. As a result, the effective coupling evolves for:

$$[t_C - t_2/2] - [\tau_C] + [t_2/2] = 0$$

From point e, back INEPT is initiated to the proton channel, which is brought in phase after sensitivity enhancement block and gets detected. This kind of analysis is possible for every triple resonance experiment.

Usually, the 3D ^{15}N NOESY-HSQC experiment is utilized, and the NH–NH connectivities, which are characteristic of α-helices, help validate the sequential assignments. For side-chain assignments, there are numerous different experiments, such as NH-detected:[70]

- 3D HBHA(CBCACO)NH;
- 3D (H)C(CO)NH;
- 3D H(CCO)NH;
- 3D 15N-edited TOCSY.

And CH-detected:[70]

- 3D (H)CCH TOCSY;
- 3D H(C)CH TOCSY.

These experiments utilize TOCSY transfer for complete side-chain assignment.

The 3D ^{15}N-edited NOESY experiment[73] and NH-NH connectivities characteristic of α-helices are successfully used for sequential residual assignment validation. Moreover, this experiment supplemented by 3D ^{13}C-edited NOESY can validate the side-chain assignments. It should be pointed out that ^{13}C-edited NOESY[74,75] is recorded separately for aliphatic and aromatic regions of the spectrum, with the $^{1}J_{CH}$ coupling constant and offset usually tuned to 140 Hz, 40 ppm and 160 Hz, 122 ppm for aliphatic and aromatic regions of the spectrum, respectively. Notably, the NOE contacts can also be successfully used for determining contacts between the residues placed in the interface of the relatively small oligomeric proteins. An example of such an application was demonstrated for a globular protein with a quaternary structure, the S100A1 protein, a dimeric protein consisting of two identical subunits forming a homodimer. As was shown by studies of a high-resolution NMR-based structure of human S100A1, and also bacterial CylR2 homodimeric proteins, the nature of the interactions within the interface of oligomeric proteins involves mainly hydrophobic interactions between the methyl-containing side chains and several aromatic residues with weak C–H···π interactions.[76] An example of a weak intermolecular C–H···π interaction is between the aromatic moiety of Trp63 and the Hγ2/3 side-chain protons of Lys48, which is observed in the structure of the CylR2 dimeric protein.[77]

1.3 Through Space Correlation

The dipole–dipole interaction between two spins enables researchers to probe through space closed by atoms even though they are far apart in the primary sequence and usually from different spin systems (Figure 1.5). This phenomenon gives special significance for magnetic resonance spectroscopy in determining 3D structures of macromolecules such as proteins. There are different techniques to measure the distance between instigated spins, such as paramagnetic relaxation enhancement, **R**otational Frame **O**verhauser **E**ffect **S**pectroscop**Y** (**ROESY**), **N**uclear **O**verhauser **E**ffect **S**pectroscop**Y** (**NOESY**) and **H**eteronuclear **O**verhauser **E**ffect **S**pectroscop**Y** (**HOESY** or hetNOE). NOESY experiments are the most common approaches in biological structure analysis, mainly for protein structure elucidation. Hence, in this section, we present the theory and applications of different NOESY experiments. As most of the through-space correlations are based on the Nuclear Overhauser Effect (NOE) phenomenon, the basic concept of NOE will be presented first.

1.3.1 Nuclear Overhauser Effect

The Nuclear Overhauser Effect (NOE) phenomenon was discovered in 1953 by Albert Overhauser, who reported spin polarization enhancements by

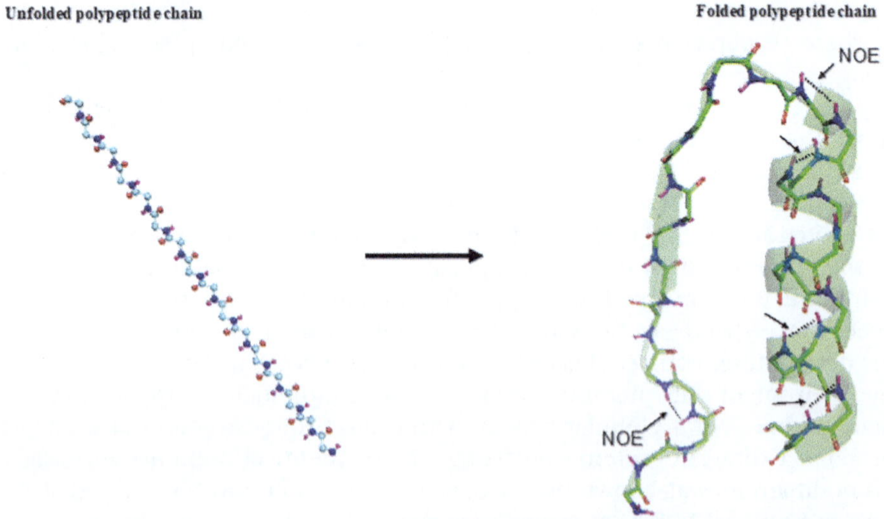

Figure 1.5 Folded and unfolded polypeptide chains highlighting the NOE effect.

microwave irradiation of conductive electrons in studied metals.[78] It can be defined as enhancing a spin's signal due to the transfer of magnetization from one nucleus to another through a cross-relaxation mechanism in a dipolar coupled system (through a space dipole–dipole relaxation effect).[79] As NOE occurs between different spins that are through-space coupled (dipole–dipole interactions), it provides important structural information.[80] Indeed, NOE has been used as a tool in NMR-based protein structural analysis for more than 20 years.[78,81] The NOE effect measures the distance between two nuclei within a short distance of less than 5 Å, therefore providing information on the identity of the atoms that are close together in space to help determine the intramolecular and intermolecular distances.[21] Such information can be obtained due to the folding of the protein (Figure 1.5), which detects the spins in proximity in space even though they are far apart in the primary sequence. As a result, the spins being irradiated will demonstrate the NOE effect independently from the direct connection of the spins by the chain of chemical bonds (scalarly or *J*-coupled). The NOE measurement is primarily used to provide information on the three-dimensional (3D) structure of biomolecules in solution by measuring the distances, and therefore calculating the global and local structures of the protein of interest. NOE has also been used to study macromolecular dynamics and is typically employed during the later stages of a structural investigation after the structure of the biomolecules has been fully defined by the use of different NMR experiments.

1.3.1.1 Basic Theory

The nuclei typically have two types of magnetic interactions: spin–spin interactions (scalar or *J*-couplings) and dipole–dipole interactions (dipolar couplings) and produce the NOE effect. The protein structure determination using NMR relies mainly on the NOE effect.[81–86] Theoretically, NOE is described as the change of signal intensity, either enhancement or decrease, of one spin when the transition of another is perturbed from its equilibrium population (Figure 1.6). This usually occurs by either inversion or saturation, eliminating a population difference between transitions that 2D-NOESY can visualize.

Dipolar relaxation is considered one of the most important sources of relaxation between the magnetic field of neighboring nuclei. For example, assume a system consisting of two spin-1/2 nuclei, *I* and *S* (*e.g.*, protons), where there is no scalar coupling (*J* for *I* and *S* is equal to zero), are spatially closed. This effect is described as cross-relaxation because spin *I* relaxes spin *S* and *vice versa*, resulting in alteration of the spin population. Therefore, this relaxation is distance-dependent, and the rate of relaxation that produces the NOE is inversely proportional to the sixth power of the distance between the interacting nuclei in space.[81]

$$I_{\text{NOE}} \propto r^{-6}$$

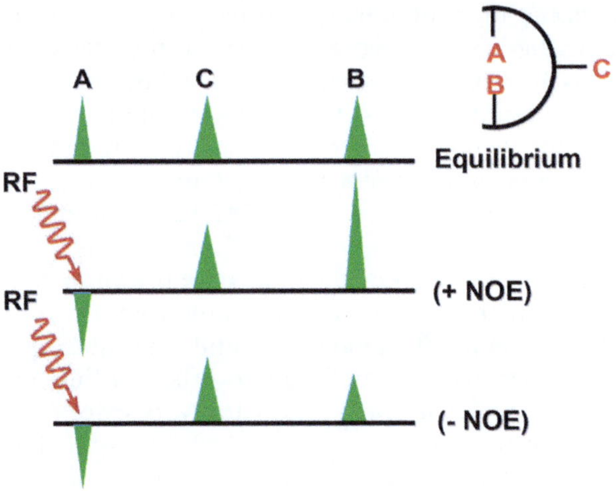

Figure 1.6 Nuclear Overhauser effect (NOE) for A and B, which are spatially close together, and A and C, which are far apart.

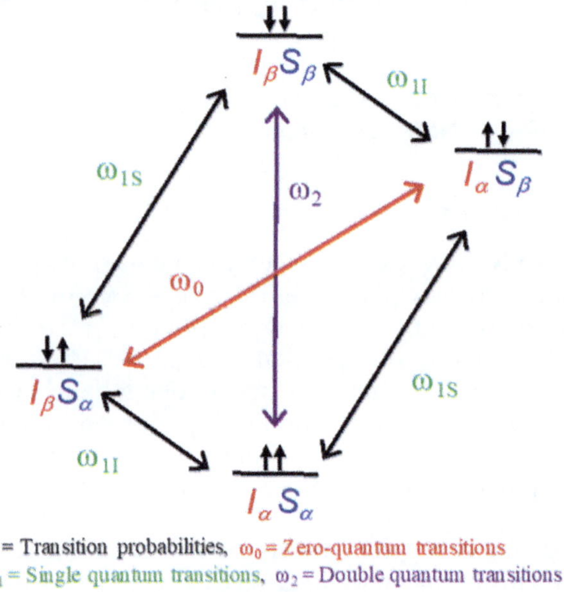

Figure 1.7 The four-spin energy level diagram and the six possible transitions of a two-spin system.

Figure 1.7 shows the energy level diagram and spin–flip transitions for the *IS* spin system, where the cross-relaxation that causes the NOE effect is represented. Such a system has four energy levels that depend on the spin-states of *I* and *S*, which correspond to αα, αβ, βα, and ββ, respectively. The dipolar interaction of the protons will cause T_1 relaxation between the spin states with the transition probabilities (ω). Notably, the ω's (ω_{1I}, ω_{1S}, ω_{0IS}, and ω_{2IS}) are the probabilities per unit time of the nuclear spin-1/2 that a transition will occur between the spin energy states. In other words, the rates at which the corresponding spin flips occur. Thus, there are two single quantum transitions, ω_{1I}, corresponding to $\alpha\alpha \rightarrow \alpha\beta$ and $\beta\alpha \rightarrow \beta\beta$, and ω_{1S}, corresponding to $\alpha\alpha \rightarrow \beta\alpha$ and $\beta\alpha \rightarrow \beta\beta$, which is the normal transition probability that gives rise to a peak in the spectrum that requires frequencies or magnetic field fluctuations near the Larmor precession frequency. In addition, there is a zero-quantum transition, ω_{0IS}, corresponding to $\beta\alpha \rightarrow \alpha\beta$, and a double quantum transition, ω_{2IS}, corresponding to $\alpha\alpha \rightarrow \beta\beta$ that are collectively referred to as cross-relaxation pathways, which are responsible for the NOE effect.

When the spin is at equilibrium, this implies that the population of $\beta\beta$ is less than that of $\alpha\beta$, and the population of $\beta\alpha$ is less than that of $\alpha\alpha$ [Figure 1.8(A)].[87] When selective irradiation of the spin S is applied, it will cause the populations of the $\alpha\alpha$ and $\alpha\beta$, and also the $\beta\alpha$ and $\beta\beta$ states, to

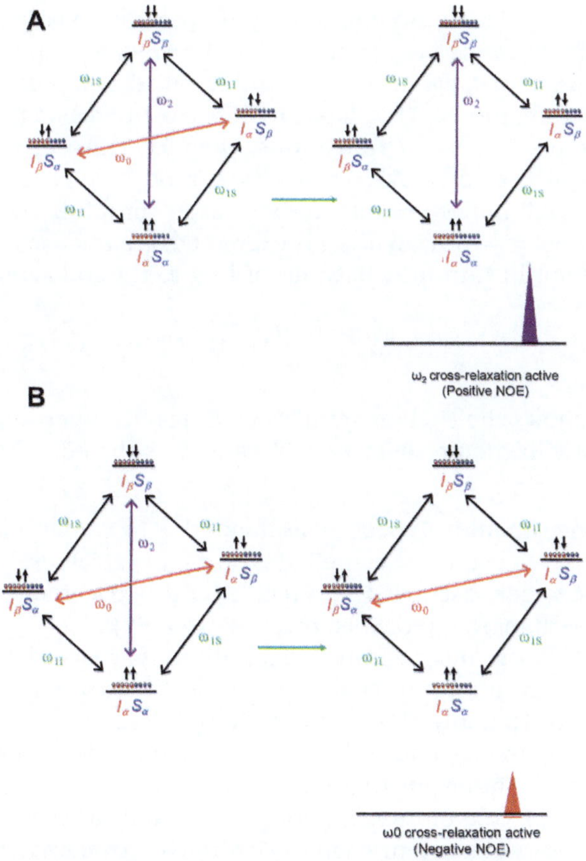

Figure 1.8 Schematic representation of population differences in energy levels of two spin system, S and I, that are dipolarly coupled. (A) after relaxation through the ω_2 process and (B) after relaxation through the ω_0 process.

become saturated (equalization of the α and β states of spin *S* through the single-quantum transition processes) [Figure 1.8(B)].[86] Then, as the relaxation takes place, the difference in these two populations depends on which possible relaxation processes dominates.[88]

First, domination by ω_1 (no NOE): upon *S* irradiation it will cause equalization of the populations of α- and β-spin states of *S*. Therefore, there will be no peak observed for *S* and because of the presence of only ω_1 relaxation ($\omega_2 = \omega_0$), spin *I* will be unaffected, which would lead to no NOE effect on *I* nuclei. Second, domination by ω_0 (negative NOE): now we consider the ω_0 process to be the only one that is operative. In the ω_0 process, when the *S* spin relaxes from the αβ to βα state, the *I* spin undergoes an α→β transition. Thus, the population will flow from the αβ to the βα state. As *S* returns to its normal population difference, it will lower the population difference for *I*, resulting in a decrease in the *I* intensity as the *S* intensity decreases. If *S* is irradiated continuously, then the signal for *I* will vanish (−100% NOE). This is a negative NOE, which is more effective for slow tumbling rate molecules (large molecules). Third, domination by ω_2 (positive NOE): for the ω_2 process, when the *S* spin relaxes from the β to the α state, *I* spin transfers from ββ states to αα states, again in order to recover the population differences across the *S* transitions. This has the effect of increasing the population difference across the two *I* transitions, therefore increasing the intensity of the *I* spin in the spectrum. This is then a positive NOE, which is more effective for small molecules due to their faster tumbling rate in a solution.

The magnitude of this transition is expressed as a relative intensity change between the equilibrium intensity, according to the following equation:

$$N_1(S) = \frac{I - I_0}{I_0} \times 100(\%)$$

where $N_1(S)$ shows the NOE of spin *I* once it has been perturbed by spin *S*, *I* is the intensity in the presence of NOE, and I_0 is the equilibrium intensity.

1.3.1.1.1 Experimental Types. Based on the NOE method, many NMR techniques have been developed for protein structure determination including the widely used **NOESY, HOESY,** and hetNOE and **ROESY.**[89]

Nuclear **O**verhauser **E**ffect **S**pectroscop**Y** **(NOESY)**

The NOESY is the most common technique that is widely used in large molecules to determine the cross-correlation between similar nuclei (*e.g.*, ^{1}H–^{1}H) that are spatially close. The basic pulse sequence of 2D NOESY is presented in Figure 1.9 and follows similar principles to the homonuclear 2D experiments. After applying a 90° pulse, the transverse magnetization evolves during a free evolution t_1 period followed by a second 90° pulse that creates longitudinal magnetization.[79] The transfer of magnetization to other spins occurs during the mixing time (τ_m), through cross-relaxation, which eventually establishes the correlations. A final 90° pulse is then applied to create transverse magnetization, thus allowing the signal to be detected.

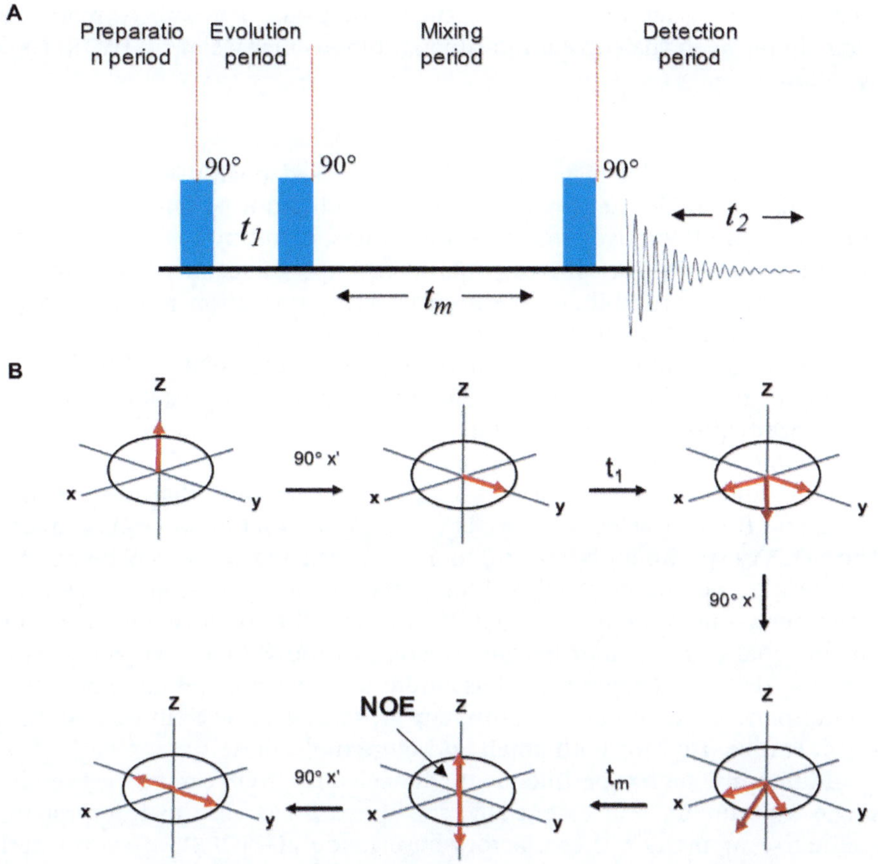

Figure 1.9 (A) The pulse sequences used to record classic 2D NOESY spectra. (B) A vector model describes ^1H spin magnetization.

In NOESY experiments, the cross-relaxation can be observed while magnetization is aligned along the *z*-axis.[74] The 2D NOESY spectrum typically shows diagonal and off-diagonal peaks (cross-peaks). The diagonal peaks usually indicate the proton peaks observed in the 1-D spectrum, whereas the cross-peaks represent the dipole–dipole interaction between nuclei that are close in space to one another showing NOE interactions between the correlated spins (<5 Å).[90] Both types of peaks can be either the same or opposite phases, depending on the magnitude of the rotational correlation time (τ_c) for the molecular size and, hence, the molecular weight.[90] The molecular correlation time (τ_c) is the average time needed for a molecule to rotate fully through an angle of 1 rad around any axis.[91] Thus, molecules with a faster tumbling rate will have a shorter correlation time, whereas slower tumbling molecules have a longer correlation time.[89] Between these extremes of molecular tumbling rates, the conventional NOE can become weak and even vanishingly small for the mid-sized molecules.[89] Therefore, the rotating-frame NOE measurements play a vital role in such situations. The approximate relationship between the correlation time and molecular weight (MW, in Da) is as follows:

$$\tau_c \approx MW \times 10^{-12}\ (S)$$

For instance, if the diagonal peaks are phased positively (and fully absorptive), the small-sized molecules that rapidly tumble have short correlation times and positive cross-relaxation rates, resulting in weakly negative 2D cross-peak intensities. In comparison, large molecules (*e.g.*, proteins) that exhibit slow tumbling have negative cross-relaxation rates, leading to intense positive cross-peaks in the 2D spectrum.[90]

Through qualitative assessment of distances, the relationship between the intensity of the cross-peaks and the internuclear distance has been categorized into three main groups based on NOE distance restraints.[92] The strong peaks correspond to the spins that are 1.8–2.7 Å apart, followed by the medium-intensity peaks, which reflect the spins that are 2.8–3.3 Å apart, and finally the weakest peaks for the spins that are 3.4–5.0 Å apart.[92] The NOESY experiment is a useful tool for identifying the stereochemistry of a molecule and the close-by residues that can be beneficial for sequence assignment in proteins.[88] NOESY can also be performed in a one-dimensional (1D) manner by the selection of individual resonances. However, the 2D NOESY experiment has a number of advantages compared to its counterpart 1D, including the simplicity for setting up, less time-consuming, and it can be used for both small and large molecules.

Additionally, such experiments are nonselective, NOE can be mapped in a single experiment, and essentially, the spectrum is easy to interpret in a qualitative manner.[90] These factors have made 2D-NOESY a favorite technique for structural chemists and biologists. However, protein structure determination by solution NMR depends on 3D NOESY spectra to provide interproton distance restraints, which are obtained from the intensity of cross-peak signals.[93] The 3D NOESY spectra provide better resolution, and therefore, it can be easier to assign and measure peak intensities (necessary for distance quantitation). The NOESY technique is applied to generate distance restraints in structure calculations, which consider the predominant source of macromolecular structural information in structural biology.

Heteronuclear **O**verhauser **E**ffect **S**pectroscop**Y** (**HOESY** or hetNOE)

The HOESY technique is the heteronuclear version of the NOESY and is used for cross-peaks between different nuclear species (*e.g.*, ^{1}H with ^{13}C) that allow the detection of heteronuclear through-space NOE connectives between nonbonded nuclei.[94] Figure 1.10(A) shows the general feature of pulse sequences used to record classic 2D HOESY, and Figure 1.10(B) shows the vector representation that describes the I spin magnetization. Let us consider a system consisting of two spins (I and S), where I represents ^{1}H and S represents ^{13}C, as an example. An initial 90° pulse on the I spin is applied to flip the I spins into the x–y plane where their components (α and β) precess during t_1. The spins precess according to the resonance frequencies of the chemically

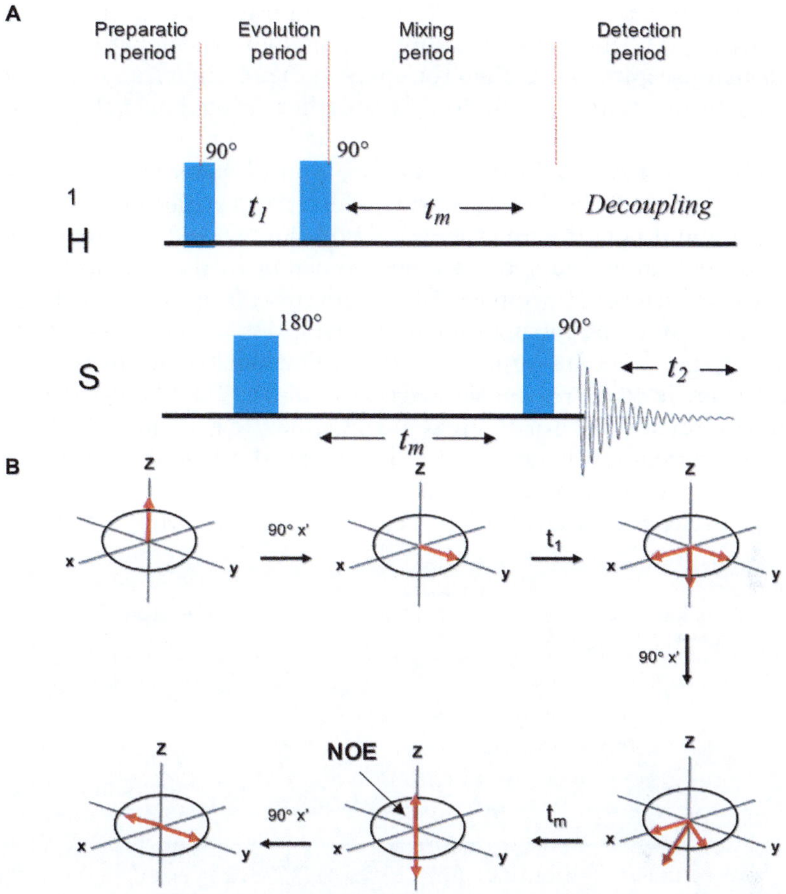

Figure 1.10 (A) The pulse sequences used to record classic 2D HOESY spectra. (B) A vector model describes I (^{1}H) spin magnetization.

distinct nuclei. Due to the coupling between the I and S spins, 180° pulse on the S spin at $t_1/2$ is applied to be removed. Then the spins precess for another $t_1/2$ period, resulting in refocusing of the two vectors. A second 90° pulse on the I spin creates longitudinal I spin magnetization followed by the mixing time (τ_m), where cross-relaxation leads to an exchange of magnetization between I and S spins through dipolar interactions. A final 90° pulse is then applied to create transverse magnetization for signal detection.

Rotational Frame **O**verhauser **E**ffect **S**pectroscop**Y** (**ROESY**)

ROESY, on the other hand, is typically used to obtain NOE information for mid-sized molecules that are approximately 1000–3000 Da in mass, such as peptides and small molecules, for which regular NOESY is not applicable. ROESY is especially suited for molecules with molecular correlation times (τ_c) such that $\omega_{\tau c} \sim 1$, where ω is the angular frequency $\omega = \gamma B$.[88] In such cases, the NOE for such molecules may be theoretically zero, so ROE (rotating frame

NOE) is often preferred since the ROE is always positive. In this technique, the homonuclear NOE effects are measured under spin-locked conditions, which involves using a spin lock to align the spins in the rotating frame where cross-relaxation occurs during the spinlock period (the mixing time, t_m).[90] For the 2D ROESY sequence, an initial 90° pulse is applied to put the net magnetization vectors (M) of the spins in the transverse (x–y) plane followed by a low-power RF spin-locking pulse for a time t_m during which magnetization transfer in the rotating frame occurs due to cross-relaxation (Figure 1.11).[95] The cross-peak generation occurs as the spins are held constant in the x–y plane and the magnetic field axis of the applied RF to prevent it from processing.[95] Under ROESY conditions, the mixing time (t_m) is typically 200–500 milliseconds depending on the T_2 of the sample, which is sufficiently long in the case of small molecules. As the ROESY pulse sequence is similar to TOCSY (Figure 1.11), it is essential to run ROESY with different mixing times. In general, the mixing time for TOCSY correlation is around 100 ms, while the ROESY ones are within 300–500 ms. Thus, running ROESY at different mixing times enables researchers to separate ROESY cross peaks from TOCSY ones. In the ROESY

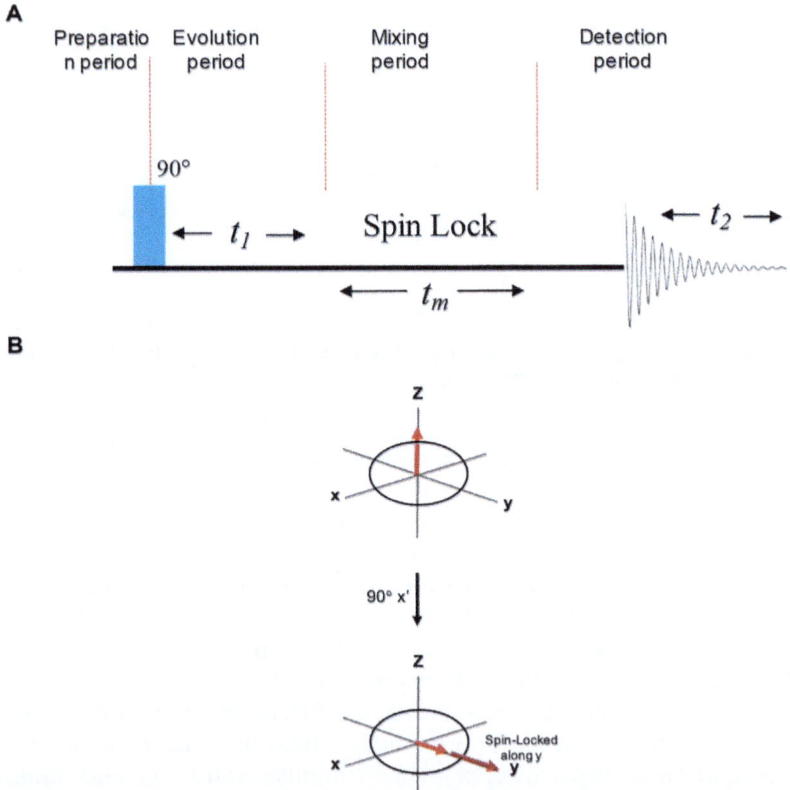

Figure 1.11 (A) The pulse sequences used to record classic 2D ROESY spectra. (B) A vector model describes ^{1}H spin magnetization.

spectrum, cross-peaks always have a sign opposite to the diagonal peaks, irrespective of the molecular weight. Such a technique is extensively used for small molecules such as saccharides, but is rarely used for proteins.

In comparison to NOESY, the main disadvantages that are associated with ROESY experiments are the lower sensitivity. Moreover, sample heating (specifically the ones in salty solutions) due to the requirement of spinlock during the mixing time can affect the molecular dynamics.[95] However, with small molecules, the trade between the negative and positive NOE effects can be comparable, leading to canceling of the NOE effects and cross-peaks of coupled spins will be undetectable, while the effect is always positive in ROESY experiments where all the cross-peaks would be detected.

1.3.1.1.2 3D NOESY Experiments. The use of solution NMR for the determination of protein structure has become a well-established technique. However, an increasing number of challenges may be encountered due to the size of the protein. Regularly, proteins with masses greater than 10–12 kDa with slower tumbling rates will display line broadening, causing a spectral congestion, which is an overcrowded rich peaks spectrum.[96] As a result, a signal overlap occurs even with the increase of the two-dimensional spectra resolution. For such proteins, both spectral crowding and rapid transverse relaxation can impair the analysis of cross-peaks in NOESY spectra that provide interproton distance restraints for structural models, therefore they can produce a spectrum that is impossible to assign.

To improve the resolution and overcome signals overlap, since it can be a major obstacle in the analysis of NMR spectra, three- and four-dimensional techniques have been developed to allow the possibility to study macromolecules up to 30 kDa in size.[97] The pulse sequence of a 3D experiment is changed to add a vertical domain, thus the information is spread into a third dimension resulting in a spectrum that can be visualized as a cube instead of a plane (usually ^{15}N or ^{13}C).[98] The 3D techniques are a natural progression of the 2D experiments so they are constructed from two 2D experiments combined, resulting in a spectrum that contains two evolution and two mixing periods that are inserted between the first mixing period and the direct data acquisition.[88,98] Each of the different indirect periods (t_1, t_2) is incremented separately.[88] Such a technique can aid in resolving the overlapped cross-peaks that show in a 2D spectrum. 3D NOESY NMR experiments can be categorized into two categories. The first is homonuclear NMR experiments where all three axes exhibit the same chemical shift of the same nucleus. They are rarely recorded due to the two transfer steps that would increase the number of possible correlations in the three-dimensional space.[98] The second is heteronuclear NMR experiments, which exhibit two or three different nuclei on the three axes and exhibit an increase in resolution due to the third dimension that is not compromised by an increased number of correlations.[98] The most widely used for structure determinations by solution NMR is heteronuclear-edited NMR experiments, such as 3D NOESY-HSQC (or HMQC), which can assist in spreading out of the signals by combining HSQC and NOESY in a single 3D experiment by addition of the HSQC step into the NOESY pulse sequence.[69] Alternatively, acquisition starts after the HSQC step rather than at the end of the NOESY mixing time. ^{1}H–^{1}H NOESY spectra are edited by ^{15}N or ^{13}C chemical shifts. ^{15}N-edited 3D NOESY can exhibit peak overlap and symmetry problems, thus uncertainties of protein calculation and structure may rise.[69] Hence, ^{13}C-edited 3D NOESY is considered to be a complementary method that can provide additional NOE constraints.

Since NMR-based structural biology depends on NOESY to obtain structure elucidation and modeling, many new techniques have been developed over the years to maximize sensitivity and data quality and, most importantly, reduce collection time. These include newly developed 3D NOESY experiments based on fast pulsing SOFAST-HMQC, 3D SOFAST-HMQC-NOESY-HMQC, and 3D SOFAST NOESY-HMQC experiments.[99]

1.4 Summary and Future Perspectives

Here we have focused on the theory and applications of basic methods of NMR spectroscopy that have been used for sequence-specific resonance assignment followed by structure elucidation. We provided some fundamental aspects of dynamics investigation using NMR for biological molecules. One can perform NMR measurements under physiological conditions, and real-time dynamics information can be derived, which is unique to NMR. We initially discussed how historically two-dimensional (2D) NMR spectroscopy was introduced, which boosted the development of multidimensional experimental possibilities combining chemical shifts of coupled homonuclear and/or heteronuclear spins. Later, the dipolar interactions between pairs of nuclear spins were utilized to derive structural information. The details of magnetization transfer, basic building blocks for polarization transfer from one spin to other spins, were discussed at great length. We first provided a comprehensive discussion of INEPT and reverse-INEPT building blocks, which form the basis of all triple resonance experiments. The nomenclature and detailed magnetization pathways for triple resonance experiments were subsequently discussed. Numerous advances have been made during the past two decades to achieve three- and four-dimensional spectroscopy in a reasonable amount of time, and the corresponding research area for FAST NMR methodologies was introduced. We have also covered the detailed theory of dipole–dipole interaction between two spins at an advanced level. In this direction, the Nuclear Overhauser Effect (NOE) phenomenon was introduced first. This is the basic concept for probing the through-space correlations by atoms and led subsequently to different techniques to measure the distance between instigated spins, such as paramagnetic relaxation enhancement (PRE), Rotational Frame Overhauser Effect Spectroscopy (ROESY), Nuclear Overhauser Effect Spectroscopy (NOESY), and Heteronuclear Overhauser Effect Spectroscopy (HOESY or hetNOE), which were also discussed.

Thus our approach has been to provide a solid foundation for the two most important aspects of NMR. The first was the foundation of building blocks, magnetization transfer, followed by details of the structure elucidation process. Recently, selective 'labeling' and 'unlabeling' have been reported as a cost-effective method for unambiguous assignment. This has been backed by simultaneous advancement in the hardware technologies, such as the availability of higher magnetic fields, cryogenic probes, pulsed-field gradients, availability of separately controllable radiofrequency channels, *etc.* New-generation pulse programs were designed to overcome spectral degeneracy and chemical shift overlap problems for different kinds of systems. To overcome the resolution barrier, and utilize the full potential of higher magnetic fields, nonuniform sampling and associated processing algorithms have been developed. The most obvious advantage of NMR spectroscopy is that it is still the only available technique to investigate molecular dynamics at different levels, such as proteins and DNA under physiological conditions.

With advancements in the last 10 years, some new avenues have appeared for NMR, making it the only tool available for addressing some biological questions. The foremost of these is understanding the dynamic picture of the molecule on a real-time scale, which is the key to its function. The residual dipolar couplings (RDCs) also provide dynamic information at a much wider time scale along with directional information. One key advantage of NMR is that it can study proteins that are amenable to crystallization. Normally, any protein with a large intrinsic disorder region or full intrinsically disordered protein cannot be crystalized.

In fact, more than 34% of eukaryotic proteins fall in this category. These proteins can undergo disorder-to-order transition upon binding to their biological targets, which are key to their functions. These systems access dynamic structural preferences on various timescales, and NMR can obtain detailed information in terms of chemical shift assignment, chemical shift index, J-couplings, paramagnetic relaxation enhancement (PRE), and RDCs. Now all chemical reactions involve intermediates, and hence there are also different short-lived confirmations that serve the purpose in the case of a biological process, which is chains of complex chemical reactions at an atomic level. However, a detailed understanding of these intermediates remains elusive as these conformers are only transiently formed and marginally populated on the energy landscape. Initially, a series of relaxation dispersion experiments were introduced to probe these intermediates, which are often less than a few percent of the total population. From these measurements, the details of this exchange process's kinetic and thermodynamic parameters can be obtained. This involves the relative population, exchange rates, and the free energy difference. However, most importantly, the minor species' chemical shifts can be obtained. Later, chemical exchange saturation transfer (CEST) was proposed for the slow exchange process, which is more straightforward and yields the chemical shift information more directly for the minor species. Like relaxation dispersion, different kinds of

CEST experiments have also been proposed, and now even the diffusion parameters of the excited states have also been reported. The future of NMR relies on such detailed exploration of this kind of structural information that is directly connected to the function of biological systems. Therefore, NMR will continue to play a dominant role in this field.

List of Abbreviations

NMR	Nuclear Magnetic Resonance
1D	one-dimensional
2D	two-dimensional
3D and 4D	three- and four-dimensional
ROESY	Rotational Frame Overhauser Effect SpectroscopY
NOESY	Nuclear Overhauser Effect SpectroscopY
HOESY or hetNOE	Heteronuclear Overhauser Effect SpectroscopY

References

1. M. Bauer, A. Bertario, G. Boccardi, X. Fontaine, R. Rao and D. Verrier, *J. Pharm. Biomed. Anal.*, 1998, **17**, 419–425.
2. M.-E. Dumas, E. C. Maibaum, C. Teague, H. Ueshima, B. Zhou, J. C. Lindon, J. K. Nicholson, J. Stamler, P. Elliott and Q. Chan, *Anal. Chem.*, 2006, **78**, 2199–2208.
3. A.-H. Emwas, R. Roy, R. T. McKay, L. Tenori, E. Saccenti, G. Gowda, D. Raftery, F. Alahmari, L. Jaremko and M. Jaremko, *Metabolites*, 2019, **9**, 123.
4. C. A. Blindauer, A. H. Emwas, A. Holý, H. Dvořáková, E. Sletten and H. Sigel, *Chem. Eur. J.*, 1997, **3**, 1526–1536.
5. B. Davaasuren, A.-H. Emwas and A. Rothenberger, *Inorg. Chem.*, 2017, **56**, 9609–9616.
6. M. Huang, S. Chen, J. Huang, R. E. Gerald II and K. Woelk, *Mater. Sci. Eng. C*, 2020, 111177.
7. E. Steimers, Y. Matviychuk, A. Friebel, K. Münnemann, E. von Harbou and D. J. Holland, *Magn. Reson. Chem.*, 2020, **59**, 221–236.
8. J. Aburabie, A. H. Emwas and K. V. Peinemann, *Macromol. Mater. Eng.*, 2019, **304**, 1800551.
9. S. Asghar, T. Shahzadi, M. Alazmi, X. Gao, A.-H. Emwas, R. S. Saleem, F. Batool and G. A. Chotana, *Synthesis*, 2018, **50**, 2211–2220.
10. A. G. A. Jameel, N. Naser, A.-H. Emwas and S. M. Sarathy, *Proc. Combust. Inst.*, 2019, **37**, 4663–4671.
11. A. G. A. Jameel, N. Naser, G. Issayev, J. Touitou, M. K. Ghosh, A.-H. Emwas, A. Farooq, S. Dooley and S. M. Sarathy, *Combust. Flame*, 2018, **192**, 250–271.
12. S. Manzoor, A. Bilal, S. Khan, R. Ullah, S. Iftikhar, A.-H. Emwas, M. Alazmi, X. Gao, A. Jawaid and R. S. Z. Saleem, *Sci. Rep.*, 2018, **8**, 1–12.

13. A. O'Rourke, S. Kremb, B. M. Duggan, S. Sioud, N. Kharbatia, M. Raji, A.-H. Emwas, W. H. Gerwick and C. R. Voolstra, *Molecules*, 2018, **23**, 1472.
14. Z. U. Rehman, S. Jeong, S. Tabatabai, A. Emwas and T. Leiknes, *Desalin. Water Treat.*, 2017, **69**, 1–11.
15. G. Zhang, A.-H. Emwas, U. F. S. Hameed, S. T. Arold, P. Yang, A. Chen, J.-F. Xiang and N. M. Khashab, *Chem.*, 2020, **6**, 1082–1096.
16. D. Hüse, L. Bíró, J. Patalenszki, A. C. Bényei and P. Buglyó, *Eur. J. Inorg. Chem.*, 2014, **2014**, 5204–5216.
17. N. H. Ahmed, G. R. Saad and M. M. Naoum, *Liq. Cryst.*, 2018, **45**, 1487–1497.
18. P. Agback, E. Woestenenk and T. Agback, *BMC Mol. Cell Biol.*, 2020, **21**, 1–15.
19. J. Deng, T. Ma, Z. Chang, W. Zhao and J. Yang, *Acta Phys.-Chim. Sin.*, 2020, **36**, 1901079.
20. J. M. Oommen, M. M. Hussain, A.-H. M. Emwas, P. Agarwal and L. A. Archer, *Electrochem. Solid-State Lett.*, 2010, **13**, K87.
21. S. Chu, S. Maltsev, A.-H. Emwas and G. A. Lorigan, *J. Magn. Reson.*, 2010, **207**, 89–94.
22. S. Li, O. Lafon, W. Wang, Q. Wang, X. Wang, Y. Li, J. Xu and F. Deng, *Adv. Mater.*, 2020, 2002879.
23. P. Rovó, *Solid State Nucl. Magn. Reson.*, 2020, 101665.
24. G. Qi, Y. Chu, Q. Wang, X. Wang, Y. Li, J. Trébosc, O. Lafon, J. Xu and F. Deng, *Angew. Chem., Int. Ed.*, 2020, **132**, 19700–19706.
25. F. Alahmari, B. Davaasuren, A.-H. Emwas and A. Rothenberger, *Inorg. Chem.*, 2018, **57**, 3713–3719.
26. S. Asghar, T. Shahzadi, M. Alazmi, X. Gao, A. H. Emwas, R. S. Z. Saleem, F. Batool and G. A. Chotana, *Synthesis*, 2018, **50**, 2211–2220.
27. T. F. Huang, G. Sheng, P. Manchanda, A. H. Emwas, Z. P. Lai, S. P. Nunes and K. V. Peinemann, *Sci. Adv.*, 2019, **5**, eaax6976.
28. A. G. A. Jameel, A. Khateeb, A. M. Elbaz, A. H. Emwas, W. Zhang, W. L. Roberts and S. M. Sarathy, *Fuel*, 2019, **253**, 950–963.
29. X. H. Qiu, D. Redwine, K. Beshah, S. Livazovic, C. G. Canlas, A. Guinov and A. H. M. Emwas, *J. Membr. Sci.*, 2019, **581**, 243–251.
30. N. Glanzmann, A. M. L. Carmo, L. M. R. Antinarelli, E. S. Coimbra, L. A. S. Costa and A. D. da Silva, *J. Mol. Model.*, 2018, **24**, 160.
31. R. A. Hill and C. P. Nicholson, *J. Chem. Educ.*, 2017, **94**, 1965–1968.
32. R. Hopson, P. Y. B. Lee and K. M. Hess, *J. Chem. Educ.*, 2018, **95**, 641–647.
33. F. Alahmari, B. Davaasuren, A. H. Emwas, P. Costa and A. Rothenberger, *Inorg. Chim. Acta*, 2019, **488**, 145–151.
34. M. Alghrably, D. Dudek, A. H. Emwas, L. Jaremko, M. Jaremko and M. Rowinska-Zyrek, *Inorg. Chem.*, 2020, **59**, 2527–2535.
35. R. A. Alsiary, M. Alghrably, A. Saoudi, S. Al-Ghamdi, L. Jaremko, M. Jaremko and A. H. Emwas, *Neurol. Sci.*, 2020, **41**, 2389–2406.
36. Z. A. Al-Talla, S. H. Akrawi and A. H. M. Emwas, *Int. J. Clin. Pharmacol. Ther.*, 2011, **49**, 469–476.

37. C. M. Davis and B. M. Dixon, *J. Chem. Educ.*, 2011, **88**, 309–310.
38. L. Ronconi and P. J. Sadler, *Coord. Chem. Rev.*, 2008, **252**, 2239–2277.
39. K. Srikanth, R. W. Schurko, I. Hung and A. Ramamoorthy, *Mater. Sci. Technol.*, 2003, **19**, 1191–1196.
40. S. Appelt, A. Kentner, S. Lehmkuhl and B. Blumich, *Prog. Nucl. Magn. Reson. Spectrosc.*, 2019, **114**, 1–32.
41. D. R. Glenn, D. B. Bucher, J. Lee, M. D. Lukin, H. Park and R. L. Walsworth, *Nature,*, 2018, **555**, 351–354.
42. M. Jiang, J. Bian, X. M. Liu, H. Y. Wang, Y. L. Ji, B. Zhang, X. H. Peng and J. F. Du, *Phys. Rev. A*, 2018, **97**, 062118.
43. L. Shen and L. M. Peng, *Chin. J. Catal.*, 2015, **36**, 1494–1504.
44. S. Ciambellotti and P. Turano, *Eur. J. Inorg. Chem.*, 2019, 569–576.
45. A. Marchanka and T. Carlomagno, in *Biological Nmr, Pt B*, 2019, ed. A. J. Wand, vol. 615, pp. 333–371.
46. N. Nishida, Y. Ito and I. Shimada, *Biochim. Biophys. Acta, Gen. Subj.*, 2020, **1864**, 129364.
47. J. Bernarding, F. Euchner, C. Bruns, R. Ringleb, D. Muller, T. Trantzschel, J. Bargon, U. Bommerich and M. Plaumann, *Chem. Phys. Chem.*, 2018, **19**, 2453–2456.
48. J. H. Du and L. M. Peng, *Chin. Chem. Lett.*, 2018, **29**, 747–751.
49. A. M. Olaru, T. B. R. Robertson, J. S. Lewis, A. Antony, W. Iali, R. E. Mewis and S. B. Duckett, *ChemistryOpen*, 2018, **7**, 97–105.
50. M. Pellecchia, I. Bertini, D. Cowburn, C. Dalvit, E. Giralt, W. Jahnke, T. L. James, S. W. Homans, H. Kessler and C. Luchinat, *Nat. Rev. Drug Discovery*, 2008, **7**, 738–745.
51. E. A. Mahrous and M. A. Farag, *J. Adv. Res.*, 2015, **6**, 3–15.
52. G. M. Clore and A. M. Gronenborn, *Annu. Rev. Biophys. Biophys. Chem.*, 1991, **20**, 29–63.
53. L. E. Kay, G. M. Clore, A. BAx and A. M. Gronenborn, *Science*, 1990, **249**, 411–414.
54. A.-H. Emwas, K. Szczepski, B. G. Poulson, K. Chandra, R. T. McKay, M. Dhahri, F. Alahmari, L. Jaremko, J. I. Lachowicz and M. Jaremko, *Molecules*, 2020, **25**, 4597.
55. J. Jeener and G. Alewaeters, *Prog. Nucl. Magn. Reson. Spectrosc.*, 2016, **94**, 75–80.
56. K. Anil, R. R. Ernst and K. Wüthrich, *Biochem. Biophys. Res. Commun.*, 1980, **95**, 1–6.
57. M. P. Williamson, T. F. Havel and K. Wüthrich, in *Nmr In Structural Biology: A Collection of Papers by Kurt Wüthrich*, World Scientific, 1995, pp. 319–339.
58. J. Ying, F. Delaglio, D. A. Torchia and A. Bax, *J. Biomol. NMR*, 2017, **68**, 101–118.
59. M. A. Thomas, R. Nagarajan, A. Huda, D. Margolis, M. K. Sarma, K. Sheng, R. E. Reiter and S. S. Raman, *NMR Biomed.*, 2014, **27**, 53–66.
60. E. Kupče and R. Freeman, *J. Biomol. NMR*, 2004, **28**, 391–395.
61. M. Zinke, P. Fricke, C. Samson, S. Hwang, J. S. Wall, S. Lange, S. Zinn-Justin and A. Lange, *Angew. Chem., Int. Ed.*, 2017, **56**, 9497–9501.

62. A.-H. Emwas, M. Alghrably, S. Al-Harthi, B. G. Poulson, K. Szczepski, K. Chandra and M. Jaremko, in *Nuclear Magnetic Resonance*, ed. N. Khaneja, IntechOpen, London, UK, 2019, ch. 5, pp. 83–106.
63. Ē. Kupče, K. R. Mote and P. K. Madhu, *J. Magn. Reson.*, 2019, **304**, 16–34.
64. R. A. Alsiary, M. Alghrably, A. Saoudi, S. Al-Ghamdi, L. Jaremko, M. Jaremko and A.-H. Emwas, *Neurol. Sci.*, 2020, 1–18.
65. G. Wagner, A. Kumar and K. Wüthrich, *Eur. J. Biochem.*, 1981, **114**, 375–384.
66. W. C. Johnson Jr, T. G. Pagano, C. T. Basson, J. A. Madri, P. Gooley and I. M. Armitage, *Biochemistry*, 1993, **32**, 268–273.
67. G. M. Clore, A. Bax, P. C. Driscoll, P. T. Wingfield and A. M. Gronenborn, *Biochemistry*, 1990, **29**, 8172–8184.
68. P. C. Driscoll, G. M. Clore, D. Marion, P. T. Wingfield and A. M. Gronenborn, *Biochemistry*, 1990, **29**, 3542–3556.
69. J. Cavanagh, W. J. Fairbrother, A. G. Palmer III and N. J. Skelton, *Protein NMR Spectroscopy: Principles and Practice*, Elsevier, 2nd edn., 2007.
70. J. Cavanagh, W. J. Fairbrother, A. G. Palmer III and N. J. Skelton, *Protein NMR Spectroscopy: Principles and Practice*, Elsevier, 1995.
71. M. P. Williamson and J. P. Waltho, *Chem. Soc. Rev.*, 1992, **21**, 227–236.
72. L. E. Kay, L. K. Nicholson, F. Delaglio, A. Bax and D. Torchia, *J. Magn. Reson. (1969)*, 1992, **97**, 359–375.
73. O. Zhang, L. E. Kay, J. P. Olivier and J. D. Forman-Kay, *J. Biomol. NMR*, 1994, **4**, 845–858.
74. N. A. Farrow, R. Muhandiram, A. U. Singer, S. M. Pascal, C. M. Kay, G. Gish, S. E. Shoelson, T. Pawson, J. D. Forman-Kay and L. E. Kay, *Biochemistry*, 1994, **33**, 5984–6003.
75. D. Muhandiram, G. Y. Xu and L. E. Kay, *J. Biomol. NMR*, 1993, **3**, 463–470.
76. M. Nowakowski, Ł. Jaremko, M. Jaremko, I. Zhukov, A. Belczyk, A. Bierzyński and A. Ejchart, *J. Struct. Biol.*, 2011, **174**, 391–399.
77. M. Jaremko, Ł. Jaremko, H.-Y. Kim, M.-K. Cho, C. D. Schwieters, K. Giller, S. Becker and M. Zweckstetter, *Nat. Chem. Biol.*, 2013, **9**, 264–270.
78. A. W. Overhauser, *Phys. Rev.*, 1953, **92**, 411–415.
79. J. Fattori, F. H. S. Rodrigues, J. G. M. Pontes, A. Paula Espíndola and L. Tasic, in *Applications of NMR Spectroscopy: Volume 3*, ed. A. ur-Rahman and M. I. Choudhary, Bentham Science Publishers, 2015, pp. 180–266.
80. N. Wang, X. Dong, L. Liu, D. Cai, Q. Cheng, J. Wang, Y. Hou, A.-H. Emwas, J. Gascon and Y. Han, *Cell Rep. Phys. Sci.*, 2021, **2**, 100309.
81. M. P. Williamson, in *Annual Reports on NMR Spectroscopy*, Academic Press, 2009, vol. 65, pp. 77–109.
82. F. Ni, *Prog. Nucl. Magn. Reson. Spectrosc.*, 1994, **26**, 517–606.
83. A. T. Brünger, P. Adams, G. M. Clore, W. L. DeLano, P. Gros, R. W. Grosse-Kunstleve, J. Jiang, J. J. Kuszewski, M. Nilges, N. S. Pannu, R. J. Read, L. L. M. Rice, T. Simonson and G. Warren, *Acta Crystallogr., Sect. D: Biol. Crystallogr.*, 1998, **54**, 905–921.

84. G. A. Morris, *Magn. Reson. Chem.*, 1990, **28**, 738.
85. V. J. H. N. U. R. E. Schirmer, Academic Press, New York, 1971.
86. R. R. Gil and A. Navarro-Vázquez, in *Modern NMR Approaches to the Structure Elucidation of Natural Products: Volume 2: Data Acquisition and Applications to Compound Classes*, The Royal Society of Chemistry, 2017, vol. 2, pp. 1–38.
87. Eds. W. M.P., Webb G. A., *Modern Magnetic Resonance*, Springer, Dordrecht, 2008.
88. C. Karunakaran, P. Santharaman and M. Balamurugan, in *Spin Resonance Spectroscopy*, ed. C. Karunakaran, Elsevier, 2018, pp. 49–110.
89. T. D. W. Claridge, in *High-Resolution NMR Techniques in Organic Chemistry*, ed. T. D. W. Claridge, Elsevier, Boston, 3rd edn, 2016, pp. 315–380.
90. R. Atta ur, M. I. Choudhary and W. Atia tul, in *Solving Problems with NMR Spectroscopy* 2nd edn, 2016, eds. R. Atta ur, M. I. Choudhary and W. Atia tul, Academic Press, Boston, pp. 227–264.
91. E. Hatzakis, *Compr. Rev. Food Sci. Food Saf.*, 2019, **18**, 189–220.
92. M. J. Howard, *Curr. Biol.*, 1998, **8**, R331–R333.
93. S. H. Mishra, B. J. Harden and D. P. Frueh, *J. Biomol. NMR*, 2014, **60**, 265–274.
94. C. Yu and G. C. Levy, *J. Am. Chem. Soc.*, 1984, **106**, 6533–6537.
95. J. H. Simpson, in *Organic Structure Determination Using 2-D NMR Spectroscopy*, ed. J. H. Simpson, Academic Press, Boston, 2nd edn, 2012, pp. 169–184.
96. M. J. Sweredoski, K. J. Donovan, B. D. Nguyen, A. J. Shaka and P. Baldi, *Bioinformatics*, 2007, **23**, 2829–2835.
97. H. Oschkinat, C. Griesinger, P. J. Kraulis, O. W. Sørensen, R. R. Ernst, A. M. Gronenborn and G. M. Clore, *Nature*, 1988, **332**, 374–376.
98. P. Schmieder, in *Encyclopedic Reference of Genomics and Proteomics in Molecular Medicine*, Springer Berlin Heidelberg, Berlin, Heidelberg, 2006, pp. 1204–1208.
99. P. Rossi, Y. Xia, N. Khanra, G. Veglia and C. Kalodimos, *J. Biomol. NMR*, 2016, **66**, 259–271.

CHAPTER 2

Multistate Structures and Dynamics at Atomic Resolution Using Exact Nuclear Overhauser Enhancements (eNOEs)

HARINDRANATH KADAVATH AND ROLAND RIEK*

Laboratory of Physical Chemistry, ETH Zurich, Switzerland
*Email: roland.riek@phys.chem.ethz.ch

2.1 Introduction

Biomolecules, such as proteins and nucleic acids, exist in a dynamic equilibrium between many structural states. However, most biomolecules that have been structurally characterized by crystallographic studies,[1] cryo electron microscopy (cryo EM) and conventional nuclear magnetic resonance (NMR)-based methods[2–5] are represented as single-state structures. The dynamics of proteins at atomic resolution can then indirectly be inferred from the presence of distinct structures due to different crystal packings, different class selection in cryo EM, the elucidated apo *versus* holo structures or/and by the measurement of dynamic-sensitive NMR probes such as spin (mostly ^{15}N)-relaxation mesaurements.[4–8] Using these indirect methods eventually in combination with molecular dynamics simulations[9,10] or/and statistical model selection techniques,[7] it has been shown that concerted molecular motions play a pivotal role in protein functions such as enzymatic activity,

New Developments in NMR No. 26
NMR Spectroscopy for Probing Functional Dynamics at Biological Interfaces
Edited by Anirban Bhunia, Hanudatta S. Atreya and Neeraj Sinha

Published by the Royal Society of Chemistry, www.rsc.org

protein–protein interactions, and protein–ligand interactions[11–14] indicating among others the presence of various mechanisms of allostery also.[15–19]

The strength of NMR is thereby on the one hand the study of the biomolecules under physiological conditions in solution and on the other hand the presence of a dense set of distinct probes throughout the molecule allowing detailed monitoring. The NMR probes are thus ensemble-averaged entities originated from the various dynamically interchanging structures of the biomolecule of interest and are sensitive to the time domain of the motion. While these probes are usually exploited for the elucidation of the dynamics of a protein (such as relaxation) they yield "only" the time order of the dynamics and the order parameters,[6,20,21] which indicate the spatial extent of the motion, but lack atomic detailed information on the dynamics. In contrast, the conventional NMR methods used for the 3D structure determination of a biomolecule at atomic resolution such as the nuclear Overhauser enhancement (NOE),[22,23] residual dipolar couplings (RDCs),[24–28] chemical shifts,[29,30] paramagnetic relaxation enhancement (PRE),[31] pseudo-contact shifts (PCS),[32–34] and *J* couplings[35] have been used by the majority of NMR structural biologists to determine an averaged conformation of the molecule.[36–38] As stated by Oleg Jardetzky, "in cases where the individual conformers contributing to the average are not known the 'average' derived from spectroscopic data represents a virtual structure devoid of physical meaning,"[39] which is in principle true as all the NMR observables are ensemble-averaged parameters.

However, there have been continuous efforts toward a holistic determination of the biomolecular structural landscape at atomic resolution and time-resolved using NMR spectroscopy. These are in addition to RDC measurements in combination with molecular dynamics simulation,[28,40] cross-correlated relaxation (CCR),[41–43] relaxation dispersion in concert with chemical-shift-based structure prediction,[44] and the exact Nuclear Overhauser Enhancement (eNOE)[45–50] approach discussed in this chapter.

As we shall see, with the use of eNOE multiple-state ensembles of biomolecule structures can be elucidated that describe the various structural states (*i.e.* usually two or three states locally) of the molecule in solution. The key to this possibility is the high accuracy of <0.1 Å of the ensemble-averaged distance restraints obtained from the NOESY measurements. To date the eNOE method has been applied to a variety of proteins (see Table 2.1) such as the WW domain of Pin1, GB3, Cyclophilin A, PDZ2 domain of human tyrosine phosphatase, ubiquitin, RNA molecules, and one very large system. After discussing

Table 2.1 List of proteins for which the eNOE structures were determined. The selection of the mixing times is proportional to the size of the molecule.

Protein	Molecular weight	Mixing times used (ms)
WW domain	4.1 kDa	20, 30, 40, 50, and 60
GB3	6.2 kDa	20, 30, 40, 50, and 60
PDZ2 domain	10.0 kDa	8, 16, 24, 32, 40, 50, and 80
Cyclophilin A	18.0 kDa	15, 20, 25, 30, 40, and 50

a detailed description of the eNOE approach to obtain the multistate structure determination of proteins we show the implications of the exact NOEs to dissect protein allostery at atomic resolution and finally discuss our current research developments and future directions.

2.2 Results and Discussion

2.2.1 Exact Nuclear Overhauser Effect Spectroscopy

In conventional NMR-based structural biology, the standard NMR structure calculation protocol has been applied by making use of the experimentally derived NOE cross-peak intensities or volumes of the NOESY spectra, which are measured in a semiquantitative manner. These NOEs are employed as the common source of distance restraints in protein structure determination. Based on the relative intensity of NOE cross-peaks they are converted to upper and lower distance restraints.[22] The reasons for this approach include the complex dependency on time and ensemble averaging of the NOE cross-relaxation rate constants (σ) that perturb the inverse proportionality between NOE rate and ensemble-averaged distance between two spins,[51] spin diffusion which is the relayed magnetization transfer between two independent spins *via* commonly coupled spins contributing to undesired cross-peaks in the NOESY spectra,[52] as well as signal perturbations in the NOESY pulse sequence. Nonetheless, this approach has been widely successful in yielding an ensemble-averaged structure.[23]

With the introduction of the eNOE, most of the above-mentioned limitations could be resolved, enabling the conversion of NOE cross-relaxation rates into nearly exact distances between two spins and explaining the name exact NOE.[45,46] In particular, it has been demonstrated that the NOE in biomolecules does not strongly depend on fast motion yielding a straight-forward relationship between the NOE rate and the distance to the power of six, that the spin diffusion can be reduced by the selection of a series of short mixing times and the remaining spin diffusion can be calculated using an available intermediate structure, and that the unbalanced pathway of magnetization during the NOESY experiment can be rescued using an optimized 3D combined [^{13}C,^{15}N]-resolved [^{1}H,^{1}H]-NOESY experiment. In addition, the rotational correlation time of the protein must be determined independently.

2.2.1.1 *From the Collection of NMR Spectra to the NOE Buildups*

A well-behaving biomolecule that can be concentrated to a concentration in the range of 1 mM or higher in addition to a high field spectrometer with a cryogenic probe is necessary for multistate structure determination by eNOEs. Standard ^{15}N relaxation measurements must be performed to determine the rotational correlation time τ_c of the biomolecule necessary for the conversion from NOE buildup curves to a distance. A series of four 3D combined [^{13}C,^{15}N]-resolved [^{1}H,^{1}H]-NOESY experiments that cover both aliphatic and aromatics must be performed because cross-peaks on both sides of the diagonal are required to determine an exact distance. The selection of the mixing times as well as the maximal mixing time depend on the size of the molecule as there is an inverse relationship between the mixing time and the rotational correlation time (τ_c) (see Table 2.1) and the theoretically suggested maximal mixing time is approximately 2.5×10^{-10} s^2/τ_c.[45,53,54] In order to compensate for potential time-dependent sample irritation during long measurements (such as precipitation, fragmentation, loss of the homogeneity of the magnetic field, *etc.*), it is highly recommended to randomize the order of the mixing time points and also to measure the whole series of spectra in the same measurement time session. Although the mixing times beyond the proposed limit will potentially lead to errors in the NOE rate (σ) determinations due to spin diffusion, NOESY spectra with longer mixing times might be valuable for unambiguous (sequential) assignment.

As in the standard procedure of NMR structure determination, the NOESY spectrum (with the longest mixing time) is assigned. In addition to the cross-peaks, the diagonal peaks must also be picked. After finishing the resonance assignment, it is further validated by removing overlapped cross-peaks and excluding overlapped diagonal peaks from further analysis. In the case of larger proteins it is important to include the buildups with overlapped diagonals and it is recommended to use them in the form of generically normalized eNOEs (gn-eNOEs, see also below).[53,55] Finally, by using the script within NMRPipe[56] or eventually other NMR spectra analysis software, the intensities of the whole diagonal and cross-peaks from the entire NOESY buildup series are extracted.

In addition, it is suggested to make a small quantity of 10% ^{13}C-labeled protein for which a high-resolution [^{13}C,^{1}H]-HSQC experiment is carried out for stereo-selective assignment of methyl resonances.[57,58]

2.2.1.2 Translating the NOE Cross-peak Intensities into Restraints Using the Software eNORA (Exact NOE by Relaxation Matrix Analysis)

The cross-peak intensities of the NOESY spectra extracted as explained above are then fitted automatically with the eNORA2[59] (**e**xact **NOE** by **R**elaxation matrix **A**nalysis) program inside CYANA[60–63] to mono-exponential curves yielding distance restraints and graphs of all the fits for manual inspection. The detailed procedure is described online at: http://www.cyana.org/wiki/index.php/ENORA_and_multi-state_structure_calculations.

As input, eNORA2 requests peak lists from several NOESY spectra with different mixing times comprising both NOE cross-peaks and diagonals, the rotational correlation time as well as a 3D structure of the protein determined with conventional NOEs (for example by using the NOESY with the longest mixing time following standard procedures) or an X-ray-based 3D structure.

eNORA2 fits the experimental data yielding from the diagonal cross-peaks, the initial magnetization at zero mixing time [$\Delta M(0)$], and the autorelaxation

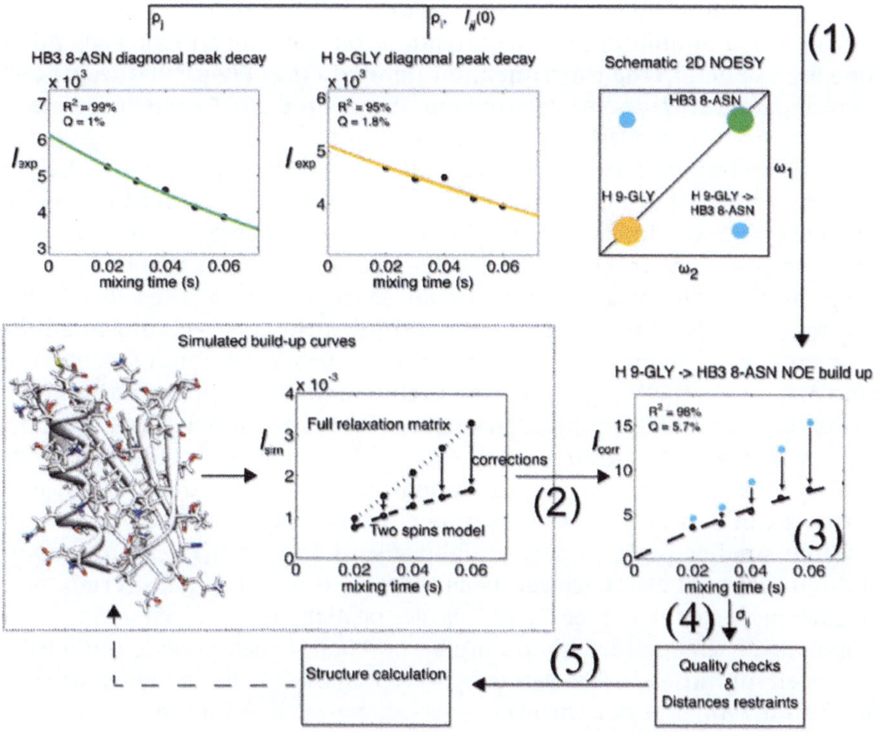

Figure 2.1 Schematic representation of the method for the determination and correction of the exact NOEs with eNORA and structure calculation with eNOEs. To demonstrate the workflow, the eNOE originating from the H (amide proton) of Gly9 (spin *i*, orange) and enhancing HB3 ($H^{\beta 3}$) of Asn8 (spin *j*, green) of protein GB3 are shown. The different steps involved in the eNOE analysis protocol are numbered 1–5. Step (1): The diagonal-peak intensities derived from the series of NOESY spectra are fitted to mono-exponential decay functions to extract the auto-relaxation rate constants ρ_i and ρ_j, and the initial magnetizations, $\Delta M_{ii}(0)$ and $\Delta M_{jj}(0)$. Step (2): Using the full relaxation matrix approach, a build-up curve is simulated by taking into account all magnetization pathways. This simulation requires a 3D structure of protein of interest as input, which can be a conventional NMR-based structure with sufficient accuracy. Step (3): Corrections for the NOE cross-peak intensities at each mixing time are applied to the experimental NOE build-ups. Step (4): The NOE build-up intensity is fitted, followed by evaluation of the quality of the fit and finally upper- and lower-bound distance restraints are generated. Step (5): A primary structure calculation is performed with the corrected new distance restraints using established software packages such as CYANA. This primary structure may be used as an input for step (2) in a new cycle, as indicated by the dashed arrow. Reproduced from ref. 89 with permission from American Chemical Society, Copyright 2012.

rate constant (ρ) (see Figure 2.1). Using the full relaxation matrix approach, an initial simulation of cross-peak build-up curves is performed[64] on the basis of the input reference structure. As the magnetization transfer pathways between all spins are simultaneously active, this approach corrects for spin diffusion. The simulation-based corrections for the intensities at each mixing time are applied to the experimental intensities (see Figure 2.1, step 1). The corrected σ rates are then converted into the desired eNOE distance restraints through the inverse sixth power relationship of σ (*i.e.* $\sigma \propto r^{-6}$). These eNOE distance restraints can be bidirectional or unidirectional (Figure 2.2), depending on whether both cross-peaks or only one cross-peak between two spins are observed in the spectra (*i.e.* NOE buildup curves are present in the provided peak list) and whether the diagonal peaks could also be fit.

The unidirectional and bidirectional eNOE restraints (Figure 2.2) have different accuracy and are thus treated differently. It is recommended (and as such set up within eNORA2) to leave ~20% error for distances from unidirectional eNOEs[65] yielding for each distance an upper and lower distance restraint that differ by 20%, whereas for bidirectional eNOEs the upper and lower distance restraints are the same. It should be noted that eNORA2 and CYANA will consider unidirectional eNOEs only if the cross-peak intensities can be normalized with respect to the diagonal peak.[65]

While smaller proteins are straightforward in classifying unidirectional and bidirectional eNOE restraints as there is minimal signal overlap, it can be challenging to extract eNOEs for larger proteins because of overlap on the diagonal. To this end a rather simple approach, called generic (gn)-eNOEs, has been proposed.[55] To extract gn-eNOEs from spin systems with unresolved diagonal peaks, the average ρ value as well as an upper limit $\Delta M(0)$ is determined on the basis of other non-overlapped values in the output lists. To further simplify the normalization it is also important to distinguish resolved methylene and methyl group protons so that the upper limit $\Delta M(0)$ values are slightly above all fitted $\Delta M(0)$ values. It is further recommended to use an averaged ρ and upper limit $\Delta M(0)$ values for each specific type of atom. Thus, eNORA2 can use these values to determine gn-eNOEs of cross-peaks associated with overlapped diagonal peaks. These gn-eNOEs yield then only upper limit restraints that should include ~20% addition (as is done in eNORA2) due to inaccuracies in their determination. It is noted that eNORA2 within CYANA carried out all the procedures described, automatically yielding two lists of upper and lower limit restraints ready for the structure calculation. Also the fitted curves are plot by eNORA for manual inspection of the calculated eNOEs in order to remove poor fits. Figure 2.3A shows a list of buildup curves.

2.2.2 Multistate Ensemble Structure Determination (Using eNOEs)

As the overall aim of the eNOE network-based structure calculation is to describe protein dynamics at atomic resolution and dissect distinct conformational states, a multistate structure calculation can be performed in CYANA using the output of the eNORA2 from eNOEs in addition to other restraints such as RDCs or J-coupling-derived angular restraints. The detailed

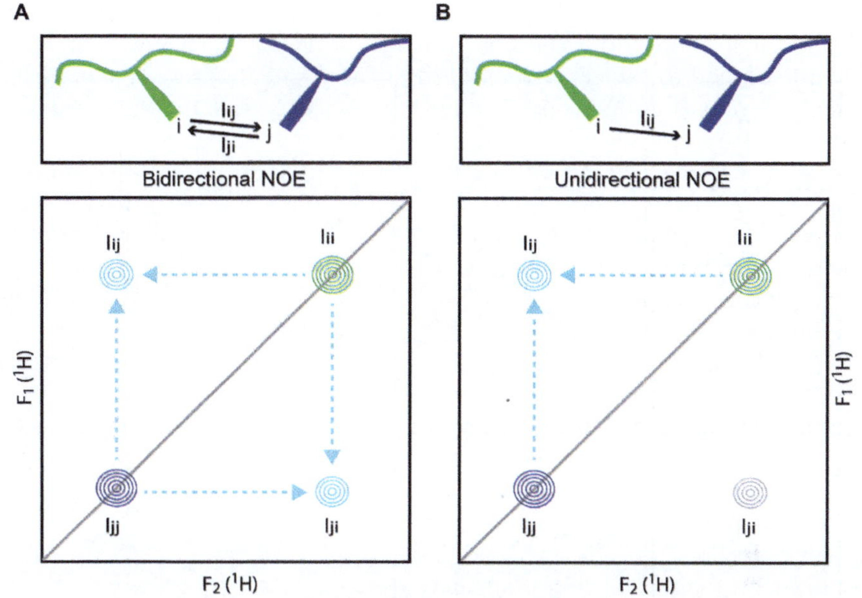

Figure 2.2 Illustration of the unidirectional and bidirectional eNOEs. eNOE distance restraints can be bidirectional or unidirectional, depending on whether only one or both cross-peaks between two spins are observed in the spectra. The dipolar interaction of two spins, *i* and *j*, are represented by arrows between two spatially close hydrogen atoms shown in green and blue. (A) The diagonal peaks of spins *i* and *j* are shown in green and blue circles, respectively, and the cross-peaks of spins *i* and *j* are shown in cyan circles (bottom). Bidirectional NOEs in the recorded ^{1}H–^{1}H NOESY spectrum (bottom) are due to the dipolar interaction of spins in both directions, as represented by arrows in both directions (top). As a result, two symmetrical and analyzable NOESY cross-peaks (I_{ij} and I_{ji}, shown in cyan) are present in the spectrum (bottom) and are called bidirectional NOEs. (B) The observed single NOESY cross-peak is attributed to the dipolar interactions in one direction (top), resulting in unidirectional NOE, I_{ij}, as shown in cyan (bottom). The theoretically possible symmetrical cross-peak, I_{ji}, is either not observed in the spectrum or is unanalyzable and is represented by the gray circles.

procedure to establish multistate structure calculations can be found online (CYANA wiki) at: http://www.cyana.org/wiki/index.php/ENORA_and_multistate_structure_calculations. Since the NMR probes are ensemble-averaged entities, a single-state structure calculation cannot fulfill the experimental distance restraints if a biomolecule exists in multiple states and a sufficiently large amount of restraints with high quality have been collected. The necessity of multistate eNOE-based structure calculation is shown as a three-atom model in Figure 2.3B.

In practice this necessity depends however on the requirement to collect more experimental distance restraints than the number of degrees of freedom. As illustrated in Figure 2.3C, in the case of the WW domain discussed

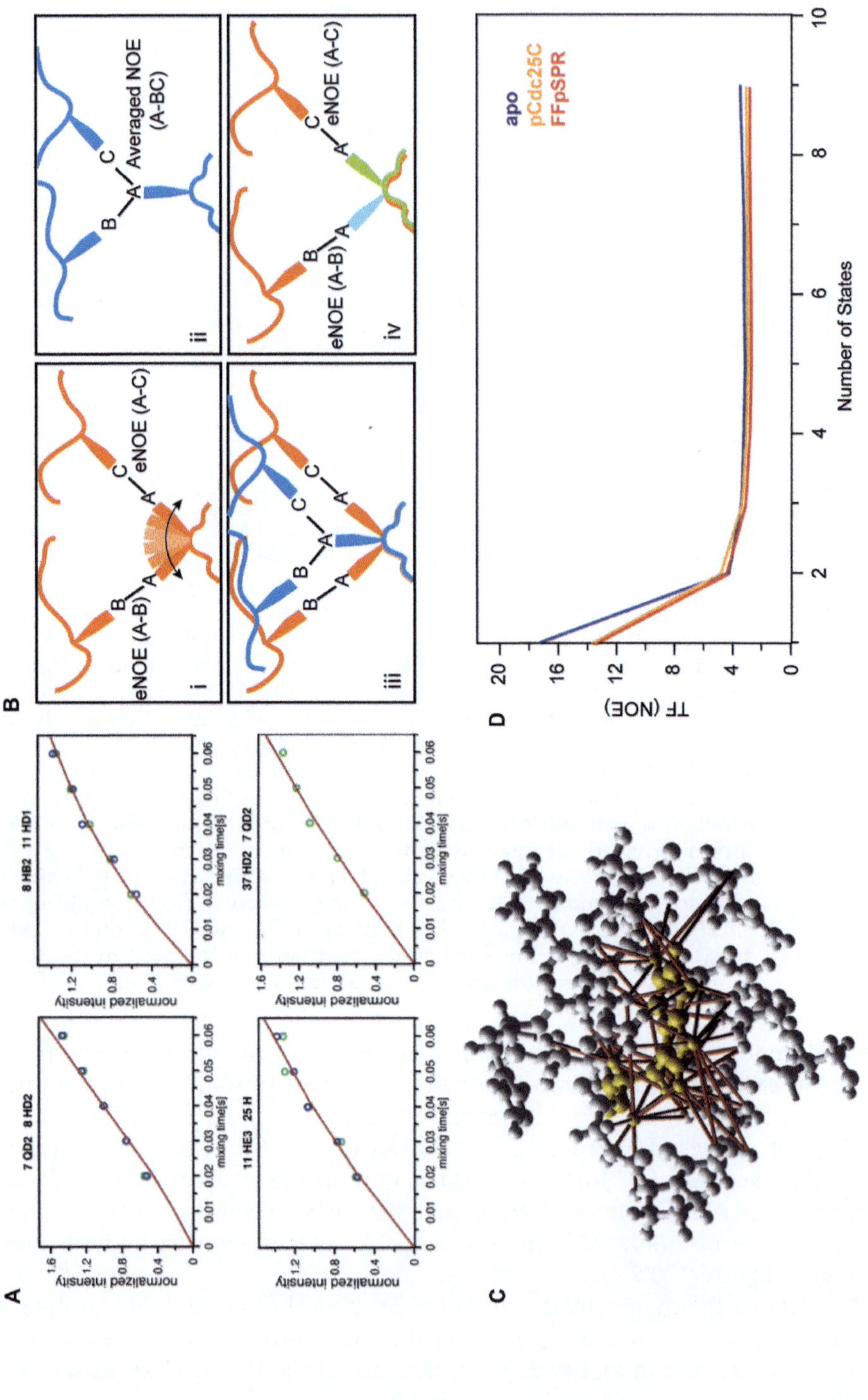
A
7 QD2 8 HD2
8 HB2 11 HD1
11 HE3 25 H
37 HD2 7 QD2
normalized intensity
mixing time[s]
B
eNOE (A-B)
eNOE (A-C)
Averaged NOE (A-BC)
i
ii
iii
iv
C
D
TF (NOE)
Number of States
apo
pCdc25C
FFpSPR

in detail below, the four degrees of freedom of the tryptophan residues Trp11 (*i.e.* four angles) are defined by more than 30 bidirectional and *circa* 30 unidirectional eNOE restraints, yielding an overdetermined system that can elucidate multistate configurations if present.

In order to determine the number of states in a structural ensemble necessary to describe the experimental data well, the CYANA target function is assessed. The CYANA target function obtained after the structure calculation protocol is a direct measure of the violation of the experimental restraints (*i.e.* it is a weighted sum of all squared violations) and it can be used to determine the appropriate number of states needed to fulfill the experimental restraints and as such indicate on the experimental requested spatial sampling of the biomolecule of interest. As shown in Figure 2.3D, the overall

Figure 2.3 Collection of eNOE distance restraints and structure calculation. (A) Five experimental eNOE distance restraints of the apo WW domain, derived from NOESY spectra measured at mixing times of 20, 30, 40, 50, and 60 ms are color coded by green and blue circles corresponding to the NOEs on both sides of the diagonal. The NOEs are between the residues and atom types indicated at the top of the buildup curve and the intensities of the NOE peaks are normalized to the average value of each buildup. The experimental data are in agreement with the back-predicted theoretical model where the predicted buildups were calculated using the eNORA2 program implemented in CYANA. (B) Cartoon representation of single-state and multistate eNOE structure calculation. (i) An ideal system of three atoms A, B, and C within a protein. The positions of atoms B and C are fixed, whereas atom A exists in dynamic equilibrium, and is in close proximity with both atoms B and C. Thus the eNOEs should be measured between atom pairs A–B and A–C. (ii) The eNOEs measured appear to be A simultaneously interacting with B and C. This assumes a single-state structure that causes B and C to be brought closer to A, resulting in distortion of the calculated structure, away from the true positions. (iii) Representation of how atoms A, B, and C in the single-state structure calculation (shown in blue) are distorted from their actual positions (shown in orange). (iv) Importance of multistate structure calculation to satisfy distinct states (cyan and green) allows atom A to occupy its original positions between atoms B and C. This multistate structure calculation satisfies the eNOE-derived distance restraints and captures the dynamics information. (C) The eNOE-based distance restraints of Trp11 of apo WW domain are mapped onto the 3D structure, which indicates the large size of the data set. The yellow and gray colors highlight Trp11 and the residues around it, respectively. Contrasting the four degrees of freedom of a Trp residue more than 60 eNOE distance restraints were collected and are highlighted as red sticks. This clearly shows the high density of information provided by eNOE-based structure calculation. (D) CYANA target function (TF) values of structure calculations against the number of simultaneously calculated states, demonstrating the importance of the ensemble-based structure. The CYANA TF for all three calculations of the apo WW domain and WW domain in complex with two ligands are shown. The decrease of the TF with increasing number of states indicates the need for multistate structure calculation to describe the experimental data. Panels A, C and D: Reproduced from ref. 67 with permission from John Wiley and Sons, Copyright © 2020 Wiley-VCH GmbH.

target function for the 20 lowest-energy conformers of the WW domain (discussed in detail below) after the single-state structure calculation is *circa* 16 $Å^2$, indicating a significant amount of violations (in order to illustrate this number, this could correspond to *circa* 64 distances that are violated by more than 0.5 Å, or 400 distance restraints that are violated by 0.2 Å). Because of the presence of two states of the WW domain this high value of the target function obtained for the single-state structure calculation decreases to approximately $\sim$3 $Å^2$ for two states.

2.2.2.1 *Structural States, Conformers, and Ensembles of a Protein*

NMR-based structural determination of biomolecules is represented by an ensemble, that is a bundle, of conformers that fulfills the experimental data. Usually the 20 lowest energy conformers having the lowest CYANA target function are used to represent the 3D structure. The root mean square deviation (RMSD) of the ensemble of conformers is the standard deviation of the calculation and as such the ensemble presents the precision of the structure determination. A conformer within the ensemble is an individual structure composed of one or more states, whereas a state is defined as a set of coordinates positioning all atoms of a molecule. Often biomolecules exist in multiple states. In such a scenario, it is important to define different states by sub-bundles of conformers. That is, there can be several sub-bundles of distinct states within an ensemble structure. A pictorial representation of the state, conformer, and ensemble is graphically illustrated in Figure 2.4.

2.2.3 Recent Applications of Exact Nuclear Overhauser Enhancement

To date we and others applied ensemble-averaged distance restraints obtained from exact NOE (eNOE) rate constants to various well-behaving model biomolecules. We measured eNOEs for an assortment of small to medium-sized systems (*i.e.* 4–18 kDa, see Table 2.1) with a high accuracy of $<$0.1 Å for the ensemble-averaged distance restraints.[54] Given the high density of the eNOE network for all the systems, we found that in all systems a multistate structure representation (usually two states) is needed to explain the experimental data. We also have demonstrated that eNOE data alone provided as much structural information as conventional NOE data provide in combination with RDC and J coupling data.[48,55] In order to grasp the potential impact of the eNOE-based multistate structures on the biomolecular structure–activity relationships, several systems are described in some detail in the following.

2.2.3.1 *eNOE-derived Protein Allostery of the Pin1 WW Domain*

The small-sized regulatory WW domain of the human peptidyl prolyl isomerase Pin1 was studied in detail by applying the eNOE-based multistate ensemble protocol.[66,67] Pin1 is an important peptidyl-prolyl *cis-trans* isomerase (PPIase) and its biological significance includes its protective

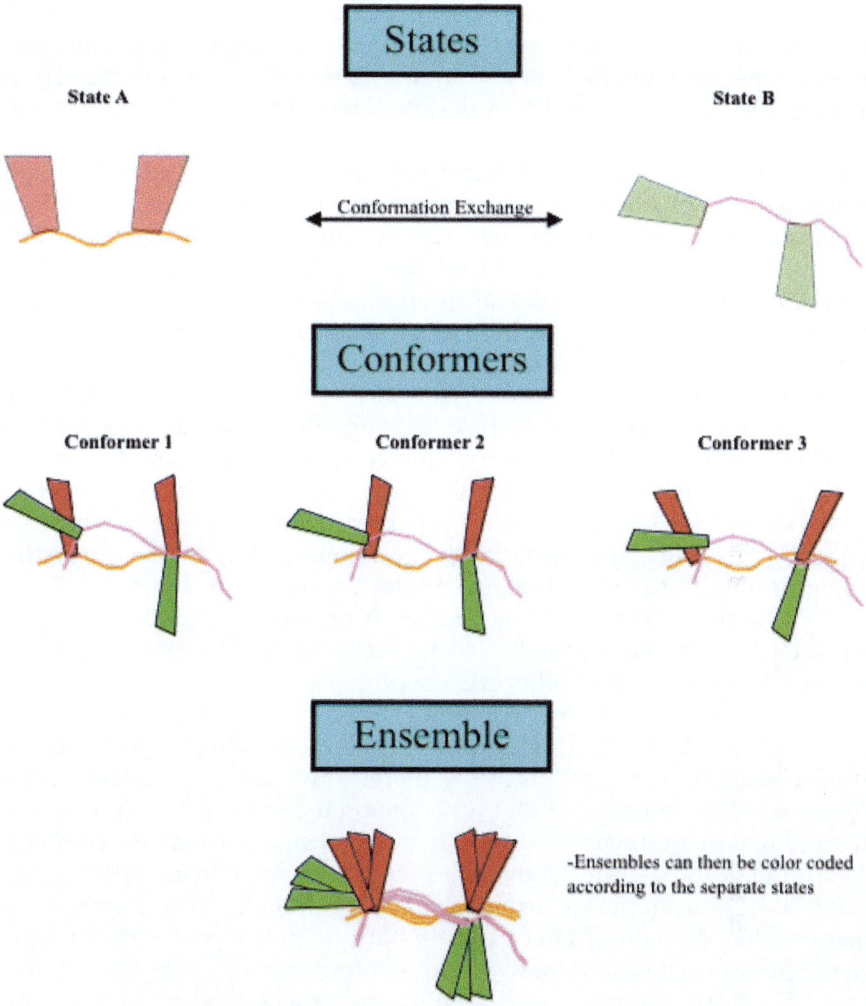

Figure 2.4 Cartoon representation of the definition of states, conformers, and ensembles. (Top) Two different states, A and B, are shown for a theoretical protein, where each state is defined as one set of coordinates for all the atoms of a molecule. Structure calculation provides a single conformer composed of one or more states that satisfy the data provided. (Middle) The structure calculation is performed multiple times and results in many conformers such as conformer 1, conformer 2, conformer 3 *etc.* (Bottom) An ensemble is the superposition of all the conformers generated from several structure calculation steps. If multiple states are available during the structure calculation, the ensemble can be divided into sub-bundles corresponding to each state, which should be identified by structural similarities in the region of interest. Reproduced from ref. 54, https://doi.org/10.3390/molecules22071176, under the terms of the CC BY 4.0 license, http://creativecommons.org/licenses/by/4.0/.

function against fibril formation in Alzheimer disease,[68] the regulation of mitosis,[69] increase of hepatitis C infection,[70] and it is also overexpressed in most human cancer cells.[71] Pin1 contains an N-terminal WW domain (due to two Trp residues) separated by a flexible linker from its C-terminal catalytic PPIase domain.[72]

Both these Pin1 domains interact with various ligands containing phosphorylated Ser/Thr-Pro motifs.[71] More interestingly, the Loop 1 (*i.e.* ligand-binding site comprising residues 15–21) and Loop 2 of the WW domain (residues 27–30) are responsible for the regulation of ligand binding and regulation of Pin1 function. The interdomain cross-talk between the WW domain and the PPIase domain is regulated by means of Loop 2 of the WW domain and residues 138 and 140–142 in the PPIase domain.[73,74] This suggests the potential allosteric communication at the interdomain interface as demonstrated by NMR-based ligand titrations and relaxation measurements.[74–76] The ligand titration NMR experiments performed by us and others with the isolated WW domain and two peptide ligands, FFpSPR and pCdc25C, showed chemical shift perturbations for those residues at the distal Loop 2 when the ligands bind at Loop 1.[67,74–77] More interestingly, upon binding to the WW domain the two ligands induce distinct chemical shift changes on Loop 2 in line with their opposing property in regulating the interdomain interaction.[67,74,75] In other words, the WW domain possesses a ligand-mediated allosteric coupling.

However, it is challenging to capture the structural aspects and dynamics nature of the Pin1 WW domain and allosteric coupling experimentally at atomic resolution. To this end we explored the nature of the structure and dynamics of the apo form of the WW domain using our eNOE approach, in a comprehensive manner.[66,67] As a first step we established a new approach to determine atomistic thermodynamic models of the allosteric WW domain using NMR data and extensive molecular simulations.[66] Our data suggested that the WW domain of Pin1 exists in both its native conformation and an energetically excited near-native state conformation.[66] This opened an avenue to perform the multistate eNOE-based structure calculation to provide a dynamic picture of the distinct states and dynamic allostery.

In order to explore the allosteric nature of the Pin1 WW domain and to further investigate the specificity of the distinct structural states proposed previously,[66] we decided to perform an in-depth analysis of the multistate structure of the WW domain at atomic resolution. As ligand-mediated allosteric coupling and conformational selection were predicted we went on to determine the eNOE ensemble structures of the apo-state of the WW domain as well as the holo-state in the presence of either of the two peptide ligands, FFpSPR and pCdc25C.

Upon eNOE restraint collection we calculated the multistate structures of all three forms of the WW domain in CYANA covering from the single state to nine states. The CYANA target function (TF) was found to be significantly dropped from one state to two states, reaching a plateau value of $\sim$3 Å^2 after three states (see Figure 2.2C), suggesting the presence of two distinct states for the WW domain in the presence and absence of ligands. The two states are still observed even in a ten-state structure calculation, as evidenced in the Ramachandran plot for residue Thr29 (see Figure 2.5) which further supported the two-state nature of the WW domain.

A closer look at the two-states ensemble of the apo WW domain shows two spatially well-separated states, which span from the ligand-binding site (Loop 1, Tyr23, and Phe25) to the distal and interdomain site (Loop 2) *via* the backbone of the second β-strand[67] (see Figure 2.6A). Despite the presence of two distinct structural states, only one set of chemical shifts is observed, suggesting the micro-millisecond timescale dynamics where the two states interchange in a concerted fashion. The backbone of the second β-strand and side chain of Asn26 thereby play significant roles in the two-state structural interchange. Another interesting observation to support the two-state ensemble structure is the relative side-chain arrangements of Loop 2

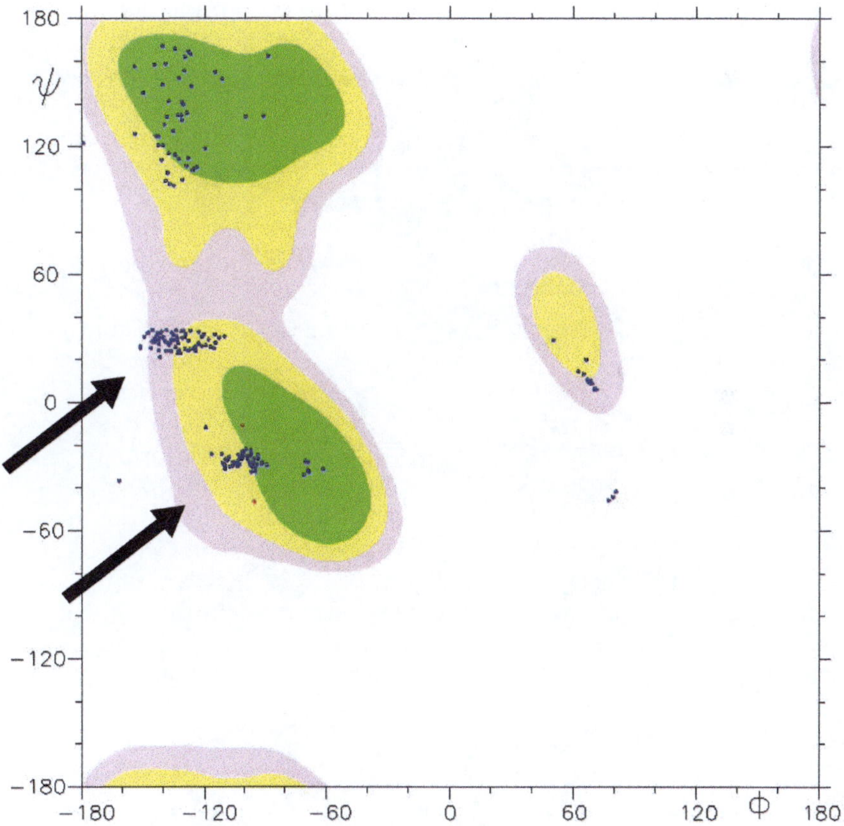

Figure 2.5 Ramachandran plot for residue Thr29. The two states are preserved in a ten-state structure calculation of apo WW as shown here. The Ramachandran plot of Thr29 of all the 20 lowest energy conformers (out of 200 conformers in total) of the ten-state structure calculation is shown. The two states of interest are still present and are highlighted by arrows.

residues, Thr29 and Ala31 (see Figure 2.6 and D), the side chains of which are arranged in the form of a seesaw sketch.

The two-state structure of the Pin 1 WW domain in complex with the ligand FFpSPR shows that the two states observed for the apo WW domain are preserved and only a population change toward the dark blue conformation of the apo WW ensemble structure has happened (see Figure 2.6A and D). The two side chains of residues Thr29 and Ala31 are still in a seesaw arrangement as in Figure 2.6F, where the yellow state of the complex matches well with the cyan state of the apo WW domain, whereas the red state of the complex superimposes with the blue state of the apo WW domain (see Figure 2.6D). Taken together it can be concluded that the FFpSPR peptide appears to select the dark blue state and this mode of positive allosteric action can be proposed as conformational selection.[78] Thus the allosteric mechanism of action of the FFpSPR ligand shifts the population of states.

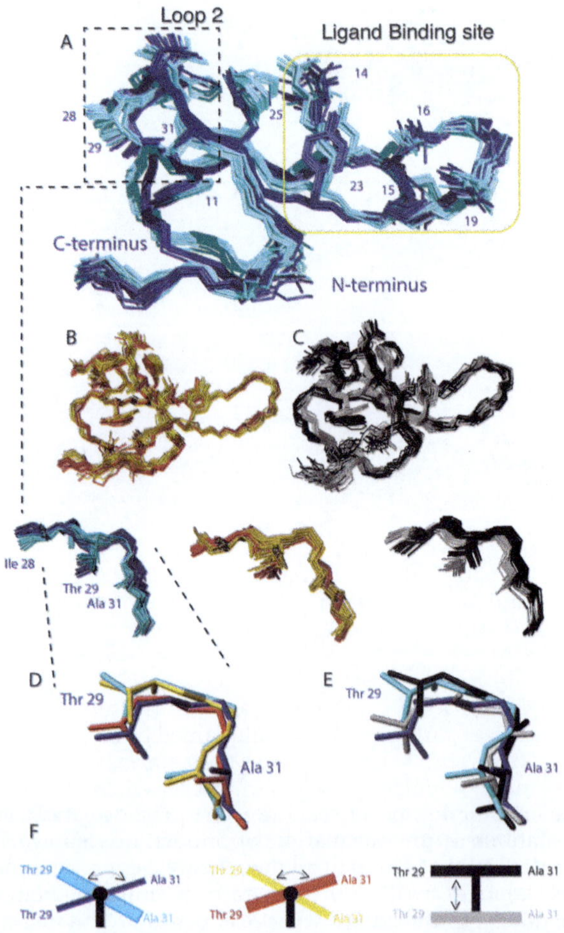

In the case of the WW–pCdc25C complex structure, the two distinct states are still observed (see Figure 2.3D). However it can be seen that the backbone and side-chain states between residues Ile28/Thr29 *versus* Ala31 present in Loop 2 are found to be anticorrelated as it is in contrast to apo and FFpSPR complex structures (see Figure 2.6C and D). As a result of the unique orientation of Loop 2, side chains are in such a way that both side chains are up or down simultaneously (see Figure 2.6D, E, and F). In short, unlike the complex with FFpSPR ligand, the two states of the apo WW structure are locally indistinguishable from the structure of the WW domain while in complex with pCdc25C. Due to the altered dynamics, the apo and holo WW domain behave differently, as it is observed in a time-resolved manner. Thus, the WW–pCdc25C complex can be regarded as an example of a dynamic allostery model. That is, the average local structure is not perturbed, however globally there is a change in the correlated motion of the free form to an anticorrelated motion in the complex (see Figure 2.6D, E, and F). Taken together, certain ligands can shift the structural states differently than the apo form and perturb the allosteric mechanism in proteins.

In summary, although various allosteric mechanisms of actions such as the population shift model and the dynamic allostery model have been proposed it was nearly impossible to explain the mechanism of action by experimental data. The most recent advances in NMR methodology

Figure 2.6 Two-state structural ensemble of the apo and holo WW domain. WW domain alone (A), WW in complex with the FFpSPR (B) and pCdc25C (C) ligands, highlighting the presence of two distinct states. Backbone traces of the 20 lowest energy structural ensembles each representing two distinct states are shown. Selected side chains are shown and labeled. The different WW states are color coded; cyan and blue for the apo WW, red and yellow for the positive allosteric ligand FFpSPR complex, and black and gray for the negative allosteric ligand pCdc25C complex. The two states at the Loop 2 present at the catalytic-domain are enlarged. (D) and (E) The two states for residues 29–31 are shown for only one conformer for clarity by superimposing the two apo states with either the two states of the positive (D) or the two states of the negative (E) allosteric-ligand–WW complex. (D) The positive allosteric ligand FFpSPR does not alter the two-state structures as the yellow state superimposes with the cyan state and the red state with the blue state for both residues Thr29 and Ala31. Meanwhile in the case of the negative allosteric ligand pCdc25C, the gry state superimposes with the blue state for residue Thr29, however for Ala31, the blue state superimposes with the black state, and the black state superimposes with the cyan state for Thr29, while the black state superimposes with the blue state of Ala31. (F) The observation from (D) and (E) is illustrated by a seesaw model. In the case of apo WW, Thr29 and Ala31 alter their states like a seesaw with the cyan state more populated (drawn thicker). When bound to the positive allosteric ligand, the seesaw states are preserved, however with different populations of the two states. In contrast, no seesaw-like two states are observed in the case of the negative allosteric ligand, shown in gray/black. Reproduced from ref. 67 with permission from John Wiley and Sons, Copyright © 2020 Wiley-VCH GmbH.

including the eNOE approach described here can, however, provide highly accurate experimental data, which in this study allowed us to elucidate the allosteric mechanism for the WW domain at atomic resolution.[67] We revealed a ligand-dependent allosteric mechanism where one of the ligands uses the population shift model, whereas the other ligand acts with the dynamic allostery model.

2.2.3.2 *Four-state Structure of GB3*

GB3 (third immunoglobulin binding domain of protein G)[46,47] is probably the protein in solution with the highest experimentally derived amount of structural and dynamical restraints known to date including, in addition to the eNOE-derived restraints, also many RDCs, J-couplings, and cross-correlated relaxation data permitting high-quality multistate structure determination. We have demonstrated that at least four structural states are required to fulfill the complete experimental data set for GB3 as shown in Figure 2.7.[47,48,54]

2.2.3.3 *Two-state Structure of the Prolyl* cis/trans *Isomerase Cyclophilin A Elucidating the Mechanism of Action*

Before our eNOE multistate structure determination, the mechanism of action of the proline *cis–trans* isomerase Cyclophilin A was proposed to be through a dynamic network on the μs–ms time scale between the enzyme's active site and nearby segments.[79–82] Cyclophilin A acts thereby with an electrostatic handle mechanism in the catalysis of proline *cis–trans* isomerization.[83] To obtain atomic resolution information of the dynamics, multistate structure determination of Cyclophilin A with eNOEs and RDCs was performed. The 165-residue-long peptidyl prolyl cis–trans isomerase Cyclophilin A[80,82,84] is thereby one of the largest proteins with which we have performed so far the eNOE-based multistate ensemble structure calculation.[55,85] Because the size of Cyclophilin A contributed to peak overlap gn-eNOEs (see Section 2.2.1.2) were also used.[55] Inclusion of gn-eNOEs doubled the number of eNOE-derived restraints, contributing significantly to the precision of the multistate structure.[55] It was demonstrated that a two-state structure calculation explains well the experimental restraints indicating that Cyclophilin A exists in multiple states. The 20 lowest energy two-state conformers are shown in Figure 2.8. The structural investigation revealed that the loop comprising residues 64–74 adjacent to the active site exists in two spatially well-separated states, indicative of large-scale correlated motion. Furthermore, 70% of all the side chains undergo the two-state correlation of the loop, including the active site (with Arg55). Due to the structural features, we named these two states the "open" (blue) and "closed" (cyan) states. Our findings can be interpreted as follows: the catalytic site of Cyclophilin A is preorganized for catalysis in the open state welcoming the

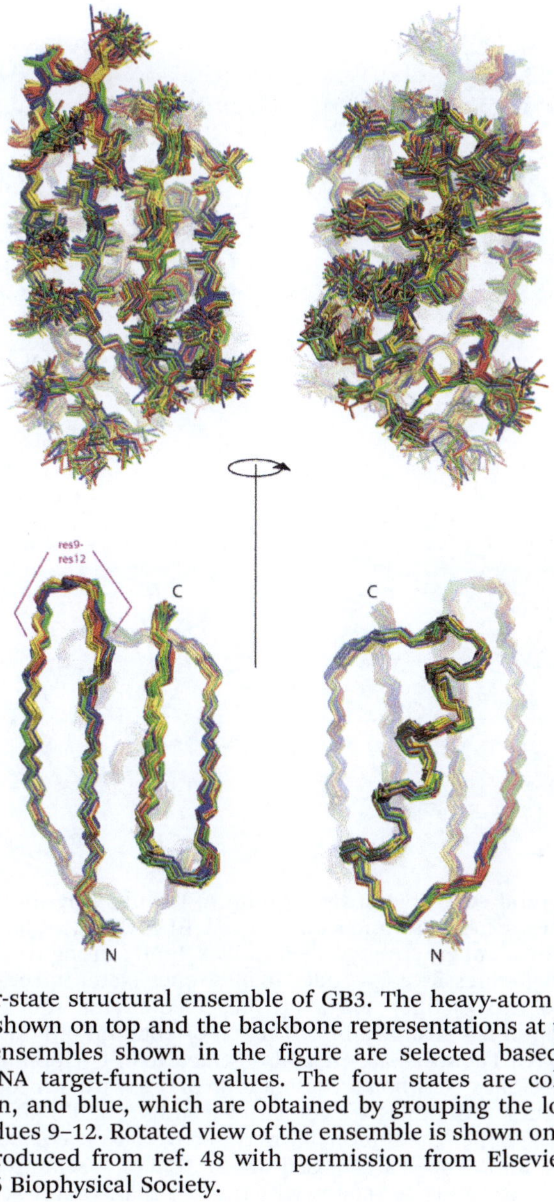

Figure 2.7 Four-state structural ensemble of GB3. The heavy-atom representations are shown on top and the backbone representations at the bottom. The 20 ensembles shown in the figure are selected based on the lowest CYANA target-function values. The four states are colored gold, red, green, and blue, which are obtained by grouping the loop constituting residues 9–12. Rotated view of the ensemble is shown on the right panel. Reproduced from ref. 48 with permission from Elsevier, Copyright © 2016 Biophysical Society.

ligand, while after catalysis the loop and with 70% of the side chains close and thereby releases the ligand. These findings were further corroborated with mutagenesis at the loop yielding, for example, mainly an open state that binds well the ligand without catalysis, or increasing the catalytic rate of Cyclophilin A by accelerated dynamics of the loop.[86] The data are in line with relaxation measurements,[80,84,87] while other interpretations appear to be also feasible.[81] Taken together, the two-state ensemble of Cyclophilin A

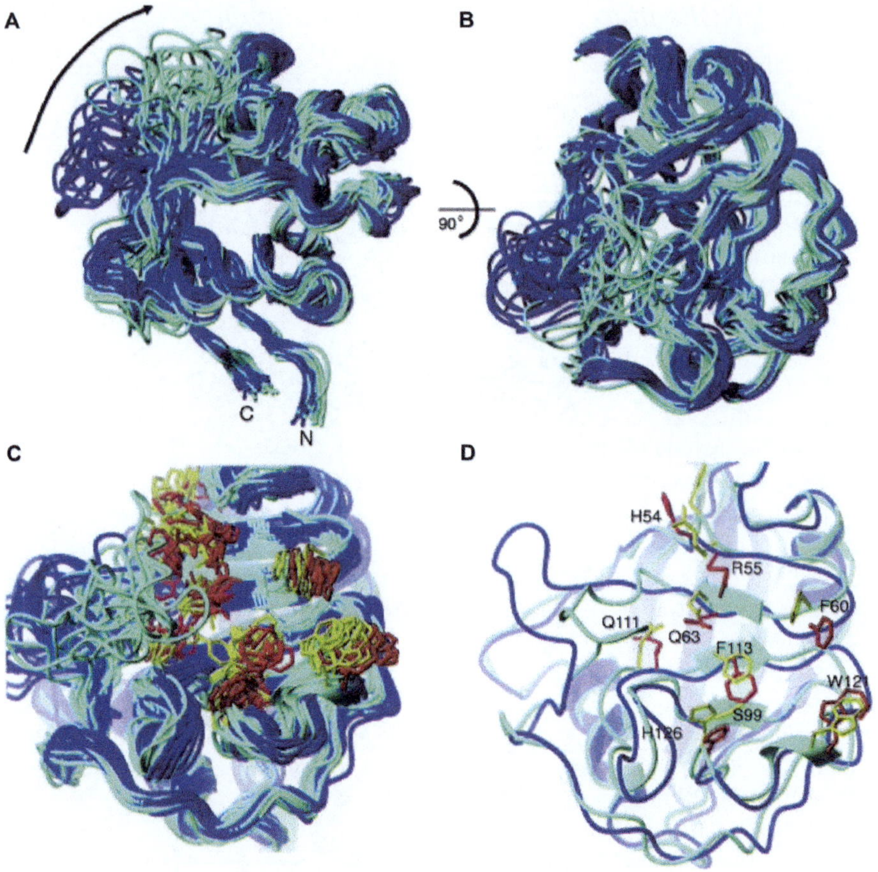

Figure 2.8 Structural ensemble of Cyclophilin A, highlighting the presence of two distinct states in its Apo form. (A) and (B) Backbone representation of the 20 lowest energy structural ensembles, representing two different states. The two states are color-coded in blue (open state) and cyan (closed state). These two distinct states are observed across the whole structure. Cyclophilin A is shown in a different orientation in B, with a 90° rotated view. (C) The two-state ensemble structure of Cyclophilin A with active-site residues. A ribbon representation of the two distinct open and closed structural ensembles are color coded in blue and cyan, respectively, for the backbone. The side chains of the active site are color coded in red and yellow for the open and closed states. (D) A representative conformer, for both closed and open states, with the side chains at the active site labeled. The lowest energy two-state conformers from both states are shown. Reproduced from ref. 85 with permission from John Wiley and Sons, Copyright © 2015 WILEY-VCH Verlag GmbH & Co. KGaA, Weinheim.

revealed a long-range conformational interchange between two states in the apo form of the protein that cover *circa* two-thirds of all the residues of the protein, highlighting the complex interplay present in biomolecules important for their action.

2.2.3.4 eNOEs in a 360-kDa Proteasome Complex

As described in the previous sections we have shown the applicability of eNOE-based multistate ensemble determination for various well-behaving proteins. Currently we are investigating several other larger proteins (up to 185 residues/20 kDa, unpublished data) following the established protocol. To further expand the applications of our eNOE method we investigated also the 360-kDa 2×7-mer half proteasome from *Thermoplasma acidophilium.*[88] We measured eNOEs in the 360 kDa 2×7-mer half proteasome. However, for larger proteins due to the overlap of NOESY diagonal peaks it is challenging to extract valuable eNOEs.[55] Another limiting factor is the spin diffusion for molecules with very large rotational correlation time. Our previous simulation study showed that spin diffusion between the methyl groups of the same Leu and Val residues can be an order of magnitude larger than the direct NOE transfer, and hence it is difficult to accurately correct spin-diffusion to derive quantitative eNOE distances.[89] To this end, we opted for the perdeuterated and selectively methyl-labeled molecule to measure eNOEs, where the methyl groups of only Ile, Leu, and Val residues will be labeled as $^{13}CH_3$ (see Figure 2.9).[90,91] Using the full relaxation matrix approach we corrected spin-diffusion for a NOESY buildup series and extracted 18 bidirectional eNOEs, 14 unidirectional eNOEs if normalized to the spin of origin, and 29 additional unidirectional eNOEs if normalized to the spin of destination (see Figure 2.9). We compared the extracted eNOE distances to two high-resolution X-ray structures of the proteasome and found that several of the eNOE distances were more than 1 Å shorter or longer than the corresponding distances from X-ray structures, but that overall a rather good correlation is obtained.[88] It is noted that the source of the differences is unknown and can range from spin diffusion correction issues to crystal packing artifacts. More important is the demonstration that rather accurate distance restraints can be collected also for very large systems because of the nature of the NOE being a relaxation-based entity.

2.2.3.5 High-resolution RNA Structures from eNOEs

RNA is an important type of biomolecule involved in a plethora of biological activities. As in the case of proteins, understanding how RNA functions demands detailed knowledge of its structure and dynamics at atomic resolution. The conformational changes and functions of RNA rely mostly on dynamic motions that occur on the picoseconds to seconds timescale,[92] which include secondary-structure transitions during RNA–ligand binding,[93–96] isomerization that regulates RNA–protein and RNA–RNA interactions[97–102] and several other functions. These types of motions can be coupled to each other to result in an overall structural rearrangement.[103] Thus, an in-depth understanding of how RNA regulates its functions requires accurate 3D structural models as well as a comprehensive description of their dynamics.

Although NMR is an outstanding method to solve RNA structures,[104] the NMR-based structural biology of RNA has several challenges. This is mainly due to the severe signal overlap and low proton density. In addition, unlike proteins, RNA is composed of only four chemically unique building blocks, in contrast to 20 amino acids for proteins. Thus, it is often difficult to carry out resonance assignment and provide high-resolution RNA structures from NOESY data. However the implementation of isotope-labeling strategies significantly increased the size of RNA structures that can be investigated by NMR spectroscopy.[105–107] In addition, there have been several other advances that allowed NMR spectroscopy to study RNA in a detailed manner.[104]

Voegeli and coworkers recently investigated the feasibility and accuracy of extracting eNOE distances from RNA.[108] These authors applied the well-developed eNOE method to the RNA 14-mer UUCG tetraloop.[104,109,110] Despite the lack of chemical-shift diversity and NOE network, that makes the structural investigation problematic, they exploited the exceptional accuracy of eNOEs to derive distances within RNA molecules.[53,111]

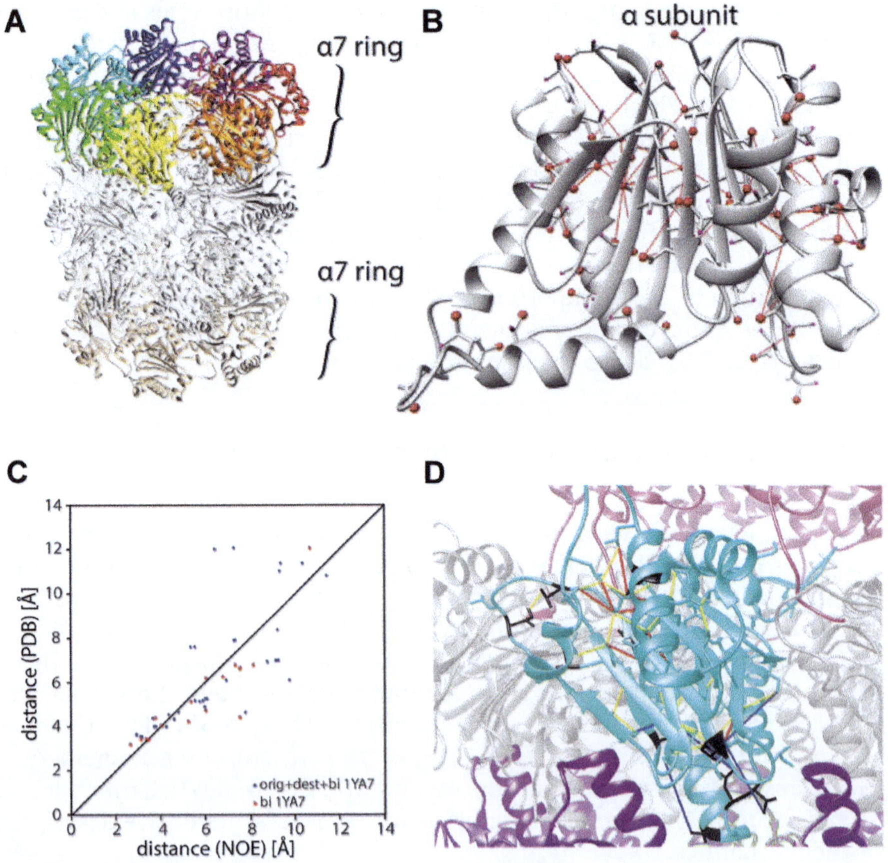

Initially, a series of 2D NOESY buildup series with various mixing times were measured. After correcting spin diffusion and excluding erroneous fits, 265 eNOEs and 88 gn-eNOEs were extracted (see Figure 2.10). The extracted eNOE distances were in agreement with those defined in the high-resolution NMR structure (PDB ID: 2KOC)[110] and the calculated ensemble structure with only eNOEs and gn-eNOEs was well defined with an RMSD of 0.44 Å.[108] The eNOE structure was significantly better than the conventional NMR structure calculated using semiquantitative NOE restraints. As in the case of the protein, a two-state structure calculation fulfilled the experimental restraints better than the single-state calculation (evaluated by the CYANA target function, see above). The obtained two-state ensemble shows minor base-pair dynamics and correlated motions within the c(UUCG)g tetraloop. These results indicate that the eNOE approach can also be used to define accurately single-state as well as multistate structures of well-behaving RNA molecules, opening an avenue for a comprehensive structure/dynamics activity relationship thereof.

Figure 2.9 eNOEs measured on a 360-kDa proteasome with the NOE-sensitive methyl groups on the α ring of 20S proteasome used for distance measurement. (A) Ribbon representation of the backbone of PDB 1PMA consisting of two heptameric α rings are colored and two heptameric β rings are shown in white. The seven α subunits in the uppermost α ring are distinguished by rainbow colors. (B) The side chains of residues Ile, Leu, and Val are plotted on an individual α subunit. The NOE-sensitive $^{1}H,^{13}C$-labeled methyl groups such as Ileδ1, Valγ2, and Leuδ2 are shown with red spheres at the carbon positions, and the NOE-insensitive $^{2}H,^{12}C$-labeled methyl groups are highlighted as sticks with pink shading. Methyl group pairs for which buildups can be measured are connected by red lines. (C) Distances extracted from the X-ray structure 1YA7 on the x-axis *versus* those from eNOEs on the y-axis. The intensities of the cross-peaks of unidirectional eNOEs were normalized with diagonal peak intensities of magnetization of origin or destination. eNOEs mentioned here were extracted from a series of NOESY buildup with mixing times of 30, 60, 90, and 120 ms. Corrections for spin diffusion were estimated with the full-relaxation matrix approach, and the resulting distances were scaled by 1.09 to account for incomplete deuteration. An overall rotational correlation time τ_c of 148 ns was estimated. (D) Clustering of deviations of experimental distances with respect to distances in X-ray structure. Distances which are at least 1 Å larger or smaller than those extracted from the X-ray structure 1YA7 are shown in red and blue respectively, whereas the distances within 1 Å are colored yellow. An α subunit A highlighted in cyan is used to map the distances and other α subunits in the same ring are highlighted in white. Black color code in α subunit A indicates those residues with methyl axis order parameters smaller than 0.33. The β ring and 11S activator are shown in purple and pink, respectively. For more details see detailed description in Chi, CN *et al.*[88] Reproduced from ref. 88 with permission from John Wiley and Sons, Copyright © 2018 Wiley-VCH Verlag GmbH & Co. KGaA, Weinheim.

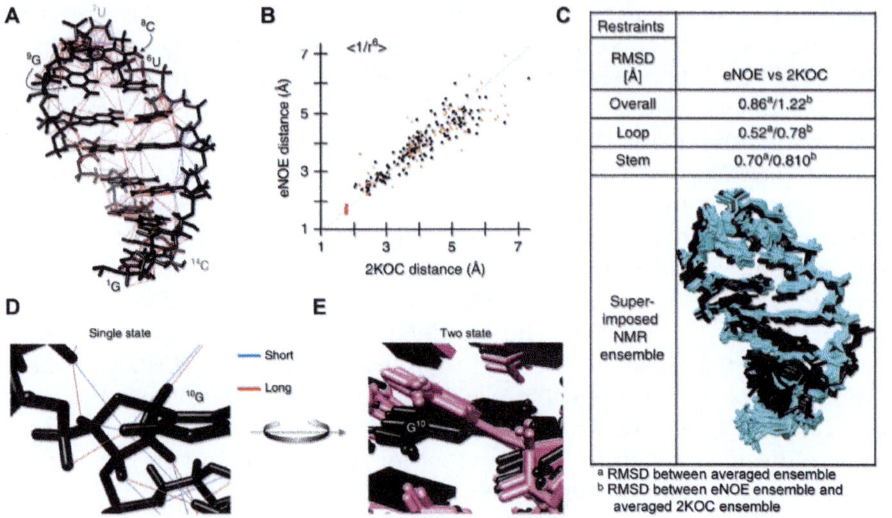

Figure 2.10 (A) The extracted 265 eNOEs (red) and 88 gn-eNOEs (blue) are mapped onto the tetraloop structure 2KOC. (B) Correlation plot between distances from 2KOC on the *x*-axis and eNOE effective distances from fits of 40–160 ms on the *y*-axis. Black circles correspond to distances between non-amino/non-methylene protons, orange diamonds correspond to distances between amino/methylene protons and non-amino/non-methylene or amino/methylene protons on a different residue, and red squares correspond to distances within amino/methylene spin pairs. (C) Superimposition of the eNOE structure (black) and 2KOC (cyan) where the RMSD values are for heavy atoms. (D) and (E) State separation enforced by distance violation in the single-state structure for G10. The distances that are too short or too long at least by 0.2 Å are shown in blue and red colors, respectively. The two-state ensemble structure is represented with five conformers for each state. Reproduced from ref. 108, https://doi.org/10.1038/s42003-018-0067-x, under the terms of the CC BY 4.0 license, http://creativecommons.org/licenses/by/4.0/.

2.3 Conclusion

In conclusion, we have presented the latest advances in the eNOE methodology and its diverse application to various biological systems. The eNOE approach is well implemented within the CYANA software package allowing for its straightforward use if a series of ^{15}N,^{13}C combined [^{1}H,^{1}H]-NOESY experiments are collected and analyzed. Further automation in both spectra analysis and structure interpretation for the elucidation of correlated motion is in progress in our group toward a fully automated multistate structure determination of well-behaving proteins. The very high accuracy and precision of the eNOE-derived distance restraints enables calculation of multistate ensembles that sample the conformational space of the biomolecule at atomic resolution. These multistate structures provide insights into the diverse mechanisms of biomolecules and highlight the multidimensional interplay between dynamics and structure.

References

1. H. Van Den Bedem and J. S. Fraser, *Nat. Methods*, 2015, **12**, 307–318.
2. R. Riek, J. Fiaux, E. B. Bertelsen, A. L. Horwich and K. Wuthrich, *J. Am. Chem. Soc.*, 2002, **124**, 12144–12153.
3. R. Riek, G. Wider, K. Pervushin and K. Wüthrich, *Proc. Natl. Acad. Sci. U. S. A.*, 1999, **96**, 4918–4923.
4. G. Wider, *Methods Enzymol.*, 2005, **394**, 382–398.
5. K. H. Gardner and L. E. Kay, *Annu. Rev. Biophys. Biomol. Struct.*, 1998, **27**, 357–406.
6. L. E. Kay, D. A. Torchia and A. Bax, *Biochemistry*, 1989, **28**, 8972–8979.
7. A. G. Palmer, 3rd and F. Massi, *Chem. Rev.*, 2006, **106**, 1700–1719.
8. Y. Xue, J. M. Ward, T. Yuwen, I. S. Podkorytov and N. R. Skrynnikov, *J. Am. Chem. Soc.*, 2012, **134**, 2555–2562.
9. D. E. Shaw, P. Maragakis, K. Lindorff-Larsen, S. Piana, R. O. Dror, M. P. Eastwood, J. A. Bank, J. M. Jumper, J. K. Salmon, Y. Shan and W. Wriggers, *Science*, 2010, **330**, 341–346.
10. K. Lindorff-Larsen, S. Piana, R. O. Dror and D. E. Shaw, *Science*, 2011, **334**, 517–520.
11. E. Papaleo, G. Saladino, M. Lambrughi, K. Lindorff-Larsen, F. L. Gervasio and R. Nussinov, *Chem. Rev.*, 2016, **116**, 6391–6423.
12. H. van den Bedem and J. S. Fraser, *Nat. Methods*, 2015, **12**, 307–318.
13. H. Kadavath, Y. Cabrales Fontela, M. Jaremko, Ł. Jaremko, K. Overkamp, J. Biernat, E. Mandelkow and M. Zweckstetter, *Angew. Chem. Int. Ed. Engl.*, 2018, **57**, 3246–3250.
14. H. Kadavath, R. V. Hofele, J. Biernat, S. Kumar, K. Tepper, H. Urlaub, E. Mandelkow and M. Zweckstetter, *Proc. Natl. Acad. Sci. U. S. A.*, 2015, **112**, 7501–7506.
15. R. B. Berlow, H. J. Dyson and P. E. Wright, *Nature*, 2017, **543**, 447–451.
16. A. B. Law, P. J. Sapienza, J. Zhang, X. Zuo and C. M. Petit, *J. Am. Chem. Soc.*, 2017, **139**, 3599–3602.
17. C. G. Kalodimos, *Ann. N. Y. Acad. Sci.*, 2012, **1260**, 81–86.
18. R. Selvaratnam, S. Chowdhury, B. VanSchouwen and G. Melacini, *Proc. Natl. Acad. Sci. U. S. A.*, 2011, **108**, 6133–6138.
19. R. Das, M. Abu-Abed and G. Melacini, *J. Am. Chem. Soc.*, 2006, **128**, 8406–8407.
20. G. Lipari and A. Szabo, *J. Am. Chem. Soc.*, 1982, **104**, 4546–4559.
21. J. Meiler, J. J. Prompers, W. Peti, C. Griesinger and R. Brüschweiler, *J. Am. Chem. Soc.*, 2001, **123**, 6098–6107.
22. K. Wüthrich, *NMR of Proteins and Nucleic Acids*, 1986.
23. A. Kumar, G. Wagner, R. R. Ernst and K. Wuthrich, *J. Am. Chem. Soc.*, 1981, **103**, 3654–3658.
24. M. Zweckstetter and A. Bax, *J. Am. Chem. Soc.*, 2001, **123**, 9490–9491.
25. M. Zweckstetter and A. Bax, *J. Biomol. NMR*, 2002, **23**, 127–137.
26. O. F. Lange, N. A. Lakomek, C. Farès, G. F. Schröder, K. F. A. Walter, S. Becker, J. Meiler, H. Grubmüller, C. Griesinger and B. L. De Groot, *Science*, 2008, **320**, 1471–1475.

27. N. A. Lakomek, T. Carlomagno, S. Becker, C. Griesinger and J. Meiler, *J. Biomol. NMR*, 2006, **34**, 101–115.
28. T. M. Sabo, C. A. Smith, D. Ban, A. Mazur, D. Lee and C. Griesinger, *J. Biomol. NMR*, 2014, **58**, 287–301.
29. P. Robustelli, K. Kohlhoff, A. Cavalli and M. Vendruscolo, *Structure*, 2010, **18**, 923–933.
30. C. Camilloni, A. Cavalli and M. Vendruscolo, *J. Phys. Chem. B*, 2013, **117**, 1838–1843.
31. V. Kocman, G. M. Di Mauro, G. Veglia and A. Ramamoorthy, *Solid State Nucl. Magn. Reson.*, 2019, **102**, 36–46.
32. L. Russo, M. Maestre-Martinez, S. Wolff, S. Becker and C. Griesinger, *J. Am. Chem. Soc.*, 2013, **135**, 17111–17120.
33. F. Peters, M. Maestre-Martinez, A. Leonov, L. Kovačič, S. Becker, R. Boelens and C. Griesinger, *J. Biomol. NMR*, 2011, **51**, 329–337.
34. D. Joss and D. Häussinger, *Prog. Nucl. Magn. Reson. Spectrosc.*, 2019, **114–115**, 284–312.
35. K. Lindorff-Larsen, R. B. Best, M. A. DePristo, C. M. Dobson and M. Vendruscolo, *Nature*, 2005, **433**, 128–132.
36. C. Mumenthaler, P. Güntert, W. Braun and K. Wüthrich, *J. Biomol. NMR*, 1997, **10**, 351–362.
37. T. Herrmann, P. Güntert and K. Wüthrich, *J. Biomol. NMR*, 2002, **24**, 171–189.
38. H. Kadavath, M. Jaremko, J. Jaremko, E. Biernat, Mandelkow and M. Zweckstetter, *Angew. Chem. Int. Ed. Engl.*, 2015, **54**, 10347–10351.
39. O. Jardetzky, *Biochim Biophys Acta*, 1980, **621**, 227–232.
40. O. F. Lange, N. A. Lakomek, C. Farès, G. F. Schröder, K. F. Walter, S. Becker, J. Meiler, H. Grubmüller, C. Griesinger and B. L. de Groot, *Science*, 2008, **320**, 1471–1475.
41. T. M. Sabo, V. Gapsys, K. F. A. Walter, R. B. Fenwick, S. Becker, X. Salvatella, B. L. de Groot, D. Lee and C. Griesinger, *Methods*, 2018, **138–139**, 85–92.
42. T. Carlomagno, W. Bermel and C. Griesinger, *J. Biomol. NMR*, 2003, **27**, 151–157.
43. B. Vogeli, *J. Biomol. NMR*, 2017, **67**, 211–232.
44. D. Ban, C. A. Smith, B. L. de Groot, C. Griesinger and D. Lee, *Arch. Biochem. Biophys.*, 2017, **628**, 81–91.
45. B. Vogeli, *Prog. Nucl. Magn. Reson. Spectrosc.*, 2014, **78**, 1–46.
46. B. Vogeli, P. Guntert and R. Riek, *Mol. Phys.*, 2013, **111**, 437–454.
47. B. Vogeli, S. Kazemi, P. Guntert and R. Riek, *Nat. Struct. Mol. Biol.*, 2012, **19**, 1053–U1110.
48. B. Vogeli, S. Olsson, P. Guntert and R. Riek, *Biophys. J.*, 2016, **110**, 113–126.
49. B. Vögeli, J. Orts, D. Strotz, C. Chi, M. Minges, M. A. Wälti, P. Güntert and R. Riek, *J. Magn. Reson.*, 2014, **241**, 53–59.
50. B. Vögeli, T. F. Segawa, D. Leitz, A. Sobol, A. Choutko, D. Trzesniak, W. van Gunsteren and R. Riek, *J. Am. Chem. Soc.*, 2009, **131**, 17215–17225.
51. I. Solomon, *Phys. Rev.*, 1955, **99**, 559–565.

52. A. Kalk and H. J. C. Berendsen, *J. Magn. Reson.*, 1976, **24**, 343–366.
53. P. J. Nichols, A. Born, M. A. Henen, D. Strotz, C. N. Celestine, P. Guntert and B. Vogeli, *ChemBioChem*, 2018, **19**, 1695–1701.
54. P. J. Nichols, A. Born, M. A. Henen, D. Strotz, J. Orts, S. Olsson, P. Guntert, C. N. Chi and B. Vogeli, *Molecules*, 2017, **22**, 1176.
55. C. N. Chi, D. Strotz, R. Riek and B. Vögeli, *J. Biomol. NMR*, 2015, **62**, 63–69.
56. F. Delaglio, S. Grzesiek, G. W. Vuister, G. Zhu, J. Pfeifer and A. Bax, *J. Biomol. NMR*, 1995, **6**, 277–293.
57. D. Neri, T. Szyperski, G. Otting, H. Senn and K. Wüthrich, *Biochemistry*, 1989, **28**, 7510–7516.
58. T. Szyperski, D. Neri, B. Leiting, G. Otting and K. Wüthrich, *J. Biomol. NMR*, 1992, **2**, 323–334.
59. D. Strotz, J. Orts, C. N. Chi, R. Riek and B. Vogeli, *J. Chem. Theory Comput.*, 2017, **13**, 4336–4346.
60. P. Guntert and L. Buchner, *J. Biomol. NMR*, 2015, **62**, 453–471.
61. P. Guntert, C. Mumenthaler and K. Wuthrich, *J. Mol. Biol.*, 1997, **273**, 283–298.
62. L. Buchner and P. Güntert, *J. Biomol. NMR*, 2015, **62**, 81–95.
63. P. Güntert, *Methods Mol. Biol.*, 2004, **278**, 353–378.
64. J. W. Keepers, *J. Magn. Reson.*, 1984, **57**, 404–426.
65. D. Strotz, J. Orts, M. Minges and B. Vogeli, *J. Magn. Reson.*, 2015, **259**, 32–46.
66. S. Olsson, D. Strotz, B. Vögeli, R. Riek and A. Cavalli, *Structure*, 2016, **24**, 1464–1475.
67. D. Strotz, J. Orts, H. Kadavath, M. Friedmann, D. Ghosh, S. Olsson, C. N. Chi, A. Pokharna, P. Güntert, B. Vögeli and R. Riek, *Angew. Chem., Int. Ed. Engl.*, 2020, **59**, 22132–22139.
68. S. L. Ma, L. Pastorino, X. Z. Zhou and K. P. Lu, *J. Biol. Chem.*, 2012, **287**, 6969–6973.
69. K. P. Lu, S. D. Hanes and T. Hunter, *Nature*, 1996, **380**, 544–547.
70. Y. S. Lim, H. T. L. Tran, S. J. Park, S. A. Yim and S. B. Hwang, *J. Virol.*, 2011, **85**, 8777–8788.
71. K. P. Lu, *Cancer Cell*, 2003, **4**, 175–180.
72. R. Ranganathan, K. P. Lu, T. Hunter and J. P. Noel, *Cell*, 1997, **89**, 875–886.
73. K. A. Wilson, J. J. Bouchard and J. W. Peng, *Biochemistry*, 2013, **52**, 6968–6981.
74. X. Wang, B. J. Mahoney, M. Zhang, J. S. Zintsmaster and J. W. Peng, *Structure*, 2015, **23**, 2224–2233.
75. J. W. Peng, *Biophys. Rev.*, 2015, **7**, 239–249.
76. J. W. Peng, B. D. Wilson and A. T. Namanja, *J. Biomol. NMR*, 2009, **45**, 171–183.
77. T. Peng, J. S. Zintsmaster, A. T. Namanja and J. W. Peng, *Nat. Struct. Mol. Biol.*, 2007, **14**, 325–331.
78. J. Monod, J. P. Changeux and F. Jacob, *J. Mol. Biol.*, 1963, **6**, 306.
79. L. D. Zydowsky, F. A. Etzkorn, H. Y. Chang, S. B. Ferguson, L. A. Stolz, S. I. Ho and C. T. Walsh, *Protein Sci.*, 1992, **1**, 1092–1099.

80. J. S. Fraser, M. W. Clarkson, S. C. Degnan, R. Erion, D. Kern and T. Alber, *Nature*, 2009, **462**, 669–673.
81. M. Wolf-Watz, V. Thai, K. Henzler-Wildman, G. Hadjipavlou, E. Z. Eisenmesser and D. Kern, *Nat. Struct. Mol. Biol.*, 2004, **11**, 945–949.
82. E. Z. Eisenmesser, O. Millet, W. Labeikovsky, D. M. Korzhnev, M. Wolf-Watz, D. A. Bosco, J. J. Skalicky, L. E. Kay and D. Kern, *Nature*, 2005, **438**, 117–121.
83. C. Camilloni, A. B. Sahakyan, M. J. Holliday, N. G. Isern, F. Zhang, E. Z. Eisenmesser and M. Vendruscolo, *Proc. Natl. Acad. Sci. U. S. A.*, 2014, **111**, 10203–10208.
84. K. Henzler-Wildman and D. Kern, *Nature*, 2007, **450**, 964–972.
85. C. N. Chi, B. Vogeli, S. Bibow, D. Strotz, J. Orts, P. Guntert and R. Riek, *Angew. Chem., Int. Ed. Engl.*, 2015, **54**, 11657–11661.
86. E. Z. Eisenmesser, D. A. Bosco, M. Akke and D. Kern, *Science*, 2002, **295**, 1520–1523.
87. S. J. Kerns, R. V. Agafonov, Y. J. Cho, F. Pontiggia, R. Otten, D. V. Pachov, S. Kutter, L. A. Phung, P. N. Murphy, V. Thai, T. Alber, M. F. Hagan and D. Kern, *Nat. Struct. Mol. Biol.*, 2015, **22**, 124–131.
88. C. N. Chi, D. Strotz, R. Riek and B. Vogeli, *Chem. – Eur. J.*, 2018, **24**, 2270–2276.
89. J. Orts, B. Vogeli and R. Riek, *J. Chem. Theory Comput.*, 2012, **8**, 3483–3492.
90. R. Kerfah, M. J. Plevin, R. Sounier, P. Gans and J. Boisbouvier, *Curr. Opin. Struct. Biol.*, 2015, **32**, 113–122.
91. R. Sounier, L. Blanchard, Z. R. Wu and J. Boisbouvier, *J. Am. Chem. Soc.*, 2007, **129**, 472–473.
92. A. M. Mustoe, C. L. Brooks and H. M. Al-Hashimi, in *Annual Review of Biochemistry, Vol. 83*, ed. R. D. Kornberg, Annual Reviews, Palo Alto, 2014, vol. 83, pp. 441–466.
93. R. R. Breaker, *Cold Spring Harbor Perspect. Biol.*, 2012, **4**, 15.
94. A. Haller, M. F. Souliere and R. Micura, *Accounts Chem Res*, 2011, **44**, 1339–1348.
95. J. Kortmann, S. Sczodrok, J. Rinnenthal, H. Schwalbe and F. Narberhaus, *Nucl. Acids Res.*, 2011, **39**, 2855–2868.
96. W. Winkler, A. Nahvi and R. R. Breaker, *Nature*, 2002, **419**, 952–956.
97. G. L. Conn, D. E. Draper, E. E. Lattman and A. G. Gittis, *Science*, 1999, **284**, 1171–1174.
98. E. A. Dethoff, K. Petzold, J. Chugh, A. Casiano-Negroni and H. M. Al-Hashimi, *Nature*, 2012, **491**, 724–728.
99. A. M. Mustoe, M. H. Bailor, R. M. Teixeira, C. L. Brooks and H. M. Al-Hashimi, *Nucleic Acids Res.*, 2012, **40**, 892–904.
100. N. Shankar, T. Xia, S. D. Kennedy, T. R. Krugh, D. H. Mathews and D. H. Turner, *Biochemistry*, 2007, **46**, 12665–12678.
101. S. A. Woodson, *RNA Biol.*, 2010, **7**, 677–686.
102. E. Jankowsky, *Trends Biochem. Sci.*, 2011, **36**, 19–29.
103. M. Wu and I. Tinoco, *Proc. Natl. Acad. Sci. U. S. A.*, 1998, **95**, 11555–11560.

104. F. H. Allain and G. Varani, *J. Mol. Biol.*, 1997, **267**, 338–351.
105. L. J. Alvarado, R. M. LeBlanc, A. P. Longhini, S. C. Keane, N. Jain, Z. F. Yildiz, B. S. Tolbert, V. M. D'Souza, M. F. Summers, C. Kreutz and T. K. Dayie, *ChemBioChem*, 2014, **15**, 1573–1577.
106. S. C. Keane, X. Heng, K. Lu, S. Kharytonchyk, V. Ramakrishnan, G. Carter, S. Barton, A. Hosic, A. Florwick, J. Santos, N. C. Bolden, S. McCowin, D. A. Case, B. A. Johnson, M. Salemi, A. Telesnitsky and M. F. Summers, *Science*, 2015, **348**, 917–921.
107. K. Lu, Y. Miyazaki and M. F. Summers, *J. Biomol. NMR*, 2010, **46**, 113–125.
108. P. J. Nichols, M. A. Henen, A. Born, D. Strotz, P. Güntert and B. Vögeli, *Commun Biol*, 2018, **1**, 61.
109. C. Cheong, G. Varani and I. Tinoco, Jr, *Nature*, 1990, **346**, 680–682.
110. S. Nozinovic, B. Fürtig, H. R. Jonker, C. Richter and H. Schwalbe, *Nucleic Acids Res.*, 2010, **38**, 683–694.
111. B. Furtig, C. Richter, J. Wohnert and H. Schwalbe, *ChemBioChem*, 2003, **4**, 936–962.

CHAPTER 3

Protein Backbone and Side-chain ^{15}N Spin Relaxation Techniques to Study Biomacromolecular Interactions

V. KHARCHENKO,[a] A. EJCHART[b] AND L. JAREMKO*[a]

[a] Division of Biological and Environmental Sciences and Engineering (BESE), King Abdullah University of Science and Technology (KAUST), Thuwal 23955-6900, Saudi Arabia; [b] Institute of Biochemistry and Biophysics, Polish Academy of Sciences, Pawinskiego 5A 02-106, Warsaw, Poland
*Email: lukasz.jaremko@kaust.edu.sa

3.1 Molecular Dynamics: Time Scales of Motions in Proteins

In complex molecular systems like proteins, structure, function, and dynamics are tightly interconnected. At present, it is widely accepted that intramolecular motions in proteins are one of the key factors determining their biological activity. Protein functioning includes interactions with ions, ligands, other proteins and nucleic acids in the course of such processes as folding, binding, allosteric regulation, recognition, signalling, and catalysis.[1–3]

Dynamic processes in proteins cover a broad range of timescales, usually quantitatively described by correlation times (τ_c) or their reciprocals – diffusion constants (D). Segmental motions and domain movements involve more extensive amplitude changes of wide-reaching regions of a protein

New Developments in NMR No. 26
NMR Spectroscopy for Probing Functional Dynamics at Biological Interfaces
Edited by Anirban Bhunia, Hanudatta S. Atreya and Neeraj Sinha

Published by the Royal Society of Chemistry, www.rsc.org

molecule. These motions falling into the 10^{-9} s$<\tau_c<10^3$ s regime can be studied using such techniques as equilibration, magnetization transfer (ZZ exchange), line-shape analysis and nuclear magnetic relaxation in the rotating and laboratory frames, from the slowest to the fastest process, respectively. Local backbone motions and side-chain reorientations (10^{-12} s$<\tau_c<10^{-9}$ s) are usually studied through relaxation in the laboratory frame. Finally, bond vibrations ($\tau_c<10^{-12}$ s) are too fast to be quantitatively probed using NMR techniques, but they do influence the values of molecular parameters, *e.g.*, dipolar coupling constants that play an essential role in the nuclear magnetic relaxation (Figure 3.1). Besides the intramolecular motions, the rotational diffusion of a whole molecule (10^{-9} s$<\tau_c<10^{-7}$ s) has to be separated from the superposed internal motions.[1,5,6]

Usually, NMR experiments are initiated from the equilibrium state. All populations of the spin system's energy levels are determined by the Boltzmann distribution (resulting in macroscopic magnetization aligned with the external magnetic field, B_0). After execution of any multiple pulse sequence generating a non-equilibrium state of the spin system (generally in which macroscopic magnetization is not collinear with the external magnetic field, B_0) the equilibrium state is finally restored. The process by which a spin system returns to the equilibrium is called nuclear magnetic relaxation. The physical phenomenon behind nuclear magnetic relaxation is a coupling of the spin system to the surrounding, often called the lattice, *via* time-dependent local magnetic fields. The lattice includes a continuum of motional degrees of freedom, creating a broad spectrum of fluctuating magnetic fields. Those that fluctuate at frequencies characteristic of the analysed spin system allow the dissipation of excess energy of the spin system absorbed during NMR experiments, hence driving it toward equilibrium.[7–10] Therefore, nuclear magnetic relaxation enables sampling of molecular motions at particular frequencies.

The first reports of relaxation measurements in proteins were published in the late 1980s.[11,12] Since then, a significant and continually increasing number of original studies on the subject, which are summarized in many reviews, have been published.[13–15]

To interpret experimental relaxation data in terms of molecular mobility, three complex tasks need to be completed. First, owing to the relatively low sensitivity of NMR spectroscopy, isotopic labelling is usually required. At present, hopefully, many efficient methods of protein isotope labelling are available.[16] Next, *trustworthy* experimental techniques allowing measurements of relaxation parameters and reliable data reduction methods have to be used. Finally, based on a realistic model of motion(s), theoretically reproduced values of experimental relaxation parameters should result in the model parameters describing molecular dynamics. The flow chart shown in Figure 3.2 presents a routine linking of the procedures used for the relaxation measurements and their interpretation to elucidate the intramolecular dynamics.

In the following text, attention is focused on the ^{15}N relaxation studies of protein mobility because spins of ^{15}N nuclei have been most often observed to monitor protein dynamics. Moreover, an efficient ^{15}N isotopic enrichment of

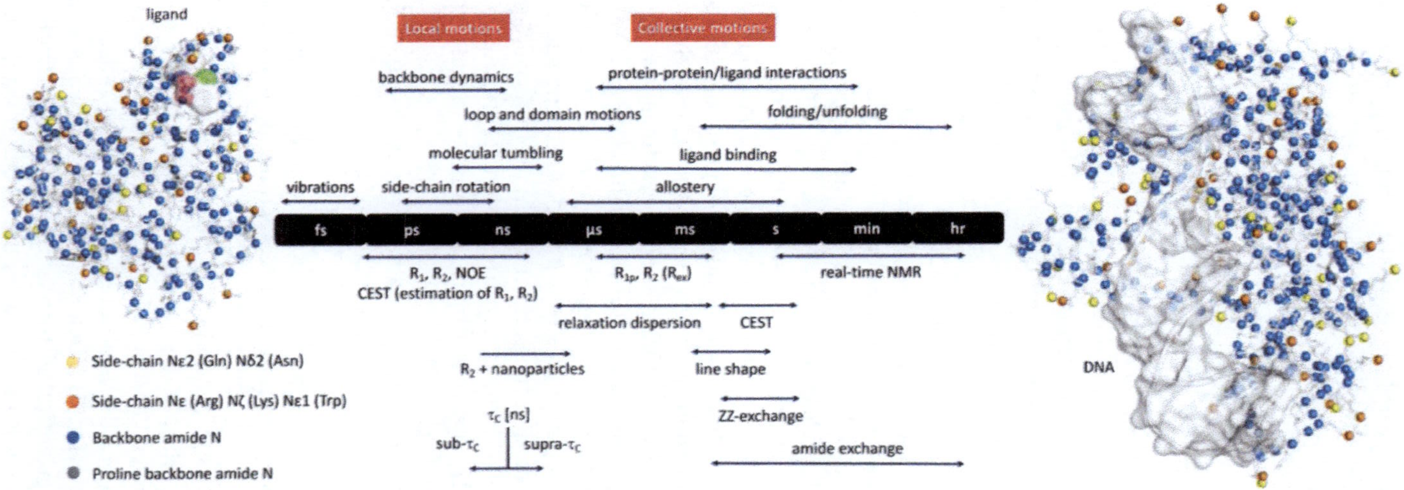

Figure 3.1 The biological motions in proteins and appropriate NMR experiments designed to probe particular timescales. Two example protein complexes comprise ubiquitin E3 ligase of RING1b/BMI1 in complex with inhibitor[4] (PDB id: 7ND1) and FEN1 endonuclease complexed with DNA substrate (PDB id: 3Q8L). Nitrogen atoms accessible *via* proton-detected NMR experiments for ^{15}N spin relaxation measurements are marked blue for main-chain, orange for side-chains of arginine and lysine and yellow for glutamine and asparagine.

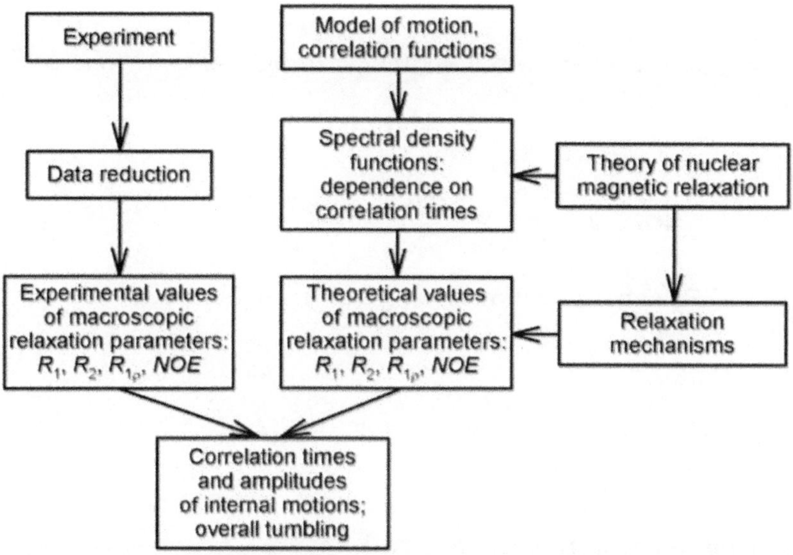

Figure 3.2 Scheme of determination of dynamics parameters in molecules from nuclear magnetic relaxation measurements.

proteins has been worked out, and methods for the determination of ^{15}N relaxation parameters are well established, delivering a wealth of information. It is important to note that uniform ^{15}N isotope labelling of proteins enables observation of the backbone amide ^{15}N nuclei and the side-chain ^{15}N resonances in several amino acid residues (Trp, Asn, Gln, Arg, His and Lys).

3.2 NMR Available Observables Dependent on Molecular Motions: Relaxation Rates

The most frequently measured protein relaxation parameters are the ^1H-detected longitudinal, R_1, and transverse, R_2, relaxation rates of ^{15}N nuclei in the laboratory frame complemented with $^{15}N\{^1H\}$ NOEs. The R_2 relaxation rates are replaced or sometimes supplemented with $R_{1\rho}$, relaxation rates in the rotating frame.[14,15]

Two basic relaxation mechanisms taking part in the relaxation of ^{15}N nuclei in protein backbone amide groups are dipolar interaction with amide ^1H spin (DD) and relaxation owing to the ^{15}N chemical shift anisotropy (CSA).[17] Additionally, the conformational exchange (EX) can contribute to the relaxation rates R_2 and $R_{1\rho}$.[18,19]

Equations describing relaxation rates of ^{15}N nuclei relaxing by dipolar and chemical shift anisotropy mechanisms in terms of spectral density functions, $J(\omega)$ at angular frequency ω, are given as:[7,20]

$$R_1 = \frac{D^2}{4}[J(\omega_H - \omega_N) + 3J(\omega_N) + 6J(\omega_H + \omega_N)] + \frac{C^2}{3}J(\omega_N) \tag{3.1}$$

$$R_2 = R_2^0 + R_{ex}^l$$
$$= \frac{D^2}{8}[4J(0) + J(\omega_H - \omega_N) + 3J(\omega_N) + 6J(\omega_H) + 6J(\omega_H + \omega_N)] \quad (3.2)$$
$$+ \frac{C^2}{18}[4J(0) + 3J(\omega_N)] + R_{ex}^l$$

$$\text{NOE} = 1 + \frac{D^2}{10}\left(\frac{\gamma_H}{\gamma_N}\right)\frac{[6J(\omega_H + \omega_N) - J(\omega_H - \omega_N)]}{R_1} \quad (3.3)$$

The $^{15}N/^{1}H$ cross-relaxation rate, σ_{NH}, is determined from $^{15}N\{^{1}H\}$ NOE[20,21]

$$\sigma_{NH} = (\text{NOE} - 1)R_1\frac{\gamma_N}{\gamma_H} \quad (3.4)$$

γ_H and γ_N are appropriate magnetogyric ratios; note that $\gamma_H > 0$ and $\gamma_N < 0$ and therefore, $\omega_H + \omega_N < \omega_H - \omega_N$. Amplitudes of DD and CSA relaxation mechanisms, D and C, respectively, are defined as: $D = -\frac{\mu_0}{4\pi}\gamma_H\gamma_N\hbar\langle r_{NH}^{-3}\rangle$ and $C = \gamma_N B_0 \Delta\sigma\sqrt{1 + \eta^2/3}$. Besides the fundamental physical constants: μ_0 and $\hbar$, there are also local, site-specific physical parameters: vibrationally averaged N–H distance $\langle r_{NH}^{-3}\rangle^{-1/3}$, and the eigenvalues of ^{15}N shielding tensor appearing as chemical shift anisotropy, $\Delta\sigma = \sigma_{zz} - (\sigma_{xx} + \sigma_{yy})/2$, and chemical shift asymmetry, $\eta = 1.5(\sigma_{xx} - \sigma_{yy})/\Delta\sigma$. It is usually assumed that backbone amide groups are characterized by $r_{NH} = 0.104$ nm,[22] $\Delta\sigma = -170$ ppm and $\eta = 0$.[23] The respective values of r_{NH} and $\Delta\sigma$ vary for the nitrogen-containing side-chains and are summarized in Table 3.1.

Table 3.1 Chemical shift anisotropy (CSA) of ^{15}N nuclei and r_{NH} distances used in the calculation of backbone and side-chain mobilities.

Residue, position	CSA (ppm)	References	r_{NH} (nm)	References
Asn, $N_{\delta 2}$	−163	24	0.103; 0.105[ab]	25
Gln, $N_{\varepsilon 2}$			0.102; 0.103[ab]	26
Arg, N_{ε}	−114	27	0.105[b]	28
His, $N_{\delta 1}$	−210	29	0.109[b]	30
Lys, $N_{\zeta}H_3^+$	15	31	0.104	32
Trp, $N_{\varepsilon 1}$	−122	33	0.102	34
Backbone N	(−169) - (−144)	35	0.104	22
	(−174) - (−177)	36	0.102 (obsolete)	
	(−148) - (−187)	37		
	(mean −166)	38		
	(−129) – (−213)			
	(mean −173)			

[a]NH interatomic distances are not equivalent; in the paper by J. Boyd[142] both NH distances in Asn and Gln were set to 0.102 nm.

[b]Data obtained from neutron diffraction were vibrational averaging corrected specifically for NMR relaxation analyses by a factor of 1.02.

R_2^0 represents relaxation originating from non-exchange mechanisms DD and CSA. The term R_{ex}^l considers the conformational exchange (EX) contribution to R_2, transverse relaxation rates in the laboratory frame, resulting from processes in the micro- to millisecond time scale often referred to as chemical exchange effects.[18,20] Such processes, slower than the molecular tumbling, but fast enough to average chemical shifts, can influence transverse relaxation rates determined using the CPMG method utilizing a train of elements [$0.5\tau_{cp}$–180°(^{15}N)–$0.5\tau_{cp}$] applied to transverse magnetization.[17,20,39] The R_{ex}^l contribution to the transverse relaxation rate is proportional to the squares of the chemical shift difference between exchanging states, $\Delta\delta$, and to ω_N, the Larmor frequency. It should be pointed out that the conformational exchange mechanism can affect the apparent transverse relaxation rate only if $\Delta\delta \neq 0$. This term can be written as $R_{ex}^l = \Phi_{\omega N}^2$.[40] The proportionality factor Φ represents the effectiveness of the conformational exchange processes, and is independent of magnetic field strength. The field-independence of Φ facilitates the direct and qualitative comparison of chemical exchange terms determined at different magnetic field strengths among other proteins.

Measurements of the relaxation rates in the presence of the transverse spin-lock radio-frequency field, applied with the carrier frequency set at ω_{SL} in the vicinity of spin ^{15}N resonances, ω, provide additional information about conformational motions. The experimentally attainable relaxation rate, $R_{1\rho}^{off}$, is:[14,20,41,42]

$$R_{1\rho}^{off} = R_1\cos^2\theta + R_{1\rho}^{on}\sin^2\theta + R_1\cos^2\theta + (R_2^0 + R_{ex}^{\rho})\sin^2\theta \tag{3.5}$$

With spin-lock field amplitude ω_1 and offset $\Delta = \omega - \omega_{SL}$, the effective field has the tilt angle $\theta = \arctan(\omega_1/\Delta)$. The measured value of $R_{1\rho}^{off}$ depends on the orientation and magnitude of the effective field, which may be changed by the spin-lock amplitude ω_1, its offset Δ, or both. An example of the $R_{1\rho}^{off}$ relaxation dispersion is shown in Figure 3.3. The term R_{ex}^{ρ} stands for the conformational exchange contribution to $R_{1\rho}^{on}$, which differs from the term R_{ex}^l in sensitivity and scale of chemical exchange effects. Complex expressions describing R_{ex}^l and R_{ex}^{ρ} at different approximation levels as functions of exchange process parameters can be found in the literature.[1,14,20,43–46]

In the fast exchange limit between two sites A and B with difference in Larmor angular frequencies $\Delta\omega_{AB}$, $\tau_{ex} \ll 1/|\Delta\omega_{AB}|$, the term R_{ex}^l can be approximated by[43,47]

$$R_{ex}^l = K\tau_{ex}\left(1 - \frac{2\tau_{ex}}{\tau_{cp}}\tanh\frac{\tau_{cp}}{2\tau_{ex}}\right) \tag{3.6}$$

where $K = p_A p_B\, \Delta\omega_{AB}^2$.

A much more general, complex expression describing the R_{ex}^l term, that comprises all conformational time scales was introduced by Jen[48] and discussed in other several papers.[1,14,43,49]

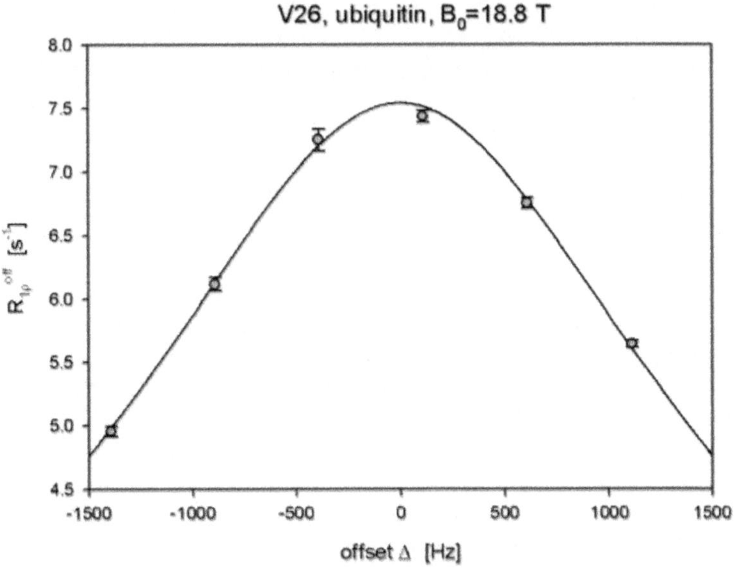

Figure 3.3 Experimental $R_{1\rho}^{\text{off}}$ determined for V26 residue of ubiquitin at 18.8 T for several offset values, Δ. The solid line represents the plot obtained for $R_{1\rho}^{\text{on}} = 7.54\ \text{s}^{-1}$ and $\omega_1/2\pi = 1582$ Hz determined from least squares fit. $R_1 = 1.63\ \text{s}^{-1}$ was determined in a separate measurement.

Two-site, fast conformational exchange in the presence of an off-resonance spin-lock field is given by[45]

$$R_{\text{ex}}^{\rho} = K \frac{\tau_{\text{ex}}}{1 + \omega_{\text{e}}^2 \tau_{\text{ex}}^2} \tag{3.7}$$

where $\omega_{\text{e}} = \sqrt{\omega_1^2 + \Delta^2}$ is the strength of the effective rotating-frame spin-lock field and its offset Δ is calculated from the population-weighted average chemical shift, $\varpi = p_{\text{A}}\omega_{\text{A}} + p_{\text{B}}\omega_{\text{B}}$.

Comparison of the exchange terms R_{ex}^{1} and R_{ex}^{ρ} calculated using experimental settings commonly applied in ^{15}N relaxation measurements is shown in Figure 3.4.

An expression describing relaxation in the rotating-frame outside of the fast-exchange limit allows more precise defining of the limit of fast exchange as $\tau_{\text{ex}}|\Delta\omega_{\text{AB}}| < 4$ and reveals that for strongly asymmetric population distribution, $p_{\text{A}} \gg p_{\text{B}}$, the maximum of R_{ex}^{ρ} moves from ϖ to ω_{B}.[46] Further improved accuracy of the R_{ex}^{ρ} description was obtained by introducing correction terms under certain limiting conditions.[41] The same problem was discussed, *i.e.*, releasing subsequent limitations as the relation between relaxation rates R_1 and R_2[50] or spin-lock amplitude.[51]

CEST (Chemical Exchange Saturation Transfer) combines the well-established method of magnetization transfer applied to slowly exchanging systems[52,53] with a selective, low-amplitude spin-lock field.[54,55]

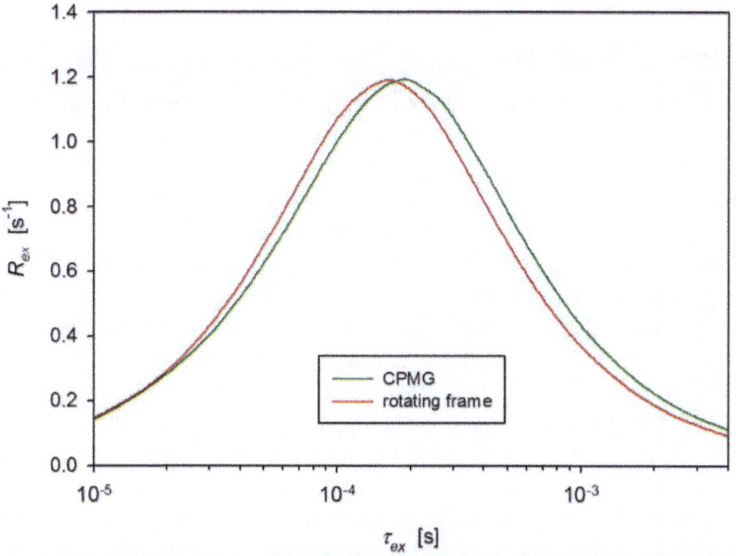

Figure 3.4 Both exchange terms were calculated assuming $K = 1.5 \times 10^4$ s^{-2}; in CPMG sequence $\tau_{cp} = 0.6$ ms, while spin-lock field with amplitude $\omega_1 = 2\pi 1000$ rad s^{-1} was on-resonance ($\Delta = 0$).

Rather than determining a profile of $R_{1\rho}$ values only one evolution time of relaxation rates is chosen, which strongly differentiates observed intensities. Such an approach apparently allows optimization of the experimental time, although one needs to densely sample the nitrogen frequencies to capture the profile with the main and the exchanging state signal extrema.

As outlined above, the ^{15}N spin relaxation techniques are predominantly used in non-modified protein solutions to probe close to native pico- to nanosecond time scale dynamics of sub-τ_C motions and detect the exchange contribution depicting slow typically μs–ms segmental motions. The sensitivity of classical longitudinal and transverse ^{15}N relaxation rates to depict high sub-τ_C motions approaching the molecular tumbling limit (*i.e.*, the τ_C itself of the system under the studied conditions) and supra-τ_C nanosecond time scale motions increases with both low[56] and ultra-high magnetic fields.[57] Recently, the ^{15}N spin relaxation measurements assisted with nanoparticles facilitated the sub-τ_C and supra-τ_C motions detection by enhancing the transverse relaxation in the presence of nanoparticles, thus shifting up the molecular tumbling τ_C limit.[58] This way the previously inaccessible intermediate nanosecond to microsecond range exceeding the molecular tumbling, the τ_C limit, became accessible to canonical ^{15}N spin relaxation rates. As follow up indicated, the protein nanoparticle interactions have marginal or negligible effects on the resulting S^2 order parameters derived from experimental ^{15}N relaxation rates.[59]

It is worth mentioning that a fair estimate of ^{15}N transverse and longitudinal rates reporting on ps–ns internal motions and molecular tumbling can be derived from ^{15}N relaxation CEST measurements.[60]

3.3 The Protein Backbone and Side-chains Accessible *via* ^{15}N Spins

There are 20 proteinogenic α-amino acid building blocks of proteins that are routinely studied by modern high-resolution structural biology techniques. Only proline, while inserted into the polypeptide chain, forms a secondary amide missing the amide proton. Proline accounts for roughly 4.8% of the amino acid composition in proteins,[61] which is not detectable, with the N-terminus $-NH_2/-NH_3^+$ moiety usually broadened beyond the detection under the regularly applied measurement conditions. Therefore, up to 95% of protein backbone dynamics through the amide signals is trackable by well-established ^1H-detected ^{15}N spin-relaxation methods.[62,63] The side-chains containing $^{15}N-^1H$ spin systems constitute 23% of the amino acid occurring in proteins. They are located mainly at the protein surface, often within the catalytic centres, ligand binding pockets, and protein interaction domains. The polar glutamine and asparagine with ($-CO-NH_2$ moiety) make up to 8% of the amino acids in proteins, while the positively charged arginine and lysine are jointly 12%. While side-chain amides of glutamine and asparagine can be directly observed in 2D [^1H,^{15}N] HSQC type experiments, there are scarce reports on their ^{15}N-spin relaxation studies.[64,65] In contrast, the dynamic studies for arginine and lysine side-chains are well supported by ^1H-detected ^{15}N-spin relaxation techniques.[29,66–68] Histidine accounts for 2.3% overall, while tryptophan is around 1.4%, although some domains or proteins miss this amino acid. It is worth mentioning here that the amino acid composition of nitrogen-containing side-chain amino acids varies with the nature of the protein. A lower occurrence of histidine of 1.2% is reported among membrane proteins. The polar side-chains containing amino acids of asparagine and glutamine occur more often in the case of membrane and intrinsically disordered proteins, while positively charged residues are more common in proteins interacting with nucleic acids.[69] In the case of folded proteins with increasing size, the hydrophobic core grows faster than the surface, indicating the importance of the N–H spin system containing side-chains in molecular recognition and interactions.

3.4 Motional Models Commonly Used in Proteins: MFA/EMFA, SDM

Experimentally available relaxation parameters are related to the spectral density functions, $J(\omega)$, characterizing the motion(s) of the $^1H-^{15}N$ bond, or shielding anisotropy tensor, which impact DD or CSA relaxation mechanisms, respectively. Analytical description of spectral density functions is an intricate task because of the complexity of motions in biomolecules. Solely for the isotropic rotational diffusion of a rigid molecule, the spectral density function takes a simple form of a single Lorentzian function[7,70]

$$J(\omega) = \frac{\tau_R}{1 + \omega^2 \tau_R^2} \tag{3.8}$$

where τ_R is the rotational correlation time characterizing overall molecular tumbling and the associated diffusion constant is equal to $D = 1/6\tau_R$. Values of relaxation parameters R_1, R_2 and NOE calculated from this model are shown in Figure 3.5 for several τ_R values as a function of the magnetic field B_0. Longitudinal relaxation rates decrease with an increase of B_0 and rotational correlation time, while changes to transverse relaxation rates are the opposite. It is noteworthy that owing to experimental limitations the determination of R_2 values exceeding *ca.* 40 s^{-1} is unfeasible at present. The R_2 data of large, slowly tumbling molecules at high magnetic fields become unavailable. This deficiency can be removed, however, by replacing the R_2 measurements with the $R_{1\rho}^{\text{off}}$ ones, which can be scaled down with the amplitude and frequency of the spin-lock field.

More complex expressions have been derived for the anisotropic rotational diffusion of rigid molecules represented by either an axially symmetric diffusion tensor or a fully asymmetric one.[71] Then, spectral density functions consist of three or five Lorentzian terms, respectively. In the latter case

$$J(\omega) = \sum_{i=1}^{5} A_i \frac{\tau_i}{1 + \omega^2 \tau_i^2} \tag{3.9}$$

Normalized coefficients A_i ($\Sigma_i A_i = 1$) are functions of three Euler angles defining the orientation of the molecular coordinate frame with respect to the laboratory frame. Adding three principal values of anisotropic diffusion tensor, the description of anisotropic tumbling is given by six parameters.

The form of a spectral density function becomes more complex when internal motions have to be taken into account. In a relatively simple case of internal threefold rotation of a DD relaxation vector superposed on the uncorrelated axially symmetric rotational diffusion, the spectral density function is composed of nine Lorentzian terms.[71,72] Moreover, it should be pointed out that, in general, coefficients A_i of spectral density functions are different for DD and CSA mechanisms. Usually it is assumed that the shielding anisotropy tensor of amide ^{15}N nuclei in proteins is approximately axially symmetric and the orientation of its unique axis makes a small angle with the ^{1}H–^{15}N bond vector.[73] Provided that DD/CSA interference is suppressed during the evolution periods of relaxation measurements, the same analytic forms of spectral density functions can be used for the description of DD and CSA mechanisms.

Owing to the complexity of internal motions of the ^{1}H–^{15}N bonds in proteins, the analysis of their ^{15}N relaxation data is most often performed using a so-called model-free approach (MFA).[74] This model is based on two assumptions imposed on the form of the autocorrelation function: firstly, internal motions and overall molecular tumbling are uncorrelated, allowing the autocorrelation function to be factorized into two components describing the overall tumbling and the internal motion, and secondly, the internal correlation function can be approximated by a single exponential. In the framework of this model, local mobility is described by two parameters: the generalized

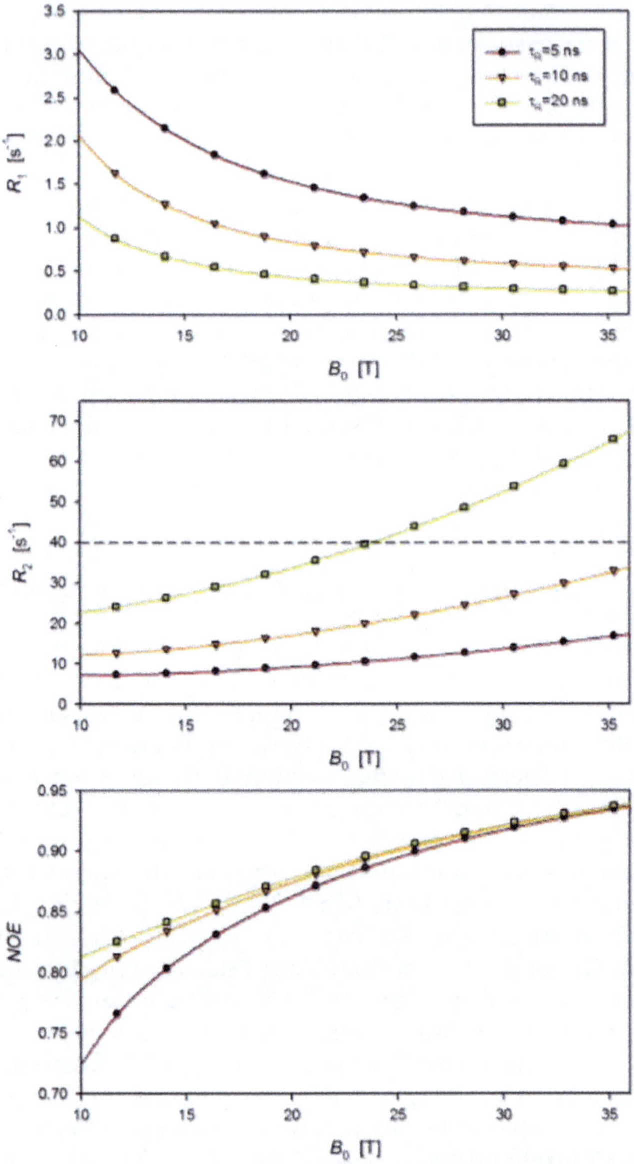

Figure 3.5 Relaxation parameters were calculated using a single Lorentzian spectral density function. Amplitudes of relaxation mechanisms, D and C, were obtained assuming $r_{NH} = 0.104$ nm and $\Delta\sigma = -170$ ppm. Overall isotropic correlation times are typical for proteins with molecular masses of 8 kDa (5 ns), 20 kDa (10 ns) and 35 kDa (20 ns) at room temperature. Symbols in figures mark the static magnetic field strengths at ^{1}H frequencies equal to 500, 600, ...,1500 MHz. The horizontal, dashed line in the central R_2 plot at the position of 40 s^{-1} shows a cut-off limit for the maximum R_2 value, which can be determined with acceptable accuracy.

order parameter, S^2, which corresponds to the spatial freedom of motion, and the internal correlation time, τ_{int}, corresponding to the rate of this motion in the 10^{-12}–10^{-9} s time scale, which is smaller than the single correlation time describing an isotropic overall molecular tumbling, τ_R. The resulting spectral density function is represented by two Lorentzian terms,

$$J^{\text{MFA}}(\omega) = \frac{2}{5}\left[\frac{S^2\tau_R}{1+\omega^2\tau_R^2} + \frac{(1-S^2)\tau}{1+\omega^2\tau^2}\right] \quad \text{where} \quad \frac{1}{\tau} = \frac{1}{\tau_R} + \frac{1}{\tau_{\text{int}}}. \tag{3.10}$$

The effect of internal motions on relaxation parameters is presented in Figure 3.6. Generally, relaxation parameters of residues exhibiting higher degrees of motional freedom (smaller values of generalized order parameter, S^2) tend to decrease.

It was quite quickly discovered that the model-free approach sometimes poorly reproduced the experimental data, especially ^{15}N{^{1}H} NOEs, which were determined together with the longitudinal, R_1, and transverse, R_2, relaxation rates of ^{15}N nuclei. The experimental NOEs tended to be larger than their values calculated based on either the relaxation rates solely or the whole set of available data. This was attributed to the additional slow internal motion outside the extreme narrowing limit pointing at motions being faster than the overall tumbling but slower than those obtained from the genuine MFA model. In order to better reproduce experimental relaxation data, the extended model-free approach (EMFA), taking into account a more complex description of internal motions characterized by two different time scales, in addition to the global motion, has been introduced at the cost of a larger number of model parameters.[75] The second motion was described by its own generalized order parameter, S_s^2, and the correlation time, $\tau_{\text{int,s}}$. In the case of isotropic overall reorientation, the spectral density comprises three Lorentzian terms:

$$J^{\text{EMFA}}(\omega) = \frac{2}{5}\left[\frac{S^2\tau_R}{1+\omega^2\tau_R^2} + \frac{(1-S_f^2)\tau_f}{1+\omega^2\tau_f^2} + \frac{S_f^2(1-S_s^2)\tau_s}{1+\omega^2\tau_s^2}\right] \quad \text{where} \quad \frac{1}{\tau_j} = \frac{1}{\tau_R} + \frac{1}{\tau_{\text{int},j}}. \tag{3.11}$$

Indices f and s correspond to the fast and slow internal motions, respectively. Please note that $S^2 = S_f^{\,2} S_s^{\,2}$. It has been shown, however, that EMFA is often unable to cope with three different time scales and fails to describe slow internal motions properly.[57]

Incorporation of the anisotropic rotational diffusion of rigid molecule[71] into spectral density functions results in more complicated formulae describing spectral density functions.[76–78] The most complex form of the $J^{\text{EMFA}}(\omega)$ function in the case of fully anisotropic tumbling comprises 15 Lorentzian terms.

An alternative to the model-free approach is direct determination of the spectral density function dependence on the frequency without using any particular model of motion. Three of the most commonly measured relaxation parameters, R_1, R_2 and NOE, depend on the spectral density functions at five angular frequencies: 0, ω_N, $\omega_H + \omega_N$, ω_H and $\omega_H - \omega_N$. Therefore, there are

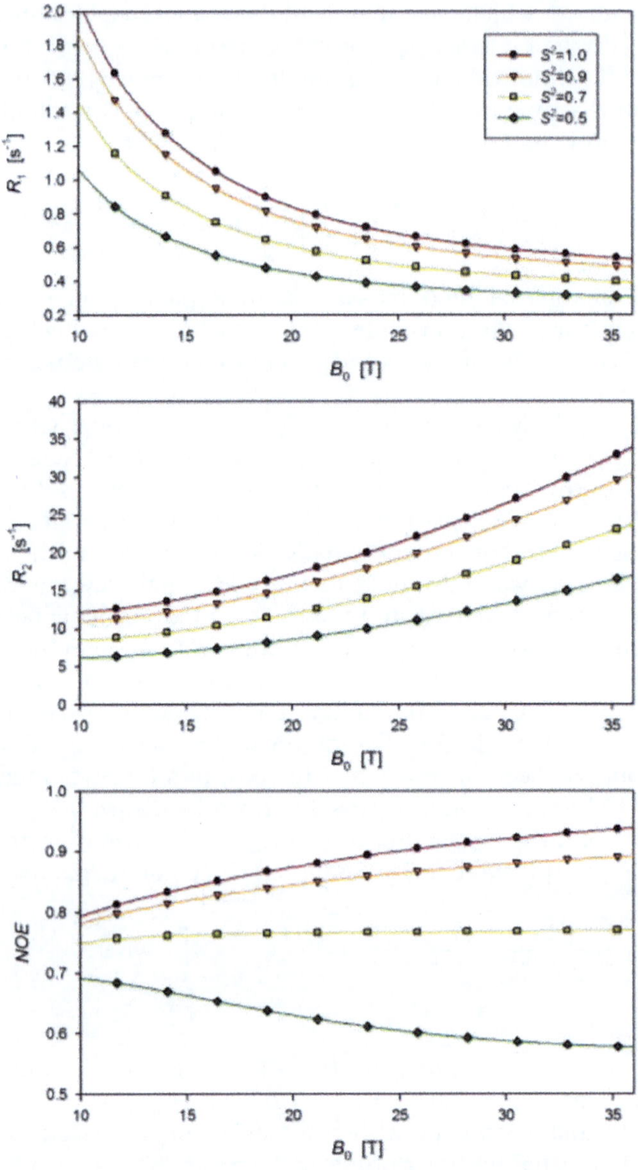

Figure 3.6 Relaxation parameters were calculated using a model-free approach (MFA) spectral density function comprising two Lorentzian terms. Overall, isotropic correlation times were assumed to be $\tau_R = 10$ ns. Increasing internal mobility with decreasing generalized order parameter S^2 and correlation time $\tau_{int} = 10$ ps display the effect of internal mobility on relaxation parameters.

insufficient experimental data for their determination. Peng and Wagner[79,80] have introduced experimental techniques allowing measurements of longitudinal two-spin order and antiphase transverse coherence relaxation rates which are described by spectral densities at the same frequencies as R_1, R_2 and NOE. This approach, called spectral density mapping (SDM), permits the calculation of spectral densities at the five frequencies and has proved to be sensitive to local intramolecular motions.[81] Extension of this method to the measurements at multiple magnetic field strengths results in accurate evaluation of spectral density functions.[82]

Despite its virtue, the SDM turned out to be experimentally cumbersome and very prone to experimental error propagation.[81] The observation that the spectral densities vary slowly at high frequencies around ω_H resulted in the modifications of genuine spectral density mapping known as the reduced spectral density mapping (rSDM) and, therefore, a number of simplifying approximations have been introduced. In one approach, each of the spectral density values $J(\omega_H+\omega_N)$, $J(\omega_H)$ and $J(\omega_H-\omega_N)$ is replaced with the averaged value $<J(\omega_H)>$,[81,82] whereas in a second approach, $J(\omega_H+\omega_N)$ is used as a replacement.[83,84] Another approach[85] relies on the replacement of the linear combinations of spectral densities with a single function according to several relations. Despite the fact that the rSDM approach does not require any knowledge concerning the analytical form of spectral density functions, the derived information about protein dynamics shows the same deficiency as the model-free approach – it requires physical interpretation. Moreover, it suffers from the lack of separation of overall and internal motions.

3.5 Determination of Model Parameters Based on Observables

Motional characterization of an anisotropically tumbling protein made up of N amino acid residues is fully described by $6+3N$ or $6+5N$ model parameters applying MFA or EMFA, respectively. On the other hand, a number of relaxation data measured at a single magnetic field can be equal to $3N$ data at most. Therefore, in order to fulfil the obvious requirement that the number of experimental data exceeds the number of model parameters, measurements at multiple magnetic fields become a necessity.

Determination of accurate information from multiple field experiments is possible only if the relaxation data share a high degree of consistency. Inconsistencies can arise from two groups of factors, sample variation (concentration, aggregation, degradation) or setup of the NMR spectrometer (temperature, handling with water magnetization, modifications of pulse sequences). Inconsistent datasets result in erroneous description of local motions and/or overall tumbling. A test of consistency based on the combination of experimental R_1, R_2 and NOE data, introduced by Morin and Gagné,[86] allows the detection of inaccurate measurements. It relies on the approach based on the calculation of the spectral density functions $J(0)$[85]

$$J(0)=\frac{1.5}{0.75D^2+\frac{1}{3}C^2}\left(R_2-\frac{R_1}{2}-0.6\sigma_{NH}\right) \tag{3.12}$$

In consistent relaxation datasets, $J(0)$ values should be uniform for a given amino acid residue, provided there is no significant participation of chemical exchange. Experimental data for residue D58 in U-[^{15}N] ubiquitin (Table 3.2) exemplify the use of the $J(0)$ approach. Relaxation data obtained at two different magnetic fields and the same temperature 298 K deliver the spectral density ratio $J(0)_{700}/J(0)_{950} = 1.013 \pm 0.019$. On the other hand, replacement of the relaxation data determined at a lower magnetic field with the data measured at higher temperature results in the ratio $J(0)_{700}/J(0)_{950} = 0.792 \pm 0.018$. At higher temperature, the molecular diffusion tumbling increases. As a consequence, $J(0)_{700}@298 > J(0)_{700}@308$.

The optimal set of model parameters is determined by a search of the global minimum of the target function χ using any convenient minimization procedure.[87,88] The global target function consists of the sum of partial target functions Γ_i characterizing the fit quality for an ith residue.

$$\chi = \sum_{i=1}^{N} \Gamma_i = \sum_{i=1}^{N} \sum_{j=1}^{M} \frac{\left(P_{ij,\mathrm{exp}} - P_{ij,\mathrm{calc}}\right)^2}{\sigma_{ij}^2} \tag{3.13}$$

where the sum is over M relaxation parameters for each of N residues, and $P_{ij,\mathrm{calc}}$ are the appropriate relaxation parameters calculated from the assumed model. The σ_{ij} values are the corresponding standard deviations of experimentally derived $P_{ij,\mathrm{exp}}$.

The combination of different types of rotational diffusion with a variable number of the model-free parameters results in many possible descriptions of protein dynamics. Appropriate selection of a model is important for the reliability of the molecular dynamics analysis. Several statistical methods have been proposed for this purpose.[88–90]

It has to be stressed that one should avoid determining rotational diffusion parameters by minimizing the difference between the experimental and the calculated R_2/R_1 ratios in separation from the local backbone dynamics. Such an approach might be applied solely to deficient relaxation data when R_2 and R_1 ratios, products and differences are explored simultaneously.[91] Local dynamics parameters are usually analysed in the framework of one of the following parameter sets: $\{S_\mathrm{f}^2, \tau_{\mathrm{int,f}}\}$, $\{S_\mathrm{f}^2, \tau_{\mathrm{int,f}}, R_\mathrm{ex}\}$ and $\{S_\mathrm{f}^2, S_\mathrm{s}^2\tau_{\mathrm{int,s}}, R_\mathrm{ex}\}$. The full five-parameter set has seldom been used.[57] The best set of local parameters can be conveniently selected by applying the Fisher – Snedecor test (F-test). Application of another statistical method, AIC (Akaike's Information Criteria), has also been suggested in the literature as an additional model selection approach.[89,90]

Table 3.2 Relaxation data determined for the amino acid residue D58 in ubiquitin.

B_0/f (^{1}H)	T (K)	$R_1 \pm \mathrm{d}R_1$	$R_2 \pm \mathrm{d}R_2$	NOE ± dNOE
16.4 T/700 MHz	308	2.273 ± 0.014	5.919 ± 0.066	0.783 ± 0.003
16.4 T/700 MHz	298	2.079 ± 0.013	7.146 ± 0.097	0.811 ± 0.002
22.3 T/950 MHz	298	1.721 ± 0.014	8.678 ± 0.060	0.865 ± 0.003

3.6 Unstructured Proteins or Their Parts

Flexible parts of structured proteins: termini, linkers, binding or catalytic loops, *etc.* should be excluded from the calculation delivering the overall diffusion parameters. These unstructured, flexible segments cannot be regarded as parts of the rigid rotor, and description of their motions in terms of a single overall correlation time seems inappropriate. Therefore, use of the J^{MFA} or J^{EMFA} function with residue-specific τ_R correlation time allows two or three local correlation times to be taken into account.[92,93] For instance, the C-terminal part of the Yfia protein comprising 22 amino acid residues displays the negative NOEs at 14.1 T magnetic field and small R_2 values that are typical for unstructured protein segments.[94] The model utilizing local correlation times describes well the dynamics of an unstructured tail attached to a fully folded fragment – the case in the Yfia protein. The longest correlation time fitted by a model-free approach, evoking two separate time scales, does not correspond to overall tumbling but rather represents a segmental motion. S^2 values are much smaller than those in the N-terminal part, pointing to the strong preference of fast local motions characterized by τ_{int} over the slower motions, the local τ_R. Since the N-domain anchors the unstructured C-domain, the systematic decrease of both correlation times when approaching the C-terminus reflects the increasing amount of conformational freedom. A similar behaviour has been reported for the isolated D2 domain of annexin I.[95] Another approach used in the analysis of the time scale dynamics is utilization of spectral density function based on a continuous distribution of correlation times. A truncated Lorentzian distribution[96] or the inverse Gaussian probability distribution[97] of correlation times have been successfully used in the characterization of highly mobile protein fragments. A number of models utilizing different distribution functions have been used for the interpretation of nuclear relaxation data in synthetic polymers[98] and polypeptide chains.[99] A comprehensive review[100] presents analyses of intrinsically disordered protein dynamics elucidated from NMR relaxation.

3.7 Biomaterial Preparation for ^{15}N Spin Relaxation Studies

The straightforward application of high-resolution ^{15}N spin relaxation techniques on proteins isolated from an organism with the natural abundance of nitrogen is unfeasible due to the inherently low percentage of NMR-active ^{15}N isotope (*ca.* 0.4%) and its low magnetogyric ratio (γ). Nevertheless, a wealth of ^{15}N labelling techniques covering host organisms from bacteria, yeasts to mammalian cells and cell-free systems is available. Among all the available stable labelling strategies, the ^{15}N labelling protocols are the cheapest and simplest, with the highest degree of success.

The primary technique available since the late 1980s is the uniform U–^{15}N labelling in M9 media in *Escherichia coli* (*E. coli*) strains.[101] By utilizing the ammonium salt – typically chloride, sulphate and nitrate – bearing the ^{15}N isotope as a sole source of any nitrogen in media, the resulting ^{15}N labelling covers all existing backbone and side-chain nitrogens in proteins and nucleic acids. Besides the uniformly ^{15}N labelled protein sample, there are cases when one needs to use sparse ^{15}N labelling. In other words, one can label or un-label selected amino acids with the ^{15}N isotope to facilitate the resonance sequence-specific backbone assignment or decrease the resonance degeneration due to poor signal dispersion.[102,103] Labelling, incorporating selected amino acids bearing ^{15}N in the *E. coli* host, and analogical un-labelling are both well understood. Few practical rules help to avoid and minimize scrambling of stable isotopes, and thus miss-incorporation, at undesired sites.[104] In multidomain proteins, chemical ligation technologies can be used to segmentally label a protein fragment of choice.[105,106] Comparably versatile labelling strategies to those available for prokaryotic *E. coli* have been developed for eukaryotic yeast systems with *Pichia pastoris*[107] and *Kluyveromyces lactis*.[108–110] In case one needs to obtain more complex and closer to native post-translational modifications with the protein of interest, *Baculovirus* with insect cells[111,112] and mammalian cell techniques,[113] including Chinese hamster ovary (CHO) and HEK 293 cells, are relatively well established and constantly developing.[114,115] A promising eukaryotic expression system that grows faster and has higher yields than mammalian cells is the slime mould.[116] In the not so rare cases when the protein is toxic or harmful to any of the host organisms, there are numerous eukaryotic and prokaryotic-based cell-free expression systems available.[117–120] Essentially, all expression systems can be supplemented with nutritional yeast, algal and bacterial hydrolysates of required ^{15}N-containing labelling boosting or, in some extreme cases, enabling protein expression. The *in vivo* ^{15}N labelling of natively folded, post-translationally modified and active plant proteins can be achieved by uniform labelling of the entire plant organism by supplementing $K^{15}NO_3$ or $^{15}NH_4{}^{15}NO_3$ salts.[121]

Besides the observation of the recombinantly expressed and purified ^{15}N-labelled proteins *in vitro* in a buffered NMR tube, it became possible mainly for the intrinsically disordered proteins to probe their dynamics *via* ^{15}N spin relaxation techniques inside the living cells, so-called in-cell NMR.[122–124]

3.8 Practical Considerations of Experimental Setup and Dedicated Pulse Programmes

The vast majority of currently used ^{15}N spin relaxation techniques rely on pulse sequences combining $^{1}H^{N}$-start and $^{1}H^{N}$-detection,[42,62,63,67,68,91,125,126] as such providing the highest sensitivity over the less numerous low-γ start and/or detected techniques.[127,128] Regardless of the gain in the signal detection, the uses of $^{1}H^{N}$-detected, and $^{1}H^{N}$-started and ^{13}C-detected[127] techniques are affected by labile hydrogen exchange with water leading to detrimental proton line broadening at low acidic, close to physiological, and basic conditions. The hydrogen-bonded and buried amides involved in intra-/intermolecular interactions can be better protected from the ^{15}N–^{1}H water-exchange broadening, remaining reasonably detectable. pH 7.5 and temperatures reaching 303 K constitute an optimal sensitivity upper limit for most globular proteins. The extensive hydrogen exchange becomes more limiting for dynamic polypeptide chains with solvent-exposed ^{15}N–^{1}H groups, including intrinsically disordered proteins (IDPs) and regions (IDRs), like loops, termini and side-chains where one is often limited to working at pH lower than 6.5 and temperatures below 288 K.[29,66,67,129,130] As a rule of thumb, the well-folded globular protein samples for $^{1}H^{N}$-detected ^{15}N spin relaxation measurements are prepared in the pH range 6.5–7.5 and temperatures between 293–303 K. IDPs are preferably measured at lower temperatures of 278–288 K and pH below 6.5, which slows down the ^{15}N–^{1}H water-exchange rate. An integral part of every spin relaxation experiment is the proper recycle delay, *i.e.*, the waiting period between the repetition of the pulse sequence that enables the return of the magnetization to the thermal equilibrium. In the case of both the backbone and side-chain labile protons, the longitudinal relaxation rates of amide/amine/guanidine ^{1}H and ^{15}N nuclei, the R_{1H} and R_{1N}, and the longitudinal relaxation rate of water protons, R_{1W}, the exchange rate between water and labile protons, k, all contribute to the efficiency of the return to the thermal equilibrium. Typically, the currently developed HSQC and TROSY-type R_1, R_2, $R_{1\rho}$ pulse programmes for backbone amides (also covering TRP side-chain and Asn, Gln, the latter two only in HSQC mode) in protonated up to 40 kDa (τ_R *ca.* 24 ns) and (per)deuterated proteins up to *ca.* 70–80 kDa ($\tau_R = 40$–50 ns) require recycle delays of 3–4 s at 298 K and 5 s at 310 K, respectively, to deliver essentially systematic and water-exchange error-free rates.[62,131,132] As reported, for a small 6 kDa protein at 298 K, a recycle delay of 1 s was also satisfactory for the R_1, and $R_{1\rho}$ rates, producing data that did not deviate above the experimental errors from the longer recycle delay of 3.5 s.[62] In the case of $^{15}N\{^{1}H\}$ NOE experiments, the situation is much more complex. One needs to optimize the ^{1}H saturation time to reach the so-called steady-state for all spin systems. The recycling delay for the reference experiment should be at least $5\times1/R_{1W}$ of water to eliminate the water-exchange effects. Importantly, the R_{1W} of water tends to increase with the magnetic field, while labile amides R_{1H} decrease, which supports the good practice of determining the R_{1W} and R_{1H} for the system at various magnetic field strengths before setting the recycle delay. It was found that recycle delays of 10 and 13 s are equally good at 298 K at 16.4–22.3 T, as verified for 8 kDa U–^{15}N ubiquitin.[63] The ^{1}H saturation times required to reach steady-state might be as long as 5–6 s or more for highly dynamic proteins or IDRs.[63] A convenient solution was presented with dynamic NOE experiments where the ^{1}H-saturation times are varied, yielding the NOE at the steady-state and a side-product of the R_1 rates for nitrogen. The addition of D_2O, usually 5–10% (v/v) to 'lock' the frequency, is slowly being recognized as a burden in ^{15}N relaxation measurements, having been reported to deteriorate the R_2 CPMG experiments,[133–135] primarily for solvent-exposed residues and flexible protein fragments, leading to the erroneous assignment of the R_{ex} contribution and erroneous estimation of total correlation time to molecules. The exchange of isotopically different species deteriorating the signal shape and relaxation rates of nitrogen is observed for lysine, glutamine and asparagine side-chains. The simplest solution is to use sealed capillaries with D_2O and to avoid adding more than 1% (v/v) of D_2O into the protein solution. Utilizing the HA-start and HA-detected pulse sequence schemes for $^{13}C,^{15}N$ labelled proteins, like HA(CACO)N-type, one can determine the ^{15}N relaxation rates of $^{15}N–^{2}H$ groups in 100% D_2O, which are shorter than the ^{15}N groups found in H_2O.[136]

Depending on the properties of the ^{15}N labelled system, one should carefully consider the pulse programmes to be applied. The most common, being the most sensitive, are the labile amide proton $^{1}H^{N}$-detected pulse programmes that are based on HSQC or TROSY schemes. The proper relaxation observable measurement is realized by implementing the appropriate relaxation block that will follow up with the proper magnetization decay over time. Usually, the nitrogen carrier is placed at 118 ppm, and the range goes from 103 up to 133 ppm. Within this range, one can readily acquire valuable spin relaxation data for the TRP side-chain and track the Asn and Gln side-chain signals.

The same pulse programmes, the HSQC type, can be used to study the IDP. However, due to the modest signal dispersion on the ^{1}H and ^{15}N axes, the indirect dimension, one might not get a reliable value due to the accompanied signal overlap, which is usually not so severe for well-folded proteins. Some improvements were proposed to enhance the IDP spectra, like the CP-HISQC experiment designed to preserve water magnetization, enabling the utilization of short recycling delays yielding enhanced resolution and sensitivity compared to proton-decoupled HSQC.[137] The situation becomes different for the side-chains, as the verified and robust pulse programme set for the ^{15}N spin relaxation is only available for Arg and Lys. Here, to specifically handle the in-phase magnetization HISQC pulse programme scheme with adjusted INEPT-transfers, one can use their apparent $^{1}J_{N\text{-}H}$ couplings, and selective ^{15}N pulse to target the ^{15}N frequencies of lysine (30–40 ppm) and arginine (70–90 ppm). Moreover, as arginine residues can be tracked under similar conditions to the remaining backbone amides, the lysine group requires a lower temperature and acidic pH.[67,138] Although the ^{13}C-detected techniques suffer from lower sensitivity than ^{1}H detection and deliver N_zC_z not pure N_z magnetization rates, they are beneficial in the case of more dynamic backbones like IDR, IDP and side-chains, as demonstrated for arginine.[139–141]

3.9 Perspectives and Future Development

Predictions of methodological developments in any field of science may be predominantly based on the evolution of existing methods and techniques. In the field of biomacromolecular dynamics based on the ^{15}N relaxation approach, there are a number of possible improvements able to enhance its value. NMR is an inherently insensitive experimental technique. Sensitivity improvements can be obtained by increasing the magnetic field strengths in spectrometers. However, at present there is an economical limit to this because high temperature superconducting materials with high critical field are not available. The highest available static magnetic field is 1.2 GHz of ^{1}H proton Larmor frequency, and the first ^{15}N relaxation reports are still to come, and will possibly reveal more imperfections and high-field challenges of the currently available techniques. Another way to enhance sensitivity would be to utilize a polarization transfer from highly polarized systems (*e.g.*, with xenon as a source). Nevertheless, these approaches are not used due to limited applicability and cumbersome technical arrangements. For sensitivity reasons, the $^{1}H^{N}$ excitation techniques ending with $^{1}H^{N}$ detection are widely used, but they are sensitive to the exchange of labile amide protons with water. The side effects of this exchange are magnetization transfer resulting from water suppression techniques and NOE mediated by water protons. Many problems might be cured by heteronuclear low γ detection, ^{15}N or ^{13}C, often providing exceptional resolution and signal dispersion, although far less sensitive when compared to ^{1}H-based techniques. An additional problem is created by the use of $^{1}H/^{2}D$ mixtures of isotopes in NMR studies, both in the form of fractionally deuterated proteins and D_2O as co-solvent for locking purposes. It could be that the use of sealed capillaries filled with D_2O, which allows the separation of a protein from the deuterium source, could be valuable. The complete perdeuteration can remove unwanted fractional isotope effects and undesirable $^{3}J_{HN-HA}$ and other J couplings among ^{1}H amide and other spin system protons. Side-chain motions in amino acid residues containing ^{15}N structural nuclei yield valuable information on interactions of proteins with other molecules of biological interest. Unfortunately, such studies remain scarce. Both, experimental methodology and analytical approaches, need developing and improving to cover all ^{15}N-spin-containing amino acid side-chains under physiological conditions of higher temperature and neutral as well as mildly basic conditions. The most commonly used ^{15}N relaxation parameters, longitudinal and transverse relaxations rates and nuclear Overhauser effect, have been measured for more than forty years. The pulse sequences used in the determination of their values are frequently improved, but new sequences are still not free of artefacts. The resulting rates are inaccurate, bearing numerous errors that might still be negligible when used for comparison among various conditions, nevertheless the exact description of motions remains a challenge. Transverse relaxation measurements, especially based on the CPMG cluster, which is prone to artefacts, should be modified or replaced with equivalent techniques, like the spin-lock-based ones.

In conclusion, ^{15}N spin relaxation is slowly entering a crisis stage of being a middle-aged technique, *i.e.*, highly successful with numerous applications, still expanding, although harried by numerous chronic imperfections coming to the surface. The coming years should be spent concentrating on expanding the applications to probe dynamic structural assemblies and their interactions, an exact *modus operandi*, thus deciphering the biological functions. At the same time, these efforts should be accompanied synergistically by extensive and necessary further improvements of the technologies.

References

1. A. G. Palmer, *Chem. Rev.*, 2004, **104**, 3623–3640.
2. L. E. Kay, *J. Magn. Reson.*, 2005, **173**, 193–207.
3. C. Micheletti, *Phys. Life Rev.*, 2013, **10**, 1–26.
4. S. Shukla, W. Ying, F. Gray, Y. Yao, M. L. Simes, Q. Zhao, H. Miao, H. J. Cho, P. González-Alonso, A. Winkler, G. Lund, T. Purohit, E. Kim, X. Zhang, J. M. Ray, S. He, C. Nikolaidis, J. Ndoj, J. Wang, M. Jaremko, R. J. H. Jaremko, M. L. Ryan, J. Guzman, Grembecka and T. Cierpicki, *Nat. Chem. Biol.*, 2021, **17**, 784–793.
5. S. Morin, *Prog. Nucl. Magn. Reson. Spectrosc.*, 2011, **59**, 245–262.
6. J. G. Kempf and J. P. Loria, *Cell Biochem. Biophys.*, 2002, **37**, 187–211.
7. A. Abragam, *The Principles of Nuclear Magnetism*, Clarendon Press, 1961.
8. R. K. Harris, *Nuclear Magnetic Resonance Spectroscopy: A Physicochemical View*, Pitman, London, 1983.
9. J. McConnell, *The Theory of Nuclear Magnetic Relaxation in Liquids*, Cambridge University Press, Cambridge, New York, 1987.
10. J. Kowalewski and L. Maler, *Nuclear Spin Relaxation in Liquids: Theory, Experiments, and Applications*, CRC press, 2017.
11. L. E. Kay, D. A. Torchia and A. Bax, *Biochemistry*, 1989, **28**, 8972–8979.
12. G. M. Clore, P. C. Driscoll, P. T. Wingfield and A. M. Gronenborn, *Biochemistry*, 1990, **29**, 7387–7401.
13. V. A. Jarymowycz and M. J. Stone, *Chem. Rev.*, 2006, **106**, 1624–1671.
14. A. G. Palmer and H. Koss, in *Methods in Enzymology*, ed. A. J. Wand, Academic Press, 2019, vol. 615, pp. 177–236.
15. M. A. Stetz, J. A. Caro, S. Kotaru, X. Yao, B. S. Marques, K. G. Valentine and A. J. Wand, in *Methods in Enzymology*, ed. A. J. Wand, Academic Press, 2019, vol. 615, pp. 237–284.
16. L.-Y. Lian and D. A. Middleton, *Prog. Nucl. Magn. Reson. Spectrosc.*, 2001, **39**, 171–190.
17. J. Cavanagh, N. J. Skelton, W. J. Fairbrother, M. Rance and A. G. Palmer, *Protein NMR Spectroscopy: Principles and Practice*, Elsevier Science, 2010.
18. M. J. Stone, W. J. Fairbrother, A. G. Palmer, J. Reizer, M. H. Saier and P. E. Wright, *Biochemistry*, 1992, **31**, 4394–4406.
19. S. Zinn-Justin, P. Berthault, M. Guenneugues and H. Desvaux, *J. Biomol. NMR*, 1997, **10**, 363–372.
20. D. M. Korzhnev, M. Billeter, A. S. Arseniev and V. Y. Orekhov, *Prog. Nucl. Magn. Reson. Spectrosc.*, 2001, **38**, 197–266.
21. D. Fushman, *Wiley Online Library*, 2002, pp. 283–308.
22. M. Ottiger and A. Bax, *J. Am. Chem. Soc.*, 1998, **120**, 12334–12341.
23. C. D. Kroenke, M. Rance and A. G. Palmer, *J. Am. Chem. Soc.*, 1999, **121**, 10119–10125.
24. J. Herzfeld, J. E. Roberts and R. G. Griffin, *J. Chem. Phys.*, 1987, **86**, 597–602.
25. J. J. Verbist, M. S. Lehmann, T. F. Koetzle and W. C. Hamilton, *Acta Crystallogr. Sect. B: Struct. Crystallogr. Crystal Chem.*, 1972, **28**, 3006–3013.

26. T. F. Koetzle, M. N. Frey, M. S. Lehmann and W. C. Hamilton, *Acta Crystallogr. Sect. B: Struct. Crystallogr. Crystal Chem.*, 1973, **29**, 2571–2575.
27. A. Esadze, C. Chen, L. Zandarashvili, S. Roy, B. M. Pettitt and J. Iwahara, *Nucleic Acids Res.*, 2016, **44**, 6961–6970.
28. M. S. Lehmann, J. J. Verbist, W. C. Hamilton and T. F. Koetzle, *J. Chem. Soc., Perkin Trans. 2*, 1973, 133–137.
29. C. A. Kemme, R. H. Luu, C. Chen, C. C. Pletka, B. M. Pettitt and J. Iwahara, *J. Phys. Chem. B*, 2019, **123**, 3706–3710.
30. H. Fuess, D. Hohlwein and S. A. Mason, *Acta Crystallogr. Sect. B: Struct. Crystallogr. Crystal Chem.*, 1977, **33**, 654–659.
31. S. K. Sarkar, Y. Hiyama, C. H. Niu, P. E. Young, J. T. Gerig and D. A. Torchia, *Biochemistry*, 1987, **26**, 6793–6800.
32. A. Esadze, D.-W. Li, T. Wang, R. Brüschweiler and J. Iwahara, *J. Am. Chem. Soc.*, 2011, **133**, 909–919.
33. R. B. Fenwick, H. van den Bedem, J. S. Fraser and P. E. Wright, *Proc. Natl. Acad. Sci. U. S. A.*, 2014, **111**, E445.
34. Y. Yuan, V. Simplaceanu, J. A. Lukin and C. Ho, *J. Mol. Biol.*, 2002, **321**, 863–878.
35. H. Saitô, I. Ando and A. Ramamoorthy, *Prog. Nucl. Magn. Reson. Spectrosc.*, 2010, **57**, 181–228.
36. J. B. Hall and D. Fushman, *J. Am. Chem. Soc.*, 2006, **128**, 7855–7870.
37. L. Yao, A. Grishaev, G. Cornilescu and A. Bax, *J. Am. Chem. Soc.*, 2010, **132**, 4295–4309.
38. D. Fushman and D. Cowburn, in *Methods in Enzymology*, ed. T. L. James, V. Dötsch and U. Schmitz, Academic Press, 2001, vol. 339, pp. 109–126.
39. S. Meiboom and D. Gill, *Rev. Sci. Instrum.*, 1958, **29**, 688–691.
40. J. W. Peng and G. Wagner, *Biochemistry*, 1995, **34**, 16733–16752.
41. V. Z. Miloushev and A. G. Palmer, *J. Magn. Reson.*, 2005, **177**, 221–227.
42. F. Massi and J. W. Peng, in *Protein NMR: Methods and Protocols*, ed. R. Ghose, Springer New York, New York, NY, 2018, pp. 205–221.
43. A. G. Palmer, C. D. Kroenke and J. Patrick Loria, in *Methods in Enzymology*, ed. T. L. James, V. Dötsch and U. Schmitz, Academic Press, 2001, vol. 339, pp. 204–238.
44. T. Yuwen, J. P. Brady and L. E. Kay, *J. Am. Chem. Soc.*, 2018, **140**, 2115–2126.
45. D. G. Davis, M. E. Perlman and R. E. London, *J. Magn. Reson., Ser. B*, 1994, **104**, 266–275.
46. O. Trott and A. G. Palmer, *J. Magn. Reson.*, 2002, **154**, 157–160.
47. Z. Luz and S. Meiboom, *J. Chem. Phys.*, 1963, **39**, 366–370.
48. J. Jen, *J. Magn. Reson. (1969)*, 1978, **30**, 111–128.
49. J. Frahm, *J. Magn. Reson. (1969)*, 1982, **47**, 209–226.
50. A. J. Baldwin and L. E. Kay, *J. Biomol. NMR*, 2013, **55**, 211–218.
51. D. M. Korzhnev, V. Y. Orekhov and L. E. Kay, *J. Am. Chem. Soc.*, 2005, **127**, 713–721.
52. S. Forsén and R. A. Hoffman, *J. Chem. Phys.*, 1963, **39**, 2892–2901.

53. J. J. Led and H. Gesmar, *J. Magn. Reson. (1969)*, 1982, **49**, 444–463.
54. P. Vallurupalli, G. Bouvignies and L. E. Kay, *J. Am. Chem. Soc.*, 2012, **134**, 8148–8161.
55. G. Bouvignies and L. E. Kay, *J. Phys. Chem. B*, 2012, **116**, 14311–14317.
56. C. Charlier, S. N. Khan, T. Marquardsen, P. Pelupessy, V. Reiss, D. Sakellariou, G. Bodenhausen, F. Engelke and F. Ferrage, *J. Am. Chem. Soc.*, 2013, **135**, 18665–18672.
57. Ł. Jaremko, M. Jaremko, M. Nowakowski and A. Ejchart, *J. Phys. Chem. B*, 2015, **119**, 11978–11987.
58. M. Xie, L. Yu, L. Bruschweiler-Li, X. Xiang, A. L. Hansen and R. Brüschweiler, *Sci. Adv.*, 2019, **5**, eaax5560.
59. S. Wardenfelt, X. Xiang, M. Xie, L. Yu, L. Bruschweiler-Li and R. Brüschweiler, *Angew. Chem., Int. Ed.*, 2021, **60**, 148–152.
60. Y. Gu, A. L. Hansen, Y. Peng and R. Brüschweiler, *Angew. Chem., Int. Ed.*, 2016, **55**, 3117–3119.
61. O. Carugo, *Protein Sci.*, 2008, **17**, 2187–2191.
62. N.-A. Lakomek, J. Ying and A. Bax, *J. Biomol. NMR*, 2012, **53**, 209–221.
63. V. Kharchenko, M. Nowakowski, M. Jaremko, A. Ejchart and Ł. Jaremko, *J. Biomol. NMR*, 2020, **74**, 707–716.
64. J. Boyd, *J. Magn. Reson., Ser. B*, 1995, **107**, 279–285.
65. M. Buck, J. Boyd, C. Redfield, D. A. MacKenzie, D. J. Jeenes, D. B. Archer and C. M. Dobson, *Biochemistry*, 1995, **34**, 4041–4055.
66. J. Iwahara and G. M. Clore, *J. Biomol. NMR*, 2006, **36**, 251–257.
67. J. Iwahara, Y.-S. Jung and G. M. Clore, *J. Am. Chem. Soc.*, 2007, **129**, 2971–2980.
68. D. Nguyen, G. L. R. Lokesh, D. E. Volk and J. Iwahara, *Molecules*, 2017, **22**, 1355.
69. M. M. Gromiha, in *Protein Bioinformatics*, ed. M. M. Gromiha, Academic Press, Singapore, 2010, pp. 29–62.
70. A. G. Redfield, in *Advances in Magnetic and Optical Resonance*, ed. J. S. Waugh, Academic Press, 1965, vol. 1, pp. 1–32.
71. D. E. Woessner, *J. Chem. Phys.*, 1962, **37**, 647–654.
72. D. E. Woessner, B. S. Snowden and G. H. Meyer, *J. Chem. Phys.*, 1969, **50**, 719–721.
73. M. Ottiger, N. Tjandra and A. Bax, *J. Am. Chem. Soc.*, 1997, **119**, 9825–9830.
74. G. Lipari and A. Szabo, *J. Am. Chem. Soc.*, 1982, **104**, 4546–4559.
75. G. M. Clore, A. Szabo, A. Bax, L. E. Kay, P. C. Driscoll and A. M. Gronenborn, *J. Am. Chem. Soc.*, 1990, **112**, 4989–4991.
76. P. Fan, M. H. Li, B. Brodsky and J. Baum, *Biochemistry*, 1993, **32**, 13299–13309.
77. N. Tjandra, S. E. Feller, R. W. Pastor and A. Bax, *J. Am. Chem. Soc.*, 1995, **117**, 12562–12566.
78. J. L. Baber, A. Szabo and N. Tjandra, *J. Am. Chem. Soc.*, 2001, **123**, 3953–3959.

79. J. W. Peng and G. Wagner, *J. Magn. Reson. (1969)*, 1992, **98**, 308–332.
80. J. W. Peng and G. Wagner, *Biochemistry*, 1992, **31**, 8571–8586.
81. J. F. Lefèvre, K. T. Dayie, J. W. Peng and G. Wagner, *Biochemistry*, 1996, **35**, 2674–2686.
82. D. F. Wyss, K. T. Dayie and G. Wagner, *Protein Sci.*, 1997, **6**, 534–542.
83. R. Ishima and K. Nagayama, *J. Magn. Reson., Ser. B*, 1995, **108**, 73–76.
84. R. Ishima and K. Nagayama, *Biochemistry*, 1995, **34**, 3162–3171.
85. N. A. Farrow, O. Zhang, A. Szabo, D. A. Torchia and L. E. Kay, *J. Biomol. NMR*, 1995, **6**, 153–162.
86. S. Morin and S. M. Gagné, *J. Biomol. NMR*, 2009, **45**, 361.
87. A. G. Palmer, M. Rance and P. E. Wright, *J. Am. Chem. Soc.*, 1991, **113**, 4371–4380.
88. A. M. Mandel, M. Akke and I. I. I. A. G. Palmer, *J. Mol. Biol.*, 1995, **246**, 144–163.
89. E. J. d'Auvergne and P. R. Gooley, *J. Biomol. NMR*, 2003, **25**, 25–39.
90. E. J. D'Auvergne, *Protein Dynamics: A Study of the Model-free Analysis of NMR Relaxation Data*, VDM Verlag Dr. Müller, Saarbrücken, Germany, 2008.
91. Ł. Jaremko, M. Jaremko, A. Ejchart and M. Nowakowski, *J. Biomol. NMR*, 2018, **70**, 219–228.
92. A. T. Alexandrescu and D. Shortlet, *J. Mol. Biol.*, 1994, **242**, 527–546.
93. B. Brutscher, R. Brüschweiler and R. R. Ernst, *Biochemistry*, 1997, **36**, 13043–13053.
94. P. B. Igor Zhukov, B. Schölermann and A. Ejchart, *Acta Biochim. Polonica*, 2007, 769–775.
95. F. Ochsenbein, R. Guerois, J. M. Neumann, A. Sanson, E. Guittet and C. van Heijenoort, *J. Biomol. NMR*, 2001, **19**, 3–18.
96. F. Ochsenbein, J.-M. Neumann, E. Guittet and C. van Heijenoort, *Protein Sci.*, 2002, **11**, 957–964.
97. A. Hsu, F. Ferrage and A. G. Palmer, *Biophys. J.*, 2018, **115**, 2301–2309.
98. I. M. K. Ward, *P. G.*, 1996, **6**, 3693–3706.
99. A. V. Buevich, U. P. Shinde, M. Inouye and J. Baum, *J. Biomol. NMR*, 2001, **20**, 233–249.
100. N. Salvi, A. Abyzov and M. Blackledge, *Prog. Nucl. Magn. Reson. Spectrosc.*, 2017, **102–103**, 43–60.
101. L. P. McIntosh and F. W. Dahlquist, *Q. Rev. Biophys.*, 1990, **23**, 1–38.
102. D. Shortle, *J. Magn. Reson., Ser. B*, 1994, **105**, 88–90.
103. P. S. C. Wu, K. Ozawa, S. Jergic, X.-C. Su, N. E. Dixon and G. Otting, *J. Biomol. NMR*, 2006, **34**, 13–21.
104. H. S. Atreya, *Isotope labeling in Biomolecular NMR*, Springer, Netherlands, 2012.
105. R. Xu, B. Ayers, D. Cowburn and T. W. Muir, *Proc. Natl. Acad. Sci. U. S. A.*, 1999, **96**, 388–393.
106. V. Muralidharan and T. W. Muir, *Nat. Methods*, 2006, **3**, 429–438.
107. A. R. Pickford and J. M. O'Leary, in *Protein NMR Techniques*, ed. A. K. Downing, Humana Press, Totowa, NJ, 2004, pp. 17–33.

108. T. Sugiki, O. Ichikawa, M. Miyazawa-Onami, I. Shimada and H. Takahashi, in *Protein NMR Techniques*, ed. A. Shekhtman and D. S. Burz, Humana Press, Totowa, NJ, 2012, pp. 19–36.
109. T. Sugiki, I. Shimada and H. Takahashi, *J. Biomol. NMR*, 2008, **42**, 159–162.
110. H. Takahashi and I. Shimada, *J. Biomol. NMR*, 2009, **46**, 3.
111. L. Skora, B. Shrestha and A. D. Gossert, in *Methods in Enzymology*, ed. and Z. Kelman, Academic Press, 2015, vol. 565, pp. 245–288.
112. A. Sitarska, L. Skora, J. Klopp, S. Roest, C. Fernández, B. Shrestha and A. D. Gossert, *J. Biomol. NMR*, 2015, **62**, 191–197.
113. T. A. Kost, J. P. Condreay and D. L. Jarvis, *Nat. Biotechnol.*, 2005, **23**, 567–575.
114. A. Dutta, K. Saxena, H. Schwalbe and J. Klein-Seetharaman, in *Protein NMR Techniques*, ed. A. Shekhtman and D. S. Burz, Humana Press, Totowa, NJ, 2012, pp. 55–69.
115. M. Sastry, C. A. Bewley and P. D. Kwong, in *Methods in Enzymology*, ed. and Z. Kelman, Academic Press, 2015, vol. 565, pp. 289–307.
116. L. Cubeddu, C. X. Moss, J. D. Swarbrick, A. A. Gooley, K. L. Williams, P. M. G. Curmi, M. B. Slade and B. C. Mabbutt, *Protein Expr. Purif.*, 2000, **19**, 335–342.
117. K. Ozawa, N. E. Dixon and G. Otting, *IUBMB Life*, 2005, **57**, 615–622.
118. K. Ozawa, P. S. C. Wu, N. E. Dixon and G. Otting, *FEBS J.*, 2006, **273**, 4154–4159.
119. K. Ozawa, M. J. Headlam, P. M. Schaeffer, B. R. Henderson, N. E. Dixon and G. Otting, *Eur. J. Biochem.*, 2004, **271**, 4084–4093.
120. M. Takeda and M. Kainosho, in *Protein NMR Techniques*, ed. A. Shekhtman and D. S. Burz, Humana Press, Totowa, NJ, 2012, pp. 71–84.
121. J. H. Ippel, L. Pouvreau, T. Kroef, H. Gruppen, G. Versteeg, P. van den Putten, P. C. Struik and C. P. M. van Mierlo, *Proteomics*, 2004, **4**, 226–234.
122. N. Sciolino, D. S. Burz and A. Shekhtman, *Proteomics*, 2019, **19**, e1800055.
123. F.-X. Theillet, A. Binolfi, B. Bekei, A. Martorana, H. M. Rose, M. Stuiver, S. Verzini, D. Lorenz, M. van Rossum, D. Goldfarb and P. Selenko, *Nature*, 2016, **530**, 45–50.
124. J. M. Plitzko, B. Schuler and P. Selenko, *Curr. Opin. Struct. Biol.*, 2017, **46**, 110–121.
125. R. Ishima, in *NMR of Proteins and Small Biomolecules*, ed. G. Zhu, Springer, Berlin, Heidelberg, 2012, pp. 99–122.
126. P. A. O'Brien and A. G. Palmer, *J. Biomol. NMR*, 2018, **70**, 205–209.
127. C. W. Lawrence and S. A. Showalter, *J. Phys. Chem. Lett.*, 2012, **3**, 1409–1413.
128. M. Bastidas, E. B. Gibbs, D. Sahu and S. A. Showalter, *Concepts Magn. Res. Part A*, 2015, **44**, 54–66.
129. B. Yu, B. M. Pettitt and J. Iwahara, *Acc. Chem. Res.*, 2020, **53**, 1802–1810.
130. D. Nguyen, C. Chen, B. M. Pettitt and J. Iwahara, in *Methods in Enzymology*, ed. and A. J. Wand, Academic Press, 2019, vol. 615, pp. 285–332.

131. M. Jaremko, S. Jaremko, C. D. Villinger, C. Schmidt, S. Griesinger, Becker and M. Zweckstetter, *Angew. Chem., Int. Ed. Engl.*, 2016, **55**, 10518–10521.
132. M. Gairí, A. Dyachenko, M. T. González, M. Feliz, M. Pons and E. Giralt, *J. Biomol. NMR*, 2015, **62**, 209–220.
133. S. Kim, K.-P. Wu and J. Baum, *J. Biomol. NMR*, 2013, **55**, 249–256.
134. P. Kumari, L. Frey, A. Sobol, N.-A. Lakomek and R. Riek, *J. Biomol. NMR*, 2018, **72**, 125–137.
135. T. Yuwen and N. R. Skrynnikov, *J. Magn. Reson.*, 2014, **241**, 155–169.
136. J. Xu, O. Millet, L. E. Kay and N. R. Skrynnikov, *J. Am. Chem. Soc.*, 2005, **127**, 3220–3229.
137. T. Yuwen and N. R. Skrynnikov, *J. Biomol. NMR*, 2014, **58**, 175–192.
138. D. Nguyen and J. Iwahara, *J. Biomol. NMR*, 2018, **71**, 45–51.
139. K. Gerecht, A. M. Figueiredo and D. F. Hansen, *Chem. Commun.*, 2017, **53**, 10062–10065.
140. H. W. Mackenzie and D. F. Hansen, *J. Biomol. NMR*, 2017, **69**, 123–132.
141. N. D. Werbeck, J. Kirkpatrick and D. F. Hansen, *Angew. Chem., Int. Ed. Engl.*, 2013, **52**, 3145–3147.
142. J. Boyd, *J. Magn. Reson., Ser. B*, 1995, **107**, 279–285.

CHAPTER 4

Lineshape Analysis as a Tool for Probing Functional Motions at Biological Interfaces

SOUMYA DE,[a] ALEXANDER I. GREENWOOD,[b]
L. ANDREA ACEVEDO,[c] NATHAN E. KORSON[d] AND
LINDA K. NICHOLSON*[d]

[a] School of Bioscience, Indian Institute of Technology Kharagpur, Kharagpur, West Bengal 721302, India; [b] University of Cincinnati, A&S-Chemistry, 301 Clifton St., Cincinnati, OH 45221-0172, USA; [c] GlaxoSmithKline, Protein, Cellular, and Structure Sciences Group, 1250 S Collegeville Rd, Collegeville, PA 19426, USA; [d] Cornell University, Department of Molecular Biology & Genetics, 526 Campus Rd, Ithaca, NY 14853, USA
*Email: lkn2@cornell.edu

4.1 Introduction

Lineshape analysis offers a powerful tool for determining the microscopic thermodynamic and kinetic parameters for multi-state reaction cycles associated with the biological functions of proteins. The shape of a peak in an NMR spectrum conveys a wealth of information about the nucleus that gave rise to the peak. For a nucleus in a rigid region of a protein, the center of the peak tells us the average chemical environment surrounding the nucleus, and the width of the peak reflects the rate of global tumbling of the whole protein molecule. A larger protein tumbles more slowly and the peak width is broader, whereas a smaller protein reorients faster and the peak width is

New Developments in NMR No. 26
NMR Spectroscopy for Probing Functional Dynamics at Biological Interfaces
Edited by Anirban Bhunia, Hanudatta S. Atreya and Neeraj Sinha

Published by the Royal Society of Chemistry, www.rsc.org

narrower. For a nucleus in a flexible region of a protein, the linewidth reflects a combination of the rates of global tumbling and internal motions. Fast (ps–ns timescale) internal motions cause narrowing of the peak width. For the case where a nucleus is in a region of a protein that undergoes an exchange process that occurs on the μs–ms timescale, the shape of the peak is further encoded with the kinetics and thermodynamics of this process (*e.g.*, conformational exchange, ligand binding and release, or enzymatic turnover). This encoding is due to the nucleus occupying different chemical environments during the exchange process that occurs during the time period in which the signal is observed. The well-established technique for modeling the lineshape to obtain the parameters that describe these dynamic processes is known as NMR lineshape analysis.

NMR lineshape analysis dates back nearly to our first awareness of nuclear spin.[1] The analysis of NMR lineshapes of proteins is particularly useful because μs–ms timescale exchange processes with barrier heights between approximately 5 and 25 kcal mol^{-1} fall in the realm of biological functions of proteins. Both 1D and 2D lineshape analysis techniques have been applied to proteins to determine the exchange scheme and kinetic and thermodynamic parameters for protein–ligand interactions,[2–5] protein folding,[6–8] and enzyme catalysis.[9–12]

In this chapter, the applications of NMR lineshape analysis to determine the detailed kinetics and thermodynamics for protein–ligand interactions and catalysis of specific substrates by prolyl isomerases are described.[2,9–11] These studies yielded fundamental insights regarding catalysis of *cis–trans* isomerization by PPIase enzymes and their role in the biological context. Determination of kinetic rates and equilibrium constants of each step helps elucidate whether the enzyme catalyzes its reaction *via* uniform binding or differential binding of the substrate and the transition state.[13] Also, determination of the internal equilibrium constant, *i.e.* $[\text{enzyme–substrate}]_{eq}/[\text{enzyme–product}]_{eq}$, helps identify whether the enzyme has evolved to work at equilibrium or far from equilibrium.[14] A powerful aspect of the quantitative model obtained from NMR experiments is that it enables the activity to be predicted at very low concentrations that are often found in cells. Thus, these studies provide valuable insights into enzyme catalysis mechanisms as well as predictive power for understanding their function at cellular concentrations.

4.2 Theoretical Background

4.2.1 NMR Lineshape is a Lorentzian Function Defined by R_2 (Peak Linewidth) and Chemical Environment (Peak Frequency)

An NMR spectrum is generated by the Fourier transform of the oscillating transverse magnetization. The time-dependence of the transverse magnetization is given by eqn (4.1), which decays with a rate constant of R_2. Fourier transform of eqn (4.1) results in a Lorentzian lineshape of the NMR spectra, eqn (4.2), where the width of the line at half height is determined by the transverse relaxation rate R_2. Thus, a narrow or broad lineshape of an NMR signal is directly influenced by the transverse relaxation of the system.

$$M(t) = M_0 e^{-R_2 t}(\cos((\omega - \omega_0)t) + i\sin((\omega - \omega_0)t)) \tag{4.1}$$

$$S(\omega) = M_0 \frac{R_2^2}{R_2^2 + (\omega - \omega_0)^2} \tag{4.2}$$

M_0 and $M(t)$ are the magnetic moments at time 0 and t; ω_0 is the Larmor frequency, ω is the reference frequency and R_2 is the transverse relaxation rate constant. $S(\omega)$ is the real component of the signal intensity of the NMR spectrum at frequency ω. As ω is varied, the intensity is largest when $\omega = \omega_0$ (*i.e.*, at the peak center), and the intensity is smaller as ω moves further away from ω_0. In eqn (4.1) and (4.2), no window function was applied prior to the Fourier transform of the time domain signal. To improve the signal-to-noise ratio, an exponential window function may be applied to $M(t)$, which will also preserve the Lorentzian lineshape of the signal.

4.2.2 Exchange Between Different Chemical Environments Alters Lineshape

The basis for lineshape analysis is that, if a given nucleus undergoes exchange processes on the μs–ms timescale during acquisition of its free induction decay (FID), the lineshape of the corresponding signal will be affected by the rates and equilibrium constants of the exchange processes, as well as by the chemical shifts and scalar couplings of the sampled states.[15] The influence of chemical exchange on NMR lineshape can be demonstrated using exchange between two sites. As explained in Box 4.1, exchange between two sites results in faster decay of the transverse magnetization, which results in broadened NMR lineshape. This dependence of NMR lineshape on k_{ex} is exploited in lineshape analysis.

4.2.3 Mathematical Framework for Simulating Lineshape

To understand how lineshape analysis works, it is useful to consider the perspective of an observed nucleus and the various ways that its magnetization can be lost during the time period in which its free induction decay is acquired. Since the magnetization signal is acquired in the transverse plane (*i.e.*, perpendicular to the static magnetic field), during the acquisition period the time-domain signal will decay according to the transverse relaxation rate, R_2, and this determines the peak width and intensity in the frequency domain. Additionally, if an exchange process occurs during the acquisition time period, the chemical shift of the observed nucleus will change as the nucleus experiences the different chemical environments of

Box 4.1 Chemical exchange results in a broader linewidth of the NMR signal.

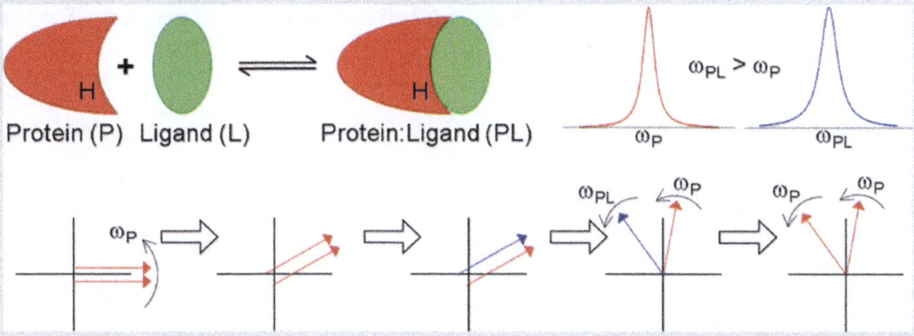

Consider binding of a ligand to a protein. Let us follow the NMR signal of particular nuclei, in this case a proton, which is on the binding interface of the protein. Thus, this particular proton should experience two distinct chemical environments in the free and ligand-bound states, leading to two distinct chemical shifts. Under experimental conditions, the protein will be in dynamic equilibrium between free and bound states. Here we follow the magnetic moment of this proton in two protein molecules. One protein molecule remains in the unbound state throughout the observation period and undergoes precession with a Larmor frequency of ω_P. The other protein molecule also starts in the free state. After some time this protein molecule binds the ligand that results in change in the Larmor frequency of the proton to ω_{PL}, which is assumed to be greater than ω_P. A little later the protein–ligand complex dissociates and the proton reverts back to ω_P. At this point, the two vectors for the proton in the free protein molecules have the same Larmor frequency but have different phases. The resulting magnetic moment will have a smaller magnitude than the sum of the two moments. Thus, for all protein molecules, ligand binding and dissociation will lead to dephasing of the magnetic moment resulting in faster decay of the signal. Since this is a random process, chemical exchange results in an exponential decay of the magnetic moment and manifests itself as an additional term in the transverse relaxation (R_{ex}). This results in a broader linewidth of the NMR signal.

the states that are visited (Box 4.1). Therefore, the chemical shifts of the different states (ϑ_i) and the rates of exchange between those states (k_{ij}) will influence the intensity at a given frequency (*i.e.*, the peak lineshape). These effects will be weighted by the thermodynamic equilibrium population of each state, p_i. We can construct an equilibrium equation for the signal in the frequency domain for each state in a reaction scheme that takes into account each of these influences. A two-state reaction scheme will have two equations, a four-state reaction scheme will have four equations, and so on.

It is instructive to begin with the construction of the lineshape simulation equations for a simple two-state protein–ligand interaction, where protein (P) and ligand (L) each exist in free and bound states (Figure 4.1a). Let us first consider the case where P is isotopically labeled. The first task is to write the equilibrium expression for the reaction from the perspective of each labeled state:

$$\frac{\mathrm{d}\left[\mathbf{P}^{\mathbf{free}}\right]}{\mathrm{d}t} = (\text{rate of formation}) - (\text{rate of loss}) = \left(k_{\text{off}}\left[\mathbf{P}^{\mathbf{bound}}\right]\right) - \left(k_{\text{on}}[\mathrm{L}]\left[\mathbf{P}^{\mathbf{free}}\right]\right)$$

$$= -(\text{rate of loss}) + (\text{rate of formation})$$

$$= -\left(k_{\text{on}}[\mathrm{L}]\left[\mathbf{P}^{\mathbf{free}}\right]\right) + \left(k_{\text{off}}\left[\mathbf{P}^{\mathbf{bound}}\right]\right)$$

$$\frac{\mathrm{d}\left[\mathbf{P}^{\mathbf{bound}}\right]}{\mathrm{d}t} = (\text{rate of formation}) - (\text{rate of loss}) = \left(k_{\text{on}}\left[\mathrm{L}^{\text{free}}\right]\left[\mathbf{P}^{\mathbf{free}}\right]\right) - \left(k_{\text{off}}\left[\mathbf{P}^{\mathbf{bound}}\right]\right)$$

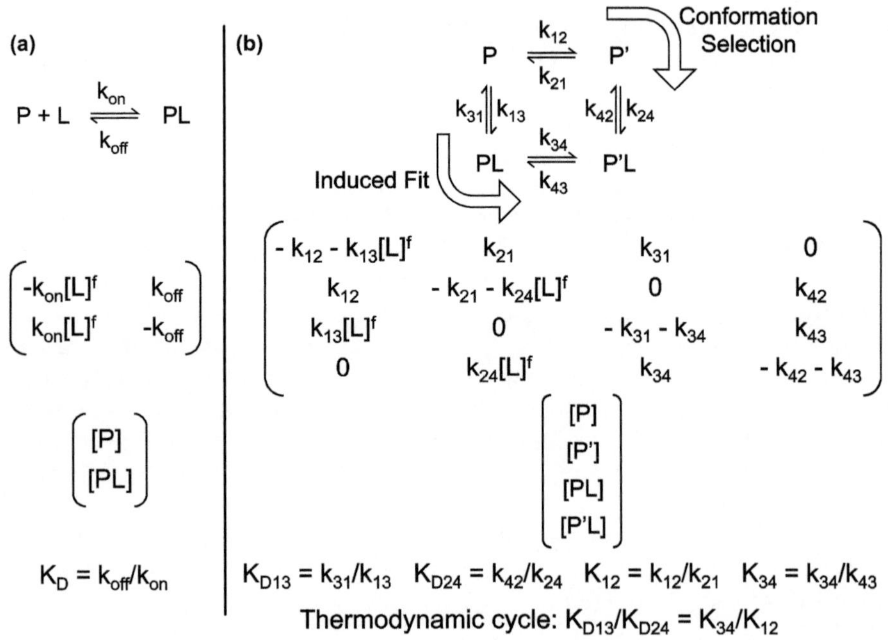

Figure 4.1 Reaction schemes for ligand binding involving two-state or four-state systems, and corresponding rate matrices and populations for lineshape simulation. Two commonly found protein–ligand interaction schemes are shown, (a) two-state interchange between free and bound protein, and (b) four-state interchange forming a thermodynamic cycle. The free protein samples two conformations, P and P′, and P has the higher population. Both conformations bind the ligand. The P′L is the most stable complex. The corresponding rate matrices for the two schemes are shown below. The relation between thermodynamic parameters such as the dissociation constant (K_D) and the rate constants are also shown.

In the above expressions, the states that contain the isotope-labeled species (P, in this case) are highlighted in bold. It is for these states that the transverse relaxation rates, chemical shifts, and exchange rates will influence the NMR lineshape. We can write the above two expressions in 2×2 matrix form as:

$$\begin{bmatrix} \dfrac{d\left[\mathbf{P}^{\text{free}}\right]}{dt} \\ \dfrac{d\left[\mathbf{P}^{\text{bound}}\right]}{dt} \end{bmatrix} = \begin{bmatrix} -k_{\text{on}}\left[\text{L}^{\text{free}}\right] & k_{\text{off}} \\ k_{\text{on}}\left[\text{L}^{\text{free}}\right] & -k_{\text{off}} \end{bmatrix} \begin{bmatrix} \left[\mathbf{P}^{\text{free}}\right] \\ \left[\mathbf{P}^{\text{bound}}\right] \end{bmatrix}$$

Notably, the matrix containing the rate terms includes elements that depend on the unlabeled free ligand concentration, $[\text{L}^{\text{free}}]$, *i.e.*, $k_{\text{on}}[\text{L}^{\text{free}}]$. The lineshape analysis experiment is performed by titrating unlabeled ligand while the labeled binding partner is held at constant concentration, and NMR spectra are collected at each titration point. The changes in lineshapes during the titration will also depend on the transverse relaxation rates and chemical shifts of the isotope-labeled free and bound states. The influence of these relaxation rates and chemical shifts will also be scaled by the populations (concentrations) of the free and bound protein states, and as illustrated above for the reaction rates these can be written in matrix form. The NMR signal can be written as:

$$S_1 = \begin{pmatrix} R_2^{\text{free}} & 0 \\ 0 & R_2^{\text{bound}} \end{pmatrix} - \begin{pmatrix} k_{\text{on}}[\text{L}^{\text{free}}] & -k_{\text{off}} \\ -k_{\text{on}}[\text{L}^{\text{free}}] & k_{\text{off}} \end{pmatrix} - 2\pi i \begin{pmatrix} \vartheta_{\text{free}} & 0 \\ 0 & \vartheta_{\text{bound}} \end{pmatrix} \tag{4.3}$$

$$S_2 = 2\pi i \vartheta \begin{pmatrix} 1 & 1 \\ 1 & 1 \end{pmatrix} \tag{4.4}$$

$$S(\vartheta) = \text{Re}\left(\begin{pmatrix} 1 & 1 \end{pmatrix} (S_1 + S_2)^{-1} \begin{pmatrix} [\mathbf{P}^{\text{free}}] \\ [\mathbf{P}^{\text{bound}}] \end{pmatrix} \right) \tag{4.5}$$

The transverse relaxation rate constants for each state in the reaction scheme (*e.g.*, free and bound protein for a two-state reaction) are in the first matrix of eqn (4.3). The second matrix of eqn (4.3) consists of the forward and reverse rate constants as described above, and the third matrix contains the chemical shifts (or frequencies) of each state. For a backbone NH group, the influence of three-bond scalar coupling with the intraresidue Hα, $J_{\text{N}\alpha}$, can be accounted for by splitting the chemical shift matrix into two matrices with a diagonal term for each state x, $2\pi i(\nu_x + eJ_{\text{N}\alpha}/2)$, where $e = (+1, -1)$ and $x = 1$ to N states. In eqn (4.4), ν represents the frequency that is varied along the spectrum. The final spectrum is calculated as the real component of eqn (4.5). For the case where the ligand is isotopically labeled, we replace $[\text{L}^{\text{free}}]$ with

[P^{free}] in eqn (4.3), and [P^{free}] and [P^{bound}] with [L^{free}] and [L^{bound}], respectively, in eqn (4.5).

By fitting the observed NMR lineshapes acquired for different titration points to the theoretical lineshape expressions above (eqn (4.3)–(4.5)), both kinetic (k_{on} and k_{off}) and thermodynamic ($K_D = k_{off}/k_{on}$) parameters can be determined. Notably, for the labeled protein case only [L^{free}] and [P^{free}] change as the titration proceeds, while the remaining parameters remain constant across all titration points. When titrations are performed with the total concentration of labeled component X held constant (where X is P or L) the bound population is not an additional unknown, since $[X^{bound}] = 1 - [X^{free}]$.

An important point to note is that the same on/off exchange reaction can be observed from the perspective of either the protein or the ligand. This can provide two different exchange regimes in which to characterize the same reaction parameters, providing higher confidence in the fitted parameters. An excellent example is when excess ligand is present such that $[L]^{free} \gg [P]^{free}$. In this case, fast exchange is often observed in HSQC spectra, while slow exchange may be detected in ROESY spectra, depending on the differences between free and bound chemical shifts and the magnetic field strength.[16]

For a generalized reaction scheme consisting of N number of states, the mathematical representation is similarly expressed in matrix form as described in more detail in Box 4.2. The equation for the theoretical NMR signal is given by:[17]

$$S_1 = \begin{pmatrix} R_2^1 & \cdots & 0 \\ \vdots & \ddots & \vdots \\ 0 & \cdots & R_2^N \end{pmatrix} - \begin{pmatrix} -k_{1N} & \cdots & k_{N1} \\ \vdots & \ddots & \vdots \\ k_{1N} & \cdots & -k_{N1} \end{pmatrix} - 2\pi i \begin{pmatrix} \vartheta_1 & \cdots & 0 \\ \vdots & \ddots & \vdots \\ 0 & \cdots & \vartheta_N \end{pmatrix} \tag{4.6}$$

$$S_2 = 2\pi i \vartheta \begin{pmatrix} 1 & \cdots & 0 \\ \vdots & \ddots & \vdots \\ 0 & \cdots & 1 \end{pmatrix} \tag{4.7}$$

$$S(\vartheta) = \mathrm{Re}\left((1 \; \ldots \; 1)(S_1 + S_2)^{-1} \begin{pmatrix} p_1 \\ \vdots \\ p_N \end{pmatrix} \right) \tag{4.8}$$

Insights regarding the derivation of this expression and, for example, why the inverse of $(S_1 + S_2)$ is required, are presented in Box 4.3.

Through NMR titration experiments and the application of eqn (4.6)–(4.8), NMR lineshape analysis can be used to study exchange processes between multiple sites. For example, a common four-state protein–ligand interaction

Box 4.2 Construction of the rate matrix.

The rate matrix in eqn (4.6) will depend explicitly on the number of states in a system (N) and on how these states are connected to each other in the reaction scheme. In other words, the rate matrix size and the form of its elements will depend on the reaction scheme. Each *row* in this matrix corresponds to the rates associated with the exchange of one state with other states in the reaction scheme to which it is directly connected. For example, let's consider a four-state reaction scheme with states A–D:

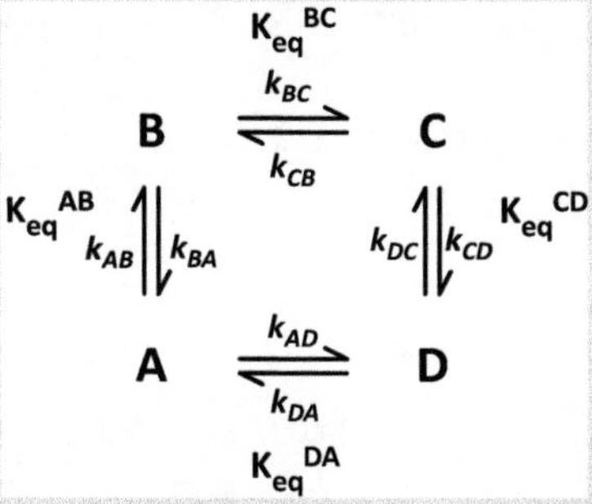

Starting with state A, we see that it exchanges directly with states B and D. We can write an expression for the rate of change of A in terms of the rate constants and concentrations of these states:

$$\frac{\mathrm{d}[\mathrm{A}]}{\mathrm{d}t} = (\text{rate of formation}) - (\text{rate of loss}) = (k_{\mathrm{BA}}[\mathrm{B}] + k_{\mathrm{DA}}[\mathrm{D}]) - (k_{\mathrm{AB}}[\mathrm{A}] + k_{\mathrm{AD}}[\mathrm{A}])$$

Collecting terms,

$$\frac{\mathrm{d}[\mathrm{A}]}{\mathrm{d}t} = -(k_{\mathrm{AB}} + k_{\mathrm{AD}})[\mathrm{A}] + k_{\mathrm{BA}}[\mathrm{B}] + 0[\mathrm{C}] + k_{\mathrm{DA}}[\mathrm{D}]$$

In preparation for constructing the rate matrix, we have conveniently ordered the above terms as

$$\frac{\mathrm{d}[\mathrm{A}]}{\mathrm{d}t} = a[\mathrm{A}] + b[\mathrm{B}] + c[\mathrm{C}] + d[\mathrm{D}]$$

where $a = -(k_{\mathrm{AB}} + k_{\mathrm{AD}})$, $b = k_{\mathrm{BA}}$, $c = 0$, and $d = \mathrm{k}_{\mathrm{DA}}$.

We can immediately construct the equivalent equation in matrix form as

$$\frac{\mathrm{d}[\mathrm{A}]}{\mathrm{d}t} = \begin{bmatrix} -(k_{\mathrm{AB}} + k_{\mathrm{AD}}) & k_{\mathrm{BA}} & 0 & k_{\mathrm{DA}} \end{bmatrix} \begin{bmatrix} [\mathrm{A}] \\ [\mathrm{B}] \\ [\mathrm{C}] \\ [\mathrm{D}] \end{bmatrix}$$

In the same way we can write expressions for each of the remaining states in our four-state reaction scheme:

$$\frac{d[B]}{dt} = k_{AB}[A] - (k_{BA} + k_{BC})[B] + k_{CB}[C] + 0[D]$$

$$\frac{d[C]}{dt} = 0[A] + k_{BC}[B] - (k_{CB} + k_{CD})[C] + k_{DC}[D]$$

$$\frac{d[D]}{dt} = k_{AD}[A] + 0[B] + k_{CD}[C] - (k_{DC} + k_{DA})[D]$$

Finally, we can now express the whole four-state reaction scheme in matrix form as:

$$\begin{bmatrix} \frac{d[A]}{dt} \\ \frac{d[B]}{dt} \\ \frac{d[C]}{dt} \\ \frac{d[D]}{dt} \end{bmatrix} = \begin{bmatrix} -(k_{AB}+k_{AD}) & k_{BA} & 0 & k_{DA} \\ k_{AB} & -(k_{BA}+k_{BC}) & k_{CB} & 0 \\ 0 & k_{BC} & -(k_{CB}+k_{CD}) & k_{DC} \\ k_{AD} & 0 & k_{CD} & -(k_{DC}+k_{DA}) \end{bmatrix} \begin{bmatrix} [A] \\ [B] \\ [C] \\ [D] \end{bmatrix}$$

For binding steps (*i.e.*, second-order reactions) in the reaction scheme, the element corresponding to the forward (association) direction will be the product of the second-order rate constant (*e.g.*, k_{on}, with units $s^{-1}M^{-1}$) multiplied by the concentration of the binding partner not being directly observed in the NMR experiment (*i.e.*, the unlabeled molecule).

scheme is shown in Figure 4.1b. The four-state exchange scheme depicts that the protein exchanges between two conformations, with P having the higher population. Both conformations bind the ligand with different affinities and the ligand-bound states also exchange among themselves, thus forming a thermodynamic cycle. The P′L is the high affinity complex. This scheme combines the classical ligand-binding models of induced fit and conformational selection.[18] This interaction scheme has eight kinetic and four thermodynamic parameters, which are dependent on each other, thus resulting in seven independent parameters. For such a complex reaction scheme, it is important to determine these parameters independently, so that the rest can be determined by lineshape analysis. Once the kinetic parameters are determined, the relative flux through each mechanism, *i.e.* induced fit and conformational selection, can be calculated and the dominant mechanism of ligand binding can be identified. Fitting the lineshape data to the two extreme models of purely induced fit (P↔PL↔P′L) and purely conformational selection (P↔P′ ↔P′L) can also provide clues

Box 4.3 Insights regarding the derivation of eqn (4.8): why take the inverse of $(S_1 + S_2)$?

In the early days before Fourier transform (FT) NMR was widely practiced, the NMR experiment was performed in a "continuous wave" (CW) mode similar to how standard types of spectroscopy work, such as ultraviolet (UV) absorption measurement. In CW NMR, the excitation radio-frequency (rf) energy was swept through a range of frequencies across a spectral window (sw), and when the excitation frequency matched the resonance frequency of a nucleus a signal would be detected as an induced voltage in a receiver coil aligned at 90° from the transmitter coil. It is useful to consider the cw mode because NMR exchange theory was originally derived in this context.[17,77–79]

For CW NMR, with the static magnetic field ($\mathbf{H}_o$) along the *z*-axis, the oscillating excitation rf magnetic field ($\mathbf{H}_1$) was generated by a transmitter coil oriented (for example) along the *x*-axis of the laboratory frame and the induced signal was detected by a receiver coil oriented along the *y*-axis. Orthogonal excitation and detection axes (or alternatively, an imbalanced transmitter coil) facilitated signal detection, since the excitation field was continuously on at the same time and same frequency as the signal being detected. Notably, because $\mathbf{H}_1$ was acting during signal observation, the Bloch equations corresponding to CW NMR include a term that accounts for $\mathbf{H}_1$ action:[17]

$$\frac{\mathrm{d}M_x}{\mathrm{d}t} = (\omega_0 - \omega)M_y$$

$$\frac{\mathrm{d}M_y}{\mathrm{d}t} = -(\omega_0 - \omega)M_x + \gamma H_1 M_z$$

$$\frac{\mathrm{d}M_z}{\mathrm{d}t} = -\gamma H_1 M_y$$

Since the signal is observed in the *x*–*y* plane, the x and y components of magnetization $\mathbf{M}$ are combined into a transverse magnetization vector $\mathbf{G} = \mathbf{M}_x + i\mathbf{M}_y$, where the orthogonal nature of these components is represented mathematically by multiplying $\mathbf{M}_y$ by i. Accounting for transverse relaxation (R_2) and considering the simple two-site exchange case between states P and PL (k_{on}[L] or k_{off}) with fractional populations p_P and p_{PL}, the Bloch equations are written as:

$$\frac{\mathrm{d}G_P}{\mathrm{d}t} = -[i(\omega_P - \omega) + R_{2P}]G_P + i\gamma H_1 M_Z^P - k_{on}[L]G_P + k_{off}G_{PL}$$

$$\frac{\mathrm{d}G_{PL}}{\mathrm{d}t} = -[i(\omega_{PL} - \omega) + R_{2PL}]G_{PL} + i\gamma H_1 M_Z^{PL} - k_{off}G_{PL} + k_{on}[L]G_P$$

Since exchange studies by NMR are typically on systems that are at equilibrium, we can apply the equilibrium conditions, where M_0 represents the total magnetization initially along z prior to application of $\mathbf{H}_1$:

$$\frac{dG_P}{dt} = \frac{dG_{PL}}{dt} = 0\,;\ M_Z^P = M_0^P = p_P M_0\,;\ M_Z^{PL} = M_0^{PL} = p_{PL} M_0$$

Defining $k = k_{on}[L]/p_{PL} = k_{off}/p_P$ and substituting this and the above expressions into the Bloch equations for $\mathbf{G_P}$ and $\mathbf{G_{PL}}$:

$$-[i(\omega_P - \omega) + R_{2P} + kp_{PL}]G_P + kp_P G_{PL} = -ip_P \gamma H_1 M_0$$

$$kp_{PL} G_P - [i(\omega_{PL} - \omega) + R_{2PL} + kp_P]G_{PL} = -ip_{PL} \gamma H_1 M_0$$

In the above expressions, we can recognize the relaxation terms (R_{2N}, where N = P, PL), exchange rates (k_N), and offset from resonance frequency that appear as matrices in S_1 (eqn (4.3)) and S_2 (eqn (4.4)) in the main text.

The above equations can be generalized into a multi-state matrix form:

$$\mathbf{AG} = -i\,C\mathbf{P}$$

In the above expression, the coefficients of the $G_\mathbf{N}$ states on the left side (*i.e.*, the R_2, chemical shift and exchange terms) are collected into a square matrix **A**, **G** is a column vector containing elements $G_\mathbf{N}$ (total transverse magnetization of each state N), **P** is a column vector containing the populations of each state N, and $C = \gamma H_1 M_0$ which functions as a scaling factor (this depends on the initial magnetization and the magnitude of H_1). For FT NMR where H_1 is not on during acquisition, the scaling factor C represents the magnitude of M_0 in the x–y-plane at the start of acquisition. The observed signal is the sum over all states in **G** (*i.e.*, the total transverse magnetization), so we multiply both sides of this expression by the inverse of **A** and multiply the right side by the Nx1 row vector $\mathbf{I}^T = [1 \ldots 1]$:

$$\mathbf{A}^{-1}\mathbf{AG} = -i\,C\,\mathbf{A}^{-1}\mathbf{P}$$

$$\mathbf{G} = -i\,C[1 \ldots 1]\mathbf{A}^{-1}\mathbf{P}$$

This expression is easily converted into eqn (4.5) in the main text, since the signal intensity $S(\omega)$ is directly proportional to the real component of the total transverse magnetization:

$$S(\omega) \propto \mathrm{Re}([1 \ldots 1]\,\mathbf{A}^{-1}\mathbf{P})$$

regarding the dominant pathway of ligand binding and help estimate some of the fitted parameters. These two models display distinctly different responses to changes in ligand concentration as described elsewhere[19–21] and are exemplified by binding of the tyrosine kinase inhibitor Gleevec to selected targets (see SI of Agafonov *et al.*[22]).

(a)

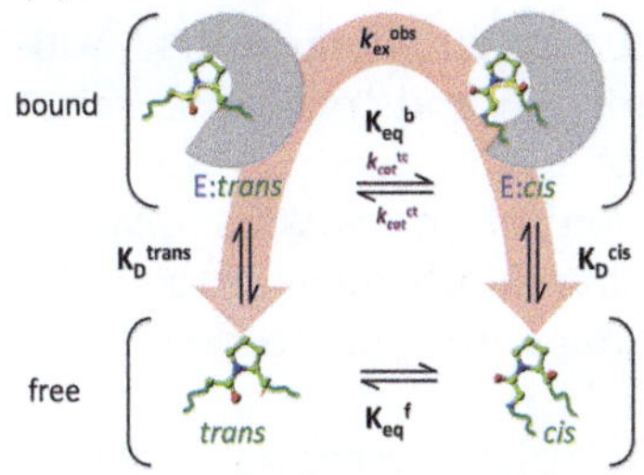

(b)

$$\begin{bmatrix} \frac{d[A]}{dt} \\ \frac{d[B]}{dt} \\ \frac{d[C]}{dt} \\ \frac{d[D]}{dt} \end{bmatrix} = \begin{bmatrix} -(k_{AB}+k_{AD}) & k_{BA} & 0 & k_{DA} \\ k_{AB} & -(k_{BA}+k_{BC}) & k_{CB} & 0 \\ 0 & k_{BC} & -(k_{CB}+k_{CD}) & k_{DC} \\ k_{AD} & 0 & k_{CD} & -(k_{DC}+k_{DA}) \end{bmatrix} \begin{bmatrix} [A] \\ [B] \\ [C] \\ [D] \end{bmatrix}$$

Figure 4.2 Four-state reaction scheme for substrate perspective of enzyme-catalyzed *cis–trans* isomerization. (a) Free substrate is in equilibrium between its *cis* and *trans* isomers (K_{eq}^{f}), each isomer is in equilibrium between free and enzyme-bound (K_{D}^{trans}, K_{D}^{cis}), and bound substrate is in equilibrium between *cis* and *trans* isomers (K_{eq}^{b}). Each equilibrium step has associated forward and reverse rate constants as shown. The overall observed enzyme-catalyzed exchange rate, k_{ex}^{obs}, reflects the overall process of isomer-specific binding and on-enzyme catalysis (double-ended pink arrow) for given enzyme and substrate concentrations. (b) The rate constant matrix, *k*, and the population matrix, *P*, for this four-state model are specified.

The reversible enzyme catalyzed prolyl *cis–trans* isomerization reaction is another important four-state exchange system (Figure 4.2). Due to its partial double bond character, the peptide bond that links amino acids together in proteins and peptides is planar, adopting most commonly the *trans* configuration ($\omega = 180°$). This configuration is destabilized in the case of the peptide bond preceding a proline residue, however, leading to the possibility of the *cis* isomer ($\omega = 0°$). In folded proteins, prolyl peptide bonds are often restricted to adopt only one of these isomers, but in unstructured peptides or disordered regions of proteins, both isomers typically exist in equilibrium, with the *cis* isomer comprising approximately between 5%–40% depending primarily on the type of residue preceding the proline.[23] The peptide bond's partial double bond character also causes the timescale of interconversion between these isomers to be unusually long, on the order of minutes. Because of its slow natural rate, prolyl isomerization can be a rate-limiting step in processes such as protein folding,[24] protein–protein binding,[25,26] or enzymatic recognition.[27] Peptidyl-prolyl isomerase enzymes (PPIases) exist that can accelerate this isomerization to faster timescales more relevant for these processes.[28] The catalytic reaction scheme includes binding of each isomer of the substrate to the enzyme, on-enzyme exchange between the two isomers, and slow isomer exchange in the free substrate (Figure 4.2a). For this four-state system, the rate constant matrix required for eqn (4.6) and the population matrix needed for eqn (4.8) are as given in Figure 4.2b.

4.2.4 Independent Measurements to Reduce the Number of Fitted Parameters in the Four-state Model

The numbers of fitted parameters increase with the increase in the number of states in a model. For example, the two-state model is described by two parameters k_{on} and k_{off} (Figure 4.1a), whereas the four-state models (Figures 4.1b and 4.2) have at least seven independent parameters. Additionally, every residue will have a chemical shift (ω) and transverse relaxation rate constant (R_2) associated with each state. Simultaneously fitting so many parameters is computationally expensive and often leads to misleading solutions. Hence, it is very important to independently measure some of these parameters and feed them into the lineshape analysis to determine the rest.

In the four-state ligand-binding model (Figure 4.1b), the exchange between the two protein conformations, P and P′, in the free state can be characterized by Z-exchange,[29,30] ZZ-exchange,[31] CPMG relaxation dispersion,[32] off-resonance rotating-frame ($R_{1\rho}$),[33] paramagnetic relaxation enhancement (PRE),[34] or chemical exchange saturation transfer (CEST) experiments.[35] For slow exchange between P and P′, with visible chemical shifts for both states, the exchange rates can be determined by Z-exchange or ZZ-exchange experiments. When the exchange rates are faster and very small population of the minor state renders it undetectable, CPMG, $R_{1\rho}$, or CEST experiments can be used to characterize the exchange process. In the case of CPMG relaxation dispersion, the ^{15}N or ^{13}C chemical shift of the minor state can be determined by a comparison of HSQC and HMQC spectra.[36] PRE experiments can also be used where the minor state represents transient long-range interactions within the protein.[37] Overall, one or more of these experiments can be used to determine the exchange rates (k_{12} and k_{21}), relative populations in the free state, and the chemical shift of the minor state (P′).

The four-state enzyme catalyzed *cis–trans* isomerization scheme (Figure 4.3) is a special case where the uncatalyzed exchange between the *cis* and *trans* isomers is very slow and, hence, the exchange rate cannot be determined by the above-mentioned experiments. These slow exchange rates can be determined by several biochemical and biophysical methods such as fluorescence spectroscopy,[2] pH- or LiCl-mediated solvent jump experiments,[38] and protease coupled assay.[39] Due to slow exchange, the chemical shift (ω) and transverse relaxation rate constant (R_2) associated with both states can be directly measured from the NMR spectra. It is common for the on-enzyme *cis–trans* isomerization exchange to be fast relative to the frequency separation between bound *cis* and *trans* peaks, in which case the binding of substrate to enzyme can be approximated as a two-state reaction with an apparent affinity, K_D^{app}, and can be measured from the perspective of the enzyme (Figure 4.3b).

Additional restraints on the fitted parameters can further be obtained from other ligand-binding experiments. Most biophysical or biochemical methods do not distinguish between the two free states (P and P′) or the two bound states (PL and PL′). Such experiments report an apparent dissociation constant (K_D^{app}) given by

$$K_D^{app} = \frac{1 + K_{12}}{1 + K_{34}} K_{D13} = \frac{(1 + K_{12})K_{34}}{(1 + K_{34})K_{12}} K_{D24} \tag{4.9}$$

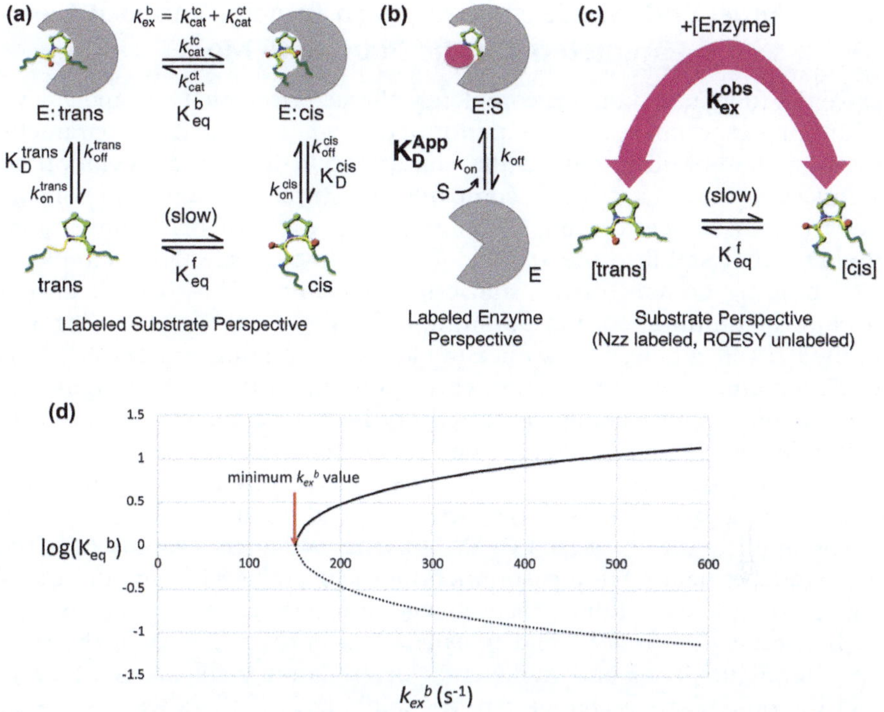

Figure 4.3 Validation of fitted parameters obtained from lineshape analysis for enzyme-catalyzed *cis–trans* isomerization. (a) From the perspective of labeled substrate, the catalytic reaction scheme is four states. (b) From the perspective of the enzyme, when on-enzyme exchange is fast relative to the chemical shift difference between bound states the binding reaction can be approximated by two states (free and bound enzyme) with a corresponding apparent binding affinity, K_D^{APP}. (c) The overall observed exchange rate for given sample conditions, k_{ex}^{obs}, can be measured from the substrate perspective using Nzz (with labeled substrate) or homonuclear ROESY (with unlabeled substrate) NMR experiments. Reproduced from ref. 9 with permission from Springer Nature, Copyright 2019. (d) Curve defined by eqn. (4.8) plotted as $\log(K_{eq}^{b})$ *vs.* k_{ex}^{b} for given sample conditions ($[E]^{tot}$, $[S]^{tot}$) and the corresponding k_{ex}^{obs} measured by ROESY or Nzz. The + (solid black) and – (dotted black) solutions of eqn (4.8) meet at $K_{eq}^{b} = 1$, and the corresponding x-coordinate is the minimum possible k_{ex}^{b} value. The minimum possible k_{ex}^{b} value occurs at the intersection of the + and – curves as indicated by a red arrow.

The parameters in the above expressions are defined in Figure 4.1b. Similarly, for the four-state enzyme-catalyzed *cis–trans* isomerization scheme (Figure 4.3), the K_D^{app} is given by

$$K_D^{app} = \frac{1 + K_{eq}^{f}}{\left(\frac{K_{eq}^{f}}{K_D^{trans}} + \frac{1}{K_D^{cis}}\right)} \tag{4.10}$$

4.2.5 Independent Measurements to Provide Validation of Fitted Parameters for the Four-state Model

Once the four-state model is completely characterized by lineshape analysis, additional experiments can be performed to validate the fitted parameters. In the ligand-binding model, the conformation-specific dissociation constants, K_{D13} and K_{D24}, can be compared to estimate dissociation constants determined by performing ligand-binding experiments using protein mutants that stabilize the P and P′ states, respectively. For example, the DNA-binding domain of the transcription factor ETV6 exists in an equilibrium of uninhibited and autoinhibited states. The DNA binding is inhibited by an α-helix that sterically blocks the binding interface.[40,41] The inhibited and uninhibited states, corresponding to P and P′ in Figure 4.1b, were stabilized by locking the inhibitory helix in place *via* an engineered disulfide bond and deleting the inhibitory helix, respectively.[42] The conformation-specific dissociation constants of ETV6 were estimated by DNA-binding experiments using these designed mutants.[40,42]

The application of either ROESY[43] or Nzz exchange spectroscopy[29] provides independent NMR measurement of the overall exchange rate, k_{ex}^{obs}, for a given sample condition. This exchange rate is the overall observed rate of exchange between free *cis* and *trans* isomers in the presence of the PPIase enzyme (Figure 4.3c), and so depends on the binding constants (K_D^{cis}, K_D^{trans}) and the on-enzyme exchange rate constants (k_{cat}^{tc}, k_{cat}^{ct}), as well as the concentrations of free *cis* and *trans* isomer forms of the substrate ($[cis]^f$, $[trans]^f$) and the total concentration of enzyme ($[E]^{tot}$):[11]

$$k_{ex}^{obs} = \left(\frac{k_{cat}^{tc}}{K_D^{trans}} + \frac{k_{cat}^{ct}}{K_D^{cis}}\right) \frac{[E]^{total}}{\left(1 + \frac{[trans]^f}{K_D^{trans}} + \frac{[cis]^f}{K_D^{cis}}\right)} \tag{4.11}$$

In the above equation, the only parameter that we know is $[E]^{total}$, which is set in the experiment. However, this equation can be rewritten in terms of only two unknowns, the on-enzyme (bound) exchange rate (k_{ex}^b) and on-enzyme equilibrium constant (K_{eq}^b), by utilizing the following relationships for these parameters and for the *cis–trans* equilibrium constant for free substrate (K_{eq}^f), the apparent binding affinity, K_D^{App}, and the isomer-specific dissociation constants (K_D^{cis}, K_D^{trans}):

$$k_{ex}^b = k_{cat}^{tc} + k_{cat}^{ct}\,;\quad K_{eq}^b = \frac{k_{cat}^{ct}}{k_{cat}^{tc}} = \frac{[trans]^b}{[cis]^b}\,;\quad K_{eq}^f = \frac{[trans]^f}{[cis]^f}$$

$$K_D^{app} = \frac{1 + K_{eq}^f}{\left(\frac{K_{eq}^f}{K_D^{trans}} + \frac{1}{K_D^{cis}}\right)}$$

$$K_D^{cis} = \frac{1 + K_{eq}^b}{1 + K_{eq}^f} K_D^{app}\ ;\ K_D^{trans} = \frac{\left(1 + K_{eq}^b\right) K_{eq}^f}{\left(1 + K_{eq}^f\right) K_{eq}^b} K_D^{app} \tag{4.12}$$

We note that the apparent binding affinity, K_D^{App}, is the two-state approximation of the dissociation constant for the enzyme–substrate binding reaction that is a weighted sum of the individual isomer dissociation constants. Using the above relationships and assuming that the total substrate concentration greatly exceeds the total enzyme concentration, eqn (4.11) can be rewritten as

$$k_{ex}^{obs} = \frac{k_{ex}^b k_{eq}^b (K_{eq}^f + 1)}{(K_{eq}^b + 1)^2 K_D^{app}} \left(1 + \frac{1}{K_{eq}^f}\right) \frac{[E]^{total}}{1 + \frac{[S]^{total}}{K_D^{app}}} \tag{4.13}$$

Although this expression is complex, it demonstrates that the observed exchange rate depends on only two parameters obtained from lineshape analysis, the on-enzyme (bound) exchange rate (k_{ex}^b) and on-enzyme equilibrium constant (K_{eq}^b). All other parameters in the equation can be independently measured or are known.

To gain important insights, it is useful to write the above equation in quadratic form in terms of the on-enzyme equilibrium constant, K_{eq}^b:

$$0 = (K_{eq}^b)^2 + (2 - k_{ex}^b C) k_{ex}^b + 1 \tag{4.14}$$

In the above expression, C is an expression that contains the independently measured values and the sample-specific concentrations:

$$C = \frac{(K_{eq}^f + 1)}{k_{ex}^{obs} K_D^{app}} \left(1 + \frac{1}{K_{eq}^f}\right) \frac{[E]^{total}}{1 + \frac{[S]^{total}}{K_D^{app}}} \tag{4.15}$$

The two possible solutions for K_{eq}^b in terms of k_{ex}^b and C are given by

$$K_{eq}^b = \frac{-\left(2 - k_{ex}^b C\right) \pm \sqrt{\left(2 - k_{ex}^b C\right)^2 - 4}}{2} \tag{4.16}$$

Both the + and – solutions are physically possible, since they correspond to $K_{eq}^b >= 1$ and $0 <= K_{eq}^b <= 1$, respectively. The above equation shows us that the value of k_{ex}^{obs} measured by ROESY or Nzz will define a curve in the two-dimensional K_{eq}^b *versus* k_{ex}^b space that shows the possible values of these parameters (Figure 4.3). Plotting the + and – solutions on a log scale, we see that these solutions converge at $\log(K_{eq}^b) = 0$ (*i.e.*, at $K_{eq}^b = 1$) and define the minimum possible value of k_{ex}^b (Figure 4.3). Moreover, the values of K_{eq}^b and k_{ex}^b derived from lineshape analysis should lie somewhere on this curve. In other words, this curve derived from k_{ex}^{obs} provides a constraining relationship between K_{eq}^{b} and k_{ex}^{b}. Notably, since K_{eq}^{b} and k_{ex}^{b} are intrinsic constants for a given enzyme and substrate and are independent of sample concentrations, ROESY or Nzz experiments can be made at different sample concentrations to obtain higher confidence in these fitted parameters.

4.2.6 Extracting Exchange Parameters (Rates, Populations) from Lineshapes

Several programs are currently available for lineshape analysis (Table 4.1) of protein–ligand interactions. Two-dimensional HSQC or HMQC spectra, collected at different titration points, are the typical input for these programs. Based on prior insight into the interaction system, the spectra are fit to one or more suitable reaction schemes. Depending on the reaction scheme, certain parameters are specific to each nucleus whereas others apply to all. For example, in a two-state binding scheme (Figure 4.1a), the k_{on} and k_{off} rate constants will be the same for all nuclei that are sensitive to the binding event. The chemical shifts (ω) and transverse relaxation rate constants (R_2) for the free and bound states will be unique to each nuclei. Lineshape analysis is performed in several iterations. At first, individual nuclei are fit to obtain their specific parameters, namely the chemical shift (ω) and transverse relaxation rate constant (R_2) for the bound state. Once several nuclei have been analyzed, a global fitting can be performed where the nuclei-specific parameters are held constant and the k_{on} and k_{off} rate constants are fit. Such iterative fitting is necessary to keep the number of fitted parameters to a minimum.

Since NMR spectra collected at different titration points are the input data for lineshape analysis, it is important to note that several factors during the titration experiment and processing of NMR spectra can affect spectral lineshapes. Mismatch in buffer conditions, such as pH and salts, between the protein and ligand solutions can systematically change the pH of the titration samples. The exchange rate of amide hydrogen is pH dependent.[44] Progressive change in pH can introduce a change in peak intensities in the HSQC or HMQC spectra. Electrostatics often plays an important role in

Table 4.1 Currently available NMR lineshape analysis programs.

	Fitting mode	Software dependency	Script or GUI	Download link
TITAN[67]	2D lineshapes	MATLAB	GUI based	https://www.nmr-titan.com/[80]
LineShapeKin	1D lineshapes	MATLAB	Script based	https://lineshapekin.net/[81]
IDAP	1D lineshapes	MATLAB	Script based	https://lineshapekin.net/#IDAP[82]
NmrLineGuru[19]	1D lineshapes	None	GUI based	https://www.purdue.edu/postlab/resources/nmrlineguru/[83]

protein–ligand interactions and hence, systematic change in salt concentration during titration can affect binding and subsequently the lineshape analysis. During processing of NMR spectra a window function is multiplied to the free induction decay (FID) to enhance the signal-to-noise ratio. For lineshape analysis, an exponential decay function should be used to preserve the Lorentzian lineshape. The same window function should be used for all spectra in the titration dataset. Using zerofilling, *i.e.* adding zeros at the end of the FID, is also helpful as it adds more points within each peak, enabling better determination of its peak center and linewidth. Finally, structure-dependent scalar couplings can be easily missed if peak splittings are not resolved. Inaccurate fitting parameters can result if scalar couplings are not included in the lineshape simulations. Taking care of these practical aspects can reduce ambiguity and help in extracting meaningful information from complex binding mechanisms by lineshape analysis.

4.3 Applications of Lineshape Analysis

4.3.1 Complete Characterization of Pin1 Catalysis of APP Substrate

Peptidyl-prolyl isomerase (PPIase) enzymes are ubiquitous in nature and play critical roles in many biological processes.[27] They can serve as chaperones for protein folding by allowing the peptide bonds preceding proline residues to rapidly sample both possible isomers until the lowest-energy folded state is reached.[45] Likewise, they can also accelerate isomer-specific protein–protein interactions, such as binding or enzyme catalysis. Some enzymes only recognize either the *cis* or *trans* isomer of a proline-containing substrate. In one example, the phosphatase Ssu72 requires the phosphorylation-specific prolyl isomerase Pin1 to recognize and dephosphorylate the C-terminal domain heptad repeats of RNA polymerase II.[46,47] Since Ssu72 only dephosphorylates the *cis* isomer of this substrate, Pin1 is necessary to convert the remaining *trans* isomers to *cis* for dephosphorylation to reach completion on an appropriate timescale (milliseconds to seconds). Pin1 is also necessary to accelerate complete dephosphorylation by *trans*-specific proline-directed phosphatases such as PP2A.[48]

Given the relevance of the rate of prolyl isomerization for numerous biological processes, the action of prolyl isomerase enzymes is of considerable interest. Assays of isomerase activity play an important role in uncovering details of isomerase substrate specificity and regulation. Early isomerase activity assays employed an isomer-specific protease (such as chymotrypsin) which would cleave the sequence Xaa-*trans*-Xaa-Yaa-Xaa where Yaa is aromatic and Xaa is any amino acid.[39,45,49] These assays would enable measurement of the *cis*-to-*trans* isomerization rate by following the slow phase of peptide bond cleavage after the initial pool of *trans* is cleaved. Protease-free assays have also been developed that take advantage of the differences in UV/Vis absorption between the *cis* and *trans* isomers of the sequence Xaa-Pro-Xaa-pNA, where pNa is *p*-nitroanilide.[50,51] Similarly, the difference in fluorescence between the *cis* and *trans* isomers of the sequence Trp-Pro has also been utilized to measure isomerization rates.[25] However, these assays all have disadvantages, including limitations on the sequence of the substrate C-terminal to the isomerized proline residue, and the need to perturb the *cis*/*trans* equilibrium in some way in order to measure the return to equilibrium.

NMR techniques such as lineshape analysis provide an alternate perspective on *cis*–*trans* isomerization. The ability to directly observe peaks corresponding to the *cis* and *trans* isomers provides numerous advantages over other biochemical techniques. For instance, any substrate sequence can be used as long as the sequence produces resolved peaks for both isomers. Additionally, NMR measurements can be performed at chemical equilibrium rather than requiring perturbation of the *cis*–*trans* equilibrium. Lastly, the sensitivity of NMR lineshapes to chemical exchange provides the potential to probe all the rates in a complex multi-state equilibrium.

The peptidyl prolyl isomerase (PPIase) enzyme Pin1 catalyzes the isomerization between the *cis* and *trans* isomers of the phosphorylated Ser-Pro or Thr-Pro (pS/T-P) peptide bond. Pin1 has a catalytic PPIase domain and a substrate binding WW domain. Pin1 is found to be involved in several diseases, including asthma, cancer, and Alzheimer's disease (AD). In AD, the amyloid precursor protein (APP) gets cleaved, leading to the formation of the neurotoxic amyloid beta peptide.[52] Pin1 interacts with the APP intracellular domain (AICD), which is intrinsically disordered,[53] and plays a neuroprotective role.[54–56] NMR spectroscopy is a very powerful technique to study intrinsically disordered regions and their interaction with partner proteins.[57] Structural characterization of AICD by NMR spectroscopy revealed a significant *cis* population for a pT-P motif within AICD.[58] The pT-P peptide bond undergoes slow exchange between the *cis* and *trans* isomers, which is catalyzed by Pin1.[56]

4.3.1.1 *Lineshape Analysis of the Pin1 PPIase Domain Acting on APP Substrate*

In Greenwood *et al.*,[11] the authors employed NMR lineshape analysis to measure the on-enzyme rates of *cis*–*trans* isomerization of an APP-derived peptide bound to the catalytic PPIase domain of Pin1. One of the purposes of this endeavor was to draw insights about the role of isomerization by Pin1 in APP biology from these microscopic rates and populations. Of particular interest was the K_{int}, or equilibrium constant describing the ratio of bound substrate to bound product at equilibrium, in this case [PPIase:*trans*]/[PPIase:*cis*].

Because the WW domain of Pin1 binds to the same phosphorylated pSer-Pro/pThr-Pro motif that the catalytic domain does, the researchers deemed it necessary to exclude the WW domain from their Pin1 construct. This simplified the reaction scheme without substantially changing the observed isomerization rate ($k_{ex} = 41 \pm 1$ s^{-1} for the PPIase domain *vs.* 28 ± 1 s^{-1} for full-length Pin1 under the conditions of 50 μM isomerase, 3 mM substrate)

(Figure 4.4). The substrate peptide, which corresponded to residues 659–682 of APP phosphorylated at T668, was ^{13}C labeled at the Proline 669 position in order to collect ^{1}H–^{13}C HSQC experiments. This labeling scheme provided multiple sensors of the catalyzed exchange process, and the resulting peaks were simultaneously fitted by lineshape analysis to increase confidence in the fitted parameters.

As Pin1-PPIase was titrated into ^{13}C-labeled peptide, the peaks corresponding to the *cis* isomer broadened severely, even at the lowest non-zero enzyme concentration of 60 μM. The *trans* peaks broadened as well, but remained detectable up to a concentration of 1 mM Pin1-PPIase. Although this substrate-perspective approach was very sensitive to the catalytic rates, the isomerization-induced broadening made it difficult to obtain spectra under conditions of saturation in order to measure the binding affinity. The apparent dissociation constant K_D^{App} was therefore determined in an independent experiment using ^{15}N-labeled Pin1-PPIase titrated with unlabeled

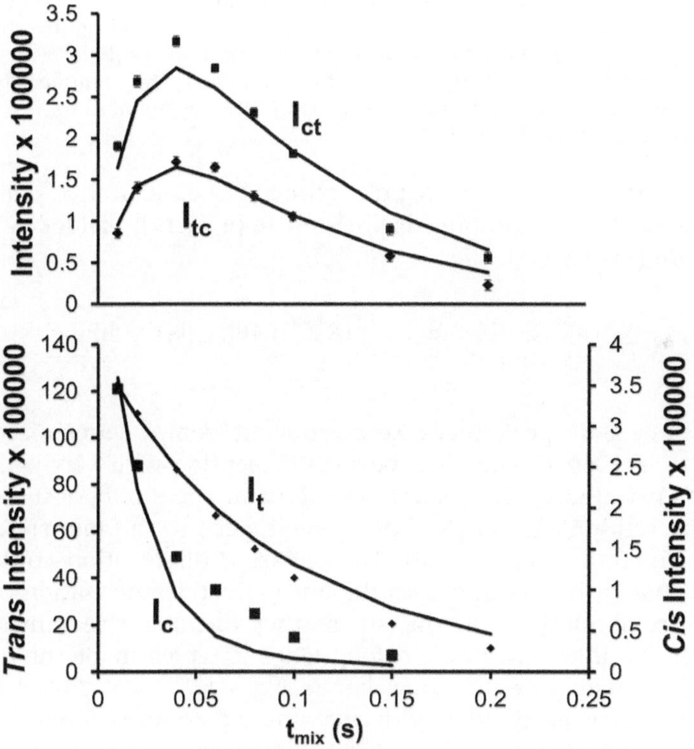

Figure 4.4 Isomerization of pAICD by Pin1-PPIase monitored by ROESY spectra. The intensity of *cis*-to-*trans* (I_{ct}) and *trans*-to-*cis* (I_{tc}) cross-peaks and *trans* (I_t) and *cis* (I_c) diagonal peaks in a series of ROESY spectra are shown as a function of ROESY mixing time. The concentration of pAICD was 3.0 mM and the concentration of Pin1-PPIase was 50 μM. Symbols represent peak intensities and lines represent a global fit to the data, which yielded a $k_{ex} = k_{cis\text{-to-}trans} + k_{trans\text{-to-}cis}$ of 41 ± 1 s^{-1}.[11]

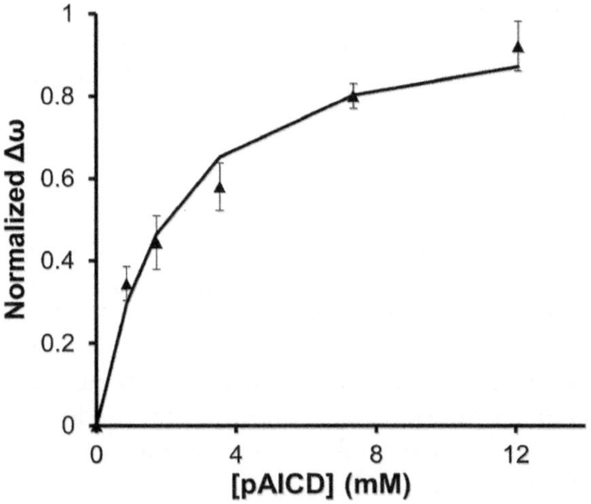

Figure 4.5 Binding curve for Pin1-PPIase binding to pAICD. The average normalized chemical shift perturbations of peaks corresponding to the amide groups of 31 Pin1-PPIase residues are plotted (symbols) as a function of pAICD concentration. Error bars represent the standard deviations across the 31 peaks and the line represents the global fit to the data used to determine the K_D^{app}.[11]

substrate peptide. Fitting the peptide-induced chemical shift changes to the fast-exchange limit-binding expression [eqn (4.17)] yielded a K_D^{App} of 1.3 ± 0.3 mM (Figure 4.5).

$$\Delta\omega = \Delta\omega^{max} \frac{K_D^{app} + [P] + [L] - \sqrt{\left(K_D^{app} + [P] + [L]\right)^2 - 4[P][L]}}{2[P]} \tag{4.17}$$

Proton dimension peak slices were extracted from a series of constant-time ^1H–^{13}C HSQCs of the ^{13}C-labeled APP peptide with varying concentrations of Pin1-PPIase. These slices were fit using the BioPhysicsLab Matlab package (currently IDAP, https://lineshapekin.net/) to the four-state model shown in Figure 4.2 (Figure 4.6). The apparent dissociation constant K_D^{App} was held constant in the fit, as was the uncatalyzed isomerization rate. The off-rates were similarly held constant as they did not have a meaningful effect on the fit unless they were below 1000 s^{-1}, which made the fit worse. The resulting on-enzyme rate of exchange, $k_{cat}^{cis} + k_{cat}^{trans}$, was 1500 ± 100 s^{-1}, which is in close agreement with the rate of conformational motions measured in Pin1 during catalysis (1200 ± 200 s^{-1}), detected by ref. 59. The equilibrium constant describing the ratio of bound *trans* to bound *cis* at equilibrium, or K_{int}, was found to be 3.9 ± 0.8. The optimal value for K_{int} for an enzyme that operates near equilibrium between its substrates and products is 1. The value of 3.9 was then taken as evidence of the possibility that Pin1 is optimized to operate under conditions where the AICD is slightly enriched in the *cis* isomer.

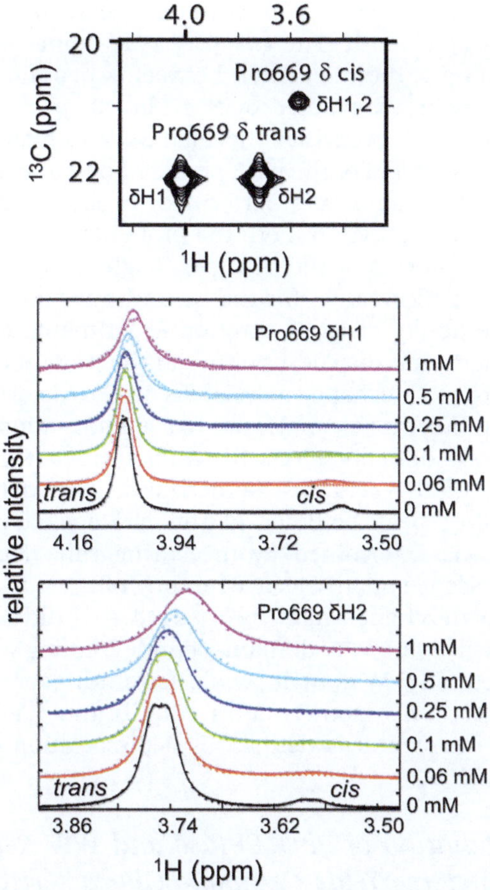

Figure 4.6 Lineshape fitting of proline delta peaks. (Top) Portion of a ^{13}C–^{1}H constant-time HSQC spectrum of ^{13}C-Pro669-pAICD showing the *cis* and *trans* Pro669 delta peaks. (Bottom) ^{1}H lineshapes of the ^{13}C-Pro669-pAICD delta protons extracted from HSQC spectra as a function of Pin1-PPIase concentration (listed on the right). Symbols represent data and lines represent fits to the data. Reproduced from ref. 11 with permission from Springer Nature, Copyright 2011.

4.3.1.2 *Lineshape Analysis of the Pin1 WW Domain Binding to APP Substrate*

The WW domains are the smallest peptide recognition modules. These domains are approximately 40 amino acids in length and have two conserved tryptophans (W). WW domains are found in many proteins and facilitate protein–protein interaction *via* recognition of proline-rich sequences.[60] Based on their specificity toward different proline-rich sequences, WW domains are classified into five groups (Verdecia 2000 *Nat. Str. Biol.*). Group I binds PPxY motif, group II binds PPLP motif, group III recognizes PGM motif, group IV WW domains bind to sequences containing pS/T-P motif, and group V binds PR motif. The group IV WW domain is present in Pin1.

In the context of AD, the interaction between a peptide derived from AICD and Pin1 WW domain was studied by NMR lineshape analysis.[2] NMR-based titration experiments showed that the Pin1 WW domain discriminates between *cis* and *trans* isomers of the pT-P peptide bond and specifically binds to the *trans* isomer. Titrations were performed to observe the protein–peptide binding from the perspectives of both the protein and the peptide by specific ^{15}N-labeling of the protein or the peptide. Neighboring residues of the pT-P showed distinct chemical shifts for the *cis* and *trans* conformers. This highlights the unique ability of NMR titration experiments to discern any conformational changes that may occur either in the protein or the ligand before or after binding. NMR lineshape analysis was used to determine the kinetics of the three-state binding model, analogous to a conformational selection model, where the Pin1 WW domain selects the *trans* isomer (Figure 4.7). The equilibrium constant of the *cis–trans* isomerization was determined from the corresponding NMR peak volumes of the ^{15}N-labeled peptide. The interconversion rate was determined by measuring time-dependent change in tryptophan fluorescence upon selective binding of the *trans* isomer to the WW domain. This slow exchange rate ($k_{ex} = 0.0144\ s^{-1}$) did not affect the NMR lineshapes and hence, the WW domain–peptide binding was fit to a two-state model (free and bound WW domain) while accounting for the thermodynamic coupling to the *cis–trans* isomerization equilibrium (Figure 4.7). The WW domain–peptide interaction is defined by a dissociation constant of 22 μM and an off-rate of 2000 s^{-1}.

4.3.1.3 Combination of Pin1 PPIase and WW Reaction Parameters Yields Complete Characterization of Pin1 Reaction Cycle

The PPIase and WW domains in Pin1 interact with the same peptide sequence, and so in the most general Pin1 reaction scheme there would be six states with eight connecting steps. These steps account for binding of each isomer to each domain, the potential transfer of substrate directly between the two domains, and *cis–trans* exchange in the free and on-enzyme substrate (Figure 4.8). Since the lineshape analysis of APP substrate binding to the WW domain showed that its binding to the *cis* isomer could be neglected, the number of distinct states in the Pin1 reaction scheme is reduced to five. Notably, the on-enzyme exchange rate determined for the PPIase domain, $1500 \pm 100\ s^{-1}$, is in the same rate regime as the off-rate of 2000 s^{-1} determined for WW interaction with substrate, suggesting a possible synchronous transfer of substrate between the two domains. If present, this transfer step would have an equilibrium constant given by the ratio of [WW:*trans*]/[PPIase:*trans*], which would be roughly 65 if the isolated domains have the same thermodynamic parameters as in the intact Pin1 protein. Moreover, the rapid binding kinetics and relatively weak affinity of

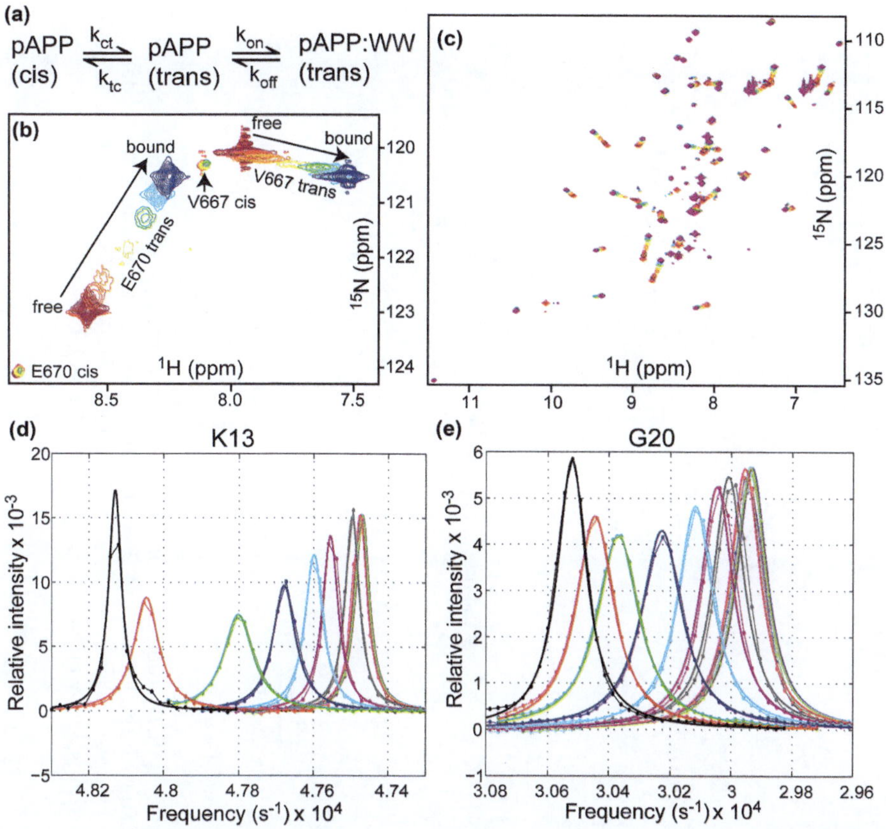

Figure 4.7 Lineshape analysis of Pin1-WW domain binding to pAICD. (a) The three-state model used for lineshape analysis. The pT-P bond in APP intracellular domain (AICD) isomerizes between *cis* and *trans* isomers with rate constants k_{ct} and k_{tc}. The Pin-WW domain specifically binds the *trans* isomer association and dissociation rate constants of k_{on} and k_{off}, respectively. (b) Residues V667 and E670, in the peptide pAPP659-679, were specifically ^{15}N-labeled and titrated with unlabeled Pin1-WW domain. Overlays of the ^{15}N-HSQC spectra collected at each titration point show chemical shift changes of the peaks corresponding to the *trans* isomer. (c) ^{15}N-labeled WW was titrated with unlabeled pAPP659-682 and the corresponding ^{15}N-HSQC spectra are overlaid. Lineshape fitting corresponding to the residues K13 (d) and G20 (e) are shown. Adapted from ref. 2 with permission from American Chemical Society, Copyright 2012.

this enzyme–substrate interaction are consistent with the APP substrate being one of many Pin1 substrates in the cellular milieu, where fast kinetics would facilitate rapid responses to phosphorylation signals. In summary, the application of lineshape analysis to each of the two Pin1 domains acting on the APP substrate yielded a complete description of the Pin1 catalytic cycle that offers insights into the role of Pin1 as a hub and timing device in the numerous biochemical pathways it is known to engage with.[2]

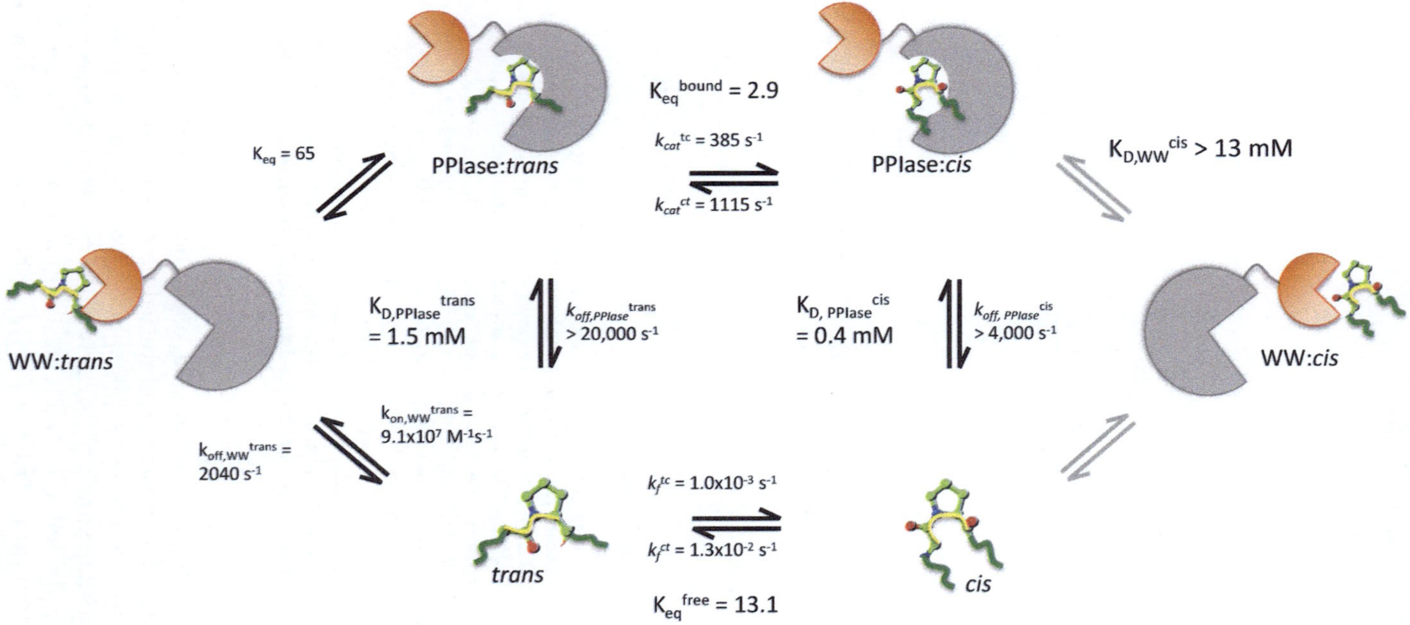

Figure 4.8 Complete reaction scheme of Pin1 interaction with pAICD. Interaction of the *cis* and *trans* isomers of the pThr-Pro motif in phosphorylated APP intracellular domain (pAICD) with the PPIase and WW domain are shown using a six-state model. The kinetic and thermodynamic parameters for each step were determined by a combination of NMR lineshape analysis and other biophysical methods. Adapted from ref. 2 with permission from American Chemical Society, Copyright 2012.

4.3.2 Characterization of the Reaction Cycle Parameters for Cyclophilin LRT2/Cyp2 Acting on Transcription Repressor OsIAA11 in Rice

The phytohormone auxin regulates many developmental and environmental response processes in plants.[61] The auxin circuit is central to this control, in which an auxin-responsive protein–protein interaction network functions as a negative-feedback gene regulation circuit (Figure 4.9). This circuit operates through auxin-triggered degradation of a given member of the Aux/IAA family of transcription repressor proteins, causing activation of the corresponding repressed genes that include the Aux/IAA repressor protein's own gene. The interaction that serves as the auxin sensor in this circuit involves the $SCF^{TIR1/AFB2}$ complex, which is an E3 ubiquitin ligase. This complex

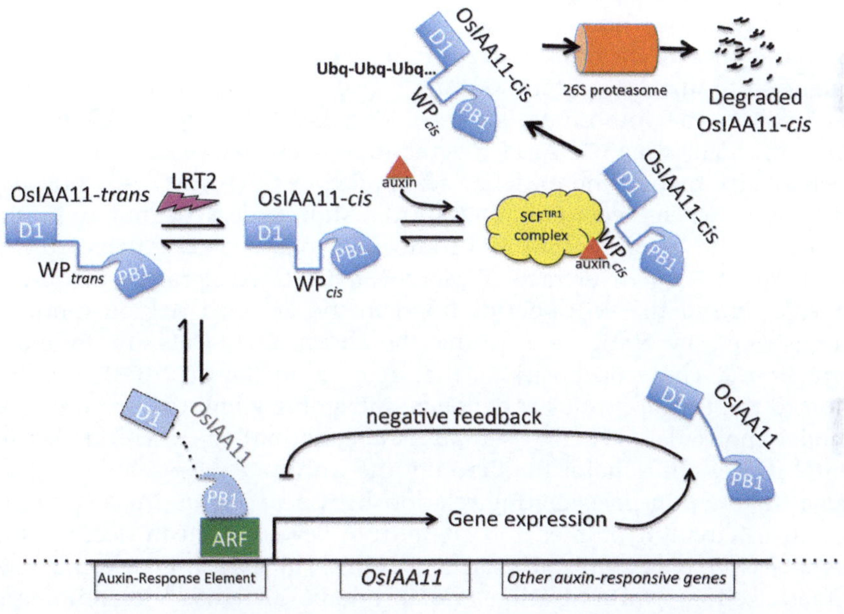

Figure 4.9 The auxin-responsive negative-feedback gene regulation circuit in rice. In the repressed or "off" state, the Aux/IAA transcription repressor protein OsIAA11 is bound to its corresponding Auxin Responsive Factor (ARF) and represses transcription. When there is an increase in auxin concentration, the *cis*-specific binding of OsIAA11 along with auxin to the $SCF^{TIR1/AFB2}$ E3 ligase complex promotes the ubiquitination and subsequent proteasomal degradation of OsIAA11. As OsIAA11 is depleted, LRT2 *cis–trans* isomerase activity replenishes the *cis* population of OsIAA11 to supply the *cis* substrate for entry into the proteasomal degradation pathway, facilitating full depletion of OsIAA11. Depletion of OsIAA11 allows ARF to activate transcription of downstream genes that include the gene for OsIAA11 and genes for proteins involved in lateral root development. Newly expressed OsIAA11 exerts negative feedback by again repressing gene transcription.

simultaneously binds to auxin and to the Aux/IAA protein, but recognition is specific for the *cis* isomer of the Trp-Pro peptide bond in the highly conserved "degron sequence" (-G**WP**PV-) of Aux/IAA protein family members.[62] In this way, only the *cis* isomer of an Aux/IAA protein is targeted for ubiquitin-mediated proteasomal degradation and only in response to auxin. As the *cis* isomer is selectively degraded, PPIase activity is required to maintain the free *cis–trans* equilibrium and continually replenish the *cis* isomer until sufficient depletion of the Aux/IAA protein is achieved. Since a given Aux/IAA protein represses transcription of a set of genes that includes its own, activation of these genes by depletion of the Aux/IAA protein leads to newly expressed AUX/IAA protein that again represses transcription, thereby forming a negative-feedback gene regulation circuit.

The essential function of the PPIase LRT2/Cyp2 (hereafter referred to as LRT2) in the auxin circuit that governs lateral root development was revealed by mutagenizing rice seedlings, screening for the lateral rootless phenotype, and identifying the mutated gene.[63,64] The same lateral rootless phenotype was found in a mutagenized rice strain where a mutation in the degron sequence of Aux/IAA protein OsIAA11 also displays the lateral rootless phenotype.[65] The importance of the LRT2:OsIAA11 enzyme–substrate reaction in facilitating OsIAA11 degradation was demonstrated in LRT2 knockout rice plants where higher accumulation of OsIAA11 was observed, and *in vitro* where recombinant OsIAA11 showed less degradation when incubated with extracts from LRT2 knockout plants as compared to incubation with WT plant extracts.[66] Moreover, LRT2 acceleration of *cis–trans* isomerization of the W-P peptide bond in the OsIAA11 degron motif was demonstrated by NMR, confirming the direct LRT2-OsIAA11 functional interaction.[66] These observations that span from the LRT2:OsIaa11 interaction to the lateral rootless phenotype intriguingly link the dynamics of a bimodal molecular switch (*cis–trans* isomerization) to a critical developmental process in a major food crop of the world.

As a first step in investigating relationships between *cis–trans* isomerization, auxin circuit dynamics and lateral root development in rice, the four-state reaction cycle for catalysis of OsIAA11 by LRT2 (Figure 4.2a) was characterized by NMR titration and lineshape analysis.[10] Recombinantly expressed ^{15}N-OsIAA11 (residues 72–125) was titrated with unlabeled LRT2 and ^{15}N–^{1}H HSQC spectra were acquired. Lineshape analysis using TITAN[67] yielded best-fit thermodynamic and kinetic parameters of the four-state reaction cycle (Figure 4.10). The free and on-enzyme equilibrium constants are nearly equal and close to 1 ($K^{f}_{eq} = 1.113$, $K^{b}_{eq}, = 1.119$), the affinities are $K^{cis}_{D} = K^{trans}_{D} = 1.55$ mM, and both binding reactions are fast (diffusion limited on-rates). The on-enzyme exchange rate, $k^{cat}_{ex} = k^{ct}_{cat} + k^{tc}_{cat}$, ~95 Hz, is roughly an order of magnitude slower than typical PPIase-catalyzed exchange rates for other substrates.[11,67,68] This slow catalysis is due to the degron substrate and not to LRT2, since human CypA shows similarly slow activity on this substrate[9] and LRT2 has higher activity on a standard cyclophilin substrate.[10]

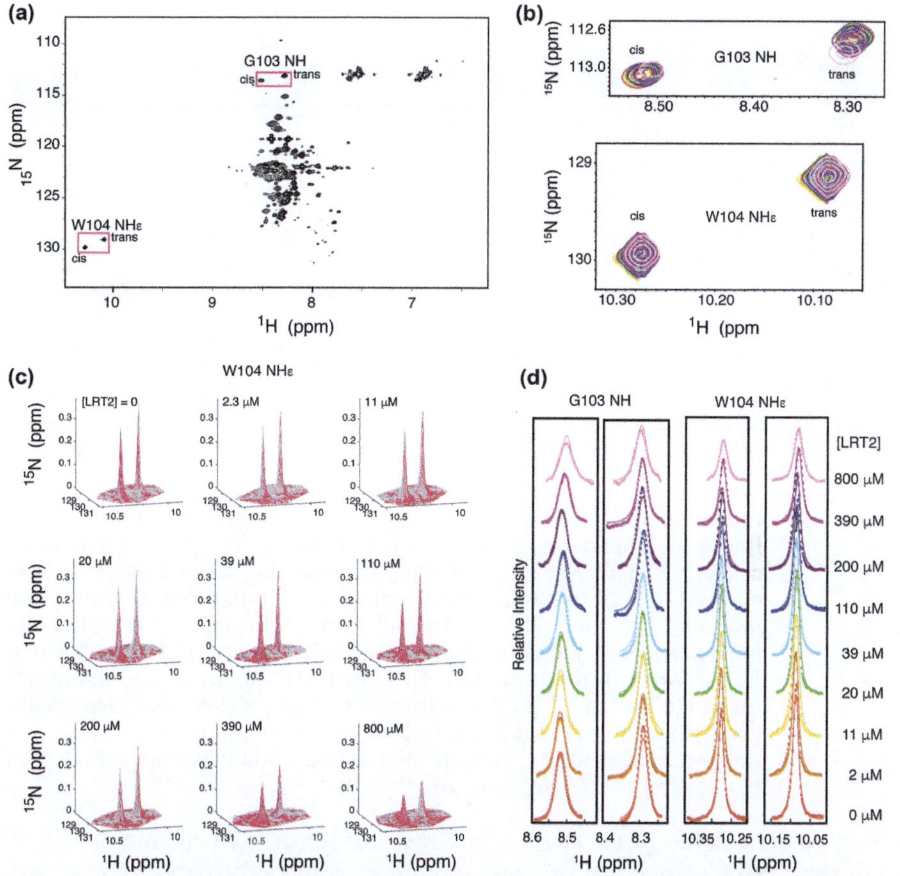

Figure 4.10 Lineshape analysis of NMR titration data for LRT2 catalysis of OsIAA11. (a) ^{15}N–^{1}H HSQC spectrum of apo ^{15}N-labeled OsIAA11 (amino acid residues 72–125), with well-resolved Gly-103 and Trp-104 *cis* and *trans* peaks indicated by red-boxed regions and labels. (b) Regions of the ^{15}N–^{1}H HSQC spectrum showing Gly-103 and Trp-104 peaks at different concentrations of LRT2. (c) Output of TITAN lineshape analysis for W104 NHe peaks at different LRT2 concentrations, showing overlays of the experimental spectra (gray) with the simulated spectra using the four-state reaction model and best-fit parameters. (d) One-dimensional slices extracted from the data shown in (c) to clearly show the goodness of fit (points are experimental data, solid lines are simulated data). Adapted from ref. 10 with permission from PNAS, Copyright 2019.

Independent validation of the parameters obtained from lineshape analysis was provided by measurement of k_{ex}^{obs} by Nzz exchange spectroscopy, and by measurement of K_D^{app} using ^{15}N-LRT2 titrated with unlabeled OsIAA11 (Figure 4.11). The experimentally measured $k_{ex}^{obs} = 0.62\ s^{-1}$, obtained using a sample containing 0.8 mM ^{15}N-OsIAA11 and 16 μM LRT2, is closely predicted by insertion of the fitted parameters and sample conditions into eqn (4.10),

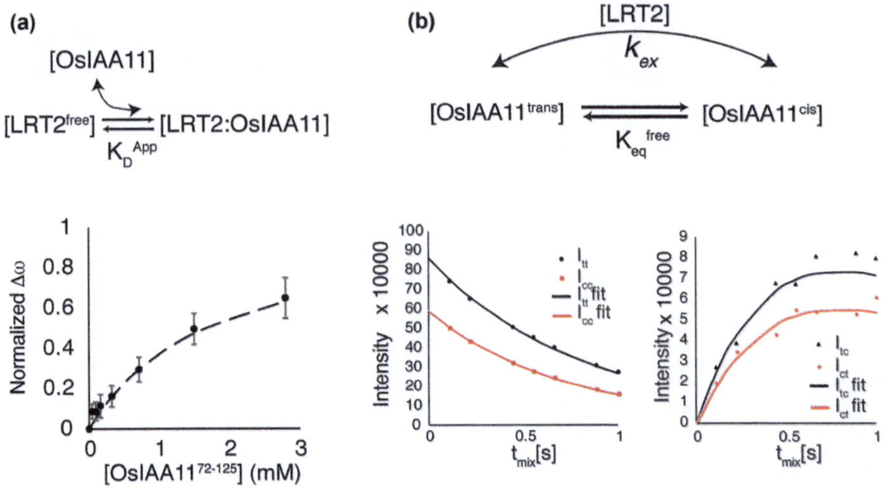

Figure 4.11 Independent measurements for validation of parameters obtained from lineshape analysis. (a) The apparent affinity (K_D^{App}) of the LRT2 interaction with OsIAA11 was measured by acquiring ^{15}N–^{1}H HSQC spectra of ^{15}N-LRT2 at different concentrations of unlabeled OsIAA11 and plotting the normalized chemical shift perturbations for 20 LRT2 amino acid residues. (b) The overall exchange rate (k_{ex}^{obs}) for 0.8 mM ^{15}N-OsIAA11 in the presence of 16 μM LRT2 was measured using Nzz exchange spectroscopy. Intensities of autopeaks (left) and cross-peaks (right) and the corresponding best-fit curves are plotted as a function of the Nzz mixing time (black and red points). Adapted from ref. 10 with permission from PNAS, Copyright 2019.

which yields a value of 0.64 s^{-1}. Similarly, independent measurement of K_D^{app} from the ^{15}N-LRT2 perspective using chemical shift perturbation and a two-state model yielded a value of 1.54 mM, in excellent agreement with the value obtained from lineshape analysis using the four-state model (1.55 mM).

These investigations revealed that LRT2 is a highly balanced enzyme, with K_{eq}^{b} close to 1. This on-enzyme equilibrium constant indicates a high degree of catalytic effectiveness in the cell for the case where the steady-state substrate/product ratio, [S]/[P], is close to the equilibrium $[S]_{eq}/[P]_{eq}$ value. Considering *cis* as both the product of LRT2 catalysis and the substrate of the SCF^{TIR1} E3 ligase, and noting that the half-life of OsIAA11 has been estimated to be on the order of 20 minutes,[69] our results indicate that LRT2 will maintain the steady-state [*trans*]/[*cis*] ratio very close to the equilibrium $[trans]_{eq}/[cis]_{eq}$ value. This means that in the non-equilibrium context of a living cell, LRT2 supplies the maximum possible *cis* isomer to the proteasomal degradation pathway. Notably, binding of LRT2 to the degron motif is weak (1.5 mM). The fact that LRT2 and OsIAA11 co-immunoprecipitate (co-IP) with SCF^{TIR1} E3 ligase is not explained by the LRT2:OsIAA11 interaction and suggests that LRT2 must interact with other components of this complex.

To explore specific LRT2 mutations that might alter the binding and/or kinetics of the reaction cycle, a series of mutants were constructed and

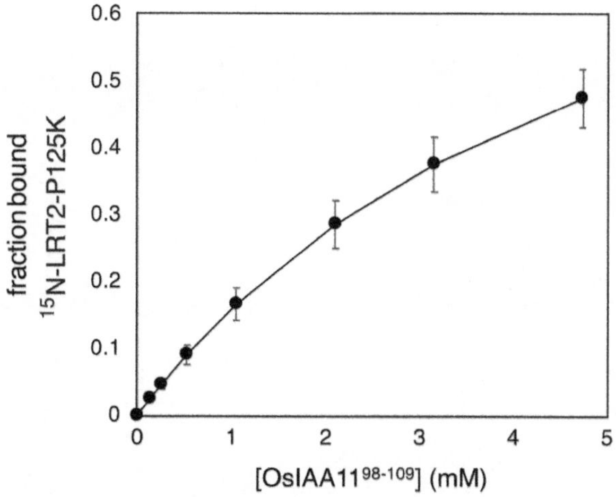

Figure 4.12 Independent measurement of K_D^{APP} for LRT2-P125K. Plot of the fraction of ^{15}N-LRT2-P125K bound (from chemical shift perturbation) as a function of OsIAA11 concentration. Fitting of experimental data (points) to the two-state binding model (line) yielded $K_D^{APP} = 5.2 \pm 0.9$ mM. Reproduced from ref. 9 with permission from Springer Nature, Copyright 2019.

screened for near-WT thermodynamic stability.[9] The LRT2-P125K mutant emerged as a variant with stability close to WT-LRT2 and with sufficient catalytic activity for characterization by lineshape analysis. For this mutant, K_D^{app} and k_{ex}^{obs} were measured first for comparison with WT LRT2. The independently measured K_D^{app} value (5.2 mM; Figure 4.12) was used as a constraint in the lineshape analysis fitting. Measurement of k_{ex}^{obs} at the same enzyme and substrate concentrations as used for WT LRT2 shows that this mutant has ~75% of WT, which will be useful for investigating relationships between LRT2 activity and lateral root development in rice plants.[9] Therefore, this mutant was subjected to NMR titration experiments and lineshape analysis to characterize the full reaction cycle.

As described above for WT LRT2, ^{15}N-OsIAA11 was titrated with unlabeled LRT2-P125K and ^{15}N–^{1}H HSQC spectra were acquired. Lineshape analysis using TITAN[67] with the independently determined K_D^{app} and R_2 values held constant yielded good fits (Figure 4.13). The on-enzyme equilibrium constant and on-enzyme exchange rate are $K_{eq}^{b} = 2.9 \pm 0.2$ and $k_{ex}^{b} = 211 \pm 5\ s^{-1}$, respectively, and again both binding reactions are diffusion-limited. To evaluate the validity of the fitted values of K_{eq}^{b} and k_{ex}^{b}, the overall observed exchange rate, k_{ex}^{obs}, was measured at two different sample conditions by ROESY ([LRT2-P125K] = 125 μM, [OsIAA11] = 1.84 mM) and Nzz ([LRT2-P125K] = 16 μM, [OsIAA11] = 0.8 mM) (Figure 4.14). The resulting k_{ex}^{obs} values, together with their corresponding enzyme and substrate concentrations and the independently measured values of K_D^{APP} and K_{eq}^{f}, define curves on which the K_{eq}^{b}, k_{ex}^{b} coordinate should fall (eqn (4.15) and (4.16)). The

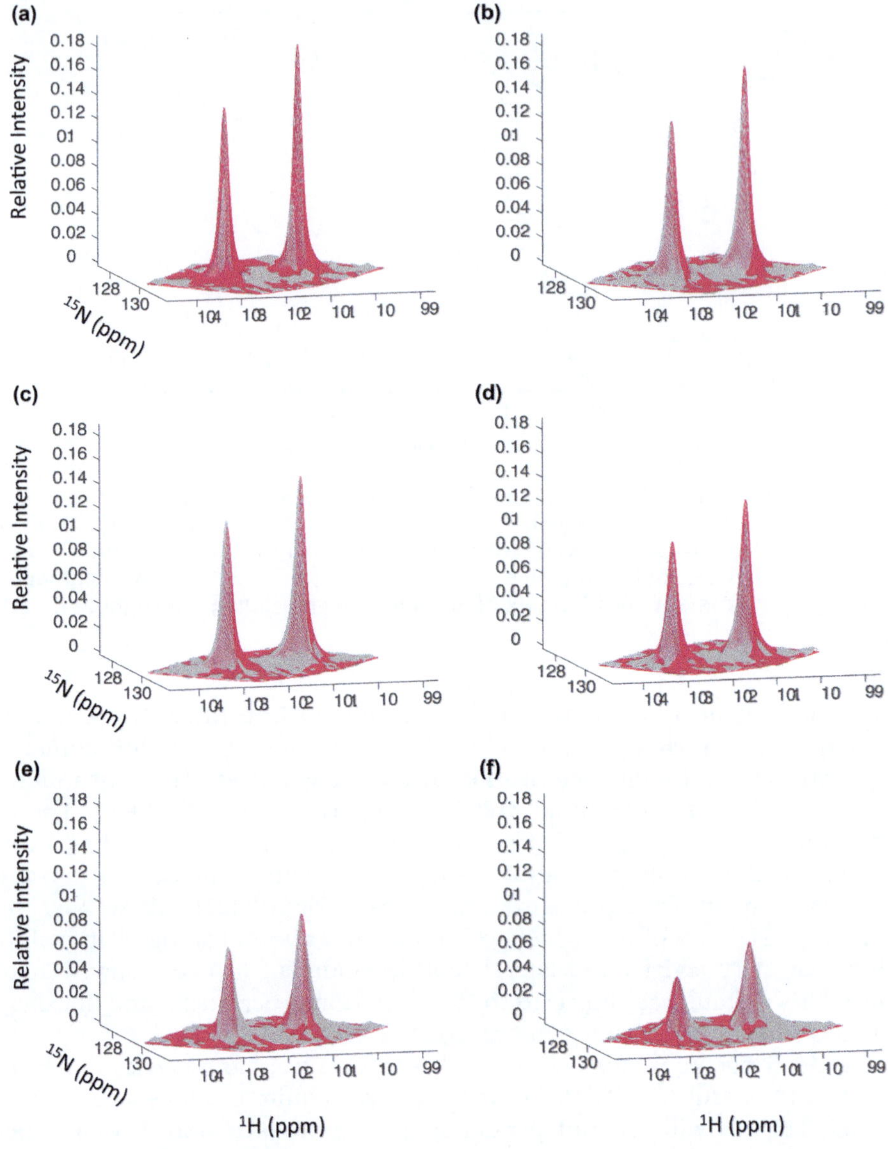

Figure 4.13 Goodness of fit for the lineshape analysis of ^{15}N-OsIAA11 titrated with LRT2-P125K. (a–f) Overlays of experimental (gray) and simulated (pink) ^{15}N-^{1}H HSQC spectra of ^{15}N-OsIAA11 showing the ^{104}W-NHε^{cis} and ^{104}W-HHε^{trans} peaks at each LRT2-P125K concentration (mM): (b) zero, (c) 0.13, (d) 0.25, (e) 0.50, (f) 1.0, and (g) 2.17. Reproduced from ref. 9 with permission from Springer Nature, Copyright 2019.

resulting curves are in close agreement with each other, and the fitted $K^{\mathrm{b}}_{\mathrm{eq}}$, $k^{\mathrm{b}}_{\mathrm{ex}}$ values define a point that indeed falls between the two curves (Figure 4.15), validating the fitted values.

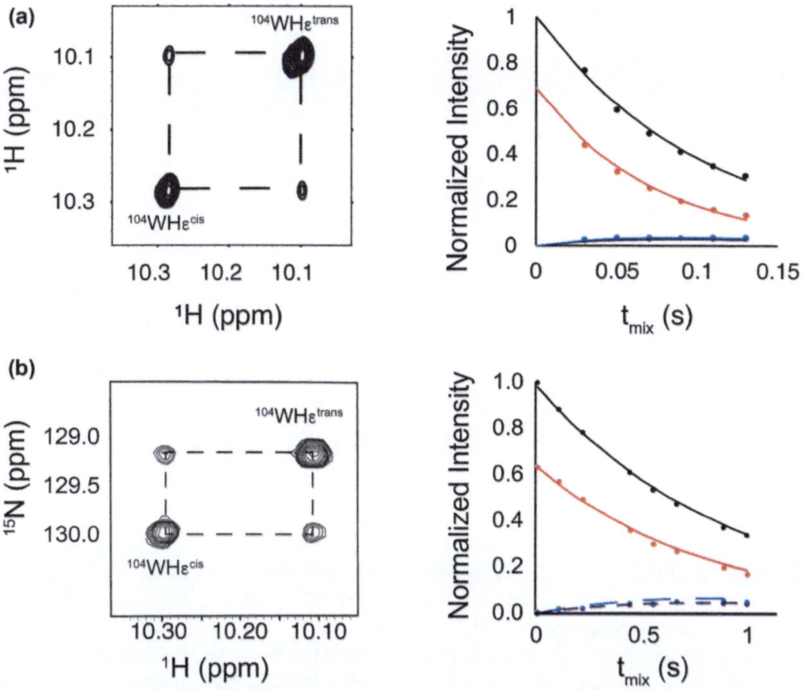

Figure 4.14 Independent measurements of k_{ex}^{obs} for LRT2-P125K. (a; left panel) Expanded region of 2D ^{1}H–^{1}H ROESY spectrum of OsIAA11 (mixing time of 120 ms) in the presence of LRT2-P125K. The ^{104}W-Hε^{cis} and ^{104}W-Hε^{trans} peaks show exchange cross-peaks, indicating detectable catalysis by this mutant. (Right panel) Fitting (lines) of the normalized peak intensities (dots) as a function of ROESY mixing time for the auto peaks (I_{tt} and I_{cc}, black and red, respectively) and exchange peaks (I_{tc} and I_{ct}, dashed blue and black, respectively) of ^{104}W-Hε yielded $k_{ex}^{obs} = 2.7 \pm 0.3$ s^{-1} for sample concentrations [LRT2-P125K] = 120 μM, [OsIAA11] = 1.84 mM. (b; left panel) Expanded region of 2D ^{1}H–^{15}N N_{ZZ}-exchange spectrum (mixing time = 0.55 s) of ^{15}N-OsIAA11 in the presence of LRT2-P125K. The ^{104}W-Hε^{cis} and ^{104}W-Hε^{trans} peaks show exchange cross-peaks. (Right panel) Fitting (lines) of normalized peak intensities (dots) for auto (I_{tt} and I_{cc}) and exchange (I_{tc} and I_{ct}) peaks [as in (a)] as a function of Nzz mixing time yielded $k_{ex}^{obs} = 0.47 \pm 0.04$ s^{-1} for sample concentrations [LRT2-P125K] = 16 μM, [OsIAA11] = 0.8 mM. Reproduced from ref. 9 with permission from Springer Nature, Copyright 2019.

For the P125K mutant, the on-enzyme equilibrium constant indicates that this mutant binds more favorably to the *trans* isomer than the *cis* ($K_{eq}^{b} = 2.9 \pm 0.2$), in contrast to WT LRT2 that binds with nearly equal affinity to the two isomers ($K_{eq}^{b} = 1.12 \pm 0.11$). Moreover, the significantly faster on-enzyme exchange rate of the P125K mutant (211 s^{-1}) compared to the WT enzyme (95 s^{-1}) indicates that either the proline at this position raises the activation energy barrier between the bound *cis* and *trans* states, or lysine at

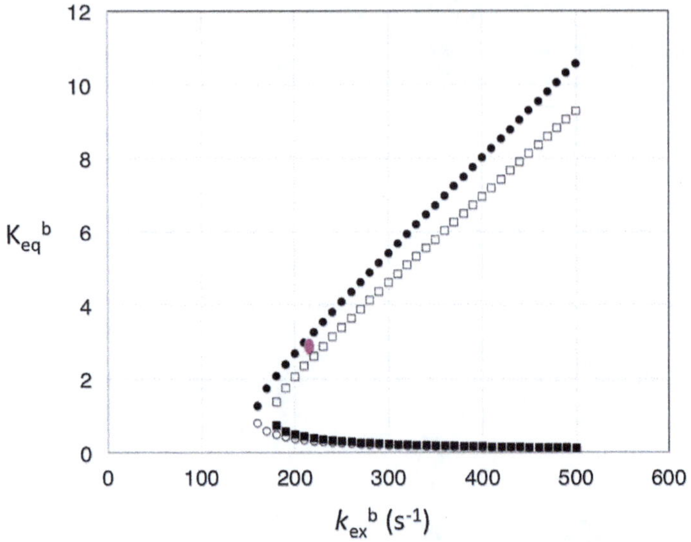

Figure 4.15 Validation of parameters from lineshape analysis for LRT2-P125K catalysis of OsIAA11. Theoretical plots of K_{eq}^{b} *versus* k_{ex}^{b} for the two independently measured k_{ex}^{obs} values for LRT2-P125K showing possible pairings of K_{eq}^{b} and k_{ex}^{b} values corresponding to the ROESY (circles) and Nzz (squares) values using the plus (filled) and minus (open) solutions of eqn (4.12) and (4.13), and the K_{eq}^{b}, k_{ex}^{b} coordinate obtained by lineshape analysis (pink ellipse). Reproduced from ref. 9 with permission from Springer Nature, Copyright 2019.

this position lowers this barrier. This result provides insight on how combining this P125K mutation with a mutation that increases substrate affinity could potentially yield a faster enzyme for this degron substrate.

Overall, these studies that characterize the full catalytic reaction scheme for LRT2 and a site-specific mutant provide groundwork for addressing the central question of the quantitative relationship between isomerization rate, auxin circuit dynamics, and lateral root development in rice plants. The full parameterization of these reaction schemes enables predictions of activity at cellular concentrations. Notably, since the binding constants for both WT and P125K LRT2 far exceed the expected cellular concentrations of OsIAA11, the overall observed exchange rate is independent of [OsIAA11] but is linearly dependent on [LRT2], as indicated by eqn (4.10). In other words, a given concentration of LRT2 will maintain a constant k_{ex}^{obs} over a wide range of [OsIAA11], facilitating efficient flux of the *cis* OsIAA11 isomer into the proteasomal degradation pathway.[9] Tuning the LRT2 reaction to slower rates as accomplished with the P125K mutant offers the opportunity to investigate how the rate of prolyl isomerization influences flux through a well-characterized gene regulation circuit and corresponding phenotypic changes at the whole organism level.

4.4 Conclusions and Perspectives

In this chapter we have presented the theoretical framework of lineshape analysis and examples of the application of lineshape analysis to protein–ligand and enzyme–substrate systems. These examples highlight the predictive power provided by a full characterization of the parameters that govern a reaction scheme. Moreover, the described systems showcase how independent measurements can be performed to reduce the number of fitted parameters and to test the validity of the parameters obtained from lineshape analysis. The two biological systems presented function as bimodal switches in protein–protein interaction networks involved in Alzheimer's disease and in lateral root development in rice. In both of these systems, the application of NMR titration coupled with lineshape analysis has yielded key insights regarding connections between the kinetics and thermodynamics of the biological interface and the role of this interaction in the context of the living cell.[2,9,11,70]

As with any process in which experimental data are theoretically fit to extract parameters, it is important to assess the confidence in the resulting fitted parameters. As described above, independent measurements can reduce the number of unknown parameters and can provide important validations of the resulting fitted parameters. Similarly, the different sensitivities to chemical exchange of single quantum (SQ), multiple quantum (MQ), zero quantum (ZQ), and double quantum (DQ) coherence experiments can be leveraged to obtain different views of the same exchange process.[71] Thus, the application of HSQC, HMQC, HZQC, and HDQC experiments to the same system with subsequent independent lineshape analysis of the resulting spectra can allow characterization of multi-state mechanisms with improved confidence in the fitted parameters.

Lineshape analysis is complementary to NMR relaxation methods and combining these approaches offers potential synergy for tackling highly complex systems. In addition to the classic lineshape analysis approach, exciting NMR-based methods have been developed during the past 30 years for the quantitative measurement of motions that occur on the millisecond to microsecond timescale. These methods include pulse sequences that employ the Carr–Purcell–Meiboom–Gill (CPMG) or $R_{1\rho}$ relaxation approaches to generate dispersion profiles that yield rates, populations, and chemical shifts associated with multi-state exchange processes.[72,73] Application of these methods to study functional motions in proteins has provided a deeper understanding of the roles that internal protein motions play in binding and catalysis.[74] The number of states in an exchange reaction scheme remains the major limitation to application of these approaches, with robust interpretations of three-state[75] and four-state[76] exchange processes reported so far. An intriguing possibility is that a combination of relaxation and lineshape analysis approaches on the same exchange reaction could enable the robust characterization of even more complex multi-state systems.

Acknowledgements

Support from the National Science Foundation (MCB-1615350, to LKN) and from Science & Engineering Research Board (SERB), India (grant CRG/2020/004984 to SD) is acknowledged.

References

1. E. L. Hahn and D. E. Maxwell, Spin Echo Measurements of Nuclear Spin Coupling in Molecules, *Phys. Rev.*, 1952, **88**(5), 1070–1084.
2. S. De, A. I. Greenwood, M. J. Rogals, E. L. Kovrigin, K. P. Lu and L. K. Nicholson, Complete Thermodynamic and Kinetic Characterization of the Isomer-Specific Interaction between Pin1-WW Domain and the Amyloid Precursor Protein Cytoplasmic Tail Phosphorylated at Thr668, *Biochemistry*, 2012, **51**(43), 8583–8596.
3. U. L. Günther and B. Schaffhausen, NMRKIN: simulating line shapes from two-dimensional spectra of proteins upon ligand binding, *J. Biomol. NMR*, 2002, **22**(3), 201–209.
4. W. F. Hawse, S. De, A. I. Greenwood, L. K. Nicholson, J. Zajicek, E. L. Kovrigin, D. M. Kranz, K. C. Garcia and B. M. Baker, TCR scanning of peptide/MHC through complementary matching of receptor and ligand molecular flexibility, *J. Immunol.*, 2014, **192**(6), 2885–2891.
5. M. Hensmann, G. W. Booker, G. Panayotou, J. Boyd, J. Linacre, M. Waterfield and I. D. Campbell, Phosphopeptide binding to the N-terminal SH2 domain of the p85 alpha subunit of PI 3'-kinase: a heteronuclear NMR study, *Protein Sci.*, 1994, **3**(7), 1020–1030.
6. P. Arora, T. G. Oas and J. K. Myers, Fast and faster: a designed variant of the B-domain of protein A folds in 3 microsec, *Protein Sci.*, 2004, **13**(4), 847–853.
7. J. Balbach, C. Steegborn, T. Schindler and F. X. Schmid, *A protein folding intermediate of ribonuclease T1 characterized at high resolution by 1D and 2D real-time NMR spectroscopy11Edited by* P. E. Wright, *J. Mol. Biol.*, 1999, **285**(2), 829–842.
8. M. Wang, Y. Tang, S. Sato, L. Vugmeyster, C. J. McKnight and D. P. Raleigh, Dynamic NMR line-shape analysis demonstrates that the villin headpiece subdomain folds on the microsecond time scale, *J. Am. Chem. Soc.*, 2003, **125**(20), 6032–6033.
9. L. A. Acevedo, N. E. Korson, J. Williams and L. K. Nicholson, Tuning a timing device that regulates lateral root development in rice, *J. Biomol. NMR*, 2019, **73**(8–9), 493–507.
10. L. A. Acevedo, J. Kwon and L. K. Nicholson, Quantification of reaction cycle parameters for an essential molecular switch in an auxin-responsive transcription circuit in rice, *Proc. Natl. Acad. Sci. U. S. A.*, 2019, **116**(7), 2589–2594.
11. A. I. Greenwood, M. J. Rogals, S. De, K. P. Lu, E. L. Kovrigin and L. K. Nicholson, Complete determination of the Pin1 catalytic domain

thermodynamic cycle by NMR lineshape analysis, *J. Biomol. NMR*, 2011, **51**(1–2), 21–34.

12. D. Kern, G. Kern, G. Scherer, G. Fischer and T. Drakenberg, Kinetic Analysis of Cyclophilin-Catalyzed Prolyl Cis/Trans Isomerization by Dynamic NMR Spectroscopy, *Biochemistry*, 1995, **34**, 13594–13602.
13. S. J. Benkovic, C. A. Fierke and A. M. Naylor, Insights into enzyme function from studies on mutants of dihydrofolate reductase, *Science*, 1988, **239**(4844), 1105–1110.
14. J. J. Burbaum, R. T. Raines, W. J. Albery and J. R. Knowles, Evolutionary optimization of the catalytic effectiveness of an enzyme, *Biochemistry*, 1989, **28**(24), 9293–9305.
15. G. Binsch, Unified theory of exchange effects on nuclear magnetic resonance line shapes, *J. Am. Chem. Soc.*, 1969, **91**(6), 1304–1309.
16. S. Boulton, M. Akimoto, S. Akbarizadeh and G. Melacini, Free energy landscape remodeling of the cardiac pacemaker channel explains the molecular basis of familial sinus bradycardia, *J. Biol. Chem.*, 2017, **292**(15), 6414–6428.
17. G. Binsch, *The Study of Intramolecular Rate Processes by Dynamic Nuclear Magnetic Resonance*, in, *Top. Stereochem.*, 1968, 97–192.
18. G. G. Hammes, Y.-C. Chang and T. G. Oas, Conformational selection or induced fit: A flux description of reaction mechanism, *Proc. Natl. Acad. Sci.*, 2009, **106**(33), 13737–13741.
19. C. Feng, E. L. Kovrigin and C. B. Post, NmrLineGuru: Standalone and User-Friendly GUIs for Fast 1D NMR Lineshape Simulation and Analysis of Multi-State Equilibrium Binding Models, *Sci. Rep.*, 2019, **9**(1), 16023.
20. E. L. Kovrigin, NMR line shapes and multi-state binding equilibria, *J. Biomol. NMR*, 2012, **53**(3), 257–270.
21. C. A. Waudby and J. Christodoulou, NMR Lineshape Analysis of Intrinsically Disordered Protein Interactions, *Methods Mol. Biol.*, 2020, **2141**, 477–504.
22. R. V. Agafonov, C. Wilson, R. Otten, V. Buosi and D. Kern, Energetic dissection of Gleevec's selectivity toward human tyrosine kinases, *Nat. Struct. Mol. Biol.*, 2014, **21**(10), 848–853.
23. U. Reimer, G. Scherer, M. Drewello, S. Kruber, M. Schutkowski and G. Fischer, *Side-chain effects on peptidyl-prolyl cis/trans isomerisation11Edited by* A. R. Fersht, *J. Mol. Biol.*, 1998, **279**(2), 449–460.
24. S. F. Göthel and M. A. Marahiel, Peptidyl-prolyl cis-trans isomerases, a superfamily of ubiquitous folding catalysts, *Cell. Mol. Life Sci.*, 1999, **55**(3), 423–436.
25. A. I. Greenwood, J. Kwon and L. K. Nicholson, Isomerase-Catalyzed Binding of Interleukin-1 Receptor-Associated Kinase 1 to the EVH1 Domain of Vasodilator-Stimulated Phosphoprotein, *Biochemistry*, 2014, **53**(22), 3593–3607.
26. L. K. Nicholson and S. De, Structural biology: The twist in Crk signaling revealed, *Nat. Chem. Biol.*, 2011, **7**(1), 5–6.

27. G. Fischer and T. Aumüller, Regulation of peptide bond cis/trans isomerization by enzyme catalysis and its implication in physiological processes, *Rev Physiol Biochem Pharmacol*, 2003, **148**, 105–150.
28. K. P. Lu, G. Finn, T. H. Lee and L. K. Nicholson, Prolyl cis-trans isomerization as a molecular timer, *Nat. Chem. Biol.*, 2007, **3**(10), 619–629.
29. N. A. Farrow, O. Zhang, J. D. Forman-Kay and L. E. Kay, A heteronuclear correlation experiment for simultaneous determination of 15N longitudinal decay and chemical exchange rates of systems in slow equilibrium, *J. Biomol. NMR*, 1994, **4**(5), 727–734.
30. J. C. Rodríguez, P. A. Jennings and G. Melacini, Using chemical exchange to assign non-covalent protein complexes in slow exchange with the free state: Enhanced resolution and efficient signal editing, *J. Biomol. NMR*, 2004, **30**(2), 155–161.
31. D. Sahu, G. M. Clore and J. Iwahara, TROSY-based z-exchange spectroscopy: application to the determination of the activation energy for intermolecular protein translocation between specific sites on different DNA molecules, *J. Am. Chem. Soc.*, 2007, **129**(43), 13232–13237.
32. A. G. Palmer, C. D. Kroenke and J. P. Loria, Nuclear magnetic resonance methods for quantifying microsecond-to-millisecond motions in biological macromolecules, *Methods Enzymol.*, 2001, **339**, 204–238.
33. O. Trott and A. G. Palmer Iii, Theoretical study of R1ρ rotating-frame and R2 free-precession relaxation in the presence of n-site chemical exchange, *J. Magn. Reson.*, 2004, **170**(1), 104–112.
34. G. M. Clore, Exploring sparsely populated states of macromolecules by diamagnetic and paramagnetic NMR relaxation, *Protein Sci.*, 2011, **20**(2), 229–246.
35. P. Vallurupalli, G. Bouvignies and L. E. Kay, Studying "Invisible" Excited Protein States in Slow Exchange with a Major State Conformation, *J. Am. Chem. Soc.*, 2012, **134**(19), 8148–8161.
36. N. R. Skrynnikov, F. W. Dahlquist and L. E. Kay, Reconstructing NMR spectra of "invisible" excited protein states using HSQC and HMQC experiments, *J. Am. Chem. Soc.*, 2002, **124**(41), 12352–12360.
37. N. J. Anthis and G. M. Clore, Sequence-specific determination of protein and peptide concentrations by absorbance at 205 nm, *Protein Sci.*, 2013, **22**(6), 851–858.
38. C. Schiene-Fischer and G. Fischer, Direct measurement indicates a slow cis/trans isomerization at the secondary amide peptide bond of glycylglycine, *J. Am. Chem. Soc.*, 2001, **123**(26), 6227–6231.
39. G. Fischer, H. Bang and C. Mech, [Determination of enzymatic catalysis for the cis-trans-isomerization of peptide binding in proline-containing peptides], *Biomed Biochim Acta*, 1984, **43**(10), 1101–1111.
40. H. J. Coyne, S. De, M. Okon, S. M. Green, N. Bhachech, B. J. Graves and L. P. McIntosh, Autoinhibition of ETV6 (TEL) DNA binding: appended helices sterically block the ETS domain, *J. Mol. Biol.*, 2012, **421**(1), 67–84.
41. S. De, A. C. Chan, H. J. Coyne, 3rd, N. Bhachech, U. Hermsdorf, M. Okon, M. E. Murphy, B. J. Graves and L. P. McIntosh, Steric mechanism of

auto-inhibitory regulation of specific and non-specific DNA binding by the ETS transcriptional repressor ETV6, *J. Mol. Biol.*, 2014, **426**(7), 1390–1406.

42. S. De, M. Okon, B. J. Graves and L. P. McIntosh, Autoinhibition of ETV6 DNA Binding Is Established by the Stability of Its Inhibitory Helix, *J. Mol. Biol.*, 2016, **428**(8), 1515–1530.
43. A. Bax and D. G. Davis, Practical aspects of two-dimensional transverse NOE spectroscopy, *J. Magn. Reson. (1969)*, 1985, **63**(1), 207–213.
44. J. B. Matthew and F. M. Richards, The pH dependence of hydrogen exchange in proteins, *J. Biol. Chem.*, 1983, **258**(5), 3039–3044.
45. L. N. Lin and J. F. Brandts, Isomer-specific proteolysis of model substrates: influence that the location of the proline residue exerts on cis/trans specificity, *Biochemistry*, 1985, **24**(23), 6533–6538.
46. A. Meinhart and P. Cramer, Recognition of RNA polymerase II carboxy-terminal domain by 3'-RNA-processing factors, *Nature*, 2004, **430**(6996), 223–226.
47. K. Xiang, T. Nagaike, S. Xiang, T. Kilic, M. M. Beh, J. L. Manley and L. Tong, Crystal structure of the human symplekin-Ssu72-CTD phosphopeptide complex, *Nature*, 2010, **467**(7316), 729–733.
48. X. Z. Zhou, O. Kops, A. Werner, P. J. Lu, M. Shen, G. Stoller, G. Küllertz, M. Stark, G. Fischer and K. P. Lu, Pin1-dependent prolyl isomerization regulates dephosphorylation of Cdc25C and tau proteins, *Mol. Cell*, 2000, **6**(4), 873–883.
49. G. Fischer, J. Heins and A. Barth, The conformation around the peptide bond between the P1- and P2-positions is important for catalytic activity of some proline-specific proteases, *Biochim Biophys Acta*, 1983, **742**(3), 452–462.
50. C. Garcia-Echeverria, J. L. Kofron, P. Kuzmic, V. Kishore and D. H. Rich, Continuous fluorimetric direct (uncoupled) assay for peptidyl prolyl cis-trans isomerases, *J. Am. Chem. Soc.*, 1992, **114**(7), 2758–2759.
51. C. Garcia-Echeverria, J. L. Kofron, P. Kuzmic and D. H. Rich, A Continuous Spectrophotometric Direct Assay for Peptidyl Prolyl cis-trans Isomerases, *Biochem. Biophys. Res. Commun.*, 1993, **191**(1), 70–75.
52. S. Li, G. M. Shankar and D. J. Selkoe, How do Soluble Oligomers of Amyloid beta-protein Impair Hippocampal Synaptic Plasticity?, *Front. Cell. Neurosci.*, 2010, **4**, 5.
53. T. A. Ramelot, L. N. Gentile and L. K. Nicholson, Transient structure of the amyloid precursor protein cytoplasmic tail indicates preordering of structure for binding to cytosolic factors, *Biochemistry*, 2000, **39**(10), 2714–2725.
54. C. L. Fisher, R. J. Resnick, S. De, L. A. Acevedo, K. P. Lu, F. C. Schroeder and L. K. Nicholson, Cyclic cis-Locked Phospho-Dipeptides Reduce Entry of AbetaPP into Amyloidogenic Processing Pathway, *J Alzheimers Dis*, 2017, **55**(1), 391–410.
55. L. Pastorino, S. L. Ma, M. Balastik, P. Huang, D. Pandya, L. Nicholson and K. P. Lu, Alzheimer's disease-related loss of Pin1 function influences

the intracellular localization and the processing of AβPP, *J Alzheimers Dis*, 2012, **30**(2), 277–297.
56. L. Pastorino, A. Sun, P. J. Lu, X. Z. Zhou, M. Balastik, G. Finn, G. Wulf, J. Lim, S. H. Li, X. Li, W. Xia, L. K. Nicholson and K. P. Lu, The prolyl isomerase Pin1 regulates amyloid precursor protein processing and amyloid-beta production, *Nature*, 2006, **440**(7083), 528–534.
57. S. Maiti, B. Acharya, V. S. Boorla, B. Manna, A. Ghosh and S. De, Dynamic Studies on Intrinsically Disordered Regions of Two Paralogous Transcription Factors Reveal Rigid Segments with Important Biological Functions, *J. Mol. Biol.*, 2019, **431**(7), 1353–1369.
58. T. A. Ramelot and L. K. Nicholson, Phosphorylation-induced structural changes in the amyloid precursor protein cytoplasmic tail detected by NMR, *J. Mol. Biol.*, 2001, **307**(3), 871–884.
59. W. Labeikovsky, E. Z. Eisenmesser, D. A. Bosco and D. Kern, Structure and dynamics of pin1 during catalysis by NMR, *J. Mol. Biol.*, 2007, **367**(5), 1370–1381.
60. M. J. Macias, S. Wiesner and M. Sudol, WW and SH3 domains, two different scaffolds to recognize proline-rich ligands, *FEBS Lett.*, 2002, **513**(1), 30–37.
61. W. D. Teale, I. A. Paponov and K. Palme, Auxin in action: signalling, transport and the control of plant growth and development, *Nat. Rev. Mol. Cell Biol.*, 2006, **7**(11), 847–859.
62. X. Tan, L. I. Calderon-Villalobos, M. Sharon, C. Zheng, C. V. Robinson, M. Estelle and N. Zheng, Mechanism of auxin perception by the TIR1 ubiquitin ligase, *Nature*, 2007, **446**(7136), 640–645.
63. B. Kang, Z. Zhang, L. Wang, L. Zheng, W. Mao, M. Li, Y. Wu, P. Wu and X. Mo, OsCYP2, a chaperone involved in degradation of auxin-responsive proteins, plays crucial roles in rice lateral root initiation, *Plant J.*, 2013, **74**(1), 86–97.
64. H. Zheng, S. Li, B. Ren, J. Zhang, M. Ichii, S. Taketa, Y. Tao, J. Zuo and H. Wang, LATERAL ROOTLESS2, a cyclophilin protein, regulates lateral root initiation and auxin signaling pathway in rice, *Mol. Plant*, 2013, **6**(5), 1719–1721.
65. Z. X. Zhu, Y. Liu, S. J. Liu, C. Z. Mao, Y. R. Wu and P. Wu, A gain-of-function mutation in OsIAA11 affects lateral root development in rice, *Mol. Plant*, 2012, **5**(1), 154–161.
66. H. Jing, X. Yang, J. Zhang, X. Liu, H. Zheng, G. Dong, J. Nian, J. Feng, B. Xia, Q. Qian, J. Li and J. Zuo, Peptidyl-prolyl isomerization targets rice Aux/IAAs for proteasomal degradation during auxin signalling, *Nat. Commun.*, 2015, **6**, 7395.
67. C. A. Waudby, A. Ramos, L. D. Cabrita and J. Christodoulou, Two-Dimensional NMR Lineshape Analysis, *Sci. Rep.*, 2016, **6**, 24826.
68. M. J. Holliday, G. S. Armstrong and E. Z. Eisenmesser, Determination of the Full Catalytic Cycle among Multiple Cyclophilin Family Members and Limitations on the Application of CPMG-RD in Reversible Catalytic Systems, *Biochemistry*, 2015, **54**(38), 5815–5827.

69. B. L. Moss, H. Mao, J. M. Guseman, T. R. Hinds, A. Hellmuth, M. Kovenock, A. Noorassa, A. Lanctot, L. I. Villalobos, N. Zheng and J. L. Nemhauser, Rate Motifs Tune Auxin/Indole-3-Acetic Acid Degradation Dynamics, *Plant Physiol.*, 2015, **169**(1), 803–813.
70. L. A. Acevedo, A. I. Greenwood and L. K. Nicholson, A Noncanonical Binding Site in the EVH1 Domain of Vasodilator-Stimulated Phosphoprotein Regulates Its Interactions with the Proline Rich Region of Zyxin, *Biochemistry*, 2017, **56**(35), 4626–4636.
71. C. A. Waudby, M. Ouvry, B. Davis and J. Christodoulou, Two-dimensional NMR lineshape analysis of single, multiple, zero and double quantum correlation experiments, *J. Biomol. NMR*, 2020, **74**(1), 95–109.
72. D. F. Hansen, P. Vallurupalli and L. E. Kay, An improved 15N relaxation dispersion experiment for the measurement of millisecond time-scale dynamics in proteins, *J. Phys. Chem. B*, 2008, **112**(19), 5898–5904.
73. F. Massi, E. Johnson, C. Wang, M. Rance and A. G. Palmer, 3rd, NMR R1 rho rotating-frame relaxation with weak radio frequency fields, *J. Am. Chem. Soc.*, 2004, **126**(7), 2247–2256.
74. A. K. Mittermaier and L. E. Kay, Observing biological dynamics at atomic resolution using NMR, *Trends Biochem. Sci.*, 2009, **34**(12), 601–611.
75. K. Sugase, H. J. Dyson and P. E. Wright, Mechanism of coupled folding and binding of an intrinsically disordered protein, *Nature*, 2007, **447**(7147), 1021–1025.
76. P. Li, I. R. Martins, G. K. Amarasinghe and M. K. Rosen, Internal dynamics control activation and activity of the autoinhibited Vav DH domain, *Nat. Struct. Mol. Biol.*, 2008, **15**(6), 613–618.
77. H. S. Gutowsky and C. H. Holm, Rate Processes and Nuclear Magnetic Resonance Spectra. II. Hindered Internal Rotation of Amides, *J. Chem. Phys.*, 1956, **25**(6), 1228–1234.
78. H. S. Gutowsky, D. W. McCall and C. P. Slichter, Nuclear Magnetic Resonance Multiplets in Liquids, *J. Chem. Phys.*, 1953, **21**(2), 279–292.
79. H. S. Gutowsky and A. Saika, Dissociation, Chemical Exchange, and the Proton Magnetic Resonance in Some Aqueous Electrolytes, *J. Chem. Phys.*, 1953, **21**(10), 1688–1694.
80. https://www.nmr-titan.com/ (accessed January 2022).
81. https://lineshapekin.net/ (accessed January 2022).
82. https://lineshapekin.net/#IDAP (accessed January 2022).
83. https://www.purdue.edu/postlab/resources/nmrlineguru/ (accessed January 2022).

CHAPTER 5

Monitoring Interfaces of Thermo- and pH-responsive Polymers Using Solvent Relaxation

IPSITA CHAKRABORTY[a] AND RANGEET BHATTACHARYYA*[b]

[a] Department of Biophysics, Bose Institute, P-1/12 C.I.T. Scheme VII-M, Kolkata 700054, India; [b] Department of Physical Sciences, Indian Institute of Science Education and Research Kolkata, Mohanpur 741246, WB, India
*Email: rangeet@iiserkol.ac.in

5.1 Introduction

The conformation of a macromolecule in a solution is controlled by the solvent–solute interaction. Among the multifarious classes of macromolecules, one class of macromolecules is found to undergo structural transition under some external stimuli, such as pH, temperature, *etc.* These polymers belong to the class of "smart polymers", *e.g.* pH-responsive polymers, thermoresponsive polymers, *etc.*[1] In general, these external factors alter the interface dynamics and also the thermodynamical parameters of a solvent–solute entity.[2,3] This drives the system to undergo conformational changes and eventually results in systematic macroscopic changes to solutes. In recent days these smart molecules have gained huge importance due to their applicability in drug-delivery systems.[4] Thus, monitoring the interface of the macromolecules is becoming one of the major challenges in the biomedical field of research.[5] A detailed analysis of the interplay between solvent and solute provides a deep insight into the understanding of the

New Developments in NMR No. 26
NMR Spectroscopy for Probing Functional Dynamics at Biological Interfaces
Edited by Anirban Bhunia, Hanudatta S. Atreya and Neeraj Sinha

Published by the Royal Society of Chemistry, www.rsc.org

physicochemical aspects of conformational or phase transitions of those macromolecules that occur under certain conditions.

It is evident that NMR spectroscopy is one of the most powerful techniques which can be applied to study the macromolecular interface. Now, as both the solvent and solute are involved in the structural transition one can study the interface from either of the two sides, solvent or solute. Some of the seminal works have shown that if we choose to probe solvent instead of solute to analyze the interface the task will be made much easier.[6] In the next section, we discuss the rationality behind selecting the solvent as a tool rather than the solute. In this context, we discuss two methods, based on two research works published previously, *i.e.* the relaxation–dispersion method and simple solvent relaxation which could be applied, along with some theoretical models to accurately monitor the interfacial dynamics.[7,8]

The remainder of this chapter is organized as follows. The rest of this section is devoted to laying the ground work on the solvent relaxation method and the specifics of the smart polymers used in the given examples. In the next two sections, we discuss the solvent relaxation method to determine the sizes of micellar aggregates and the mechanism of the coil-to-globule transition, respectively. In the final section, we summarize our findings and discuss how to broaden the scope of the solvent relaxation method.

5.1.1 Solvent Relaxation Method

Consider an NMR active solvent having some solute molecules. These solutes can have detectable effects on the relaxation time of the solvent in a variety of ways. Proton exchange between the solute and the solvent is known to affect the transverse relaxation time of the solvent. The extent of the proton exchange could be controlled using the temperature or pH of the solution.[9] If one measures the transverse relaxation time of the solvent, then it would also change as a function of these parameters. By reversing the problem, one may estimate the extent of the exchange by monitoring the relaxation time of the solvent.[7] Similarly, if the presence of the solute gives rise to multiple thermal regions in the solvent (*e.g.* the presence of thermotropic polymers or large macromolecules), this may induce the presence of the bulk *hot* water and the close-to-surface *cold* water. The continuous thermal mixing of the waters of the two types influences the relaxation time of the water. As such, monitoring the water relaxation time yields valuable information about the extent of water having different thermal properties.[8]

In general, the solvent relaxation method is employed to measure the longitudinal relaxation times (T_1) or transverse relaxation times (T_2) of the protons of the solvent molecules within dispersions or suspensions as a function of extrinsic factors such as temperature, pH, time, *etc.* This is one of the simple techniques in NMR spectroscopy which has been used to study the interactions of a variety of polymers, surfactants, and colloidal particles with solvents.[7,8] Hence, this technique has wide applicability in industrial research where it has been mainly used for studying the relative efficiency of various compounds.[6,10] In recent decades, a variety of homogeneous and inhomogeneous systems such as colloidal suspensions, surfactants, polymers gel, silica gel, *etc.* in aqueous and non-aqueous solvents have been extensively studied using solvent relaxation.[11–14]

Often, the solvent spectrum shows a single sharp Lorentzian peak, and it has a very high signal-to-noise ratio compared to the solute peak. As a result, performing any relaxation experiment on the solvent peak requires a lower number of scans and hence less time. Moreover, in the case of macromolecules some specific techniques like solvent peak suppression or deuteration of samples are required for masking the solvent signal. However, these are not required while probing the solvent. As such, the solvent relaxation method is less costly and can be applied efficiently in a low- and high-field NMR spectrometer.

5.1.2 Interfaces of Smart Polymers

Here we discuss the interfacial dynamics of two types of smart polymers, pH-responsive polymers and thermoresponsive polymers, as a function of pH and temperature, respectively.

5.1.2.1 pH-responsive

pH-responsive polymers are characterized by multiple ionizable functional groups, such as carboxylic acid, phosphoric acid, amines, *etc.* Such functional groups play key roles in ascribing hydrophilic or hydrophobic properties at various ends of the polymer molecule, depending on the nature of abundance of ionizable groups on the polymer molecules.[15] As such, pH-responsive polymers are usually amphiphilic in nature. Such amphiphilic molecules form a macrostructure in the aqueous solvent where the hydrophobic or hydrophilic parts form clusters in space. The resulting macrostructure could be a micelle, bicelle, vesicle, *etc.*, depending on the precise geometry of the participating polymer molecule. We note that the concentration of the micelles must be greater than the critical micelle concentration (CMC) for the micellization to be a thermodynamically favorable process and for the micelles to be stable.

Currently, there are several known shapes and structures of micelles. The relative lengths of the hydrophobic and hydrophilic regions of the polymers play a key role in the formation of these structures. For the hydrophobic and hydrophilic regions of comparable lengths, one usually obtains spherical micelles. For di-block copolymer of AB type, which has been used for the study of the interface, the spherical micelles are common and, depending on the relative sizes of A and B, one encounters 'hairy', 'starlike' or 'core–shell' macrostructures. On the other hand, the ABA type tri-block copolymers usually form 'flowerlike' or 'starlike' spherical micelles. Such a triblock copolymer can also give rise to a 'core–shell–corona' architecture or cylindrical worm-like micelles.[16,17]

5.1.2.2 *Thermoresponsive*

Thermoresponsive polymers undergo morphological phase transitions from free coil-like form to a collapsed globule-like form and *vice versa*, in an aqueous solution as the temperature of the solution is changed. A subset of thermoresponsive polymers switch from coil structure form to globular structure form as the temperature crosses a set threshold (known as the lower critical solution temperature or LCST).[1] Among the many thermoresponsive polymers, some of the well-investigated examples include *N*-isopropylacrylamide (NIPAM)-based polymer (PNIPAM) and its derivative copolymers. LCST of PNIPAM homopolymer (∼32 °C) is close to the ambient temperature.[18] The precise value of LCST could be tuned by copolymerization with other co-monomers.[19–21] PNIPAM and its derivatives are soluble in an aqueous solvent at temperatures lower than their LCST owing to the presence of multiple hydrophilic groups.

Since the early 1990s, some research groups have monitored the coil-globule transition of PNIPAM and some of its derivatives using solution-state NMR spectroscopy.[22–28] These groups have found that the longitudinal and transverse relaxation time (T_1 and T_2, respectively) of the solvent, *i.e.* water, bear signatures of this morphological transition of the solute as a function of temperature. It has been found that as the temperature is swiped across LCST from the lower side, the relaxation times show an *S*-shaped behavior. Therefore, the relaxation time rises linearly with the increase of temperature, while it is below the LCST. At the LCST, the relaxation time drops abruptly and then starts rising again with an increase in temperature.[22,24–26,29] While this behavior has been observed by multiple groups, no clear explanation for it has been offered.

We note that the solvent dynamics of the macromolecular solutions have been extensively studied using mainly proteins and polymers.[27,30–34] Overall, these studies established that in a macromolecular aqueous solution, the water molecules could be broadly classified into two categories depending on their timescale of motion. It has been found that the waters within the solvation shell of the macromolecules or a biomolecule move slowly compared to the water present in the bulk (far away from the macromolecule).[27,30,32–34] A macromolecule moves slowly compared to a solvent water molecule. The water molecules close to the macromolecule form hydrogen bonds with it in a stochastic manner (forms the bond, stays for a random duration and then breaks away). As such, these water molecules are also slowly moving and have low rotational kinetic energy. These groups have studied numerically the dielectric relaxation constants and found that the *slow* water close to the macromolecules moves with a typical nanosecond timescale, whereas, the *hot* water in the bulk moves much faster with a typical timescale of picoseconds.[27,31]

Interestingly, the hydrogen bonds (H-bonds) between the solvent water molecules and the polymer play an important role in the coil-globule transition.[27,32,34–36] At temperatures lower than the LCST, water molecules are intermolecularly H-bonded to the hydrophilic ends (C=O and N–H) of NIPAM and stabilize the coil structure. A small number of C=O and N–H sites also have intramolecular H-bonds, as revealed by FTIR and other spectroscopic methods. At the LCST, a large number of water–polymer H-bonds break and intramolecular H-bonds are formed. As a result, the polymer is collapsed from a coil-like free form to a clustered globular form.[34–38]

We use the thermotropic polymer PNIPAM to demonstrate how solvent relaxation can provide greater insight into the mechanism of the LCST with the help of a phenomenological model. The model has been constructed based on numerical studies on the H-bond numbers between the water molecule and the hydrophilic end of the polymer. In Section 5.1.3, we show that it is possible to accurately fit the observed relaxation data with the help of this model.

5.1.3 Relaxation Dispersion and Chemical Exchange

We show that tracking the solvent relaxation is a viable method to monitor the morphological changes of solute particles.[7,8] As discussed in the preceding subsections, the relaxation time of the solvent depends (i) on the chemical exchange and also (ii) on the extent of the slow-moving water in the aqueous solution of the polymers. As such, there are two methods to monitor the relaxation time. In the first method, we use relaxation dispersion (RD), where the aim is to measure the chemical exchange rate.

Measuring the chemical exchange from the relaxation rate of at least one of the species taking part in the exchange is a well-established method proposed and demonstrated originally by Luz and Meiboom.[9] Let us suppose that an atom bearing a spin is being exchanged intermittently between two sites. A simple example is the H^+ being exchanged between a hydronium ion and an amine ($NH_3{}^+$). In such a case, for slow exchange, the water peak and the amine peak would be visible in the proton NMR spectrum, but for a very fast exchange one would observe a single peak at the population-weighted average of the chemical shifts of the two sites (water and amine). In any case, the measured transverse relaxation rate (of either site) would bear the information of the exchange. When the proton (the H^+) is at the hydronium form, it would have a relaxation time typical of solvent water, whereas, when the proton is at the amine, then it would relax like the solute. Now, the exchange takes place intermittently and we measure this by using a train of π pulses (details given below). If the proton moves between the two sites then the final spin-echo obtained at the end of the train of pulses is expected to be a function of the probability of the proton to be at either site (*i.e.*, the population), the rate with which the proton jumps from site to site (*i.e.*, the exchange rate), and of course on the speed with which we apply the pi pulses (frequency of the pulse train). Luz and Meiboom proposed a simple theory for the exchange rate-dependent relaxation which we summarize below.

Before we proceed, we note that our method requires the following conditions:

1. Chemical exchange rate should lie within the fast exchange limit.
2. Population difference between two sites should be large and correspond to asymmetric exchange.

When the above conditions are met, a single resonance is observed, reflecting the population-averaged chemical shift for fast exchange, or the highly populated species for slow exchange.[9] Carr–Purcell–Meiboom–Gill (CPMG) is one of the common sequences used in RD experiments. The experiment works in the following way.[39] A 90°_y pulse brings the equilibrium magnetization along the x-axis. Over time the transverse magnetization acquires a phase with respect to the x-axis which can be inverted by applying a π pulse. If we consider protons with two different chemical shifts ω_A and ω_B that do not undergo chemical exchange, the transverse magnetization will be defocused with time as the components of magnetization with different chemical shifts rotate with different frequencies. This is a similar case where the spin isochromats are getting out of phase under a static magnetic field inhomogeneity. Therefore, after the evolution period $\tau/2$ if a π pulse would be applied, at time τ, the phase difference between magnetization from two sites, A and B, will be zero.

Consider a proton undergoing chemical exchange from site A to site B. This causes the precession frequency of a proton to vary stochastically between ω_A and ω_B. As a result, the acquired phases of magnetization originating on different molecules would be non-identical. Thus the NMR signal which is basically the vector sum of the magnetization from all of the molecules present in the sample broadens in the frequency domain and decreases in intensity. Hence, application of a π_x pulse at time τ, *i.e.* in the middle of the evolution period of spins, cannot refocus the exchange-induced magnetization. However, as more and more π pulses are applied during τ, refocusing improves progressively. The efficient refocusing continues as long as the rate of application of pulses would be on the order of or faster than the exchange rate.

Depending on the time-scale of exchange, multiple analytic solutions have been derived to extract the exchange rate from the RD method using the CPMG sequence. In this context, only the two-site exchange equation derived by Luz and Meiboom will be discussed.[9,40] Luz and Meiboom proposed that the measured relaxation rate $R(g)$ by CPMG with inter-pulse frequency g is given by,[9]

$$R(g) = R_\circ + [1 - 2\tau g \tanh(2\tau g)^{-1}]\tau \sum_i p_i \delta_i^2. \tag{5.1}$$

where τ is the inter-π pulse gap, δ_i is the frequency difference between the ith site and the frequency of the applied rf field. The exchange is assumed to be randomly distributed between the different components, *i.e.* the probability of transfer of a proton (P_{ij}) from the jth site to the ith site is independent of j and proportional to the relative intensity of the ith line, *i.e.* $P_{ij} = p_i$.

For chemical exchange between two sites, the above equation takes the form,

$$R_2(f_{cp}) = R_{2,\circ} + p_1 p_2 \frac{\delta\omega^2}{k_{ex}} \left[1 - \frac{2f_{cp}}{k_{ex}} \tanh\left(\frac{k_{ex}}{2f_{cp}}\right)\right], \tag{5.2}$$

where g in eqn (5.1) indicates the frequency of the applied CPMG train f_{cp} in eqn (5.2) and $1/\tau = k_{ex}$ and $\delta\omega^2$ is δ_i^2. p_1 and p_2 are the populations of the two involving sites. We have used eqn (5.2) for estimating the proton exchange rate between solute and solvent. By measuring the solvent peak intensity as a function of inter π pulse delay τ, we can easily measure the exchange rate and also the populations of two sites.

5.2 Exchanges at the Interface of Micellar Aggregates

In this section, we show how one can monitor the dynamic changes of the micellar aggregates. For this study, we use a diblock AB-type copolymer which forms spherical micelles above critical micellar concentration.[41,42] With the change in the pH of the solution, these micelles adhere to each other to form larger aggregates. We intend to monitor the interface of these aggregates with the solvent. At the interface, hydrogen ions (protons) undergo exchanges from the amine groups present at the micellar surface and the hydronium ions in the solvent.

The formation of micellar aggregates is characterized by a competition between two processes.[43,44] The first process is diffusion-limited aggregation (DLA). Micelles undergo thermal diffusion in the aqueous solvent and when two micelles are physically close, they then form an aggregate through the van der Waal's forces. More and more micelles can diffuse toward the aggregate and the aggregate grows in size. Therefore, DLA assists in the growth of the aggregates. The second process is repulsion between the micelles due to the accumulated surface charges. Micelles have labile protons on the surface which are intermittently exchanged between the solvent water. These labile protons render a positive charge on the surface of the micelles. As a result, micelles repulse each other. At lower pH, with the availability of more protons, the surface charge density is high and hence micellar aggregation is inhibited. At higher pH, one encounters an opposite effect. Micellar aggregation is favored, since the DLA dominates over the surface charge repulsion. The aggregation often leads to precipitation since such large structures can no longer remain suspended. The formation of aggregates is usually measured and characterized by fluorescence spectrometry, UV spectroscopy, dynamic light scattering (DLS), atomic force microscopy (AFM), Fourier transform infrared spectroscopy (FTIR), *etc.*[41,45,46]

Since the micellar surface has labile protons which undergo chemical exchange, one can, in principle, monitor the interface by tracking the chemical exchange; the latter requires a standard relaxation dispersion experiment of the solvent. We provide a more microscopic or molecular view of the process. The polymer strands have ionizable groups. These strands form the micelles and the ionizable groups buried deep inside the micelles do not exchange protons. Thus, the exchange remains a surface phenomenon for micelles. At low pH, one has many small micelles (no aggregation due to high surface charge), each of which takes part in the proton exchange from the surface. As the pH increases, the aggregation takes place and one encounters a loss of surface area. The change in the surface area indicates a change in the population of one of the chemical exchange sites (water). We capture this information by analyzing the relaxation dispersion data of the solvent water. We note that this is not the first time that these data have been measured,[47–49] but it is the first time the data have been successfully analyzed to gain an insight into the aggregation process.

5.2.1 The Model and the Method

We have chosen a pH-responsive diblock copolymer having n repeating units with RNH_2-type side chains; the amine groups serve as the ionizable group. For this polymer, the chemical exchange processes are the following:[50,51]

$$RNH_2 + H_3O^+ \rightleftharpoons RNH_3{}^+ + H_2O \tag{5.3}$$

$$H_3O^+ + D_2O \rightleftharpoons H_2O + D_2H^+O \tag{5.4}$$

The second chemical process is expected to have a very small effect on the exchange time-scale compared to the first, and so we monitor the first reaction from the relaxation dispersion data of the solvent peak.[52]

Using the exchange eqn (5.3), the product of the populations of the two sites $p_A p_B$ as well as k_{ex} can be experimentally determined since all other parameters are known.[7] The product $p_A p_B$ is given by,

$$p_A p_B = \frac{N_{NH_2}^{eff} N_{H_2O}}{(N_{NH_2}^{eff} + N_{H_2O})^2}, \tag{5.5}$$

where, $N_{NH_2}^{eff}$ is the number of NH_2 groups which undergo exchange, *i.e.* $N_{NH_2}^{eff}$ denotes the number of amine groups at the interface and $N_{NH_2}^{all}$ indicates the total number of amine groups in the solution. Since for this system, $N_{H_2O} \gg N_{NH_2}^{eff}$, we have,

$$p_A p_B = \left(\frac{N_{NH_2}^{eff}}{N_{NH_2}^{all}}\right)\left(\frac{N_{NH_2}^{all}}{N_{H_2O}}\right) = \alpha\beta, \tag{5.6}$$

where α denotes the ratio of $N_{NH_2}^{eff}$ to $N_{NH_2}^{all}$ and β denotes the ratio of $N_{NH_2}^{all}$ to N_{H_2O}. Since some of the amine groups are buried within the aggregates and thus do not actively take part in the exchange process, this number is expected to be smaller than the total number of amine groups in the solution $N_{NH_2}^{all}$. Since the ratio β is known *a priori*, α can be easily determined using eqn (5.6) if $p_A p_B$ is known.

We construct a simple spherical model of micelles by assuming that the polymer strands are stacked in a spherical arrangement, with the hydrophobic parts inside such that a fraction of the ionizable groups protrudes outside the surface from which the exchange takes place. For such a model, the average number of polymers packed into a micelle of diameter, d, is given by,

$$N_{\rm agg} = \frac{\pi d^3}{6aL}, \tag{5.7}$$

where a is the cross-section of the hydrophobic end. Therefore, the number of polymers on the surface of the sphere is

$$N_{\rm surface} = \frac{\pi d^2}{a}. \tag{5.8}$$

Now, the total number of NH_2, *i.e*, $N^{\rm all}_{NH_2} = nN_{\rm agg}$ and $N^{\rm eff}_{NH_2} = n_1 N_{\rm surface}$, where, n_1 is the average number of NH_2 groups participating in the exchange from a single polymer molecule at the surface of a spherical aggregate. Hence, one obtains,

$$d = \frac{6n_1 L\beta}{np_{\rm A}p_{\rm B}}. \tag{5.9}$$

Since all parameters in the right-hand side of the above equation can be extracted from NMR experiments, and if n_1 is known, then d can be easily evaluated using eqn (5.9). For example, a set of CPMG experiments at various $f_{\rm cp}$ values can be carried out which will yield the value of $p_{\rm A}p_{\rm B}$. One can estimate the values of n and L from the geometry of the molecule and β can be calculated from the concentration of the polymer solution.

To determine the value of n_1, we adopt the following strategy: using an alternate spectroscopy the value of d can be estimated at a particular pH value for which NMR measurements yield $p_{\rm A}p_{\rm B}$. Then using eqn (5.9), n_1 can be estimated since the other parameters are known. Once the estimated n_1 is available, the diameter can be easily evaluated from NMR measurements at any other value of pH. UV–visible spectroscopy provides the required alternative. It is evident that the transmittance in UV–visible spectroscopy undergoes a sharp transition when the mean wavelength of the laser is of the order of the size of the suspension. Hence an estimation of the aggregate size is possible.

5.2.2 Results

The example of the determination of the diameter is given with the experimental details in our earlier work.[7] Here, we describe the final outcome of this study, which exemplifies the method described above. Figure 5.1 shows the calculated average diameters of the micellar aggregates as a function of the pH of the solution.

This figure shows that the diameters of the aggregates increase linearly with the increase in pH. This behavior agrees with the previously mentioned dominance of DLA over surface charge repulsion with the increase of pH. Between pH 3 to pH 5.5, the aggregation results in the loss of the surface

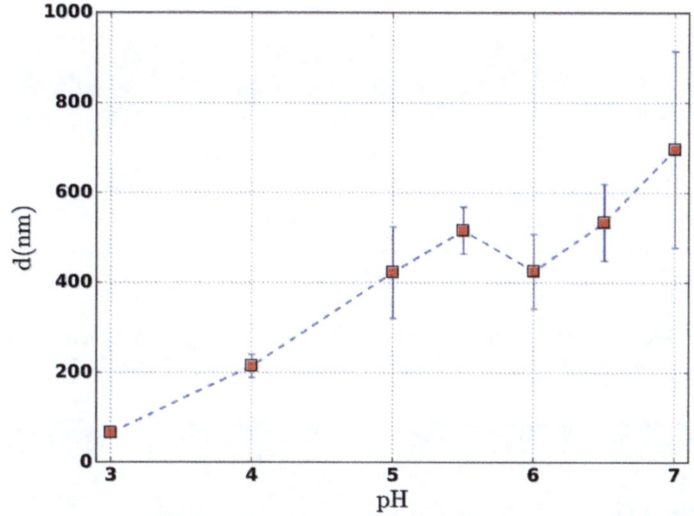

Figure 5.1 Experimentally determined diameter of the aggregates using NMR, as a function of pH as determined from eqn (5.9). Reproduced from ref. 7 with permission from the Royal Society of Chemistry.

area and, as a result, the determined diameter (approximately inverse of the volume to surface ratio with volume being constant) increases. The trend continues from pH 6 to pH 7 (at which we have the onset of precipitation). Experiments at pH 7 have been conducted with freshly prepared samples to avoid precipitation. We note that the errorbar of pH 7 is higher than at other pH values. The major source of the error at this pH is precipitation. Precipitation results in irregularity of the measured T_2, which in turn affects the results. However, we note that the most surprising feature is a drop at pH 6.

The drop has been explained using the following argument. At low pH, *i.e.*, high concentration of labile protons, the surface charge would be high and the micellar aggregation would be limited. One would expect the aggregates to have an irregular shape. At higher pH, the loss of surface charge may lead to an uncontrolled aggregation which may result in long rod-like or cylindrical structures. But, with higher pH, these structures can undergo compactification. Consider a long cylinder being reshaped into a spherical one. Naturally, one expects the former to have a longer lengthscale compared to the latter. Post-compactification, the aggregation continues until the precipitation takes over. The argument is depicted here with a cartoon (Figure 5.2).

The above argument is only conjecture and requires further confirmation. We have collected the SEM images of the dried aggregates obtained from solutions of various pH values and these images completely corroborate the arguments outlined above (Figure 5.3). The SEM images show that at around pH 5.8 the aggregates indeed have long rod-like shapes, whereas for pH 6.5, the shape is spherical. Naturally, the hypothesis of the compactification is vindicated.

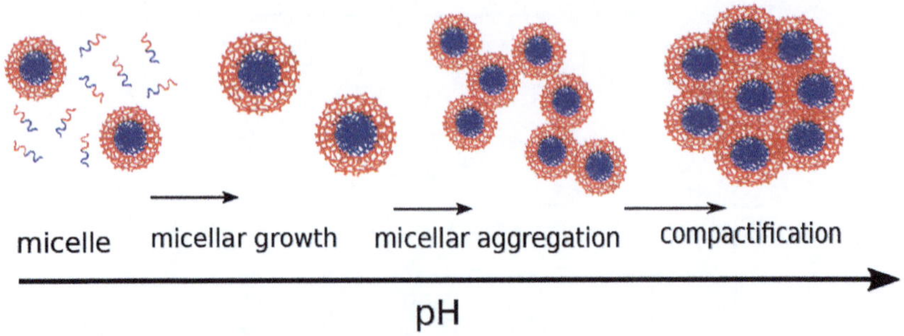

Figure 5.2 Pictorial representation of aggregation of a pH-responsive polymer as a function of pH. Reproduced from ref. 53 with permission from the authors.

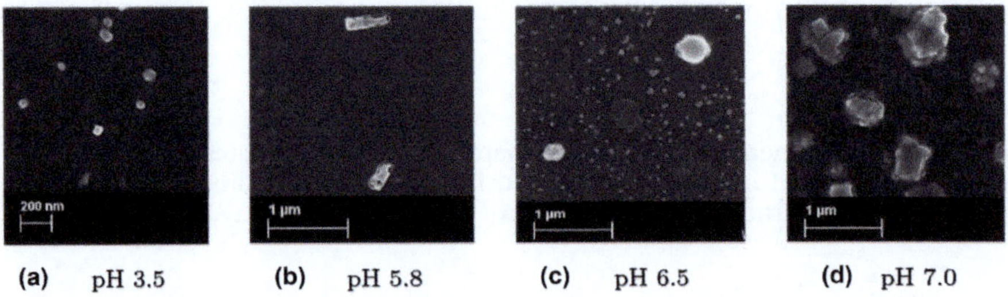

Figure 5.3 SEM images of the dried micellar aggregates at different pH values. Reproduced from ref. 7 with permission from the Royal Society of Chemistry.

Using this method, one can also study the exchange rate as a function of the pH of the solution.[7] As noted in this work, the exchange rate seems to be strongly correlated to the electrokinetic potential or the zeta potential. Both curves (not shown here), behave identically. The rate (or potential) remains a slow linear function of pH until pH 5.5 and then undergoes a sharp drop. The behavior has been speculated to be connected to the surface charge density, but the exact mechanism remains unknown.

Finally, we have demonstrated that the interface of micellar aggregates could be monitored using NMR spectroscopy. This method relies on the relaxation dispersion experiments on the solvent. The solvent, by virtue of its abundance, is easily observed and such an experiment is not limited by the sensitivity of the solutes. Therefore, the method is widely applicable. The present model assumes that the solute has micelles of a single variety. However, if the solution contains polymers of more than one variety then the model needs to be augmented appropriately.

5.3 Thermal Mixing: Coil-to-globule Transitions

The other method that we describe here also relies on solvent relaxation and is able to monitor the coil-to-globule transition of thermoresponsive polymers. We consider a solution where the solute is a macromolecule (*e.g.* proteins, polymers, *etc.*). It is known that the dynamics of solvent molecules depends on their proximity with the dissolved macromolecule. Depending on the mobility, the solvent molecules could be classified as bound or free.[27,30,31] The *bound* solvent molecules close to the macromolecule move slowly and have a long dielectric relaxation time (of the order of microseconds to nanoseconds), as opposed to the *free* ones, far away from the macromolecule, which move fast and have short dielectric relaxation time (of the order of picoseconds). Let us suppose the solution to be an aqueous solution of a polymer and hence the solvent molecules are water molecules. The above dynamics can also be explained in terms of the hydrogen bonds (H-bonds). The water molecules close to the polymer undergo an exchange-like process, where these molecules are intermittently H-bonded to the polymer. When these molecules break free from the polymer, they have slow dynamics commensurate with the motional timescale of the polymer molecule. We name these molecules *cold* water molecules, as opposed to the *hot* water molecules far away from the polymer. Therefore, at any given temperature the relaxation time of the water of this polymer solution is but a weighted average of the relaxation times of the *cold* and *hot* water.

Now, for certain thermotropic polymers in aqueous solutions, there exists a lower critical solution temperature (LCST), below which the polymer remains in the coil form. On the other hand, above the LCST the polymer collapses to a globule form. The underlying mechanism has a close relation with the H-bond process with the solvent molecule. The thermoresponsive polymers have sites where solvent water can form H-bonds with the polymer. Below the LCST, most such sites in the polymer have H-bonded water and this water sheath helps to prevent the polymer from collapsing to a globular form. On the other hand, above the LCST, most such sites form an intramolecular H-bond and the polymer remains in a stable globular form. As such a temperature run from below the LCST to above the LCST results in the polymer collapsing to a globular form aided by a temperature-dependent H-bond process. A pictorial depiction of the process described above is given in Figure 5.4.

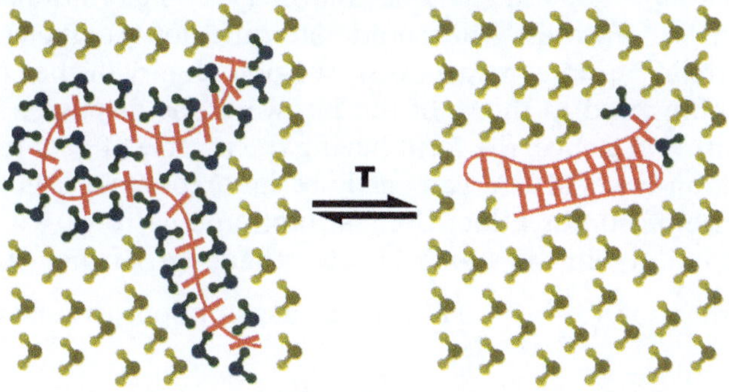

Figure 5.4 The *cold* water molecules (dark) and the *hot* water molecules (light) in a dilute solution of a polymer strand. Below the LCST, many *cold* water molecules are present due to the large number of polymer–water H-bonds. The amount of *cold* water reduces drastically the post coil-globule transition. Reproduced from ref. 8 with permission from American Chemical Society, Copyright 2018.

It is clear from the above that the number of *cold* water molecules would be different for below and above the LCST cases. Below the LCST, there would be more *cold* water molecules which bind to the polymer because of the availability of the H-bond sites. But, above the LCST, this number would be much less since the binding sites are unavailable. Since the relaxation time of water would be a function of the weighted average of the two types of water, in principle, the change in the *cold* to *hot* fraction should be reflected in the relaxation time of the solvent water. Such studies have been undertaken earlier, but the experimental data could not be analyzed due to the lack of a suitable model.[8] We provide a model and exemplify the usefulness of the solvent relaxation method for the extraction of H-bond information from the experimental data.

5.3.1 The Model and the Method

We recognize that the average energy of a water molecule in this polymer–water solution would be a weighted average of the energy of the *cold* and *hot* water. Here we refer to the rotational kinetic energy, since the NMR relaxation times are closely connected to the rotational correlation time. Thus we can write:

$$\left[N_{\mathrm{H}}^{\circ} - N_{\mathrm{H}}(T)\right]\frac{1}{2}I\langle\omega^2\rangle_{\mathrm{cold}} + N_{\mathrm{bulk}}\frac{1}{2}I\langle\omega^2\rangle_{\mathrm{hot}} = \left[N_{\mathrm{H}}^{\circ} - N_{\mathrm{H}}(T) + N_{\mathrm{bulk}}\right]\frac{1}{2}I\langle\omega^2\rangle_{\mathrm{mean}} \tag{5.10}$$

where N_{H}° and $N_{\mathrm{H}}(T)$ indicate the total number of amine sites where water molecules can form an intermolecular H-bond with the polymer and the number of active hydrophilic sites at a given temperature, respectively. N_{bulk} indicates the number of water molecules in the bulk. I denotes the moment of inertia of the water molecules. $\langle\omega^2\rangle_{\mathrm{cold}}$ and $\langle\omega^2\rangle_{\mathrm{hot}}$ represent the variance of the angular velocity of *cold* and *hot* water molecules, respectively, and $\langle\omega^2\rangle_{\mathrm{mean}}$ represents the mean variance of the angular velocity at the steady state.

We note that $\langle\omega^2\rangle \propto D_{\mathrm{r}}$ and $\tau_{\mathrm{c}} = 1/6D_r$ for the isotropic rotational diffusion, where D_{r} and τ_{c} represent the rotational diffusion coefficient and the rotational correlation time, respectively. We note that τ_{c} of the cold water is much larger compared to the τ_{c} of the hot water (as discussed in the preceding section).[30,31] Hence, the term having the inverse of τ_{c} of *cold* water in the last equation is small and can easily be ignored. Also, it is well known that in the case of liquids under the fast motional narrowing limit, the rotational correlation time is much shorter than the Larmor time period, i.e., $\omega\tau_c \ll 1$, therefore, $\frac{1}{T_2} \propto \tau_{\mathrm{c}}$. From these assumptions, we obtain,[8]

$$T_2^{\mathrm{mean}} = \frac{N_{\mathrm{bulk}} T_2^{\mathrm{hot}}}{N_{\mathrm{H}}^{\circ} - N_{\mathrm{H}}(T) + N_{\mathrm{bulk}}} \tag{5.11}$$

Further, we note that T_2^{hot} is expected to have a linear dependence on temperature, since $T_2 \propto 1/\tau_{\mathrm{c}} \propto D_{\mathrm{rot}} \propto T$, the later proportionality follows from the Einstein relation. Hence one may tentatively write $T_2^{\mathrm{hot}} = aT + b$, where a

and b are two different physical parameters, which could be estimated from the experimental results.

In recent years, molecular dynamics simulations[32–35] and IP spectroscopy[38,54,55] have suggested that the fraction of intramolecular H-bonds in a thermoresponsive polymer having LCST may have an inverted tanh-like nature. As such, we phenomenologically assume that the temperature dependence of $N_H(T)$ is a tanh-like behavior. At temperatures much above T°, *i.e.* in the post-coil-globule transition, the number of H-bonded water molecules is small and remains almost constant.[35] This number is denoted here by N_H^{∞}. Using the above assumptions, from eqn (5.11) we obtain (see for details the work of Chakraborty and others[8]):

$$T_2^{\text{mean}} = \frac{aT + b}{\frac{\beta\kappa}{2}[1 + \tanh\alpha(T - T_{\circ})] + 1.0} \tag{5.12}$$

where α is a measure of the steepness of the transition; smaller α indicates that a transition takes place over a larger range of temperature. $\kappa = \frac{N_H^{\circ} - N_H^{\infty}}{N_H^{\circ}}$ and $\beta = \frac{N_H^{\circ}}{N_{\text{bulk}}}$. Therefore, κ denotes the fractional change of H-bonded water molecules across LCST and β measures the ratio of active intermolecular H-bond sites to the total number of bulk water molecules. Note that β is known as N_H° and N_{bulk} are known *a priori*.

5.3.2 Results

The efficacy of this method has been shown with thermoresponsive polymer of more than one type and also with a polymer dissolved in various solvents. In fact, a particularly interesting aspect of this work was the study of the effect of isotopic substitution on H-bonds. In this section, we discuss the latter topic with a brief review of the former aspects.

Some of the experimental results are shown in Figure 5.5. This figure shows the transverse relaxation time (T_2) of water as a function of the temperature for PNIPAM solution made with different isotopic compositions, *i.e.*, the ratio of H_2O and D_2O has been varied. First, we note that there is a generic behavior of the water relaxation time. Below the LCST, it rises linearly with temperature and then undergoes an *S*-like bend at the LCST. Then, post-LCST, it keeps on rising linearly with temperature. This behavior is in total agreement with the previously reported values (see our works in[8] and references therein). The behavior is identical in nature for different solvents (as shown with different markers). The continuous, dashed, and dash-dotted lines are the fit of experimental data with the final formula stated in the previous section. It is obvious that the proposed model has satisfactorily fit the recorded data. We have not shown the case for different solvents here, but the behavior and the fit are again quite good.[8]

Table 5.1 shows the extracted parameters along with the confirmatory LCST values measured by UV visible spectroscopy. It is clear from the plots in

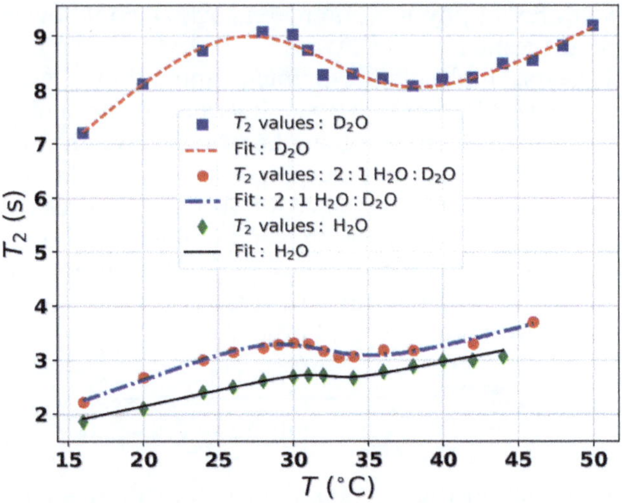

Figure 5.5 T_2 of protons of the solvent *versus* temperature, for different PNIPAM solutions along with the fit of the experimental data with eqn (5.12). Marker-size is chosen to reflect the error in T_2 data. Reproduced from ref. 8 with permission from American Chemical Society, Copyright 2018.

Table 5.1 Numerical values of the parameters extracted from the experimental data using our model. Reproduced from ref. 8 with permission from American Chemical Society, Copyright 2018.

Solvent	K	α (°C^{-1})	T (°C)[a]	a (s °C^{-1})	b (s)	T (°C)[b]
D_2O	0.4 ± 0.2	0.14 ± 0.04	33.0 ± 0.5	0.25 ± 0.09	3.0 ± 2.0	33.0
$H_2O:D_2O$ (2:1)	0.2 ± 0.1	0.3 ± 0.1	32.3 ± 0.6	0.09 ± 0.02	0.7 ± 0.4	32.7
H_2O	0.12 ± 0.06	0.5 ± 0.06	32.3 ± 0.9	0.06 ± 0.01	1.0 ± 0.3	32.4

[a]From the fitting of experimental data using eqn (5.12).
[b]From UV–visible spectroscopy.

Figure 5.5 that the transition is sharper for D_2O compared to H_2O. The extracted α values in Table 5.1 also confirm this (α increases from 0.14 to 0.5 as one moves from D_2O to H_2O). On the other hand, the drop of the *S*-bend, which is characterized by κ, shows a completely opposite behavior. κ decreases as we move from D_2O to H_2O. This behavior is explained by protons forming an H-bond with a wider range of strength. The latter fact explains why one needs to move to higher temperature to break all H-bonds in H_2O compared to a narrow range in D_2O. Therefore, heavy water forms stronger H-bonds with a narrow range of bond strengths. This observation agrees well with earlier reports.[56,57] Naturally, one needs higher temperature to break the heavy water H-bonds, which leads to a slightly larger value of LCST (confirmed also by UV visible spectroscopy).

This work clearly shows that the ratio of the *cold* and *hot* waters in a macromolecular solution can be determined by monitoring the transverse relaxation time (T_2) of the solvent water. The method is easy to implement since solvent NMR does not suffer from sensitivity issues. The method may also be extended to the case of non-H-bonded water. The molecular dynamics simulations suggest that the presence of such *cold* and *hot* water could also be due to simple van der Waal's interactions between the solute and the solvent molecules along with the crowding effects.[27,30,31] We note that the above method, with suitable modifications, may also be applied to study such cases. If the H-bond is not the dominant mechanism then one needs to replace the tanh-like dependence of H-bonds on the temperature by something more appropriate. We note that the phase transition of polymers in a mixture of solvents could also be monitored using this method.

5.4 Summary and Outlook

Solvent relaxation is a powerful technique that helps uncover the dynamics of the interfaces. In this chapter, we have exemplified two of the important usages of this technique involving smart polymers. The biggest advantage of the solvent relaxation in this example is the use of standard relaxation measurement schemes which are available with all NMR spectrometers, and acquiring the solvent spectrum is one of the simplest of the NMR spectrum, owing to its highly abundant nature.

In the first example, we show that one can monitor the micellar aggregation of the pH-responsive polymers by looking at the interfacial dynamics. The interface is, in turn, probed by using the proton exchange occurring from the micellar interface. In this method, one records the transverse relaxation time (T_2) of the solvent water containing the micellar aggregates. The T_2 of the water strongly depends on the proton exchange taking place between the solvent and the interface of the micellar aggregates. From a simple model, one can easily perform spin-counting, and hence estimate the size of the interface. Since the volume of the aggregate is known, from the surface-to-volume ratio the size of the aggregate can also be estimated.

This method also has been proven to be advantageous over other methods for determining the micellar size. A performance comparison of this technique with dynamic light scattering shows that while both methods are capable of providing the size and shape of the aggregates, the solvent relaxation in this case is more accurate. The underlying reason for this accuracy is that the solvent relaxation monitors the true surface from which the exchange takes place, whereas DLS tracks the hydrodynamic radii. Naturally, the recorded size from the solvent relaxation is smaller compared to that of DLS and closer to the actual value. One can also approximate the size of the aggregates by using the UV–visible spectroscopy from the irradiation wavelength at which the transmitted intensity drops. We note that the pH of the solution needs to be constant during the monitoring of the relaxation. Since pH is quite sensitive to the environmental conditions, the latter limits the accuracy of the method. However, this limitation equally affects other methods of size distribution such as the light-scattering methods. Moreover, the analysis of this method is most accurate for the micellar aggregates having spherical shapes, which may not always be the case.

For the thermoresponsive polymers, the solvent relaxation method proves to be equally advantageous. There are several parameters which could be accurately estimated using our proposed model. The precise value of LCST may be estimated using other spectroscopic methods, but the strengths of the H-bond or its dispersion can only be estimated using our model. Moreover, this model could be easily used for a mixture of solvents and possibly also a mixture of solute polymers if they have well-separated LCST values. The model has been exemplified using transverse relaxation but could be used with longitudinal relaxation as well, particularly in the fast motional narrowing regime. Finally, the versatility of the solvent relaxation method is exemplified by applications in the pH-responsive and thermoresponsive polymers. It is envisaged that this method could be easily extended to more complex situations involving mixtures of polymers and solvents.

References

1. D. Schmaljohann, Thermo- and pH-responsive polymers in drug delivery, *Adv. Drug Delivery Rev.*, 2006, **58**(15), 1655–1670.
2. P. R. Sperry, *et al.*, Role of water in particle deformation and compaction in latex film formation, *Langmuir*, 1994, **10**(8), 2619–2628.
3. S. Hirotsu, Phase transition of a polymer gel in pure and mixed solvent media, *J. Phys. Soc. Jpn.*, 1987, **560**(1), 233–242.
4. D. Schmaljohann, Thermo-and pH-responsive polymers in drug delivery, *Adv. Drug Delivery Rev.*, 2006, **58**, 1655–1670.
5. H. Zhang, Y. Tian, J. Hou, X. Hou, G. Hou, R. Ou, H. Wang and L. Jiang, Bioinspired smart gate-location-controllable single nanochannels: experiment and theoretical simulation, *ACS Nano*, 2015, **90**(12), 12264–12273.
6. C. L. Cooper, T. Cosgrove, J. S. van Duijneveldt, M. Murray and S. W. Prescott, The use of solvent relaxation NMR to study colloidal suspensions, *Soft Matter*, 2013a, **9**, 7211–7228.
7. I. Chakraborty, I. Mukherjee, U. Haldar, P. De and R. Bhattacharyya, Monitoring aggregation of a pH-responsive polymer via proton exchange, *Phys. Chem. Chem. Phys.*, 2017, **190**(26), 17360–17365.
8. I. Chakraborty, K. Mukherjee, P. De and R. Bhattacharyya, Monitoring Coil-Globule Transitions of Thermoresponsive Polymers by Using NMR Solvent Relaxation, *J. Phys. Chem. B*, 2018, **1220**(22), 6094–6100.
9. Z. Luz and S. Meiboom, Nuclear magnetic resonance study of the protolysis of trimethylammonium ion in aqueous solution–order of the reaction with respect to solvent, *J. Chem. Phys.*, 1963, **32**(0), 366–370.
10. S. J. Mears, T. Cosgrove, L. Thompson and I. Howell, Solvent relaxation NMR measurements on polymer, particle, surfactant systems, *Langmuir*, 1998, **14**, 997–1001.
11. G. P. Van der Beek, M. A. Cohen, Stuart and T. Cosgrove, Polymer adsorption and desorption studies via proton NMR relaxation of the solvent, *Langmuir*, 1991, **7**, 327–334.
12. N. Doshi. Colloidal *Mixtures* of *Spheres* and *Plates*, PhD thesis, 2011.

13. M. G. Brereton, NMR transverse relaxation function calculated for constrained polymer chains: application to entanglements and networks, *Macromolecules*, 1990, **23**, 1119–1131.
14. M. G. Brereton, I. M. Ward, N. Boden and P. Wright, Nature of the proton NMR transverse relaxation function of polyethylene melts. 1. monodispersed polyethylenes, *Macromolecules*, 1991, **24**, 2068–2074.
15. G. Kocak, C. Tuncer and V. Bütün, pH-responsive polymers, *Polym. Chem.*, 2017, **8**, 144–176.
16. M. Karayianni and S. Pispas. *Self-assembly of Amphiphilic Block Copolymers in Selective Solvents*, Springer, 2016.
17. A. Chakrabarti and R. Toral, Computer simulation of the aggregation process in self-associating polymer and surfactant systems, *J. Chem. Phys.*, 1989, **91**, 5687–5693.
18. R. Umapathi, P. M. Reddy, A. Kumar, P. Venkatesu and C.-J. Chang, The biological stimuli for governing the phase transition temperature of the "smart" polymer pnipam in water, *Colloids Surf., B*, 2015, **135**, 588–595.
19. A. J. Convertine, B. S. Lokitz, Y. Vasileva, L. J. Myrick, C. W. Scales, A. B. Lowe and C. L. McCormick, Direct synthesis of thermally responsive dma/nipam diblock and dma/nipam/dma triblock copolymers via aqueous, room temperature raft polymerization, *Macromolecules*, 2006, **39**(0), 1724–1730.
20. P. De and B. S. Sumerlin, Precision control of temperature response by copolymerization of di (ethylene glycol) acrylate and an acrylamide comonomer, *Macromol. Chem. Phys.*, 2013, **214**, 272–279.
21. K. Bauri, S. Roy, S. Pant and P. De, Controlled synthesis of amino acid-based pH-responsive chiral polymers and self-assembly of their block copolymers, *Langmuir*, 2013, **29**, 2764–2774.
22. H. Ohta, I. Ando, S. Fujishige and K. Kubota, Molecular motion and ^{1}H NMR relaxation of aqueous poly (n-isopropylacrylamide) solution under high pressure, *J. Polym. Sci. B*, 1991, **29**(0), 963–968.
23. J. Spěvácek and J. Dybal, Temperature-induced phase separation and hydration in aqueous polymer solutions studied by NMR and ir spectroscopy: Comparison of poly(n-vinylcaprolactam) and acrylamide-based polymers, *Macromol. Symp.*, 2014, **336**(0), 39–46.
24. H. Kouřilová, J. Šťastná, L. Hanyková, Z. Sedláková and J. Spěváček, ^{1}H NMR study of temperature-induced phase separation in solutions of poly (n-isopropylmethacrylamide-co-acrylamide) copolymers, *Eur. Polym. J.*, 2010, **46**, 1299–1306.
25. J. Spěváček, L. Hanyková and L. Starovoytova, ^{1}H NMR relaxation study of thermotropic phase transition in poly (vinyl methyl ether)/d^{2}o solutions, *Macromolecules*, 2004, **37**, 7710–7718.
26. J. Spěváček, NMR investigations of temperature-induced phase transition in aqueous polymer solutions, *Macromol. Symp.*, 2011, **305**, 18–25.
27. T. Tokuhiro, T. Amiya, A. Mamada and T. Tanaka, NMR study of poly (n-isopropylacrylamide) gels near phase transition, *Macromolecules*, 1991, **24**, 2936–2943.

28. B. Sierra-Martín, M. S. Romero-Cano, T. Cosgrove, B. Vincent and A. Fernández-Barbero, Solvent relaxation of swelling pnipam microgels by NMR, *Colloids Surf., A*, 2005, **270**, 296–300.
29. C. L. Cooper, T. Cosgrove, J. S. van Duijneveldt, M. Murray and S. W. Prescott, The use of solvent relaxation NMR to study colloidal suspensions, *Soft Matter*, 2013, **9**, 7211–7228.
30. N. Nandi and B. Bagchi, Dielectric relaxation of biological water, *J. Phys. Chem. B*, 1997, **101**, 10954–10961.
31. K. Bhattacharyya, Solvation dynamics and proton transfer in supramolecular assemblies, *Acc. Chem. Res.*, 2003, **36**, 95–101.
32. X. J. Zhao and Z. F. Gao, Role of hydrogen bonding in solubility of poly (n-isopropylacrylamide) brushes in sodium halide solutions, *Chin. Phys. B*, 2016, **25**, 074703.
33. M. S. Liu, C. Taylor, B. Chong, L. Liu, A. Bilic, N. S. Terefe, R. Stockmann, S. H. Thang and K. D. Silva, Conformational transitions and dynamics of thermal responsive poly (n-isopropylacrylamide) polymers as revealed by molecular simulation, *Eur. Polym. J.*, 2014, **55**, 153–159.
34. E. Dormidontova and G. T. Brinke, Phase behavior of hydrogen-bonding polymer- oligomer mixtures, *Macromolecules*, 1998, **31**, 2649–2660.
35. S. A. Deshmukh, S. K. R. S. Sankaranarayanan, K. Suthar and D. C. Mancini, Role of solvation dynamics and local ordering of water in inducing conformational transitions in poly (n-isopropylacrylamide) oligomers through the lcst, *J. Phys. Chem. B*, 2012, **116**, 2651–2663.
36. S. A. Deshmukh, Z. Li, G. Kamath, K. J. Suthar, S. K. R. S. Sankaranarayanan and D. C. Mancini, Atomistic insights into solvation dynamics and conformational transformation in thermo-sensitive and non-thermo-sensitive oligomers, *Polymer*, 2013, **54**, 210–222.
37. Y. Katsumoto, T. Tanaka, K. Ihara, M. Koyama and Y. Ozaki, Contribution of intramolecular co–h- n hydrogen bonding to the solvent-induced reentrant phase separation of poly (n-isopropylacrylamide), *J. Phys. Chem. B*, 2007, **111**, 12730–12737.
38. B. Sun, Y. Lin, P. Wu and H. W. Siesler, A ftir and 2d-ir spectroscopic study on the microdynamics phase separation mechanism of the poly (n-isopropylacrylamide) aqueous solution, *Macromolecules*, 2008, **41**, 1512–1520.
39. P. Neudecker, P. Lundström and L. E. Kay, Relaxation dispersion NMR spectroscopy as a tool for detailed studies of protein folding, *J. Magn. Reson.*, 2009, **96**, 2045–2054.
40. J. P. Carver and R. E. Richards, A general two-site solution for the chemical exchange produced dependence of t^2 upon the carr-purcell pulse separation, *J. Magn. Reson.*, 1972, **6**, 89–105.
41. B. P. Bastakoti, S. H. Liao, M. Imura, M. Inoue, S. I. Yusa, K. Nakashima, K. C. Wu and Y. Yamauchi, pH-responsive polymeric micelles with core–shell–corona architectures as intracellular anti-cancer drug carriers, *Sci. Technol. Adv. Mater.*, 2013, **14**, 044402.
42. D. Lombardo, M. A Kiselev, S. Magazù and P. Calandra, Amphiphiles self-assembly: basic concepts and future perspectives of supramolecular approaches, *Adv. Condens. Matter Phys.*, 2015, 2015.

43. K. Bauri, S. Ghosh Roy, S. Pant and P. De, Controlled synthesis of amino acid-based pH-responsive chiral polymers and self-assembly of their block copolymers, *Langmuir*, 2013, **29**, 2764–2774.
44. T. A. Witten Jr and L. M. Sander, Diffusion-limited aggregation, a kinetic critical phenomenon, *Phys. Rev. Lett.*, 1981, **47**, 1400.
45. J. Humpolícková, M. Štěpánek, K. Procházka and M. Hof, Solvent relaxation study of pH-dependent hydration of poly (oxyethylene) shells in polystyrene-block-poly (2-vinylpyridine)-block-poly (oxyethylene) micelles in aqueous solutions, *J. Phys. Chem. A*, 2005, **109**, 10803–10812.
46. M. Karayianni and S. Pispas, *Self-assembly of Amphiphilic Block Copolymers in Selective Solvents*, Springer, 2016b.
47. H. Takahashi, T. Nakanishi, K. Kamia, Y. Arata and I. Shimada, A novel NMR method for determining the interfaces of large protein–protein complexes, *Nat. Struct. Biol.*, 2000, **7**, 220.
48. R. Bhattacharyyaa, B. Key, H. Chen, A. S. Best, A. F. Hollenkamp and C. P. Grey, In situ NMR observation of the formation of metallic lithium microstructures in lithium batteries, *Nat. Mater.*, 2010, **9**, 504.
49. F. Berti, P. Costantino, M. Fragai and C. Luchinat, Water accessibility, aggregation, and motional features of polysaccharide-protein conjugate vaccines, *Biophys. J.*, 2004, **86**, 3–9.
50. J. P. Carver and R. E. Richards, A general two-site solution for the chemical exchange produced dependence of t^2 upon the carr-purcell pulse separation, *J. Magn. Reson.*, 1972, **6**, 89–105.
51. E. Grunwald, A. Loewenstein and S. Meiboom, Rates and mechanisms of protolysis of methylammonium ion in aqueous solution studied by proton magnetic resonance, *J. Phys. Chem.*, 1957, **27**(0), 630–640.
52. Z. Luz and S. Meiboom, Nuclear magnetic resonance study of the protolysis of trimethylammonium ion in aqueous solutionorder of the reaction with respect to solvent, *J. Chem. Phys.*, 1963, **39**, 366–370.
53. Ipsita Chakraborty. PhD thesis, Department of Chemical Sciences, Indian Institute of Science Education and Research Kolkata, India, 2018.
54. S. Sun and P. Wu, Role of water/methanol clustering dynamics on thermosensitivity of poly (n-isopropylacrylamide) from spectral and calorimetric insights, *Macromolecules*, 2010, **43**, 9501–9510.
55. Y. Maeda, T. Higuchi and I. Ikeda, Change in hydration state during the coil- globule transition of aqueous solutions of poly (n-isopropylacrylamide) as evidenced by ftir spectroscopy, *Langmuir*, 2000, **16**, 7503–7509.
56. M. Philipp, K. Kyriakos, L. Silvi, W. Lohstroh, W. Petry, C. M. Papadakis, J. K. Krüger and P. Müller-Buschbaum, From molecular dehydration to excess volumes of phase-separating pnipam solutions, *J. Phys. Chem. B*, 2014, **118**, 4253–4260.
57. K. Kyriakos, M. Philipp, L. Silvi, W. Lohstroh, W. Petry, P. Müller-Buschbaum and C. M. Papadakis, Solvent dynamics in solutions of pnipam in water/methanol mixture: A quasi-elastic neutron scattering study, *J. Phys. Chem. B*, 2016, **120**, 4679–4688.

CHAPTER 6

NMR-based Ligand–Receptor Interaction Studies under Conventional and Unconventional Conditions

C. CIARAMELLI, A. PALMIOLI AND C. AIROLDI*

BioOrgNMR Lab, Department of Biotechnology and Biosciences, University of Milano-Bicocca, Piazza della Scienza 2, 20126, Milan, Italy
*Email: cristina.airoldi@unimib.it

6.1 Introduction

With the biological function of biomacromolecules being dependent on their interactions with ligand molecules, the characterization of the molecular recognition events is critical to understanding the biological processes.

The ligand of a biomacromolecule, such as a protein, can be represented by small molecules, for example second messengers, cofactors, inhibitors and drugs, or by another macromolecule, for example activator and effector proteins, DNA or RNA and lipids. Among the variety of biophysical methods allowing the characterization of macromolecule binding processes, Nuclear Magnetic Resonance (NMR) has found wide application, because of its ability to characterize highly dynamic, partially inhomogeneous molecules and heterogeneous complexes in a very highly reliable and feasible manner.

The aim of this chapter is to provide a description of the most common NMR experiments based on ligand resonance observation, that have been

New Developments in NMR No. 26
NMR Spectroscopy for Probing Functional Dynamics at Biological Interfaces
Edited by Anirban Bhunia, Hanudatta S. Atreya and Neeraj Sinha

Published by the Royal Society of Chemistry, www.rsc.org

increasingly widespread and employed in the last two decades, together with specific examples of their applications in unconventional samples, such as those containing complex compound mixtures, multivalent ligands, liposomes, nanoparticles and/or cells.

6.2 NMR Molecular Recognition Techniques Based on Ligand Resonance Observation

The NMR approach for the study of a specific receptor–ligand pair needs to be chosen based on the kind of information required. If the final goal of the study is represented by the screening of compounds for the identification of new ligand(s) of a specific receptor of interest, a pool of experiments based on the direct observation of ligand resonances can be applied. When the ligand is a low- or medium-molecular-weight molecule, the same experiments also provide information about the "ligand-binding mode", such as the ligand-binding epitope and/or the ligand bioactive conformation.

Many different experiments can be performed, by monitoring several NMR variables that are related to the ligand behaviour. As a matter of fact, ligand-based methods observe and compare one or more NMR parameters of the ligand resonances in the presence and the absence of the receptor. As they are based on the observation of the ligand resonances, they can be exploited only for ligands at the equilibrium between the free and bound states, often assuming K_D values in the high nanomolar to high millimolar range.

Among the NMR variables that can be monitored in this screening the most significant are (1) line broadening, (2) changes to the Nuclear Overhauser Effect (NOE), as in Saturation Transfer Difference (STD) spectroscopy,[1] Water Ligand Observed *via* Gradient SpectroscopY (WaterLOGSY),[2] transferred-NOESY (trNOESY)[3] and NOE pumping experiments,[4,5] (3) changes in ligand diffusion coefficients measured by diffusion-edited NMR spectroscopy,[6] and (4) changes in the transverse and longitudinal relaxation rates of the ligand due to interaction with the receptor and/or spin labels.[7]

For all these ligand-based approaches, the exchange-mediated transfer of bound state information to the free state is then a fundamental assumption. This requirement implies that the formation of the complex under investigation is reversible and the free and bound ligand states are in rapid exchange.

To perform such a kind of experiment, a high ligand molar excess is adopted. One of the main advantages of this relies on the low requirement in terms of quantity of the molecular receptor. Less than 0.5 mg of unlabelled protein is required for these experiments. In this way, new ligands can be evaluated very rapidly, and the experiments can be performed on a time scale useful for chemistry and high-throughput screens.

6.2.1 Brief Description of the Main Experiments Based on Ligand-resonance Observation

The main ligand-based techniques, whose unconventional applications are briefly and qualitatively described in the following subsections.

6.2.1.1 Saturation Transfer Difference (STD)

The STD experiment was firstly described by Mayer and Meyer in 1999 and has since been applied to a large series of ligand–receptor pairs.[1] STD spectra come from the subtraction of two experiments (see Figure 6.1).

In the first experiment, the "on-resonance" experiment, several receptor magnetizations are selectively saturated *via* a train of frequency-selective radio frequency (rf) pulses. The rf train is applied to a frequency window containing the sole receptor resonances, and where compound resonances are not present. The saturation propagates from the selectively irradiated protons to other receptor protons *via* the large intramolecular ^1H–^1H cross-relaxation pathways (spin-diffusion) that typically is quite efficient, for high-molecular-weight receptors. Bound ligands receive this saturation *via* intermolecular ^1H–^1H cross-relaxation at the ligand–receptor interface. When they dissociate back into free solution, the relative saturation of their resonances persists and can be detected, while new ligand molecules bind the receptor receiving saturation energy as long as the application of the selective rf continues. This process allows the "amplification" of the signal. For the on-resonance irradiation, frequency values between −2 and 0 ppm are very frequently employed because no ligand nuclei resonances are found in this spectral region, whereas proteins present many resonances in this spectral region. Another common choice, when the putative ligand(s) show

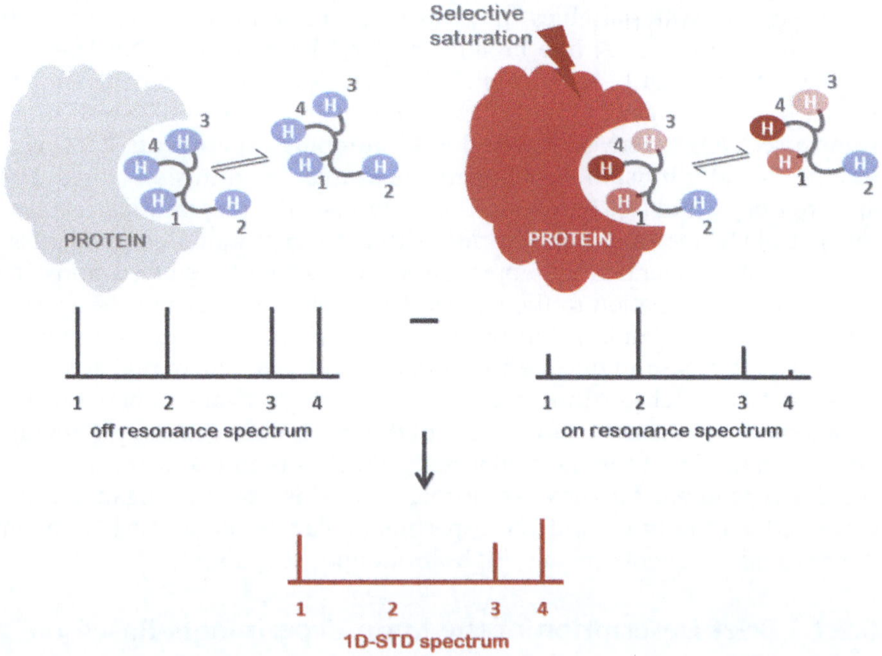

Figure 6.1 STD NMR experiment. The STD spectrum is obtained by subtraction of the "on-resonance" spectrum from the "off-resonance" spectrum.

no resonance signals in the aromatic proton spectral region, consists in placing the saturation frequency at 7–8 ppm. The second experiment, the "off-resonance" experiment, consists in the application of an identical rf train for off-resonance (at frequencies where neither receptor nor putative ligand resonances are present); the spectrum yields a normal NMR spectrum of the ligand–protein mixture.

The "on-resonance" and "off-resonance" experiments are recorded in an interleaved manner. After the subtraction of the "on-resonance" spectrum from the "off-resonance" spectrum, the corresponding STD difference spectrum displays only the signals of the ligand that experienced the transfer of saturation from the protein. The receptor resonances will not be visible, or very low, as a consequence of their minimal concentration or because of the application of R_2 relaxation-filtering.

STD experiments can be performed for 50 nM$<K_D<$10 mM. When the binding is very tight, the off rates are in the range of 0.1–0.01 Hz. In this case, the saturation transfer to ligand molecules is not very efficient. This is usually the case for K_D values less than 1 nanomolar. If the K_D values are 50 nanomolar, or larger, fast exchange of free and bound ligands leads to a very efficient build-up of saturation of the ligand molecules in solution. However, when the binding is very weak, the probability that the ligand is in the receptor site becomes very low, affording very weak STD signals.

STD signal intensity depends on the irradiation time/saturation time and on the excess of ligand molecules used.[8] The more ligand is used and the longer the irradiation time, the stronger the STD signal is. A commonly used irradiation time is 2 s and until a 100-fold excess of ligand is used to achieve good-quality STD spectra. In fact, upon dissociation, the saturation of the ligand is transported into solution, where it accumulates due to the slow decline of the saturation by relaxation processes in solution. Moreover, working with high ligand:receptor ratios, very low amounts of receptor are required. The ligand saturation is higher for those protons that are in closer contact with the receptor. For those ligand protons that interact with protein protons through an intermolecular NOE, a decrease in intensity is observed in the "on-resonance" spectrum, while no intensity variations take place in the "off-resonance" spectrum. Subtraction of the "on-resonance" from the "off-resonance" spectra leads to a difference spectrum, in which only the signals of the protons that were attenuated by saturation transfer are visible (Figure 6.1).

Thus, STD spectra not only allow the identification of ligand(s), but also provide the ligand epitope mapping, that is the map of the ligand regions involved in the interaction with the receptor. This information is fundamental for pharmacophore group identification for the rational drug design of new receptor ligands.

6.2.2 Solvent Saturation Transfer

A variant of STD consists in a process of saturation transfer that involves bound water instead of the receptor. This is based on the selective saturation (or inversion) of water resonance and on the following transfer of magnetization to the ligand. The main technique exploiting this principle is the Water-Ligand Observed *via* Gradient SpectroscopY (WaterLOGSY).[2]

The magnetization transfer from bulk water, often found at the receptor–ligand interface where a ternary complex is formed, to ligand takes place *via* labile receptor protons that can exchange with the solvent. They can be within but also distant from the ligand-binding pocket. In the presence of small ligands, the intermolecular water–ligand NOE becomes negative only if the ligand is bound to the receptor (see Figure 6.2).

In fact, the correlation time is much longer for the ternary water–ligand–receptor complex than for the water–ligand complex. Therefore, if the residence time of water molecules is long enough, negative NOE effects are observed between water and the ligand protons. In this experiment, the resonances of non-binding compounds appear with opposite sign and are weaker than those of the interacting ligands.

Because of the large reservoir of bulk water magnetization, the Water-LOGSY experiment is very sensitive and allows for the use of protein concentrations as low as 100 nanomolars. A detailed discussion of many practical aspects and the range of applicability of this experiment is reported in Dalvit *et al.*[9]

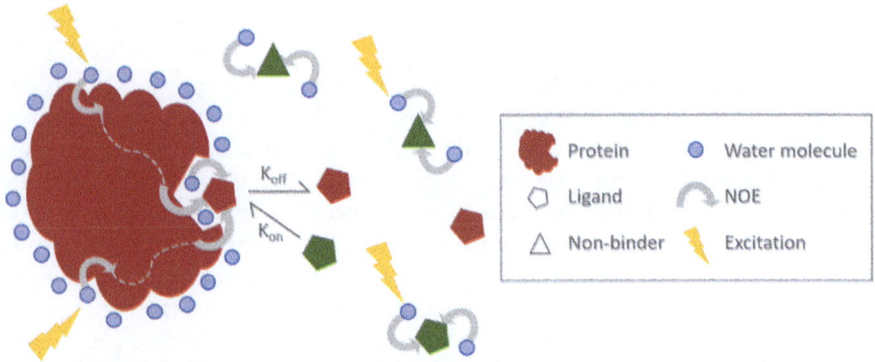

Figure 6.2 Water-LOGSY experiment. Magnetization transfer pathways between water molecules, small molecules, and proteins in waterLOGSY after the initial excitation of the bulk water resonance. Molecules that do not bind the receptor (triangles) display positive NOE (green) with water. Proteins display negative NOE (red) with water. Ligands (pentagons) that are in fast exchange between free and protein-bound forms are in two populations: the first is made by the molecule receptor that exhibits negative NOE (red), and the second is composed by free ligands that exhibit positive NOE (green) with water. The overall ligand signal is an average of the two-opposite magnetization.

6.2.3 Transferred NOESY (trNOESY)

NOE effects (NOEs), traditionally exploited for the determination of 3D structure of molecules in solution,[10] have found large applications also for the characterization of molecular recognition events.

Low- or medium-molecular-weight molecules (MW < 1000 Da) have short correlation times τ_c and, consequently, such molecules show positive NOEs. On the contrary, biomacromolecules exhibit strongly negative NOEs. Thus, when low-molecular-weight ligands bind to a large receptor molecule, their NOEs undergo drastic changes leading to the observation of transferred NOEs (trNOEs). After the binding to a high-molecular-weight receptor, the small ligand behaves as a part of the large molecule and adopts the corresponding NOE behaviour, that is, it shows strong negative NOEs, so-called trNOEs. The interaction of a ligand with a receptor protein can thus be easily detected by looking at the sign and size of the observed NOEs (see Figure 6.3).

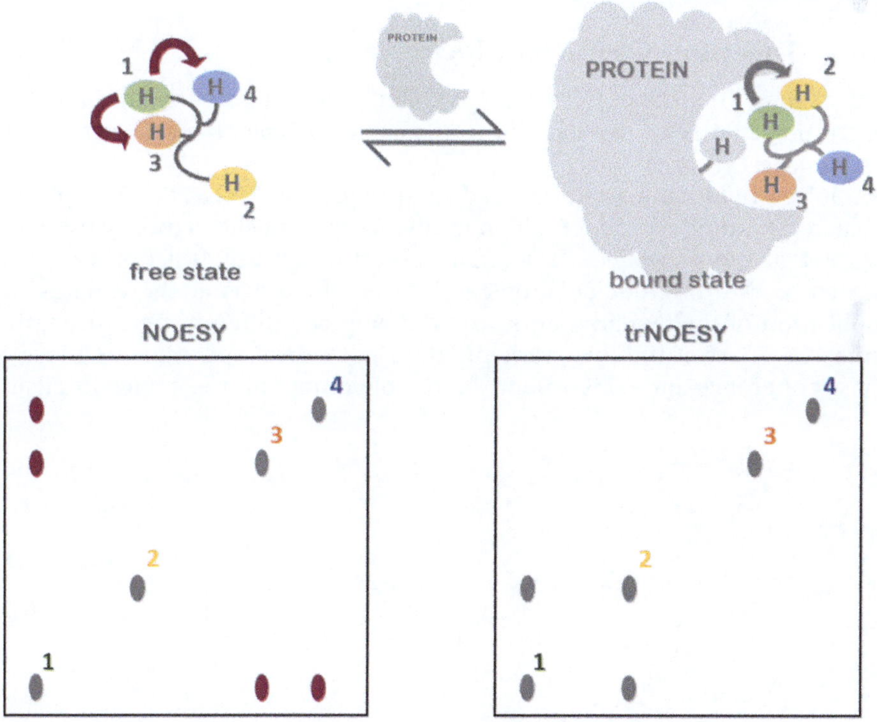

Figure 6.3 The tr-NOESY experiment. (Left) Schematic representation of a NOESY spectrum for a free low-molecular-weight ligand. Cross-peaks and diagonal peaks have different signs. (Right) Schematic representation of a trNOESY spectrum acquired for an exchanging ligand–protein system. Cross-peaks and diagonal peaks have the same sign, as expected for a large molecule, thus providing evidence of molecular interaction.

Moreover, trNOEs depend on the bound conformation of the ligand, allowing determination of the ligand bioactive conformation, that provides fundamental structural information.

However, due to spin diffusion effects, which are typical for large molecules, these data must be interpreted very carefully. In this case, receptor spins may mediate the exchange of magnetization producing negative cross-peaks between protons far apart in the macromolecule. Thus, protein-mediated, indirect trNOE effects among ligand protons can appear in trNOESY spectra, leading to errors in the analysis of the ligand bound conformation. However, the use of trNOEs in the rotating frame (trROESY) experiments[11] allows to distinguish which cross-peaks are dominated by an indirect effect, from those due to a direct enhancement. trNOESY allows to probe binding affinities in the range 100 nanomolar $\leq K_D \leq$ 1 millimolar.[12]

Thus, the trNOE method allows fast screening of putative ligands of a specific target receptor and, at the same time, provides information on the ligand conformation able to recognize and bind the receptor. These data are fundamental for the rational structure-based drug design.

6.2.4 Diffusion Coefficient Measurements

The change in diffusion rate of a small ligand upon its binding to a macromolecule is at the base of a diffusion coefficient measurement approach (see Figure 6.4).

A pulsed-field gradient stimulated echo sequence (PFG-STE)[13] is exploited in such a kind of experiment. Two gradients of opposite sign, separated by an interval t, are applied. As a small molecule diffuses fast during t, it is located at two different positions along the z-field axis at the time as the application of the two gradients, experiencing two different field strengths. Thus, the effect of the two gradients does not cancel out, causing a loss of phase coherence and consequently a loss of compound resonance intensity.

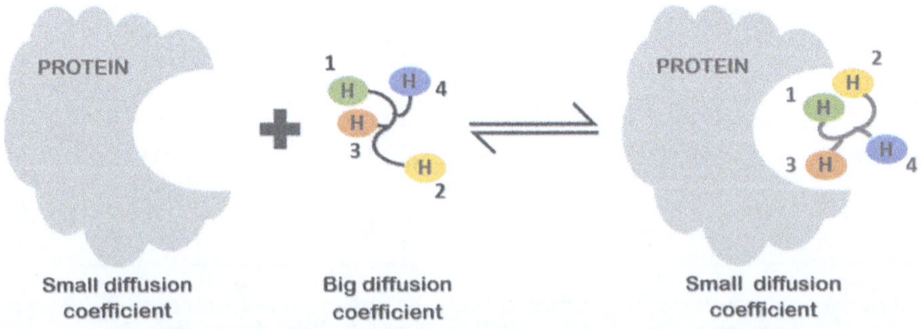

Figure 6.4 The diffusion filtering. A macromolecular receptor shows a small diffusion coefficient (in contrast, a low-molecular-weight compound diffuses fast in solution); when the small compound interacts with the receptor, the corresponding diffusion coefficient becomes smaller.

After its binding to a macromolecule, the small molecule diffuses along the *z*-field axis more slowly. Therefore, the intensities of the transiently bound ligand signals are attenuated, assuming lower values than those of the free ligand. Differences in diffusion coefficients are the observable NMR parameter exploited in this kind of experiment, and there is no high-molecular-weight limit for the receptor. This approach does not provide specific structural information; it is exclusively adopted for ligand screening.

6.2.5 Relaxation Time Measurements

Small molecules are characterized by low relaxation rates $R_1 = 1/T_1$ and $R_2 = 1/T_2$. When the same compounds bind a receptor with high molecular weight, their NMR relaxation properties can change significantly, becoming larger $R_1 = 1/T_1$ and $R_2 = 1/T_2$ values (see Figure 6.5).

The basic procedure for ligand screening using R_1- and R_2-edited experiments requires the following steps: (1) the R_1 or R_2 experiment for the free compound and for a set transverse relaxation delay is recorded; and (2) a second R_1 or R_2 experiment on the putative ligand is recorded in the presence of the receptor. If there is an interaction between the compound and receptor, T_1 and T_2 values for the ligand decrease.

R_1-edited experiments need to rely on selective R_1 measurements, that must be distinguished from non-selective ones. To achieve selective R_1 values the magnetizations must be differentially perturbed, either by selective inversions (as in 1D-NOE experiments) or at different times (as in 2D-NOESY experiments).

R_2 relaxation can be determined by a spin-lock such as in the Carr–Purcell–Meiboom–Gill (CPMG) pulse train[14] or continuous wave irradiation as in a $R_{1\rho} = 1/T_{1\rho}$ experiment.[15]

In the following sections, unconventional applications of these experiments are illustrated and discussed.

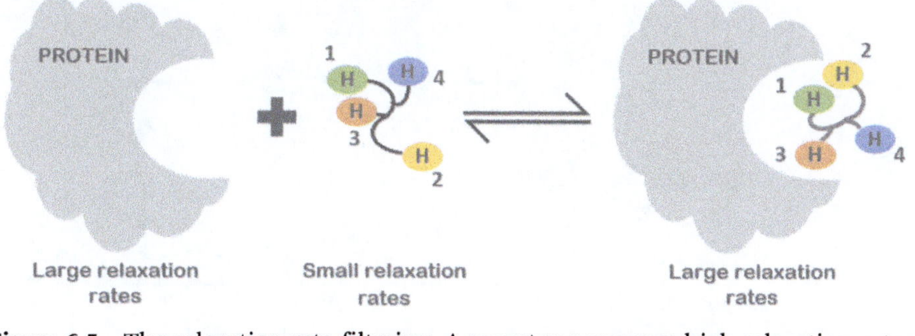

Figure 6.5 The relaxation rate filtering. A receptor possesses high relaxation rates; by contrast, a low-molecular-weight compound shows low relaxation rates in solution; when the small compound interacts with the receptor, its relaxation rates become higher.

6.3 NMR Interaction Studies on Complex Compound Mixtures

NMR is a powerful technique to study the interaction of a biomacromolecule with complex mixtures of molecules in a single step, providing the identification of receptor ligand(s), if any. This result can be achieved by performing even only one simple and very fast experiment.

The most common experiment used for this purpose is indeed STD NMR,[1] that can be used to screen different molecules at once through the analysis of a sample containing the receptor and a mixture of putative ligands.[16] The ligands can readily be distinguished from nonbinders in the STD NMR spectrum (see Figure 6.6).

An advantage of the STD technique is the possible combination with any other NMR pulse sequence generating a whole suite of STD NMR experiments such as STD TOCSY (TOtal Correlation SpectroscopY), STD HSQC (Heteronuclear Single Quantum Coherence) and STD DOSY (Diffusion Ordered SpectroscopY).[8]

Other NMR techniques that enable the rapid and comprehensive screening of active components in compound mixtures include trNOESY (see Figure 6.7), WaterLOGSY, $T_{1\rho}$ and T_2 relaxation experiments, that are described in the next examples.

Thus, by exploiting these approaches, or their combination, NMR can be successfully used for the screening of a particular activity both in cocktails of structurally known molecules, as libraries of compounds, generally represented by a panel of synthetic molecules, and in mixtures of natural compounds such as those contained in an extract obtained from natural sources.

6.3.1 NMR-based Screening of Libraries of Synthetic Compounds

The use of NMR for the screening of a library of putative ligands has been primarily developed in the context of drug discovery, representing a

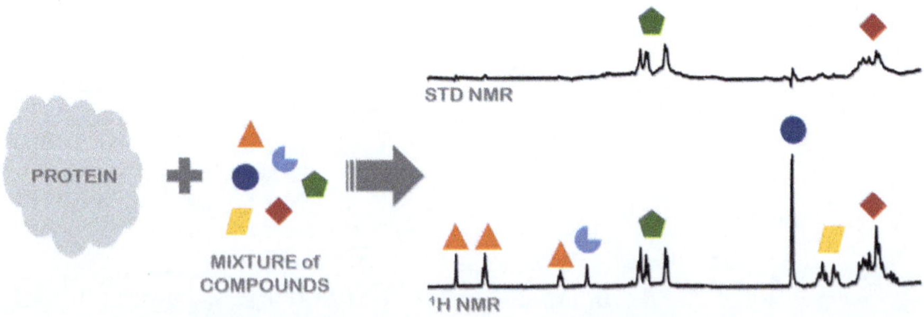

Figure 6.6 STD NMR experiment to screen a mixture of compounds for the binding to a target macromolecule. The ^{1}H NMR spectrum of the mixture is at bottom and the corresponding STD spectrum in the presence of the macromolecule is on top.

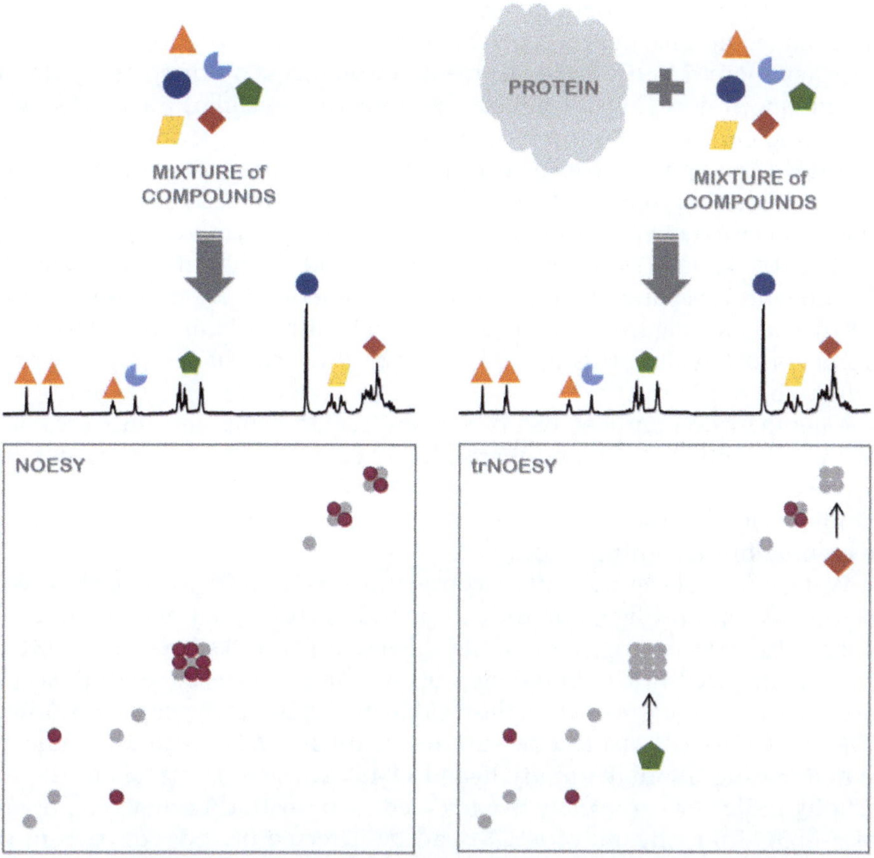

Figure 6.7 2D NOESY and trNOESY experiments of a mixture of compounds in the absence and presence of a target macromolecule, respectively. Positive cross-peaks of the small molecules are in purple and negative cross-peaks in grey. The change in the cross-peak sign of the ligands (green pentagon and red rhombus) in trNOESY is indicated by arrows.

high-throughput screening method of synthetic compound libraries. The identification of biologically active compounds within a library of known compounds has been largely reported.[17]

An excellent example of this application is the work of Vogtherr and Peters (2000), where, by using STD NMR and trNOESY experiments, compound libraries of randomly methylated *O*-methyl glycopyranosides were screened for binding activity towards *Sambucus nigra* agglutinin (SNA).[18] Even if the compounds have identical molecular weights and rather similar physico-chemical properties, they were able to identify the active compounds directly from the mixture. The results obtained with this approach were consistent with earlier biochemical work, demonstrating the reliability of the method applied.

The maximum number of compounds contained in the mixture that can be analysed at once depends on the number and type of protons present in each compound and on the degree of resonances' overlapping in the different spectral regions. In fact, an issue to consider, especially when screening cocktails of similar fragments in a single experiment, is the resonance overlapping between ligands that can prevent the correct identification of the binding molecules.[19] If, in addition to the standard STD, the Diffusion Ordered SpectroscopY (DOSY) element is introduced, the resulting 2D STD DOSY spectrum shows all resonances of the binding molecules, but each ligand is separated along the second dimension based on its diffusion coefficient (see Figure 6.8).[20] If all ligands in the cocktail possess different diffusion behaviours this approach allows the issue of resonance overlapping to be overcome.

A competitive approach can be used to screen compound mixtures, both in the case of STD NMR and WaterLOGSY experiments. This approach may provide information on the specificity of two or more ligands for the same binding site on the target and it is particularly useful for the indirect screening of high-affinity ligands.

Wang and colleagues[21] used competitive STD NMR to identify a competing high-affinity ligand in the compound mixture. They demonstrated the possibility of detecting high-affinity ligands for the β-secretase BACE1, an enzyme involved in the initial step of Alzheimer's disease (AD). First, they proved the binding of the eight-residue transition-state peptide inhibitor (OM99-2) to BACE1, using a low-affinity ligand of BACE1 as an STD indicator. When adding the high-affinity ligand OM99-2, the STD signals of the low-affinity indicator completely disappeared, demonstrating that the presence of a high-affinity ligand of BACE1 can be detected by the reduction or disappearance of the STD signals of the weak ligand. Then they demonstrated that this competition strategy can be applied also for the screening of

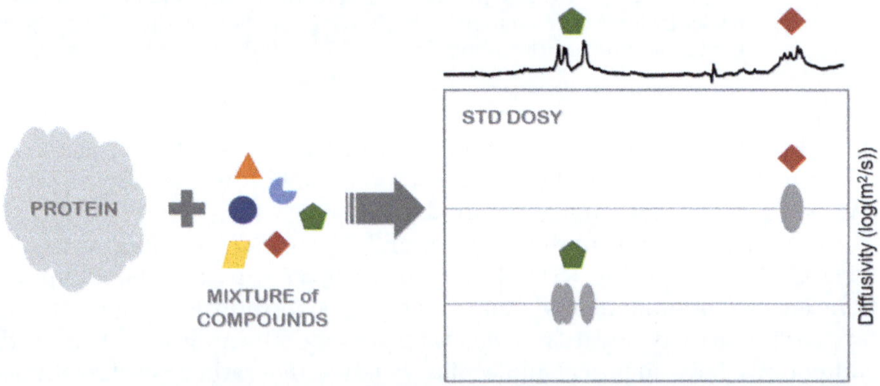

Figure 6.8 STD DOSY experiment of a mixture of compounds with a target macromolecule. The peaks from the two ligands, that show different diffusive properties, are indicated in the spectrum with the corresponding symbols (green pentagon and red rhombus).

compound mixtures including high-affinity ligands that would be missed by the standard STD NMR method. They used the human serum albumin (HSA) as the model system, 6-CH_3-D,L-Trp as the weak ligand, diazepam as the high-affinity ligand and 7-CH_3-D,L-Trp as the non-binding compound (control). The STD NMR spectrum of the mixture was compared with that of the indicator alone. The reduction in the STD signal intensities observed for 6-D,L-CH_3-Trp indicated the presence of an active ligand in the mixture and the new STD signal at 3.45 ppm (the methyl signal of diazepam) identified diazepam as a ligand and no STD signals were detected for the control compound 7-CH_3-D,L-Trp. This approach allows the screening of compound libraries against a broad range of drug targets to identify both high- and low-affinity ligands.

Dalvit *et al.*[22] used a similar strategy to perform competitive WaterLOGSY (c-WaterLOGSY) experiments, taking advantage of a compound identified as a weak affinity ligand of the target of interest. Indeed, if another molecule of the testing mixture (containing sucrose, 7-CH_3-tryptophan and diazepam, in this work) associates with the target, the human serum albumin (HSA), employed as test case, displacing the reference molecule (6-CH_3-tryptophan), an intense reduction or disappearance of the reference resonances will be observed. If this is not the case, the other molecules are not competitors, or they do not bind the target at all. In this study they observed a drastic reduction of 6-CH_3-tryptophan resonances when diazepam was tested, enabling its identification as a high-affinity ligand of HSA.

6.3.2 NMR-based Screening of Complex Mixtures of Natural Compounds

Drugs obtained from natural sources and used in traditional herbal medicines are currently attracting great interest. However, the screening of natural extracts to identify active substances remains a challenging topic, due to their intrinsic complexity. Drug discovery from natural products can start from the isolation of compounds from the natural source and the assay for their biological activity using *in vitro*/*in vivo* experiments. Then, the synthetic structural modification of natural bioactive compounds can be exploited to improve compound efficacy and reduce toxicity when present. Thus, the entire process requires huge effort in terms of time and resources.

For this reason, the classical phytochemical approaches have usually avoided the study of such complex mixtures. However, NMR-based approaches make it possible to also reveal the simultaneous interaction of different ligands with the same receptor, highlighting possible synergic effects often involved in the mechanism of action of several herbal remedies.

Moreover, these approaches allow the screening of natural extracts for the presence of ligands and modulators of a pharmacological target without requiring the separation and purification of the single components. These processes are time consuming and, in addition, some natural extracts are often sensitive towards solvents, light or some isolation methods. For all these reasons, NMR-based approaches allowing the identification of potential bioactive compounds through a one-step process seem to be an ideal choice that offers the economically and environmentally healthy conditions required for the screening of these complex mixtures.

One of the first works that demonstrated the possibility to use NMR as a fast and simple method for the screening of mixtures of natural extracts was reported in 2005.[23] The authors used NMR experiments to study the water extract of a mushroom, the *Pleurotus ostreatus* hot water extract, and the lentil lectin from *Lens culinaris*. From the screening *via* STD NMR of the extract, it was evident that one or more compounds, with resonances in the carbohydrate region, were ligands of the lectin. Then, trNOESY and STD TOCSY experiments provided more information on the ligand, the bound epitope and the bound conformation, indicating that lentil lectin binds the double *exo*-anomeric conformation of trehalose in solution.

Tanoli *et al.* used STD NMR, trNOESY and STD TOCSY experiments to obtain a rapid and comprehensive screening of the active components present in the hydroethanolic fraction of the crude extract from *Rauia resinous* for their interaction with Bovine Serum Albumin (BSA) protein and their structural characterization.[24] Isovitexin-2-rhamnoside, vitexin-2-rhamnoside and neoschaftoside were identified as BSA ligands, with aromatic moieties of the ligands involved in stronger interactions with the protein. The application of liquid chromatography (LC)-solid-phase extraction (SPE)-NMR separation followed by structure characterization helped to obtain the epitope mapping of these compounds, indicating isovitexin-2-rhamnoside as the best ligand.

In a subsequent work, the same authors studied the natural herbal extract of *Stryphnodendron polyphyllum* through STD NMR, hyphenations (LC-SPE-NMR) and docking simulations.[25] In particular, they applied STD NMR, trNOESY and STD TOCSY experiments to evaluate the potential binding of the natural extract of *Stryphnodendron polyphyllum*, used in herbal medicine in Brazil, towards HSA. Then, a hyphenated system comprising LC-SPE-NMR was used to obtain a full structural characterization and information about the exact number of the compounds involved in the binding. The results obtained from the combination of these techniques indicated myricetin-3-*O*-rhamnopyranoside, quercetin-3-*O*-glucopyranoside, quercetin-3-*O*-xylopyranoside and quercetin-3-*O*-rhamnopyranoside as the ligands of the protein.

A recent example of an STD NMR-based interaction study regards the interaction of coffee extracts components with oncogenic Ras Proteins wt H-RasGDP or H-RasGDP G13D. It was based on an experimental protocol that, combining NMR spectroscopy, molecular docking, surface plasmon resonance and *ex vivo* assays, contributed to the elucidation of the molecular basis of the activity of natural extracts from green and roasted coffee beans and their main polyphenolic components (chlorogenic acids, CGAs) as chemoprotective dietary supplements.[26] CGAs, and in particular 5-CQA (5-caffeoylquinic acid), were identified as ligands. However, the STD NMR spectra revealed the presence not only of 5-CQA, but also of caffeine resonances, which suggested an interaction between this molecule and the target protein. Caffeine signals were also visible in blank STD NMR spectra recorded in the absence of the protein, which indicated its interaction with other macromolecules present in the extracts and not with Ras proteins.

This example brings out the importance of preforming a blank STD NMR experiment to exclude the non-specific interaction of some molecules of the extracts with other components, taking into account that the natural extracts are complex mixtures containing different classes of molecules.

Our group applied the NMR-based approaches to the screening of several edible matrices to identify the potential ligand of amyloidogenic proteins. In these works, STD NMR was used to study the interaction of natural extract components with oligomers of amyloidogenic peptides as Aβ1–42 peptides that oligomerize under specific experimental conditions and are in this way different from the above-described conventional studies on proteins. Aβ1–42 oligomers were used as the model for neurotoxic amyloid oligomers, being the soluble Aβ oligomers involved in the aetiology of Alzheimer's disease (AD) the most neurotoxic among the other Aβ aggregates.[27]

In the first work, the NMR characterization of the interaction of *Salvia sclareolide* extract with Aβ1–42 peptide was based on STD NMR and trNOESY experiments.[28] After obtaining from STD experiments the evidence of the interactions of some extract components, and in particular of rosmarinic acid, with the Aβ1–42 oligomers, the existence of interaction was further confirmed by trNOESY experiments acquired on the same ligand/peptide mixture. A blank NOESY spectrum of the extract solution was recorded in the absence of the Aβ1–42 peptide (see Figure 6.9).

An inversion of the sign of the NOESY cross-peaks of the main component present in the extract was observed between both spectra, passing from positive (red), in the absence of Aβ1–42 peptide, to negative (blue) in the

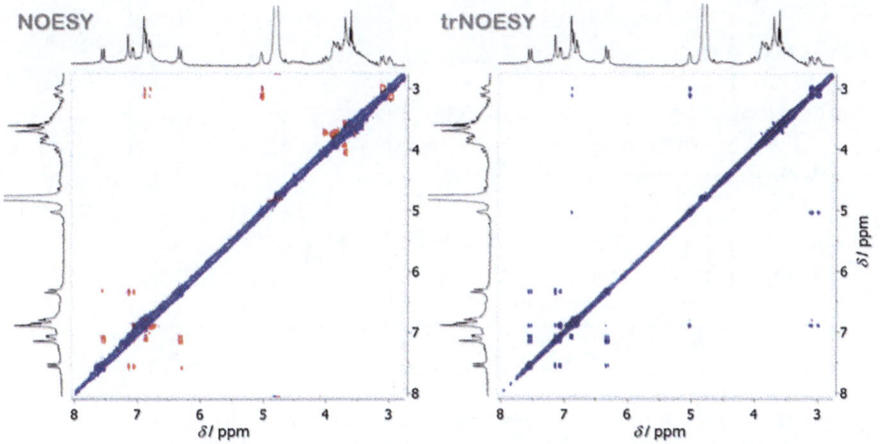

Figure 6.9 2D-NOESY spectrum of the butanol extract of *Salvia sclareolide* and trNOESY spectrum of the same *Salvia sclareolide* mixture and Aβ1–42. Positive cross-peaks are shown in red, and negative ones in blue.

presence of Aβ1–42 peptide (see Figure 6.9). This change is due to an increase in the effective rotational correlation time of the molecule in the presence of the Aβ oligomers, thereby strongly supporting the existence of binding between the small molecules present in the extract to a large molecular entity. In this context, the macromolecule corresponds to the Aβ peptide oligomeric state.

The same approach, combining NMR spectroscopy with atomic force microscopy, *in vitro* biochemical and cell assays, was used to investigate the potential anti-amyloidogenic properties of coffee and its molecular constituents.[29] STD NMR spectra demonstrated the binding of CGAs, and in particular of 5-CQA, contained in coffee extracts to the Aβ-oligomers. Also in this case, caffeine signals appear in STD spectra recorded in the presence of Aβ oligomers, but, as already mentioned above, blank STD spectra acquired on coffee extracts in the absence of Aβ peptide contained the same resonances, suggesting an interaction between caffeine and other components of the extracts (see Figure 6.10).

An interesting outcome of this work is the possibility of using STD NMR to monitor the interaction of different isomers with a target protein over time. It is known that 5-CQA isomerization to regioisomers 3-CQA and 4-CQA occurs over time. This isomerization was monitored for 7 days and the interaction of the isomeric mixtures with Aβ-oligomers was verified through STD NMR, confirming that all three isomers are Aβ ligands.

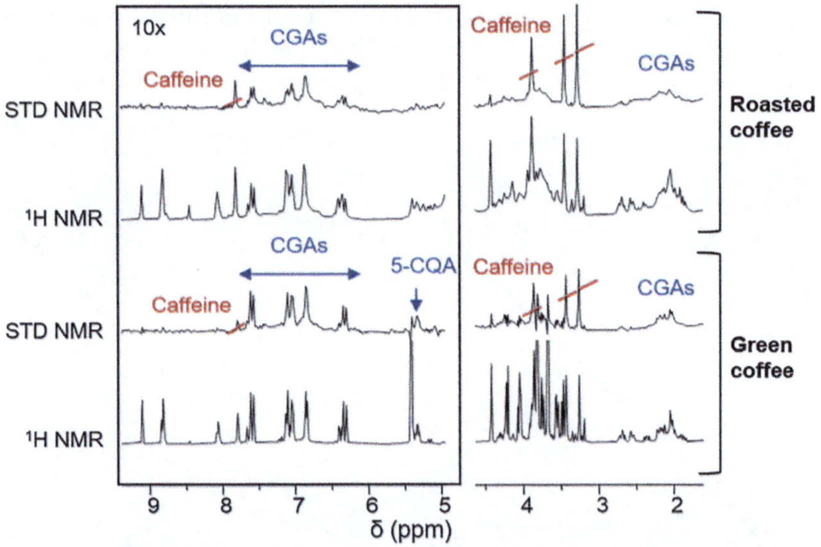

Figure 6.10 ^{1}H NMR and STD NMR spectra acquired on a sample containing green or roasted coffee extracts and Aβ1–42 oligomers. The resonances of the ligands (CGAs and 5-CQA in particular) are indicated in STD spectra. Caffeine signals in STD spectra should not be considered since it is a non-binder of Aβ1–42 oligomers, as demonstrated by the blank STD.

As is well known, STD NMR experiments allow revealing of the interaction between small molecules (MW under 1000 Da) and a high-molecular-weight biomolecule, such as a protein or, as in this case, peptide/protein oligomers. However, the presence in roasted coffee extracts of melanoidins, the heterogeneous macromolecules formed during roasting because of pyrolysis and the Maillard reaction, and the speculation that melanoidins might be able, at least in part, to sequester Aβ1–42 peptide, prompted the authors to investigate these macromolecules. In this case STD NMR is not feasible because of the high molecular weight of the potential ligands (3–100 kDa). Thus, another NMR-based approach was used to overcome this issue. Samples of extracts enriched in high-molecular-weight melanoidins (HMW, MW > 10 kDa) obtained by ultrafiltration were used to see whether melanoidins affected the oligomerization equilibria of Aβ1–42: ^{1}H NMR spectra of the Aβ1–42 were acquired in the absence and presence of the HMW sample at increasing concentrations. When melanoidins were added to the Aβ1–42 peptide, a reduction of peptide signal intensities occurred and their line broadening increased in a concentration-dependent manner, validating the hypothesis that melanoidins interact with the Aβ1–42 peptide.

A similar approach, based on STD NMR experiments, was employed also on *Peucedanum ostruthium*, a plant traditionally employed in Austria and Italy for its several potential therapeutic applications.[30] STD NMR experiments revealed that CQAs, and in particular di-substituted CQAs, are the best ligands of Aβ1–42 oligomers contained in *Peucedanum ostruthium* rhizome extract, and also glycosylated flavonoids and furanocoumarins are able to bind Aβ1–42 oligomers. The co-presence of these compounds in the same extract allows the obtaining of a significant biological activity.

In a very recent work, this approach was used to investigate and compare the potential anti-amyloidogenic activity of different cocoa extracts, combining NMR spectroscopy, preparative reversed-phase (RP) chromatography, atomic force microscopy, *in vitro* biochemical and cell assays on a human neuroblastoma cell line.[31] In this work, a separation procedure was used to obtain fractions of extracts enriched in different classes of molecules, with reversed-phase chromatography. The binding of the ligands to Aβ oligomers was demonstrated through saturation transfer double difference (STDD) experiments (see Figure 6.11).

The difference between STD spectra of extract/fraction and protein and STD spectra of the only extract/fraction (control) were calculated to estimate the STD effect due to the extract and fractions' component(s) directly binding to Aβ oligomers. The subtraction of the control spectrum permits, indeed, the suppression of possible contributions from non-specific interactions within the extract. Therefore, the STDD spectrum contains only the signals related to the molecular interaction with Aβ oligomers. The flavanols catechin and epicatechin and the methylxanthine theobromine were identified as ligands of Aβ-oligomers. The separation step allowed the performance of the biological assays also on the enriched fractions, confirming the fractions containing mainly catechins and theobromine as the most active

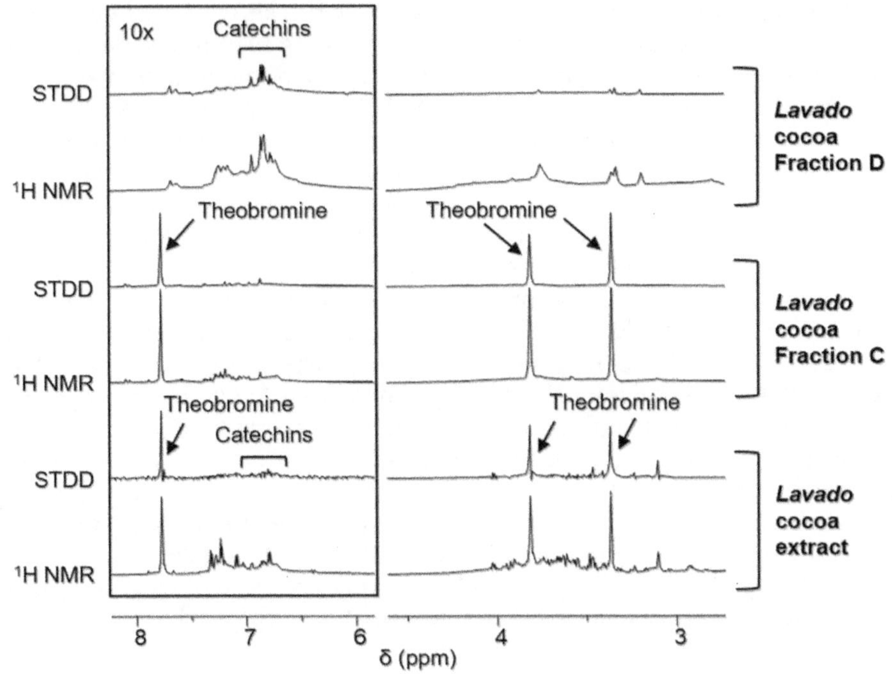

Figure 6.11 ^{1}H NMR and STDD NMR spectra acquired on a mixture of cocoa extract/enriched fractions and Aβ1–42 oligomers. The resonances of the ligands (catechins and theobromine) are indicated in the STDD spectra.

ones, thus correlating the evidence of the interaction at an atomic level with the biological affect. Interestingly, theobromine, that is structurally very similar to caffeine, interacts with Aβ peptides, both in the extract and in the purified fraction, conversely to what is observed for the caffeine contained in coffee, as reported in the previous work. STD NMR proved to be able to shed light on the different behaviours of these two methylxanthines contained in different natural extracts with Aβ-oligomers at the atomic level. The approach used to screen natural extracts to identify ligands of Aβ1–42 oligomers was also extended to other amyloidogenic proteins: PrP106–126 and ataxin-3 oligomers.[32] In fact, NMR spectroscopy contributed to the identification of the structural determinants involved in the interaction of green tea catechins with Aβ1–42, PrP106–126 and ataxin-3 oligomers. The data provided important information for the rational design of new compounds with anti-amyloidogenic activity, giving information at an atomic level on the interaction, showing that the flavan-3-ol unit of catechins is essential for the interaction and that the gallate moiety, when present, seems to increase the affinity for the target oligomers.

A very recent NMR-based drug-screening strategy for discovering active substances from the radix *Polygoni multiflori*, a herb radix used in traditional Chinese medicine, was applied, gaining information on the mechanism of action at an atomic level. Since *Polygoni multiflori* is known for its benefits to the glycolipid metabolism, the authors chose as the target protein the human Fatty Acid Binding Protein 4 (FABP4), which represents a potential therapeutic target for metabolic diseases such as diabetes and atherosclerosis.[33] They propose a four-step NMR-based protocol to implement the screening: (1) preparation of libraries of pre-purified subfraction of the extract; (2) NMR-based screening of subtractions to identify ligands of target protein (^{1}H NMR line broadening or chemical shift perturbations measurements in combination with STD NMR); (3) identification and structural characterization of the binder/s coupling HPLC-SPE-NMR; and (4) optimization of the molecular structure using the structural information from STD NMR. This strategy allowed them to identify 2,4-dihydroxy-6-[(1*E*)-2-(4-hydroxyphenyl)ethenyl]phenyl-ß-D-glucopyranoside (TSG) contained in one of the subfractions as the ligand of the target protein FABP4.

A particular application of STD NMR for the screening of mixtures was presented in a work published in 2016. It regarded the possible binding of metabolites to proteins present in the cellular environment. The identification of possible interactions of endogenous metabolite with protein can represent a key step in the understanding of the regulation of numerous cellular processes. It has been proved that NMR spectroscopy can be successfully used in this field, compared to the few available studies, largely mass spectrometry-based, that are restricted to specific metabolite classes, such as lipids. This work represents a proof of concept: they identified the binders of four well-characterized bacterial and mammalian proteins (AroG, Eno, PfkA and bovine serum albumin) and in complex mixtures of up to 33 metabolites.

They compared different NMR methods for ligand-detected primary screening of unlabelled components: STD, diffusion editing, WaterLOGSY, $T_{1\rho}$ and T_2 relaxation. They reported the results from WaterLOGSY and $T_{1\rho}$ relaxation experiments in interaction maps, where the colour intensity in the boxes correlates with the signal intensity. Both methods showed generally good agreement about the detection of protein – metabolite interactions, suggesting that using only one method should be sufficient to detect such interactions.

In this work, WaterLOGSY was slightly more sensitive. However, the authors would prefer $T_{1\rho}$ relaxation experiments for high-throughput screening because $T_{1\rho}$ relaxation should be less biased towards the chemical properties of the ligand, depending primarily on the macroscopic properties of the target – ligand complex and because $T_{1\rho}$ relaxation experiments can be performed in D_2O ($T_{1\rho}$ does not depend on H_2O for magnetization transfer as does WaterLOGSY), increasing the robustness of the experimental setup.

6.3.3 A Brief Resume of the Advantages and Disadvantages (and How to Overcome Them) of Studies on Ligand Mixtures

The examples reported previously pointed out the advantages and disadvantages of NMR-based identification of ligands of a target macromolecule within a mixture of several compounds. Here we want to summarize the key advantages and consider the major disadvantages and, when possible, give some advice to overcome them.

The first advantage of ligand-detected NMR over classical activity-based assays is the possibility of testing several molecules in a mixture simultaneously. The issue of how many molecules can be screened in the same testing mixtures is difficult to address. Of course, an aspect that limits the applicability of this approach is the overlapping of NMR resonances in the spectra. Testing a simple mixture containing few molecules greatly simplifies the data analysis and peak assignment at the expense of limited throughput. On the other hand, increasing the complexity of the mixture will inevitably lead to ambiguous assignment because of spectral overlapping. Obviously, the degree of signal overlapping is due to the chemical structure of the molecules contained in the mixture: more similar molecules will probably have resonances in the same spectral regions, while the different chemical natures of the compounds will reduce the overlapping of the signals. The trade-off between the simplicity of data interpretation and the level of throughput obtained should be evaluated for each different mixture analysed.[34]

A key advantage of this approach is represented by the NMR experiments on which it is based that are simple, fast and robust, and allow the screening of compounds in mixtures in a very short time, when performing even only one experiment. Moreover, NMR binding experiments allow identification of the specific molecule(s) involved in the interaction with the macromolecular target, with an unequivocal structural characterization of the ligand(s) and they give information on the binding at an atomic level, providing the binding epitope.

A fundamental point, especially when working on mixtures of natural extracts, is that NMR screening does not require any prior purification step that could be not only time consuming but could also alter the mixture composition and properties.

Another point to be considered, together with the amount of macromolecule, is the ratio between the protein and the ligand. When dealing with mixtures of potential ligands, this evaluation is not that straightforward. For a successful STD NMR experiment, for example, the ligand needs to be in a large excess. When the concentration of the mixture component is very different this can result in a diverse probability to detect their putative binding.

The possible non-specific binding of molecules of the mixtures can represent a confounding factor. When dealing with natural extracts, in which the heterogeneous composition of the mixture probably contains also some high-molecular-weight species difficult to characterize, it is possible that the potential ligands interact with these components or with each other. Some possible non-specific binding can be interpreted as evidence of the binding with the macromolecule of interest in the STD NMR spectra. The acquisition of a blank STD NMR spectrum of the extract in the absence of the macromolecular target will clarify this possible effect, as reported in some of the examples cited previously.

When screening possible ligands through STD NMR, another limitation regarding the ligands is their molecular weight that generally should be under 1000 Da. However, some alternative strategies can be applied to study the interaction of molecules with higher molecular weight with the target; these are discussed in the following section.

6.4 NMR Interaction Studies Involving Multivalent Ligands, Liposomes, Nanoparticles and Peptide/Protein Aggregates

Hyperbranched structures (dendrimers or/and dendrons), nanoparticles (NPs) and, more generally, supramolecular aggregates (micelles, liposomes, *etc.*) are emerging as bioactive key tools for the developments of innovative biopharmaceuticals and biotechnological applications.[35] Probing the interactions of these complex structures at the biological interfaces is a crucial step for a deep insight into the elucidation of their mechanism of action with atomic resolution and this knowledge is essential in the rational drug-discovery process. Moreover, the results of the characterization of interactions occurring at the surfaces of nanoparticles are fundamental for the development of novel biocompatible and sustainable nanomaterials.

Unfortunately, this task is often very challenging, mainly due to molecular size of the interactors and further optimization of the analytical approaches is still required. Large molecules, supramolecular adducts and heterogeneous structures imply relatively slow molecular tumbling that leads to short nuclear transverse spin relaxation time (T_2) and results in broad or even undetectable signals. For these reasons, for example, the direct investigation of protein–NP interaction results is very difficult with standard NMR experiments. Moreover, some difficulties lie in the unfavourable hydrodynamic properties of many nanoscale or supramolecular assemblies, and, in addition, agglutination properties of multivalent interaction networks, such as in the case of carbohydrate–lectin complexes, prevent the NMR sample from remaining homogeneously suspended during NMR data acquisition. On the other hand, an advantage of these approaches is the possibility to characterize simultaneously the binding of multiple ligands, as long as their signals do not overlap.

Although there remain many challenges to address, Martins and coworkers have shown that STD NMR was a valuable additional toolbox for the characterization of colloidal dispersant particle interactions,[36,37] paving the way for the use of STD NMR experiments for studying interactions at the nanoparticle surface. As the first examples of STD NMR applied to detect small ligand interactions with organic nanoparticles widely used for biological application, we report the works of Casabianca and coworkers.[38–41] The authors observed that ethanol and isopropanol interact with the carboxylate-modified surface of polystyrene nanobeads, whereas other molecules such as glucose and glycine do not interact with the surfaces of the beads. In addition, treating water as ligands in a mixture of $H_2O:D_2O$, they quantitatively characterized for the first time the interaction of water molecules with the surface of a nanoparticle.[38] Then, they screened the binding of single amino acids to the carboxylate surface of polystyrene nanobeads, probing the driving force in terms of electrostatic and hydrophobic effects.[39,40] Moreover, more recently they explored the feasibility of an approach based on the saturation of protons in the receptor molecule followed by the observation of the STD effect on nearby carbon nuclei. Polarization is transferred from protons to carbons using the insensitive nuclei enhanced by a polarization transfer pulse sequence (^1H–^{13}C INEPT STD NMR). This allowed for examination of the binding events between small molecules and polystyrene nanoparticles.[41] This approach can be extremely useful for ligand screening and epitope mapping in systems characterized by a significant overlap of ^1H signals, such as for natural or synthetic complex mixtures of compounds.

Carbohydrate–protein interactions are fundamental events in a plethora of biological processes among which host-pathogen recognition, immune response and cell–cell communication promote proliferation and differentiation. Since most of these networks of interactions have a multivalent nature, in the last few decades several multivalent hyperbranched glycoconjugates have been designed to study or interfere with many biologically relevant recognition events. STD and trNOESY experiments were powerful tools to understand these interaction networks, but some technical and practical issues have limited their use to a few examples in the literature.

The majority of mammalian and bacterial carbohydrate-binding proteins (also called lectins) are membrane integral protein receptors or oligomeric circulating receptors, and this makes their expression and purification for binding studies very challenging. At the same time, the multivalent lectins–glycoconjugates clustering often results in the agglutination and precipitation of the sample, greatly limiting the acquisition of good-quality NMR spectra.

Arce *et al.* reported the STD interaction studies of mannosylated Boltorn H20 and H30 hyperbranched dendritic structures with the commercially available mannose-binding *Lens culinaris* lectin, as a model protein.[42] Giri *et al.* reported the characterization of human serum albumin (HSA) binding to PAMAM dendrimers. They combined ^1H NMR, STD NMR and DOSY experiments to validate the use of PAMAM scaffold as biocompatible nanomaterials.[43]

Ponader *et al.* evaluated the binding of a series of heteromultivalent glycooligomers presenting different combinations of Man, Glc and Gal residues, toward Concanavalin A (*Canavalia ensiformis*) as a model of mannose-binding lectin, revealing that non-binders seem to promote steric shielding, suppressing the receptor clustering with an overall increase in affinity.[44]

Mirabella *et al.* reported STD NMR and molecular modelling-based studies that supported a multivalent mechanism with specific interactions of pyrrolidine-based multivalent iminosugars with two lysosomal α-mannosidases (GMIIb and LManII) from *Drosophila melanogaster*. These proteins represent two models of target enzymes for anticancer therapy and the treatment of lysosomal storage disorders, respectively.[45]

As a proof of concept, our group recently reported the first example of STD NMR exploited to characterize the binding of a glycodendrimer bearing 18 copies of galactose (Gal18) and the pathogenic lectin LecA, a *Pseudomonas aeruginosa* virulence factor.[46] The binding epitope map revealed that both the sugar entities and the dendrimer scaffold are involved in the molecular recognition process, suggesting that this specific scaffold could contribute to stabilizing of the interaction with the target. As expected, STD NMR experiments performed on the scaffold alone or on the glycodendrimer functionalized with another sugar (Man18) did not show an interaction with the target protein (see Figure 6.12). Moreover, the quantitative evaluation of the STD effect on Gal_{18} allowed determination of the B_{max} and K_D values of 11 and 40 μm, respectively.[46]

The investigation of the initial stage of the oligomerization process of amyloidogenic peptides represents a crucial point to understand the evolution and aetiopathogenesis of several neurodegenerative diseases, such as Alzheimer disease (AD) and other amyloidosis pathologies. Due to the high molecular weight and heterogenicity of the species involved, the application of NMR-based interaction studies is extremely difficult and challenging.

Interestingly, STD NMR combined with non-selective off-resonance relaxation (ORR) experiments have been exploited to understand the molecular basis of the inhibition of Alzheimer's Aβ-peptide oligomerization by Human Serum Albumin (HSA).[47] This combined approach, applied by Melacini and coworkers, showed that HSA can recognize selectively hydrophobic loci exposed on the surface of soluble high-MW prefibrillar oligomers and inhibits their growth to fibrillar species. These important results revealed a general defence mechanism in physiological conditions and lay the foundations for the design of new inhibition strategies against Aβ-peptide oligomerization for the development of preventive anti-amyloid therapies.

Our group, for example, has been involved for several years in the identification of inhibitors of the process of oligomerization of amyloid peptides through STD NMR and some examples have been reported in the previous section. A possible approach in the formulation of anti-Alzheimer therapy is use of the nanoliposome decorated at the surface with Aβ-peptide ligands to target and inhibit their oligomerization. In this context, we report the characterization of the liposome surface functionalization and the assessment of their ability to bind Aβ1–42 oligomers by mean of NMR experiments. In particular, to face the atypical condition in which the putative ligands (functionalized liposome) were potentially larger than the receptors (Aβ oligomers), we applied the Water-LOGSY experiment for the characterization of the interaction between a liposome/nanoparticle and Aβ1–42 oligomers.[48]

Another interesting example of STD NMR and trNOESY experiments in unconventional conditions concerns the study of the interaction between lipopolysaccharide (LPS) micelles and potential ligands. LPS is the major constituent of the outer membrane of Gram-negative bacteria and represents a valid target for the development of antimicrobial agents. In addition, LPS is responsible for fatal septic shock in humans, due to its toxic, inflammatory

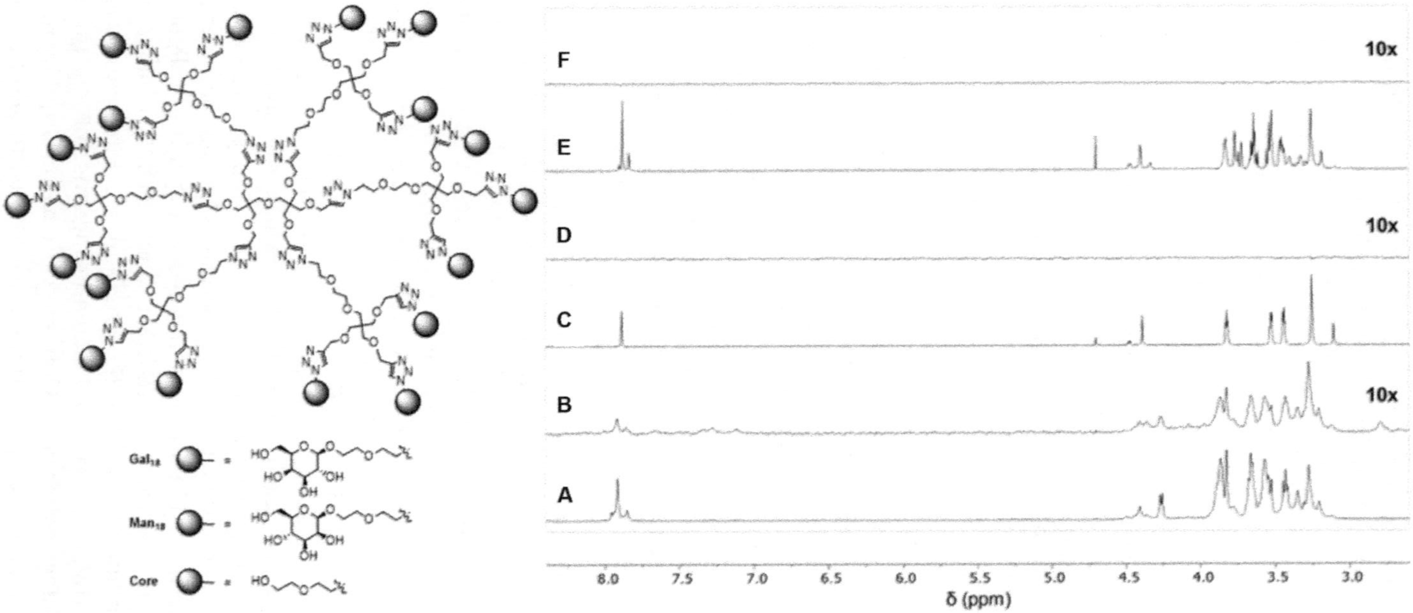

Figure 6.12 Structure of glycodendrimer Gal18, Man18 and core scaffold. ^{1}H (A) and STD NMR (B) spectra recorded on a mixture containing Gal18 250 µM and LecA 20 µM dissolved in PB 10 mM, pH 7.4, $CaCl_2$ 1 mM, 25 °C. ^{1}H (C) and STD NMR (D) spectra recorded on a mixture containing compound 8 250 µM and LecA 20 µM dissolved in PB 10 mM, pH 7.4, $CaCl_2$ 1 mM, 25 °C. ^{1}H (E) and STD NMR (F) spectra recorded on a mixture containing Man18 250 µM and LecA 20 µM dissolved in PB 10 mM, pH 7.4, $CaCl_2$ 1 mM, 25 °C. STD spectra were acquired with 256 scans and 2 s saturation time at 600 MHz.

properties. LPS micelles have a molecular weight of about 100 kDa, which apparently places it outside the range of STD NMR applications. Despite this, Bhunia and coworkers reported several examples of the application of this technique in the screening of antimicrobial peptides.[49–54] Moreover, they reported the investigation of the role of LPS micelles, released by gut microbiota, in the modulation of α-synuclein amyloidogenesis, providing new insights into an alternative pathway for Parkinson disease progression.[55] In this pioneering work they combined a multitude of NMR experiments to map the epitope of α-Syn–LPS interaction and demonstrated an overall increase of α-synuclein aggregation kinetics under an LPS dose-dependent modulation.

Undoubtedly, the study of the transient interactions between proteins and nanoparticles remains the most challenging goal to be pursued by NMR spectroscopy. Recently, this task has been critically reviewed for the first time.[56] The authors briefly described the most promising techniques to address experimental issues concerning protein–NP interactions, including Heteronuclear Single-Quantum Correlation protein fingerprint, Transverse-Relaxation Optimized SpectroscopY (TROSY),[57,58] Hydrogen/Deuterium eXchange (HDX)[59,60] and Dark-state Exchange Saturation Transfer experiments (DEST).[61,62]

The combination of these techniques was extensively used, for example, to study the partitioning of intracellular lipid-binding proteins, such as bile acid- and fatty acid-binding proteins (BABP and FABP) between free and lipid vesicle-bound states.[63–66]

Interestingly, DEST experiments have been introduced for the characterization, at the atomic level, of the chemical exchange equilibrium between monomers of amyloid-β (Aβ) peptides or amyloid inhibitors protein and polydisperse NMR-invisible ("dark") protofibril aggregates.[61,67–70] The 2D ^{15}N-DEST experiment consists of the ^{15}N longitudinal magnetization transfer by chemical exchange from an observable species (free monomer) to a corresponding partially saturated "invisible" state (the high-molecular-weight binding complex) and transferred back to the dissociated free state where the partial perturbation was recorded as cross-peak attenuation over the spectrum obtained without saturation (see Figure 6.13).[61]

Overall, the combination of DEST and ^{1}H–^{15}N HSQC experiments represents a powerful tool to investigate the dynamic exchange processes on the millisecond–second timescale of high-molecular-weight interactors, including large supramolecular machines such as the chaperonin GroEL[71–74] and protein–, peptide– or ligand–nanoparticle complexes.[63,75–82] For example, the DEST approach has been exploited by Cecconi *et al.* to detect, at the atomic level, the bound-state dynamics of the human liver fatty acid binding protein (LFABP), a globular polypeptide playing a key role in intracellular lipid trafficking, and liposomal nanoparticles.[63] The results of quantitative analysis showed that larger ^{15}N-$\Delta R_2^{\text{bound}}$ values were displayed by specific residues located at the so-called "anti-portal region", providing epitope mapping of the interaction. Also, Egner *et al.* recently investigated the mechanistic insights of absorption/desorption of ligands to the surface of nanoparticles through a combination of DEST and relaxation dispersion

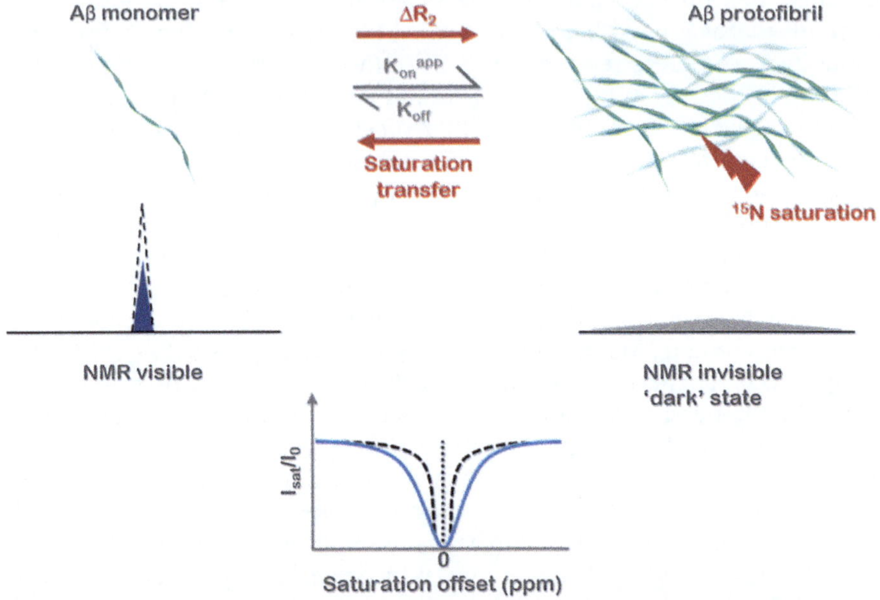

Figure 6.13 Principles of DEST experiments. The chemical exchange equilibrium between a monomeric (NMR visible) and a partially saturated "dark" oligomeric state (NMR invisible) was recorded as attenuation of the signals of the visible species.

(RD) NMR experiments.[81] They showed that the preparation of NP samples in aqueous agarose-based gel prevents NP sedimentation for extended periods of time (up to several months), allowing extensive NMR data acquisition for the investigation of the dynamics and kinetics of NP–ligand interactions.

6.5 On-cell NMR Molecular Recognition Studies

As reported in the previous sections, modern ligand–receptor NMR experiments, including STD NMR, allow the characterization of ligand–receptor binding events using very low amounts of receptor molecules. However, in some cases, the availability of the receptor might represent a great limitation. This can occur in the case of membrane receptors, such as integral membrane proteins. They can present inherent problems associated with the extreme difficulty of isolating and maintaining these proteins in solution with the correct folding and the proper functionality. These points are particularly critical for proteins presenting many transmembrane fragments.

In 2005, Mari *et al.*[83] demonstrated that this problem can be overcome by performing STD NMR experiments on samples containing living cells expressing the receptor of interest on their surface (see Figure 6.14).

In particular, they characterized the binding of the dendritic cell-specific ICAM-3-grabbing non-integrin (DC-SIGN) receptor, mediating the binding and internalization of a large array of pathogens, along with HIV, Ebola virus,

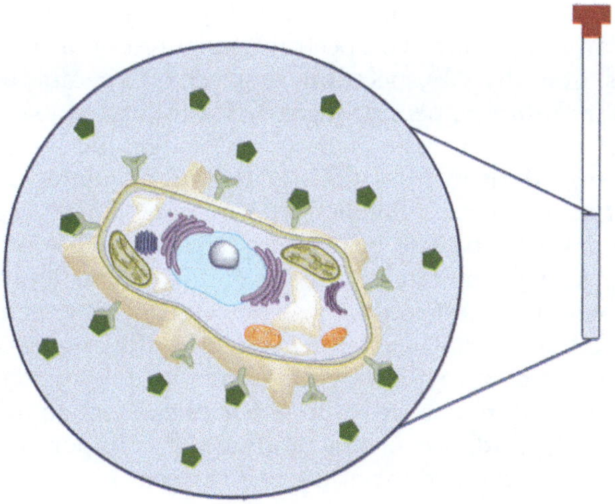

Figure 6.14 On-cell STD NMR experiments allow the detection of binding events among ligands and receptors bound to cell membrane, avoiding the otherwise necessary purification and resuspension of the isolated receptor in solution, which in the case of many integral membrane proteins is not achievable.

SARS-CoV, *Leishmania amastigotes*, *Mycobacterium tuberculosis*, in a Man- or Fuc-dependent manner,[84] with *S. cerevisiae* mannan, previously shown to compete with most of the DC-SIGN-dependent pathogen recognition events. They used two different cell lines, K562 transfected with the prototype DC-SIGN, namely K562-CD209 (DC-SIGN+), and mock-transfected K562 (DC-SIGN−), differing only for the expression of the DC-SIGN receptor on the cell surface.

A very critical aspect of the on-cell STD experiment, indeed, is the availability of a negative control that needs to be constituted by a cell line differing from the positive control only for the expression of the receptor of interest. This control allows to assess whether any ligand interaction observed with cells representing the positive control takes place with the receptor of interest.

With this experimental set-up they demonstrated, for the first time, that receptor–ligand interactions involving integral membrane proteins may be investigated by 1D STD NMR experiment, performed on samples containing living cells.

In the same year, Claasen and coworkers[85] described STDD (Saturation Transfer Double Difference) NMR experiments to detect the binding of the pentapeptide cyclo(RGDfV) to the surface glycoprotein integrin αIIbβ3 expressed by intact human blood platelets. In this paper they substituted the application of a $T_{1\rho}$-filter (spin-lock), an established NMR method to suppress NMR signals of large molecules, in this case cellular macromolecules, with a double difference spectrum. In their example, the spin lock field was not enough to erase completely the signals originating from the biological components contained in platelets. Thus, they set up an STDD filter, through the subtraction of a blank STD spectrum, acquired on a sample containing only platelets, from the STD spectrum, acquired on a sample containing both platelets and the inhibitor peptide (ligand). This subtraction results in a STDD spectrum that shows only resonances from the ligand.

The advantage derived from the STDD method mainly consists in the possibility of achieving very detailed information on the membrane-bound receptor – ligand epitope also by working in the presence of living cells.

Interestingly, they compared both the binding mode and the affinity for the receptor determined through these on-cell experiments and by working with αIIbβ3 in liposomes. They found that the binding mode of cyclo(RGDfV) in the arginine glycine region was different in the two systems and a significantly higher affinity for αIIbβ3 in native platelets than in liposomes was determined. This evidence clearly outlined the importance of studying membrane protein interactions at the cell surface.

Potenza *et al.*[86] exploited both STD and trNOESY experiments on platelets to study the interaction of integrin αvβ3, highly expressed on their surface, and of two synthetic cyclic RGD mimics. By combining these two experiments they were able not only to determine ligands' binding epitopes, but also their bioactive conformations, being able to justify their different biological activities in terms of different binding modes.

They exploited the same approach to study the binding of RGD peptidomimetics with integrins αvβ3 and αIIbβ3, expressed on the surface of ECV304 bladder cancer cells and human platelets,[87] and integrin α5β1, expressed on MDA-MB-231 breast cancer cells.[88] Such an approach was very useful for the rational design of new bioactive compounds interacting with membrane receptors.

Fattorusso and coworkers used NMR experiments both in solution, on the purified receptor, and in the presence of intact endothelial cells to characterize the binding between the peptide QK, a vascular endothelial growth factor (VEGF)-mimetic molecule with significant proangiogenic activity, activating VEGF receptors, and the membrane receptor VEGFR2, expressed on the surface of endothelial cells.[89] This approach allowed them to obtain a detailed description of this molecular interaction, deriving a structural model of QK/VEGFR1D2 complex and defining QK residues interacting with VEGFR2. This structural information can be exploited for the development of potent VEGF-mimetic peptides that can find potential applications for the targeting of angiogenesis in tumour cells.

The same authors exploited the combination of STD NMR, WaterLOGSY and trNOESY experiments to study the interaction of peptide heptamers, derived from the C-X-C Motif Chemokine 12 (CXCL12), and the C-X-C Chemokine Receptor Type 4 (CXCR4), expressed on the surface of human T-leukaemia cells (CCRF-CEM cells).[90] To assess that the observed interactions occur through the peptide binding to the CXCR4 receptor, they performed competition experiments using plerixafor, a compound reported to be a potent and selective CXCR4 antagonist. A significant decrease in ligand signal intensities in STD spectra after plerixafor addition to the ligand/CCRF-CEM cells mixture guaranteed the specificity of binding to CXCR4 receptor. This approach is an alternative to the use of a negative control represented by a second cell line not expressing the receptor of interest; nevertheless, this strategy remains, in our experience, the most reliable choice.

Another example of an on-cell STD NMR experiment applied to the elucidation of molecular interaction involving tumour cells was reported by our group. We characterized the binding of Gastrin Releasing Peptide Receptor (GRP-R), overexpressed on the surface of different human tumour cell lines, and its natural ligand bombesin,[91] the 14-residue peptide that hormonally functions as autocrine or paracrine growth factors in a variety of cells. Being overexpressed on the surface of different human cancer cell lines, GRP-R is an ideal target for the selective delivery of anticancer drugs, and also for diagnostic device delivery to tumour cells. Nevertheless, detailed structural information on the receptor interaction with natural ligands is required for the design of new GRPR binders.

For this purpose, an experimental protocol, based on on-cell STD NMR, was set up. Briefly, three cell lines were used as positive controls, *i.e.*, PC3, MCF-7 and HEK-GRPR cell lines. Both PC3 and MCF-7 cell lines overexpress GRP-R and their proliferation is increased after treatment with BN.[92–94] Moreover, GRP-R is essential for their survival and the corresponding gene cannot be silenced or deleted. On the other hand, GRP-R expression is negligible in the HEK cells used in this study as a negative control. Hence, HEK-GRPR cells, transfected with the gene encoding for GRP-R, were used as further positive controls directly comparable with the negative control made by wt HEK cells. Figure 6.15A reports the STD spectra acquired in presence of bombesin and each of the four cell lines.

The relative STD intensities of spectra acquired in the presence of the three positive controls are comparable, as can be deduced from their comparison [see Figure 6.15A4 (PC3 cells), Figure 6.15A6 (MCF-7 cells) and A10 (HEK-GRPR cells)]. Instead, no evidence of binding was obtained in the presence of wt HEK cell; this result demonstrated the specificity of the binding observed with cell lines expressing the receptor of interest.

An STDD spectrum for each positive control was calculated by subtraction of a blank STD spectrum, recorded on a sample containing cells alone, from the respective STD spectrum, acquired in the presence of the ligand (see Figure 6.15B). As already mentioned, through this procedure resonances from the different cell components are erased and the ligand epitope can be determined. Trp8, His12, Met14 and a Leu residue were identified as the amino acids fundamental for bombesin interaction with GRP-R, *i.e.* bombesin-binding epitope. These structural data were exploited for the design of new synthetic GRP-R ligands, leading to the identification of small non-peptide compounds able to antagonize bombesin in the nanomolar range of concentration.[95]

Moreover, our group demonstrated the possibility to perform on-cell STD NMR experiments also under high-resolution magic angle spinning (HR-MAS) conditions.[96] The applications of on-cell STD NMR described above concern the use of cells that remain in suspension for a time long enough to allow the acquisition of STD spectra (at least 30 minutes). However, many cell lines deriving from solid tissues show a strong tendency to aggregate and precipitate. In many cases, this prevents the application of good-quality STD NMR experiments.

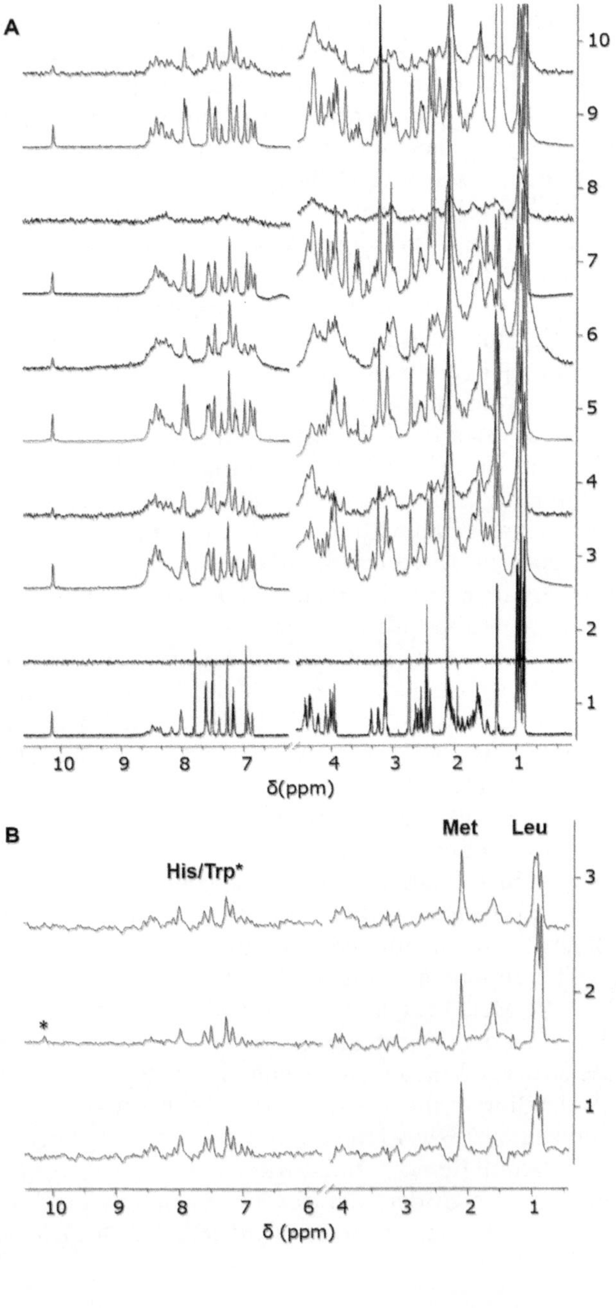

We explored the HR-MAS NMR techniques to use the sample rotation at high speeds to maintain cells in the sample "active window" (see Figure 6.16). In this way, cells do not sediment at the bottom of the sample, as occurs in the liquid state, but distribute along the internal walls of the MAS rotor, forming a homogeneous layer that mimics cell disposition at the tissue surface.[97] As the rotor walls fit into the detection region of an HR-MAS probe, the ligand that dissociates from the receptor is immediately available for detection in the cell-free central core of the rotor,[98] meanwhile, when the sample is filled into an NMR tube, the ligand has to spread from the bottom of the tube, where the pellet sediments, to the sample active window (see Figure 6.16).

The feasibility of this approach was demonstrated using the SGLT1 co-transporter and two of its ligands, phlorizin and naphthyl-β-D-galactoside, chosen as a model system. HEK cells express hSGLT1 on their surface. As SGLT1 expression is essential for HEK cell survival, its expression cannot be inhibited by silencing or deleting the encoding genes. In this case, to verify the specificity of the observed binding, we made STD NMR experiments on two different cell lines differing for the expression level of hSGLT1. In these conditions, we attributed the differences between the STD spectra obtained with the two cell lines solely to the phlorizin or naphthyl-β-D-galactoside–hSGLT1 interaction. To this aim, tsA 201 cells were transfected with a vector containing a hSGLT1 encoding sequence (pEYFP-hSGLT1), to increase the protein expression level with respect to wt HEK cells. For each ligand, the STD spectrum acquired in the presence of wt HEK cells was subtracted from the STD spectrum acquired in the presence of cells overexpressing hSGLT1, obtaining the corresponding STDD spectra. The STDD spectra enabled suppression of the possible contributions deriving from phlorizin or

---

Figure 6.15 (A) (1) ^{1}H spectrum of a sample containing BN (2 mM), PBS; (2) 1D STD spectrum of a sample containing BN (2 mM), PBS; (3) ^{1}H spectrum of a sample containing BN (2 mM) and 12×10^{6} PC3 cells, PBS; (4) 1D STD spectrum of a sample BN (2 mM) and 12×10^{6} PC3 cells, PBS; (5) ^{1}H spectrum of a sample containing BN (2 mM) and 12×10^{6} MCF-7 cells, PBS; (6) 1D STD spectrum of a sample containing BN (2 mM) and 12×10^{6} MCF-7 cells, PBS; (7) ^{1}H spectrum of a sample containing BN (2 mM) and 12×10^{6} wt HEK-293 cells, PBS; (8) 1D STD spectrum of a sample containing BN (2 mM) and 12×10^{6} wt HEK-293 cells, PBS; (9) ^{1}H spectrum of a sample containing BN (2 mM) and 12×10^{6} HEK-GRPR cells, PBS; (10) 1D STD spectrum of a sample containing BN (2 mM) and 12×10^{6} HEK-GRPR cells, PBS. (B) (1) STDD spectrum obtained subtracting 1D STD spectrum of a sample containing 12×10^{6} PC3 cells from spectrum A4; (2) STDD spectrum obtained subtracting 1D STD spectrum of a sample containing 12×10^{6} MCF-7 cells from spectrum A6; (3) STDD spectrum obtained subtracting 1D STD spectrum of a sample containing 12×10^{6} HEK-GRPR cells from spectrum A10. All spectra were acquired at 600 MHz and 25 °C. ^{1}H NMR spectra were acquired with 16 scans; STD NMR spectra were acquired with 256 scans, selective irradiation frequency −1.0 ppm, off-resonance irradiation frequency 30 ppm, and reported with a 5× signal enhancement.

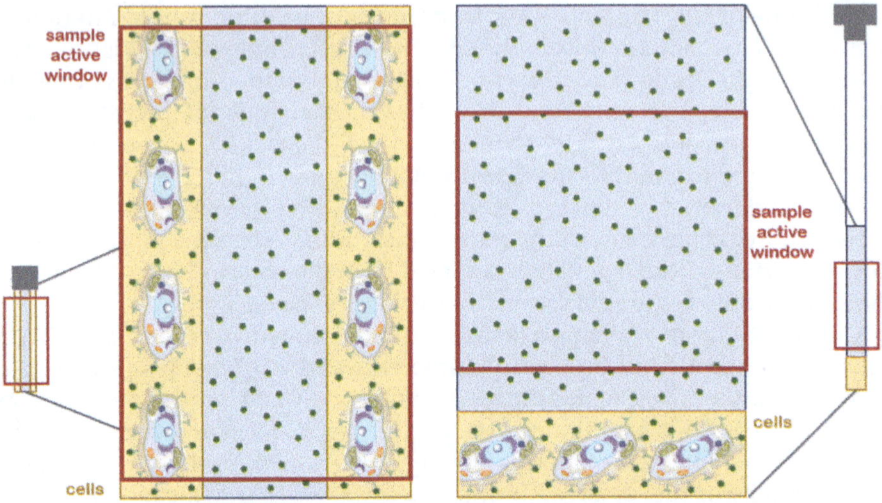

Figure 6.16 Comparison of cell distributions into an NMR sample when an HR-MAS rotor (left) or a liquid state NMR tube (right) is used. In the first case, due to centrifugal force, cells remain in the sample active window and the ligand is available for detection immediately after its dissociation from the receptor. In the second case, cells precipitate at the tube bottom and, after its dissociation from the receptor, the ligand spreads along the sample before being detected.

naphthyl-β-D-galactoside additional specific and non-specific interactions with cellular membrane components different from hSGLT1.

The binding specificity was further confirmed by an STD NMR competition experiment. Naphthyl-β-D-galactoside and phlorizin were added at a 1:1 molar ratio to a sample of HEK cells overexpressing hSGLT1. As expected, because of the significantly higher phlorizin affinity for hSGLT1, naphthyl-β-D-galactoside signals, appearing in the STD spectrum achieved in the absence of phlorizin, disappeared from the STD spectrum acquired in the presence of both ligands. This experiment also demonstrated the feasibility of competitive on-cell STD NMR experiments under HR-MAS conditions.

Finally, very recently we also reported the application of on-cell STD NMR for characterization of the bacterial adhesin FimH binding to multivalent ligands.[99] FimH is a mannose-binding bacterial adhesin expressed at the apical end of type 1 pili of uropathogenic bacterial strains (UPEC) and responsible for their D-mannose-sensitive adhesion to host mammalian epithelial cells. Due to these properties, FimH is a key virulence factor and an attractive therapeutic target for urinary tract infection. We synthesized glycodendrimers decorated with 4, 6 or 18 D-mannose units. They were tested as FimH-mediated yeast agglutination inhibitors and the most effective were used to set up STD NMR experiments on UPEC (*E. coli* CFT073) cells. We obtained very high-quality STD NMR spectra and the reliability of our experimental data was confirmed by using, as a negative control for blank STD experiments, a CFT073Δ*fimH* mutant.

This methodology can be a powerful tool for the screening and epitope mapping of FimH ligands, allowing verification and identification of the structural determinants of the binding of new potential FimH ligands in a very fast and reliable way and, most importantly, under physiologically relevant conditions.

All these examples highlight the significant versatility of ligand-based NMR experiments, in particular STD NMR, made on samples containing living cells. They can be performed on a large variety of ligand–receptor pairs, as well as on a large panel of both procaryotic and eukaryotic cell lines.

These approaches allow for the performance of interaction studies involving membrane receptors, that represent a high percentage of cellular proteins, also when the purification and/or solubilization of the same biomacromolecules cannot be achieved. Moreover, and most importantly, they allow performance of these studies under experimental conditions closed to physiological ones, at the cell surface, where the studied interactions take place *in vivo*, increasing the reliability of the gained structural data.

However, these experiments require a careful choice of the experimental conditions, both in terms of sample preparation and spectrum acquisition, to assure cell integrity during NMR measurements, and a proper use of control experiment(s) to confirm that the molecular interaction observed actually involves the biological receptor of interest.

6.6 Final Remarks

NMR spectroscopy is one of the most powerful tools to achieve the detailed characterization of molecular recognition events involving biomacromolecules.

Notably, NMR allows the investigation of ligand–receptor interactions, deepening both their structural and dynamic features. Moreover, many experimental conditions permit work under near-physiological conditions.

In this chapter, we have reviewed different approaches mainly based on observation of the ligand resonances and their modulation as a consequence of the equilibrium between the free and receptor-bound forms.

Initially described for the study of interactions between purified receptors and ligands constituted by low-molecular weight molecules, these techniques have rapidly evolved, allowing the study of these events on increasingly complex samples. Emphasis was therefore dedicated to the discussion of these applications, in an attempt to highlight the main advantages, compared to conventional applications, but also the critical points to be considered carefully in order to achieve reproducible and reliable experimental results.

Knowledge and in-depth analysis of the current state of the art of these methodologies certainly represent the best starting point for their further development and implementation.

Acknowledgements

The authors acknowledge MIUR for funding PRIN-2017 project 2017E44A9P – BacHounds: Supramolecular nanostructures for bacteria detection.

References

1. M. Mayer and B. Meyer, *Angew. Chem., Int. Ed.*, 1999, **38**, 1784–1788.
2. C. Dalvit, P. Pevarello, M. Tatò, M. Veronesi, A. Vulpetti and M. Sundström, *J. Biomol. NMR*, 2000, **18**, 65–68.
3. F. Ni, *Prog. Nucl. Magn. Reson. Spectrosc.*, 1994, **26**, 517–606.
4. A. Chen and M. J. Shapiro, *J. Am. Chem. Soc.*, 2000, **122**, 414–415.
5. A. Chen and M. J. Shapiro, *J. Am. Chem. Soc.*, 1998, **120**, 10258–10259.
6. M. Lin, M. J. Shapiro and J. R. Wareing, *J. Am. Chem. Soc.*, 1997, **119**, 5249–5250.
7. W. Jahnke, S. Rüdisser and M. Zurini, *J. Am. Chem. Soc.*, 2001, **123**, 3149–3150.
8. B. Meyer and T. Peters, *Angew. Chem., Int. Ed. Eng.*, 2003, **42**, 864–890.
9. C. Dalvit, G. Fogliatto, A. Stewart, M. Veronesi and B. Stockman, *J. Biomol. NMR*, 2001, **21**, 349–359.
10. D. Neuhaus and M. P. Williamson, *The Nuclear Overhauser Effect in Structural and Conformational Analysis*, VCH, 1989.
11. J. L. Asensio, F. J. Canada, M. Bruix, A. Rodriguez-Romero and J. Jimenez-Barbero, *Eur. J. Biochem.*, 1995, **230**, 621–633.
12. M. Mayer and B. Meyer, *J. Med. Chem.*, 2000, **43**, 2093–2099.
13. P. Stilbs, *Prog. Nucl. Magn. Reson. Spectrosc.*, 1987, **19**, 1–45.
14. H. Y. Carr and E. M. Purcell, *Phys. Rev.*, 1954, **94**, 630–638.
15. S. Meiboom and D. Gill, *Rev. Sci. Instrum.*, 1958, **29**, 688–691.
16. S. Walpole, S. Monaco, R. Nepravishta and J. Angulo, *Methods Enzymol.*, 2019, **615**, 423–451.
17. M. Pellecchia, I. Bertini, D. Cowburn, C. Dalvit, E. Giralt, W. Jahnke, T. L. James, S. W. Homans, H. Kessler, C. Luchinat, B. Meyer, H. Oschkinat, J. Peng, H. Schwalbe and G. Siegal, *Nat. Rev. Drug Discovery*, 2008, **7**, 738–745.
18. M. Vogtherr and T. Peters, *J. Am. Chem. Soc.*, 2000, **122**, 6093–6099.
19. J. L. Wagstaff, S. L. Taylor and M. J. Howard, *Mol. BioSyst.*, 2013, **9**, 571–577.
20. M. Kramer and E. Kleinpeter, *J. Magn. Reson.*, 2010, **202**, 245–249.
21. Y. S. Wang, D. Liu and D. F. Wyss, *Magn. Reson. Chem.*, 2004, **42**, 485–489.
22. C. Dalvit, M. Fasolini, M. Flocco, S. Knapp, P. Pevarello and M. Veronesi, *J. Med. Chem.*, 2002, **45**, 2610–2614.
23. M. Politi, M. I. Chávez, F. J. Cañada and J. Jiménez-Barbero, *Eur. J. Org. Chem.*, 2005, 1392–1396.
24. S. A. K. Tanoli, N. U. Tanoli, T. M. Bondancia, S. Usmani, R. Kerssebaum, A. G. Ferreira, J. B. Fernandes and Z. Ul-Haq, *Analyst*, 2013, **138**, 5137–5145.
25. S. A. K. Tanoli, N. U. Tanoli, T. M. Bondancia, S. Usmani, Z. Ul-Haq, J. B. Fernandes, S. S. Thomasi and A. G. Ferreira, *RSC Adv.*, 2015, **5**, 23431–23442.
26. A. Palmioli, C. Ciaramelli, R. Tisi, M. Spinelli, G. De Sanctis, E. Sacco and C. Airoldi, *Chem. – Asian J.*, 2017, **12**, 2457–2466.
27. C. Haass and D. J. Selkoe, *Nat. Rev. Mol. Cell Biol.*, 2007, **8**, 101–112.

28. C. Airoldi, E. Sironi, C. Dias, F. Marcelo, A. Martins, A. P. Rauter, F. Nicotra and J. Jimenez-Barbero, *Chem. - Asian J.*, 2013, **8**, 596–602.
29. C. Ciaramelli, A. Palmioli, A. De Luigi, L. Colombo, G. Sala, C. Riva, C. P. Zoia, M. Salmona and C. Airoldi, *Food Chem.*, 2018, **252**, 171–180.
30. A. Palmioli, S. Bertuzzi, A. De Luigi, L. Colombo, B. La Ferla, M. Salmona, I. De Noni and C. Airoldi, *Bioorg. Chem.*, 2019, **83**, 76–86.
31. C. Ciaramelli, A. Palmioli, A. De Luigi, L. Colombo, G. Sala, M. Salmona and C. Airoldi, *Food Chem.*, 2021, **341**, 128249.
32. E. Sironi, L. Colombo, A. Lompo, M. Messa, M. Bonanomi, M. E. Regonesi, M. Salmona and C. Airoldi, *Chemistry*, 2014, **20**, 13793–13800.
33. H. Wan, Y. Tian, H. Jiang, X. Zhang and X. Ju, *J. Ethnopharmacol.*, 2020, **254**, 112712.
34. Y. V. Nikolaev, K. Kochanowski, H. Link, U. Sauer and F. H. Allain, *Biochemistry*, 2016, **55**, 2590–2600.
35. B. Pelaz, C. Alexiou, R. A. Alvarez-Puebla, F. Alves, A. M. Andrews, S. Ashraf, L. P. Balogh, L. Ballerini, A. Bestetti, C. Brendel, S. Bosi, M. Carril, W. C. W. Chan, C. Chen, X. Chen, X. Chen, Z. Cheng, D. Cui, J. Du, C. Dullin, A. Escudero, N. Feliu, M. Gao, M. George, Y. Gogotsi, A. Grünweller, Z. Gu, N. J. Halas, N. Hampp, R. K. Hartmann, M. C. Hersam, P. Hunziker, J. Jian, X. Jiang, P. Jungebluth, P. Kadhiresan, K. Kataoka, A. Khademhosseini, J. Kopeček, N. A. Kotov, H. F. Krug, D. S. Lee, C.-M. Lehr, K. W. Leong, X.-J. Liang, M. Ling Lim, L. M. Liz-Marzán, X. Ma, P. Macchiarini, H. Meng, H. Möhwald, P. Mulvaney, A. E. Nel, S. Nie, P. Nordlander, T. Okano, J. Oliveira, T. H. Park, R. M. Penner, M. Prato, V. Puntes, V. M. Rotello, A. Samarakoon, R. E. Schaak, Y. Shen, S. Sjöqvist, A. G. Skirtach, M. G. Soliman, M. M. Stevens, H.-W. Sung, B. Z. Tang, R. Tietze, B. N. Udugama, J. S. VanEpps, T. Weil, P. S. Weiss, I. Willner, Y. Wu, L. Yang, Z. Yue, Q. Zhang, Q. Zhang, X.-E. Zhang, Y. Zhao, X. Zhou and W. J. Parak, *ACS Nano*, 2017, **11**, 2313–2381.
36. A. Szczygiel, L. Timmermans, B. Fritzinger and J. C. Martins, *J. Am. Chem. Soc.*, 2009, **131**, 17756–17758.
37. Z. Hens and J. C. Martins, *Chem. Mater.*, 2013, **25**, 1211–1221.
38. Y. Zhang, H. Xu, A. M. Parsons and L. B. Casabianca, *J. Phys. Chem. C*, 2017, **121**, 24678–24686.
39. Y. Zhang and L. B. Casabianca, *J. Phys. Chem. Lett.*, 2018, **9**, 6921–6925.
40. H. Xu and L. B. Casabianca, *Sci. Rep.*, 2020, **10**, 12351.
41. H. Xu, D. Lustig and L. B. Casabianca, *Appl. Magn. Reson.*, 2020, **51**, 277–286.
42. E. Arce, P. M. Nieto, V. Díaz, R. G. Castro, A. Bernad and J. Rojo, *Bioconjugate Chem.*, 2003, **14**, 817–823.
43. J. Giri, M. S. Diallo, A. J. Simpson, Y. Liu, W. A. Goddard, R. Kumar and G. C. Woods, *ACS Nano*, 2011, **5**, 3456–3468.
44. D. Ponader, P. Maffre, J. Aretz, D. Pussak, N. M. Ninnemann, S. Schmidt, P. H. Seeberger, C. Rademacher, G. U. Nienhaus and L. Hartmann, *J. Am. Chem. Soc.*, 2014, **136**, 2008–2016.

45. S. Mirabella, G. D'Adamio, C. Matassini, A. Goti, S. Delgado, A. Gimeno, I. Robina, A. J. Moreno-Vargas, S. Šesták, J. Jiménez-Barbero and F. Cardona, *Chem. – Eur. J.*, 2017, **23**, 14585–14596.
46. A. Palmioli, P. Sperandeo, A. Polissi and C. Airoldi, *ChemBioChem*, 2019, **20**, 2911–2915.
47. J. Milojevic, V. Esposito, R. Das and G. Melacini, *J. Am. Chem. Soc.*, 2007, **129**, 4282–4290.
48. C. Airoldi, S. Mourtas, F. Cardona, C. Zona, E. Sironi, G. D'Orazio, E. Markoutsa, F. Nicotra, S. G. Antimisiaris and B. La Ferla, *Eur. J. Med. Chem.*, 2014, **85**, 43–50.
49. A. Bhunia, P. N. Domadia and S. Bhattacharjya, *Biochim. Biophys. Acta*, 2007, **1768**, 3282–3291.
50. A. Bhunia, H. Mohanram and S. Bhattacharjya, *Pept. Sci.*, 2009, **92**, 9–22.
51. A. Bhunia, A. Ramamoorthy and S. Bhattacharjya, *Chem. – Eur. J.*, 2009, **15**, 2036–2040.
52. A. Bhunia, P. N. Domadia, J. Torres, K. J. Hallock, A. Ramamoorthy and S. Bhattacharjya, *J. Biol. Chem.*, 2010, **285**, 3883–3895.
53. A. Bhunia and S. Bhattacharjya, *Biopolymers*, 2011, **96**, 273–287.
54. A. Bhunia, R. Saravanan, H. Mohanram, M. L. Mangoni and S. Bhattacharjya, *J. Biol. Chem.*, 2011, **286**, 24394–24406.
55. D. Bhattacharyya, G. M. Mohite, J. Krishnamoorthy, N. Gayen, S. Mehra, A. Navalkar, S. A. Kotler, B. N. Ratha, A. Ghosh, R. Kumar, K. Garai, A. K. Mandal, S. K. Maji and A. Bhunia, *ACS Chem. Neurosci.*, 2019, **10**, 2229–2236.
56. M. Assfalg, L. Ragona, K. Pagano, M. D'Onofrio, S. Zanzoni, S. Tomaselli and H. Molinari, *Biochim. Biophys. Acta*, 2016, **1864**, 102–114.
57. K. Pervushin, *Q. Rev. Biophys.*, 2000, **33**, 161–197.
58. Y. Xu and S. Matthews, in *Modern NMR Methodology*, ed. H. Heise and S. Matthews, Springer, Berlin, Heidelberg, 2013, pp. 97–119.
59. R. Zahn, A. Liu, T. Lührs, R. Riek, C. von Schroetter, F. López García, M. Billeter, L. Calzolai, G. Wider and K. Wüthrich, *Proc. Natl. Acad. Sci. U. S. A.*, 2000, **97**, 145–150.
60. V. Chevelkov, Y. Xue, D. K. Rao, J. D. Forman-Kay and N. R. Skrynnikov, *J. Biomol. NMR*, 2010, **46**, 227–244.
61. N. L. Fawzi, J. Ying, R. Ghirlando, D. A. Torchia and G. M. Clore, *Nature*, 2011, **480**, 268–272.
62. V. Tugarinov and G. M. Clore, *J. Biomol. NMR*, 2019, **73**, 461–469.
63. A. Ceccon, M. Lelli, M. D'Onofrio, H. Molinari and M. Assfalg, *J. Am. Chem. Soc.*, 2014, **136**, 13158–13161.
64. M. Pedò, F. Löhr, M. D'Onofrio, M. Assfalg, V. Dötsch and H. Molinari, *J. Mol. Biol.*, 2009, **394**, 852–863.
65. A. Ceccon, M. D'Onofrio, S. Zanzoni, D. L. Longo, S. Aime, H. Molinari and M. Assfalg, *Proteins*, 2013, **81**, 1776–1791.
66. D. L. Longo, E. Di Gregorio, R. Abategiovanni, A. Ceccon, M. Assfalg, H. Molinari and S. Aime, *Analyst*, 2014, **139**, 2687–2690.

67. N. L. Fawzi, J. Ying, D. A. Torchia and G. M. Clore, *J. Am. Chem. Soc.*, 2010, **132**, 9948–9951.
68. N. L. Fawzi, J. Ying, D. A. Torchia and G. M. Clore, *Nat. Protoc.*, 2012, **7**, 1523–1533.
69. N. L. Fawzi, D. S. Libich, J. Ying, V. Tugarinov and G. M. Clore, *Angew. Chem., Int. Ed.*, 2014, **53**, 10345–10349.
70. M. Algamal, R. Ahmed, N. Jafari, B. Ahsan, J. Ortega and G. Melacini, *J. Biol. Chem.*, 2017, **292**, 17158–17168.
71. D. S. Libich, N. L. Fawzi, J. Ying and G. M. Clore, *Proc. Natl. Acad. Sci. U. S. A.*, 2013, **110**, 11361–11366.
72. D. S. Libich, V. Tugarinov and G. M. Clore, *Proc. Natl. Acad. Sci.*, 2015, **112**, 8817.
73. D. S. Libich, V. Tugarinov, R. Ghirlando and G. M. Clore, *Biochemistry*, 2017, **56**, 903–906.
74. M. A. Wälti, J. Steiner, F. Meng, H. S. Chung, J. M. Louis, R. Ghirlando, V. Tugarinov, A. Nath and G. M. Clore, *Proc. Natl. Acad. Sci.*, 2018, **115**, E11924–E11932.
75. A. Ceccon, G. Marius Clore and V. Tugarinov, *J. Biomol. NMR*, 2016, **66**, 1–7.
76. A. Ceccon, V. Tugarinov, A. Bax and G. M. Clore, *J. Am. Chem. Soc.*, 2016, **138**, 5789–5792.
77. A. Ceccon, V. Tugarinov, A. J. Boughton, D. Fushman and G. M. Clore, *J. Phys. Chem. Lett.*, 2017, **8**, 2535–2540.
78. A. Ceccon, G. M. Clore and V. Tugarinov, *J. Phys. Chem. B*, 2018, **122**, 11271–11278.
79. A. Ceccon, T. Schmidt, V. Tugarinov, S. A. Kotler, C. D. Schwieters and G. M. Clore, *J. Am. Chem. Soc.*, 2018, **140**, 6199–6202.
80. A. Ceccon, V. Tugarinov and G. M. Clore, *J. Am. Chem. Soc.*, 2019, **141**, 94–97.
81. T. K. Egner, P. Naik, N. C. Nelson, I. I. Slowing and V. Venditti, *Angew. Chem., Int. Ed.*, 2017, **56**, 9802–9806.
82. S. Zanzoni, A. Ceccon, M. Assfalg, R. K. Singh, D. Fushman and M. D'Onofrio, *Nanoscale*, 2015, **7**, 7197–7205.
83. S. Mari, D. Serrano-Gómez, F. J. Cañada, A. L. Corbí and J. Jiménez-Barbero, *Angew. Chem., Int. Ed.*, 2005, **44**, 296–298.
84. Y. van Kooyk and T. B. H. Geijtenbeek, *Nat. Rev. Immunol.*, 2003, **3**, 697–709.
85. B. Claasen, M. Axmann, R. Meinecke and B. Meyer, *J. Am. Chem. Soc.*, 2005, **127**, 916–919.
86. D. Potenza, F. Vasile, L. Belvisi, M. Civera and E. M. V. Araldi, *ChemBioChem*, 2011, **12**, 695–699.
87. I. Guzzetti, M. Civera, F. Vasile, E. M. Araldi, L. Belvisi, C. Gennari, D. Potenza, R. Fanelli and U. Piarulli, *Org. Biomol. Chem.*, 2013, **11**, 3886–3893.
88. I. Guzzetti, M. Civera, F. Vasile, D. Arosio, C. Tringali, U. Piarulli, C. Gennari, L. Pignataro, L. Belvisi and D. Potenza, *ChemistryOpen*, 2017, **6**, 128–136.

89. R. Di Stasi, D. Diana, D. Capasso, S. Di Gaetano, L. De Rosa, V. Celentano, C. Isernia, R. Fattorusso and L. D. D'Andrea, *Chem. Eur. J.*, 2018, **24**, 11461–11466.
90. D. Brancaccio, D. Diana, S. Di Maro, F. S. Di Leva, S. Tomassi, R. Fattorusso, L. Russo, S. Scala, A. M. Trotta, L. Portella, E. Novellino, L. Marinelli and A. Carotenuto, *J. Med. Chem.*, 2018, **61**, 2910–2923.
91. A. Palmioli, C. Ceresa, F. Tripodi, B. La Ferla, G. Nicolini and C. Airoldi, *Bioorg. Chem.*, 2020, **99**, 103861.
92. H. Reile, P. E. Armatis and A. V. Schally, *Prostate*, 1994, **25**, 29–38.
93. W. J. Wasilenko, J. Cooper, A. J. Palad, K. D. Somers, P. F. Blackmore, J. S. Rhim, G. L. Wright Jr and P. F. Schellhammer, *Prostate*, 1997, **30**, 167–173.
94. M. Gugger and J. C. Reubi, *Am. J. Pathol.*, 1999, **155**, 2067–2076.
95. A. Palmioli, G. Nicolini, F. Tripodi, A. Orsato, C. Ceresa, E. Donzelli, M. Arici, P. Coccetti, M. Rocchetti, B. La Ferla and C. Airoldi, *Bioorg. Chem.*, 2021, **109**, 104739.
96. C. Airoldi, S. Giovannardi, B. La Ferla, J. Jimenez-Barbero and F. Nicotra, *Chemistry*, 2011, **17**, 13395–13399.
97. E. Bruno, G. Digilio, C. Cabella, A. de Reggi, S. Baroni, V. Mainero and S. Aime, *Magn. Reson. Med.*, 2006, **56**, 978–985.
98. O. Beckonert, M. Coen, H. C. Keun, Y. Wang, T. M. Ebbels, E. Holmes, J. C. Lindon and J. K. Nicholson, *Nat. Protoc.*, 2010, **5**, 1019–1032.
99. A. Palmioli, P. Sperandeo, S. Bertuzzi, A. Polissi and C. Airoldi, *Bioorg. Chem.*, 2021, **112**, 104876.

CHAPTER 7

Multi-frequency Saturation Transfer Difference NMR to Characterize Weak Protein–Ligand Complexes

SERENA MONACO*[a] AND JESÚS ANGULO*[a,b,c]

[a] School of Pharmacy, University of East Anglia, Norwich Research Park, NR4 7TJ, Norwich, UK; [b] Department of Organic Chemistry, Faculty of Chemistry, University of Seville, 41012 Seville, Spain; [c] Instituto de Investigaciones Químicas (CSIC-US), Avda. Américo Vespucio, 49. 41092 Sevilla, Spain
*Emails: jangulo@us.es; jesus@iiq.csic.es

7.1 Introduction

7.1.1 Weak Protein–Ligand Interactions

Intermolecular interactions regulate the dynamic behaviour of many multi-molecular systems. In biological systems, the specific intermolecular interactions between the constituent biomolecules are eventually responsible for triggering, shutting down, or controlling processes of signal transduction leading to activation of biological functions. Importantly, the 3D molecular structure of a biomolecule determines with which partners it will be able to interact in a specific way and, consequently, which biological function it will be able to trigger. To gain a profound knowledge of the molecular basis of

New Developments in NMR No. 26
NMR Spectroscopy for Probing Functional Dynamics at Biological Interfaces
Edited by Anirban Bhunia, Hanudatta S. Atreya and Neeraj Sinha

Published by the Royal Society of Chemistry, www.rsc.org

biological processes of interest it is necessary to unveil the 3D structures of the participant biomolecules and their complexes (structural biology).

Most structural biology studies on protein–ligand interactions focus on strong or high-affinity interactions ($K_D < 10^{-6}$ M), which are readily accessible by many of the available biophysical experimental techniques. In particular, X-ray crystallography and very recently cryo-electron microscopy (cryo-EM) are two of the most powerful techniques to characterize strong biomolecular complexes at high resolution. In recent years interest has been growing on gaining structural details of weak protein–ligand interactions ($K_D > 10^{-4}$ M). Our increasing understanding of the interconnectivity, dynamics and complexity of biological systems (coined with the broad term "systems biology"), has revealed many examples where weak protein interactions play crucial roles, *e.g.*, when quick cellular responses are required after temporary stimuli (signal transduction, reversible cell–cell contacts, transient interactions in host/pathogen recognition, leucocyte rolling, *etc.*).[1] It is clear that to reach a complete understanding of life processes it is necessary to investigate both strong and weak protein–ligand interactions. However, for weak biomolecular interactions many conventional experimental approaches fail or become unreliable as, due to their transient nature, the complexes are mostly dissociated with major populations of the interacting partners in the free state, rendering detailed structural studies by most common biophysical techniques challenging.

A good number of physical techniques have been demonstrated to be well suited to studying intermolecular interactions, [*e.g.*, X-ray crystallography,[2] surface plasmon resonance (SPR),[3] NMR spectroscopy,[4] cryo-electron microscopy,[5] small-angle X-ray scattering (SAXS),[6] mass spectroscopy (MS),[7] isothermal titration calorimetry, *etc.*]. In particular, nuclear magnetic resonance (NMR) spectroscopy is still considered to be one of the most useful spectroscopic techniques for the scope of investigating weak protein–ligand interactions. The advantages of NMR spectroscopy include its robustness, versatility, and ability to provide high-resolution structural details as well as dynamics properties of biomolecules in solution, in a physicochemical context much closer to the natural biological system than in the case of solid-state techniques, such as X-ray diffraction.[8]

7.1.2 The Role of NMR Spectroscopy

In the context of the structural analysis of biological complexes by NMR, a *ligand-based* or *receptor-based* approach can be employed, depending on the point of view (NMR signals) from which the system is observed. While receptor-based approaches are very powerful in mapping interaction interfaces (*e.g.*, through chemical shift perturbations,[9] or intermolecular cross-saturation), their downside is the need for protein isotopic labelling and total protein assignment for data interpretation. These studies can be carried out with the protein being in a stoichiometric ratio with the interacting partner (*e.g.* a small molecule or another protein). Ligand-based techniques, in contrast, rely on the detection of signals from the small molecule, which must be in a large excess relative to the receptor. Favourably, the receptor does not need to be either labelled or assigned/resolved, and its only requirement is to be of a much larger size than the small molecule and formally with no upper limit for its size as long as it is soluble, so that it can range from protein to DNA, micelles or live cells.[10]

Among the many ligand-based techniques, the focus of this chapter is on saturation transfer difference (STD) NMR, a robust *ligand-based* tool for the detection and structural elucidation of ligand–macromolecule interactions.[11] STD NMR has become very popular both in drug discovery and fundamental structural biology due to (i) its ability to provide atomic-level structural details on the interactions (*e.g.*, relative to other ligand-based techniques which are only applicable for screening for binding), and (ii) its ease of implementation.[12]

In the first place, STD NMR can be used to screen for binding. Potential ligands are typically analysed in a pool or cocktail, in the presence of the receptor, and the hits are then identified simply by detecting the appearance of the corresponding STD NMR signals.[12] Unlike other ligand-based techniques which are mainly used for binding screening (*e.g.*, CMPG or waterLOGSY), quantitative STD NMR has the great advantage of allowing a deep level of structural elucidation. This is possible because the relative intensities of different ligand STD NMR signals can be used to construct a so-called ligand-binding epitope map, which gives information about the mode of binding of the ligand and some insights about the relative orientation of the ligand with respect to the binding site.[13,14] This approach is particularly powerful when combined with molecular modelling approaches, such as *molecular docking*, because the generated 3D model complexes can then be validated by comparing experimental STD NMR data with the data predicted from the model.[15] Structural STD NMR can be employed at many levels, namely: (i) qualitative; (ii) semi-quantitative; and (iii) quantitative, and these will be explored in their respective sections. As a rule of thumb, STD NMR is generally applicable for interactions with dissociation constants in the millimolar to micromolar range ($K_D > 10^{-6}$ M), making it optimal to investigate weak protein–ligand interactions of different natures. Indeed, STD NMR has been successfully applied to highly relevant weak protein–ligand interactions like protein–carbohydrate interactions,[16,17] interactions in amyloid fibres,[18] and hydrogels,[19] to name but a few.

In recent years, advances have been focusing on expanding the potentiality of STD NMR. In particular, the recently developed differential epitope mapping by the STD NMR (DEEP-STD NMR) approach, allows further structural information beyond the ligand-binding epitope to be gained, by identifying the nature of the amino acid side chains (polar, aliphatic, or aromatic) that are in close proximity to different regions of the ligand. This information, that is related to the pharmacophore of the receptor under study, allows also to orientate the ligand molecule in the binding pocket if the protein pocket geometry is known, significantly aiding structural studies of the complexes.[20]

This technique relies on the same extremely simple experimental set up, just by introducing the concept of *multi-frequency irradiation*. Remarkably, in 2006, the group of Gomez-Paloma had exploited a similar idea based on *differential irradiation* in STD NMR to screen the binding of small molecules to the minor or major groove of DNA, in DNA–ligand complexes (notoriously more challenging, and therefore less studied, than protein–ligand complexes).[21] In this chapter, we first provide some brief theoretical and practical concepts on the traditional STD NMR approach, and then we will illustrate the huge structural potential of *multi-frequency* STD NMR for ligand-based characterization of protein–ligand interactions.

7.2 Saturation Transfer Difference NMR

7.2.1 Theoretical Concepts

Saturation transfer difference (STD) NMR is based on the nuclear Overhauser effect (NOE) and chemical exchange under fast kinetics conditions. NOE is a phenomenon that manifests as the perturbation of the signal intensity of a particular nuclear spin (proton, typically, in STD NMR studies), due to the presence of a previously perturbed spin nearby in space. The STD NMR experiment relies on the generation of that type of perturbation or non-equilibrium magnetization (generated by a selective irradiation) on a small group of protons of the receptor, followed by transfer of that non-equilibrium population all over the protons of the protein, and eventually leading to intermolecular NOEs with the closest protons of the ligand in the bound state.[22]

We can briefly outline the STD NMR experiment in five processes, as illustrated in Figure 7.1a: (i) a selective shaped pulse of variable length is applied to a given set of protein protons to perturb their equilibrium, long enough to induce a state of no proton magnetization (saturation); (ii) by dipole–dipole direct NOE, the irradiated protons transfer the perturbed magnetization state to *adjacent* protons (note that they can belong either to the protein or to the ligand);

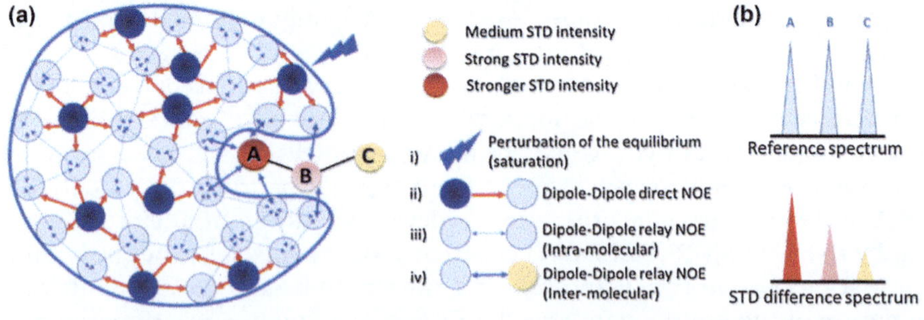

Figure 7.1 (a) Illustration of the sequence of events taking place during an STD NMR experiment. The outcome is the binding epitope mapping of the ligand constituted by the three different moieties A–B–C, reporting on its binding mode. (b) Illustration of the reference spectrum (top) and STD difference spectrum (bottom) for the A–B–C ligand in part (a).

(iii) by dipole–dipole relayed NOE, the protein protons spread the saturation to the whole protein, in an intramolecular process (so-called spin diffusion) that is very efficient in large molecules with a dense network of proton–proton distances appropriate for NOE propagation; (iv) when a small molecule binds to the surface of the protein, the magnetization is transferred to the interacting ligand protons *via* an intermolecular protein–ligand process; and (v) the fast exchange between the bound and free states of the ligand transfer that perturbed magnetization to the bulk (free state) where it is accumulated during the saturation time due to the slow relaxation properties of the ligand proton signals in the free state. This finally leads to an observable reduction in the intensities of the ligand signals in the ^{1}H NMR spectrum under saturation of the protein, in comparison to the equilibrium spectrum (without protein saturation) as shown in Figure 7.1b. Importantly, those ligand protons that are in closest contact with the macromolecule in the binding pocket receive the greatest saturation.

The STD intensity (I_{STD}) is calculated as:

$$I_{STD} = \frac{I_0 - I_{sat}}{I_0} \tag{7.1}$$

where I_0 is the intensity of a signal in the reference spectrum and I_{sat} is the intensity of the same signal in the saturated spectrum. To observe a difference between the reference and saturated spectra, the saturated ligand must be able to accumulate in solution. For this reason, the ligand must fulfil the conditions of fast exchange between the free and bound states on the NMR timescale (with K_D ranging from mM to μM), since relaxation is less efficient in small molecules, leading to retention of the perturbed magnetization for longer than in the bound state. On the contrary, if binding is too tight, three things can occur: (i) the efficient spin diffusion process can reach all the protons of the ligand, precluding any discrimination in terms of contacts in the bound state, (ii) extremely low accumulation in the free state, due to the slow turnover during the saturation time, and (iii) the ligand perturbations can relax too fast in the bound state even before it can accumulate in solution. In this way, STD NMR is an optimal technique to gain structural information only for weak protein–ligand interactions.

7.2.2 STD NMR Sample Preparation and Pulse Sequences

Experimentally, STD NMR is rather straightforward. An STD NMR sample is composed of a protein and a ligand in large excess (typically 1 : 20–1 : 100 protein : ligand ratio), in a suitable buffer. For screening purposes, a cocktail of small molecules constitutes the sample, where signal overlapping must be taken into consideration.

The standard STD NMR pulse sequence consists typically of a train of low-power Gaussian-shaped pulses, that are applied to a selected frequency as to selectively saturate specific protein protons, followed by a 90° pulse to convert the signal into an observable state (Figure 7.2, orange square). The power of the shaped pulse is set to cover a bandwidth of about 100 Hz, while

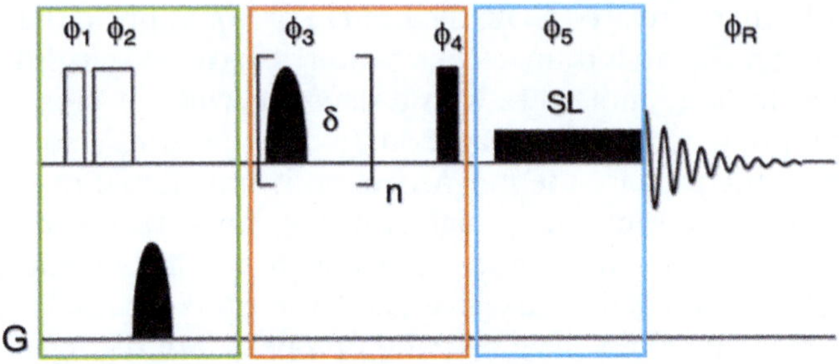

Figure 7.2 Pulse sequence for STD NMR. The narrow and wide-open bars correspond to trim pulses with lengths of 2.5 and 5 ms, respectively, performed at the power level for spin-lock. The narrow-filled bar is a 90° hard pulse. A series of selective pulses (typically 50 ms duration, 1 ms inter-pulse delay, frequency switched for on-resonance and off-resonance spectrum) is applied for saturation. The first two pulses as well as the gradient are applied to destroy the residual magnetization. Phase cycling: $\varphi 1 = x$; $\varphi 2 = y$; $\varphi 3 = x$; $\varphi 4 = \varphi R = x, 2(-x), x, y, 2(-y), y$; $\varphi 5 = 4(y), 4(-x), 4(-y), 4(x)$.

the number of pulses in the train (n) being adjusted to obtain the desired length of total saturation time (the usual saturation time for STD NMR in screening experiments is 2–3 s, as a compromise between sensitivity and reasonable experimental time).

Other selective shaped saturation pulses have been proposed,[23] though their advantages besides differences in irradiation power are not yet clear. The experiment is run in an interleaved fashion, meaning that the reference and saturated spectra can be acquired simultaneously. In practice, the reference spectrum is acquired by applying the saturating pulse far from any signals (*e.g.* 40 ppm) and is termed the off-resonance spectrum. The saturated spectrum (on-resonance spectrum) is acquired by applying the saturating pulse to a frequency where the receptor signals can be hit, but far from any ligand signal.[24] Typically, this is around 0 or 7 ppm (where the aliphatic and aromatic residues of the protein generally resonate, respectively). Choosing this frequency is a crucial feature of the *multi-frequency* STD approach, as we will see in the dedicated section.

Optionally, the introduction of a trim pulse plus a gradient, introduced between each scan, ensures that any unwanted magnetization is destroyed before the next acquisition and it has been demonstrated to be useful for STD NMR experiments in high-field spectrometers >700 MHz[24] (Figure 7.2, green square). Also, the presence of a spin-lock filter prior to acquisition removes signals originating from the protein from the spectrum (Figure 7.2, turquoise square).[24]

7.2.3 Qualitative and Semi-quantitative STD NMR: Ligand-binding Epitope Mapping

It is important to note that, unlike other NOE-based techniques, absolute distance measurements based on STD NMR data cannot be made, since, first, no assignment of the protein protons is known, precluding a pairwise interpretation of intensities, and second, chemical exchange contributes to the signal intensity. Nevertheless, theoretically, the magnetization transferred from receptor to ligand protons by intermolecular NOE in the bound state depends on the inverse of the sixth power of their distances.[25] Thus, the observed differences in saturation transferred among the different ligand protons can be correlated with different spatial proximities between the distinct parts of the ligand molecule and the protein in the bound state. Qualitatively, within a ligand STD NMR spectrum, stronger STD intensities reflect closer inter-proton distances between the corresponding ligand protons and the receptor surface in the bound state.

This allows mapping of the protons of the ligand that are closer to the protein surface, and hence to obtain *qualitative* structural information about the binding, by comparing the relative STD intensities of the different protons of the ligand at a given saturation time. This is done by normalizing all the measured STD factors ($I_0 - I_{sat}/I_0$) against the most intense one within the ligand, which arbitrarily is assumed to be 100%. Thus, the resulting percentage STD values represent a map of the ligand–protein contacts delineating which ligand moieties are key for recognition by the receptor. This is what we call ligand-binding epitope mapping, an example of which is shown in the left panel of Figure 7.3.

In choosing the saturation time, it is crucial to consider that the STD intensity of any given proton of a ligand typically increases at increasing protein saturation time, until reaching a plateau, the STD_{max}, which depends on the relaxation properties of the ligand (for fast-tumbling small molecules

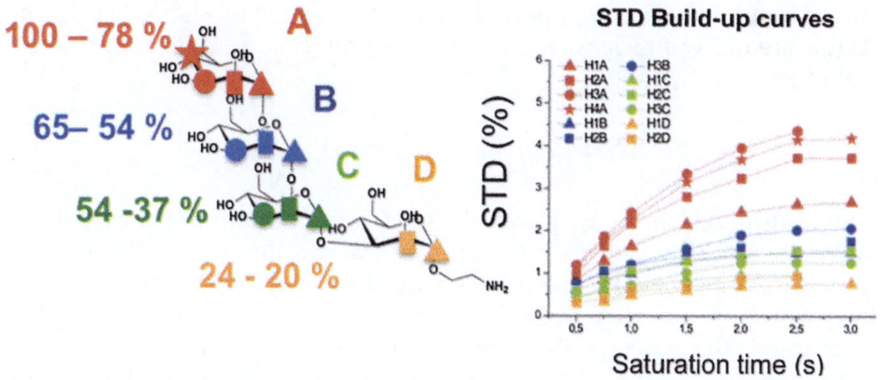

Figure 7.3 Ligand-binding epitope mapping by STD NMR. (Left) Group epitope mapping of a tetramannoside binding to the human anti-HIV-1 antibody 2G12 by normalization of the STD initial slopes from the experimental curves (right). In this example, the distribution of saturation transfer indicates binding through the non-reducing mannose residue (mannose A). Adapted from ref. 16 with permission from John Wiley and Sons, Copyright © 2011 WILEY-VCH Verlag GmbH & Co. KGaA, Weinheim.

we usually expect STD_{max} to be reached within 5–6 s). The plateau is affected by the longitudinal relaxation rate (R_z) (of the ligand in the free state, *i.e.* R_z plays a more relevant role the closer we get to the STD_{max}. Ideally, the group epitope mapping information should not depend on the chosen saturation time. However, significantly different R_z relaxation rates of the different protons of the ligand can produce artefacts in the epitope definition, leading to wrong structural conclusions.

Specifically, protons with small R_z will accumulate more efficiently saturation in solution. Thus, their relative STD intensity will be enhanced in comparison to other ligand protons with larger R_z values, so that their proximity to the binding pocket may be largely overestimated at long saturation times. Since these artefacts are a consequence of differences in the ability to accumulate saturation in the free state, a way to cancel them out is to derive STD intensities for extremely short saturation times (tending to zero), when virtually no accumulation of saturated ligand takes place. Then, the problem arises of extremely low intensities that are difficult to distinguish from the baseline noise. To solve this difficulty, Mayer and James proposed measuring STD intensities at increasing saturation times, as opposed as to a single saturation time experiment, and fitting the experimental build-up curves to the mono-exponential function:[26]

$$\mathrm{STD}(t_{\mathrm{sat}}) = \mathrm{STD}_{\mathrm{max}}\left(1 - e^{k_{\mathrm{sat}} * t_{\mathrm{sat}}}\right) \tag{7.2}$$

where $\mathrm{STD}(t_{sat})$ is the observed STD intensity at a given saturation time, t_{sat}, STD_{max} is the asymptotic maximum of the build-up curve, and k_{sat} is a rate constant that measures the speed of STD build-up. An example of the resulting build-up curves is reported in the right panel of Figure 7.3. k_{sat} and STD_{max} are derived by least-squares fitting, and the initial slope of the curve is obtained as:

$$\mathrm{dSTD}(t_{\mathrm{sat}})/\mathrm{d}t_{\mathrm{sat}}\big|_{t_{sat}\to 0} = \mathrm{STD}_{\mathrm{max}} k_{\mathrm{sat}} \tag{7.3}$$

Now, although the slope physically does not represent STD intensity, but rather an intensity growth rate, we can assume that at the limit of zero saturation time the STD intensities will grow linearly with the saturation time, which means that those intensity values just contain a multiplying factor that is the same for all the protons (the saturation time). As this factor is cancelled out during the normalization process, the ligand-binding epitope mapping can be then obtained by normalization of the set of initial slopes obtained, relative to the value of an arbitrarily chosen proton, usually the one with the highest initial slope. With such an analysis of the initial slopes obtained from the build-up curves, the contribution of differential relaxation effects is removed, and this approach is considered as the *semi-quantitative* analysis of STD NMR experiments.

7.2.4 Quantitative STD NMR: CORCEMA-ST Approach; Analysis of Multi-modal Binding

A theoretical framework, CORCEMA-ST (Complete Relaxation and Conformational Exchange Matrix Analysis of Saturation Transfer), was developed by Rama Krishna in 2002, which allows computational prediction of the STD intensities starting with the atomic Cartesian coordinates of 3D molecular models of the receptor and the ligand in the free state, as well as that of the ligand–receptor complex.[15]

CORCEMA-ST is a valuable tool for the *quantitative* interpretation of STD NMR data, allowing predictions from a proposed 3D model of the complex to be compared and validated against experimental STD build-up curves. Typically, the 3D structure of the protein–ligand complex is obtained from either X-ray crystallography, NMR, or molecular modelling (like docking or molecular dynamics simulations). For the matrix calculations in CORCEMA-ST, the Cartesian coordinates of all ligand protons and those protein protons within a given cut-off distance from the ligand are considered. How well the molecular model reproduces the experimental NMR data can be quantified by the so-called NOE R-factor:

$$\text{NOE R-factor} = \sqrt{\frac{\sum(\text{STD}_{\text{exp},i} - \text{STD}_{\text{calc},i})^2}{\sum(\text{STD}_{\text{exp},i})^2}} \tag{7.4}$$

where $\text{STD}_{\text{exp},k}$ and $\text{STD}_{\text{cal},k}$ are the experimental and calculated STD intensities of each proton k, respectively. Lower NOE R-factor values indicate better fit between experimental and theoretical data and, thus, better pertinence of the chosen structural model of the complex. In this way, different theoretical structural models derived, *e.g.*, from different docking runs, can be ranked according to how well they explain the experimental STD NMR data. To validate the 3D model as *quantitatively* representative of the experimental STD NMR build-up curves, NOE R-factors below 0.30 would be desirable.

The approach can also be pushed further as far as to detecting processes of multi-modal binding of a ligand in the same binding pocket (a not that uncommon feature in weak interactions). When a ligand binds to a binding pocket in a multi-modal fashion (*e.g.*, head to tail *vs.* tail to head binding or, in the case of protein–carbohydrate interactions, if two arms of a branched oligosaccharide both bind in the same binding pocket), the detected STD intensities report on an *average* resulting from the different saturation transfer steps occurring in each binding mode. If the two binding modes are equimolar, then the STD intensities will be the exact average of the saturation transfer received in each of the binding orientations. If, instead, one of the two binding mode is preferential (accounting for a population higher than 50%), then the STD intensities will account for a weighted average of the two binding modes. In these scenarios, CORCEMA-ST is an invaluable tool to analyse multimodal binding and to properly quantify the occurrence of each of them.[16,27–29]

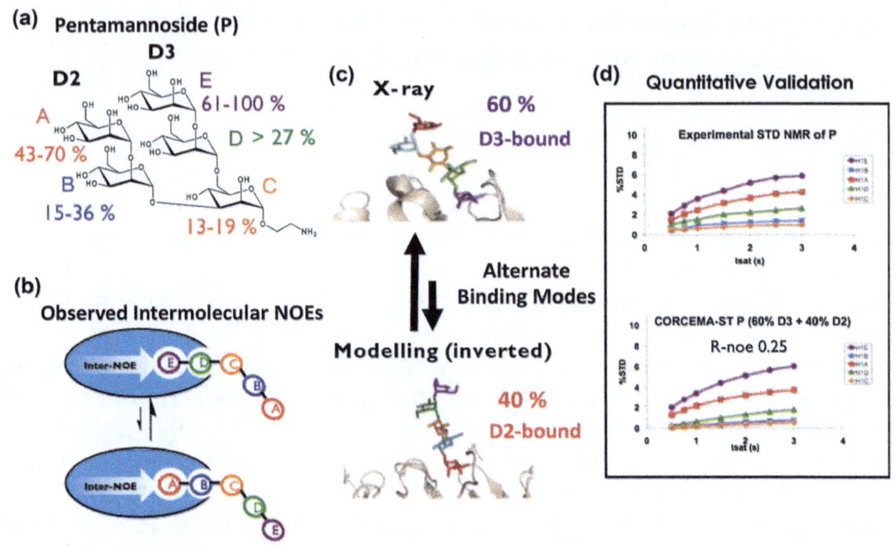

Figure 7.4 Multi-modal binding studied by STD NMR. (a) Biantennary pentamannoside with two α1,2-ending mannose disaccharides (arms D2 and D3). The obtained binding epitope mapping indicates both arms contact the protein, in contrast to the X-ray structure which indicated binding through D3-arm. (b) Intermolecular NOEs suggest that both terminal mannoses A and E interact with the same protein pocket. (c and d) CORCEMA-ST calculations validate a 60 : 40 equilibrium of D3- and D2-bound, respectively. Adapted from ref. 16 with permission from John Wiley and Sons, Copyright © 2011 WILEY-VCH Verlag GmbH & Co. KGaA, Weinheim.

As shown in Figure 7.4, indeed, we can combine the theoretical STD NMR data calculated by CORCEMA-ST on the main binding modes identified in a weighted fashion and identify for which combination the NOE R-factor is lowest, *i.e.* which combination represents best the multi-modal binding event taking place in solution.

7.3 Multi-frequency STD NMR

In the previous section, we have shown how STD NMR, in combination with computational tools, can provide 3D models of protein–ligand complexes in solution. However, this approach fails to provide *direct information* about the ligand location or its orientation in the binding pocket. To this aim, additional experiments are required: to characterize the location of binding, competitive STD NMR experiments with "spy" ligands of known binding location must be carried out,[30,31] while to identify the orientation of binding in relation to other bound fragments, ILOE tr-NOESY experiments must be performed.[32,33]

Importantly, as seen previously, STD NMR relies on the efficacy of spin diffusion to homogeneously spread the saturation over the whole receptor.

This, however, is not always ensured and, indeed, differences in ligand-binding epitope mappings can be observed if STD experiments are acquired at different saturating frequencies.[15] As explained in this section, this is the main idea behind *multi-frequency* STD NMR approaches. Under an appropriate experimental setup, small inhomogeneities in saturation can be detected, and their analysis delivers specific information on the intermolecular interactions beyond just the ligand epitope, as they allow unveiling of the nature of the protein amino acids *from where* the saturation is transferred to the ligand protons. Formally, the approach can be formulated in the statement: *Ligand protons receive larger amounts of saturation if protein protons in their proximity are "directly irradiated" instead of all of them being "relayed-NOE" saturated* (Figure 7.5a).

Based on this, *differential epitope mapping STD NMR* (DEEP-STD NMR) was developed, with multi-frequency STD NMR experiments tailored to select certain types of protein (or adjacent ligand) protons being directly irradiated, taking advantage of the inhomogeneity of the spin diffusion process within the protein.

7.3.1 Differential Epitope Mapping STD NMR (DEEP-STD NMR)

DEEP-STD NMR aims at detecting and quantifying the differences in the binding epitope maps of a given ligand, when distinct STD NMR experiments irradiating different protein residues are carried out. Experimental

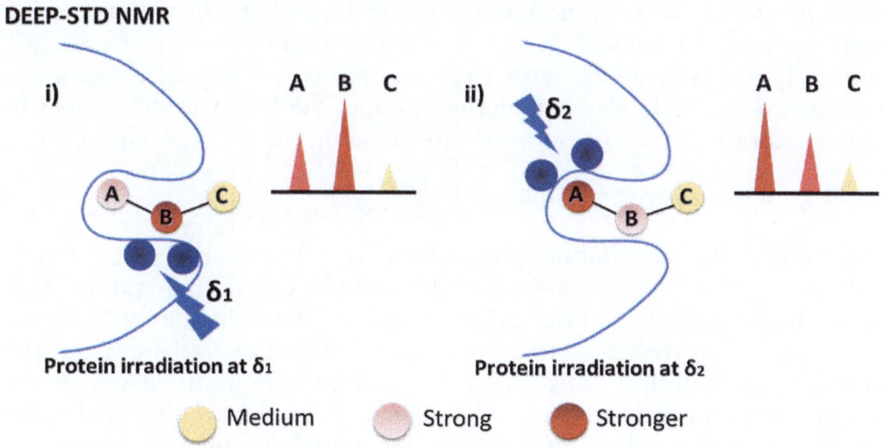

Figure 7.5 Illustration representing the implementation of the *multi-frequency* approach termed DEEP-STD NMR. In (i) selective irradiation is on protein protons at $\delta 1$ (chemical shift) and (ii) selective irradiation is on protein protons at $\delta 2$. Assuming spin diffusion cannot spread homogeneously the saturation on the whole protein, ligand protons on residue B will receive greater saturation by irradiating at $\delta 1$, in comparison to the experiment irradiating at $\delta 2$. The discussion is the opposite for ligand protons on residue A.

determination of such a "differential epitope map" of a ligand obtained at two different frequencies can inform us on the main "source" of magnetization received by different protons of the ligand. As the set of protein residues directly irradiated in each different experiment can be known (either experimentally or theoretically), the results report on the nature of the protein residues lining the binding pocket and, more generally, on its architecture.

The outcome is a map of the differences in relative ligand STDs (*i.e.*, epitope differences) from each pair of STD NMR experiments (reflecting contacts to specific types of amino acids). The "differential epitope map" of a ligand at two different frequencies ($\delta1$ and $\delta2$) is determined through the DEEP-STD factor ($\Delta\mathrm{STD}_i$) for each proton i as:

$$\Delta\mathrm{STD}_i = \frac{\mathrm{STD}_{\delta1,i}}{\mathrm{STD}_{\delta2,i}} - \frac{1}{n}\sum_i^n \left(\frac{\mathrm{STD}_{\delta1,i}}{\mathrm{STD}_{\delta2,i}}\right) \tag{7.5}$$

The first step in the analysis of DEEP-STD NMR data consists in determining which of the two experimental conditions, $\delta1$ or $\delta2$, produces stronger STD intensities (comparing their total sum of STD values). The experiment with the largest total sum will be called the experiment at frequency $\delta1$ in the equation. Next, the ratio of STD intensities $\left(\frac{\mathrm{STD}\,\%_{\delta1,i}}{\mathrm{STD}\,\%_{\delta2,i}}\right)$ is calculated for each proton of the ligand. These ratios report not only on differences in the epitopes but also on the different global levels of protein saturation achieved in both experiments (*e.g.*, protein saturation on the aromatics leads to reduced global saturation, as the number of aromatic residues is normally much lower than the number of aliphatic ones). For that reason, to obtain the differential epitope, the intrinsic differences in protein saturation must be removed. To that aim, the average ratio of STDs over all protons must be calculated: $\left[\frac{1}{n}\sum_i^n\left(\frac{\mathrm{STD}\,\%_{\delta1,i}}{\mathrm{STD}\,\%_{\delta2,i}}\right)\right]$.

To remove the contribution from differences in saturation level, and to establish a "zero" in the scale for analysis of the differential epitope, that average factor must be subtracted from each of the individual STD ratios. This assures that ($\Delta\mathrm{STD}_i$)s reveal only those differences arising from the distinct types of amino acids hit by the saturating radio-frequency. If the change in conditions is affecting the residues lining the binding pocket, we will observe positive values for protons experiencing a relative increase in STD values at $\delta1$ and negative values for protons experiencing a relative increase in STD values at $\delta2$. It is important to highlight that the increase/decrease of STD values is "relative", as it comes from the comparison of the binding epitopes under the two conditions, as expressed in eqn (7.5), and not from the comparison of absolute STD values.

The protocol has been first tested by analysing the effect of different irradiation frequencies on the binding of 2,7-anhydro-Neu5Ac to GH33,[34] the glycosyl hydrolase domain of the intramolecular (IT) *trans*-sialidase from *Ruminococcus gnavus* (IT-sialidase from *Rg*NanH, PDB ID: 4X4A[34]). The compound 2,7-anhydro-Neu5Ac is a Neu5Ac derivative resulting from the intramolecular *trans*-glycosylating activity of GH33. Two STD NMR experiments were carried out irradiating at 0.60 ppm ($\delta 1$) and 6.55 ppm ($\delta 2$), for 0.5 s. These frequencies are known to be centred in the aliphatic and aromatic (protein) proton spectral regions, respectively.[35] We discuss the criteria for selection of irradiation frequencies in the following sub-section.

The DEEP-STD factors (ΔSTD_i) (0.60 ppm/6.55 ppm) are shown in Figure 7.6a. Positive DEEP-STD factors report relative STD increases when irradiation was at 0.60 ppm (aliphatics); negative ones indicate increases when irradiation was at 6.55 ppm (aromatics). The resulting differential epitope map is shown in Figure 7.6b. The results clearly show how different protons of the ligand occupy distinct areas of the GH33 binding pocket lined by either aliphatic or aromatic residues. Positive ΔSTDs for CH3 and H3a suggest vicinity to aliphatic side chains, whereas negative ΔSTDs for H8, H9 and H9′ suggest vicinity to aromatic protons.

The DEEP-STD NMR pattern in Figure 7.6a is in perfect agreement with the crystal structure of the GH33/2,7-anhydro-sialic acid complex (Figure 7.6c), where the ligand is surrounded by an aliphatic pocket (Ile258, Ile338, and Val502) and an opposite aromatic patch (Tyr667 and Trp698). Protons CH_3 and H3a from the ligand point towards the aliphatic residues, while H8, H9, and H9′ point towards the aromatic side chains. Protons H3e, H4, H5, H6, and H7 sit between these two regions, again in agreement with their small ΔSTDs.

Further, the novel protocol was also applied to the complex of cholera toxin subunit B (CTB), a larger receptor (65 kDa), with 3-nitrophenyl-α-D-galactopyranoside (3NPG).[36] 3NPG is a well-known CTB inhibitor that occupies the galactose-binding subsite of the GM1 pocket, mimicking the galactose of the natural binder[36] and is frequently chosen as a reference (and/or starting material) in CTB inhibitor drug design projects.[37–40]

The presence of aromatic moieties in the ligand precludes protein irradiation in this spectral region. In *multi-frequency* DEEP-STD NMR, other groups of protein protons can be chosen for selective irradiation, providing that they are in spectral regions with no ligand signals. For this system, protein resonances were irradiated at 2.25 ppm, where, in CTB, protein signals are available for saturation and no ligand proton signals are present. The prediction of chemical shifts of CTB protons within 4 Å of the ligand in the X-ray structure showed Glu51 and Gln56 protons as those likely to be directly irradiated at this frequency (Figure 7.6f). The *multi-frequency* STD NMR experiments were then run at 2.25 ppm and 0.60 ppm, giving positive ΔSTD values for protons H4, H5, H6, and H6′ on the galactose, indicating an increase in relative STDs when the irradiation was at 2.25 ppm (Figure 7.6d). In contrast, negligible ΔSTD factors were observed for H1, H2, H3, and the aromatic protons at the opposite end of the molecule.

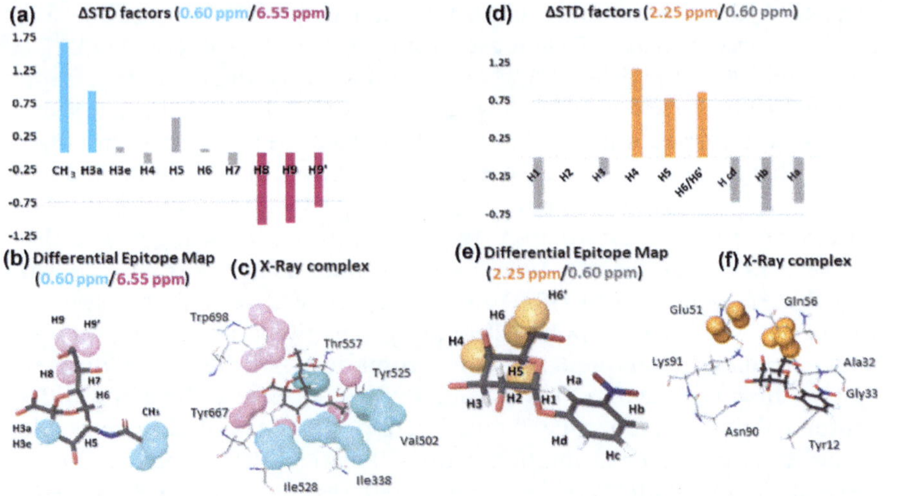

Figure 7.6 (Left panel) Differential Epitope Mapping (0.60 ppm/6.55 ppm) of 2,7-anhydro-Neu5Ac in complex with GH33. (a) ΔSTD histogram: positive ΔSTD (above the limit of +0.75) after aliphatic irradiation (0.60 ppm) are in cyan, and negative ΔSTD (below −0.75) after aromatic irradiation (6.55 ppm) are in magenta. (b) Differential Epitope: ΔSTD map of the ligand. Cyan surfaces highlight ligand contacts with aliphatic side chains; magenta, contacts with aromatic side chains. The ligand polar protons have been omitted. (c) Crystal structure of the complex (PDB ID: 4X4A[34]). Protein protons are coloured as a function of their NMR frequencies: directly irradiated at 0.60 ppm (aliphatics) in cyan and directly irradiated at 6.55 ppm (aromatics) in magenta. (Right panel) Differential Epitope Mapping (2.25 ppm/0.60 ppm) of 3-nitrophenyl-α-D-galactopyranoside (3NPG) in complex with Cholera Toxin Subunit B (CTB). (d) ΔSTD histogram: protons with positive ΔSTD (above the limit of +0.75) after irradiation at 2.25 ppm are in orange. (e) Differential Epitope: ΔSTD map of the ligand. Orange surfaces indicate ligand contacts with protein side chains directly irradiated at 2.25 ppm. Ligand polar protons have been omitted. (f) Crystal structure of the complex (PDB ID: 1EEI[36]). Protein protons directly irradiated at 2.25 ppm are enclosed in an orange surface. Comparison of (b) and (c), and (e) and (f), highlights the excellent match of the differential epitope map of the ligand with the distribution of distinct types of residues in the binding pocket. Reproduced from ref. 20, https://doi.org/10.1002/anie.201707682, under the terms of the CC BY 4.0 license, https://creativecommons.org/licenses/by/4.0/.

The ΔSTD pattern of 3NPG (Figure 7.6e) was in perfect agreement with the published crystal structure of the CTB/3NPG complex (Figure 7.6f),[36] where 3NPG's H4 to H6 (on the galactose ring) are surrounded by the side chains of Glu51 and Gln56. On the other hand, H1, H2, H3, and aromatic carbons are pointing far from those side chains (Figure 7.6f).

The results obtained on these two complexes confirmed that the *multi-frequency* DEEP-STD NMR approach allows identification of the nature of the ligand–receptor contacts and ligand orientation in the binding pocket.

7.3.1.1 Selection of Irradiation Frequencies: A Theoretical and an Experimental Approach

For proteins of known 3D structure, directly irradiated protons can be easily identified in the binding pocket if their chemical shift assignments are known. In many cases, however, the protein assignment is not available. Then, statistical averages of chemical shifts of the residues present in the binding pocket (from existing NMR databases) can be used. Alternatively, the 3D structure of the protein can be used to predict the ^{1}H chemical shifts using existing software. For the studies reported above (Figure 7.6), the irradiation frequencies were selected based on predictions of chemical shifts using the available 3D structures of the proteins and ShiftX2 (http://www.shiftx2.ca35). Another option is to check the chemical shifts of selected protons on the proteins against average chemical shift histograms from the Biological Magnetic Resonance Data Bank (http://www.bmrb.wisc.edu).

Alternatively, a more general method has been proposed to *experimentally* identify the chemical shifts of the residues present in binding hot-spots at the surface of the receptor by using 2D NMR experiments combined with a paramagnetic probe.[41] This is most relevant when the full NMR assignment is not possible, and/or the predicted chemical shifts from the structure are not in line with the experimental data (*e.g.*, due to the dynamics of the protein, not accounted for in calculations on a static X-ray structure).

Protein hot-spots can be mapped by adding paramagnetic probes such as 4-hydroxy-2,2,6,6-tetramethylpiperidin-1-oxyl (TEMPOL) to a protein sample and can be readily detected by means of 2D ^{1}H,^{1}H TOCSY experiments.[42] Due to the paramagnetic relaxation enhancement (PRE) effect, a decrease in the intensity of specific protein TOCSY cross-peaks, in comparison with the spectrum recorded without the paramagnetic probe, allows the signals of residues in contact with the probe to be identified (Figure 7.7a–c).[43,44] The greater accessibility of the TEMPOL probe to protein-specific binding sites (rather than surface regions) has already been demonstrated by previous PRE studies with TEMPOL, and also is shown by MixMD simulations (Figure 7.7d).[42,45,46] As a result of these experiments, the identified hot-spot resonances can then be considered as experimentally derived input frequencies for the DEEP-STD NMR experiments.

7.3.1.2 Application to Challenging Systems: Quest for WWP2 Ubiquitin Ligase Inhibitors

DEEP-STD NMR has been applied to a set of small molecules that are potential inhibitors of the ubiquitin ligase WWP2.[46] WWP2 ubiquitin ligases (together with WWP1 and Nedd4) are frequently mis-expressed in cancer,[47–49] and WWP2 is responsible for ubiquitin-dependent degradation of some tumour suppressor proteins found in many types of cancer. Initially, a large-scale high-throughput screening (HTS) single-shot enzymatic assay was used to identify the first generation of WWP2 ligase inhibitors from the *NCI Diversity Set V* and

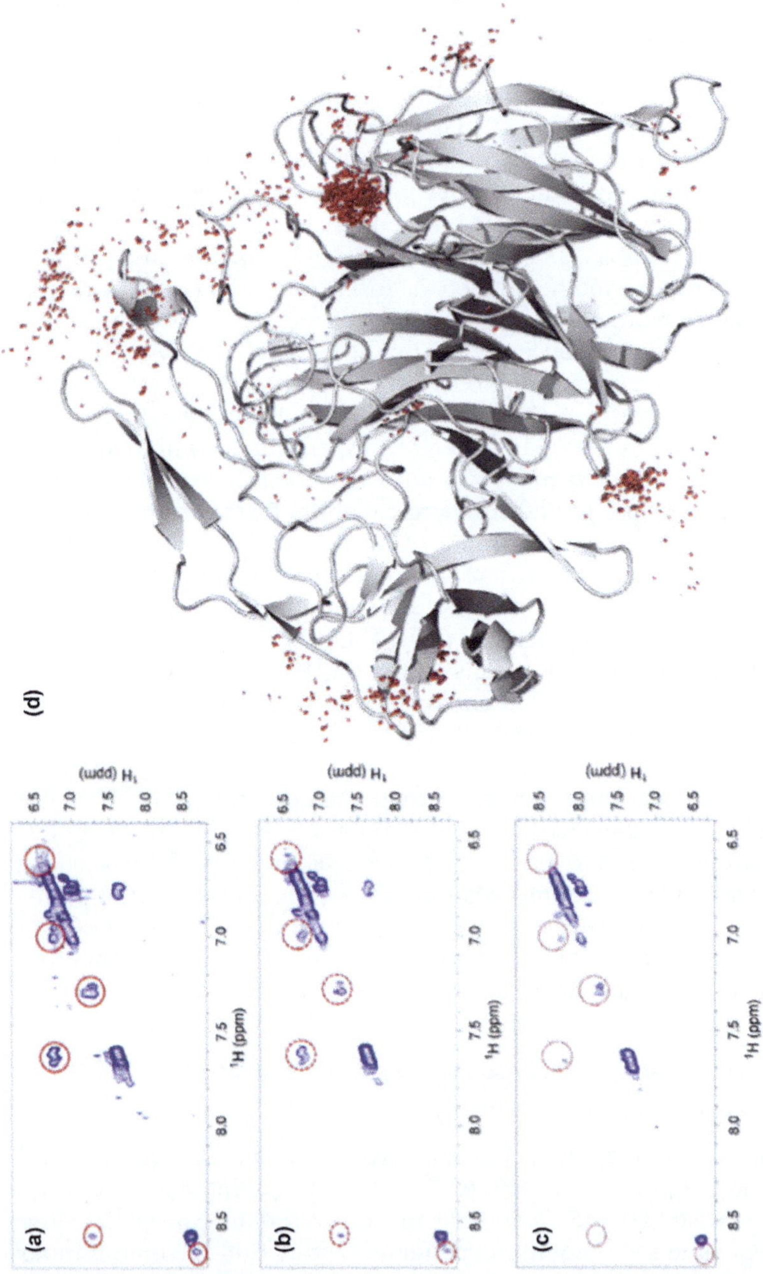

Figure 7.7 (Left panel) Expansion of the aromatic spectral region of the 2D ^{1}H,^{1}H TOCSY spectra of GH33 1.2 mM alone (a), and in the presence of TEMPOL (b) 2 mM and (c) 12 mM. Red circles highlight some resonances affected by the presence of TEMPOL. (Right panel) (d) Distribution of TEMPOL (red) around GH33 (grey) as determined by MixMD simulations. The distribution reveals a single large hot-spot at the known 2,7-anhydro-Neu5Ac binding site, thereby revealing that TEMPOL can act as a probe to selectively target residues of the binding site. Reproduced from ref. 41, https://doi.org/10.1002/cbic.201800568, under the terms of the CC BY 4.0 license, https://creativecommons.org/licenses/by/4.0/.

NCI Approved Oncology Set V, for a total of more than 2000 compounds. Twenty-four potential hits were identified and more stringent assays, coupled with rigorous control experiments allowed identification of 10 hit compounds, which were then screened singularly by ligand-based NMR, by means of (i) STD NMR and (ii) Carr–Purcell–Meiboom–Gill (CPMG) NMR experiments. Thus, six of the hits were confirmed as WWP2 binders, and one of them, compound **20** (NSC288387), was chosen for its optimal STD signal intensities to carry out a deeper structural investigation. WWP2 is a rather challenging protein to work with: due to its highly hydrophobic nature it is very challenging to purify and keep in solution, as proven by the fact that very few X-ray structures are available, and then only in its apo form (PDB IDs: 4Y07,[50] 5TJ7/8[51]). This is a remarkable showcase of how, when the binding site is unknown and no other structural information could be extracted by XRD, the STD NMR approach is perfectly suited to fill the experimental gap.

STD NMR provided the binding epitope map of compound **20** (from the build-up curve initial slope approach, Figure 7.8a) and then the *multi-frequency*

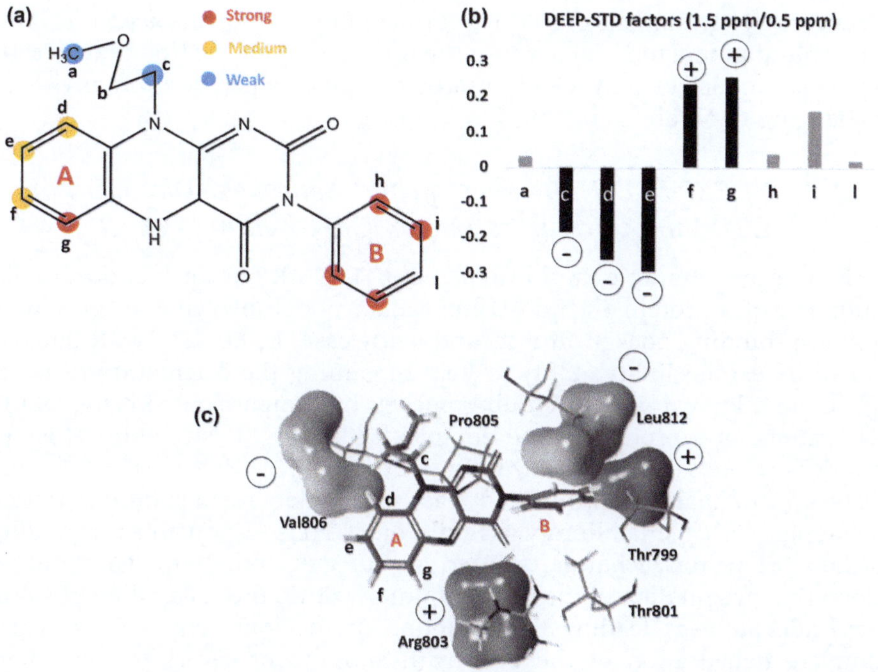

Figure 7.8 Differential epitope mapping of compound **20** (NSC288387) in complex with the WWP2-HECT domain. (a) STD NMR binding epitope mapping, (b) DEEP-STD factors (1.5 ppm/0.5 ppm), and (c) details of the binding pocket for NSC288387. In (c), protein residues within 5 Å from the ligand are shown, where protein protons resonating at 0.5 ppm and 1.5 ppm are enclosed in a surface marked as " – " (negative ΔSTD values) or in a surface marked as " + " (positive ΔSTD values), respectively. Reproduced from ref. 41, https://doi.org/10.1002/chem.201804169, under the terms of the CC BY 4.0 license, https://creativecommons.org/licenses/by/4.0/.

DEEP-STD NMR approach was used to investigate the nature of the amino acid in the binding pocket (Figure 7.8c). Computational techniques were also employed to identify the most likely putative binding site of WWP2, and compound **20** was docked into it. The lowest energy-docking pose was in perfect agreement with the experimental data, showing that the aromatic ring B of compound **20** is more deeply buried in the identified binding cleft relative to ring A, in agreement with the STD binding epitope data (Figure 7.8a and b). Importantly, DEEP-STD NMR validated the model. By visual analysis of the putative binding pocket, two aliphatic residues can be identified (one leucine and one valine) resonating at about 0.5 ppm and two polar residues (one arginine and one threonine), whose β and γ protons resonate at about 1.5 ppm. Thus, DEEP-STD NMR was performed using 1.5 ppm/0.5 ppm frequencies, and positive ΔSTD factors were observed for the f and g protons of ring A, which in the model are pointing towards the more polar Arg803. In contrast, the protons c, d, and e showed negative ΔSTD factors, as they point towards Val806 instead. Interestingly, the protons of ring B showed neglectable ΔSTD factors, in agreement with the close proximity to both Leu812 and Thr799 (Figure 7.8c and d).

Here, the combination of STD NMR (including *multi-frequency* STD NMR) and computational tools allowed for the proposition for the first time of a 3D molecular model of a WWP2-inhibitor complex, which was otherwise so challenging to obtain.

7.3.1.3 "DEEP-STD NMR Fingerprint" Approach: Discovery of CTB Inhibitors in Its Known GM1-binding Pocket

In the previous example, multi-frequency STD NMR has been shown to aid validation of a protein–ligand 3D molecular model involving an otherwise unknown binding pocket. In a second study case, DEEP-STD NMR demonstrated its extraordinary ability to help in gaining the orientation of novel ligands in a known protein-binding pocket, by comparison with the DEEP-STD pattern of a protein–ligand complex of known 3D structure. This approach is called "DEEP-STD NMR fingerprinting".

The system analysed in this study case was cholera toxin subunit B (CTB), in complex with a small library of inhibitors. CTB is responsible for binding to the GM1 pentasaccharide exposed on our gut epithelium. The binding allows the invagination of the toxic subunit A within the cells, where it starts a cascade pathway leading to disruption of the ionic channels, causing complete dehydration of the cells with subsequent severe to fatal diarrhoea.[52] While some strains of *Vibrio cholera* have started to show antibiotic resistance, drug discovery groups have put a great deal of effort into designing CTB inhibitors which could prevent the binding of CTB to GM1 and therefore stopping the onset of the infection.[53] Based on the high-resolution structural knowledge of the GM1-binding pocket, with a deep and specific galactose-binding subsite and a shallow and less specific galactose-binding subsite, most inhibitors are designed to mimic the two galactose and sialic acid moieties, to fill the binding site entirely.[36]

In this case, a small library composed of decorated thio-galactoside, linked to an aromatic moiety (known to be able to bind the sialic acid subsite) through a furoate (Figure 7.9a) was studied.[54] As STD NMR implies that only the ligand is observed, if we want to be sure of the location of binding on the protein surface we have to resort to indirect methods. It is very common to use STD competition experiments between novel ligands and "probe-ligands" known to bind in the binding pocket under investigation, to verify whether the novel compounds compete for the same binding pocket, confirming the location of binding.

However, DEEP-STD NMR can provide similar information about the location of binding, simultaneously shedding light on the orientation of the unknown ligand in the binding pocket, something that cannot be inferred by STD competition experiments.

This *multi-frequency* approach implies comparing the differential epitope map of a complex with known 3D structure and the differential epitope maps of the novel ligands, just as if we were comparing fingerprints. If the novel ligands respond to the differential conditions similarly to the ligand with known 3D structure (*i.e.* with comparable DEEP-STD maps), the results strongly support them being exposed to the same type of residues in the bound state and hence sustaining the likelihood that they most likely occupy the same binding pocket.

This approach has been applied to elucidate the location of binding of four CTB ligands **4–7**, by comparing the differential irradiation DEEP-STD maps (2.25 ppm/0.60 ppm) with the DEEP-STD map (2.25 ppm/0.60ppm) of the 3NPG/CTB complex, for which the 3D structure from X-ray diffraction is known. The DEEP-STD maps (2.25 ppm/0.60 ppm, Figure 7.9c) showed a stark increase in the STD intensities of the galactose protons (from H3 to H6) upon irradiation at 2.25 ppm, similar to what was observed for 3NPG.

As shown in Figure 7.6d–f, 2.25 ppm is the predicted frequency around which the β and γ protons of Gln56 and Glu51 resonate. The proximity of Gln56 and Glu51 to H4, H5, and H6 of the galactose moiety is appreciable in the crystal structure of the 3NPG/CTB complex (PDB ID: 1EEI,[36] Figure 7.9d and e). Overall, the positive ΔSTDs observed for the galactose protons of the four ligands confirm that the thio-galactose moiety occupies the galactose binding site, in a quicker way (no need for the preparation of double samples) and at a greater degree of detail about ligand orientation, confirming that the galactose moieties in the novel ligands indeed occupy the galactose subsite, and not, for example, the aromatic moiety.

7.4 Summary and Conclusions

The examples of applications given in this chapter showcase the potentiality of the novel *multi-frequency* STD NMR approach, and its powerful applicability in combination with classical STD NMR binding epitope mapping studies and molecular modelling. This original approach provides access to an extra layer of structural information previously inaccessible through

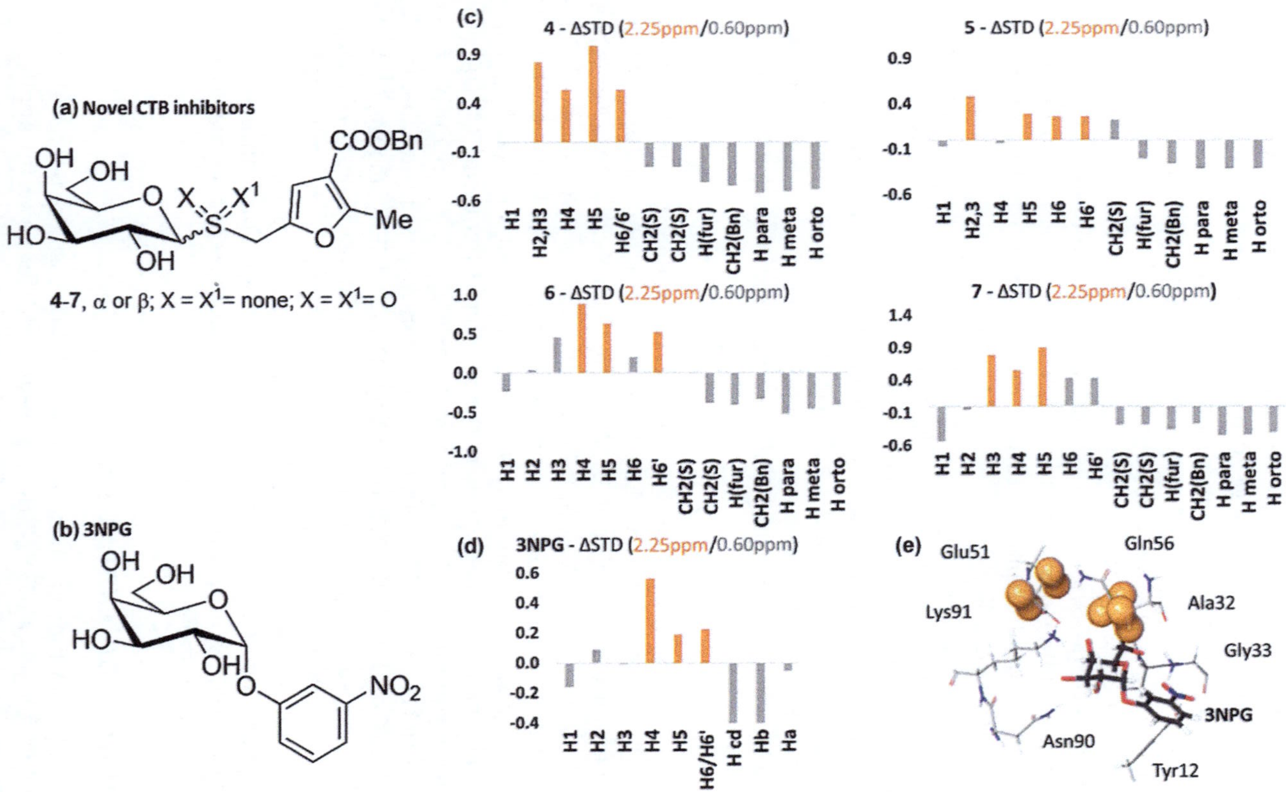

Figure 7.9 (Left panel) Chemical structures of (a) the novel set of thio-galactoside CTB ligands 4–7 described in ref. 54, and (b) 3-nitrophenyl-α-D-galactopyranoside (3NPG). (Right panel) Differential Epitope Mapping ΔSTD (2.25 ppm/0.60 ppm) of (c) ligands **4–7** and (d) 3NPG in complex with CTB (reference sample), obtained at 2 s saturation time. For each ligand, the three largest positive ΔSTD values are shown as orange bars. (e) Crystal structure of the complex 3NPG/CTB (PDB ID: 1EEI[36]). Protein protons directly irradiated at 2.25 ppm are highlighted in orange surface. Reproduced from ref. 54, https://doi.org/10.1002/chem.202001723, under the terms of the CC BY 4.0 license, https://creativecommons.org/licenses/by/4.0/.

ligand-based NMR techniques, by simply selecting appropriately protein direct irradiation at different frequencies. Indeed, it allows exploration of the nature of the residues lining the binding sites, indirectly, on the ligand signals, and hence allows assessment of the pharmacophore of the protein. Indeed, given the amphiphilic or diverse nature of many ligand–receptor interactions (with both synthetic and natural ligands), this approach is advantageous both in fragment-based drug discovery and in fundamental biological research.

As we have demonstrated in this chapter, if the 3D structure of the protein is known, novel *multi-frequency* STD NMR protocols allow for the first time gaining information from STD NMR on the orientation of the ligand. Noticeably, DEEP-STD NMR is applicable even when the 3D structure of the protein is unknown, as shown in the case of the ubiquitin ligase WWP2, for which the nature and architecture of the binding site were unveiled for the first time. In conclusion, *multi-frequency* STD NMR should be regarded as a versatile tool providing structural information at the atomic level on the binding pocket environment in which the ligand resides.

References

1. J. Qin and A. M. Gronenborn, Weak protein complexes: challenging to study but essential for life, *FEBS J.*, 2014, **281**(8), 1948–1949.
2. S. Katen and A. Zlotnick, The thermodynamics of virus capsid assembly, *Methods Enzymol.*, 2009, **455**, 395–417.
3. C. C. Moser, *et al.*, Nature of biological electron transfer, *Nature*, 1992, **355**(6363), 796–802.
4. J. Qin, O. Vinogradova and A. M. Gronenborn, Protein-protein interactions probed by nuclear magnetic resonance spectroscopy, *Methods Enzymol.*, 2001, **339**, 377–389.
5. K. Murata and M. Wolf, Cryo-electron microscopy for structural analysis of dynamic biological macromolecules, *Biochim. Biophys. Acta*, 2018, **1862**(2), 324–334.
6. A. T. Tuukkanen and D. I. Svergun, Weak protein–ligand interactions studied by small-angle X-ray scattering, *FEBS J.*, 2014, **281**(8), 1974–1987.
7. C. Schmidt and C. V. Robinson, Dynamic protein ligand interactions–insights from MS, *FEBS J.*, 2014, **281**(8), 1950–1964.
8. I. Bertini, K. S. McGreevy, and and G. Parigi, *NMR of Biomolecules: Towards Mechanistic Systems Biology*: John Wiley & Sons. 2012.
9. M. P. Williamson, Using chemical shift perturbation to characterise ligand binding, *Prog. Nucl. Magn. Reson. Spectrosc.*, 2013, **73**, 1–16.
10. C. Ludwig and U. L. Guenther, Ligand based NMR methods for drug discovery, *Front. Biosci.*, 2009, **14**(4565–4574), 24.
11. M. Mayer and B. Meyer, Characterization of ligand binding by saturation transfer difference NMR spectroscopy, *Angew. Chem., Int. Ed.*, 1999, **38**(12), 1784–1788.

12. A. Viegas, *et al.*, Saturation-transfer difference (STD) NMR: a simple and fast method for ligand screening and characterization of protein binding, *J. Chem. Educ.*, 2011, **88**(7), 990–994.
13. M. Mayer and B. Meyer, Group epitope mapping by saturation transfer difference NMR to identify segments of a ligand in direct contact with a protein receptor, *J. Am. Chem. Soc.*, 2001, **123**(25), 6108–6117.
14. J. Angulo and P. M. Nieto, STD-NMR: application to transient interactions between biomolecules—a quantitative approach, *Eur. Biophys. J.*, 2011, **40**(12), 1357–1369.
15. V. Jayalakshmi and N. R. Krishna, Complete relaxation and conformational exchange matrix (CORCEMA) analysis of intermolecular saturation transfer effects in reversibly forming ligand–receptor complexes, *J. Magn. Reson.*, 2002, **155**(1), 106–118.
16. P. M. Enríquez-Navas, *et al.*, A solution NMR study of the interactions of oligomannosides and the anti-HIV-1 2G12 antibody reveals distinct binding modes for branched ligands, *Chem. – Eur. J.*, 2011, **17**(5), 1547–1560.
17. C. D. Owen, *et al.*, Unravelling the specificity and mechanism of sialic acid recognition by the gut symbiont Ruminococcus gnavus, *Nat. Commun.*, 2017, 2196.
18. S. Narayanan and B. Reif, Characterization of chemical exchange between soluble and aggregated states of β-amyloid by solution-state NMR upon variation of salt conditions, *Biochemistry*, 2005, **44**(5), 1444–1452.
19. S. M. Ramalhete, *et al.*, Supramolecular amino acid based hydrogels: probing the contribution of additive molecules using NMR spectroscopy, *Chem. – Eur. J.*, 2017, **23**(33), 8014–8024.
20. S. Monaco, *et al.*, Differential Epitope Mapping by STD NMR Spectroscopy To Reveal the Nature of Protein–Ligand Contacts, *Angew. Chem., Int. Ed.*, 2017, **56**(48), 15289–15293.
21. S. Di Micco, *et al.*, Differential-Frequency Saturation Transfer Difference NMR Spectroscopy Allows the Detection of Different Ligand–DNA Binding Modes, *Angew. Chem., Int. Ed.*, 2006, **45**(2), 224–228.
22. P. J. Hore, *Nuclear Magnetic Resonance*, Oxford University Press, USA, 2015.
23. B. Cutting, *et al.*, Sensitivity enhancement in saturation transfer difference (STD) experiments through optimized excitation schemes, *Magn. Reson. Chem.*, 2007, **45**(9), 720–724.
24. T. Brand, E. J. Cabrita and S. Berger, Intermolecular interaction as investigated by NOE and diffusion studies, *Prog. Nucl. Magn. Reson. Spectrosc.*, 2005, **46**(4), 159–196.
25. D. Neuhaus, *Nuclear Overhauser Effect*, eMagRes, 2000.
26. M. Mayer and T. L. James, NMR-based characterization of phenothiazines as a RNA binding scaffold, *J. Am. Chem. Soc.*, 2004, **126**(13), 4453–4460.
27. J. D. Martínez, *et al.*, The Interaction of Fluorinated Glycomimetics with DC-SIGN: Multiple Binding Modes Disentangled by the Combination of NMR Methods and MD Simulations, *Pharmaceuticals*, 2020, **13**(8), 179.

28. C. Guzzi, *et al.*, Detection and quantitative analysis of two independent binding modes of a small ligand responsible for DC-SIGN clustering, *Org. Biomol. Chem.*, 2016, **14**(1), 335–344.
29. J. Angulo, *et al.*, Saturation Transfer Difference (STD) NMR Spectroscopy Characterization of Dual Binding Mode of a Mannose Disaccharide to DC-SIGN, *ChemBioChem*, 2008, **9**(14), 2225–2227.
30. I. K. Leung, *et al.*, Reporter ligand NMR screening method for 2-oxoglutarate oxygenase inhibitors, *J. Med. Chem.*, 2013, **56**(2), 547–555.
31. Y. S. Wang, D. Liu and D. F. Wyss, Competition STD NMR for the detection of high-affinity ligands and NMR-based screening, *Magn. Reson. Chem.*, 2004, **42**(6), 485–489.
32. B. Becattini, *et al.*, Structure–activity relationships by interligand NOE-based design and synthesis of antiapoptotic compounds targeting Bid, *Proc. Natl. Acad. Sci. U. S. A.*, 2006, **103**(33), 12602–12606.
33. B. Becattini and M. Pellecchia, SAR by ILOEs: An NMR-Based Approach to Reverse Chemical Genetics, *Chem. – Eur. J.*, 2006, **12**(10), 2658–2662.
34. L. E. Tailford, *et al.*, Discovery of intramolecular trans-sialidases in human gut microbiota suggests novel mechanisms of mucosal adaptation, *Nat. Commun.*, 2015, **6**, 7624.
35. E. Ulrich, *et al.*, Biological magnetic resonance data bank, *Nucleic Acids Res.*, 2007, **36**, 402–408.
36. E. Fan, *et al.*, Exploration of the GM1 receptor-binding site of heat-labile enterotoxin and cholera toxin by phenyl-ring-containing galactose derivatives, *Acta Crystallogr. Sect. D: Biol. Crystallogr.*, 2001, **57**(2), 201–212.
37. E. A. Merritt, *et al.*, Characterization and Crystal Structure of a High-Affinity Pentavalent Receptor-Binding Inhibitor for Cholera Toxin and E. c oli Heat-Labile Enterotoxin, *J. Am. Chem. Soc.*, 2002, **124**(30), 8818–8824.
38. P. Cheshev, *et al.*, Synthesis and affinity evaluation of a small library of bidentate cholera toxin ligands: towards nonhydrolyzable ganglioside mimics, *Chem. – Eur. J.*, 2010, **16**(6), 1951–1967.
39. M. Bergström, *et al.*, Cholera Toxin Inhibitors Studied with High-Performance Liquid Affinity Chromatography: A Robust Method to Evaluate Receptor–Ligand Interactions, *Chem. Biol. Drug Des.*, 2009, **73**(1), 132–141.
40. J. Ramos-Soriano, *et al.*, Synthesis, Biological Evaluation, WAC and NMR Studies of S-Galactosides and Non-Carbohydrate Ligands of Cholera Toxin Based on Polyhydroxyalkylfuroate Moieties, *Chem. – Eur. J.*, 2013, **19**(52), 17989–18003.
41. R. Nepravishta, *et al.*, Deriving Ligand Orientation in Weak Protein–Ligand Complexes by DEEP-STD NMR Spectroscopy in the Absence of Protein Chemical-Shift Assignment, *ChemBioChem*, 2019, **20**(3), 340–344.
42. N. Niccolai, *et al.*, Hot spot mapping of protein surfaces with TEMPOL: Bovine pancreatic RNase A as a model system, *Biochim. Biophys. Acta, Proteins Proteomics*, 2017, **1865**(2), 201–207.

43. I. Solomon, Relaxation processes in a system of two spins, *Phys. Rev.*, 1955, **99**(2), 559.
44. G. M. Clore and J. Iwahara, Theory, practice, and applications of paramagnetic relaxation enhancement for the characterization of transient low-population states of biological macromolecules and their complexes, *Chem. Rev.*, 2009, **109**(9), 4108–4139.
45. S. Fesik, *et al.*, Identification of solvent-exposed regions of enzyme-bound ligands by nuclear magnetic resonance, *J. Am. Chem. Soc.*, 1991, **113**(18), 7080–7081.
46. A. M. Petros, L. Mueller and K. D. Kopple, NMR identification of protein surfaces using paramagnetic probes, *Biochemistry*, 1990, **29**(43), 10041–10048.
47. E. Cerami, *et al.*, The cBio cancer genomics portal: an open platform for exploring multidimensional cancer genomics data, *Cancer Discovery*, 2012, **2**, 401. AACR.
48. S. M. Soond, *et al.*, Novel WWP2 ubiquitin ligase isoforms as potential prognostic markers and molecular targets in cancer, *Biochim. Biophys. Acta, Mol. Basis Dis.*, 2013, **1832**(12), 2127–2135.
49. X. Zou, G. Levy-Cohen and M. Blank, Molecular functions of NEDD4 E3 ubiquitin ligases in cancer, *Biochim. Biophys. Acta, Rev. Cancer*, 2015, **1856**(1), 91–106.
50. W. Gong, *et al.*, Structure of the HECT domain of human WWP2, *Acta Crystallogr. Sect. F: Struct. Biol. Commun.*, 2015, **71**(10), 1251–1257.
51. Z. Chen, *et al.*, A tunable brake for HECT ubiquitin ligases, *Mol. Cell*, 2017, **66**(3), 345–357, e6.
52. S. Mandal, M. D. Mandal and N. K. Pal, Cholera: a great global concern, *Asian Pac. J. Trop. Med.*, 2011, **4**(7), 573–580.
53. K. J. Baldauf, *et al.*, Cholera toxin B: one subunit with many pharmaceutical applications, *Toxins*, 2015, **7**(3), 974–996.
54. S. Monaco, *et al.*, Exploring Multi-Subsite Binding Pockets in Proteins: DEEP-STD NMR Fingerprinting and Molecular Dynamics Unveil a Cryptic Subsite at the GM1 Binding Pocket of Cholera Toxin B, *Chem. – Eur. J.*, 2020, **26**(44), 10024–10034.

CHAPTER 8

Structural Insight into the Slowly Exchanging Dark States at the Functional Interaction Interface

DIPITA BHATTACHARYYA,*[a]
JANARTHANAN KRISHNAMOORTHY*[b] AND
ANIRBAN BHUNIA*[a]

[a] Department of Biophysics, Bose Institute, 700054 Kolkata, India;
[b] Department of Biomedical Engineering, Jimma Institute of Technology, Jimma University, Jimma, Ethiopia
*Emails: anirbanbhunia@gmail.com; dipita15791@gmail.com; jana.jk2006@gmail.com

8.1 Introduction

Biomolecular interactions underlie all vital physiological functions concerned with the maintenance of homeostasis.[1] The precise knowledge of the three-dimensional structure of biomolecules is crucial for understanding their structure–function correlation. Recent developments in high-resolution Nuclear Magnetic Resonance (NMR) spectroscopy have enabled us to gain an insight into the three-dimensional structure of the biologically active peptides and proteins in relevance to their function.[2] Intriguingly, a significant fraction of our entire proteome exhibits intrinsically disordered regions. These highly dynamic protein segments are responsible for imparting the protein its functions—encompassing their interactome and physiological quality control.[3] However, the lack of a defined, rigid structural

New Developments in NMR No. 26
NMR Spectroscopy for Probing Functional Dynamics at Biological Interfaces
Edited by Anirban Bhunia, Hanudatta S. Atreya and Neeraj Sinha

Published by the Royal Society of Chemistry, www.rsc.org

conformation has stymied much of our understanding of the structure–function correlation for these segments. Nevertheless, intrinsically disordered proteins (IDPs) or regions (IDRs) have attracted much scientific attention over the last few years. Their conformational multiplicity and functional associativity to a wide range of biomolecules have been particularly interesting.[4,5] In fact, their involvement has been suggested to be very tightly controlled in the cellular homeostatic machinery.

Structural dynamics is essential for the protein to perform multiple functions.[6] It paves the way for the evolution of newer functionality with external environmental changes.[7,8] The numerous conformational possibilities of these disordered regions in IDPs make them prone to non-specific interactions, often leading to misfolding or diseased pathophysiology.[9–11] IDPs or IDRs have been typically implicated in diseased conditions and metabolic disorders. In fact, several inter-protein complexes defining the biological functioning or the folding/aggregation kinetics of the protein[12,13] are associated with the transient dynamic intermediate structures. Therefore, these studies have highlighted the importance of understanding the dynamics involved in protein misfolding and aggregation.

Correlating the structural disorder of proteins with their functions has been one of the major focuses in many recent scientific works.[14,15] Atomic-resolution structural insight into the protein dynamics would, thus, help in probing the interactions at the functional interface.[16–18] This would eventually help in defining newer epitopes for translation of effective targeted therapeutic intervention.[19–22] However, these meta-stable dynamic intermediates comprise only a small fraction of the system when at equilibrium,[23–26] thus rendering them "dark" or invisible to the conventional techniques used for structure determination.

8.2 Probing the Dynamicity of Functional Conformers

All proteins exhibit dynamics accompanied by structural changes at distinct timescales, which are experimentally observed through corresponding changes in the atomic coordinates.[22] However, the structural elucidations at the dynamic interfaces of the biomolecular interactions have been largely restricted, given the low population of these conformers and the associated exchange rates. The current techniques for the efficient probing of dynamic protein structures include advanced X-ray crystallography[27] (by generating electron density maps[28] and B-factors[29]) and single-particle cryo-EM techniques,[30–32] both of which are based on population-averaged sample analysis and therefore fail to provide essential insights into the dynamics of the low populated exchanging conformers.

Solution-state NMR has served as an effective tool in determining the atomic-resolution structure of biomacromolecules.[22,33,34] However, conventional NMR experiments are largely based on distance constraints and dihedral angle restraints obtained from J coupling measurements.[35] Despite the long-range orientation constraints obtained from residual dipolar couplings (RDCs) and chemical shift anisotropy (CSA) values that improve the resolution of the elucidated atomic structure,[36] these approaches still report an ensemble of the major conformer in the sample space. Moreover, the dynamic intermolecular interactions often result in high-molecular-weight (HMW) conformers at the interaction milieu, making it particularly difficult to gain NMR-based high-resolution insight into the functionally important interaction domain. Thus, owing to their transient and dynamic nature, the low population in the cellular pool and comparatively higher molecular weight these functional conformers have remained "dark" or invisible to the conventional high-resolution NMR experiments.

However, recent advancements in high-resolution NMR have made the structure of these states more realizable. NMR relaxation-based methods have been useful in probing local motions in the picoseconds–nanoseconds (ps to ns) timescales. The hetero-nuclear experiments[37] further enabled the efficient characterization of the backbone dynamics of proteins and other macromolecules based on the 'model-free formalism'.[38,39] While these experiments are efficient in probing the dynamics, they still have limitations in characterizing the minor state. Therefore, several approaches have been undertaken to extend the application of NMR further into describing slower motions on the ms to μs and μs to ns timescales—comparable to the dynamic rates of inter-molecular interaction and the molecular reorientation, respectively. In fact, paramagnetic NMR,[40] relaxation dispersion,[41] saturation transfer, lifetime line broadening, and hydrogen exchange have enabled the characterization of the low-populated, invisible states in chemical exchange with the NMR-visible species under different timescales.[42]

8.3 Relaxation in Studying Protein Dynamics

Solution-state NMR-based dynamic studies focus mainly on isotropic populations with equally probable conformational orientations—in constant random motion and exchange. The transitions between these orientations are very fast, *i.e.*, in the range of a few ps to ns. Molecules in solution essentially represent two components of motion at the atomic scale: translational and rotational. The overall rotational correlation time of a protein is often in the ns range, while for water molecules in solution, the reorientation has been observed to be in the ps timescale. This rapid rotational motion results in averaging any intramolecular orientation-dependent interactions, resulting in a series of sharp lines or peaks. However, the molecular motion at the atomic scale results in fluctuations in the local magnetic field of nuclei in a spin system on the ps to ns timescales. This can be probed as a gradual and irreversible spin system evolution during the pre-steady-state phase that contributes to the relaxation profiles before reaching equilibrium. The vertical component or the longitudinal relaxation rate—denoted as R_1 is calculated from the time constant, T_1 as $R_1 = 1/T_1$, and is used to define the distribution of the spin-up and spin-down states along the magnetic field (z-axis).[37] On the other hand, the transverse relaxation rate, R_2, is calculated from the time constant, T_2 as $R_2 = 1/T_2$, and defines the decay of magnetization along the *XY*-plane, perpendicular to the external magnetic field.[37] In fact, it is this coherent *XY* signal that is probed in the NMR experiments, which modulates the height and line width of each peak (refer to Chapter 4). The line width of NMR signals is directly proportional to R_2, with larger R_2 resulting in broader peaks and lower intensities.[43]

Many types of internal interactions result in molecular motions that contribute to different relaxation mechanisms observed in proteins and protein segments. The dominant motion in a macromolecular system is the rotational diffusion of the entire molecule/domain—quantitatively expressed as the hydrodynamic radius and its 'correlation period' (τ). In addition, specific segregated domains in proteins, such as random coil/loops or the IDRs, exhibit large-amplitude internal motion.[3,44–46] In fact, these segments have been correlated to the plasticity of the protein and conformational transitions or reorientations involved with the folding or kinetics. Some conformational transitions involve hydrogen bonds or salt bridges that define the threshold for crossing energy barriers among these states. These transitions take place on timescales ranging from hundreds of ns to seconds. In other disordered proteins, the rotational diffusion of a short segment or stretch of a few consecutive residues contributes to the transient structural reorientation.

Typically, the relaxation is significantly faster in proteins than in small molecules, and the contributions from translational motions are usually considered negligible. Translational motions are restricted in the rigid secondary structural elements of well-folded proteins, while the flexible loops undergo significant motions.[47] Thus, the conformational space exploration in disordered proteins is principally similar to translations of parts of the polypeptide chain.[48] Chemical bond vibrations are small, pm-scale translations, which are ubiquitous in all molecules. While such small motions are often innocuous to long-range interactions, they can significantly alter the effective dipolar couplings present between two nuclei involved in the chemical bond.[43] These motions are fast, on low ps to high fs timescales, and rather uniform. For instance, negligible variations of these motions will be observed between two different N–HN pairs in a protein.[43] Thus, for IDP segments, the separation of the timescales of segmental (translation and rotational) diffusion and conformational transitions, or their interplay presents itself as a challenge to the research in this field.[49,50]

The conformational transitions in a protein have attracted substantial attention over recent years, owing to their close association with the protein's functional attribute.[49] Interestingly, recent studies have been increasingly focusing on the conformational dynamics of small or large segments of the protein that contribute to a protein's function as opposed to the rigid folded domains. Over the past decades, many experimental approaches have been developed and defined for the complete characterization of ms to s timescale motions in biomolecules. This has been made possible by the quantification and control of chemical exchange contributions to relaxation.

8.4 Chemical Exchange at the Functional Interface

Most biomolecular systems are in stable dynamic equilibrium within the experimental timeframe. This includes proton exchange with the solvent for acidic or basic moieties, conversion between conformational states, and exchange between a free form and a bound-state complex. Thus, in the dynamic equilibrium state, some nuclei may experience two or more chemical environments in the course of an NMR observation due to a chemical reaction or conformational changes. During these processes, the nuclei exchange between different chemical environments leads to this phenomenon being called "chemical exchange" in the NMR literature.[51,52] The chemical shift of a nucleus, such as 'a' in Figure 8.1, is extremely sensitive to the surrounding chemical environment. Hence, when the chemical environment around it changes, the chemical shift changes to a different state, 'b' in Figure 8.1. As a solution, NMR usually records the chemical shifts of the various nuclei in a molecule, chemical exchange or inter-conversion of two functional groups with different environments can affect the resultant spectrum. Figure 8.1 represents the chemical exchange between the two sites $a \leftrightarrow b$ with varying rates of exchange—fast, intermediate, or slow.

The four main characteristics of a peak, such as the offset (frequency), line width (width at half height), amplitude (height), and the phase (absorption or dispersive), get modulated by the chemical exchange as well as relaxation (Figure 8.1, i). The dynamic line shape analysis is often used to determine the characteristics of the invisible or dark state from the known aspects of the visible states. For example, the differences in the angular frequency of the chemical shifts ($\Delta\omega_{a\leftrightarrow b}$), the apparent relaxation rates and the population of the exchanging species can be obtained through the Carr–Purcell–Meiboom–Gill pulse-train (CPMG)-relaxation dispersion experiments.[33,53] In general, any change in the chemical environment that prompts exchange, manifests in the gradual line broadening and eventual merging of the individual peaks varying with respect to the rate of chemical exchange (or inter-conversion of the different states) (Figure 8.1). Alternatively, molecular interactions underlying chemical exchange between two states can largely affect the transverse magnetic relaxation rates of distinct individual states at equilibrium through cross-relaxation (Figure 8.1, i). The appearance of the spectrum, however, depends on the individual populations, P_a, P_b corresponding to the states, 'a' and 'b', respectively. If one of the states were more populated in comparison to the other, then the lowly populated species forms the minor fraction, and often its peak remains invisible due to low amplitude, in addition to chemical exchange and relaxation dependent line broadening. Figure 8.1, ii, exemplifies the effect of varying the population of the exchanging species by changing their relative molar ratio $P_a : P_b$ to 75 : 25 or 50 : 50. Additionally, the effect of the rate of exchange on the peaks of the visible species is presented row-wise in Figure 8.1, ii. The sensitivity of the peak characteristics to an underlying chemical exchange makes NMR a very powerful method for studying the kinetics and dynamics of the molecules in solution.

When the system undergoes a fast chemical exchange, the rate of the chemical exchange, k_{ex}, the sum of the effective first-order forward and backward rates for the reaction: $k_{ex} = k_{a \to b} + k_{b \to a}$, is much greater than the difference in chemical shift, $k_{ex} \gg \Delta\omega_{a \to b}$. This results in a single signal appearing at the population-weighted average chemical shift (Figure 8.1, ii, bottom row). However, when k_{ex} is comparable to the difference in the chemical shift between the two sites measured, $k_{ex} \simeq \Delta\omega_{a \to b}$, then the two peaks merge or coalesce to form an extremely broad signal (Figure 8.1, ii, middle row). In the case of slow exchange regimes, where, k_{ex} is much lower than the chemical shift difference between the two sites, *i.e.*, $k_{ex} \ll \Delta\omega_{a \to b}$, the individual signals corresponding to the two chemical environments are

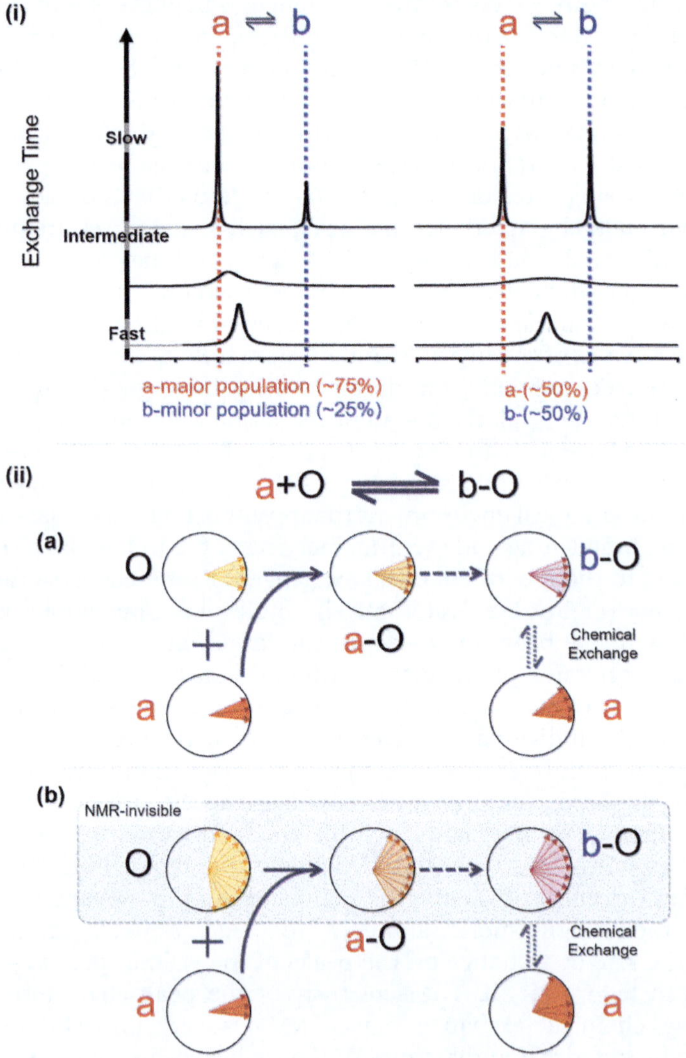

visible (Figure 8.1, ii, top row). Such conditions, however, are comparable to sample conditions where any inter-conversion/exchange is non-existent and the NMR signal is obtained as sharp peaks corresponding to the chemically distinct nuclear resonances. Thus, a direct correlation of the apparent signals with the actual chemical exchange regime is neither obvious nor easily distinguishable. In fact, a single peak in an exchanging system is not a certain indication of the presence of fast exchange, as the peak of a minor state in the slow exchange timescales may be broadened beyond detection. Such weakly populated excited states in slow exchange are often called the

Figure 8.1 Chemical exchange regimes between two states, a and b. (i) The chemical exchange between states "a" and "b" can be categorized into slow, intermediate, and fast. The representative simulation peaks demonstrate the chemical exchange profiles between "a" and "b" with varying populations of the two species seen column-wise (left: 75 : 25; right: 50 : 50) and with varying exchange rates seen row-wise (top: slow; middle: intermediate; bottom: fast). Differences in the corresponding peak intensities are observed as a function of the population (molar fraction) of each species. In the slow exchange regime (top row), states "a" and "b" give rise to two separate peaks. A low signal/noise ratio is observed for state "b" (when it forms the minor fraction ~25%; top row, left column), resulting in a peak with low intensity compared to "a" (with the major population of ~75%; top row, left column). However, comparable intensities are obtained for equimolar concentrations (top row, right column). In the intermediate exchange regime, the peaks undergo extreme chemical shift broadening (middle row). A single peak is observed in a fast exchange regime at the population-averaged chemical shift value (bottom row). (ii) Vector model demonstrating the effect of exchange on the magnetization in the transverse plane, resulting in peak broadening. Considering an intermediate to slow exchange regime, the monomeric conformer of the protein in state "a" (represented in coral red) interacts with an oligomeric conformer, O (represented in yellow chrome), resulting in a transient intermediate complex, a-O (represented in brown). This eventually forms the state b in association with the oligomer, *i.e.*, b-O (represented in magenta). Because of the chemical exchange between "a" and "b," both these states are in a dynamic equilibrium. The spanning of the arrows with solid color-fill represents the transverse magnetization after time "t" following the initial excitation. The larger the span, the greater the observed apparent transverse relaxation rate (R_2^*) and *vice versa*. Two cases of the chemical exchange between states "a" and O are considered: (a) where the oligomeric conformer (O) is comparable in molecular weight (*e.g.*, dimer, trimer, *etc.*) to that of both the exchanging conformers, "a" and "b." Meanwhile in (b), the oligomeric conformer (O) has a higher molecular weight and larger transverse relaxation rates (referred to here as the NMR-invisible or dark species). In case (a), the relaxation rates of both the a-O and b-O are not significantly different from that of species "a." Hence, the indirect effect of the relaxation of "b" on the line width of "a," mediated through chemical exchange, is less prominent. On the other hand, in case (b), the relaxation rates of both the a-O and b-O are significantly higher than that of the species "a." Hence, the broadening effect of the relaxation of "b" upon the line width of "a" is quite significant.

"invisible states". To further complicate matters, the transverse relaxation of the individual states prompt rapid line broadening to the point beyond detection when it exchanges with higher molecular weight conformers.

Thus efforts have been in full swing over the last few years to gain better insight into the quantification of the chemical exchange process, the kinetic rate, and the thermodynamic parameters, along with the atomic-resolution structural insight into the interface of the molecular binding process.

In the case of a two-site a$\leftrightarrow$b fast exchange, the contribution of chemical exchange to relaxation, R_{ex}, can be calculated using the following equation:

$$R_{ex} = \frac{p_a p_b \Delta\omega^2_{a\leftrightarrow b}}{k_{a\rightarrow b} + k_{b\rightarrow a}} \tag{8.1}$$

where p_a and p_b are the populations of states a and b; $k_{a\rightarrow b}$ and $k_{b\rightarrow a}$, are the effective rates of the first-order exchange and their sum, $k_{ex} = k_{a\rightarrow b} + k_{b\rightarrow a}$, quantifies the number of stochastic events per unit time (HZ or s^{-1}). $\Delta\omega_{a\leftrightarrow b}$ is the difference in the chemical shift in angular frequency; and is linearly proportional to the external magnetic field strength (B_0). Increasing the strength of B_0 results in a decrease in the relative timescale of exchange. This equation provides quantitative insight into the kinetic parameters and chemical structures of the two systems in exchange.[22] Due to its simplicity, this two-state model is often effectively employed to interpret and analyze the solution-NMR spectrum. However, more complex dynamics have been simulated using the Bloch–McConnell equations.[54]

In a sample undergoing slow exchange, *i.e.*, $k_{ex} \ll \Delta\omega_{a\rightarrow b}$, distinct peaks corresponding to the chemical shift of either state are observed in the absence of significant molecular exchange during the experimental timeframe (Figure 8.1, ii, top row). The individual peaks represent the total population (peak intensity) of a particular conformer with distinct chemical shift values. The line width (λ), on the other hand, is the full peak width at half maximum height, reports on the dynamic properties of the molecule in solution, including ~ns molecular rotation and ps–ms internal motion. λ is directly related to the transverse relaxation rate, R_2 (Hz) = $1/T_2$ (s) = $\lambda/2$ (Hz), and hence λ, R_2, and T_2 are used interchangeably to interpret an NMR spectrum (refer to Chapter 4). For example, if the exchange rate of the state, k_{ex}, is greater than ~10 Hz, the signal line width will increase due to the depopulation of the state during a period of 1/10 Hz = 100 ms of the detection period. This process directly contributes to the exponential decay of the free induction decay (FID) intensity and increases the observed transverse relaxation rate *via* $R_2^{obs} = R_2^0 + k_{ex}$, where R_2^0 is the relaxation rate in the absence of any exchange.[22]

In such a scenario, despite the two distinct chemical environments, one of the two peaks remains "invisible" or "dark" to the direct probing techniques. The line broadening effect beyond the point of detection due to the difference in resonance frequencies has already been reported for chemical exchanges occurring in the intermediate timescale (100 μs–10 ms), allowing detection by relaxation dispersion.[13,55] However, for the slower exchanging regimes (1 ms–1 s), the source of any increase in the relaxation rate beyond detection can vary. The major limiting cause can be the molecular weight enhancement—prompting rapid dissipation of the magnetization in the *XY* plane. This, in corroboration with the low population of one of the states, often contributes to the challenges of probing these "dark" states.

Nevertheless, various experimental strategies have been undertaken to characterize and define the chemical exchange. In fact, several of these have been extensively studied for decades, including paramagnetic relaxation enhancement (PRE),[56] CPMG,[57,58] saturation transfer,[59] *etc.* However, the recent breakthrough in methodology and technical aspects of experiments has extended the capability to characterize the "dark" states.

8.5 Characterizing Chemical Exchange Between an NMR-visible and Invisible State

The NMR-based relaxation studies serve as an excellent probe for the atomic-level characterization of biomolecular dynamics. Detection and characterization of an otherwise NMR-invisible species have been made possible by experimentally amplifying NMR observables through the chemical exchange. Interestingly, chemical exchange with the "dark" state often results in an overall increase in the apparent R_2 of the visible state manifested *via* line broadening (Figure 8.1, i). This effect indirectly enables the detection and characterization of the otherwise invisible state. In many cases, the minor state is detected through an experimental enhancement of spin relaxation. The chemical exchange between NMR-visible, low-molecular-weight (LMW), fast-tumbling conformers with known resonances and the sparsely populated or slowly tumbling HMW "dark" conformers present a unique situation where the chemical shifts and the apparent R_2 (Line width, $\text{LW} = R_2\pi = (1/T_2)\pi$), are well defined for these two states. This difference in the magnetic environment of the two exchanging species creates differential profiles mediated by spin diffusion. Resonances in the "dark" state can be selectively saturated by applying off-resonance radio frequencies (RF) that are sufficiently different from the NMR-observable state. The resonances of less populated, HMW-invisible or "dark" states exhibit significant line broadening. The application of off-resonance RF saturates the "dark" species while the NMR-observable resonances remain essentially unperturbed. This saturation can then be transferred to the visible species by chemical exchange. Based on this principle, Saturation Transfer Difference (STD) experiments are an excellent tool for detecting chemical exchange.

8.6 Basic Principle of Applying Saturation Transfer: Characterizing the "Dark" Species

The main parameter in an STD NMR experiment is the saturation time (t_{sat}) in which intermolecular transfer of magnetization occurs from the saturated large protein to the small interacting ligand.[60,61] The degree of saturation is dependent mainly on the "residence time" (τ_{res}) of the NMR-observable species in the bound-state with the saturated macromolecule (as explained already in Chapter 7). Despite the immense possibilities of STD experiments in studying biomolecular interaction interfaces, its application in the characterization of self-oligomerization events associated with amyloids has been largely affected by the non-selective cross-relaxation that can occur due to the high concentration of the monomers in the sample space. Furthermore, the dynamic nature of the exchanging conformers limits the efficient diffusion of the saturation, as a fraction of oligomers reverts to the monomeric pool before being fully saturated. Nevertheless, introducing a range of different RF saturation frequencies for the sample, spanning the methyl and aromatic regions of the spectra, can diminish much of the spin-diffusion of the saturation in the oligomers.[62] The use of different filtration states (physically separated/purified) also helps remove any contributions from the monomers. The normalization of the average STD values for the different RF frequencies with respect to the highest value recorded serves as a critical parameter for defining the self-aggregation prone sites.[62]

The principles of saturation transfer have been used in several biomolecular applications, with a recent one being used for probing the "dark" states that are chemically exchanging. Chemical Exchange in Saturation Transfer (CEST) and the "Dark-state Exchange Saturation Transfer" (DEST) are two such experiments used to study the "dark" or invisible states. Both experiments use off-resonance RF to saturate the ^{15}N nucleus, incorporated into a 2D hetero-nuclear coherence experiment—providing residue-specific information of the protein's backbone. Both these experimental strategies rely largely on the decay of RF-induced NMR spin polarization. In fact, the use of relaxation-based experiments in combination with STD NMR was employed for studying oligomerization events in the early 2000s. In 2005, Melacini *et al.* proposed off-resonance relaxation (ORR) as an efficient probe to map the epitope of self-aggregation.[63] The introduction of an adiabatic off-resonance spin-lock prior to the 90° pulse allows determination of the decay rate along the effective field tilted by an angle $\theta = 35.5°$.[63] Generally, the relaxation rates, R_1 or R_2, are modeled as a sum of "direct or self" relaxation component (originating from the spin "*i*" itself) and "cross-relaxation" component (originating from dipolar interaction with the adjacent spin "*j*"). The study showed that the non-selective relaxation rate ($R_{\theta,\mathrm{ns}}$) of a spin *i*, in the presence of a spinlock field tilted to 35.5°, was predominately contributed by the self or direct relaxation[64] component (ρ_i) with negligible contributions from the cross relaxation component ($\sigma_{i,j}$). The hard 90° pulse prior to off-resonance spin lock excites not only the spin "*i*" but also its neighboring spins such as "*j*," "*k*" *etc.*, which are correlated through dipolar coupling—thus allowing the measurement of the "non-selective" relaxation rates, as opposed to the "selective" experiments, where only the spin "*i*" would be excited through a selective pulse at its offset value and the relaxation rate measured.

The measured off-resonance relaxation rate, $R_{35.5,\mathrm{ns}}$, is thus directly correlated to the molecular weight of the species containing the spin, *i*, being interrogated. The angle was thus chosen because of the fact that the effect of cross-relaxation between the spins is removed efficiently and the "self" relaxation component of the R_2 contributes predominately compared to the non-selective effect of $R_{1,\mathrm{ns}}$. Furthermore, the spin-lock relaxation block helped to obliterate any R_2 associated artifacts (*J*-coupling effect), thus making it an excellent method to probe the residue-specific susceptibilities of self-aggregating proteins.[63] In a subsequent paper, the efficient collaboration between STD and the ORR experimental strategies was shown to characterize the segment-specific inhibitory mechanisms of HSA against Aβ amyloids, reinstating the possibilities of extending the principle into studying amyloidogenesis.[65]

DEST and CEST NMR experiments exploit the principle of RF irradiation of the exchanging nucleus to probe the amide–proton exchange rates in dynamic protein samples at equilibrium. Essentially similar, both these experiments employ a series of weak offset RF, off-resonance from the major state of the monomeric conformers. The CEST experiment was initially designed to improve magnetic resonance imagining (MRI) by reducing the need to include high concentrations of the contrast agents. Instead, RF irradiation of the low-concentration exchangeable solute protons helped magnify the visualization in the subsequent imaging. The basic principle involved a prolonged period of saturation time in comparison to the rate of exchange. Thus, the signals from the invisible fraction of metabolites were significantly amplified and, through chemical exchange, saturated the bulk water molecules that were being probed.[66] Extending the application of studying the exchanging protein systems in biomolecular NMR, the CEST pulse program is used to effectively characterize an exchange with a lifetime of about 2–100 ms between a minor and a major state, with distinct chemical shifts but comparable rotational correlation time.[67] Thus, CEST relies on the chemical shift differences between the visible major state and the lowly populated minor state. DEST, on the other hand, has been effective in probing intermediate-to-slow chemical exchange (500 μs^{-1}) between two states that have significant differences in their transverse relaxation rates, majorly owing to the formation of higher-molecular-weight aggregates or conformers in the process.[68–70] Given the fundamental difference between the two experiments, CEST employs weak B_1 fields (~5–50 Hz) at a specific offset from the resonances of the major state in the visible spectrum, while DEST employs a saturation field far-off from the monomeric resonance frequencies (~150–350 Hz). Both these experiments yield a plot of signal intensity *versus* the chemical shift offset of the B_1 saturating field for each resonance being monitored. These profiles can then be fitted to a model (based on the McConnell equations) to quantify the population and rates of exchange between the different conformers. In addition to the kinetic parameters, the DEST experiment also provides information on R_2 of the resonances in the dark state.[69,71]

8.7 Dark-state Exchange Saturation Transfer: The Standing Principle

The DEST pulse program employs weak radio frequencies, far off-resonance to the frequencies of the rapidly tumbling NMR-observables with narrow signals in the absence of the larger dark state. In the event of chemical exchange with the slowly tumbling "dark" species, with greater R_2, the RF results in their saturation (Figure 8.2). Suppose the invisible species converts to the visible species at a rate that is of the order of or faster than its longitudinal relaxation rate. In that case, saturation will be transferred to the NMR-visible species and observed as a loss of intensity of cross-peaks for the visible species (Figure 8.2). Thus, it becomes possible to qualitatively detect and characterize large aggregates and macromolecular assemblies that would otherwise be invisible. Alternatively, the hetero-nuclei-based DEST experiments eradicate the practical complications arising from spin diffusion, thus making it an indispensable technique for the quantitative characterization of chemical exchange.

Typically, DEST is possible when there is a slow exchange between the reference sample, containing only the NMR-observable, and an additional sample containing an NMR-invisible "dark" state whose chemical shifts are identical or comparable to those of the visible state but with substantial differences in their R_2 values. By convention, the observation of measurable $\Delta R_2 > 1\ s^{-1}$, between two samples (where species a alone exists, and where both a + b, exist) serves as a valuable indicator for the efficient applicability of DEST for the exchanging system. For Aβ40 peptides, it was experimentally shown that the differences in the R_2 values measured at two different concentrations (300–60 μM or 150–60 μM), *i.e.*, ΔR_2 values, were independent of both the magnetic field (600 MHz and 900 MHz) and the nucleus (^{15}N or ^{1}H).[70] Thus, the increase in R_2 is assumed to be independent of the chemical shift timescale (line broadening observed for intermediate rates of exchange) and results from chemical interaction with a larger aggregating or slowly tumbling specimen with substantially high transverse relaxation rates (Figure 8.1, i). Thus, alternatively, the maximum increase in the apparent R_2 is a direct indication of the apparent first-order rate constant, k_{app}^{on} for the conversion of the NMR-visible to the NMR-invisible species.

8.8 Description of the 2D ^{15}N-DEST Experiment and Pulse Sequence

Figure 8.3 represents the pulse sequence proposed by Fawzi *et al.*[69,71] for the ^{1}H–^{15}N HSQC-based DEST experiment to study the molecular aggregation of proteins underlied by chemical exchange: (NMR-observable) monomers ↔ oligomers ↔ fibrils (invisible). The pulse-sequence efficiently probes the resonances of the NMR-observable monomeric resonances. Broadly, the sequence employs INEPT transfer of magnetization involving the ^{1}H–^{15}N

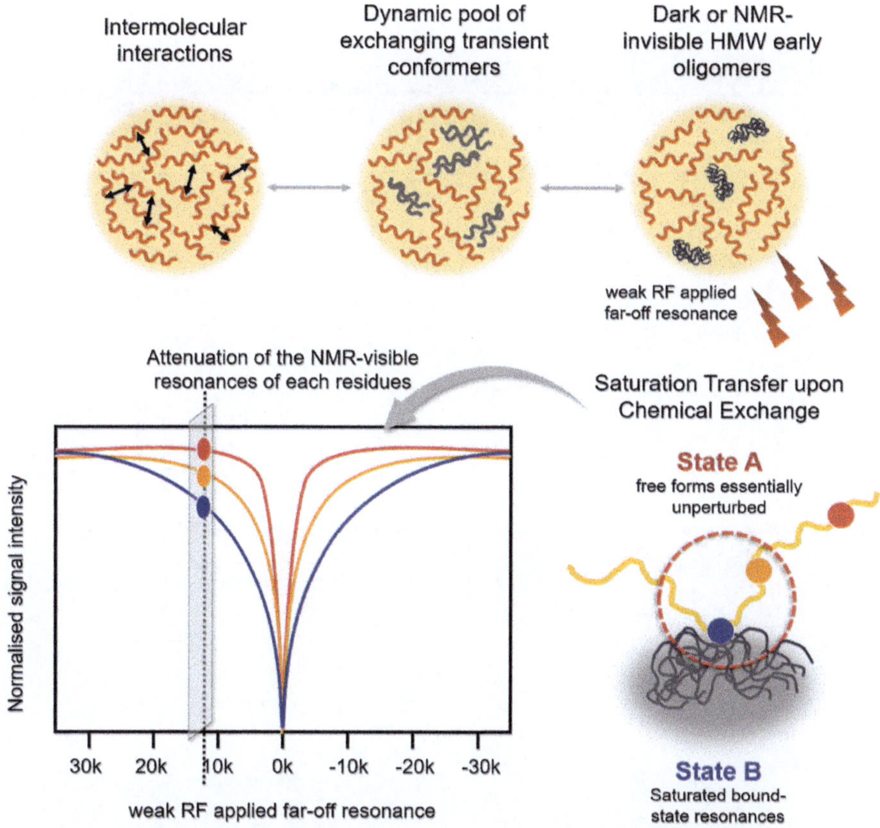

Figure 8.2 DEST NMR, the standing principle. (i) Sample condition promoting increased intermolecular interactions, resulting in the dynamic pool of exchanging transient conformers. (ii) The experiment: Applying a weak RF field to the sample in (i) applied far off-resonance leaves resonances of the free form of the protein essentially unperturbed. However, the bound state resonances are selectively saturated as the small transverse magnetization created by an RF field is efficiently relaxed by the large R_2 in the bound state. The saturation of the bound resonances is transferred back to the resonances of the free species by chemical exchange and subsequently measured as attenuation of the easily observable resonances of the NMR-visible species. The residue-specific attenuation profile—as a function of the offset frequencies—enables probing of the residue-specific dynamics in the bound state. The representative normalized signal intensities as a function of the applied RF offsets (left) for three different residues (color-coded: red, green, blue), depending on the proximity to the dark species (correlate to the colored dots on the right, *i.e.*, in-contact: blue; tethered: green; free: red). The spatial molecular proximity can be correlated to the attenuations, with greater attenuation observed for the residues closer (blue) to the dark specimen for a given off-resonance saturation offset value. The residue least perturbed by the chemical exchange results in minimal attenuation upon RF application (red).

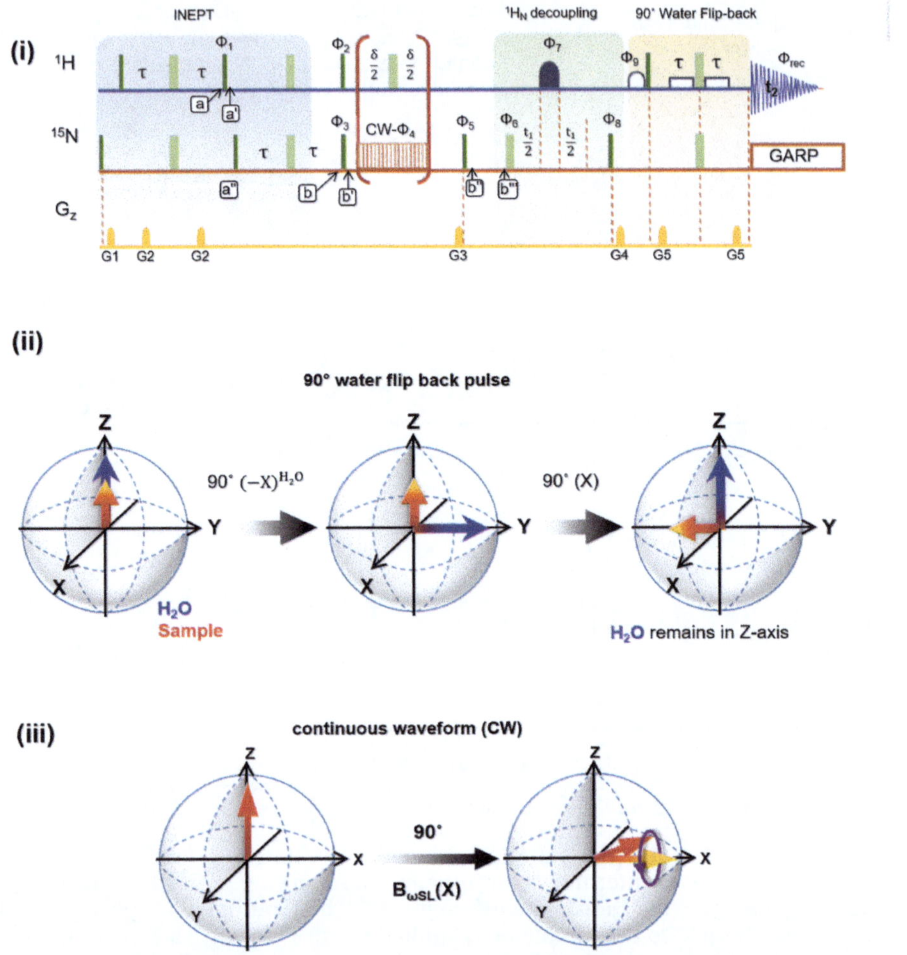

Figure 8.3 The DEST pulse program. (i) The pulse scheme, as described by Fawzi *et al.*,[71] showing the major blocks to delineate the basic segments of the experimental design. (ii) The vector representation delineating the water flip-back pulse to remove the residual effect of the water signal in the DEST profile. (iii) The vector representation of the CW pulse sequence, an electromagnetic wave of constant amplitude and frequency, almost always a sine wave, is considered to be of infinite duration for mathematical analysis. Once the magnetization is in the *XY* plane, the CW proportional to the spin-lock system holds it there and keeps it precessing around a new magnetic field aligned with *X* or *Y*.

resonances and a weak continuous wave (CW) saturation pulse at variable RF offsets.

As demonstrated in Figure 8.3i, the initial magnetization of the amide proton (H_z) is converted to its anti-phase form, $2H_xN_z$ at (a), $-2H_zN_z$ at (a′), and $2H_zN_y$ at (a″) at the end of the first INEPT segment. The subsequent refocusing set of pulses generates the in-phase magnetization of ^{15}N, $-N_x$ at (b), and N_z at (b′). This magnetization is converted to $-N_y$ at (b″), which evolves under the influence of the *J*-couplings to generate the anti-phase, $2H_zN_x$ at (b‴). In the following segment, the frequency is encoded during the "t_1" delay. Subsequently, the reverse INEPT segment transfers back the detectable magnetization from ^{15}N to ^{1}H and gets converted to an in-phase signal.

Notably, the 90° ^{1}H pulse-ϕ_2 applied at (b) helps in flipping back the water signal to the $+z$ axis, also eliminating the residual anti-phase magnetization, $2H_zN_y$, which can introduce artifacts in the spectrum (Figure 8.3, ii). A weak CW saturation pulse is then applied onto the ^{15}N nuclei (b′) for a period of ~0.9 s – a realizable timeframe in which the chemical exchange between the NMR-observable and the "dark" state can take place (Figure 8.3, iii). The offset frequency of the CW pulse, with reference to the ^{15}N carrier frequency (~119 ppm), is varied between +35 to −35 kHz and separate experiments are recorded. The experiments are also repeated by varying the RF field strengths, B_1 (generally 170 or 350 Hz) of saturation pulses. The offset of the CW pulse is set such that it remains far away from the offset of the monomer resonances/peaks; hence, only the excitation of spins of the "dark" state resonances occurs. The extremely high R_2 values of these "dark" states contribute to the efficient and uniform magnetization transfer among all its spins through the spin diffusion mechanism and allow for partial or complete saturation.[72] In the event of chemical exchange, the transfer of the saturation occurs between the dark state to the interacting NMR-observable monomer (Figure 8.2), manifesting as a reduction of its N_z magnetization by a multiplicative factor, *k*. In contrast, a reference experiment is performed without any saturation pulse, where the NMR-observable resonance remains essentially unperturbed. Thus, the reduction in intensity is directly correlated with the efficiency of the saturation transfer resulting from the effect of chemical exchange and the CW pulse applied at a particular offset frequency and RF strength. The modulated signal provides essential information on the rate and kinetics of the exchange process occurring within the experimental timeframe of the applied saturation pulse. Upon the dissociation of the NMR-observable monomer from the bound conformation with the "dark" species, the reduced magnetization, kN_z, is detected from the heteronuclear single quantum coherence (HSQC) spectra.

The phases of the 90 ^{1}H pulse, ϕ_1, and the receiver phase are varied so that the N_z magnetizations at both the $+z$ and $-z$ axes are obtained. Sometimes, the non-zero residual magnetization present at the outset of the experiment (non-Boltzmann equilibrium) also contributes to the intensity and modulates the signal in a complex manner. Hence, summation of the two transients, *i.e.*, N_z and N_{-z} at b′, ensures that such non-exponential contributions to the longitudinal relaxation are eliminated through phase cycling. The resulting signal at the end of the CW saturation pulse has contribution only from R_1 relaxation and is represented as a simple scaling factor, e^{-R_1t} pre-multiplied to (N_z+N_{-z}). The R_1 dependence can then be effectively removed from the DEST effect when the cross-peak intensities are normalized using the intensities from the reference spectrum recorded without saturation. Additionally, the interference between the two relaxation mechanisms, such as the ^{1}H–^{15}N dipolar interaction and the ^{15}N chemical shift anisotropy, is eliminated by the application of the 180° ^{1}H pulse train on the proton channel during the CW ^{15}N saturation pulses (Figure 8.3). The WATERGATE scheme is incorporated during the refocusing pulses of the reverse INEPT segment to selectively suppress the residual water signal prior to detecting the monomeric peptide amide proton mag netization (Figure 8.3, iii).

8.9 Quantifying the Chemical Exchange

The ^{15}N-DEST profiles obtained through the HSQC-based experiment contain useful information on the interaction dynamics involving an NMR invisible at the slow-exchange regime. A complete line shape analysis of the individual peaks can provide quantitative dynamic information about the system under consideration. The dynamic line shape analysis requires an assumption for the kinetic chemical exchange equilibrium to formulate the Bloch–McConnell equations. Mathematically, the Bloch–McConnell equations represent an inhomogeneous system of first-order differential equations for a system undergoing a chemical exchange process. The chemical exchange present among a set of nuclei in a magnetic field can be simulated by incorporating its kinetics into the Bloch equations.[73] The resultant Bloch–McConnell equations serve as the classical description of the chemical kinetic equilibrium in the magnetic field.

To further clarify, let us consider a simple spin system undergoing a chemical exchange between two states as:

$$\mathrm{a} \underset{k_{\mathrm{ba}}}{\overset{k_{\mathrm{ab}}}{\rightleftharpoons}} \mathrm{b} \tag{8.2}$$

where a and b are the two conformational states of the nuclei present in a molecule and k_{ab} and k_{ba} are the forward and reverse kinetic rates, respectively, for the conformational exchange. The Bloch equation for such a two-state system in a rotating frame of reference is written as:

$$\frac{\mathrm{d}M^{\mathrm{a,b}}}{\mathrm{d}t} = \left(M^{\mathrm{a,b}} \times \gamma B^{\mathrm{a,b}}\right) - R^{\mathrm{a,b}}\left(M^{\mathrm{a,b}} - M_0^{\mathrm{a,b}}\right) \tag{8.3}$$

where $M^{a,b}$ represents the time-dependent components of the magnetization along the x, y, z, directions for the species, a or b. Similarly, $M_0^{a,b}$ represents the initial magnetization present along the x, y, z, directions for either a or b. The relaxation matrix for a or b is given by $R^{a,b}$. In the above expression, $M^{a,b}{\times}\gamma B^{a,b}$ represents the cross-product between the magnetization vector, $M^{a,b}$ and the magnetic-field vector, $B^{a,b}$, multiplied with the gyromagnetic ratio (γ). Thus, the Bloch equation for species a alone can be explicitly written[73] as:

$$\underbrace{\begin{bmatrix} \frac{dM_x^a}{dt} \\ \frac{dM_y^a}{dt} \\ \frac{dM_z^a}{dt} \end{bmatrix}}_{M^{a'}} = \underbrace{\begin{vmatrix} \vec{i} & \vec{j} & \vec{k} \\ M_x^a & M_y^a & M_z^a \\ \gamma B_x^a & \gamma B_y^a & \gamma B_z^a \end{vmatrix}}_{(M^a \times B^a)} - \underbrace{\begin{bmatrix} R_2^a & 0 & 0 \\ 0 & R_2^a & 0 \\ 0 & 0 & R_1^a \end{bmatrix}}_{R^a} \left(\underbrace{\begin{bmatrix} M_x^a \\ M_y^a \\ M_z^a \end{bmatrix}}_{M^a} - \underbrace{\begin{bmatrix} 0 \\ 0 \\ M_{z_0}^a \end{bmatrix}}_{M_0^a} \right) \tag{8.4}$$

After expanding the cross-product ($M^a \times B^a$) and rewriting the resultant coefficients of $\vec{i}$, $\vec{j}$, and $\vec{k}$, (the unit vectors along, x, y, z, directions, respectively) along a column vector, we obtain:

$$\underbrace{\begin{bmatrix} \frac{dM_x^a}{dt} \\ \frac{dM_y^a}{dt} \\ \frac{dM_z^a}{dt} \end{bmatrix}}_{M^{a'}} = \underbrace{\begin{bmatrix} -R_2^a & 0 & 0 \\ 0 & -R_2^a & 0 \\ 0 & 0 & -R_1^a \end{bmatrix}}_{(R^a)} + \gamma \underbrace{\begin{bmatrix} 0 & +B_z^a & -B_y^a \\ -B_z^a & 0 & +B_x^a \\ +B_y^a & -B_x^a & 0 \end{bmatrix}}_{L} \left(\underbrace{\begin{bmatrix} M_x^a \\ M_y^a \\ M_z^a \end{bmatrix}}_{M^a} + \begin{bmatrix} 0 \\ 0 \\ R_1^a * M_{z_0}^a \end{bmatrix} \right) \tag{8.5}$$

In the above equation, the relaxation and the frequency components (the classical equivalent of the Liouville matrix) are resolved and written explicitly as R^a and L^a. The magnetic field B_x^a, B_y^a, B_z^a, multiplied with the γ, represent the angular frequencies, ω_x^a, ω_y^a, ω_z^a. Since the RF pulses are applied along either the x or y directions, the field strengths of these pulses are represented through ω_x^a or ω_y^a—specific for the species a, in eqn (8.5). On the other hand, the "offset" frequency (Ω^a) of the precessing nuclei's frequency (ω_i^a with respect to the rotating frame (ω_{ref}) is represented through $\omega_z^a = \Omega^a = (\omega_i^a - \omega_{ref})$. The combined Bloch equation for both the species *a* and *b*, can, therefore, be written as:

$$\begin{bmatrix} M_x^{a'} \\ M_y^{a'} \\ M_z^{a'} \\ M_x^{b'} \\ M_y^{b'} \\ M_z^{b'} \end{bmatrix} = (\bar{R} + \bar{L}) \begin{bmatrix} M_x^a \\ M_y^a \\ M_z^a \\ M_x^b \\ M_y^b \\ M_z^b \end{bmatrix} + \begin{bmatrix} 0 \\ 0 \\ M_{z_0}^a \\ 0 \\ 0 \\ M_{z_0}^b \end{bmatrix} \tag{8.6}$$

The composite relaxation ($\bar{R}$) and Liouville matrix ($\bar{L}$) are independently constructed by taking the Kronecker or tensor products ($\otimes$) between the composite matrix (C_a or C_b) and the relaxation or Liouvillian matrix specific to each species. For example, the relaxation matrix, $\bar{R}=\bar{R}_a+\bar{R}_b$, is constructed as follows:

$$\bar{R}=\underbrace{\underbrace{\begin{bmatrix}1 & 0\\ 0 & 0\end{bmatrix}}_{C_a}\otimes\begin{bmatrix}R_2^a & 0 & 0\\ 0 & R_2^a & 0\\ 0 & 0 & R_1^a\end{bmatrix}}_{\bar{R}_a}+\underbrace{\begin{bmatrix}0 & 0\\ 0 & 1\end{bmatrix}}_{C_b}\otimes\underbrace{\begin{bmatrix}R_2^b & 0 & 0\\ 0 & R_2^b & 0\\ 0 & 0 & R_1^b\end{bmatrix}}_{\bar{R}_b} \tag{8.7}$$

On expanding the tensor product, we have,

$$\bar{R}=\begin{bmatrix}1* & \begin{bmatrix}R_2^a & 0 & 0\\ 0 & R_2^a & 0\\ 0 & 0 & R_1^a\end{bmatrix} & 0* & \begin{bmatrix}R_2^a & 0 & 0\\ 0 & R_2^a & 0\\ 0 & 0 & R_1^a\end{bmatrix}\\ 0* & \begin{bmatrix}R_2^a & 0 & 0\\ 0 & R_2^a & 0\\ 0 & 0 & R_1^a\end{bmatrix} & 0* & \begin{bmatrix}R_2^a & 0 & 0\\ 0 & R_2^a & 0\\ 0 & 0 & R_1^a\end{bmatrix}\end{bmatrix}+\begin{bmatrix}0* & \begin{bmatrix}R_2^b & 0 & 0\\ 0 & R_2^b & 0\\ 0 & 0 & R_1^b\end{bmatrix} & 0* & \begin{bmatrix}R_2^b & 0 & 0\\ 0 & R_2^b & 0\\ 0 & 0 & R_1^b\end{bmatrix}\\ 0* & \begin{bmatrix}R_2^b & 0 & 0\\ 0 & R_2^b & 0\\ 0 & 0 & R_1^b\end{bmatrix} & 1* & \begin{bmatrix}R_2^b & 0 & 0\\ 0 & R_2^b & 0\\ 0 & 0 & R_1^b\end{bmatrix}\end{bmatrix} \tag{8.8}$$

Thus, the composite relaxation matrix is given by,

$$\begin{bmatrix}R_2^a & 0 & 0 & 0 & 0 & 0\\ 0 & R_2^a & 0 & 0 & 0 & 1\\ 0 & 0 & R_1^a & 0 & 0 & 0\\ 0 & 0 & 0 & R_2^b & 0 & 1\\ 0 & 0 & 0 & 0 & R_2^b & 0\\ 0 & 0 & 0 & 0 & 0 & R_1^b\end{bmatrix} \tag{8.9}$$

The Liouville matrix ($\bar{L}$) is also constructed similarly. Finally, to account for the chemical exchange between the two species, a, and b, we start with the kinetic rate equation for each species as:

$$\begin{bmatrix}\dfrac{da}{dt}\\ \dfrac{db}{dt}\end{bmatrix}=\underbrace{\begin{bmatrix}-k_{ab} & +k_{ba}\\ +k_{ab} & -k_{ba}\end{bmatrix}}_{K}\begin{bmatrix}a\\ b\end{bmatrix} \tag{8.10}$$

To write the composite kinetic matrix, $\bar{K}$, we assume that the chemical exchange exists only between the following magnetizations: $M_x^a \underset{k_{ba}}{\overset{k_{ab}}{\rightleftharpoons}} M_x^b$, $M_y^a \underset{k_{ba}}{\overset{k_{ab}}{\rightleftharpoons}} M_y^b$ and $M_z^a \underset{k_{ba}}{\overset{k_{ab}}{\rightleftharpoons}} M_z^b$. The required $\bar{K}$ matrix can, therefore, be expressed as:

$$\bar{K} = K \otimes \begin{bmatrix} 1 & 0 & 0 \\ 0 & 1 & 0 \\ 0 & 0 & 1 \end{bmatrix} = \begin{bmatrix} -k_{ab} & 0 & 0 & +k_{ba} & 0 & 0 \\ 0 & -k_{ab} & 0 & 0 & +k_{ba} & 0 \\ 0 & 0 & -k_{ab} & 0 & 0 & +k_{ba} \\ +k_{ab} & 0 & 0 & -k_{ba} & 0 & 0 \\ 0 & +k_{ab} & 0 & 0 & -k_{ba} & 0 \\ 0 & 0 & 0 & 0 & 0 & -k_{ba} \end{bmatrix} \quad (8.11)$$

Thus, for a two-state mechanism, a↔b considering the relaxation and chemical exchange, the final Bloch–McConnell equation can be written as:

$$\begin{bmatrix} M_x^{a'} \\ M_y^{a'} \\ M_z^{a'} \\ M_x^{b'} \\ M_y^{b'} \\ M_z^{b'} \end{bmatrix} = \left(\underbrace{\begin{bmatrix} R_2^a & 0 & 0 & 0 & 0 & 0 \\ 0 & R_2^a & 0 & 0 & 0 & 0 \\ 0 & 0 & R_1^a & 0 & 0 & 0 \\ 0 & 0 & 0 & R_2^b & 0 & 0 \\ 0 & 0 & 0 & 0 & R_2^b & 0 \\ 0 & 0 & 0 & 0 & 0 & R_1^b \end{bmatrix}}_{\bar{R}} \underbrace{\begin{bmatrix} 0 & +\Omega^a & -\omega_y^a & 0 & 0 & 0 \\ -\Omega^a & 0 & +\omega_x^a & 0 & 0 & 0 \\ +\omega_y^a & -\omega_x^a & 0 & 0 & 0 & 0 \\ 0 & 0 & 0 & 0 & +\Omega^b & -\omega_y^b \\ 0 & 0 & 0 & -\Omega^b & 0 & +\omega_x^b \\ 0 & 0 & 0 & +\omega_y^B & -\omega_x^b & 0 \end{bmatrix}}_{\bar{L}} \right.$$

$$\left. - \underbrace{\begin{bmatrix} -k_{ab} & 0 & 0 & +k_{ba} & 0 & 0 \\ 0 & -k_{ab} & 0 & 0 & +k_{ba} & 0 \\ 0 & 0 & -k_{ab} & 0 & 0 & +k_{ba} \\ +k_{ab} & 0 & 0 & -k_{ba} & 0 & 0 \\ 0 & +k_{ab} & 0 & 0 & -k_{ba} & 0 \\ 0 & 0 & +k_{ab} & 0 & 0 & -k_{ba} \end{bmatrix}}_{\bar{K}} \right) \begin{bmatrix} M_x^a \\ M_y^a \\ M_z^a \\ M_x^b \\ M_y^b \\ M_z^b \end{bmatrix} + \begin{bmatrix} 0 \\ 0 \\ M_{z_0}^a \\ 0 \\ 0 \\ M_{z_0}^b \end{bmatrix}$$

(8.12)

Summing up the elements as $\bar{R}+\bar{L}-\bar{K}$, we get:

$$\underbrace{\begin{bmatrix} M_x^{a'} \\ M_y^{a'} \\ M_z^{a'} \\ M_x^{b'} \\ M_y^{b'} \\ M_z^{b'} \end{bmatrix}}_{M'} = \begin{bmatrix} R_2^a + k_{ab} & +\Omega^a & -\omega_y^a & -k_{ba} & 0 & 0 \\ -\Omega^a & R_2^a + k_{ab} & +\omega_x^a & 0 & -k_{ba} & 0 \\ +\omega_y^a & -\omega_x^a & R_1^a + k_{ab} & 0 & 0 & -k_{ba} \\ -k_{ab} & 0 & 0 & R_2^b + k_{ba} & +\Omega^b & -\omega_y^b \\ 0 & -k_{ab} & 0 & -\Omega^b & R_2^b + k_{ba} & +\omega_x^b \\ 0 & 0 & -k_{ab} & +\omega_y^b & -\omega_x^b & R_1^b + k_{ba} \end{bmatrix} \times \begin{bmatrix} M_x^a \\ M_y^a \\ M_z^a \\ M_x^b \\ M_y^b \\ M_z^b \end{bmatrix} + \underbrace{\begin{bmatrix} 0 \\ 0 \\ M_{z_0}^a \\ 0 \\ 0 \\ M_{z_0}^b \end{bmatrix}}_{M_0} \quad (8.13)$$

The above inhomogeneous system of linear differential equation $[M' = (\bar{R}+\bar{L}-\bar{K})M + M_0]$ can be made homogeneous $[M' = (\bar{R}+\bar{L}-\bar{K}+M_0)M]$ by introducing the identity element (E) as one of the variables as follows,

$$\underbrace{\begin{bmatrix} E' \\ M_x^{a'} \\ M_y^{a'} \\ M_z^{a'} \\ M_x^{b'} \\ M_y^{b'} \\ M_z^{b'} \end{bmatrix}}_{M'} = \begin{bmatrix} 0 & 0 & 0 & 0 & 0 & 0 & 0 \\ 0 & R_2^a + k_{ab} & +\Omega^a & -\omega_y^a & -k_{ba} & 0 & 0 \\ 0 & -\Omega^a & R_2^a + k_{ab} & +\omega_x^a & 0 & -k_{ba} & 0 \\ M_{z_0}^a & +\omega_y^a & -\omega_x^a & R_1^a + k_{ab} & 0 & 0 & -k_{ba} \\ 0 & -k_{ab} & 0 & 0 & R_2^b + k_{ba} & +\Omega^b & -\omega_y^b \\ 0 & 0 & -k_{ab} & 0 & -\Omega^b & R_2^b + k_{ba} & +\omega_x^b \\ M_{z_0}^b & 0 & 0 & -k_{ab} & +\omega_y^b & -\omega_x^b & R_1^b + k_{ba} \end{bmatrix} \times \underbrace{\begin{bmatrix} E \\ M_x^a \\ M_y^a \\ M_z^a \\ M_x^b \\ M_y^b \\ M_z^b \end{bmatrix}}_{M} \quad (8.14)$$

This concise matrix formulation of the McConnell–Bloch equation system, first presented by Reewes *et al.*, has been beneficial and extensively reported for the quantitative analysis of chemical exchange for an array of NMR experiments.[73] The above set of homogeneous ordinary differential equations (ODEs) in matrix form, $M' = PM$, can be easily solved, and its solution is represented as $M = e^{(P\Delta t)}M_0$, where $e^{(P\Delta t)}$ is the exponentiation of the matrix, P, multiplied with a short time interval, Δt; and M_0 represents the initial magnetization vector.

In eqn (8.13 and 8.14), the $(n \times n)$ dimensional matrix, $\bar{R} - \bar{K}$ is called the dynamic matrix (D), which is responsible for the decrease of the total magnetization by relaxation and exchange. If we consider the transverse relaxation elements of D its diagonal (D_{ii}) and off-diagonal elements (D_{ij}) are defined as:

$$D_{ij} = \begin{cases} \left(1/T_{2i} + \sum\limits_{\substack{j=1 \\ j \neq i}}^{n} k_{ij} \right) & \text{for}, i = j \\ -k_{ji} & i \neq j \end{cases} \tag{8.15}$$

where T_{2i}, is the transverse relaxation time constant of the ith species, and k_{ij} and k_{ji} are the pseudo-first-order rate constants of magnetization transfer from site i to j and backward. Given the system in equilibrium, the correct fulfillment of the condition requires:

(i) The reversibility of the two states in equilibrium, *i.e.*, $k_{ij} \times p_i = k_{ji} \times p_j$ (where p_i and p_j are the molar fractions of each state).
(ii) The sum of elements of any j^{th} column of the D matrix minus R_{2i} must be zero—thus establishing balance.

The ultimate goal of the dynamic NMR experiments is to determine the rate constants (k elements) in the matrix eqn (8.14). Since the kinetic rates are significantly less than the relaxation rate (k_{ij} or $k_{ji} \ll R_{2i}$) in slow exchange, the diagonal members of the D matrix are dominant, while the off-diagonal elements are significant in the intermediate or fast exchange regimes.

However, earlier, because of limitations in computation, this method was applied only to two or three exchanging sites. The recent advancements in the simulation and fitting programs have enabled us to characterize and quantify the exchange parameters directly from the 2D experimental spectra. Several tools are available that allow for the exploration of common binding models from the line shape analyses.[74–77] Nevertheless, the pre-processing steps often incorporate additional artifacts since the input spectrum has already undergone phase correction, baseline correction, along with apodization and numeric Fourier transformation. This limits the accurate interpretations of the functional aspects of the protein dynamics from the analyses. Furthermore, the line shape analysis is sensitive to the instrumental settings, such as the homogeneity of the magnetic field (shimming), which affects the non-exchange line width broadening, which is often difficult to estimate accurately. There are limitations also in the time scale.[78] Thus, most of the rate constants obtained through dynamic analysis are rather an estimate of the accurate experimentally determinable value. Nevertheless, in most cases, a visual comparison of the experimental and calculated spectra and observing the line widths and the positions of all peaks often provides invaluable insights into the chemical exchange phenomenon involving the dark species.

Simulation and fitting of the DEST experimental data for Aβ amyloidogenesis was described by Fawzi *et al.* considering a two-state model (eqn (8.16)) or three-state model [eqn (8.17)] as follows,

$$M + D_\mathrm{e} \underset{k_\mathrm{off}}{\overset{k_\mathrm{on}}{\rightleftarrows}} D_\mathrm{f} \tag{8.16}$$

$$M(i) + D_\mathrm{e} \begin{cases} \underset{k_\mathrm{off}}{\overset{k_{\mathrm{on}_1}(i)}{\rightleftarrows}} & D_\mathrm{t}(i) \\ \underset{k_\mathrm{off}}{\overset{k_{\mathrm{on}_2}(i)}{\rightleftarrows}} & D_\mathrm{c}(i) \end{cases} \tag{8.17}$$

In eqn (8.16), M, D_e, and D_f, represent the free monomer, "empty" or free dark state, and "filled" dark state species, respectively. Assuming the concentration of D_e to be constant and abundant (relative to D_f); the forward rate constant, k_on, can be combined with D_e and represented as an apparent rate constant, $k_\mathrm{on}^\mathrm{off} = k_\mathrm{on}[D_\mathrm{e}]$ Similarly, for the three-state system, eqn (8.17), D_t, D_c represent the "tethered" monomer, and the "contact" monomer bound to the dark state (proto-fibrils), respectively. Here, Fawzi *et al.* assumed that each residue (i) of the monomer behaved differently (*i.e.*, either tethered or in direct contact); hence, the rate constants, $k_{\mathrm{on}_1}^\mathrm{app}(i)$, $k_{\mathrm{on}_2}^\mathrm{app}(i)$, were also allowed to vary in a residue-specific manner. In this mechanism, tethering of a particular residue (i) refers to its loci being adjacent (but not physically bound) to residues that are in direct "contact" (or bound) with proto-fibrils. Given the CW saturation pulse of defined pulse length (Δt) and the RF parameters (ω_x or ω_y), the "DESTfit" program takes the longitudinal (R_1) and transverse (R_2) relaxation parameters of the monomer (M) as inputs and determines the rate constants, $k_{\mathrm{on}_1}^\mathrm{app}(i)$, $k_{\mathrm{on}_2}^\mathrm{app}(i)$, for each residue, and the relaxation parameters concerned with the dark states. DEST analysis provides residue-specific mechanistic insight into the oligomer formation mediated through monomer–proto-fibril interaction.

For a given saturation pulse (of defined time, $\Delta t = \sim 50$ ms) and RF strength ($\omega_y^\mathrm{a} = 100$ or 200 Hz) in a DEST experiment, the spectral intensities are proportional to the M_z^a component of the ^{15}N magnetization present along the *z* axis. To quantify the experimentally observed, residue-specific DEST intensity data (normalized with respect to the reference non-saturated data) are compared with the simulation. The ^{15}N magnetization is computed along both the $+z$ axis $[M^a_{+z}(\Delta t)]$, and $-z$ axis $[M^a_{-z}(\Delta t)]$, at each saturation offset frequency and RF field strength. Subsequently, its difference, $[M^a_{+z}(\Delta t) - M^a_{-z}(\Delta t)$, is calculated and normalized with respect to the non-saturation condition ($\Delta t = 0$), as $\frac{M^a_{+z}(\Delta t) - M^a_{-z}(\Delta t)}{M^a_{+z}(\Delta t = 0) - M^a_{-z}(\Delta t = 0)}$ In an ideal case, the initial magnetization along the $-z$ axis is identical to that of the $+z$ axis except for the sign reversal, *i.e.*, $M^a_{-z}(\Delta t = 0) = M^a_{+z}(\Delta t = 0)$ and, experimentally, it represents the magnetization which is alternated through 180° phase cycling of the ^1H pulse, ϕ_1. This protocol, as described by Fawzi *et al.* for DEST analysis, allowed the determination of R_2 for the backbone ^{15}N nuclei positions in the plausible dark states, along with the kinetic rates connecting them to the visible states by fitting the different kinetic models.[69]

8.10 Applications

8.10.1 Probing the Atomic-resolution Nucleation Events in Aβ Amyloidogenesis

DEST NMR was applied by Fawzi *et al.*[71] to study the exchange dynamics of Aβ monomer (M or $M_{monomer}$) with the nucleating protofibrillar (D_f or $M_{oligomer}$) states that lead to Aβ fibrillation. The residue-specific relaxation data were fitted using the Bloch–McConnell equations to quantify the apparent global first-order association (k_{on}^{app}) and dissociation (k_{off}) rate constants. The k_{on}^{app} was equated to ΔR_2, *i.e.*, the difference between the $^{15}N - R_2$ values of the reference, M sample, and the monomer sample associated with the bound dark state ($M + D_f$) (Figure 8.4, i). The assumption was based on the fact that the k_{on}^{app} of Aβ fibrillation during the nucleation step is slow relative to the (k_{off}) and that the rate is limited by the concentration of "empty" dark state [D_e]. On the other hand, the experimentally observed ΔR_2 is proportional to D_e; hence the maximal ΔR_2 value observed among all the residues is taken to be a reasonable estimate of the k_{on}^{app}. Fawzi *et al.* found that the residue-specific Aβ dynamics could not be explained either through the classical two-state model or the three-state sequential model with an intermediate present. Instead, they identified the co-existence of an ensemble of two distinct subsets of dark states, D_c or $I_{contact}$ and D_t or $I_{tethered}$, by analyzing the experimental data. Thus defining two functionally distinct subsets of $M_{oligoner}$, where a given residue is either in direct contact ($I_{contact}$) or is just tethered ($I_{tethered}$) *via* residues in direct contact with the oligomers. Hence, for each residue, *i*, there exists a partitioning factor that defines a fraction of it being in the tethered ($I_{tethered}(i)$) state while the remnant is in the contact ($I_{contact}(i)$)

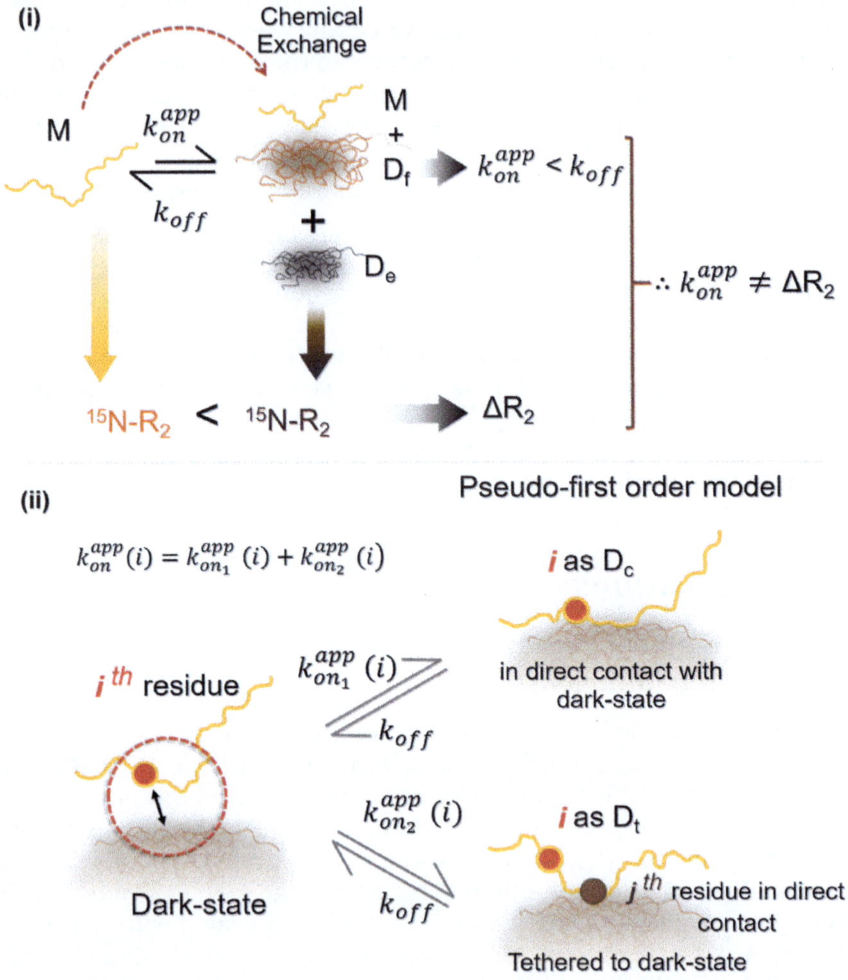

Figure 8.4 High-resolution insight into the interface of Aβ fibrillation using ^{15}N DEST and ΔR_2 as described by Fawzi *et al.*[71] (i) The apparent global first-order association (k_{on}^{app}) was initially equated to the difference between the $^{15}N - R_2$ values for the reference, monomeric (M) protein sample and the corresponding dark state nucleating fibrillary species ($M + D_f$). However, the k_{on}^{app} and the dissociation (k_{off}) constants obtained from the DESTfit program failed to provide good fits to the experimentally observed ΔR_2 values. This suggested a slower overall k_{on}^{app} value owing to the presence of the empty dark states [D_e]. (ii) The data required the assumption of a pseudo-first-order model for chemical exchange that essentially is comparable with the classical two-state exchange, except that the high-molecular-weight oligomeric protein conformers get partitioned into two states, tethered (D_t or state 1) and in direct contact (D_c or state 2) to the nucleating proto-fibrillar state. Thus, for a particular residue, i, the resultant k_{on}^{app} is given by $k_{on}^{app} = k_{on_1}^{app}(i)$, $k_{on_2}^{app}(i)$.

state. This partitioning was varied as an optimization parameter for each residue, ignoring the state of the remnant residues—whether tethered, bound, or in between these states. Nevertheless, this "pseudo-first-order" model for chemical exchange represents a simple modification of the overall two-state association/dissociation process with an apparent global association rate constant, $k_{on}^{app} = k_{on_1}^{app}(i)$, $k_{on_2}^{app}(i)$, constituted by residue-specific, $k_{on_1}^{app}(i)$, $k_{on_2}^{app}(i)$, are the partitioning factors or rate constants concerned with the conversion of a residue i from the monomeric state ($I_{monomer}(i)$) to $I_{tethered}(i)$) and $I_{contact}(i)$, respectively (Figure 8.4, ii). Here, $k_{on}^{app}(i)$ is considered to be a global kinetic rate, identical for all residues alike and representing an averaged k_{on} value across all residues. Similarly, the global dissociation rate constant, k_{off}, represents the conversion of protofibril-bound states, $I_{tethered}(i)$ or $I_{contact}(i)$ to the monomer ($I_{monomer}(i)$) and is considered to be identical for all residues alike [*i.e.*, $k_{off} = k_{-1}(i) = k_{-2}(i)$)]. In principle, this model is equivalent to that of a conventional two-state equilibrium model by having ensemble-averaged rates, k_{on}^{app} and k_{off}, for all residues alike, but differing only by the partitioned rates, $k_{on_1}^{app}(i)$, $k_{on_2}^{app}(i)$, whose sum should add up to k_{on}^{app} for each residue separately.

An extensive description of their work, the details of the pulse program, and the DESTfit script for the quantitative analysis using the MATLAB software suite has been reported as part of nature protocols for the easy comprehension and deployment of this technique.[69] A detailed description of the usage of the DESTfit script, including the set of commands used for calculating the residue-specific equilibrium constants, is provided therein.

The DEST pulse program, in particular, is effective for intrinsically disordered protein segments of the NMR-visible states. However, the situation is slightly different for a globular protein, where the kinetic rates of each residue are more or less similar to that of the folded domain itself, thereby restricting the residue-specific information on the binding interactions. Nevertheless, the same pulse program can still be used to gain high-resolution insight on the ordering or folding of the flexible, unstructured disordered regions or loops. Efforts are underway to develop extensions to the application of the DEST methodology in understanding the side-chain dynamics observed at the functional interfaces.[79,80]

8.10.2 Qualitative Insight into the Interaction Interfaces

Alternatively, the ^{15}N-DEST experiments can also be analyzed in a simplified qualitative manner. The multiple offset frequencies for the ^{15}N-selective saturation often require a long acquisition time, resulting in a biased interpretation of the data, especially for the dynamic exchanging systems undergoing aggregation.[81] Nevertheless, studies by Melacini *et al.* focusing on the modulatory effect of human serum albumin (HSA) on the Aβ oligomerization equilibrium showed the possibility of evading the obvious setbacks of such a long experimental time.[81] They employed the principles of traditional STD NMR analysis in calculating the DEST difference (Θ) or "single-point DEST" values, calculated from the average peak intensities at the positive and negative "near off-resonance" offset frequencies (I_{st}) and those at "far off-resonance" offset frequencies or reference (I_0) as: $\Theta = \frac{\Sigma \pm I_0 - \Sigma \pm I_{st}}{\Sigma \pm I_0}$. Here, the symbol $\Sigma \pm$ represents the averaging of the DEST signal acquired at both $\pm\omega$, where ω is the offset frequency corresponding to either ± 2 kHz, or ± 4 kHz, or ± 8 kHz, in the case of I_{st} and ± 35 kHz in the case of I_0. The chosen offset frequencies were shown to be a "balanced compromise" that efficiently saturates the proto-fibrillar resonances without significantly attenuating the monomeric conformers. The calculated Θ value can effectively probe the width of the individual peak intensities as a function of

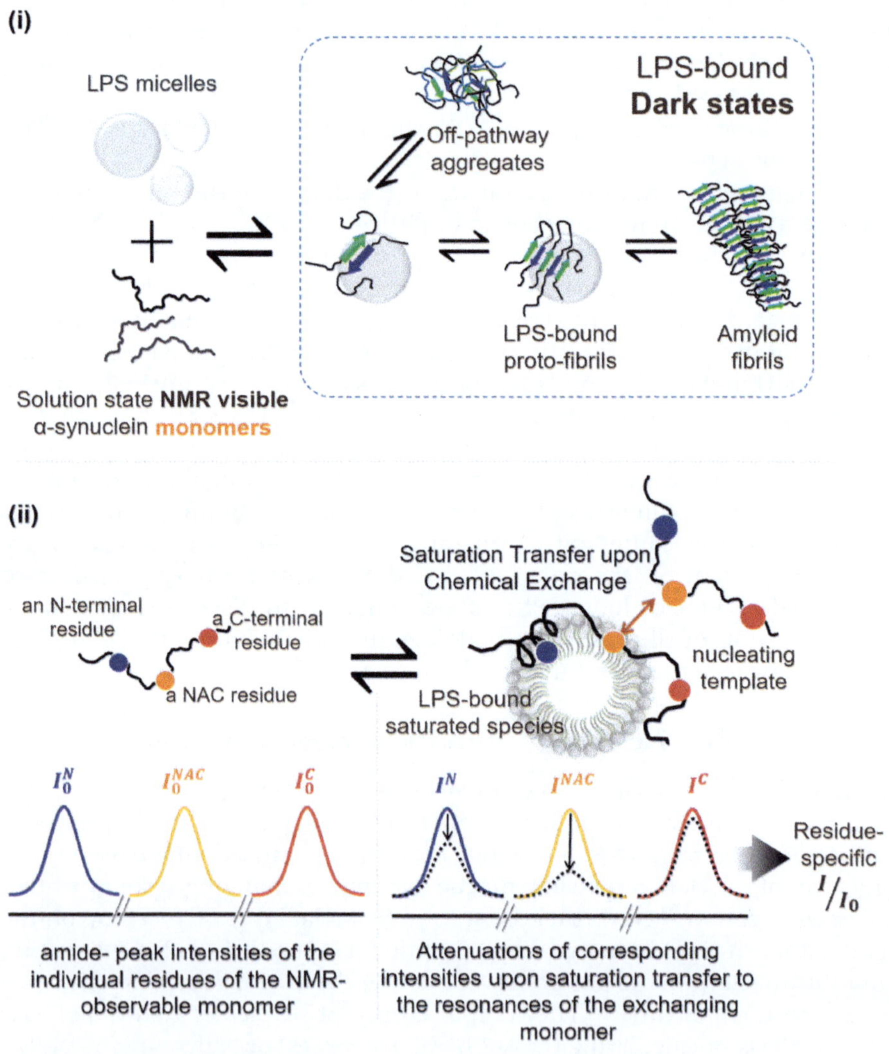

^{15}N-saturation offset profile that serves as an optimum alternative to elucidate the otherwise elaborately calculated global fit parameters from a full DEST dataset (see above).[81] Their subsequent studies have demonstrated the use of the residue-specific Θ values from DEST experiments to assess the propensity of individual residues of the Aβ monomer participating in the oligomerization events and the membrane-interacting functional domains.[82] Thus, by applying CW saturation pulses at fewer offset values in the range of kHz, it is possible to efficiently map the residue-specific epitope of the interaction by measuring the DEST-modulated change in the peak intensities.

Alternatively, many earlier studies have often exploited the qualitative analysis of the line broadening in response to the chemical exchange to define the epitope concerned with the interaction. Brender *et al.* demonstrated the possibilities of extending the applications of "DEST-like" experiments to study the residues directly involved in self-interaction events of human islet polypeptide (hIAPP) amyloidogenesis.[83] Further, an early amyloidogenic epitope for α-synuclein was delineated and compared with both the positive and negative familial mutant forms of the protein.[84]

Recently, DEST-like experiments were employed by Bhunia *et al.* to study the membrane–protein interaction interface, such as the association of α-synuclein with bacterial lipopolysaccharide (LPS).[85] This study delineated the molecular events of Parkinson's disease in a residue-specific manner by linking the events originating within the gut to the onset of the symptoms in the central nervous system decades ahead.[85] α-Synuclein exhibits greater

Figure 8.5 DEST experiments were performed to probe the domain-specific interactions of α-synuclein with the LPS micelles by Bhunia *et al.*[85] (i) The hetero-molecular interactions resulted in a complex ensemble of high-molecular-weight LPS-associated protein conformers that were in dynamic equilibrium with the resultant dark-state sample. Nevertheless, an overall increase in ΔR_2 ($\Delta R_2 = R_2$ [α-synuclein + LPS] − R_2 [α-synuclein monomer]) for all the residues indicated the applicability of the DEST pulse program for this system. However, due to the lack of a prominent DEST profile, the data allowed only for qualitative analysis of α-synuclein's interaction with LPS. (ii) Segment-specific attenuation profiles even at 16 kHz offset frequency, compared to the reference spectrum, showed the direct contact of the N-terminal residues and the central hydrophobic region of the protein in the heteromolecular LPS complex. In the reference spectrum (left), the peak intensities for the individual residues from the three segments, N-terminal (blue), central non-Aβ component (NAC) region (green), and the C-terminal (red), were shown for the free monomeric protein alone. The single representative residues from each segment and their corresponding peak intensities are kept arbitrarily the same for all three residues for simplicity. In the experimental sample (right), the NMR-invisible LPS-bound protein conformers are saturated, and in the event of chemical exchange, the saturation is transferred to the resonances of the monomeric protein that are participating in the nucleation events. This results in the subsequent attenuation of those respective peaks. The corresponding intensity ratios with respect to the reference (I_0) are obtained to provide a qualitative interpretation of the underlying chemical exchange.

aggregation propensity in the presence of LPS, rapidly forming invisible oligomeric species that nucleate distinct amyloidogenic pathways in the gut–brain axis of Parkinson's disease pathogenesis. The DEST experiments were performed on freshly prepared ^{15}N-labeled monomeric protein samples and highly concentrated protein samples directly interacting with bacterial LPS resembling the LPS-modulated amyloidogenic pathway (Figure 8.5, i). The small fraction of LPS-bound oligomers in the resultant sample was applied with weak RF, far off-resonance from the monomeric species. The presence of any small population of oligomers that were invisible to the NMR was wholly or partially saturated. As a consequence of chemical exchange between the monomers and the oligomers during the nucleation process, the saturation transfer occurred from the LPS-bound dark species to the interacting monomers. This was observed as signal attenuations of the specific residues participating in the early nucleation events. The analysis of the individual residue intensities obtained from the HSQC profiles upon employing RF (I) with respect to the reference spectra (I_0) provided qualitative insights into the nucleating residues present at the interaction interface of bacterial LPS (Figure 8.5, ii). Further, the qualitative analysis of the residue-specific DEST profiles obtained as a function of the offset frequencies can itself provide essential insight into the biomolecular interaction interfaces. A similar analysis was used in yet another study by Melacini *et al.* that probed the region-specific interactions of HSA as a chaperone in modulating the overall membrane availability of the α-synuclein conformers.[86] This laid the foundation for the possibilities of extending the application of DEST NMR in mapping the interactions of IDPs with chaperones and how chaperones modulate the interactions of IDPs with lipids.

Previous studies by Fusco *et al.* used dark-state NMR principles to probe the segment-specific functional membrane-interaction affinity for α-synuclein. They employed CW saturation pulses for a range of offset frequencies far off-resonance from the monomeric conformers at two different saturation fields (350 and 170 Hz). The resultant intensity ratios between the near on-resonance and off-resonance frequencies enabled them to characterize the α-synuclein interaction with the small unilamellar vesicles that mimic the synaptic vesicles.[87]

8.11 Experimental Sample Conditions are Critical for DEST Analysis

The experimental conditions that best represent the dynamic equilibrium in the protein samples are essential for the successful probing of the chemical exchange during DEST NMR. The long acquisition time of DEST experiments has an impact on the samples that undergo aggregation and thereby limits their immense capability of characterizing the biomolecular interaction interfaces. For this purpose, the sample preparation that effectively stabilizes the dynamic conformers of interest is an absolute prerequisite. The concentration of the exchanging HMW conformers should be sufficient for effective saturation transfer to the monomers without causing rapid line broadening due to high R_2 values. An approximation approach would be to monitor the ΔR_2 of the sample containing HMW with respect to a reference sample comprised essentially of monomers alone. This partially allows for the separation of the effect of any intrinsic dynamics of the monomers as opposed to the dynamics involving the HMW conformers.

Lowering the protein concentration often results in a stable LMW sample space that can remain at equilibrium—facilitating a longer experimental time frame. Studies by Fawzi *et al.*, focusing on the Aβ amyloidogenesis, employed two different protein concentrations, $\leq$50 μM and $\geq$150 μM for the essentially monomeric sample and the sample containing monomeric species in pseudo-equilibrium with the chemically exchanging HMW conformers, respectively.[71,88] These, however, are very specific characteristics of the system as the equilibrium properties can vary, and therefore, optimization is essential based on the best sample condition to reproduce the prerequisite difference in the R_2 values (ΔR_2). This is true for both studying the exchange dynamics involving self-oligomers/protofibrils as well as the heteromolecular complex HMW conformers. Alternatively, several strategies such as changing the buffer salt concentration and temperature conditions can modulate the overall exchanging equilibrium to enable the effective recording of DEST experiments. The use of strong desalting columns is critical to trap oligomers that are in dynamic exchange with Aβ monomers for a sufficiently long timeframe to enable the acquisition of a complete DEST profile, which requires several hours of NMR acquisition time.[70,71]

Temperature, too, plays a significant role in determining the dynamics of the chemical exchange for any system. Maintaining low-temperature conditions was critical for the DEST application to Aβ; however, to determine the best temperature conditions for other systems, the use of temperature gradients could be helpful. The Self-Assembly monitored by Temperature-Induced Relaxation Enhancement (SATIRE) approach helped monitor the ΔR_2 values as a function of temperature.[89] Employing this approach showed that in the absence of oligomerization, the accelerated internal dynamics of the free monomers undergo a steady decay in the R_2 rates while self-assembly results in local deviations of the R_2 rates. Thus, SATIRE allows for a reliable means to determine the most effective temperature conditions for the system of interest. In-depth knowledge of the optimum sample conditions is essential for the efficient characterization of chemical exchange in the sample space.

8.12 Conclusions

Thus DEST or DEST-like experiments have become an indispensable tool for probing high-resolution structural information of the protein dynamics at the interaction interface, especially for intrinsically disordered proteins. Tuning the experimental parameters to obtain quantitative DEST profiles is essential to understanding the dynamics of challenging protein interaction systems.

Acknowledgements

This work was partly supported by the Department of Biotechnology (BT/PR29978/MED/30/2037/2018 to AB), the Government of India, and partly by the Council of Scientific and Industrial Research (02(0292)/17/EMR-II to AB), Government of India, and the intramural, extramural research fund by Bose Institute (R/16/19/1615 to AB). DB thanks CSIR-UGC, Govt. of India, for providing the research fellowship.

References

1. W. D. Wilson, *Science*, 2002, **295**, 2103–2105.
2. K. Wüthrich, *Nat. Struct. Biol.*, 1998, **5 Suppl**, 492–495.
3. V. N. Uversky, *Front. Phys.*, 2019, **7**, 10.
4. A. Wallmann and C. Kesten, *Int. J. Mol. Sci.*, 2020, **21**(6), 2105.
5. U. B. Choi, H. Sanabria, T. Smirnova, M. E. Bowen and K. R. Weninger, *Biomolecules*, 2019, **9**, 114.
6. O. Khersonsky and D. S. Tawfik, *Annu. Rev. Biochem.*, 2010, **79**, 471–505.
7. A. P. Valente, C. A. Miyamoto and F. C. Almeida, *Curr. Med. Chem.*, 2006, **13**, 3697–3703.
8. N. Tokuriki and D. S. Tawfik, *Science*, 2009, **324**, 203–207.
9. C. J. Oldfield and A. K. Dunker, *Annu. Rev. Biochem.*, 2014, **83**, 553–584.
10. M. M. Babu, R. van der Lee, N. S. de Groot and J. Gsponer, *Curr. Opin. Struct. Biol.*, 2011, **21**, 432–440.
11. V. N. Uversky, *Biochim. Biophys. Acta*, 2011, **1814**, 693–712.
12. G. M. Clore, *Protein Sci.*, 2011, **20**, 229–246.
13. A. Sekhar and L. E. Kay, *Proc. Natl. Acad. Sci. U. S. A.*, 2013, **110**, 12867–12874.
14. P. Lieutaud, F. Ferron, A. V. Uversky, L. Kurgan, V. N. Uversky and S. Longhi, *Intrinsically Disord Proteins*, 2016, **4**, e1259708.
15. V. N. Uversky, *F1000Res earch*, 2020, **9**, DOI: 10.12688/f1000research.20867.1.
16. P. Tompa and M. Fuxreiter, *Trends Biochem. Sci.*, 2008, **33**, 2–8.
17. D. D. Boehr and P. E. Wright, *Science*, 2008, **320**, 1429–1430.
18. D. D. Boehr, R. Nussinov and P. E. Wright, *Nat. Chem. Biol.*, 2009, **5**, 789–796.
19. J. W. Peng, *Structure*, 2009, **17**, 319–320.
20. G. M. Lee and C. S. Craik, *Science*, 2009, **324**, 213–215.
21. M. Pellecchia, I. Bertini, D. Cowburn, C. Dalvit, E. Giralt, W. Jahnke, T. L. James, S. W. Homans, H. Kessler, C. Luchinat, B. Meyer, H. Oschkinat, J. Peng, H. Schwalbe and G. Siegal, *Nat. Rev. Drug Discovery*, 2008, **7**, 738–745.
22. I. R. Kleckner and M. P. Foster, *Biochim. Biophys. Acta*, 2011, **1814**, 942–968.
23. P. G. Wolynes, *Q. Rev. Biophys.*, 2005, **38**, 405–410.
24. J. D. Bryngelson and P. G. Wolynes, *Proc. Natl. Acad. Sci. U. S. A.*, 1987, **84**, 7524–7528.
25. J. D. Bryngelson, J. N. Onuchic, N. D. Socci and P. G. Wolynes, *Proteins*, 1995, **21**, 167–195.

26. O. Miyashita, P. G. Wolynes and J. N. Onuchic, *J. Phys. Chem. B*, 2005, **109**, 1959–1969.
27. B. T. Burnley, P. V. Afonine, P. D. Adams and P. Gros, *eLife*, 2012, **1**, e00311.
28. J. S. Fraser, M. W. Clarkson, S. C. Degnan, R. Erion, D. Kern and T. Alber, *Nature*, 2009, **462**, 669–673.
29. M. Rueda, C. Ferrer-Costa, T. Meyer, A. Pérez, J. Camps, A. Hospital, J. L. Gelpí and M. Orozco, *Proc. Natl. Acad. Sci. U. S. A.*, 2007, **104**, 796–801.
30. M. L. Baker, C. F. Hryc, Q. Zhang, W. Wu, J. Jakana, C. Haase-Pettingell, P. V. Afonine, P. D. Adams, J. A. King, W. Jiang and W. Chiu, *Proc. Natl. Acad. Sci. U. S. A.*, 2013, **110**, 12301–12306.
31. M. van Heel, B. Gowen, R. Matadeen, E. V. Orlova, R. Finn, T. Pape, D. Cohen, H. Stark, R. Schmidt, M. Schatz and A. Patwardhan, *Q. Rev. Biophys.*, 2000, **33**, 307–369.
32. Z. H. Zhou, *Curr. Opin. Struct. Biol.*, 2008, **18**, 218–228.
33. P. Neudecker, P. Lundström and L. E. Kay, *Biophys. J.*, 2009, **96**, 2045–2054.
34. H. J. Dyson and P. E. Wright, *Methods Enzymol.*, 2005, **394**, 299–321.
35. G. M. Clore and A. M. Gronenborn, *Proc. Natl. Acad. Sci. U. S. A.*, 1998, **95**, 5891–5898.
36. N. Tjandra, J. Marquardt and G. M. Clore, *J. Magn. Reson.*, 2000, **142**, 393–396.
37. L. E. Kay, D. A. Torchia and A. Bax, *Biochemistry*, 1989, **28**, 8972–8979.
38. G. Lipari and A. Szabo, 1981.
39. G. Lipari and A. Szabo, *J. Am. Chem. Soc.*, 1982, **104**, 4546–4559.
40. G. M. Clore and J. Iwahara, *Chem. Rev.*, 2009, **109**, 4108–4139.
41. N. Tjandra and A. Bax, *Science*, 1997, **278**, 1111–1114.
42. N. J. Anthis and G. M. Clore, *Q. Rev. Biophys.*, 2015, **48**, 35–116.
43. C. Charlier, S. F. Cousin and F. Ferrage, *Chem. Soc. Rev.*, 2016, **45**, 2410–2422.
44. B. He, K. Wang, Y. Liu, B. Xue, V. N. Uversky and A. K. Dunker, *Cell Res.*, 2009, **19**, 929–949.
45. R. van der Lee, M. Buljan, B. Lang, R. J. Weatheritt, G. W. Daughdrill, A. K. Dunker, M. Fuxreiter, J. Gough, J. Gsponer, D. T. Jones, P. M. Kim, R. W. Kriwacki, C. J. Oldfield, R. V. Pappu, P. Tompa, V. N. Uversky, P. E. Wright and M. M. Babu, *Chem. Rev.*, 2014, **114**, 6589–6631.
46. P. E. Wright and H. J. Dyson, *Nat. Rev. Mol. Cell Biol.*, 2015, **16**, 18–29.
47. S. Fukuchi, K. Homma, Y. Minezaki and K. Nishikawa, *J. Mol. Biol.*, 2006, **355**, 845–857.
48. M. R. Jensen, M. Zweckstetter, J. R. Huang and M. Blackledge, *Chem. Rev.*, 2014, **114**, 6632–6660.
49. N. Salvi, A. Abyzov and M. Blackledge, *Angew. Chem., Int. Ed. Engl.*, 2017, **56**, 14020–14024.
50. N. Salvi, A. Abyzov and M. Blackledge, *Prog. Nucl. Magn. Reson. Spectrosc.*, 2017, **102–103**, 43–60.
51. A. D. Bain, *Biochem. Cell Biol.*, 1998, **76**, 171–176.

52. A. D. Bain, *Prog. Nucl. Magn. Reson. Spectrosc.*, 2003, **43**, 63–103.
53. A. G. Palmer, *J. Magn. Reson.*, 2014, **241**, 3–17.
54. R. Ishima and D. A. Torchia, *J. Biomol. NMR*, 1999, **14**, 369–372.
55. D. M. Korzhnev and L. E. Kay, *Acc. Chem. Res.*, 2008, **41**, 442–451.
56. I. Solomon, *Phys. Rev.*, 1955, **99**, 559.
57. H. Y. Carr and E. M. Purcell, *Phys. Rev.*, 1954, **94**, 630.
58. S. Meiboom and D. Gill, *Rev. Sci. Instrum.*, 1958, **29**, 688–691.
59. S. Forsén and R. A. Hoffman, *J. Chem. Phys.*, 1963, **39**, 2892–2901.
60. A. Bhunia, S. Bhattacharjya and S. Chatterjee, *Drug Discovery Today*, 2012, **17**, 505–513.
61. J. R. Brender, J. Krishnamoorthy, A. Ghosh and A. Bhunia, in *Rational Drug Design*, Springer, 2018, pp. 49–65.
62. H. Huang, J. Milojevic and G. Melacini, *J. Phys. Chem. B*, 2008, **112**, 5795–5802.
63. V. Esposito, R. Das and G. Melacini, *J. Am. Chem. Soc.*, 2005, **127**, 9358–9359.
64. A. Kumar, R. C. R. Grace and P. Madhu, *Prog. Nucl. Magn. Reson. Spectrosc.*, 2000, **37**, 191–319.
65. J. Milojevic, V. Esposito, R. Das and G. Melacini, *J. Am. Chem. Soc.*, 2007, **129**, 4282–4290.
66. P. C. van Zijl and N. N. Yadav, *Magn. Reson. Med.*, 2011, **65**, 927–948.
67. P. Vallurupalli, G. Bouvignies and L. E. Kay, *J. Am. Chem. Soc.*, 2012, **134**, 8148–8161.
68. D. S. Libich, N. L. Fawzi, J. Ying and G. M. Clore, *Proc. Natl. Acad. Sci. U. S. A.*, 2013, **110**, 11361–11366.
69. N. L. Fawzi, J. Ying, D. A. Torchia and G. M. Clore, *Nat. Protoc.*, 2012, **7**, 1523.
70. N. L. Fawzi, J. Ying, D. A. Torchia and G. M. Clore, *J. Am. Chem. Soc.*, 2010, **132**, 9948–9951.
71. N. L. Fawzi, J. Ying, R. Ghirlando, D. A. Torchia and G. M. Clore, *Nature*, 2011, **480**, 268–272.
72. P. Hore, 2010.
73. M. Helgstrand, T. Härd and P. Allard, *J. Biomol. NMR*, 2000, **18**, 49–63.
74. M. Niklasson, R. Otten, A. Ahlner, C. Andresen, J. Schlagnitweit, K. Petzold and P. Lundström, *J. Biomol. NMR*, 2017, **69**, 93–99.
75. D. A. Cheshkov, K. F. Sheberstov, D. O. Sinitsyn and V. A. Chertkov, *Magn. Reson. Chem.*, 2018, **56**, 449–457.
76. C. A. Waudby, A. Ramos, L. D. Cabrita and J. Christodoulou, *Sci. Rep.*, 2016, **6**, 24826.
77. C. A. Waudby and J. Christodoulou, in *Intrinsically Disordered Proteins*, Springer, 2020, pp. 477–504.
78. I. Bányai, *New J. Chem.*, 2018, **42**, 7569–7581.
79. N. L. Fawzi, D. S. Libich, J. Ying, V. Tugarinov and G. M. Clore, *Angew. Chem. Int. Ed. Engl.*, 2014, **53**, 10345–10349.
80. J. M. Edwards, J. E. Bramham, A. Podmore, S. M. Bishop, C. F. van der Walle and A. P. Golovanov, *Anal. Chem.*, 2019, **91**, 4702–4708.

81. M. Algamal, R. Ahmed, N. Jafari, B. Ahsan, J. Ortega and G. Melacini, *J. Biol. Chem.*, 2017, **292**, 17158–17168.
82. R. Ahmed, M. Akcan, A. Khondker, M. C. Rheinstädter, J. C. Bozelli, R. M. Epand, V. Huynh, R. G. Wylie, S. Boulton, J. Huang, C. P. Verschoor and G. Melacini, *Chem. Sci.*, 2019, **10**, 6072–6082.
83. J. R. Brender, J. Krishnamoorthy, M. F. Sciacca, S. Vivekanandan, L. D'Urso, J. Chen, C. La Rosa and A. Ramamoorthy, *J. Phys. Chem. B*, 2015, **119**, 2886–2896.
84. D. Bhattacharyya, R. Kumar, S. Mehra, A. Ghosh, S. K. Maji and A. Bhunia, *Chem. Commun.*, 2018, **54**, 3605–3608.
85. D. Bhattacharyya, G. M. Mohite, J. Krishnamoorthy, N. Gayen, S. Mehra, A. Navalkar, S. A. Kotler, B. N. Ratha, A. Ghosh, R. Kumar, K. Garai, A. K. Mandal, S. K. Maji and A. Bhunia, *ACS Chem. Neurosci.*, 2019, **10**, 2229–2236.
86. R. Ahmed, J. Huang, D. K. Weber, T. Gopinath, G. Veglia, M. Akimoto, A. Khondker, M. C. Rheinstädter, V. Huynh, R. G. Wylie, J. C. Bozelli, R. M. Epand and G. Melacini, *J. Am. Chem. Soc.*, 2020, **142**, 9686–9699.
87. G. Fusco, T. Pape, A. D. Stephens, P. Mahou, A. R. Costa, C. F. Kaminski, G. S. Kaminski Schierle, M. Vendruscolo, G. Veglia, C. M. Dobson and A. De Simone, *Nat. Commun.*, 2016, **7**, 12563.
88. N. L. Fawzi, J. Ying, D. A. Torchia and G. M. Clore, *Nat. Protoc.*, 2012, **7**, 1523–1533.
89. R. Ahmed, J. Huang, M. Akimoto, T. Shi and G. Melacini, *J. Am. Chem. Soc.*, 2021, **143**, 4668–4679.

CHAPTER 9

Dynamics of Protein–Nanoparticle Interactions Using NMR

VARSHA BRAHMKHATRI* AND HANUDATTA S. ATREYA†

Centre for Nano and Material Sciences, Jain University, Jain Global Campus, Bengaluru 562112Karnataka, India
*Email: b.varsha@jainuniversity.ac.in

9.1 Introduction

The interaction of nanoparticles with proteins or biomolecules is the important aspect of nanoparticle bio-reactivity that decides their fate in biological milieu. This interaction results in the formation of a dynamic nanoparticle–protein corona.[1–3] The protein corona may impact the cellular uptake, accumulation, inflammation, degradation and clearance of the nanoparticles.[4] Additionally, conformational changes induced in adsorbed protein molecules post interaction with the nanoparticle surfaces may affect the overall bio-reactivity of the nanoparticle.[5] Comprehensive knowledge of such interactions may lead to creating bio-compatible nanomaterials with controlled surface characteristics in a biological environment. Proteins are the polypeptides with amino acids as their building blocks, with defined conformation that carry a net surface charge depending on the pH of the surrounding medium. Adsorption of proteins at the nano–bio interface is aided by several forces such as hydrogen bonds, solvation

†This book chapter is dedicated to Prof. Hanudatta S. Atreya, NMR Research Centre, IISc Bangalore, India, who passed away on July 30, 2020.

New Developments in NMR No. 26
NMR Spectroscopy for Probing Functional Dynamics at Biological Interfaces
Edited by Anirban Bhunia, Hanudatta S. Atreya and Neeraj Sinha

Published by the Royal Society of Chemistry, www.rsc.org

forces, van der Waals interactions and electrostatic interactions. The overall nanoparticle–protein corona formation is a multifactorial process and not only depends on the characteristics of the nanoparticles (NP), but also on the interacting proteins and the medium. Specific association and dissociation rates for each protein decide the durability of their interaction with the NP surface. Irreversible binding of proteins on the NP leads to formation of a "hard corona", whereas quickly reversible binding of proteins that have faster exchange rates defines a "soft corona". The analysis of proteins bound to the nanoparticle surface is the first step towards understanding the true nature of the nanoparticle-mediated biological effects.[6] The research done so far highlights that the size, shape and surface characteristics of NP affect protein adsorption, and NP also have the capability to perturb the structure of the adsorbed protein molecules.[7] A thorough understanding of NP–protein interactions might lead to strategic manipulation of NP surfaces to adsorb specific functional proteins or small drug molecules intended for delivery *in vivo*.[8] The protein corona or protein NP interaction can be characterized by various spectroscopies such as surface plasmon resonance (SPR),[9] fluorescence correlation spectroscopy (FCS),[10,11] circular dichroism (CD) spectroscopy,[12] Fourier-transform infrared spectroscopy (FTIR)[13] and nuclear magnetic resonance (NMR) spectroscopy, and other techniques such as transmission electron microscopy (TEM),[14] mass spectrometry (MS),[15] electrophoresis, dynamic light scattering (DLS)[16] and isothermal titration calorimetry (ITC).[17] These techniques have been used to understand the kinds of structural and conformational changes proteins experience on the NP surface. Nuclear magnetic resonance (NMR) is the most commanding technique to study biomolecular complexes in aqueous solutions mimicking biological media, owing to its high sensitivity and resolution to transient protein–protein interactions.[18–20] In this chapter, we intend to cover the basic fundamentals understating protein–NP interactions and how solution NMR spectroscopy can be used to understand the details of NP–protein interactions (Figure 9.1). However, for magnetic NPs, NMR may fail to provide reliable characterization of the protein–NP corona. Hence, for magnetic NP systems other techniques may give exciting and more reliable results.[21,22]

9.2 Characterization of Protein–NP Interactions by Various Techniques

In a biological medium, NPs may interact with biomolecules such as proteins, nucleic acids, lipids and even biological metabolites due to their nano size and large surface-to-mass ratio. Specifically, the adsorption of proteins on the NP surface is the most significant. So far, the NP–protein corona has been explored experimentally using circular dichroism, fluorescence, infrared spectroscopy and various other methods that are limited to monitoring changes in the secondary structure of proteins.[23] Finding comprehensive details of the structural changes occurring in the protein at the atomic level upon binding to NP remains a challenge.

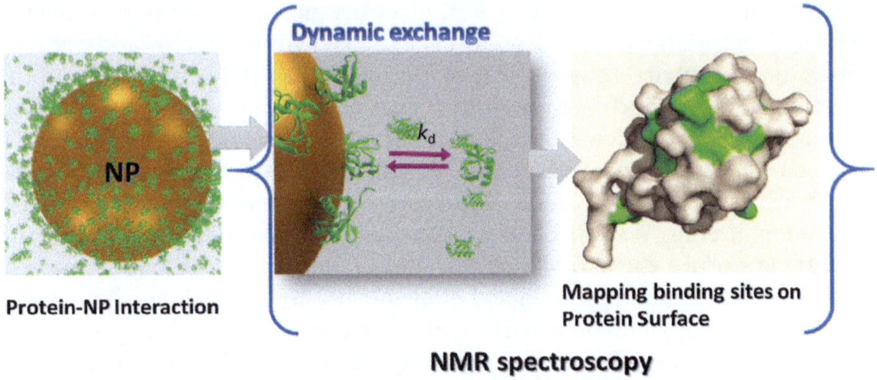

Figure 9.1 Probing the protein–nanoparticle interaction by solution NMR.

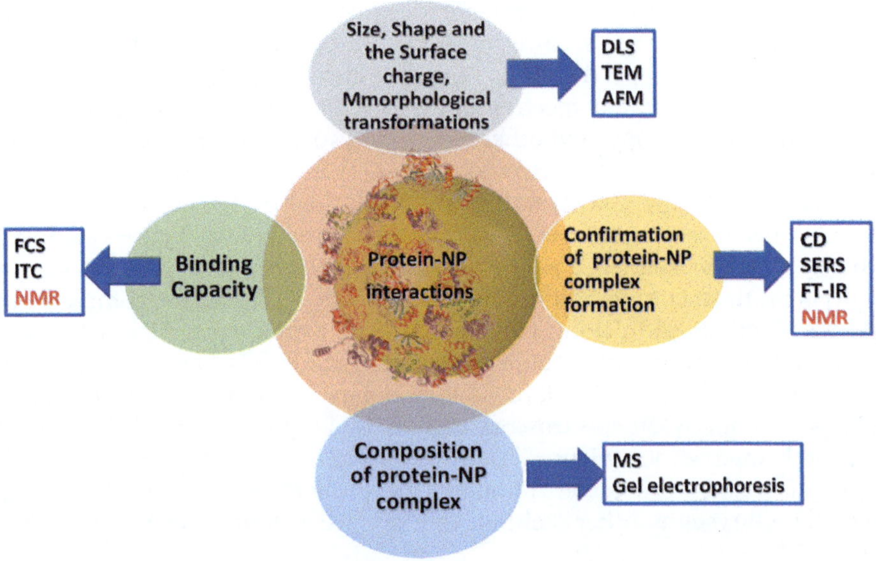

Figure 9.2 Technique to understand protein–nanoparticle interactions.

When nanoparticles are exposed to a biological environment, they are surrounded by various biological molecules, hence to understand the interaction of NP with biomolecules, for example, proteins, it is important to probe the change in inherent properties of NPs and the structural changes in the protein which illustrate the overall configuration of the protein–NP complex. The various analytical techniques that have been used to study NP-–protein interactions include UV–vis, CD, FT-IR, MS, DLS and ITC.[24] Each technique provides unique information (Figure 9.2) about the protein–NP complex and each also has its own advantages and drawbacks while exploring NP–protein interactions.

UV–Visible spectroscopy is used on a routine basis for the characterization of protein–NP interactions due to its ease of operation and low cost. The binding of proteins to nanomaterials results in some changes in the UV–visible absorption spectra such as the small spectral shifts and minute spectral intensity changes, especially in the case of metallic nanoparticles like AuNPs and AgNPs. Although the UV–vis technique requires very simple sample preparations and binding constant can be determined, it is highly influenced by the type of solvent used, pH and concentration. Vaishanav et.al. reported the interaction of gold nanoparticles with bovine serum albumin (BSA) protein using UV–vis.[25] They studied various parameters influencing BSA and AuNP binding such as concentration, pH, contact time and temperature. The binding kinetics was found to follow the pseudo-second-order rate. The Freundlich isotherm model was fitted at equilibrium adsorption and the maximum adsorption capacity for BSA protein adsorbed on AuNPs was reported to be 58.71 $mg\,g^{-1}$.

The CD spectroscopy is used to determine structural changes in proteins upon interacting with NP. It gives information about the secondary structure of proteins and how their structure changes upon binding to the NP surface. In a previous study carried out by Satzer *et al.* the proteins like myoglobin and BSA undergo conformational changes upon adsorption onto silica nanoparticles[25]. Such confirmational changes in proteins were dependent on the size of the silica nanoparticles. They evaluated the phenomenon of the NP size-dependent conformational change in proteins (BSA and myoglobin) using a circular dichroism protocol.

Fluorescence spectroscopy can be also used to characterize protein–NP interactions if there is a natural fluorophore such as tyrosine or tryptophan present in the proteins. If these are absent in the protein of interest, in such cases cysteine or amines can be labelled with fluorescent probes, like fluorescein, to study their structural and dynamic properties upon protein binding. Bing *et al.* studied TiO_2 nanoparticles adsorbing plant-derived proteins such as glutenin, gliadin, zein and soy protein that form coronas.[26] The interactions between the TiO_2 nanoparticles and plant proteins altered the secondary structures of the proteins as confirmed by fluorescence quenching, circular dichroism and infrared spectroscopy.

Raman spectroscopy is preferable to fluorescence spectroscopy as it is label-free, non-invasive and very specific, and can manage both solid and liquid samples. Surface-enhanced Raman spectroscopy (SERS) exhibits higher sensitivity and higher selectivity of chemical groups as compared to Raman spectroscopy. When metal NP surfaces are involved in protein corona formation, SERS can be used to determine the structural and conformational changes of proteins. Laruinaz *et al.* reported the interactions of AgNPs with different proteins (lysozyme, bovine serum albumin, cytochrome-C and haemoglobin) using SERS.[27] They also report chemisorption of the proteins on Ag-NP surfaces based on SERS, which demonstrates a sharp peak in the low-frequency range that is attributed to Ag–N vibrations.

The main advantage of using optical spectroscopies such as fluorescence, Raman and CD spectroscopy is that they can analyse any protein–NP system irrespective of the presence of a plasmonic NP system. Apart from spectroscopic techniques, other methods such as dynamic light scattering, chromatography, mass spectrometry and isothermal titration calorimetry are also important in gaining an idea about the effects of shape, size and surface charges of NP on protein binding.

Dynamic light scattering (DLS) is the most conventional method to determine the hydrodynamic size of NPs before and after protein interactions, based on their Brownian motion.[28] The time-dependent scattering intensity fluctuations that are caused by Brownian motion are measured by DLS using the Stokes–Einstein equation and relate the diffusion coefficient to the NP size.[29] DLS is very sensitive towards biomolecules and hence it is appropriate to measure the size changes in NPs after protein adsorption which eventually gives the size of the protein–NP corona.[30] It has been reported by various research groups that the NP diameter increases after protein corona formation[31]. The various characterization methods give unique information about protein–NP interactions, as shown in Table 9.1. The information about the protein–NP complex size and surface charge can be obtained from DLS and zeta potential measurements, whereas AFM and TEM can be used to obtain the exact shape and size. From MS and gel electrophoresis the exact composition of the protein–NP complex will be found. Techniques like FCS, ITC and NMR enable calculation of the binding capacity. The structural, confirmational changes and the stability of the protein bound to the NP surface can be investigated by NMR spectroscopy and CD.

9.3 Solution NMR Spectroscopy to Understand Protein–Nanoparticle Interactions

Solution NMR spectroscopy is a well-known technique for the high-resolution characterization of biomolecules and structural assignments of proteins, which gives important information about the structure and dynamics. Using NMR, it is possible to study the local mobility occurring in a short time window from picoseconds to hours. The solution NMR spectroscopy is the only technique that can locate the exact binding sites on proteins at the residue level upon interacting with the NP. The information on protein–NP binding kinetics, stability, rotational and diffusion coefficient can be acquired by performing specific NMR experiments.[39] However, NMR spectroscopy also has some limitations regarding molecular size. Therefore, characterizing protein–NP assemblies using solution NMR spectroscopy is highly challenging.

The use of solution NMR spectroscopy helps to understand various questions regarding protein–NP interactions by performing several diverse NMR experiments.[40,41] In the following we cover the more extensively used NMR experiment to understand the dynamics of protein–NP interactions

Table 9.1 Protein–nanoparticle interaction studies using different methods.

Nanoparticle	Protein	Technique	Findings	References
AuNP nanospheres (AuNSPs), nanorods (AuNRs), nanostars (AuNSs)	Fibrinogen (FIB) and trypsin (Try)	UV–vis, FT-IR, TEM, DLS and microscale thermophoresis (MST)	Shape of NP matters, AuNRs and AuNSs induce greater changes in the secondary structures of the proteins as compared to AuNPs	32
AgNPs	HSA	UV–visible spectroscopy	Fabrication of nano-biosensor	33
	Albumin and α-1-acid glycoprotein	TEM, DLS, ELS, ICP-MS, CD, FCS	Binding affinities	34
Magnetic hydroxyapatite (MHAp)	HEWL and Pepsin A	DLS, UV and FCS	Protein unfolding in the presence of specific functional groups on the NP surface	35
Hydroxyapatite (HAp)	Pepsin	UV–visible spectroscopy and CD	Effect of morphology of nanoparticles on protein structure	36
Carbon nanotubes (CNTs)	Cytochrome c	CD- spectroscopy	CNT surface allows more hydrophilic residues such as Lys to get into close contact with the CNT	37
Selenium nanoparticles	HSA, HHb and Cyt c	UV–vis, FT-IR, SEM-EDX, TEM, DLS and zeta potential	Biological activity of HSA, HHb and Cyt c was retained post nanoparticle interaction	30
Iron oxide magnetic nanoparticles (MNPs)	Immunoglobulin G (IgG)	Fluorescence spectroscopy	Binding affinity and thermodynamic parameters	38

such as 1D proton (1H) NMR, 2D NMR methods, hydrogen–deuterium exchange, saturation transfer and relaxation experiments.

In 1D proton (^{1}H) NMR of the proteins, the amide and aliphatic regions are clearly distinct. To investigate protein–NP interactions by solution NMR spectroscopy, a protein–NP titration is carried out, wherein the protein concentration is kept fixed and the NP concentration is varied or *vice-versa*. At each titration point, 1D ^{1}H NMR spectra are recorded. The intensity of the 1D ^{1}H signal decreases with an increase in the NP concentration. By plotting the concentration of the protein bound against the total NP concentration added during titration, the binding capacity can be calculated (typically $K_d < 1$ μM).[42] Wang *et al.* reported the interaction of five different proteins, namely GB3, human Pin 1, glutathione (GSH), bovine carbonic anhydrase (BCA), and bovine serum albumin (BSA) with AuNPs of 15 nm size using solution NMR spectroscopy.[43] The titration of GB3 with AuNPs using 1D ^{1}H spectra is shown in Figure 9.3A. In the case of GB3 protein they reported an adsorption capacity of 177 ± 20 per NP.

Two-dimensional NMR spectroscopy (2D NMR) helps to recognize a change in backbone amide nitrogen, proton chemical shifts and the intensities of proteins upon interaction with NPs.[45] 2D NMR methods can be used to identify the exact binding site and map the residues on the protein surface

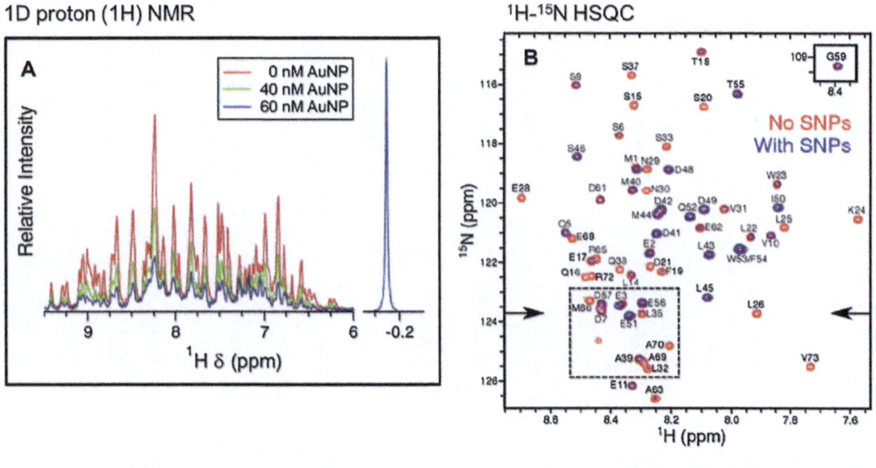

Figure 9.3 (A) Behaviour of GB3 in the presence of AuNPs. Amide proton NMR region in GB3 with various AuNP concentrations. GB3 without AuNPs (red) is overlaid with spectra of 40 nM and 60 nM AuNPs. The internal reference peak for TMSP is shown on the right. Reproduced from ref. 43 with permission from American Chemical Society, Copyright 2014. (B) Effect of interactions of WT p53TAD with silica nanoparticles on the ^{1}H–^{15}N HSQC NMR spectrum. (a) Superimposed spectra of 200 μM protein samples in the absence (red) and presence (blue) of 4.7% (9.5 μM) SNPs by weight. The ^{1}H cross-sections of the two HSQC spectra, indicated by the arrows, are displayed at the top, showing the effect of SNPs on the relative cross-peak intensities. Reproduced from ref. 44 with permission from American Chemical Society, Copyright 2016.

that are binding to the NP surface. Additionally, the confirmational changes of protein can be also studied by 2D NMR methods. ^{1}H–^{15}N HSQC (Heteronuclear Single-Quantum coherence spectrum) is extensively used to study proteins as the fingerprint because each protein exhibits a unique pattern of signal positions in the 2D spectrum. Another 2D experiment TROSY (Transverse Relaxation Optimized SpectroscopY)[46] can be used for larger proteins with high molecular weight. Even with the TROSY method it is highly challenging to study protein–NP interactions due to relaxation issues. The dependence of line broadening on different relaxation mechanisms, mainly dipolar coupling between proton and nitrogen and chemical shift anisotropy (CSA), is utilized in TROSY experiments. The CSA relaxation component strongly depends on the field strength and thus TROSY-type experiments are only effective at high magnetic fields (higher than 600 MHz).[40] Xie *et al.* reported human tumour suppressor p53TAD as a model system to study its interactions with anionic silica nanoparticles.[44] The HSQC NMR spectra of WT p53TAD samples in the absence and presence of silica nanoparticles are shown in Figure 9.3B. They have demonstrated Arg, Lys and His as the binding residues with positive charge that specify electrostatic interactions with negatively charged silica nanoparticles.

To confirm the chemical shift perturbations observed in 2D experiments for proteins are actually arising from interaction with NP and not by a change in the pH of the medium, it is advisable to have the protein and the NP in the same pH and solvent. In the case of commercially obtained NPs, knowledge about solvent or buffer pH in which NPs are suspended is very significant.

Large NP–protein supramolecular complexes exhibit broadened and undetectable signals due to slow rotational diffusion and the fast transverse spin relaxation ($R2$). Therefore, the direct inspection of molecular motions of proteins adsorbed to NP surfaces is difficult with standard 2D NMR experiments.[47] Recently the dark-state exchange saturation transfer (DEST) experiments have been used prominently to study supramolecular protein–NP assemblies.[48] In a DEST experiment there is transfer of ^{15}N magnetization of the observable species by chemical exchange to the invisible or dark state which is then transferred back after dissociation from the high-molecular-weight species. The supramolecular entity prevents direct observation of transverse relaxation rates ($R2$) but allows the efficient partial saturation of the magnetization by a weak radio frequency at a large offset where the signals of the visible species are least affected. When ^{1}H–^{15}N HSQC experiments are used in combination with DEST, the dynamic NMR invisible states of the NP-bound protein can be obtained at a single residue level in the form of $^{15}NR2$ values.[49]

Hydrogen–Deuterium Exchange (HDX) NMR-based experiments are mostly used to monitor the amide protons in the protein structure as they exchange with deuterated solvent.[50–52] This exchange (protons with deuterium) is highly sensitive to structural changes in the protein upon binding with NPs and other environmental factors like the pH of the solvent and the temperature.[53] For the protein–NP complexes, the protein molecules that are bound on the NP surface are in constant exchange with unbound protein molecules in the vicinity, where the exchange dynamics are very fast on the scale of seconds. In such systems if the chemical environment changes, the observable perturbations in HDX rate are expected when protein attaches to the NP surface. In the case of hard protein corona with slower timescale binding, it is difficult to get observable changes in the HDX rates.[54]

9.4 Protein–NP Systems Investigated by Solution NMR Spectroscopy

This section includes various protein–NP systems studied by solution NMR spectroscopy such as noble metal nanoparticles (AuNPs and AgNPs), silica nanoparticles, carbon-based nanomaterials like graphene and fullerene and many others including polymer and lipid nanoparticles.

9.4.1 Noble Metal Nanoparticles (AuNPs and AgNPs)

Noble metal nanoparticles like gold and silver nanoparticles (AuNPs and AgNPs) are gaining tremendous interest due to their size and shape and unique plasmonic properties. They have been used for various biomedical applications as they are easy to synthesize and surface functionalization is feasible also. Amongst AuNPs and AgNPs, NMR investigations on interactions of the former with proteins are abundant in the literature because, unlike AgNPs, they are more stable. Different types of proteins such as Ubiquitin (Ubq), FGF1, BSA, BCA, GSH and GB3 adsorbed onto the AuNP surface have been studied in detail using solution NMR spectroscopy. It has been possible to quantify proteins adsorbed on the AuNP surface and to get binding affinity and structural changes in the protein. The interactions of proteins with nanoparticles are also affected by the size, shape and charge on the NP. In one report, AuNPs of different sizes (14 nm, 30 nm, 43 nm and 86 nm) and three different proteins (GB3, BCA and drkN SH3 domain) were chosen to evaluate the effect of NP size on these proteins using solution NMR spectroscopy along with other biophysical methods.[55] It was observed that GB3 and BCA retained their globular conformation on the surface of AuNP of all size, whereas drkN SH3 exhibits protein deformation on both 14 nm and 30 nm AuNPs as the folding stability of the protein itself is sufficiently low, as shown in Figure 9.4.

Other noble metal nanoparticles that are commercially important as well as finding a wide variety of applications in the nanotechnology field are silver nanoparticles (AgNPs). Nevertheless, these are not as stable as AuNPs and are highly susceptible to oxidation. Recently Atreya's group has extensively studied the interaction of AgNPs with different proteins like ubiquitin,[56] intrinsically disordered proteins (IDP)[57] and antimicrobial peptide Odorranain-A-OA1[58] using solution NMR methods. The protein–NP titration was performed using 2D NMR experiments like sfHMQC. The observed chemical shift perturbations (CSP) were used to identify the binding site on Ubq upon binding with AgNP. They have demonstrated that there is a

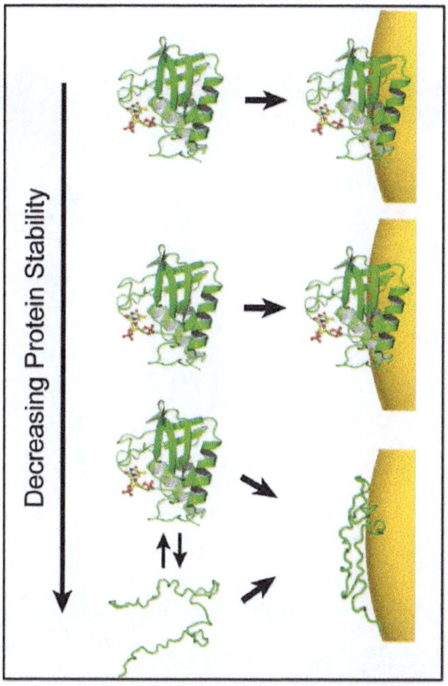

Figure 9.4 Relationship between protein stability upon AuNP adsorption. Stable proteins appear to remain globular when adsorbed, but the unstable drkN SH3 domain exhibits anomalous binding. Reproduced from ref. 55 with permission from American Chemical Society, Copyright 2016.

dynamic and reversible association and dissociation of Ubq from the surface of AgNPs, and that this exchange occurs at a rate much greater than 25 s^{-1}. They have reported the binding constant of 55 μM from NMR titration studies as shown in Figure 9.5, and have also demonstrated the number of protein molecules per AgNP particle using eqn (9.1),[56] where n indicates the number of ubq per one AgNP, and R is the radius of the nanoparticle/protein.

$$n_{\text{ubq}} = 0.65^{*}(R^{3}_{\text{AgNP-Ubq}} - R^{3}_{\text{AgNP}})/R^{3}_{\text{Ubq}} \tag{9.1}$$

The same group carried out similar studies to unravel the structural and dynamic changes of antimicrobial peptide Odorranain-A-OA1 (OA1) upon binding to AgNP.[58] For this peptide, 1D 1H NMR experiments were carried out to monitor the titration of peptide with AgNPs. The observable chemical shift perturbations in peptide upon binding with AgNP confirm the interactions. Another 2D set of NMR experiments, specifically ^{13}C–^{1}H HSQC spectra, was useful to find residues interacting with AgNPs. Interestingly, it was the Cys 15 residue of the peptide that was perturbed due to a direct interaction with AgNP.

In another study by the same group, 10 nm citrate-capped AgNPs were conjugated with unstructured linker domain of human insulin-like growth factor binding protein-2 (L-hIGFBP2). The 2D NMR studies revealed the

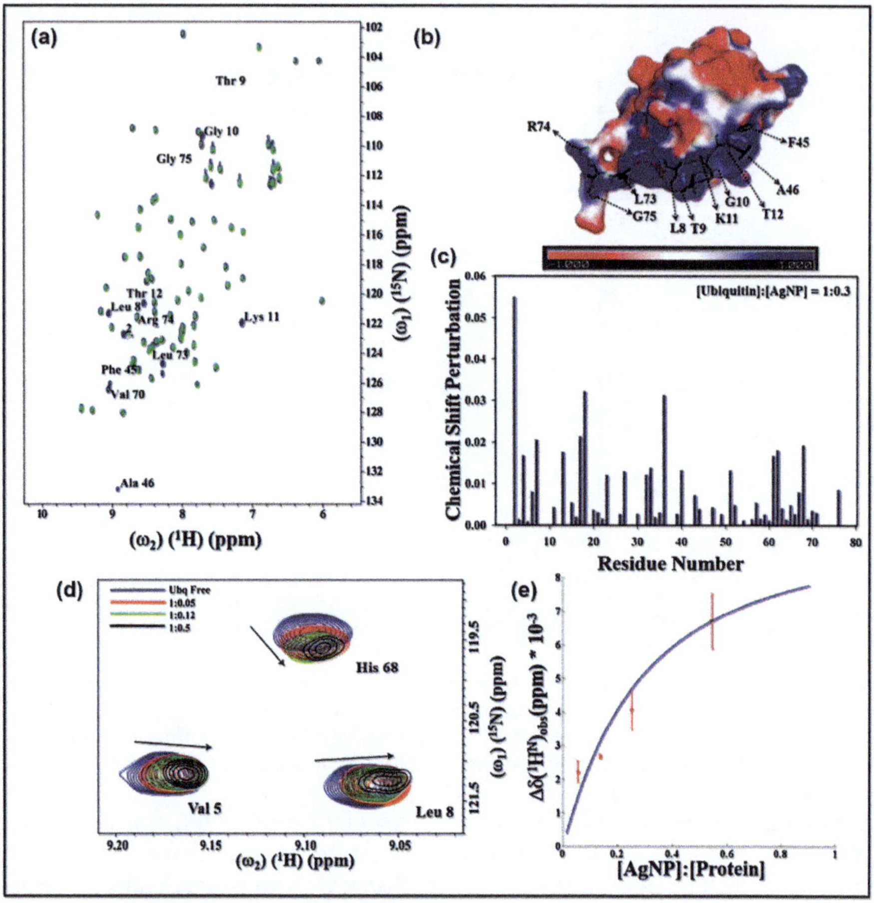

Figure 9.5 (a) Overlay of free (blue) ubiquitin and AgNP–Ubq (green) at a protein: AgNP. (b) The residues broadened out upon interaction with AgNP are mapped on the electrostatic surface of the protein. (c) Chemical shift perturbation (CSP) between the free and bound forms (at a ubiquitin:AgNP ratio of 1 : 0.3) plotted in a residue-wise manner. (d) Overlay of three residues at different protein:AgNP ratios from the reverse titration. (e) A plot of change in ^{1}H chemical shifts of a few residues at different AgNP : protein ratios to estimate the dissociation constant, KD. Reproduced from ref. 56 with permission from the Royal Society of Chemistry.

binding site on IDP upon interaction with AgNPs. Post interaction with AgNPs, the reported IDP becomes more stable as the proteolysis-prone sites were masked after interaction with nanoparticles.[57]

9.4.2 Carbon-based Nanoparticles

Carbon-based nanomaterials (CNMs) like graphene oxide, fullerenes, carbon nanotubes, graphene quantum dots and nanodiamonds are widely used in biomedical applications.[59,60] Therefore, it will be important to investigate

CNMPs at bio interfaces. Calvaresi *et al.* reported the interaction of lysozyme with C_{60} using NMR. They observed the stoichiometric 1 : 1 complex of lysozyme with C_{60}, in which the protein retains its tridimensional structure upon interaction with C_{60}.[61] In another study by Malik *et al.*, the interaction of graphene oxide (GO) with Ubq was reported.[62] Their finding suggests that the protein–GO interactions were electrostatic in nature and by changing the degree of oxidation in GO, protein–GO binding is affected, as shown in Figure 9.6.

9.4.3 Silica Nanoparticles

Silica nanoparticles (SNPs) are the most common nanocarriers for therapeutics and drug-delivery applications because of their tuneable size and shape, ease of functionalization and exceptional biocompatibility. There are several different proteins have been reported to interact with SNPs, such as BSA,[63] RNaseA,[64] cytochrome c,[65] human carbonic anhydrase and lysozyme.[66] 2D NMR methods have been applied to discover the interactions of human carbonic anhydrase (HCAI and HCAII) with 6 and 9 nm SNPs[45,67,68] and assess the interactions of 15 nm SNPs with acylphosphatase (AcP), respectively.[16] The use of solution NMR spectroscopy to investigate various protein–NP complexes in combination with other biophysical techniques is summarized in Table 9.2.

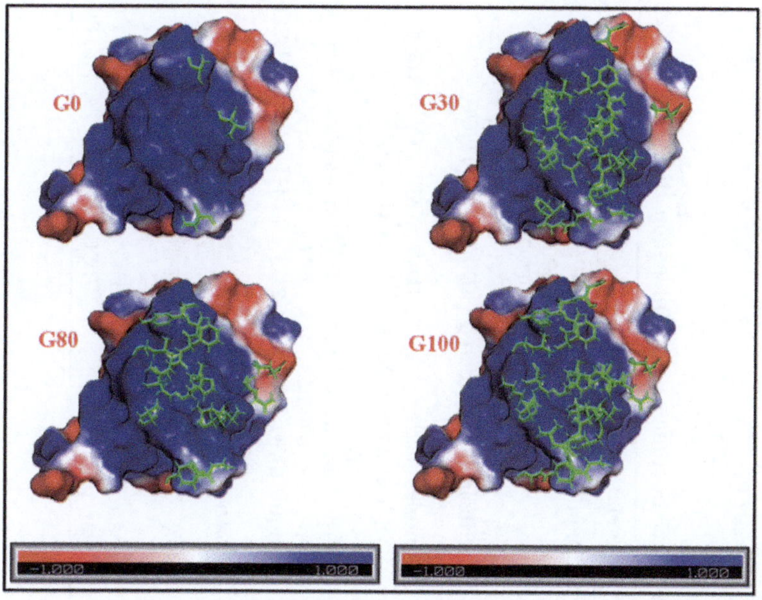

Figure 9.6 Variations in the ubiquitin binding site upon interacting with GO as the degree of oxidation of GO changes; residues marked in green represent the protein backbone residues involved in the interaction with different GO samples. Reproduced from ref. 62 with permission from the Royal Society of Chemistry.

Table 9.2 Solution NMR studies of protein–nanoparticle interactions.

Nanoparticle	Protein	NMR experiments	Findings	References
Gold nanoparticles (AuNPs)	Ubiquitin	CSP; ^{1}H–^{15}N HSQC	Identification of binding site	69
	GB3	CSP and peak intensity; ^{1}H–^{15}N TROSY–HSQC	Three-step model on protein–NP binding was proposed	43
Silver nanoparticles (AgNPs)	Ubiquitin	HMCQ, CSP, $T2$ (^{1}H) relaxation	Highly stable Ubq-AgNP complex	56
	Intrinsically disordered proteins (IDP)	2D [^{15}N, ^{1}H] HSQC	IDP retained stability and becomes proteolysis resistant upon binding with AgNP	57
	Peptides	1D ^{1}H NMR ^{13}C–^{1}H HSQC	Identification of binding site	58
Silica nanoparticles (SNPs)	Acylphosphatase (AcP)	[^{15}N – ^{1}H]-HSQC	Protein undergoes specific and size-dependent orientation on nanoparticle surfaces.	16
	Intrinsically disordered proteins (IDP)p53	^{1}H–^{15}N cross-peaks of enhanced-HSQC-type, CPMG	Long-range electrostatic and short-range hydrophobic interactions	44
Carbon-based nanoparticles	Ubiquitin-GO	2D [^{15}N, ^{1}H] HSQC and CSP	Interaction of the protein with GO can thus be modulated by tuning the degree of oxidation	62
	Lysozyme-C60	2D ^{1}H,^{15}N HSQC and CSP	Stoichiometric 1 : 1 complex of lysozyme with C60	61
Polymer nanoparticles	BLA	^{1}H–TOCSY; RT-HDX	Helix C, calcium binding site, andpart of its domain	70
Lipid	β-barrel protein OprHΔL1ΔL4	^{1}H–^{13}C HMQC and H(C)(CC)-TOCSY-(CO)- [^{15}N,^{1}H]-TROSY	Binding sites and dissociation constants	71

9.5 Therapeutic Approaches to Protein–NP Interactions

Protein–NP interactions are essentially important to study because when the NPs are subjected to a biological medium, they form a protein–NP complex, the so-called "protein corona", which defines the true identity of the NPs.[72] Once the NPs are administered as therapeutics or drug-delivery agents their circulation, biodistribution, bioavailability, cellular uptake and immune response are affected by corona formation.[73–76] The uncontrolled protein corona may lead to negative effects in targeted therapies, rapid clearance of NPs and induced toxicity.[77] Meanwhile, the protein corona formation is controlled in two ways: (1) through controlled protein adsorption by modifying NPs with various functional proteins and (2) by tailoring appropriate NPs which favour selective peptides and protein adsorption. By systematic engineering of protein corona and by modulation of protein–NP interactions, approaches to promising therapeutics and targeted cancer treatments and other nanomedicines can be designed.[78–80]

9.6 Conclusion and Future Prospects

It is well established now that "protein corona" is the true identity of NPs when exposed to a biological medium as a consequence of protein adsorption. Therefore, it is vital to understand protein–NP interactions. The combinations of various techniques such as UV–vis, CD, FT-IR, MS, DLS, ITC and solution NMR have been used to understand protein–NP interactions. Nonetheless by solution NMR spectroscopy, it is possible to obtain high-resolution structural information of proteins at the specificity of amino acid level on protein–NP complex formation. There remains wider scope to explore the involvement of amino acid side chains in protein–NP interactions and therefore advanced NMR methods and experiments based on isotope labelling, analysis of bound state and dynamics at nano bio-interfaces need to be developed. The accessibility of this detailed information on protein–NP interactions using NMR will open up new opportunities in the design of NP-based therapeutics and nanomedicines to achieve the desired biological and therapeutic results using nanoparticles.

Acknowledgements

We are thankful to Jain University, Bangalore, India, for providing facilities. VB also acknowledges TARE-SERB File No.: TAR/2018/000547.

References

1. D. Docter, D. Westmeier, M. Markiewicz, S. Stolte, S. K. Knauer and R. H. Stauber, *Chem. Soc. Rev.*, 2015, **44**, 6094–6121.
2. P. C. Ke, S. Lin, W. J. Parak, T. P. Davis and F. Caruso, *ACS Nano*, 2017, **11**, 11773–11776.

3. E. Casals, T. Pfaller, A. Duschl, G. J. Oostingh and V. Puntes, *ACS Nano*, 2010, **4**, 3623–3632.
4. C. Gunawan, M. Lim, C. P. Marquis and R. Amal, *J. Mat. Chem. B*, 2014, **2**, 2060–2083.
5. S. Juling, A. Niedzwiecka, L. Böhmert, D. Lichtenstein, S. Selve, A. Braeuning, A. F. Thünemann, E. Krause and A. Lampen, *J. Proteome Res.*, 2017, **16**, 4020–4034.
6. M. Lundqvist, J. Stigler, T. Cedervall, T. Berggård, M. B. Flanagan, I. Lynch, G. Elia and K. Dawson, *ACS Nano*, 2011, **5**, 7503–7509.
7. R. L. Pinals, L. Chio, F. Ledesma and M. P. Landry, *Analyst*, 2020, **145**, 5090–5112.
8. S. Shahabi, L. Treccani, R. Dringen and K. Rezwan, *Nanoscale*, 2015, **7**, 16251–16265.
9. M. Vitali, E. Casals, F. Canals, N. Colomé and V. Puntes, *Nanoscale*, 2020, **12**, 15832–15844.
10. F. Ahmad, Y. Zhou, Z. Ling, Q. Xiang and X. Zhou, *RSC Adv.*, 2016, **6**, 35719–35730.
11. L. Shang and G. U. Nienhaus, *Acc. Chem. Res.*, 2017, **50**, 387–395.
12. D. T. Jayaram, S. M. Pustulka, R. G. Mannino, W. A. Lam and C. K. Payne, *Biophys. J.*, 2018, **115**, 209–216.
13. L. Shang, Y. Wang, J. Jiang and S. Dong, *Langmuir*, 2007, **23**, 2714–2721.
14. V. Pareek, A. Bhargava, V. Bhanot, R. Gupta, N. Jain and J. Panwar, *J. Nanosci. Nanotechnol.*, 2018, **18**, 6653–6670.
15. C. Wang, B. Chen, M. He and B. Hu, *ACS Nano*, 2021, **15**, 3108–3122.
16. S. Shrivastava, S. A. McCallum, J. H. Nuffer, X. Qian, R. W. Siegel and J. S. Dordick, *Langmuir*, 2013, **29**, 10841–10849.
17. M.-M. Yin, W.-Q. Chen, Y.-Q. Lu, J.-Y. Han, Y. Liu and F.-L. Jiang, *Nanoscale*, 2020, **12**, 4573–4585.
18. E. R. P. Zuiderweg, *Biochemistry*, 2002, **41**, 1–7.
19. R. E. Luna, S. R. Akabayov, J. J. Ziarek and G. Wagner, *FEBS J.*, 2014, **281**, 1965–1973.
20. O. Vinogradova and J. Qin, *Top. Curr. Chem.*, 2012, **326**, 35–45.
21. U. Sakulkhu, M. Mahmoudi, L. Maurizi, J. Salaklang and H. Hofmann, *Sci. Rep.*, 2014, **4**, 5020.
22. M. Jansch, P. Stumpf, C. Graf, E. Rühl and R. H. Müller, *Int. J. Pharm.*, 2012, **428**, 125–133.
23. L. Zeng, J. Gao, Y. Liu, J. Gao, L. Yao, X. Yang, X. Liu, B. He, L. Hu, J. Shi, M. Song, G. Qu and G. Jiang, *TrAC, Trends Anal. Chem.*, 2019, **118**, 303–314.
24. R. Rampado, S. Crotti, P. Caliceti, S. Pucciarelli and M. Agostini, *Front. Bioeng. Biotechnol.*, 2020, **8**, 166.
25. P. Satzer, F. Svec, G. Sekot and A. Jungbauer, *Eng. Life Sci.*, 2016, **16**, 238–246.
26. J. Bing, X. Xiao, D. J. McClements, Y. Biao and C. Chongjiang, *Food Hydrocolloids*, 2021, **115**, 106594.
27. S. Reymond-Laruinaz, L. Saviot, V. Potin and M. d. C. Marco de Lucas, *Appl. Surf. Sci.*, 2016, **389**, 17–24.

28. Z. Varga, B. Fehér, D. Kitka, A. Wacha, A. Bóta, S. Berényi, V. Pipich and J.-L. Fraikin, *Colloids Surf., B*, 2020, **192**, 111053.
29. C. Gillespie, P. Halling and D. Edwards, *Colloids Surf., A*, 2011, **384**, 233–239.
30. N. Shahabadi, S. Zendehcheshm and F. Khademi, *Biotechnol. Rep.*, 2021, **30**, e00615.
31. N. William, F. Bamidoro, P. A. Beales, R. Drummond-Brydson, N. Hondow, S. Key, A. Kulak, A. C. Walsh, S. Winter and L. A. Nelson, *J. Colloid Interface Sci.*, 2021, **594**, 101–112.
32. G. Wang, W. Wang, E. Shangguan, S. Gao and Y. Liu, *Mater. Sci. Eng. C*, 2020, **111**, 110830.
33. Y. Wang and Y. Ni, *Analyst*, 2014, **139**, 416–424.
34. R. Barbir, I. Capjak, T. Crnković, Ž. Debeljak, D. Domazet Jurašin, M. Ćurlin, G. Šinko, T. Weitner and I. Vinković Vrček, *Chem. - Biol. Int.*, 2021, **335**, 109364.
35. K. Kadu, G. Ghosh, L. Panicker, M. Kowshik and S. Roy Ramanan, *Colloids Surf., B*, 2019, **177**, 362–369.
36. K. Kadu, M. Kowshik and S. Roy Ramanan, *Mater. Today Commun.*, 2021, **26**, 102172.
37. C. Zhang, X. Li, Z. Wang, X. Huang, Z. Ge and B. Hu, *J. Phys. Chem. B*, 2020, **124**, 684–694.
38. A. V. Bychkova, M. V. Lopukhova, L. A. Wasserman, P. G. Pronkin, Y. N. Degtyarev, A. I. Shalupov, A. D. Vasilyeva, L. V. Yurina, A. L. Kovarski, A. S. Kononikhin and E. N. Nikolaev, *Biochim. Biophys. Acta*, 2020, **1868**, 140300.
39. M. Assfalg, L. Ragona, K. Pagano, M. D'Onofrio, S. Zanzoni, S. Tomaselli and H. Molinari, *Biochim. Biophys. Acta*, 2016, **1864**, 102–114.
40. Y. R. Perera, J. X. Xu, D. L. Amarasekara, A. C. Hughes, I. Abbood and N. C. Fitzkee, *Molecules*, 2021, **26**, 5788.
41. A. Bortot, S. Zanzoni, M. D'Onofrio and M. Assfalg, *Chemistry*, 2018, **24**, 5911–5919.
42. Y. R. Perera, T. M. South, A. C. Hughes, A. N. Parkhurst, O. C. Williams, M. B. Davidson, C. A. Wilks, D. A. Mlsna and N. C. Fitzkee, *J. Chem. Educ.*, 2020, **97**, 820–824.
43. A. Wang, K. Vangala, T. Vo, D. Zhang and N. C. Fitzkee, *J. Phys. Chem. C*, 2014, **118**, 8134–8142.
44. M. Xie, A. L. Hansen, J. Yuan and R. Brüschweiler, *J. Phys. Chem. C*, 2016, **120**, 24463–24468.
45. M. Lundqvist, I. Sethson and B.-H. Jonsson, *Langmuir*, 2005, **21**, 5974–5979.
46. M. Salzmann, K. Pervushin, G. Wider, H. Senn and K. Wüthrich, *Proc. Natl. Acad. Sci. U S A*, 1998, **95**, 13585–13590.
47. A. Ceccon, M. Lelli, M. D'Onofrio, H. Molinari and M. Assfalg, *J. Am. Chem. Soc.*, 2014, **136**, 13158–13161.
48. T. Das, D. Acosta and D. Eliezer, *Methods Mol. Biol.*, 2020, **2141**, 585–608.
49. N. L. Fawzi, J. Ying, D. A. Torchia and G. M. Clore, *Nat. Protoc.*, 2012, **7**, 1523–1533.

50. L. Ragona, F. Fogolari, S. Romagnoli, L. Zetta, J. L. Maubois and H. Molinari, *J. Mol. Biol.*, 1999, **293**, 953–969.
51. V. Chevelkov, Y. Xue, D. K. Rao, J. D. Forman-Kay and N. R. Skrynnikov, *J. Biomol. NMR*, 2010, **46**, 227–244.
52. J. Abildgaard, P. E. Hansen, M. N. Manalo and A. LiWang, *J. Biomol. NMR*, 2009, **44**, 119–126.
53. A. Wang, T. Vo, V. Le and N. C. Fitzkee, *J. Phys. Chem. B*, 2014, **118**, 14148–14156.
54. D. Baimanov, R. Cai and C. Chen, *Bioconjugate Chem.*, 2019, **30**, 1923–1937.
55. K. E. Woods, Y. R. Perera, M. B. Davidson, C. A. Wilks, D. K. Yadav and N. C. Fitzkee, *J. Phys. Chem. C*, 2016, **120**, 27944–27953.
56. V. P. Brahmkhatri, K. Chandra, A. Dubey and H. S. Atreya, *Nanoscale*, 2015, **7**, 12921–12931.
57. S. A. Malik, S. Mondal and H. S. Atreya, *RSC Adv.*, 2019, **9**, 28746–28753.
58. I. Pal, V. P. Brahmkhatri, S. Bera, D. Bhattacharyya, Y. Quirishi, A. Bhunia and H. S. Atreya, *J. Colloid Interface Sci.*, 2016, **483**, 385–393.
59. C. Cha, S. R. Shin, N. Annabi, M. R. Dokmeci and A. Khademhosseini, *ACS Nano*, 2013, **7**, 2891–2897.
60. H. Wang, Q. Chen and S. Zhou, *Chem. Soc. Rev.*, 2018, **47**, 4198–4232.
61. M. Calvaresi, F. Arnesano, S. Bonacchi, A. Bottoni, V. Calò, S. Conte, G. Falini, S. Fermani, M. Losacco, M. Montalti, G. Natile, L. Prodi, F. Sparla and F. Zerbetto, *ACS Nano*, 2014, **8**, 1871–1877.
62. S. A. Malik, Z. Mohanta, C. Srivastava and H. S. Atreya, *Nanoscale Adv.*, 2020, **2**, 1904–1912.
63. P. Roach, D. Farrar and C. C. Perry, *J. Am. Chem. Soc.*, 2006, **128**, 3939–3945.
64. W. Shang, J. H. Nuffer, J. S. Dordick and R. W. Siegel, *Nano Lett.*, 2007, **7**, 1991–1995.
65. W. Shang, J. H. Nuffer, V. A. Muñiz-Papandrea, W. Colón, R. W. Siegel and J. S. Dordick, *Small*, 2009, **5**, 470–476.
66. M. Lundqvist, I. Sethson and B. H. Jonsson, *Langmuir*, 2004, **20**, 10639–10647.
67. M. Lundqvist, I. Sethson and B.-H. Jonsson, *Langmuir*, 2004, **20**, 10639–10647.
68. M. Lundqvist, I. Sethson and B.-H. Jonsson, *Biochemistry*, 2005, **44**, 10093–10099.
69. L. Calzolai, F. Franchini, D. Gilliland and F. Rossi, *Nano Lett.*, 2010, **10**, 3101–3105.
70. D. Bhowmik, K. R. Mote, C. M. MacLaughlin, N. Biswas, B. Chandra, J. K. Basu, G. C. Walker, P. K. Madhu and S. Maiti, *ACS Nano*, 2015, **9**, 9070–9077.
71. I. Kucharska and L. K. Tamm, *Biochemistry*, 2017, **56**, 4291–4292.
72. S. J. Park, *Int. J. Nanomed.*, 2020, **15**, 5783–5802.
73. F. S. M. Tekie, M. Hajiramezanali, P. Geramifar, M. Raoufi, R. Dinarvand, M. Soleimani and F. Atyabi, *Sci. Rep.*, 2020, **10**, 9664.

74. L. Ding, C. Yao, X. Yin, C. Li, Y. Huang, M. Wu, B. Wang, X. Guo, Y. Wang and M. Wu, *Small*, 2018, **14**, 1801451.
75. M. Barbalinardo, F. Caicci, M. Cavallini and D. Gentili, *Small*, 2018, **14**, 1801219.
76. R. Cai, J. Ren, Y. Ji, Y. Wang, Y. Liu, Z. Chen, Z. Farhadi Sabet, X. Wu, I. Lynch and C. Chen, *ACS Appl. Mater. Interfaces*, 2020, **12**, 1997–2008.
77. M. M. Yallapu, N. Chauhan, S. F. Othman, V. Khalilzad-Sharghi, M. C. Ebeling, S. Khan, M. Jaggi and S. C. Chauhan, *Biomaterials*, 2015, **46**, 1–12.
78. J. Y. Oh, H. S. Kim, L. Palanikumar, E. M. Go, B. Jana, S. A. Park, H. Y. Kim, K. Kim, J. K. Seo, S. K. Kwak, C. Kim, S. Kang and J.-H. Ryu, *Nat. Commun.*, 2018, **9**, 4548.
79. V. H. Nguyen and B. J. Lee, *Int. J. Nanomed.*, 2017, **12**, 3137–3151.
80. D. H. Jo, J. H. Kim, J. G. Son, K. S. Dan, S. H. Song, T. G. Lee and J. H. Kim, *Biomaterials*, 2016, **109**, 23–31.

CHAPTER 10

NMR Structures, Dynamics and Interactions of Protein Complexes in β2 Integrins

SURAJIT BHATTACHARJYA

School of Biological Sciences, Nanyang Technological University, 60 Nanyang Drive, Singapore 637551, Singapore
Email: surajit@ntu.edu.sg

10.1 Integrins: An Overview

Integrins are master regulators of cell adhesion and migration in multicellular eukaryotes or metazoans.[1] In addition, integrins establish transmembrane linkage with cytoskeleton filaments and activate multiple cell signalling cascades.[1] Integrins and cognate ligand molecules are pivotal in key physiological processes, including leucocyte trafficking, immune regulation, haemostasis and development. The advancement of effective therapeutics against cancer, auto-immune, inflammation and blood disorders is targeted at integrins.[2–4] Integrin molecules can be described as type I transmembrane protein receptors consisting of two subunits, α and β, of a functional heterodimer. To date, eight such β subunits are known to be present with 18 α subunits representing the 24-integrin family of receptors (Figure 10.1). It is noteworthy that all of 24 integrin receptors are functionally nonredundant as disclosed from mouse gene knockout studies.[5–7] Each subunit of integrin, based on cellular localization, can be divided into three domains, a large extracellular domain, a transmembrane (TM) helix and a

New Developments in NMR No. 26
NMR Spectroscopy for Probing Functional Dynamics at Biological Interfaces
Edited by Anirban Bhunia, Hanudatta S. Atreya and Neeraj Sinha

Published by the Royal Society of Chemistry, www.rsc.org

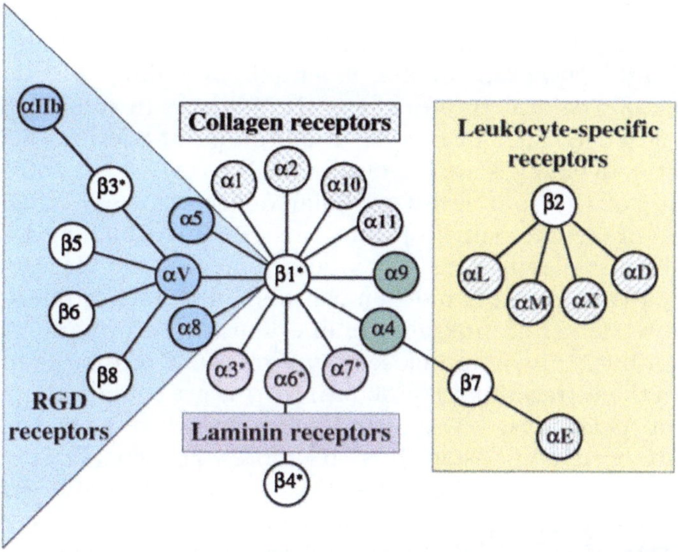

Figure 10.1 Integrin receptor family in mammals. Twenty-four heterodimeric αβ integrins are formed from eight β subunits (β1, β2, β3, β4, β5, β6, β7 and β8) and 18 α subunits (α1 to α10, αIIb, αL, αM, αX, αD and αE). Asterisks indicate the presence of alternatively spliced cytoplasmic domains in β1, β3, β4, α3, α6 and α7 subunits. α subunits marked with hatching contain inserted I/A domains. These α subunits, also α4, α9 (green) and β2–β8 are found in chordates. α subunits, observed across all metazoans, restricted to RGD and laminin are highlighted in blue and purple, respectively. Leucocyte-specific β2 integrins (αLβ2, αMβ2, αDβ2, αXβ2) and sole β7αE are indicated in yellow. Reproduced from ref. 1 with permission from Elsevier, Copyright 2002.

relatively short cytoplasmic tail (CT) (20–50 residues long) except for β4 (~1000 amino acids). The α and β subunits are held together by numerous non-covalent interactions contributed by all the three domains. The inter-subunit contacts are better realized for the resting state of the receptor compared to its activated state. Bidirectional (inside-out and outside-in) signalling is a remarkable feature observed in all members of the integrin family.[1,2,8–10] Inside-out signalling of integrins occurs because of either agonist stimulation, signalling *via* G-protein-coupled receptors and/or growth factor receptors. These signalling processes can generate cytosolic binding proteins which are necessary for the activation of integrins. On the other hand, in the case of the outside-in signalling pathway, ligand molecules directly bind to the extracellular domains of integrins resulting in activation for signal transduction across the receptor. In the absence of any signal, residing on the cell surface integrins maintain an "OFF" or resting state.

The resting or inactive state of integrins does not bind to ligands and does not transmit a signal. Constitutively active integrins are known to be responsible for serious pathological consequences, *e.g.* bleeding disorder in platelet αIIbβ3 integrin and LAD (leucocyte adhesion deficiency) in β2 leucocyte integrins.[11,12] Apparently, large allosteric structural changes to integrins are proposed to be essential for bidirectional signalling (Figure 10.2). The structure of the entire integrin in full-length either at inactive or active states at atomic resolution is unknown. The models of conformational changes are largely assumed based on x-ray structures and EM images of the soluble extracellular domains of integrins. The x-ray structures of ectodomains of integrins αVβ3, αIIbβ3 and αXβ2 have revealed bent conformations that would represent the resting states.[13–17] EM images coupled with biochemical and functional studies evidenced that activated states of integrins are in extended conformations with facile ligand binding.[17] Open and closed conformations of the independent I/A domain in the extracellular region of integrins are stabilized upon binding to divalent metal ions.[18] The ligand-bound structure of the headpiece of αIIbβ3 has been demonstrated to be more open compared to the free counterpart.[14] Similarly to ectodomains, the TMs of α and β subunits of integrins are packed against each other in resting receptors. NMR structures of TMs of αIIbβ3 integrins are determined in bicelles (Figure 10.3) and in isotropic organic solvent (Figure 10.3).[19,20] Two parallel TM α-helices are intertwined coiled-coil, whereby the sidechains of hydrophobic residues are interacted as a knob into a hole pattern where Gly residues are involved in maintaining such close interactions. TM helices are well packed at the central hydrophobic region with reduced sidechain–sidechain interactions towards the C-termini. The extensive TM–TM packing interactions of α and β subunits are necessary for the resting state of the receptors which are thought to be disrupted during activation stabilizing the elongated conformation (Figure 10.2).[21,22] Amino acid sequences of TM domains are well conserved across several integrins, indicating similar structural characteristics of TM–TM association. Two cytosolic tails or α CT and β CT of integrins, although short and flexible (in the absence of membrane), can indeed form transient CT complexes. CTs and molecular interactions between them are of high significance in bidirectional signalling of integrins. Constitutive activation of integrins is observed upon deletion of one of the CTs or mutations of key residues that destabilized CT–CT interactions.[21–23] Therefore, allosteric changes of heterodimeric integrin from the bent conformation to upright or straight conformation would necessitate cooperative disruption of several interactions involving ectodomains, TMs and CTs.[7–9] The structural and dynamical views of the bent to straight conformational transitions and potential intermediates in the pathway in an intact or full-length integrin are largely elusive. Structural determinants and protein/ligand interactions which are central to bidirectional signalling are investigated with independent domains. Regardless, the uses of various structural tools, NMR, X-ray, cryo-EM, MD simulations, in conjunction with functional analyses are successful in deciphering much of the mechanistic principles involved in integrin functional regulations.

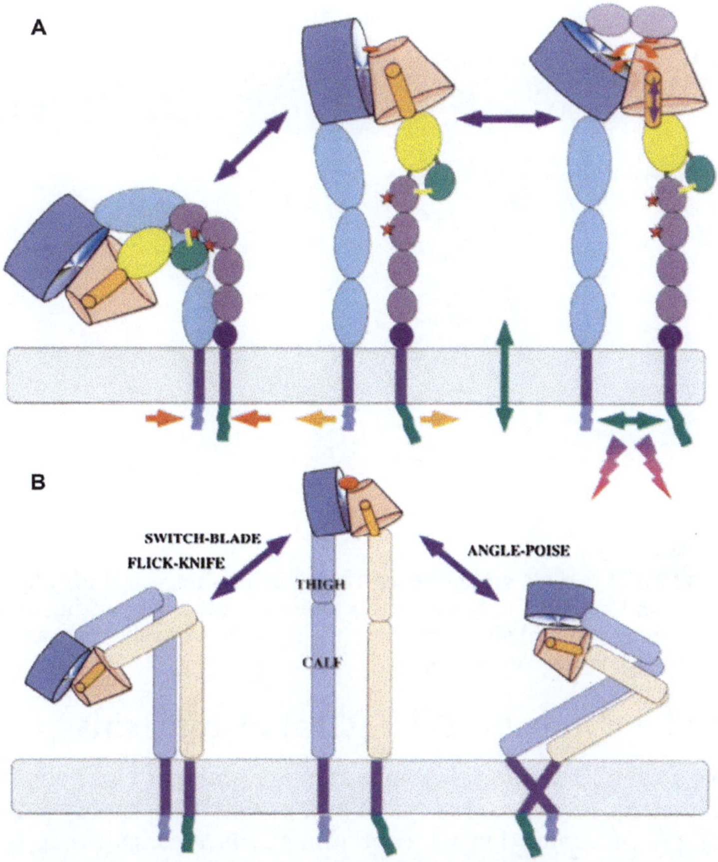

Figure 10.2 Models of activation of integrins bidirectional signalling. (A) Resting integrins assume bent conformations stabilized by multiple intersubunit interactions (left). During activation, with ligand binding either to the extracellular domains or recruitment of proteins to cytosolic tails, inter-subunit interactions are diminished resulting in upright or straight conformations. Conversion of inactive bent to straight conformations may occur *via* more than one step (middle and right). Maximum separation is likely to require signalling. (B) Two proposed models for the straightening up of integrins, namely the switch blade or flick knife and angle poised, differ on the possible orientation of the cytosolic tails with respect to the TMs. The angle poised model suggests that long TMs are dynamic with potential movement in or out of the membrane plane during activation. Domains of α and β subunits are differently colour coded. α subunit: from extracellular N-terminus, β-propeller (blue), thigh and calf (1 and 2) domains (light blue), TM (purple, thick straight line), cytosolic tail (light blue, jagged line). β subunit: from extracellular N-terminus, βIA (pink), hybrid domain (yellow), C-terminus helix, PSI domain (green), I-EGF domain (purple), disulfide bond between the first I-EGF and PSI (yellow stick), β-TD (purple, small circle), TM (purple, thick straight line), cytosolic tail (green, jagged line), activation epitopes in I-EGFs in star. Inward and outward arrows indicate interactions and separation between the subunits, respectively, with lightening-type arrows representing signalling. Membrane is shown as a thin rectangle (light grey). Reproduced from ref. 1 with permission from Elsevier, Copyright 2002.

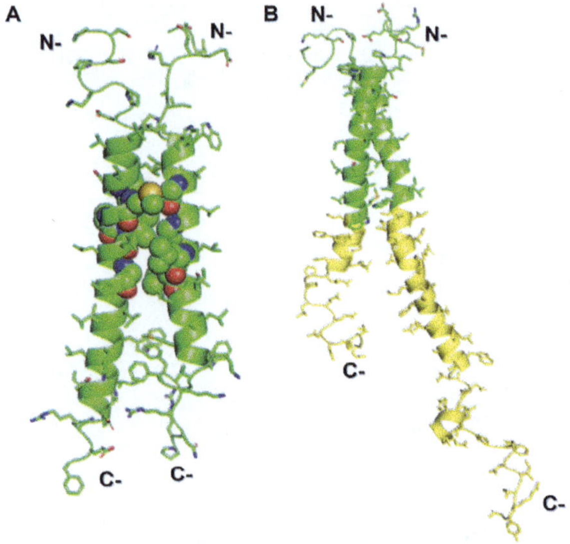

Figure 10.3 NMR-derived structures of TMs of αIIbβ3 integrin in (A) lipid bicelles and in (B) organic solvent acetonitrile. Sidechain packing of the knob–hole is highlighted as space-fill (A). The cytosolic tails are indicated with yellow ribbon (B).

10.2 The Cytosolic Tails (CTs) of Integrins

The CTs of α and β subunits of integrins are eminent components for receptor signalling functions. Except β4 and β8 CTs, the primary structures of all β CTs are of similar length, containing conserved sequence motifs for intracellular protein interactions (Figure 10.4). The β4 CT is unusually long (>1000 amino acids) with the occurrence of multiple fibronectin-like domains. Meanwhile the β8 CT is devoid of canonical protein interaction motifs and is believed to be regulated by alternative mechanisms (Figure 10.4). By contrast, the primary structures of α CTs demonstrate considerable variations in length, sequence and composition (Figure 10.5). A plethora of intracellular proteins have been identified to interact with β CTs.[24–26] On the other hand, a relatively reduced set of cytosolic proteins are known to be associated with α CTs.[24–28] Specific conserved motifs in β CTs are identified for the recruitment of integrin regulatory proteins (Figure 10.4). For example, membrane proximal NPxY/F or NPxxY motif is central for binding to the head domain of talin which is known to activate integrins (Figure 10.4). The large cytoskeletal protein filamin binds to the membrane distal (MD) segment of β CTs and maintains the resting state of integrins.[29–33] Other pivotal proteins, namely Dok1, 14-3-3ζ, kindlins and ICAP1, regulate integrin activations. The conserved MP GFFKR motif in α CTs confers interactions of integrin-inactivating proteins (Figure 10.5). Interactions of cytosolic proteins with CTs, in particular with β CTs, play major roles in integrin bidirectional signalling, including maintaining the β1: ^{752}KLLMIIHDRREFAKFEKEKMNAKWDTGENPIYKSAVTTVVNPKYEGK^{798}

β2: ^{724}KALIHLSDLREYRRFEKEKLKSQWNNDNPLFKSATTTVMNPKFAES^{769}

β3: ^{742}KLLITIHDRKEFAKFEEERARAKWDTANNPLYKEATSTFTNITYRGT^{788}

β4: ^{734}KYCACCKACLALLPCCNRG----FQT^{1182}

β5: ^{743}KLLVTIHDRREFAKFQSERSRARYEMASNPLYRKPISTHTVDFTFNKFNKSYNGTVD^{799}

β6: ^{731}KLLVSFHDRKEVAKFEAERSKAKWQTGTNPLYRGSTSTFKNVTYKHREKQKVDLSTDC^{788}

β7: ^{747}RLSVEIYDRREYSRFEKEQQQLNWKQDSNPLYKSAITTTINPRFQEADSPTL^{798}

β8: ^{705}RQVILQWNSNKIKSSSDYRVSASKKDKLILQSVCTRAVTYRREKPEEIKMDISKLNAHETFRCNF^{769}

Figure 10.4 Amino acid sequences of β CTs of integrins. Protein interaction motifs, *i.e.* membrane proximal NPxY/F and membrane distal NxxY/F, are highlighted in red. Phosphorylation TTT motif and Ser phosphorylation of β2 CT are indicated in blue and green, respectively. The β4 CT is long and so the entire sequence is not shown here. The β8 CT does not contain canonical protein-binding motifs.

αL:^{1113}KVGFFKRNLKEKMEAGRGVPNGIPAEDSEQLASGQEAGDPGCLKPLHEKDSE SGGGKD^{1170}

αM: ^{1129}KLGFFKRQYKDMMSEGGPPGAEPQ^{1152}

αX: ^{1129}KVGFFKRQYKEM MEEANGQIAPENGTQTPSPPSEK^{1163}

αD: ^{1124}KLGFFKRHYKEMLEDKPEDTATFSGDDFSCVAPNVPLS^{1161}

αV: ^{1017}RMGFFKRVRPPQEEQEREQLQPHENGEGNSET^{1048}

αE: ^{1148}KCGFFKRKYQQLNLESIRKAQLKSENLLEEEN^{1179}

αIIb: ^{1020}KVGFFKRNRPPLEEDDEEGE^{1039}

α1: ^{1165}KIGFFKRPLKKKMEK^{1179}

α2: ^{1155}KLGFFKRKYEKMTKNPDEIDETTELSS^{1181}

α3: ^{1015}KCGFFKRARTRALYEAKRQKAEMKSQPSETERLTDDY^{1051}

α4: ^{1001}KAGFFKRQYKSILQEENRRDSWSYINSKSNDD^{1032}

α5: ^{1022}KLGFFKRSLPYGTAMEKAQLKPPATSDA^{1049}

α6: ^{1077}KCGFFKRSRYDDSVPRYHAVRIRKEEREIKDEKYIDNLEKKQWITKWNENESYS^{1130}

α7: ^{1105}KMGFFKRAKHPEATVPQYHAVKIPREDRQQFKEEKTGTILRNNWGSPRREGPDAHPILAADGHPELGPDGHPGPGTA^{1181}

α8: ^{1035}KCGFFDRARPPQEDMTDREQLTNDKTPEA^{1036}

α9: ^{1003}KMGFFRRRYKEIIEAEKNRKENEDSWDWVQKNQ^{1035}

α10: ^{1146}KLGFFAHKKIPEEEKREEKLEQ^{1167}

α11: ^{1165}KLGFFRSARRRREPGLDPTPKVLE^{1188}

Figure 10.5 Amino acid sequences of α CTs of integrins. The conserved membrane proximal GFFRR motif is highlighted in red. Note that the α8 CT contains residue Asp in place of Lys. Also, in the α10 CT the motif is shorter.

resting states, activation and clustering of receptors (see below). Notably, binding interactions of CTs/proteins or CT/CT are often of low affinity. NMR spectroscopy has provided much needed atomic resolution information of these transient complexes as discussed in the following sections.

10.3 The Cytosolic Tails of β2 Integrins: Structures and Interactions

The family of leucocyte-specific integrins consists of four members: αLβ2, αMβ2, αXβ2 and αDβ2 (Figure 10.1).[1,8] β2 integrins are essential for leucocyte migration, adhesion and functional immune systems. Members of β2 integrins demonstrate differences in cellular expression and extracellular ligand-binding specificity. All leucocytes express αLβ2 integrin, whereas the expressions of the other three integrins (αMβ2, αXβ2 and αDβ2) are cell specific.[1,8] The extracellular domain of αLβ2 predominantly binds to the ICAM group of ligands. By contrast, αMβ2, αXβ2 and αDβ2 integrins are more promiscuous in recognizing extracellular ligands. Structures of β2, αL, αM and αX CTs are solved together with interactions between cognate CTs through the use of heteronuclear NMR methods.[34–36] The very first structures and interactions were reported for β2 and αL CTs.[34] The CT of αL is uniquely longer compared to other α CTs (Figure 10.5). Recombinant expression of full-length CTs as soluble proteins was found to be difficult due to undesired proteolytic digestion. Therefore, CTs were recombinantly expressed as fusion proteins of ketosteroid isomerase (KSI) as inclusion bodies in *E. coli* cells. The expressed protein constructs KSI-M-CT-His_6 or KSI-His_6-D-P-CT, isotopically labelled (^{15}N, $^{15}N/^{13}C$) for NMR, were purified under denaturing conditions using histidine-tag affinity binding with Ni-NTA agarose column. Further, KSI and His-tag were removed from fusion proteins either by treatment with cyanogen bromide or formic acid followed by repurification in HPLC. This procedure has yielded soluble CT proteins in large quantities for NMR structural studies. Well-resolved ^{15}N–^{1}H HSQC spectra were obtained and most of the amino acid residues can be assigned for β2 and αL CTs (Figure 10.6). The limited dispersion of amide proton chemical shift (~0.8 ppm) observed in ^{15}N–^{1}H HSQC spectra indicates the absence of highly populated well-folded conformations of both CTs in free solutions. However, analyses of secondary $^{13}C_{\alpha}$ chemical shift values and nuclear Overhauser effect spectra (NOSEY) indicated that residues of the MP region of β2 CT assumed helical conformations. Residues belonging to the C-terminus MD part appeared to lack any detectable regular secondary structures. Interestingly, αL CT demarcated stretches of discretely folded short helical structures at the N- and C-termini. Several long-range NOEs indicated that the helices are in proximity. An ensemble of atomic-resolution structures of αL CT was determined using NOE-driven distance constraints (Figure 10.7).

The compact structure of αL CT is defined by the presence of three folded helices, *i.e.* N-terminal helix 1 or MP helix (residues K6–I13), helix 2 or mid helix (residues A25–E36) and C-terminal helix 3 (residues K44–G55) (Figure 10.7). The topology of the global fold of three helices describes by folding back of the C-terminal helix showing simultaneous interactions with the N-terminal MP helix 1 and mid-helix (Figure 10.7). Notably, there are potential ionic and/or hydrogen-bonding interactions involving sidechains of helices (Figure 10.7). In other words, the spatial orientations of three helices of αL CT are predominantly maintained by polar interactions rather than packing among hydrophobic residues as observed in structures of well-folded proteins. In HSQC titrations and ITC experiments, αL CT demonstrated coordination with divalent Ca^{2+} ions. The cation-binding sites could be mapped onto a well-formed negatively charged surface embraced by

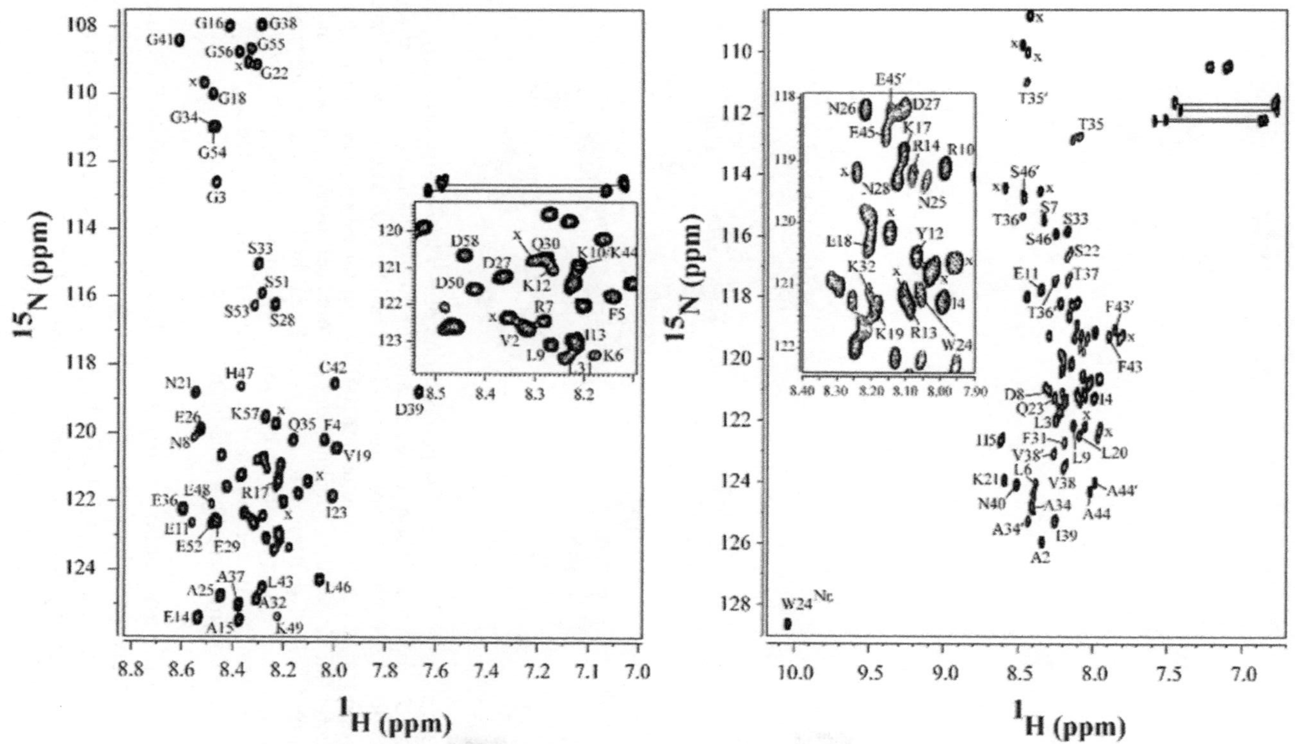

Figure 10.6 ^{15}N–^{1}H heteronuclear single quantum coherence (HSQC) spectra of α CT (left panel) and β2 CT (right panel) showing identification of ^{15}N–^{1}H correlations of each amino acid residue. The 58-residue α CT (residues K1088–D1145) and 46-residue β2 CT (residues K724–S769) are numbered 1–58 and 1–46, respectively. Reproduced from ref. 34, https://doi.org/10.1074/jbc.M807236200, under the terms of the CC BY 4.0 license, https://creativecommons.org/licenses/by/4.0/.

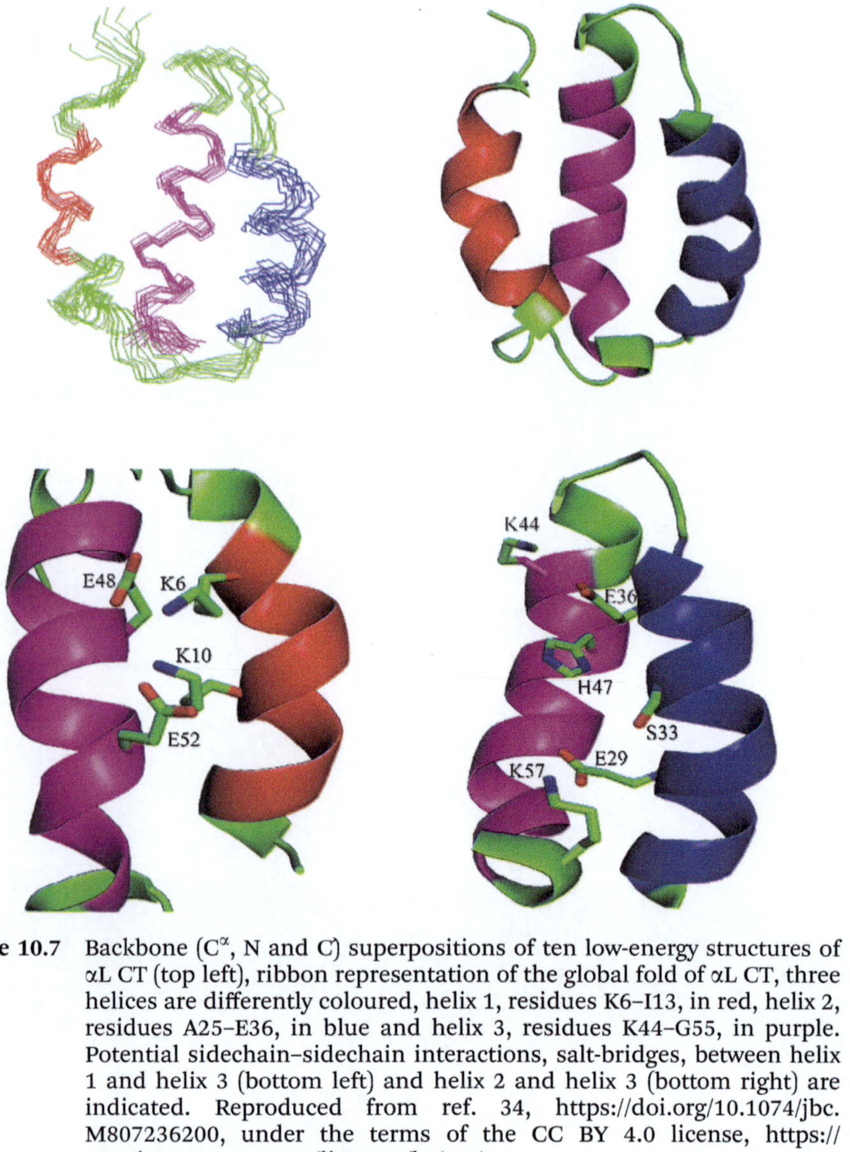

Figure 10.7 Backbone (C^{α}, N and C′) superpositions of ten low-energy structures of αL CT (top left), ribbon representation of the global fold of αL CT, three helices are differently coloured, helix 1, residues K6–I13, in red, helix 2, residues A25–E36, in blue and helix 3, residues K44–G55, in purple. Potential sidechain–sidechain interactions, salt-bridges, between helix 1 and helix 3 (bottom left) and helix 2 and helix 3 (bottom right) are indicated. Reproduced from ref. 34, https://doi.org/10.1074/jbc.M807236200, under the terms of the CC BY 4.0 license, https://creativecommons.org/licenses/by/4.0/.

sidechains of anionic and polar residues of helix 2, C-terminal helix 3 and residues of the intervening loop between the two helices. Notably, residues E26, D27, S28, E29, Q35 and E36 of mid-helix 2, residues G38, D39, G41 and C42 from the connecting loop, and residues H47, E48, K49, D50, S51, E52 and S53 of helix 3 are in the ion-binding surface. Ca^{2+} binding to αL CT is driven by favourable changes in entropy, suggesting the release of water molecules from the binding sites. Further, rigidity of the compact structure of αL CT was assessed by measuring ^{15}N relaxations (R_1, R_2) and $\{^1H\}$–^{15}N

heteronuclear NOE (Figure 10.8). The relaxation parameters (R_1, R_2) and {^{1}H}–^{15}N NOE generally describe N–H bond motion in the ps–ns time scale (Figure 10.8). The lower {^{1}H}–^{15}N NOE values (≤0.25) for most of the residues indicated that the global fold of αL CT is dynamically flexible (Figure 10.8C). In particular, residues of MP helix 1 exhibited relatively higher R_2 and low {^{1}H}–^{15}N NOE values compared to helix 2 and helix 3, implying more dynamic characteristics of helix 1. The relaxation studies support the notion that the populated helical fold of αL CT is in dynamic equilibrium, plausibly with more open states. However, at present it remains unclear whether the three helical folds of αL CT deduced in free solution would assume a stable conformation in continuation with the membrane-anchored TM domain.

^{15}N spin relaxation indicated that the MP helix and MD region of β2 CT are also dynamic in nature (unpublished results). Observation of negative values

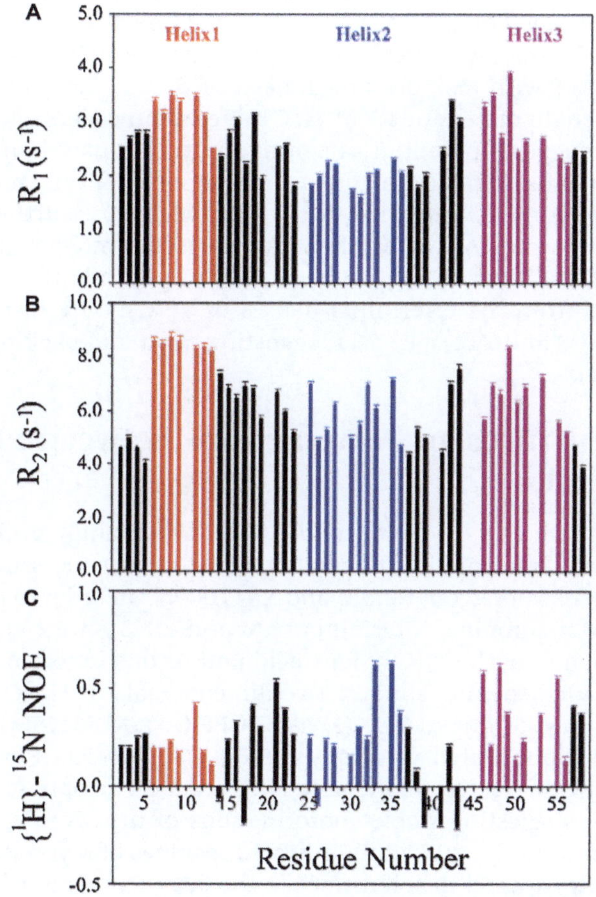

Figure 10.8 Bar diagram showing R_1, longitudinal relaxation (A), R_2, transverse relaxation (B) and {^{1}H}–^{15}N het-NOE (C) of αL CT (our unpublished data). Residues of helix 1, helix 2 and helix 3 are colour coded.

of {^{1}H}–^{15}N het-NOE for several residues of β2 CT including segments of MP and MD suggested a lack of stable conformations. Interactions between αL and β2 CTs were probed by ^{15}N–^{1}H HSQC titrations and ITC measurements.[34] ^{15}N–^{1}H HSQC spectra of ^{15}N-labelled samples of either αL or β2 CT were acquired in the presence of unlabelled CTs at higher molar ratios. The binding-induced changes could be detected in ^{15}N–^{1}H HSQC spectra in terms of chemical shift changes and/or resonance broadening. As can be seen, several residues of the three helices of αL CT experienced binding-induced chemical shift perturbations (Figure 10.9A). Meanwhile, residues belonging to the MP helix of β2 CT demonstrated changes upon interactions with αL CT (Figure 10.9B). The CT/CT complex formation is exothermic in nature, signifying involvement of polar (ionic and/or hydrogen bonding) interactions (Figure 10.9C). The interaction affinity (K_d) between the two CTs was estimated from ITC to be 2.63 μM. Based on chemical shift perturbations, a docked model of the complex of αL/β2 CTs was proposed (Figure 10.10). The complex of αL/β2 CTs could be sustained by several potential polar interactions involving residues of MP helix 1 and C-terminal helix 3 of αL CT with residues of MP helix of β2 CT. Close proximity of the cationic sidechain of residue R7 of αL CT and anionic sidechains of residues D8 and E11 suggests potential salt bridge interactions (Figure 10.10). Notably, functional data showed plausible interactions between residues R7 and D8 in αLβ2 and αIIbβ3 integrins in the inactive states. Further, more ionic interactions between the two MP helices may occur among residues, R14 (β2 CT)/E11, E14(αL CT) and E18 (β2 CT)/R17 (αL CT). In addition, residues D39, K44 and H47 from the C-terminal helix 3 of αL CT were in proximity with residues H5, D8 and Y12 of β2 CT, suggesting polar and packing interactions (Figure 10.10).

10.4 Structures and Interactions between αM, αX and β2 CTs

The three other α CTs, αM, αX and αD CTs, in the family of β2 integrin are much shorter in sequence in comparison to αL CT. Structures and interactions with β2 CT were worked out for αM and αX CTs.[35,36] In order to mimic TM-like insertion into membrane, N-termini of αM and αX CTs were conjugated with myristic acid, a long-chain (C_{14}) fatty acid and reconstituted into solutions of DPC micelle for structural studies. Two-dimensional ^{1}H-^{1}H NOESY spectra of myristoylated αM or Myr-αM CT (Myr-^{1}KLGFFKRQYKDMMSEGGPPGAEPQ)[24] and a Ser phosphorylated version Myr-PαM CT (Myr-^{1}KLGFFKRQYKDMM**p**-**S**EGGPPGAEPQ)[24] yielded many more NOE contacts as opposed to αM CT in free solution, suggesting stable conformations of the Myr-αM and Myr-PαM CTs in DPC micelles.[35] PRE NMR indicated residues of Myr-αM and Myr-PαM CTs are largely exposed to solvent, while the fatty acyl chain of Myr moiety is inserted to the hydrophobic core of the DPC micelle, mimicking localization of TM attached CT in membrane. The micelle insertion strategy has yielded high-resolution structures of Myr-CTs (Figure 10.11). The backbone RMSDs of the

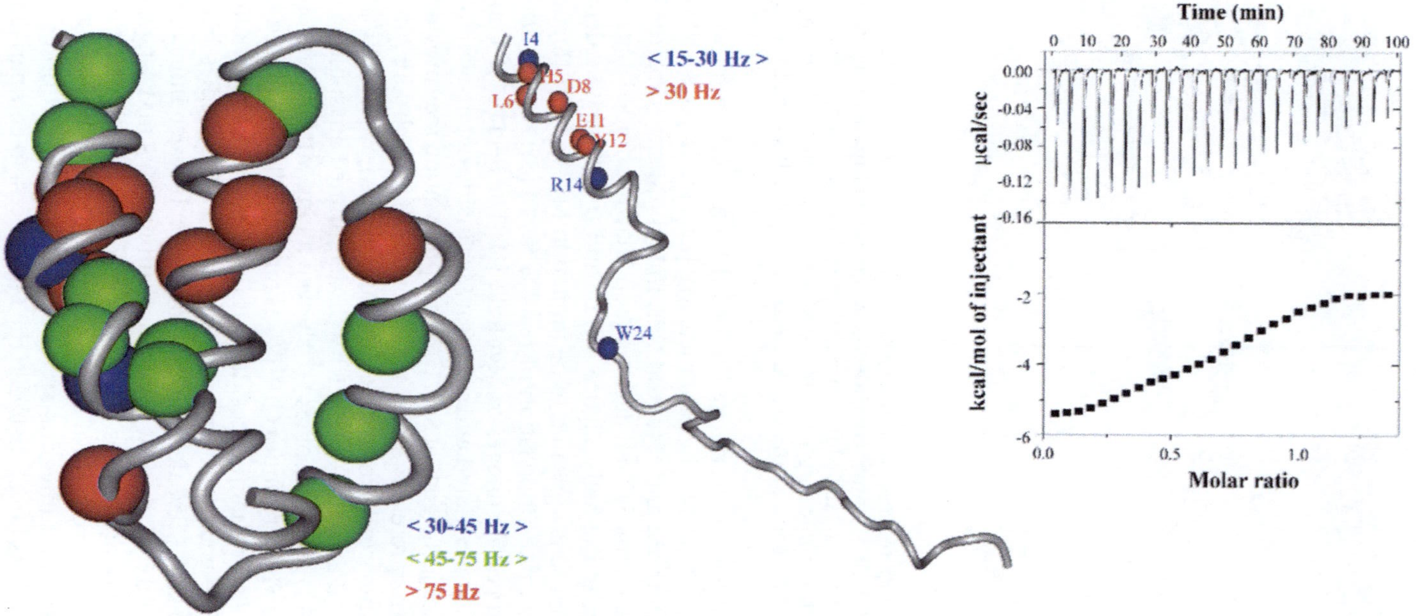

Figure 10.9 Ribbon representation of (A) αL CT and (B) β2 CT showing residues that experienced chemical shift changes upon binding to cognate CT. (C) ITC thermograph of αL and β2 CTs interactions. Reproduced from ref. 34, https://doi.org/10.1074/jbc.M807236200, under the terms of the CC BY 4.0 license, https://creativecommons.org/licenses/by/4.0/.

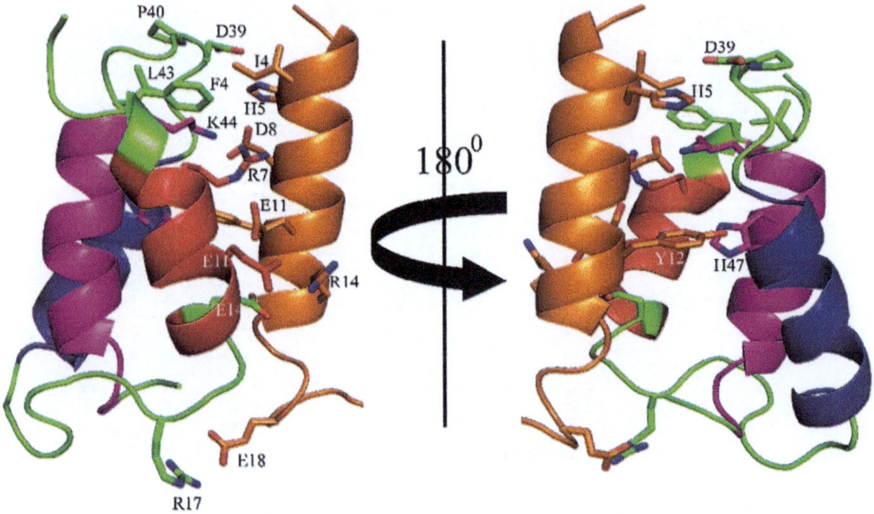

Figure 10.10 Ribbon representation of the docked complex of αL/β2 CTs in two different orientations. Sidechain–sidechain interactions among potential interfacial residues are indicated by sticks. Reproduced from ref. 34, https://doi.org/10.1074/jbc.M807236200, under the terms of the CC BY 4.0 license, https://creativecommons.org/licenses/by/4.0/.

structural ensembles were estimated to be 0.14 Å and 0.39 Å for Myr-αM and Myr-PαM CTs, respectively. The canonical MP helix (residues F4–S14) are well defined in both Myr-αM and Myr-PαM CTs. In Myr-αM structure, the MP helix is followed by a relatively well-defined loop that has folded back onto the hydrophobic face of the helix (Figure 10.11C and D). The spatial definition of the loop conformations appeared to be limited in Myr-PαM CT (Figure 10.11E and F). Therefore, phosphorylation of residue Ser14 (Ser1127 in αM) did not impart many structural changes to Myr-PαM CT. The Ser1127 is known to be constitutively phosphorylated and necessary for inside-out activation of αMβ2 integrin. It was conjectured that Ser1127 phosphorylation may provide binding interactions with proteins like 14-3-3ζ or Dok1 which are involved in regulations of integrins. The NMR structure of Myr-αX CT (KVGFFKRQY-KEMMEEANGQIAPENGTQTPSPPSEK) was determined in solutions of DPC micelles.[36] The 3-D structure of αX CT showed a long MP helix encompassing residues F4-N17, whilst residues Q19-T26 assumed well-defined fold-back loop conformations. The remaining residues at the C-terminus (Q27-K35) did not adopt any definite conformations. NMR ^{15}N–^{1}H HSQC studies between αM and αX CTs with β2 CT revealed complex formation between independent CTs. However, CTs exhibited very low interaction affinity, $K_d \sim$ mM range, based on limited HSQC spectral changes. The docked complexes of αM/β2 and αX/β2 could be obtained by chemical shift perturbation data (Figure 10.12). The CT/CT complexes of β2 integrins are to be sustained by salt-bridge, H-bonds and polar interactions largely between the MP helices. In comparison to αL/β2 CTs, the complexes of αM/β2 and αX/β2 CTs contain limited interfacial

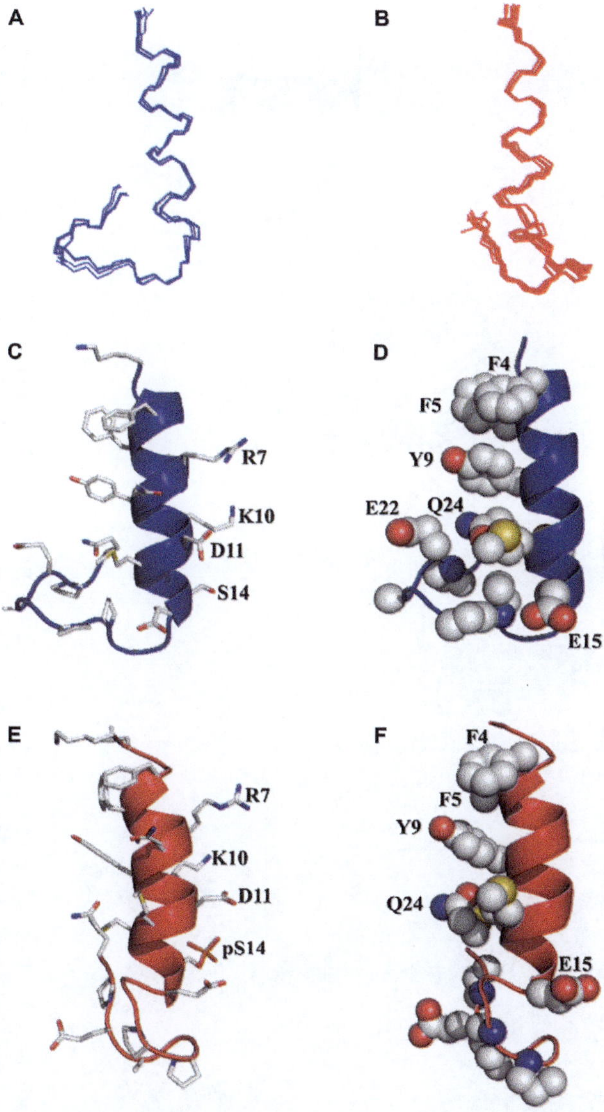

Figure 10.11 (A and B) Superposition of backbone atoms of 20 low-energy structures of (A) Myr-αM CT and (B) Myr-PαM CT determined by CYANA. (C and D) Ribbon representation of a representative structure of Myr-αM CT showing sidechains (C) as stick and (D) space-fill. (E and F) Ribbon representation of a representative structure of Myr-PαM CT showing sidechains as (E) stick and (F) space-fill. Reproduced from ref. 35, https://doi.org/10.1074/jbc.M111.280164, under the terms of the CC BY 4.0 license, https://creativecommons.org/licenses/by/4.0/.

contacts akin to the CT/CT structure of αIIβ and β3 platelet integrin.[37] We have surmised that in the absence of TMs CT/CT interactions could be dynamic and transient in characteristics. Notably, determining native interactions between

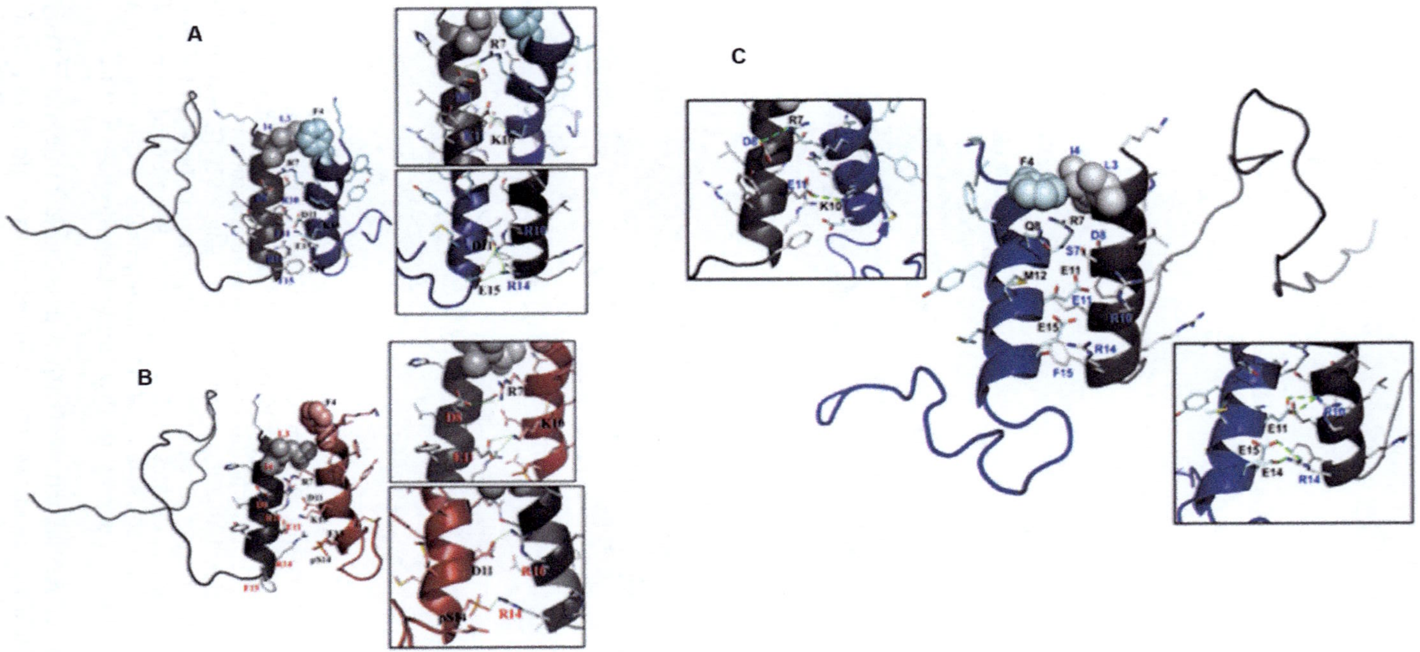

Figure 10.12 Ribbon representations of the docked complexes of (A) αM/β2 CTs, (B) PαM/β2 CTs and (C) αX/β2 CTs. Sidechain–sidechain interactions among potential interfacial residues are indicated by sticks. Ribbon of β2 CT is in black, whereas αM and αX are in red and blue, respectively. Panels A and B: Reproduced from ref. 35, https://doi.org/10.1074/jbc.M111.280164, under the terms of the CC BY 4.0 license, https://creativecommons.org/licenses/by/4.0/. Panel C: Reproduced from ref. 36, https://doi.org/10.1371/journal.pone.0041924, under the terms of the CC BY 4.0 license, https://creativecommons.org/licenses/by/4.0/.

CTs while they are covalently attached to TMs in membrane could be highly challenging. The structure of the αIIb/β3 TMs complex was determined in a lipid bicelle without the CTs.[19] In the presence of CTs, NMR spectra of long TM-CT constructs appeared to be of limited resolution for structural analyses. Much more resolved NMR spectra were obtained for TM-CT in isotropic organic solvent (CH_3CN) medium.[20] However, interactions between CTs were largely diminished under such a non-native solution environment. We designed fluorescence resonance energy transfer (FRET) experiments to evaluate binding interactions between dansylated αM CT or Dan-αM CT and myristoylated β2 CT in membrane mimic environments.[35] In order to detect FRET unambiguously, residue H5 in MP helix of β2 CT (Myr-β2H5W) was substituted with residue Trp acting as a donor fluorophore whereas the N-terminal dansyl group of αM CT served as a fluorescence acceptor moiety. Myr-β2H5W was reconstituted in 1.2 mM DPC solutions for FRET analysis. A dramatic quenching of Trp fluorescence emission of Myr-β2H5W was observed upon the additions of graded concentrations of either Dan-αM CT or Ser phosphorylated Dan-PαM CT evidencing efficient FRET between the CTs. Analyses of binding isotherms demonstrated β2 CT binds to both αM CT ($K_d \sim 5.5$ μM) and PαM CT ($K_d \sim 4.2$ μM) at higher affinity while inserted into micelles of DPC. Taken together, it is tempting to speculate that the interactions between CTs may be stronger while they are attached to TM in membrane. However, more work is needed, in particular using native membrane or close membrane mimics (bicelles, lipid nano-disc), to assess structures and interactions of integrin CTs.

10.5 Structures and Interactions of β2 Cytosolic Tails with Effector Proteins

β CTs of integrins, being well conserved, bind to several cytosolic proteins modulating bidirectional signalling processes (see above). The FERM domain containing proteins talins (talin 1 and talin 2) and kindlins-1,2,3 are the activator and coactivator of integrins, respectively.[38–46] Structure of the FERM domain of the talin head consists of four well-folded subdomains, F0, F1, F2 and F3, which are arranged in a clover leaf-like fashion.[42,43,46] The F3 sub-domain provides most of the binding interactions with β CTs of integrins.[40–43] Obtaining atomic-resolution structures of the complexes of β CTs either with the F3 sub-domain or the entire talin head has met with limited success largely due to transient binding affinity. As noted, β3 CT of platelet integrin binds to the F3 domain of talin1 and talin2 with estimated K_d values of 223 μM and 438 μM, respectively.[43,47] Similar observations were also made for β1A/talin1 ($K_d \sim 491$ μM) and β1A/talin2 ($K_d \sim 652$ μM).[43,47] Our work on talin1 F3/β2 CT and F3/Thr758 phosphorylated pT^{756}β2 CT has determined K_d values 244 μM and 578 μM, respectively.[48] Initial studies gaining the atomic resolution information of this key complex have had to use an optimized chimeric peptide construct (MP region of β3 CT and C-terminal PIPKIγ) that binds to the F3 subdomain with higher affinity.[40] The NMR-derived structure of the F3/peptide complex revealed important intermolecular contacts between talin F3 and the MP region of β3 CT (Figure 10.13). A later work found high-affinity interactions between talin2 head and integrin β1D CT, $K_d \sim 36$ μM and solved the X-ray structure of the F2F3/β1D CT complex (Figure 10.13).[43] A complex structure, generated using chemical shift changes and HADDOCK docking, of talin1 F3/full-length β2 CT could be determined (Figure 10.13).[48]

The talin F3/β CTs structures revealed a set of conserved interactions despite the wide-ranging differences in binding affinity. The MP helix of CTs exhibits several interactions with the PTB fold of the F3 domain. By contrast, conformations of the MD region of β CTs are less well established in these complexes. In the 2.2 Å X-ray structure of talin2 F2F3/β1D the last 13 residues of the MD segment were not resolved and assumed to be unstructured. The MP NPxY/F motif of β CTs is recognized to be critical for talin binding. In β1D CT, the $N^{780}PIY^{783}$ motif delineates close contacts with F3 residues, *e.g.* Y^{783}/F3 $T^{357}TVK^{360}$, N^{780}-P^{781}/F3 W361. However, intermolecular contacts involving the NpxY motif in the talin F3/β3 CT complex could not be obtained as key residues were replaced by C-terminal PIPKIγ in the chimeric tail construct. In β2 CT, the non-canonical N^{751}PxF motif demonstrated

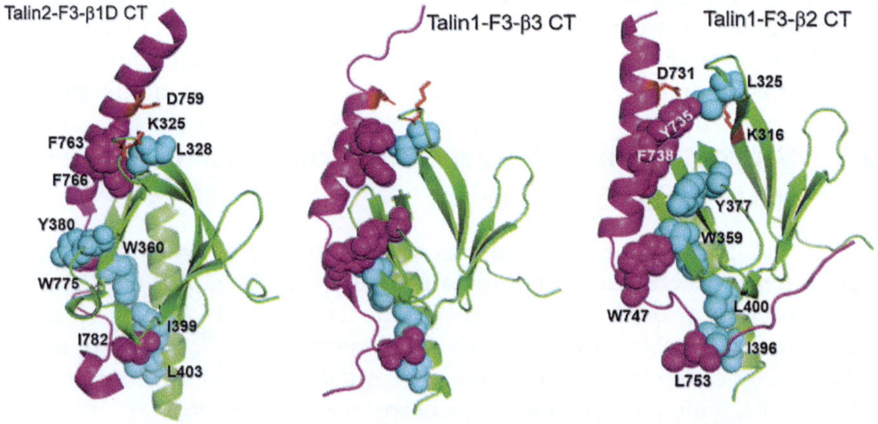

Figure 10.13 Structures of talin1/talin2 F3 domain with β CTs. (Right panel) NMR-based docked complex of β2 CT with F3 domain of talin 1. Sidechains of residues of β2 CT and F3 involved in mutual non-polar packing interactions are shown as a space-fill representation. Sidechains of residue D731 of β2 CT and residue K316 of F3 domain are shown as sticks. (Centre panel) Representative NMR structure of talin 1 F3 with β3 CT chimeric peptide (pdb:2h7e) (see text for more detail). Interacting residues of F3 and β3 CT are similar to the β2 complex and hence are not marked. X-ray structure of the high-affinity complex of talin 2 F2F3/β1D CT (pdb:3g9w). The F2 domain has been omitted for more clarity. Sidechains of residues of β1D CT and F3 involved in mutual non-polar packing interactions are shown as a space-fill representation. Sidechains of residue D759 of β1D CT and residue K325 of F3 domain are shown in stick. In all structures, talin F3 domain and integrin β CTs are shown as green and magenta ribbons, respectively.

limited interactions with the F3 domain. Only the sidechain of residue L400 of F3 sustained packing interactions with residue F754 in the motif. Meanwhile, the MP helix of β CTs established most of the interfacial contacts with the F3 domain of talin (Figure 10.13). The conserved Asp/Lys salt bridge (*e.g.* D731 β2 CT/K361 F3) observed in all three structures is known to be important in regulating integrin activation. Residue Asp of the respective β CTs is engaged in ionic interactions with the MP helix of α CTs maintaining the resting state of the receptor (see above). Talin-mediated activation of integrins has been correlated with binding of the F3 domain with the MP helix of β CT that may essentially disrupt interfacial contacts between the α and β CTs, including the critical ionic interactions.

A number of cytosolic proteins are involved in maintaining the resting state of integrins. Docking protein 1 or Dok-1, an adaptor protein, has been implicated in various signalling pathways in immune cells including cell migration.[49–51] The 481-residue long Dok-1 consists of two distinct folded domains, an N-terminus PH domain and an adjacent phosphotyrosine-binding domain (PTB) domain.[51] The long C-terminus of Dok-1 is largely unstructured and binds to SH2 domain of p120 rasGap upon phosphorylation of specific Tyr residues.[51] Cell based co-immuno-precipitation and pull-down studies detected association of integrins β1, β2, β3 and β7 CTs with the PTB domain of Dok-1.[52] A mechanism of negative regulation of integrin by Dok-1 has been proposed based on binding of the PTB domain with β3 CT of platelet integrin αIIbβ3. Seminal NMR studies demonstrated that the Dok-1 PTB domain can bind tightly with Tyr phosphorylated (pY) β3 CT at the $NPLY^{747}$ motif.[53,54] The talin F3 domain, also a PTB fold, binds to unphosphorylated β3 CT with estimated $K_d \sim 223$ μM, whereas interactions of unphosphorylated β3 CT with Dok-1 PTB were deduced to be significantly weaker, $K_d \gg 1$ mM.[53] On the other hand, an opposite binding scenario was observed with pY747 β3 CT. The phosphorylated (pY747) β3 CT binds to the PTB domain of Dok-1 with higher affinity, $K_d \sim 37.1$ μM, compared to the talin F3 domain, $K_d \sim 6.53$ mM. Taken together, a model of integrin regulation was put forward, whereby binding of Dok-1 at the phosphorylated β3 CT is likely to compete with binding of talin for activation. Also, unlike talin F3, Dok-1 PTB does not interact with the MP helix of β3 CT, suggesting a plausible resting state of the receptor in the complex.[53,54] However, Dok-1 binding to β2 integrin using a similar mechanism appeared to be unfeasible since β2 CT contains a non-phosphorylatable $NPLF^{754}$ motif at the canonical binding site. Phosphorylations of two Ser residues at positions 745 and 756 of β2 CT, flanking the NPLF motif, have been demonstrated using cellular studies.[55] It was conjectured that phosphorylation of these Ser (pS) residues may induce Dok-1 PTB binding to β2 CT using an alternative mode of action.[56] It may be pertinent to note that β CTs of β3, β5 and β6 integrins which contain an NPxY motif do not contain Ser residues at the corresponding positions. Extensive NMR studies have demonstrated that full-length β2 CT phosphorylated either at pS756 or pS745 binds to Dok-1 PTB with K_d of 98 μM and 215 μM, respectively.[56] Furthermore, structural analyses of Dok-1 and β2 CT phosphorylated peptide fragment or pSer756-KT15 (KSQWNNDNPLFKpSAT) revealed that the phosphorylated β2 CT peptide shares the same binding pocket of the PTB domain akin to β3 CT (Figure 10.14). In addition, saturation transfer difference (STD) NMR experiments showed an intimate association of pSer756-KT15 with Dok-1 PTB. In addition, ^{31}P NMR studies delineated direct interaction of the phosphate group of pSer756-KT15 with PTB binding residues. A low-energy docked structure of the complex of pSer756-KT15/ Dok-1 PTB was generated based on NMR data and a tr-NOESY-derived structure (Figure 10.14). The phosphate group of the pSer756-KT15 β2 CT demonstrates ionic/hydrogen bonding interactions with the Arg residues R54, R69 and R70 of Dok-1 PTB (Figure 10.4).[54] These Arg residues situated at the binding pocket of the PTB domain also bind to phosphorylated peptide ligands from other proteins. In addition, complexes of pSer756-KT15/Dok-1 PTB perhaps can be stabilized by other interactions (Figure 10.14).[54] Binding interactions of the phosphorylated Ser756 β2 CT with Dok-1, $K_d \sim 98$ μM, can essentially inactivate β2 integrin-mediated signalling by competing with talin binding to the β2 CT. The F3 domain of talin has been deduced to bind to non-phosphorylated β2 CT at a much lower affinity with an estimated $K_d \sim 244$ μM. Therefore, the binary complexes of talin, Dok-1 with integrin CTs described here play dominant roles in regulating bidirectional signalling of integrins. The CTs/proteins interfacial contacts could be exploited in designing drugs against diseases such as inflammation, cancer and thrombosis.

10.6 Concluding Remarks

Elucidation of the molecular mechanisms of integrin bidirectional signalling is of great interest for understanding the cellular physiology and development of novel drugs. Deciphering the sequence of events of multiple protein interactions with integrin CTs is one of the major aspects of integrin research. In this regard, NMR spectroscopy has been found to be an eminently suited technique. The transient nature of many of the CTs–CTs or CTs–proteins interactions can only be determined by NMR methods with atomic resolution. Moving forward, NMR relaxation dispersion can be utilized for further analyses of the functional dynamics of CT–protein complexes. Also, at present, conformational states of integrin CTs remain elusive while covalently bonded to the respective TMs. NMR studies of TM–CT constructs in membrane environments, *e.g.* bicelles, lipid nano-disc and how such constructs interact with cytosolic proteins would significantly improve mechanistic insights into integrin-mediated cell signalling.

Acknowledgements

Integrin research in the author's laboratory has been funded by the Biomedical Research Council (BMRC), Singapore and Ministry of Education (MOE), Singapore.

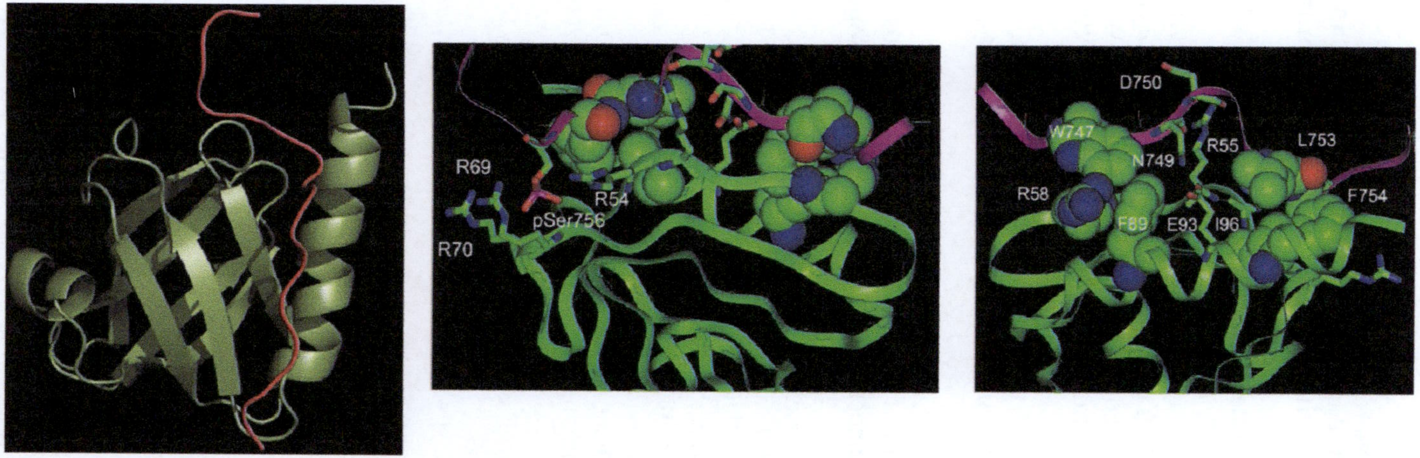

Figure 10.14 Structure of the Dok-1 PTB domain in complex with phosphorylated β2 CT peptide pSer756-KT15. (Left panel) NMR-based docked complex of β2 CT pSer756-KT15 with PTB domain of Dok-1. β2 CT pSer756-KT15 peptide occupies the canonical peptide binding grove of the PTB domain. (Centre panel) Potential ionic and hydrogen boding interactions between the negatively charged phosphate group of pSer756-KT15 peptide with cationic guanidium sidechains of residues Arg 54, R69 and R70 of the PTB domain. (Right panel) The probable additional interactions between pSer756-KT15 and the Dok-1 PTB domain, shown are aromatic–aromatic stacking and cation–π interactions residues Phe89 and Arg58 (Dok-1) with Trp747 of pSer756-KT15, non-polar packing interactions between residues Phe754 and Leu753 of the NPLF motif of pSer756-KT15 peptide with Ile96 of the PTB domain and polar and ionic interactions: Asp749 (K15) with Glu93 (Dok1) and Asp750 (K15) with Arg55 (Dok1). Reproduced from ref. 56, https://doi.org/10.1038/srep11630, under the terms of the CC BY 4.0 license, http://creativecommons.org/licenses/by/4.0/.

References

1. R. O. Hynes, *Cell*, 2002, **110**, 673–687.
2. D. S. Harburger and D. A. Calderwood, *J. Cell Sci.*, 2009, **122**, 159–163.
3. U. K. Marelli, F. Rechenmacher, T. R. Sobahi, C. Mas-Moruno and H. Kessler, *Front. Oncol.*, 2013, **3**, 222.
4. R. J. D. Hatley, S. J. F. Macdonald, R. J. Slack, J. Le, S. B. Ludbrook and P. T. Lukey, *Angew. Chem., Int. Ed. Engl.*, 2018, **57**, 3298–3321.
5. A. R. Rosenkranz and T. N. Mayadas, *Exp. Nephrol.*, 1999, **7**, 125–136.
6. K. Scharffetter-Kochanek, H. Lu, K. Norman, N. van Nood, F. Munoz, S. Grabbe, M. McArthur, I. Lorenzo, S. Kaplan, K. Ley, C. W. Smith, C. A. Montgomery, S. Rich and A. L. Beaudet, *J. Exp. Med.*, 1998, **188**, 119–131.
7. A. Coxon, P. Rieu, F. J Barkalow, S. Askari, A. H. Sharpe, U. H. von Andrian, M. A. Arnaout and T. N. Mayadas, *Immunity*, 1996, **6**, 653–666.
8. S. M. Tan, *Biosci. Rep.*, 2012, **32**, 241–269.
9. J. Qin, O. Vinogradova and E. F. Plow, *PLoS Biol.*, 2004, **6**, e169.
10. F. Ye, C. Kim and M. H. Ginsberg, *J. Thromb. Haemost.*, 2011, **Suppl 1**(1), 20–25.
11. E. van de Vijver, A. Maddalena, O. Sanal, S. M. Holland, G. Uzel, M. Madkaikar, M. de Boer, K. van Leeuwen, M. Y. Koker, N. Parvaneh, A. Fischer, S. K. Law, N. Klein, F. I. Tezcan, E. Unal, T. Patiroglu, B. H. Belohradsky, K. Schwartz, R. Somech, T. W. Kuijpers and D. Roos, *Blood Cells, Mol., Dis.*, 2012, **48**, 53–61.
12. A. Kato, *Crit. Rev. Oncol. Hematol.*, 1997, **26**, 1–23.
13. J. P. Xiong, T. Stehle, B. Diefenbach, R. Zhang, R. Dunker, D. L. Scott, A. Joachimiak, S. L. Goodman and M. A. Arnaout, *Science*, 2001, **294**, 339–345.
14. J. Zhu, B. H. Luo, P. Barth, J. Schonbrun, D. Baker and T. A. Springer, *Mol. Cell*, 2009, **34**, 234–249.
15. C. Xie, J. Zhu, X. Chen, L. Mi, N. Nishida and T. A. Springer, *EMBO J.*, 2010, **29**, 666–679.
16. J. Zhu, B. H. Luo, T. Xiao, C. Zhang, N. Nishida and T. A. Springer, *Mol. Cell*, 2008, **32**, 849–861.
17. T. Xiao, J. Takagi, B. S. Coller, J. H. Wang and T. A. Springer, *Nature*, 2004, **432**, 59–67.
18. M. Shimaoka, T. Xiao, J. H. Liu, Y. Yang, Y. Dong, C. D. Jun, A. McCormack, R. Zhang, A. Joachimiak and J. Takagi, *et al.*, *Cell*, 2003, **112**, 99–111.
19. T. L. Lau, C. Kim, M. H. Ginsberg and T. S. Ulmer, *EMBO J.*, 2009, **28**, 1351–1361.
20. J. Yang, Y. Q. Ma, R. C. Page, S. Misra, E. F. Plow and J. Qin, *Proc. Natl. Acad. Sci. U. S. A.*, 2009, **106**, 17729–17734.
21. P. E. Hughes, F. DiazGonzalez, L. Leong, C. Y. Wu, J. A. McDonald, S. J. Shattil and M. H. Ginsberg, *J. Biol. Chem.*, 1996, **271**, 6571–6574.

22. M. Kim, C. V. Carman and T. A. Springer, *Science*, 2003, **301**, 1720–1725.
23. Y. F. Li, R. H. Tang, K. J. Puan, S. K. Law and S. M. Tan, *J. Biol. Chem.*, 2007, **282**, 24310–24319.
24. E. J. Brown, *Curr. Opin. Cell Biol.*, 2002, **14**, 603–607.
25. M. Moser, K. R. Legate, R. Zent and R. Fässler, *Science*, 2009, **324**, 895–899.
26. E. M. Morse, N. N. Brahme and D. A. Calderwood, *Biochemistry*, 2014, **53**, 810–820.
27. M. Das, S. S. Ithychanda, J. Qin and E. F. Plow, *Biochim. Biophys. Acta*, 2014, **1838**, 579–588.
28. K. R. Legate and R. Fässler, *J. Cell Sci.*, 2009, **122**, 187–198.
29. M. Das, S. S. Ithychanda, J. Qin and E. F. Plow, *PLoS One*, 2011, **6**, e26355.
30. T. Kiema, Y. Lad, P. Jiang, C. L. Oxley, M. Baldassarre, K. L. Wegener, I. D. Campbell, J. Ylanne and D. A. Calderwood, *Mol. Cell*, 2006, **21**, 337–347.
31. H. Takala, E. Nurminen, S. M. Nurmi, M. Aatonen, T. Strandin, M. Takatalo, T. Kiema, C. G. Gahmberg, J. Ylanne and S. C. Fagerholm, *Blood*, 2008, **112**, 1853–1862.
32. D. A. Calderwood, A. Huttenlocher, W. B. Kiosses, D. M. Rose, D. G. Woodside, M. A. Schwartz and M. H. Ginsberg, *Nat. Cell Biol.*, 2001, **3**, 1060–1068.
33. D. Chatterjee, L. L. Zhiping, S. M. Tan and S. Bhattacharjya, *Sci. Rep.*, 2018, **8**, 5490.
34. A. Bhunia, X. Y. Tang, H. Mohanram, S. M. Tan and S. Bhattacharjya, *J. Biol. Chem.*, 2009, **284**, 3873–3884.
35. G. L. Chua, X. Y. Tang, M. Amalraj, S. M. Tan and S. Bhattacharjya, *J. Biol. Chem.*, 2011, **286**, 43842–43854.
36. G. L. Chua, X. Y. Tang, A. T. Patra, S. M. Tan and S. Bhattacharjya, *PLoS One*, 2012, **7**, e41924.
37. O. Vinogradova, A. Velyvis, A. Velyviene, B. Hu, T. Haas, E. Plow and J. Qin, *Cell*, 2002, **110**, 587–597.
38. H. Li, Y. Deng, K. Sun, H. Yang, J. Liu, M. Wang, Z. Zhang, J. Lin, C. Wu, Z. Wei and C. Yu, *Proc. Natl. Acad. Sci. U. S. A.*, 2017, **114**, 9349–9354.
39. M. Moser, K. R. Legate, R. Zent and R. Fassler, *Science*, 2009, **324**, 895–899.
40. K. L. Wegener, A. W. Partridge, J. Han, A. R. Pickford, R. C. Liddington, M. H. Ginsberg and I. D. Campbell, *Cell*, 2007, **128**, 171–182.
41. S. J. Shattil, C. Kim and M. H. Ginsberg, *Nat. Rev. Mol. Cell Biol.*, 2010, **11**, 288–300.
42. D. R. Critchley and A. R. Gingras, *J. Cell Sci.*, 2008, **121**, 1345–1347.
43. N. J. Anthis, K. L. Wegener, F. Ye, C. Kim, B. T. Goult, E. D. Lowe, I. Vakonakis, N. Bate, D. R. Critchley, M. H. Ginsberg and I. D. Campbell, *EMBO J.*, 2009, **284**, 36700–36710.
44. W. Bu, Z. Levitskaya, Z. Y. Loh, S. Jin, S. Basu, R. Ero, X. Yan, M. Wang, S. F. C. Ngan, S. K. Sze, S. M. Tan and Y. G. Gao, *PLoS Biol.*, 2020, **18**, e3000755.

45. F. Ye, G. Hu, D. Taylor, B. Ratnikov, A. A. Bobkov, M. A. McLean, S. G. Sligar, K. A. Taylor and M. H. Ginsberg, *J. Cell Biol.*, 2010, **188**, 157–173.
46. E. Goksoy, Y. Q. Ma, X. Wang, X. Kong, D. Perera, E. F. Plow and J. Qin, *Mol. Cell*, 2008, **31**, 124–133.
47. N. J. Anthis, K. L. Wegener, D. R. Critchley and I. D. Campbell, *Structure*, 2010, **8**, 1654–1666.
48. D. Chatterjee, L. L. Zhiping, S. M. Tan and S. Bhattacharjya, *J. Mol. Biol.*, 2016, **428**, 4129–4142.
49. T. Yasuda, K. Bundo, A. Hino, K. Honda, A. Inoue, M. Shirakata, M. Osawa, T. Tamura, H. Nariuchi, H. Oda, T. Yamamoto and Y. Yamanashi, *Int. Immunol.*, 2007, **19**, 487–495.
50. T. Yasuda, M. Shirakata, A. Iwama, A. Ishii, Y. Ebihara, M. Osawa, K. Honda, H. Shinohara, K. Sudo, K. Tsuji, H. Nakauchi, Y. Iwakura, H. Hirai, H. Oda, T. Yamamoto and Y. Yamanashi, *J. Exp. Med.*, 2004, **200**, 1681–1687.
51. R. Mashima, Y. Hishida, T. Tezuka and Y. Yamanashi, *Immunol. Rev.*, 2009, **232**, 273–285.
52. D. A. Calderwood, Y. Fujioka, J. M. de Pereda, B. García-Alvarez, T. Nakamoto, B. Margolis, C. J. McGlade and R. C. Liddington, *Proc. Natl. Acad. Sci. U. S. A.*, 2003, **100**, 2272–2277.
53. C. L. Oxley, N. J. Anthis, E. D. Lowe, I. Vakonakis, I. D. Campbell and K. L. Wegener, *J. Biol. Chem.*, 2008, **283**, 5420–5426.
54. N. J. Anthis, J. R. Haling, C. L. Oxley, M. Memo, K. L. Wegener, C. J. Lim, M. H. Ginsberg and I. D. Campbell, *J. Biol. Chem.*, 2009, **284**, 36700–36710.
55. S. Fagerholm, N. Morrice, C. G. Gahmberg and P. Cohen, *J. Biol. Chem.*, 2002, **277**, 1728–1738.
56. S. Gupta, J. C. Chit, C. Feng, A. Bhunia, S. M. Tan and S. Bhattacharjya, *Sci. Rep.*, 2015, **5**, 11630.

CHAPTER 11

Describing Dynamic Chaperone–Client Complexes by Solution NMR Spectroscopy

S. HILLER*[a] AND B. M. BURMANN*[b,c]

[a] Biozentrum, University of Basel, Klingelbergstrasse 70, 4056 Basel, Switzerland; [b] Wallenberg Centre for Molecular and Translational Medicine, University of Gothenburg, Gothenburg 40530, Sweden; [c] Department of Chemistry and Molecular Biology, University of Gothenburg, Gothenburg 40530, Sweden
*Emails: sebastian.hiller@unibas.ch; bjorn.marcus.burmann@gu.se

11.1 Introduction

The beauty of life is directly based on the gigantic variety of protein structures that can be generated from linear chains of 20 different amino acids[1]. Not surprisingly, this richness of possible structures is connected to conformational landscapes of the polypeptides that are non-trivial.[2,3] They are highly complex, rugged, and full of local traps. In order to navigate polypeptide chains across these conformational spaces, nature has evolved a special class of helper proteins, the chaperones.[4–6] Molecular chaperones interact with polypeptide chains in folded, partially folded, as well as misfolded conformations and can thus directly influence their path on the

New Developments in NMR No. 26
NMR Spectroscopy for Probing Functional Dynamics at Biological Interfaces
Edited by Anirban Bhunia, Hanudatta S. Atreya and Neeraj Sinha

Published by the Royal Society of Chemistry, www.rsc.org

folding landscape.[7] To achieve this goal, the chaperones interact dynamically with their respective client proteins, which are themselves in highly dynamic ensemble states.[8–10] Chaperones thus denature clients or are able to bind them in unfolded states to transport and release them for proper spatial distribution in the cell.[11,12]

The most prominent chaperone systems are the ATP-dependent Hsp60, Hsp70, and Hsp90, which are found in prokaryotes and eukaryotes alike.[13–15] These systems come along with a plethora of regulating factors, co-chaperones, and additional chaperones, all in all resulting in a large network tasked with maintaining the cellular proteome.[16] Hsp60 forms a large, football-like double cavity that can incorporate and subsequently process clients at its interior. Hsp70 is a large two-domain chaperone comprised of one domain hydrolyzing ATP and the other domain interacting with client proteins. Hsp90 is a large dimer, which can undergo structural rearrangements as part of its ATP-dependent functional cycle, thus modulating its client interaction sites. Additional factors regulate the functional cycles of the chaperones in a cell-state-dependent manner, recruit the clients, or differentiate the functional cycle toward specific applications.[17,18] The three ATP-dependent core machineries are supported by a plethora of ATP-independent chaperones, including the eukaryotic prefoldin, or bacterial Trigger Factor, SecB, Spy, SurA, or Skp.[19–26] These interact with client proteins in partially unfolded states, denaturing them and keeping them unfolded while bound.

Understanding molecular chaperones properly requires biophysical descriptions of their client-interacting states, in particular including a characterization of the underlying dynamics. To this end, solution NMR spectroscopy is the ideal technique, as it can resolve these parameters at the atomic level, even for complex sizes of several hundreds of kilodalton (kDa). Consequently, the past decade has brought tremendous insight into chaperone function from solution NMR studies of such complexes. In this chapter, we aim at providing a comprehensive overview of the NMR experiments that have proven suitable to study chaperone–client complexes at the atomic level. We characterize the individual experiments systematically according to their application purpose and discuss for each of them the specific aspects that are encountered in their application to molecular chaperones and their clients. We distinguish in particular two types of clients, those that are water soluble in the absence of the chaperones at comparable concentrations (soluble clients) and those that are not soluble in aqueous solution on their own (insoluble clients). For the soluble clients the thermodynamic equilibrium condition encountered in NMR spectroscopy results in averaged data over the bound and unbound forms. Insoluble clients, in turn, have their equilibrium shifted almost completely to the chaperone-bound and with negligible free-form. This may require special preparation schemes, while at the same time yielding homogeneous data sets.

11.2 NMR Techniques for Chaperone–Client Complexes

11.2.1 Determination of Spatial Contacts Between Client and Chaperone

11.2.1.1 Chemical Shift Perturbation

Chemical shift perturbation (CSP) is a well-established technique to characterize interfaces and to derive dissociation constants of classical biomolecular interactions, such as protein–protein, protein–nucleic acid, and protein–ligand complexes.[27,28] CSPs are typically monitored by suitable 2D or 3D NMR fingerprint spectra, such as the 2D [^{15}N,^{1}H]-TROSY experiment for the amide group or the 2D [^{13}C,^{1}H]-HMQC experiment for specific methyl groups on a deuterated background. The CSPs are then calculated as the difference between the free and bound forms for the proton as $\Delta\delta(^1\text{H}) = \delta(^1\text{H})_{\text{bound}} - \delta(^1\text{H})_{\text{free}}$, and analogously for the heavy atoms, $\Delta\delta(\text{X}) = \Delta\delta(^{15}\text{N})$ or $\Delta\delta(^{13}\text{C})$. For atoms in chemical moieties that represent a suitable single probe for structural changes, such as amide or methyl groups, the individual CSPs are typically combined toward a quadratic sum, weighted relative to each other by an empiric factor α, typically reflecting on the difference in chemical shift dispersion of the different nuclei,

$$\Delta\delta(\text{HX}) = \sqrt{\Delta\delta(^1\text{H})^2 + \alpha \cdot \Delta\delta(\text{X})^2} \qquad (11.1)$$

For the experimental determination of CSPs in chaperone–client systems, two fundamentally different cases need to be distinguished:

- For soluble clients, CSPs upon binding can be determined for both the chaperone and the client. In one experiment, the unlabeled client is added to the isotope-labeled chaperone to determine the CSPs of the chaperone. In a second experiment, the unlabeled chaperone is added to the isotope-labeled client to determine the CSPs of the client. If the exchange kinetics between the free and the bound forms are in the fast or intermediate time regime on the NMR time scale, a titration in small steps is possible, which allows transfer of the assignment between the free and bound forms and can yield the dissociation constant. If the exchange kinetics are slow, a partial or complete re-assignment of the resonance peaks in the client-bound state might be necessary.
- For insoluble clients, the only accessible CSP is to compare the free state of the chaperone with its client-bound form. Because the client is insoluble in the absence of the chaperone, these two forms must be in slow exchange on the NMR time scale. Depending on the magnitude of the CSPs, a partial or complete re-assignment of the resonance peaks in the client-bound state might be necessary.

Data interpretation: The CSPs are typically determined in a residue-specific manner and can therefore be plotted subsequently against the amino acid sequence or on a structural model of the protein (see Figure 11.1a). Furthermore, a statistical analysis yields the level of noise and, above these, identifies the significantly non-zero CSPs, which can be clustered. Generally, the CSPs comprise effects both from the direct interaction of the molecules and from indirect (allosteric) effects induced by the binding. Unlike for classical rigid body interactions, the direct effects of binding are often comparably weak on the chaperone, because they result from interactions with an ensemble of bound conformations and are accordingly averaged out. The CSP can thus be dominated by the allosteric effects that reflect on the structural switches of the chaperone. Therefore, the determination of the direct interaction sites must thus be made with caution. For example, for the two clients OmpX and tOmpA binding to Skp, the CSPs were basically identical, thus reflecting mostly allosteric effects and hardly the direct binding.[29]

In contrast, allostery is typically less prominent on a dynamic client protein, and the CSPs are dominated by the direct interaction with the chaperone. Still, also under these conditions indirect effects could be contributing, in cases that the chaperones dissolved a pre-formed long-range cluster within the unfolded client.

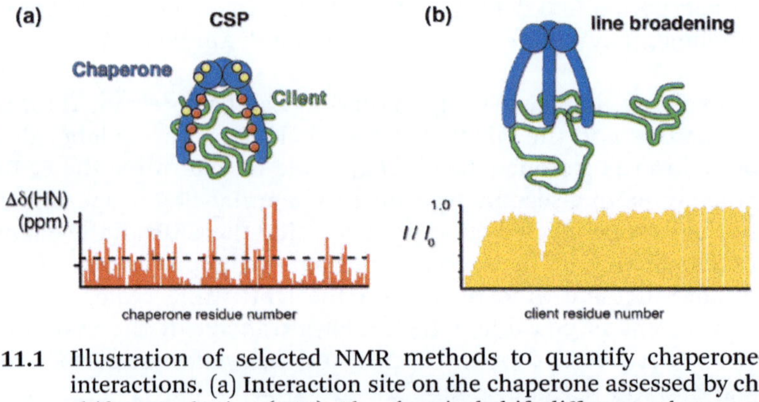

Figure 11.1 Illustration of selected NMR methods to quantify chaperone–client interactions. (a) Interaction site on the chaperone assessed by chemical shift perturbation (CSP). The chemical shift differences between client-free and client-bound forms are plotted against the sequence (orange bars). A statistical significance line (dashed line) identifies the strongest interactions. The residues with the strongest CSPs comprise both direct interactions (orange points) as well as allosteric effects (yellow points). (b) Interaction site on a client protein identified by resonance line broadening. The intensity changes between the free and chaperone-bound forms are determined as the intensity ratio I/I_0. Segments with a reduced intensity ratio correspond to chaperone interactions.

11.2.1.2 Line Broadening

Many chaperone–client systems do not solely undergo chemical shift perturbations upon binding, but also signal intensity changes. These result from changes in the transverse relaxation times, either by a difference between the relaxation properties in the bound and the free states, as well as by contributions of fast-to-intermediate exchange processes that occur within the bound state or as part of the binding/dissociation kinetics. The contributions of the latter are particularly important for chaperone–client systems, because the client interaction as well as the client state itself are highly dynamic.

The signal intensity reduction due to line broadening can be monitored by the same 2D fingerprint spectra that are used for the CSP analysis. For a two-state system, the intensity change related to the modulation in R_2 can be approximated by the equation

$$\frac{I}{I_0} = \frac{R_{2,0}}{R_2} \cdot \exp\left((R_2 - R_{2,0}) \cdot \tau\right) \tag{11.2}$$

where I and I_0 are the NMR signal intensities, and R_2 and $R_{2,0}$ are the transverse relaxation rate constants of the detected proton in the bound and free forms, respectively. τ is the time the signal spends in the transverse plane during the pulse sequence. For systems that exchange between more than two states, the corresponding equation comprises additional terms that relate to the differential relaxation rates as well as to contributions from the kinetic exchange. The system becomes under-determined and it is thus most convenient to record the I/I_0 values and interpret them directly.[30] The method is applicable for soluble and insoluble clients with the same considerations as the CSP.

Data interpretation: The interpretation of intensity changes follows the same rationale as the CSPs and may thus provide information on the direct interaction site as well as allosteric effects (see Figure 11.1b). Notably, intensity changes are mostly a monodirectional parameter, while the CSPs are a bidirectional parameter, and therefore, these two parameters react differently to conformational averaging. The averaging over many conformations with different chemical shifts might result in a perturbation close to zero, but a non-zero effect on the intensities. Hence, the intensity changes can in many cases be more sensitive. In particular, this effect is observed for unfolded clients binding to folded chaperones, where the intensity changes can be very strong and the interpretation is thus straightforward, directly indicating the interacting sites.

11.2.1.3 Intermolecular NOE

The nuclear Overhauser effect (NOE) is caused by the dipolar interaction, leading to longitudinal cross-relaxation between spins that are in close spatial contact for times longer than the rotational correlation time.[31] When combined with multidimensional correlation spectroscopy, the NOE can resolve both the originating and the receiving spin site-specifically and thus yields a highly sensitive and highly precise measure of close intermolecular contacts between chaperone and client, up to at least 6 Å and in favorable cases even further in deuterated systems.[32,33] Suitable pairs of spin probes have to be chosen and prepared with suitable isotope labeling and corresponding experiment selection. For the characterization of chaperone–client complexes, methyl groups of amino acid side chains are the most suitable spin probes, because they are in direct spatial contact at the interface and at the same time provide the highest experimental sensitivity. Specific isotope labeling schemes for methyl groups are available for isoleucine, leucine, valine, methionine, alanine, and threonine residues, with the ILV labeling scheme being most commonly used, although amide groups are also a suitable addition for smaller chaperone–client systems. The NOESY experiment is recorded in chemical equilibrium on the chaperone–client complex and therefore generally is suited for both soluble and insoluble clients.

- The highest information content is obtained in 4D NOESY experiments that are able to resolve the receiving and originating spin in two dimensions and thus in an essentially unambiguous manner.[34] Due to the otherwise prohibitively long recording times, four-dimensional NOESY NMR spectroscopy typically requires the use of non-uniform time-domain sampling.[35] So far, however, 4D NOESYs have not been reported for chaperone–client complexes.
- 3D NOESY experiments provide similar information, but somewhat reduced information content compared to the 4D experiment, while at the same time having increased sensitivity. For optimal sensitivity, 3D NOESY experiments are chosen that resolve the receiving spin moiety in two dimensions and the originating spin moiety in only the proton dimension. Despite the lower dimensionality, by considering transposed peaks, the NOE assignments can for most peaks become unambiguous, at least in the case of rigid structures.

Data interpretation: Site-specific intermolecular NOEs can in principle be converted directly into semi-quantitative structural distance constraints and used in classical structure calculation protocols. This is, however, only reasonable in cases of chaperone–client complexes where the bound client occupies a single conformation (see Figure 11.2). In the more typical case of dynamic multi-conformational complexes, the observed NOE signal intensities result from different conformations and in addition may, due to the narrow chemical shift dispersion, also originate from superimposition of multiple spins.[36] The conversion of NOEs into meaningful distance constraints is under these conditions no longer possible and the data are then typically not able to overdetermine the conformational space occupied by the polypeptide.

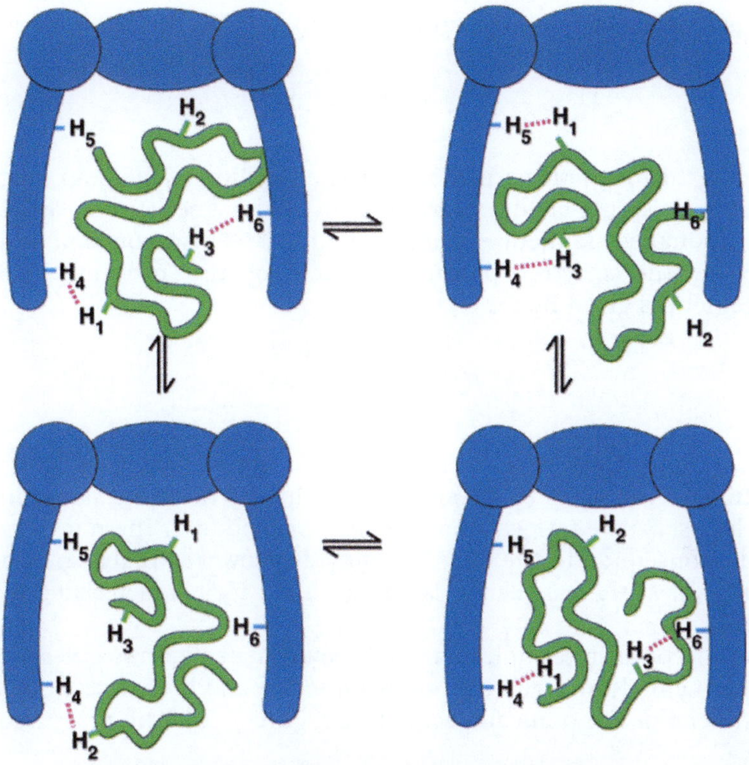

Figure 11.2 Conformational averaging of intermolecular NOEs in a chaperone–client complex. Schematic model of four conformations of a chaperone protein (blue) in complex with its unfolded client (green) in fast chemical equilibrium. Six protons, H_1–H_6, are highlighted. For each conformation, pairs of protons involved in short-range contacts, corresponding to strong intermolecular NOEs, are connected with dashed magenta lines.

In such situations of dynamic complexes, the NOEs can nonetheless still be fruitfully interpreted as an ensemble-averaged contact parameter, similar to CSPs or line broadening. An elegant application of this approach has been presented in combination with an orthogonal methyl-labeling scheme that allowed the protein-specific assignment of the peaks directly from their chemical shift.[36]

11.2.1.4 Paramagnetic Relaxation Enhancement (PRE)

Paramagnetic relaxation enhancement (PRE) results from the dipolar interaction between a free electron spin and a nuclear spin, with the notable difference to the NOE that the interaction is substantially stronger due to large gyromagnetic ratio (*g*-factor) of the electron.[37–40] The PRE is a transverse relaxation term Γ_2 of the nuclear spin that is added to the diamagnetic transverse relaxation rate constants R_2^{dia} to result in the paramagnetic transverse relaxation rate constant $R_2^{\text{para}} = R_2^{\text{dia}} + \Gamma_2$. The PRE is given by

$$\Gamma_2 = \frac{1}{20}\left(\frac{\mu_0}{4\pi}\right)^2 \mu_B^2 g^2 \gamma_H^2 (4J(0) + 3J(\omega_H)) \tag{11.3}$$

where μ_0 is the magnetic permeability of the vacuum, μ_B the Bohr magneton, g the electron spin g-factor, γ_H the proton gyromagnetic ratio, and ω_H the proton resonance frequency in angular units.[41] For a constant inter-spin distance r, the spectral density function of the resulting relaxation enhancement is given by

$$J(\omega) = \frac{1}{r^6}\frac{\tau_r}{1 + \omega^2\tau_r^2} \tag{11.4}$$

where τ_r is the effective rotational correlation time of the inter-spin vector. The PRE is sensitive up to distances of around 30 Å. At the same time, the PRE is less sensitive to local changes in the position of the nuclear spin of the NOE. Consequently, the PRE does provide a lower spatial resolution than the NOE, but at the same time is better suited for characterizing dynamic ensembles.

Recording of an intermolecular PRE experiment requires site-specific labeling of one molecule with a paramagnetic tag, and then detecting the PRE on the nuclei of the other molecule (see Figure 11.3). The most frequently

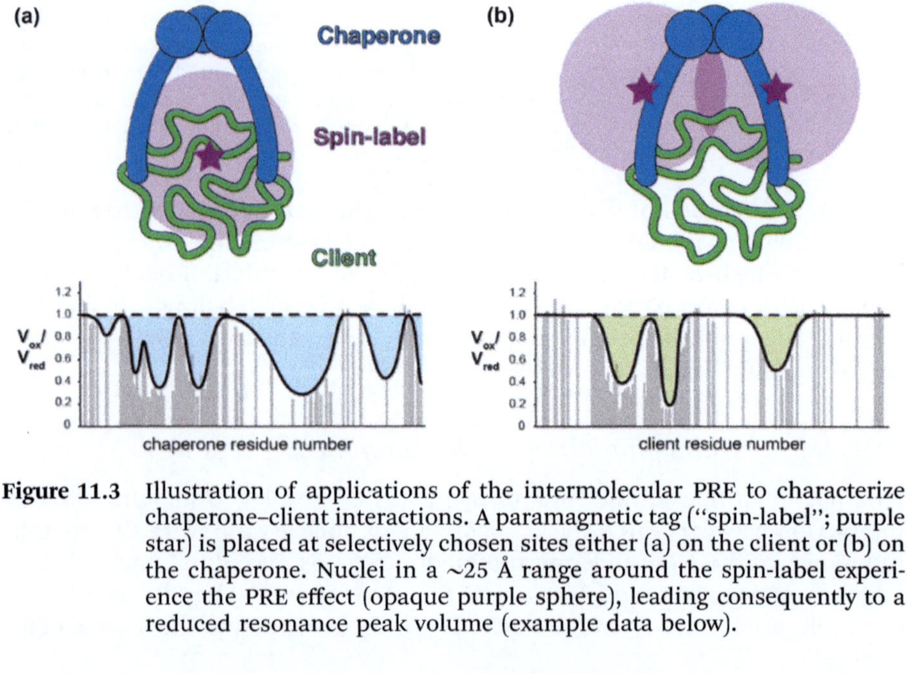

Figure 11.3 Illustration of applications of the intermolecular PRE to characterize chaperone–client interactions. A paramagnetic tag ("spin-label"; purple star) is placed at selectively chosen sites either (a) on the client or (b) on the chaperone. Nuclei in a ~25 Å range around the spin-label experience the PRE effect (opaque purple sphere), leading consequently to a reduced resonance peak volume (example data below).

used tag is the MTSL spin-label, a nitroxide spin-label. The paramagnetic tag is typically attached covalently to cysteine residues that are introduced by site-directed mutagenesis into suitable positions at the protein of interest.[42] The tag is added in excess and the unreacted free radicals need to be removed after the reaction by subsequent purification.

In the most basic version, the PRE can be determined as a 1-point measurement by determining peak volumes in a suitable fingerprint spectrum, such as the [^{15}N,^{1}H]-TROSY for amide groups or the [^{13}C,^{1}H]-HMQC for methyl groups. The PRE rate can then be obtained *via* the relationship

$$\frac{V_{\mathrm{ox}}}{V_{\mathrm{red}}} = \frac{V_{\mathrm{para}}}{V_{\mathrm{dia}}} = \exp(-\Gamma_2 \cdot \tau_{\mathrm{I}}) \tag{11.5}$$

where τ_{I} is the total duration during which the nuclear spin magnetization is transverse in the pulse sequence.[43] The experiments are run twice, once in the oxidized, paramagnetic state of the sample and a second time, after a reducing agent, such as ascorbic acid, has been added to the sample. If increased data precision is needed, a multi-point approximation can be recorded, where an additional relaxation delay is included in the pulse sequence, to vary τ_{I}. The data are then fitted to multiple experimental data points, resulting in increased precision and accuracy of the result.[41]

Data interpretation: Using the equations outlined above, the PRE can be related to intermolecular distances. For an inter-spin vector with constant distance and distances between 20 and 30 Å, exact values can theoretically be determined.[44] For distances below 20 Å, the PRE is so strong that it leads to a complete bleaching of the nuclear resonance and thus is effectively not quantifiable anymore. Therefore, in most practical cases, only upper distance limits can be obtained. For the use of PRE constraints in dynamic ensembles, different structure calculation protocols have been developed, which are discussed in Section 11.2.6.

11.2.2 Intramolecular Client Contacts

11.2.2.1 PRE

The PRE is not only applicable to determine intermolecular contacts but also highly suitable to quantify intramolecular contacts in random-coil-like ensembles. This application was initially introduced for unfolded polypeptides[45,46] and subsequently adapted to dynamic clients bound to chaperones.[29,47] To determine intramolecular PREs in a dynamic polypeptide, a paramagnetic spin-label is attached at a suitable site and the PREs are measured on nuclei within the same polypeptide, using the procedures as described above for the intermolecular PRE.

Data interpretation: PRE data in unfolded ensembles can be interpreted using the Gillespie–Shortle model for dynamic polypeptides.[43,48] In this motional model, the spectral density function of the relaxation enhancement is given by

$$J(\omega) = \left\langle \frac{1}{r^6} \right\rangle \frac{\tau_r}{1 + \omega^2 \tau_r^2} \tag{11.6}$$

where r is the inter-spin distance and τ_r is an effective correlation time. The cornered brackets denote a time and ensemble average. The model differs from motional models used for the interpretation of rigid molecules in two ways. First, the correlation time used is an effective local correlation time that takes into account the local flexibility of the polypeptide chain which is different from the global rotational correlation time of a rigid molecule. A suitable value of $\tau_r = 4.0$ ns was empirically established by fitting model calculations to experimental PRE data of a urea-denatured polypeptide.[29] Second, because the polypeptide chain samples multiple conformations, the inter-spin distance r is not a constant value but results from a suitably chosen ensemble description.

One possible approach for this description is polymer theory of a random flight chain.[29,49,50] This is applicable to completely disordered random coils. In these calculations, r is given by the expected value for a random flight chain $\langle r^2 \rangle = Nl^2$, with N being the distance in residues from the spin-label, and l being the length of an amino acid residue backbone, 3.8 Å.[49] For clients bound to chaperones, however, the polypeptide is constrained in its motion and consequently, another approach needs to be chosen. If the spin-label and the nucleus are sufficiently far separated in the amino acid sequence, the two sites can be assumed to diffuse independently in a constrained volume. Restricted co-diffusion in a spatial volume is given by

$$\left\langle \frac{1}{r^6} \right\rangle = \int \mathrm{d}\vec{x}_1 \mathrm{d}\vec{x}_2\, f(\vec{x}_1 - \vec{x}_2) \tag{11.7}$$

where the integrals run over the volume. The distance dependence is modeled by the function

$$f(y) = \begin{cases} \frac{1}{|y|^6} & \text{if} |y| > D \\ 0 & |y| \leq D \end{cases} \tag{11.8}$$

where D is the minimal inter-spin distance that is possible regarding steric hindrance. The volume can in first-order approximation be approximated by a spherical volume. By calculating the expected PRE using eqn (11.6) for spherical volumes with different radii R and comparing with the experimentally determined PRE value, it is thus possible to determine the expansion of the polypeptide in first-order approximation.

This approach has been fruitfully used for several client proteins bound in an unfolded state to molecular holdase chaperones.[29,47] The intramolecular PRE was used to assess the degree of chaperone-dependent compactness of

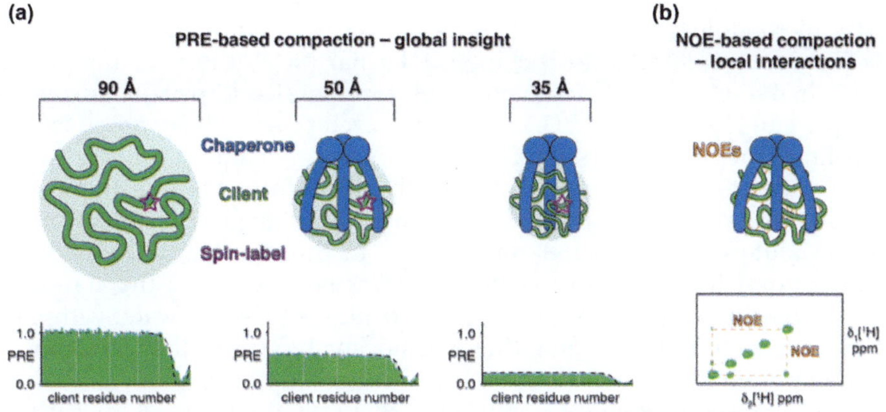

Figure 11.4 Intramolecular contacts client. (a) Dependence of the magnitude of the intramolecular global PRE effect of the client provides insight into its compactness. (b) Intramolecular NOEs could eventually provide detailed insights into the local environment and packaging of the chaperone-bound client.

the client, and to compare the completely unrestrained case in the presence of chaotropic denaturants.[29] For the highly dynamic client OmpX bound to the molecular chaperone Skp, an average intramolecular PREs of 0.26 between an electron spin located at position 140 and individual nuclei on the polypeptide chain was observed.[29] This value corresponds to random diffusion in a sphere with radius $R = 21$ Å (see Figure 11.4a). Compared to the smallest theoretically possible volume for a polypeptide of such size of 17 Å the resulting ensemble thus contains about 50% polypeptide and 50% internal water molecules within the spherical volume encapsulated within the chaperone interior.[29] Notably, these results are in very good agreement with a subsequent study using small angle neutron scattering (SANS) to determine the related eight-stranded β-barrel OmpW bound to Skp in an ellipsoidal volume with a radius of gyration of 24 Å.[51]

The intramolecular PRE approach was also applied to the Spy protein and its client $Im7_U$,[47,52] which adopted a rapidly exchanging conformational ensemble as evidenced by the narrow chemical shift dispersion.[47] Therefore, here also the first-order approximation and the Gillespie–Shortle model were employed, revealing a spherical volume of 23 Å while bound to Spy compared to 35 Å in the urea denatured state.[48,53] This represents a chaperone-dependent compaction to about 66%.[22]

11.2.2.2 NOE

For the determination of intramolecular distance information within the dynamic client, the NOE can readily be employed. The NOE provides a way to assess short-distance contacts, which can result in a detailed description of the "interior" of the bound client (see Figure 11.4b). The intramolecular NOE methodology has however so far not yet been employed in chaperone-bound clients. Similar to the intermolecular NOE, the presence of fast-interchanging conformational ensembles would also impair the analysis of the intramolecular NOE.[36] Therefore, the developments of advanced isotope-labeling schemes such as specific methyl group labeling[54] and site-specific aromatic residues[55] will hopefully set the stage for exploiting these possibilities. Potentially, fluorinated amino acids[56] might further contribute to deconvolute observed NOESY cross-peaks in the near future. Besides providing novel insights into the spatial arrangements within the compacted client state, it would also provide access to study the bound water within this compacted ensemble by adapting and advancing established methods for folded proteins.[57]

Data interpretation: Intramolecular NOEs could be used to calculate structures of local clusters of non-random behavior within the ensembles, following established protocols. Alternatively, they could also be used to map spatial contacts within the unfolded client as a function of different conditions and external parameters.

11.2.3 Lifetime of Chaperone–Client Complexes

To assess the stability of the complexes of molecular chaperones with their client proteins, the measurement of the global lifetime by NMR spectroscopy, the "GLIMPSE" experiment [Global Lifetime Measurements by (PRE spin-label) exchange], was introduced by Hiller and Burmann.[29] The experimental setup consists of an assembled chaperone–client complex that is mixed at $t = 0$ with the corresponding free chaperone in a way that the newly added free chaperone is spectroscopically different from the chaperone that has client bound. The client will equilibrate over time into the free chaperone, limited by the dissociation or off-rate of the chaperone–client complex. This transfer of the client from the initial chaperone to the free chaperone is then detected in real time by suitable NMR methods and fitting of the data yields the off-rate and hence the lifetime of the complex. The GLIMPSE experiment can be performed with different NMR readouts, as outlined in the following sections (see Figure 11.5).

11.2.3.1 GLIMPSE by Chemical Shifts

In the first implementation of the GLIMPSE experiment, the NMR spectral differences of apo-Skp and holo-SKP were directly used. A complex of unlabeled Skp–OmpX was mixed in a 1 : 1 ratio with client-free [U-^{2}H,^{15}N]-Skp and a set of 2D [^{15}N,^{1}H]-TROSY spectra were recorded.[29] The appearance of peaks distinctive for client-bound Skp indicated the transfer of the client from the unlabeled chaperone to the isotopically labeled one. Exponential fitting of the peak intensities under the assumption of off-rate-limited first-order kinetics yielded an average lifetime for the Skp–OmpX complex of

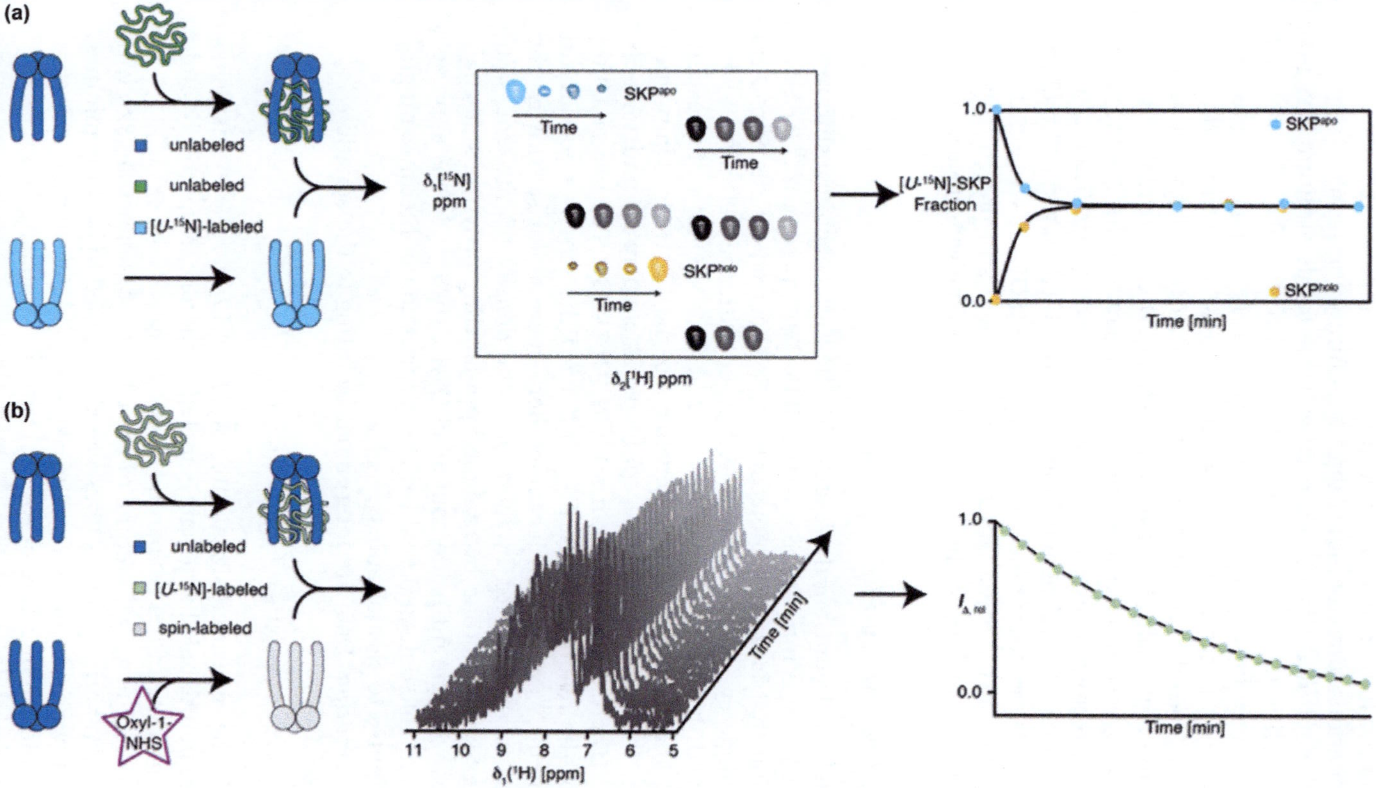

Figure 11.5 The GLIMPSE approach. (a) Initial implementation of the GLIMPSE approach, using [U-^{15}N]-labeled apo-chaperones. At $t = 0$ holo- and apo-chaperones are mixed in equimolar amounts and real-time 2D NMR experiments are measured. Analysis of the resulting intensities of unique signals (blue, yellow) over the time course enables determination of the respective fractions of holo- and apo-state, providing access to the global lifetime in the higher minutes to hours range. (b) Implementation of the GLIMPSE approach using [U-^{15}N]-labeled client and spin-labeled chaperone. At $t = 0$, holo- and apo-chaperones (spin-labeled) are mixed in equimolar amounts and real-time 1D NMR experiments are measured. Due to the PRE effect the signal intensity of the client will be reduced over time, providing direct access to the global lifetime in the minutes range.

2.6 ± 0.9 hours. Remarkably, this global lifetime was found to be seven orders of magnitude longer than the lifetimes of individual chaperone-bound client conformations, directly indicating high-affinity complexes.[29]

Although the outlined approach proved suitable for determining the global lifetime, specific requirements prevented its direct application to other systems: (i) the need for complete sequence-specific resonance assignments together with easily accessible peaks unambiguously representing the apo- as well as the holo-state for either the chaperone or the client and (ii) the time-consuming approach of measuring a whole 2D dataset for each individual timepoint, resulting in low time sampling rate. Whereas the latter aspect can be directly addressed by using instead a standard TROSY HSQC taking about 2 hours experimental time in the outlined example,[29] making use of fast pulsing techniques employing the SOFAST or BEST approaches,[58,59] the spectral assignments needed would still prevent this approach for screening purposes in the early stages of a given project.

11.2.3.2 GLIMPSE by PRE

As an approach to circumvent the issues with the initial chemical shift readout, the PRE effect was introduced into the GLIMPSE experiment.[60] Based on the observation that large intermolecular PREs occurred for the Skp–Omp complexes, single cysteine Skp mutants labeled with the MTSL spin-label were mixed with holo-complexes containing [U-^{2}H,^{15}N]-Omp at the experiment onset. The PRE was thus observable on the [U-^{2}H,^{15}N]-Omp upon transfer from the unlabeled to the spin-labeled chaperone, and read-out using δ_1[^{15}N]-1D cross-sections from 2D [^{15}N,^{1}H]-TROSY. The use of 1D experiments for the readout significantly enhanced the sample frequency of the time domain.[60] Subsequently, for higher experimental sensitivity, the approach was further refined by increasing the density of nitroxide spin-labels on the Skp chaperone by random spin-labeling of lysine side-chain amino-groups with the nitroxide spin-label Oxyl-1-NHS.[61–63] This approach yielded reproducible data and enabled characterization of lifetimes ranging from ~10 minutes to several hours.

Data interpretation: Data are interpreted by fitting to the equation

$$I_{\text{rel}}(t) = I_{\text{U}} + 0.5(I_{\text{L}} - I_{\text{U}})\text{e}^{\left(-2\frac{t}{\tau}\right)} \tag{11.9}$$

where I_{rel} is the measured signal intensity, I_{U} and I_{L} are the intensities of the substrates in unlabeled and (spin)-labeled chaperone, respectively, and τ is the global lifetime of the chaperone–client complex ($\tau = k_{\text{off}}^{-1}$).[60]

As an example, the global lifetime of the complex of chaperone Skp with the 22-stranded β-barrel proteins FhuA was determined to be ~100 minutes, which is about two-fold smaller than for Skp–OmpX, likely directly reflecting the more complex architecture of two Skp trimers stabilizing FhuA.[29,60,64,65] In contrast, the periplasmic chaperone SurA showed a one order of magnitude lower stability with a global lifetime of 12 minutes.[60]

The usefulness of the random spin-labeling approach was also instrumental in the study of a different bacterial molecular chaperone, Trigger Factor.[66] Trigger Factor, the first molecular chaperone to directly interact with newly synthesized polypeptides is known to exist in an equilibrium between a monomeric and dimeric form with a dissociation constant in the low micromolar range.[66–68] For modeling the monomer–dimer equilibrium, consider mixing an unlabeled Trigger Factor dimer (N) with an equivalent amount of spin-labeled Trigger Factor dimer (S), resulting in the following equilibrium reactions:

$$\begin{aligned} &2\mathrm{N} \rightleftarrows \mathrm{N_2} \\ &\mathrm{N} + \mathrm{S} \rightleftarrows \mathrm{NS} \\ &2\mathrm{S} \rightleftarrows \mathrm{S} \end{aligned} \tag{11.10}$$

Importantly, the three reactions [see eqn (11.10)] have the same k_{off}, k_{on}, and K_{d} values. Taking the detailed considerations outlined in Morgado *et al.*[66] into account, the observed NMR signal in the lifetime experiment is given by

$$I(t) = \alpha_1[\mathrm{N_2}](t) + \alpha_2[\mathrm{NS}](t) = \left(\frac{1}{3}\alpha_1 - \frac{2}{3}\alpha_2\right)c\mathrm{e}^{(-k_{\text{off}}t)} + (\alpha_1 + \alpha_2)\frac{2}{3}c \tag{11.11}$$

where α_1 and α_2 are the specific molar NMR intensities of the N_2 and NS species, respectively, and c is the used concentration. Since α_1, α_2, and c are constants, the equation can be rewritten as:

$$I(t) = (I_0 - I_\infty)\mathrm{e}^{(-k_{\text{off}}t)} + I_\infty \tag{11.12}$$

where the constants I_0 and I_∞ are the signal intensities at $t=0$ and $t=\infty$, respectively. The off-rate constant can thus be determined from non-linear fits of the experimental data. Analysis of the global lifetimes of the Trigger Factor dimer as a function of temperature revealed that the lifetimes were modulated slightly by temperature ranging from ~16 min at 25 °C to ~2.5 min as the estimated upper limit at 35 °C.[66] Although in this particular application the limit of the experimental set-up due to manual mixing was evidently reached, even shorter mixing times down to the millisecond regime could be accessible, when employing a fast-mixing device directly in the NMR magnet, as described by Brutscher and co-workers.[69]

11.2.3.3 GLIMPSE by Line Broadening

The GLIMPSE experiment[60] was subsequently adapted in an elegant way by the Schanda group in their study of mitochondrial holdase chaperones.[70] The authors used the high sensitivity of methyl resonances as the readout of the ensuing transfer kinetics. Unlabeled chaperone loaded with

[U-^{2}H,^{15}N]-client was mixed with equal amounts of [Ala, Leu, Val]-methyl-labeled chaperone. This labeling scheme enabled the extraction of the global lifetime while having at the same time access to specific internal controls. Due to the isotopically labeled client, its state could be monitored, ensuring that no signal loss occurred due to precipitation. Then, by assessing the lifetime *via* 2D [^{13}C,^{1}H]-SOFAST-HMQC experiments, the stability of the chaperone was also ensured, because resonances away from the binding site were non-affected.[70] The numeric analysis of the resulting intensity drop of the methyl resonances in the binding cleft revealed a global lifetime of ~4 hours,[70] which is remarkably similar to the related bacterial Skp system with 2.6 hours.[29] The difference between individual conformation and global lifetime, *i.e.* in both cases ~7 orders of magnitude difference, indicated similar governing principles for these evolutionary non-related holdase chaperones.[29,70]

As these different examples show, the determination of global lifetimes in the seconds–hours timescale with the GLIMPSE experiment is a highly informative and useful approach, that can be easily adapted to the needs of a specific project by combining different tools from the ever-growing NMR-toolbox: specific isotope labeling, usage of PREs, with and without sequence specific resonance assignment, *etc.*

11.2.3.4 Biophysical Methods

For completeness, we would like to mention here also the multitude of biophysical techniques beyond NMR experiments that are generally available to study kinetics and thermodynamics of protein–protein complexes. These techniques include fluorescence spectroscopy, surface plasmon resonance (SPR), isothermal titration calorimetry (ITC), microscale thermophoresis, and several more. These methods can be directly applied in studies including soluble client proteins, providing thermodynamical and kinetic parameters characterizing the respective complexes.[71–74] For insoluble clients however, most of these techniques have a crucial disadvantage that they rely on an equilibrium between a bound and unbound form of the client, a feat that cannot be achieved for these proteins.[75]

11.2.4 Dynamics of the Client

11.2.4.1 NMR Relaxation Measurements

The dynamic properties of chaperone-bound clients can be characterized over a range of time scales by NMR backbone relaxation measurements.[76,77] In order to quantify the backbone motions of proteins, the standard set of three NMR relaxation parameters, R_1(^{15}N), R_2(^{15}N), and ^{15}N{^{1}H} NOE (het-NOE), is most commonly recorded. Unlike for globular proteins, their interpretation is however not straightforward for chaperone-bound clients, because the frameworks of the Lipari–Szabo approach or other motional models[78–80] are not applicable to fast-interchanging conformer ensembles. Rather, reduced spectral density mapping, an approach initially developed for intrinsically disordered polypeptides, proved to be extremely useful.[81,82] Reduced spectral density mapping directly converts the three experimentally determined spin relaxation rate constants into the amplitudes of motions at three specific frequencies, in a site-specific manner[81–84] (see Figure 11.6).

$$\sigma_{NH} = R_1(\text{hetNOE} - 1)\gamma_N/\gamma_H \tag{11.13}$$

$$J(0) = (6R_2 - 3R_1 - 2.72\sigma_{NH})/(3d^2 + 4c^2) \tag{11.14}$$

$$J(\omega N) = (4R_1 - 5\sigma_{NH})/(3d^2 + 4c^2) \tag{11.15}$$

$$J(0.87\omega_H) = 4\sigma_{NH}/(5d^2) \tag{11.16}$$

where σ_{NH} is the NH^N cross-relaxation rate, and J the respective spectral densities at the indicated frequencies. γ_N and γ_H are the nitrogen and hydrogen gyromagnetic ratios, respectively. $d = (\mu_0 h\ \gamma_N\gamma_H/8\pi^2)(r_{NH}^{-3})$, with μ_0 the vacuum permeability, h is Planck's constant, and r_{NH} the amide bond length (1.02 Å). $c = \omega_N\Delta\sigma/\sqrt{3}$, with $\Delta\sigma$ the ^{15}N chemical shift anisotropy in the amide moiety (160 ppm).

Data interpretation: Reduced spectral density mapping was used to benchmark the chaperone-bound states of the client protein OmpX directly against the respective urea-denatured state.[29,60] The analysis revealed that the chaperone-bound polypeptide exhibits backbone dynamics with uniform amplitude over the entire chain on all assessed timescales, confirming the interpretation of an essentially featureless polypeptide. The amplitude of motions on the nano- and sub-nanosecond timescale, $J(0.87\omega_H)$ and $J(\omega_N)$, were similar in the chaperone-bound and urea-denatured forms, indicating a locally flexible polypeptide in both cases. The zero-frequency motions, $J(0)$,

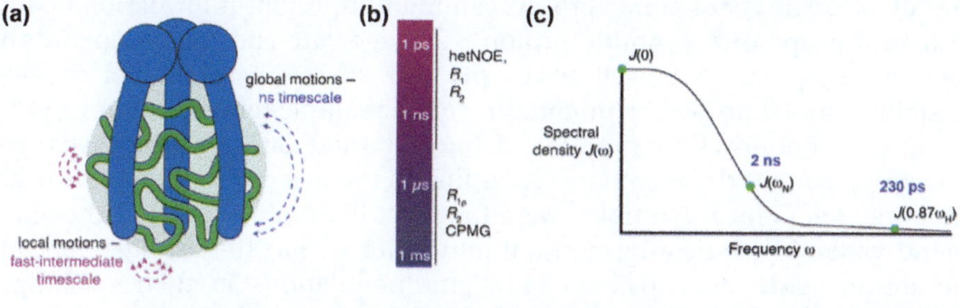

Figure 11.6 Client dynamics. (a) Local and global motions of a chaperone-bound client can occur on different time scales. (b) Characteristic timeline and different suitable NMR experiments to assess motions on the different timescales. (c) The spectral density function $J(\omega)$ is the Fourier transform of the autocorrelation function of the spin system, describing the dynamics over all the accessible timescales. While the spectral density function is inferred in model-free approaches, in the reduced spectral density mapping approach, three distinct points (green points) are directly determined.

were however significantly different and larger in the chaperone-bound form, reflecting the motional coupling of the client to the global tumbling of the chaperone[1]. The same features of a restricted overall tumbling, while preserving the inherent dynamics, were later also confirmed for the larger substrate FhuA bound to Skp.[60]

Reduced spectral density mapping was subsequently also applied to characterize client dynamics in the Spy–Im7 complex.[47] Unlike the membrane protein OmpX, the client Im7 is a partially folded, soluble protein, and its reference state is therefore the free, soluble form of the client. $J(0)$ was found to be globally increased in the chaperone-bound form, in perfect agreement with the earlier observations for the Skp–OmpX system.[29,47] Notably, a sub-set of resonances showed enhanced $J(0)$ contributions, indicative of conformational exchange. Furthermore, the amplitude of motions on the pico- to nanosecond time scales, $J(0.87\omega_H)$ and $J(\omega_N)$, increased for most parts of the protein when Im7 bound to Spy.[47] Also, a direct comparison of the hetNOE in the chaperone-bound and free forms revealed that the chaperone-bound Im7 was overall more dynamic on the nanosecond time scale than the free form.[47] The Spy chaperone thus modulated the dynamics of its client toward unfolding, a feature lately also observed for other holdase chaperones.[47,73] In fact, this ability might be an inherent biophysical property of at least the sub-class of holdase chaperones, enabling them to act on their clients in a manner similar to chaotropic reagents.[11] In this context, future research might well benefit from recently developed techniques to describe the dynamics of structural ensembles of intrinsically disordered proteins.[85,86]

11.2.4.2 Amide Protection Factors

For a refined characterization of the chaperone-bound client ensemble, the solvent accessibility of amide groups can hold important information about structural propensities. Amide proton exchange rate constants depend on both the temperature as well as the pH and, therefore, amide solvent accessibility can be probed by modulating either sample temperature or pH.[87] For the Skp-bound client proteins, using this modulation of the exchange rates revealed that the segments remaining in the fast-exchange regime over the whole temperature sampled were the most hydrophilic parts of the Skp-bound ensemble indicating non-random order within the arrangement of the amino acids, as evidenced by slight modulations in their exchange kinetics, within the highly dynamic fluid globule state.[29,88]

Hydrogen/deuterium exchange experiments can be employed to directly quantify amide exchange kinetics. To perform this experiment, the protein is held in a fully protonated state and then rapidly diluted into deuterium solution, such that exchange of accessible amides to deuterons starts to take place. The exchange process is monitored in real time by 2D [^{15}N,^{1}H]-TROSY NMR experiments and the observed amide signal intensities are fitted to obtain the amide exchange rates in a sequence-specific manner. For the

Spy–Im7 complex amide, such hydrogen/deuterium exchange measurements revealed that the interaction with the chaperone accelerated the exchange kinetics dramatically, showing that the chaperone denatures the client Im7.[47] This increase in conformational dynamics contributes to the entropy of the bound state, contributing to the affinity between client and chaperone, while at the same time due to partial denaturation of the client the binding to the chaperone is strengthened, and as such avoiding misfolding or aggregation.[89]

11.2.5 Dynamics of the Chaperone

The dynamics of the chaperone in chaperone–client complexes can readily be quantified following the above-mentioned classical approaches for determining protein dynamics of folded proteins. For the characterization of backbone dynamics, pulse sequences for amide nitrogen $R_1(^{15}N)$, $R_2(^{15}N)$, as well as hetNOE in the amide group, are well established and can be used in their routine way. The data can then be interpreted either by a motional model, in its most simple form by model-free analysis, resulting in an order parameter and the frequency of internal motion. This method parameterizes the contributions of the global and internal motions to the autocorrelation function in a residue-specific manner.[79] An alternative is provided by reduced spectral density mapping, which directly assesses the underlying spectral density at three different frequencies, as described above. The use of spectral density mapping can have the advantage that the same method can also be used for the client protein in the same setup, thus creating internal consistency for the chaperone–client system.

For the characterization of side-chain dynamics, methyl groups are the preferrable probe, due to their superior sensitivity. Methyl groups feature fast local rearrangement that may be of interest, and they are involved in motions ranging from the pico- to nanosecond timescale that can be assessed by "forbidden" double-quantum/triple quantum transitions, to micro- to milliseconds where motions on this time scale can readily be assessed by either Carr–Purcell–Meiboom–Gill (CPMG)-type or chemical exchange saturation transfer (CEST)-type experiments (reviewed in ref. 90 and 91). With these last types of experiments, events on the microsecond timescale can be resolved and these motions might be directly coupled to chaperone function.

Data interpretation: Two main contributing factors need to be distinguished for the interpretation of dynamics in chaperone–client interactions. One part is the inherent dynamics of the chaperone molecules. These may be obtained separately by comparing the client-free and client-bound forms. The other parts are effects caused by the interaction of the client with the chaperones. Those might likely be observed on the side chains but may also occur on the backbone. Note that these dynamics are connected to the line-broadening experiments listed above. Overall, the assignment of observed dynamics to a particular molecular motion is a non-trivial task that, depending on the claims that are being made, requires additional control experiments and validation.

11.2.6 Structural Modeling

We have now seen that a variety of different NMR parameters are available to characterize chaperone–client systems. In contrast to well-folded globular proteins, the dynamic chaperone–client systems are hampered by sparse data as well as substantial technical challenges and therefore the protocols to integrate the available data into structural models have to consider the available information thoroughly. Most notably, the traditional determination protocol for protein structures based on NOEs breaks down for most of the chaperone–client systems, because the inherent dynamical nature of the clients does not lead to a single coherent set of NOEs. For most chaperone systems, the structures of the client-free forms of the chaperones are known from crystallographic or cryo-EM studies and the protocols for determining structural models of chaperone–client complexes thus consist in their core of docking the client protein onto the chaperone structure by the available data as constraints. Here we discuss and compare three protocols that have been fruitfully used.

1. **Protocol for docking an unfolded client into a folded chaperone by intra- and intermolecular PRE** (see Figure 11.7a). This three-step protocol was used to determine a structural model of the client OmpX bound to the chaperone Skp.[29] In a first step, the spatial volume occupied by the client protein OmpX was determined in first-order approximation as outlined in Section 11.2.2.1, resulting in a sphere of radius 21 Å. In a second step, the intermolecular PRE was determined on the Skp chaperone, originating from a paramagnetic spin-label on the client protein. The observed PRE effects were spatially averaged, to result in the determination of the center of the conformational substrate ensemble by geometric averaging. In a third step, the resulting model was cross-validated by intermolecular PREs from a paramagnetic spin on Skp on the amides of the OMP substrate. This protocol should be generally applicable whenever a fully unfolded client is bound to a chaperone.
2. **Protocol for determining a dynamic ensemble of two rigid bodies from CSP and PRE data** (see Figure 11.7b). This protocol was used to calculate the dimeric structure of the chaperone Trigger Factor.[66] Using advanced high-resolution NMR spectroscopy together with specific isotope labeling of methyl groups and aromatic amino acids and PRE distance restraints revealed the intertwined structure of the Trigger Factor dimer.[66,74] Since this arrangement is highly dynamic, a single structure determination protocol was not feasible and the following approach was developed to calculate representative snapshots of the conformational ensemble of dimers.

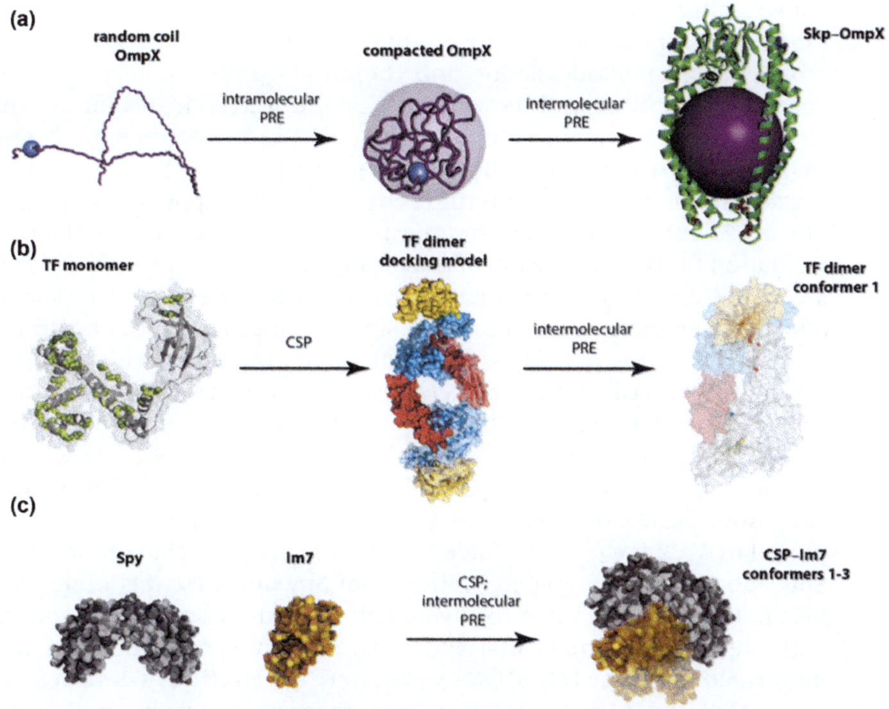

Figure 11.7 Illustration of protocols for determination of chaperone client complexes. (a) Determination of the Skp–OmpX complex. Reproduced from ref. 29 with permission from Springer Nature, Copyright © 2013, Nature Publishing Group, a division of Macmillan Publishers Limited. (b) Trigger Factor (TF) dimer. Reproduced from ref. 66, https://doi.org/10.1038/s41467-017-02196-7, under the terms of the CC BY 4.0 license, http://creativecommons.org/licenses/by/4.0/. (c) Spy–Im7 complex. Client and chaperone proteins and their complexes are shown as labeled. The arrows correspond to individual NMR methods yielding information to connect the protein conformations to structural models. See text for details.

The models were calculated in two steps: In a first step, the two monomers were docked using CSP data, with the monomeric structures based on an existing crystal structure. CSP docking was done with the HADDOCK protocol.[92] In the second calculation step, rearrangement of the individual domains was allowed using PRE distance restraints. Here, the results from HADDOCK together with the PRE data were used as input for the structure calculation software XPLOR-NIH.[93] Based on the experimental confirmation that the individual domains feature the secondary structure elements known from the crystal structure, a set of distance restraints was introduced to maintain the geometry of these elements, referred to as elastic fold network (EFN) constraints. The EFN restraints maintain the domains folded by constraining distances and leave the linkers between the domains flexible. PRE restraints were introduced as distances between the Cβ atom of the mutated residue and the amide proton detected in the spectra. The PRE restraints were divided into three classes, according to the ratio between the volume of the peaks in the paramagnetic and diamagnetic samples. The PRE restraints were used according to their classification as intra- or intermolecular. Thereby, approximately 75% of the 521 PRE restraints were intermolecular, providing sufficient information for the calculation of the arrangement of the protomers. Two structural conformers resulted that share the same overall topology, and each fits the experimental data similarly. Each conformer fulfilled a majority of, but not all, experimental constraints. These two structures thus describe two representative conformers of the conformational equilibrium of dimeric Trigger Factor, in an overall underdetermined system.

3. **Protocol for a model of a partially folded client interacting with a chaperone based on PREs** (see Figure 11.7c). This protocol was employed to dock the partially folded client Im7 onto the chaperone Spy.[47] Spin labeling of the cysteine mutants of Spy and Im7 was done with MTSL. PRE distance restraints were determined according to standard methods. For intermolecular PRE restraints, molecular docking was performed using the HADDOCK webserver. The starting PDB structures were 3O39 and 1AYI for Spy and Im7, respectively. Ambiguous Interaction Restraints (AIRs) were defined between Spy residues 53, 85, 99, and 123, and Im7 residue ranges 23–37 and 50–65, according to the observed PRE values. The HADDOCK sampling protocol resulted in 200 structures that were subjected to further semi-flexible refinement and the final structures were refined in an explicit water model. The docking results were clustered and the three best clusters with very similar HADDOCK scores were selected to represent the final solution. This protocol is similar to the outlined protocol 2 and is suitable when no internal domain motion is possible.

11.3 Conclusion and Outlook

Determining structures of chaperone–client complexes and characterizing their dynamic properties is key to resolving the underlying chaperone biophysics. The methods discussed here have allowed for many systems the establishment of structural models and will prove useful for many additional systems. Molecular chaperones are essentially connected to evolution and their study will arguably reveal many future fundamental biophysical principles that teach us the molecular origin of life.

References

1. C. B. Anfinsen, *Science*, 1973, **181**, 223–230.

2. P. G. Wolynes, J. N. Onuchic and D. Thirumalai, *Science*, 1995, **267**, 1619–1620.
3. D. J. Brockwell and S. E. Radford, *Curr. Opin. Struct. Biol.*, 2007, **17**, 30–37.
4. R. J. Ellis, *Nature*, 1987, **328**, 378–379.
5. D. Balchin, M. Hayer-Hartl and F. U. Hartl, *Science*, 2016, **353**, aac4354.
6. S. Walter and J. Buchner, *Angew. Chem., Int. Ed. Engl.*, 2002, **41**, 1098–1113.
7. F. U. Hartl, A. Bracher and M. Hayer-Hartl, *Nature*, 2011, **475**, 324–332.
8. B. M. Burmann and S. Hiller, *Prog. Nucl. Magn. Reson. Spectrosc.*, 2015, **86–87**, 41–64.
9. S. Hiller, *Trends Biochem. Sci.*, 2019, **44**, 517–527.
10. S. Hiller and B. M. Burmann, *J. Magn. Reson.*, 2018, **289**, 142–155.
11. S. Hiller, *J. Biomol. NMR*, 2021, **75**, 1–8.
12. A. Finka, R. U. Mattoo and P. Goloubinoff, *Annu. Rev. Biochem.*, 2016, **85**, 715–742.
13. M. P. Mayer and L. M. Gierasch, *J. Biol. Chem.*, 2019, **294**, 2085–2097.
14. M. Hayer-Hartl, A. Bracher and F. U. Hartl, *Trends Biochem. Sci.*, 2016, **41**, 62–76.
15. H. Wiech, J. Buchner, R. Zimmermann and U. Jakob, *Nature*, 1992, **358**, 169–170.
16. Y. E. Kim, M. S. Hipp, A. Bracher, M. Hayer-Hartl and F. U. Hartl, *Annu. Rev. Biochem.*, 2013, **82**, 323–355.
17. R. Zhao, M. Davey, Y. C. Hsu, P. Kaplanek, A. Tong, A. B. Parsons, N. Krogan, G. Cagney, D. Mai, J. Greenblatt, C. Boone, A. Emili and W. A. Houry, *Cell*, 2005, **120**, 715–727.
18. M. Taipale, G. Tucker, J. Peng, I. Krykbaeva, Z. Y. Lin, B. Larsen, H. Choi, B. Berger, A. C. Gingras and S. Lindquist, *Cell*, 2014, **158**, 434–448.
19. L. Ferbitz, T. Maier, H. Patzelt, B. Bukau, E. Deuerling and N. Ban, *Nature*, 2004, **431**, 590–596.
20. T. Hesterkamp and B. Bukau, *FEBS Lett.*, 1996, **385**, 67–71.
21. T. A. Walton, C. M. Sandoval, C. A. Fowler, A. Pardi and M. C. Sousa, *Proc. Natl. Acad. Sci. U. S. A.*, 2009, **106**, 1772–1777.
22. S. Quan, P. Koldewey, T. Tapley, N. Kirsch, K. M. Ruane, J. Pfizenmaier, R. Shi, S. Hofmann, L. Foit, G. Ren, U. Jakob, Z. Xu, M. Cygler and J. C. A. Bardwell, *Nat. Struct. Mol. Biol.*, 2011, **18**, 262–269.
23. P. Bechtluft, N. Nouwen, S. J. Tans and A. J. Driessen, *Mol. BioSyst.*, 2010, **6**, 620–627.
24. A. N. Calabrese, B. Schiffrin, M. Watson, T. K. Karamanos, M. Walko, J. R. Humes, J. E. Horne, P. White, A. J. Wilson, A. C. Kalli, R. Tuma, A. E. Ashcroft, D. J. Brockwell and S. E. Radford, *Nat. Commun.*, 2020, **11**, 2155.
25. I. P. Korndörfer, M. K. Dommel and A. Skerra, *Nat. Struct. Mol. Biol.*, 2004, **11**, 1015–1020.
26. T. A. Walton and M. C. Sousa, *Mol. Cell*, 2004, **15**, 367–374.
27. M. P. Williamson, *Prog. Nucl. Magn. Reson. Spectrosc.*, 2013, **73**, 1–16.
28. D. S. Wishart, *Prog. Nucl. Magn. Reson. Spectrosc.*, 2011, **58**, 62–87.

29. B. M. Burmann, C. Wang and S. Hiller, *Nat. Struct. Mol. Biol.*, 2013, **20**, 1265–1272.
30. B. M. Burmann, J. A. Gerez, I. Matečko-Burmann, S. Campioni, P. Kumari, D. Ghosh, A. Mazur, E. E. Aspholm, D. Šulskis, M. Wawrzyniuk, T. Bock, A. Schmidt, S. G. D. Rüdiger, R. Riek and S. Hiller, *Nature*, 2020, **577**, 127–132.
31. K. Wüthrich, *NMR of Proteins and Nucleic Acids*, Wiley, New York, 1986.
32. R. Kerfah, M. J. Plevin, O. Pessey, O. Hamelin, P. Gans and J. Boisbouvier, *J. Biomol. NMR*, 2015, **61**, 73–82.
33. G. M. Clore, in *Encyclopedia of Cell Biology*, ed. R. A. Bradshaw and P. D. Stahl, Academic Press, Waltham, 2016, pp. 98–107.
34. S. Hiller, I. Ibraghimov, G. Wagner and V. Y. Orekhov, *J. Am. Chem. Soc.*, 2009, **131**, 12970–12978.
35. M. Mobli and J. C. Hoch, *Prog. Nucl. Magn. Reson. Spectrosc.*, 2014, **83**, 21–41.
36. M. Callon, B. M. Burmann and S. Hiller, *Angew. Chem., Int. Ed. Engl.*, 2014, **53**, 5069–5072.
37. G. Otting, *Annu. Rev. Biophys.*, 2010, **39**, 387–405.
38. G. Pintacuda, M. John, X. C. Su and G. Otting, *Acc. Chem. Res.*, 2007, **40**, 206–212.
39. I. Bertini, C. Luchinat and G. Parigi, *Prog. Nucl. Magn. Reson. Spectrosc.*, 2002, **40**, 249–273.
40. I. Bertini, M. B. Janik, Y. M. Lee, C. Luchinat and A. Rosato, *J. Am. Chem. Soc.*, 2001, **123**, 4181–4188.
41. G. M. Clore and J. Iwahara, *Chem. Rev.*, 2009, **109**, 4108–4139.
42. P. H. Keizers and M. Ubbink, *Prog. Nucl. Magn. Reson. Spectrosc.*, 2011, **58**, 88–96.
43. Y. Xue, I. S. Podkorytov, D. K. Rao, N. Benjamin, H. Sun and N. R. Skrynnikov, *Protein Sci.*, 2009, **18**, 1401–1424.
44. J. L. Battiste and G. Wagner, *Biochemistry*, 2000, **39**, 5355–5365.
45. M. A. Lietzow, M. Jamin, H. J. Dyson and P. E. Wright, *J. Mol. Biol.*, 2002, **322**, 655–662.
46. D. J. Felitsky, M. A. Lietzow, H. J. Dyson and P. E. Wright, *Proc. Natl. Acad. Sci. U. S. A.*, 2008, **105**, 6278–6283.
47. L. He, T. Sharpe, A. Mazur and S. Hiller, *Sci. Adv.*, 2016, **2**, e1601625.
48. J. R. Gillespie and D. Shortle, *J. Mol. Biol.*, 1997, **268**, 158–169.
49. P. J. Flory, *Statistical Mechanics of Chain Molecules*, Hanser Publishers, New York, 1989.
50. H. Hofmann, A. Soranno, A. Borgia, K. Gast, D. Nettels and B. Schuler, *Proc. Natl. Acad. Sci. U. S. A.*, 2012, 109, 16155–16160.
51. N. R. Zaccai, C. W. Sandlin, J. T. Hoopes, J. E. Curtis, P. J. Fleming, K. G. Fleming and S. Krueger, *Methods Enzymol.*, 2016, **566**, 159–210.
52. C. L. Pashley, G. J. Morgan, A. P. Kalverda, G. S. Thompson, C. Kleanthous and S. E. Radford, *J. Mol. Biol.*, 2012, **416**, 300–318.
53. L. J. Smith, K. M. Fiebig, H. Schwalbe and C. M. Dobson, *Fold. Des.*, 1996, **1**, 95–106.

54. R. Kerfah, M. J. Plevin, R. Sounier, P. Gans and J. Boisbouvier, *Curr. Opin. Struct. Biol.*, 2015, **32**, 113–122.
55. D. F. Gauto, P. Macek, A. Barducci, H. Fraga, A. Hessel, T. Terauchi, D. Gajan, Y. Miyanoiri, J. Boisbouvier, R. Lichtenecker, M. Kainosho and P. Schanda, *J. Am. Chem. Soc.*, 2019, **141**, 11183–11195.
56. A. Boeszoermenyi, S. Chhabra, A. Dubey, D. L. Radeva, N. T. Burdzhiev, C. D. Chanev, O. I. Petrov, V. M. Gelev, M. Zhang, C. Anklin, H. Kovacs, G. Wagner, I. Kuprov, K. Takeuchi and H. Arthanari, *Nat. Methods*, 2019, **16**, 333–340.
57. N. V. Nucci, B. Fuglestad, E. A. Athanasoula and A. J. Wand, *Proc. Natl. Acad. Sci. U. S. A.*, 2014, **111**, 13846–13851.
58. P. Schanda, E. Kupče and B. Brutscher, *J. Biomol. NMR*, 2005, **33**, 199–211.
59. P. Schanda, H. Van Melckebeke and B. Brutscher, *J. Am. Chem. Soc.*, 2006, **128**, 9042–9043.
60. J. Thoma, B. M. Burmann, S. Hiller and D. J. Müller, *Nat. Struct. Mol. Biol.*, 2015, **22**, 795–802.
61. M. J. Hartl, B. M. Burmann, S. J. Prasch, C. Schwarzinger, K. Schweimer, B. M. Wöhrl, P. Rösch and S. Schwarzinger, *J. Biomol. Struct. Dyn.*, 2012, **29**, 793–798.
62. B. M. Burmann, U. Scheckenhofer, K. Schweimer and P. Rösch, *Biochem. J.*, 2011, **435**, 783–789.
63. M. J. Hartl, K. Schweimer, M. H. Reger, S. Schwarzinger, J. Bodem, P. Rösch and B. M. Wöhrl, *Biochem. J.*, 2010, **427**, 197–203.
64. B. Schiffrin, A. N. Calabrese, P. W. A. Devine, S. A. Harris, A. E. Ashcroft, D. J. Brockwell and S. E. Radford, *Nat. Struct. Mol. Biol.*, 2016, **23**, 786–793.
65. D. A. Holdbrook, B. M. Burmann, R. G. Huber, M. V. Petoukhov, D. I. Svergun, S. Hiller and P. J. Bond, *Structure*, 2017, **25**, 1079–1088.
66. L. Morgado, B. M. Burmann, T. Sharpe, A. Mazur and S. Hiller, *Nat. Commun.*, 2017, **8**, 1992.
67. C. M. Kaiser, H. C. Chang, V. R. Agashe, S. K. Lakshmipathy, S. A. Etchells, M. Hayer-Hartl, F. U. Hartl and J. M. Barral, *Nature*, 2006, **444**, 455–460.
68. H. Patzelt, G. Kramer, T. Rauch, H. J. Schönfeld, B. Bukau and E. Deuerling, *Biol. Chem.*, 2002, **383**, 1611–1619.
69. R. Franco, A. Favier, P. Schanda and B. Brutscher, *J. Magn. Reson.*, 2017, **281**, 125–129.
70. K. Weinhäupl, C. Lindau, A. Hessel, Y. Wang, C. Schütze, T. Jores, L. Melchionda, B. Schönfisch, H. Kalbacher, B. Bersch, D. Rapaport, M. Brennich, K. Lindorff-Larsen, N. Wiedemann and P. Schanda, *Cell*, 2018, **175**, 1365–1379.
71. L. He and S. Hiller, *Angew. Chem., Int. Ed. Engl.*, 2018, **57**, 5921–5924.
72. T. Saio, X. Guan, P. Rossi, A. Economou and C. G. Kalodimos, *Science*, 2014, **344**, 1250494.
73. C. Huang, P. Rossi, T. Saio and C. G. Kalodimos, *Nature*, 2016, **537**, 202–206.

74. T. Saio, S. Kawagoe, K. Ishimori and C. G. Kalodimos, *eLife*, 2018, **7**, e35731.
75. B. M. Burmann, D. A. Holdbrook, M. Callon, P. J. Bond and S. Hiller, *Biophys. J.*, 2015, **108**, 1516–1526.
76. A. G. Palmer 3rd and F. Massi, *Chem. Rev.*, 2006, **106**, 1700–1719.
77. A. G. Palmer 3rd, C. D. Kroenke and J. P. Loria, *Methods Enzymol.*, 2001, **339**, 204–238.
78. G. Lipari and A. Szabo, *J. Am. Chem. Soc.*, 1982, **104**, 4559–4570.
79. G. Lipari and A. Szabo, *J. Am. Chem. Soc.*, 1982, **104**, 4546–4559.
80. G. M. Clore, A. Szabo, A. Bax, L. E. Kay, P. C. Driscoll and A. M. Gronenborn, *J. Am. Chem. Soc.*, 1990, **112**, 4989–4991.
81. J. W. Peng and G. Wagner, *Biochemistry*, 1992, **31**, 8571–8586.
82. H. J. Dyson and P. E. Wright, *Methods Enzymol.*, 2001, **339**, 258–270.
83. C. Bracken, P. A. Carr, J. Cavanagh and A. G. Palmer 3rd, *J. Mol. Biol.*, 1999, **285**, 2133–2146.
84. N. A. Farrow, O. Zhang, J. D. Forman-Kay and L. E. Kay, *Biochemistry*, 1995, **34**, 868–878.
85. N. Salvi, A. Abyzov and M. Blackledge, *Prog. Nucl. Magn. Reson. Spectrosc.*, 2017, **102 – 103**, 43–60.
86. W. Adamski, N. Salvi, D. Maurin, J. Magnat, S. Milles, M. R. Jensen, A. Abyzov, C. J. Moreau and M. Blackledge, *J. Am. Chem. Soc.*, 2019, **141**, 17817–17829.
87. K. Wüthrich, *Curr. Opin. Struct. Biol.*, 1994, **4**, 93–99.
88. B. M. Burmann and S. Hiller, *Chimia*, 2012, **66**, 759–763.
89. L. He and S. Hiller, *ChemBioChem*, 2019, **20**, 2803–2806.
90. Z. K. Boswell and M. P. Latham, *Biochemistry*, 2019, **58**, 144–155.
91. S. Schütz and R. Sprangers, *Prog. Nucl. Magn. Reson. Spectrosc.*, 2020, **116**, 56–84.
92. C. Dominguez, R. Boelens and A. M. Bonvin, *J. Am. Chem. Soc.*, 2003, **125**, 1731–1737.
93. C. D. Schwieters, J. J. Kuszewski, N. Tjandra and G. M. Clore, *J. Magn. Reson.*, 2003, **160**, 65–73.

CHAPTER 12

Methyl-TROSY NMR Spectroscopy in the Investigation of Allosteric Cooperativity in Large Biomolecular Complexes

RUI HUANG

University of Guelph, Department of Chemistry, 482 Gordon Street, Guelph, ON N1G 1Y4, Canada
Email: rhuang08@uoguelph.ca

12.1 Introduction

Cooperativity is an important and universal phenomenon in biological systems.[1] In a broad sense, it describes ecological interactions between species or groups of organisms which result in evolutionary advantages, while it may also refer to metabolic interconnectivity between cells and signaling pathways.[1,2] In biochemistry and enzymology, cooperativity arises from interactions at the molecular level which render the entire molecular system of interest to behave differently from what is expected based on the properties of the individual building blocks. Two typical examples introduced in biochemistry textbooks are oxygen binding to hemoglobin and protein folding. Cooperativity observed in the former originates from allosteric communication between different binding sites of the ligand—one in

New Developments in NMR No. 26
NMR Spectroscopy for Probing Functional Dynamics at Biological Interfaces
Edited by Anirban Bhunia, Hanudatta S. Atreya and Neeraj Sinha

Published by the Royal Society of Chemistry, www.rsc.org

each subunit of the multimeric protein, with binding of a ligand to one site affecting the affinity at a different site; hence this type of cooperativity is also termed 'allosteric cooperativity'.[1–3] On the other hand, cooperativity in the folding of biopolymers, such as proteins and nucleic acids, is largely due to an entropic advantage in which an initial folding event greatly reduces the conformational space that the rest of the molecule needs to search in order to reach a minimum in the free energy landscape. Cooperative phenomenon of such an origin is sometimes referred to as 'configurational cooperativity'[1] or 'chelating cooperativity,'[3] which derives from a somewhat distinct mechanism from allostery. With the various levels at which cooperativity manifests in biological systems, in what follows we focus on 'allosteric co-operativity' in the context of molecular complexes containing multiple identical subunits.

Since the discovery of cooperativity by Bohr, Hasselbach and Krogh in hemoglobin,[4] a large number of enzymes have been shown to display positive or negative cooperativity, or both.[2] Conventionally, cooperativity is indicated by the phenomenological sigmoidal response curve of ligand binding or substrate catalysis as a function of ligand or substrate concentration. The curve can be fitted either with an empirical Hill equation or with explicit cooperativity models to extract more insights into the system.[5] Despite the effectiveness of these models in describing the data, they provide limited structural insights into how the binding of a ligand/substrate leads to the observed allosteric effect and cooperativity.[5] With the progress in structural biology, in particular, the emergence of cryo-electron microscopy (cryo-EM)[6–9] and the development of modern nuclear magnetic resonance (NMR) spectroscopy techniques for high-molecular-weight assemblies,[10–12] it has become increasingly possible to understand allostery and cooperativity from a fundamental and structural perspective, even for large biomolecular complexes.

Amongst the structural biology techniques, NMR spectroscopy is a unique and powerful tool to study allostery and cooperativity of a biomacromolecule in a near-physiological environment. The resolving power of NMR allows the occupancies of the binding sites to be monitored in a site-specific manner in a titration experiment.[13] Importantly, NMR is an excellent technique for characterizing conformational interconversion which frequently occurs in a cooperative system. Thermodynamic and kinetic parameters of the interconversion, as well as structural information of the conformers, can be extracted from the NMR experiments to provide a detailed description of the energy landscape of the molecule.[14–19] The NMR chemical shift is extremely sensitive to changes in the local environment; hence, subtle structural rearrangements, which propagate from the ligand-binding site toward distal areas in the molecule leading to allostery and cooperativity can be monitored by chemical shift perturbation.[13,20–23] Notably, due to the special capability of NMR to probe molecular motions over a wide range of timescales,[24] it is particularly useful to decipher mechanisms of allostery and cooperativity that are mediated exclusively or predominantly *via* changes in protein dynamics.[25,26] In such studies, NMR is often used in conjunction with other biochemical and biophysical techniques, such as functional assays and isothermal titration calorimetry, to paint a more complete picture of the system of interest.

Conventional solution NMR techniques are limited to biomolecules and complexes under 40 kDa due to rapid transverse relaxation rates and spectra overcrowding with increased molecular size. Recent decades have witnessed the expansion of NMR studies into the realm of large biological systems thanks to developments in isotope labeling strategies, such as deuteration and selective labeling schemes, and the exploitation of transverse relaxation optimized spectroscopy (TROSY) effect.[27] A particular TROSY technique tailored for $^{13}CH_3$-labeled methyl groups, termed methyl-TROSY,[28] has increased the size limit of biological complexes studied by solution NMR to over 1 MDa.[29] High-quality methyl-TROSY spectra have enabled structural and dynamic studies of a number of high-molecular-weight complexes by solution NMR,[11,30–32] which are otherwise not amenable to this technique due to fast signal decay.

In this chapter, we provide an overview of the application of methyl-TROSY NMR spectroscopy in studying cooperativity in high-molecular-weight (>100 kDa) assemblies. First, we give an introduction of cooperativity and the classical models. We then delineate the concept of ligand-free conformational cooperativity and discuss the general approaches to investigate it, including protomer specific isotope labeling strategies. Subsequently, we provide a brief review of the methyl-TROSY technique, focusing on the practical aspects of isotopic labeling strategies as well as the current approaches to obtain resonance assignments for methyl groups. Finally, we focus on several specific applications in which methyl-TROSY NMR is utilized to probe allosteric cooperativity in different molecular assemblies strategies as case studies. This chapter illustrates the unique insights that methyl-TROSY NMR can provide in the investigation of cooperativity in large molecular complexes, and hopefully provides readers with essential background, general strategies, and examples to carry out independent studies in this direction.

12.2 Cooperativity

12.2.1 What Is Cooperativity?

The term ‘cooperativity’ gives an intuitive description of the phenomenon at a microscopic level. Take the classical case of hemoglobin[4] as an example, the sigmoidal binding curve of O_2 can be explained by a simple idea that the binding of the first O_2 molecule renders the binding of the next one easier, and hence the binding events are ‘cooperative’ toward a fully occupied enzyme. This very idea can be extended to give a general description of cooperativity in multimeric proteins: after the initial event occurs in one of the subunits in the complex, *e.g.* ligand binding, conformational conversion, change in dynamics, *etc.*, the effect is transmitted allosterically to other

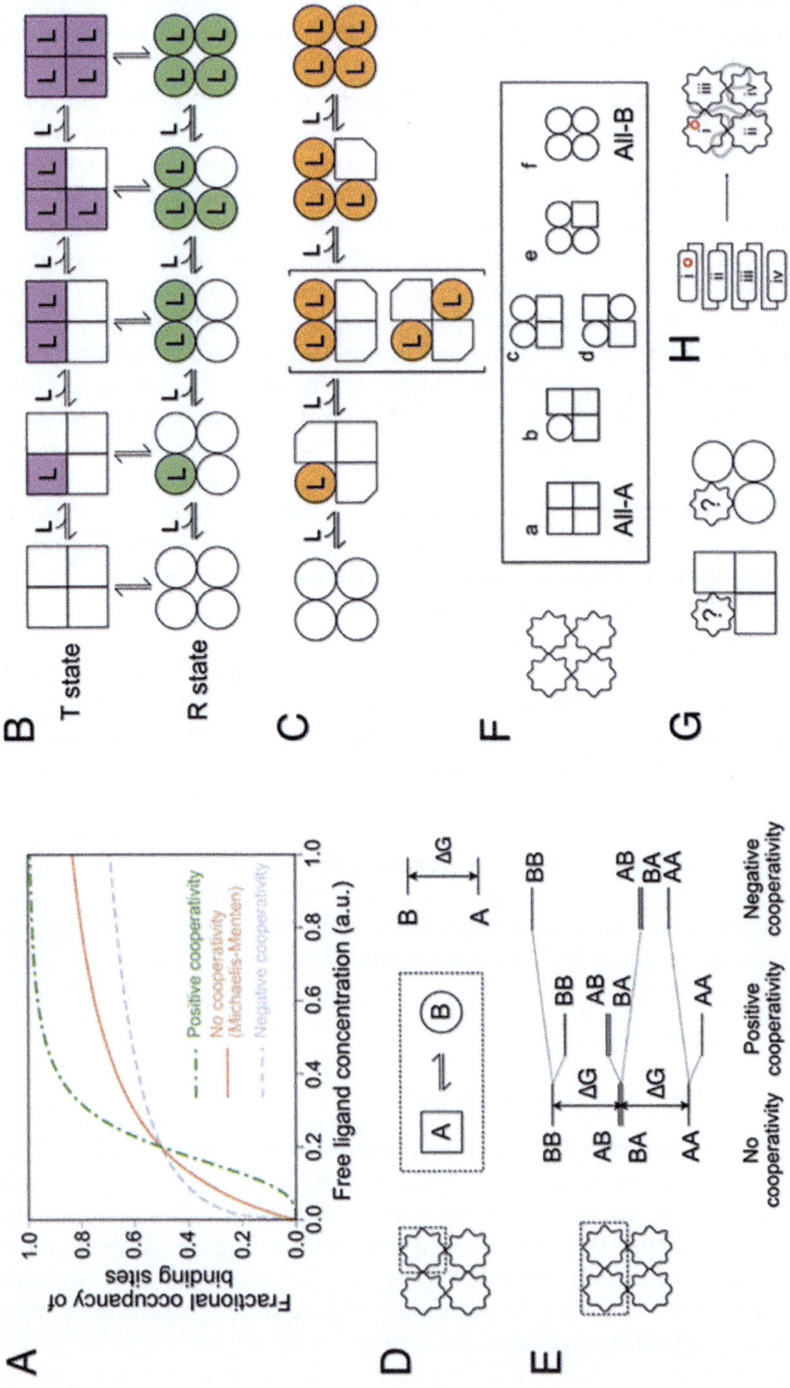
A
Fractional occupancy of binding sites
1.0
0.8
0.6
0.4
0.2
0.0
Free ligand concentration (a.u.)
Positive cooperativity
No cooperativity (Michaelis-Menten)
Negative cooperativity
B
T state
R state
L
C
D
A
B
ΔG
E
BB
AB
BA
AA
No cooperativity
Positive cooperativity
Negative cooperativity
F
a
b
c
d
e
f
All-A
All-B
G
?
H
i
ii
iii
iv

subunits, making it (i) easier, (ii) more difficult, or (iii) with no effect for the same event to occur to other subunits. The three scenarios aforementioned correspond to (i) positive, (ii) negative, and (iii) no cooperativity, respectively. In the case of strong positive cooperativity, it can be expected that any individual complex is likely to be either ligand-free or fully bound, with the partially bound state lowly populated, while for strong negative cooperativity, the fully bound state is much less probable than the intermediate state.

At the macroscopic level, a sigmoidal binding curve or rate of catalysis as a function of substrate concentration is characteristic of positive cooperativity (Figure 12.1A, green). The sigmoidal shape indicates that the ensemble of enzyme molecules switches from mostly free to mostly bound over a small change of ligand concentration. The binding curve with negative cooperativity has a sharp increase at the low concentration of the ligand, yet the binding sites remain unsaturated over a wide range of ligand

Figure 12.1 (A) Characteristic binding curves of ligands with positive, negative, and no cooperativity. The curves are generated with a Hill–Langmuir equation with the 50% saturation concentration set to be the same (0.2) for all cases and with Hill coefficients of 3 (green broken line), 0.5 (lavender dashed line), and 1 (orange solid line). (B) Illustration of the MWC model for a square tetramer (*e.g.* hemoglobin). The complex is proposed to interconvert between two symmetric quaternary states (T and R). Ligands bind to both states yet with different affinities. High binding affinity of the ligand/substrate to the R state relative to the T state leads to highly cooperative binding. (C) Illustration of the KNF model. Ligand binding induces allosteric conformational change in the neighboring subunits and hence affects the affinities for the subsequent binding events. The two binding configurations in the brackets illustrates that both positive (top) and negative (bottom) cooperativity can be accounted for by the KNF model. (D–E) A tetrameric complex with each subunit interconverting between conformations A and B. Under an 'uncoupled' scenario, the conformational equilibrium depends solely on the free energies of state A and B (panel D). In the complex, the energy landscape could be affected by inter-subunit interactions (panel E). For example, focusing on two neighboring subunits (E, in the dotted-line box), the free energies of the combined states of the two subunits (AA, AB, BA, and BB) are illustrated in cases of positive, negative, and no cooperativity. (F) A total of six microstates (a–f) exist for a tetramer with C4 symmetry in which each subunit exchanges between A (square) and B (circle) states. (G) Hetero-tetramers consisting of WT (star shape) and mutant subunits, with the conformation of the mutants constrained in either A (left, square) or B (right, circle) state. Comparison of the conformational equilibrium of the WT subunits in these two species reveals inter-subunit allostery and cooperativity in the complex. The illustration also describes the conditional probability of P(X|AAA) (left) and P(X|BBB) (right) (see text), where X can be either the A or B state. (H) Concatenation of all four subunits (i– iv) of a tetramer on a single polypeptide chain encoded in one DNA sequence (left) allows a mutation (red circle) to be introduced to a specific subunit *via* site-directed mutagenesis.

concentrations (Figure 12.1A, purple). In both cases, the curves deviate from a typical Michaelis–Menten model (Figure 12.1A, orange) and can be fit with the Hill–Langmuir equation to obtain a qualitative description of co-operativity.[33] Fitting the data with a model that precisely describes the properties of the system grants further insights into the mechanism of the cooperativity, which is discussed in detail later.

12.2.2 Biological Significance of Allosteric Cooperativity

For many enzymes, allosteric cooperativity is evolutionarily selected, perhaps to regulate the metabolic responses in the cell. Compared with a Michaelis–Menten system, an enzyme with positive cooperativity has an amplified sensitivity to substrate concentration, converting a small change in concentration to a larger effect on the output, such as ligand binding and substrate turnover. On the other hand, negative cooperativity decreases the sensitivity of the response to substrate concentration while broadening the range of concentration that gives some response. In a multimeric protein, allosteric cooperativity originates from delicate inter-subunit communications that can be easily altered by amino acid mutations. In this regard, in order to fine-tune the sensitivity range of an enzyme, it is evolutionarily easier to select mutations on the allosteric pathways that impact co-operativity of the enzyme than to change the precise structure and geometry of the active site.[2]

The importance of cooperativity is also highlighted in large molecular motors and degradation machines, in which concerted motions or coordinated conformational interconversion between subunits are often required for function. Examples of such complexes include the family of the AAA+ (ATPases associated with diverse cel-lular activities) proteins, which convert the chemical energy of ATP hydrolysis into mechanical forces for substrate remodeling. Recent cryo-EM studies of a number of AAA+ proteins revealed highly coordinated conformational changes and motions between subunits during the process of substrate translocation, suggesting intricate allosteric communications and cooperation in these molecular machines.[34] Extensive allosteric network and/or cooperativity has also been discovered in large protein degradation machines and chaperones, such as the proteasome core particle, ClpP and GroEL.[20,35–37] Cooperativity, in these cases, plays an essential role in the biological functions of the enzymes.

12.2.3 Classical Models of Cooperativity: MWC and KNF Models

Two classical theories—the MWC[38] and KNF[39,40] models—have conventionally been used to describe the mechanism of cooperativity and to fit the experimental data. The MWC model (also known as the 'concerted' or 'symmetry' model), formulated by Monod, Wyman, and Changeux in 1965,[38] presents an elegant mathematical approach to describe cooperative binding. The key hypothesis involves a concerted conformational switch of all subunits in the complex which retains the symmetry of the whole molecule. That is to say, the entire complex interconverts between (at least) two quaternary states—tense (T) or relaxed (R), which have distinct binding affinities to a ligand/substrate (Figure 12.1B). An important feature of the model is that the microscopic affinity of the ligand is the same for all the available sites in the complex in a particular state (T or R), which is independent of the number of already-bound ligands and the configuration of the ligand–protein complex (or microstate). The energetic drive behind the T to R conversion with bound ligands originates from a higher affinity of the ligand toward the R state of the enzyme.

The beauty of the MWC model is partially manifested in the fact that for homotropic cases (with only one type of ligand involved) the fractional saturation (Y_s) can be expressed by a simple equation (eqn (12.1)) of only three independent variables α, L, and c, as shown below.

$$Y_S = \frac{\alpha(1+\alpha)^{n-1} + Lc\alpha(1+c\alpha)^{n-1}}{(1+\alpha)^n + L(1+c\alpha)^n} \tag{12.1}$$

where $\alpha = [L]/K_R$, $L = [T_0]/[R_0]$ and $c = K_R/K_T$. L is the equilibrium constant of the enzyme between the T and R states in the absence of ligand. K_R and K_T are the microscopic dissociation constants of the ligand for the R and T states, respectively. And [L] is the total concentration of the ligand. Based on the framework of the MWC model, more sophisticated models can be constructed with relative ease to account for cases when binding of multiple ligands, such as substrates, activators, and inhibitors, are involved (heterotopic interactions), or when more precisely defined structural states are needed.[5,36] Although the assumptions of obligated symmetry and quaternary structural change sometimes deviate from reality, the MWC model has proved to be a very effective approximation to describe many cooperative systems, and has gained wide usage throughout the years due to its mathematical succinctness.[41]

On the other hand, the KNF model,[40] by Koshland, Némethy, and Filmer, postulated that binding of ligand on one subunit induces allosteric conformational changes in the neighboring subunits, which in turn alters their ligand-binding affinities. In contrast to the MWC assumption in which the affinity of the ligand solely depends on the distinct quaternary states of the complex, the KNF model explicitly describes the allosteric effects between the neighboring subunits with the affinity of the ligand varying sequentially with the number of bound ligands as well as their positions in the complex (Figure 12.1C). In this regard, the KNF model is also known as the 'sequential' or 'induced fit' model. Derivation of the KNF model usually requires all structural configurations (or microstates) of the ligand–protein complex to be explicitly considered. For instance, a symmetric square tetramer has a total of six unique configurations with zero to four bound ligands (Figure 12.1C), while the total number of configurations goes up to 20 in the case of a C7 symmetric heptamer. In the original KNF paper,[40] an interaction constant, *K*, was defined to describe the likelihood of occurrences of certain combinational states. For example, consider two neighboring promoters, each of which can adopt either a high-affinity (A) or a low-affinity binding state (B). K_{AB} is defined to be the likelihood of the case in which one of the two subunits is in state A and the other is in state B, divided by the likelihood of both being in state A. By this definition, $K_{AB} > 1$ would suggest a higher occurrence of AB neighbors than the AA neighbors, implying a case of negative cooperativity. Notably, negative cooperativity cannot be accounted by the MWC model due to the restriction of symmetry. Following the sample principle, the KNF model can be modified to include interaction constants specific to a certain combinational state—beyond just the nearest-neighbor effect—based on the hypothesis of the mechanism of cooperativity.[42] However, due to the requirement of explicit description, the KNF model becomes rather cumbersome to derive in cases where binding of multiple ligands is involved, considering the many existing combinatorial configurations; convergence of data fitting, in this case, can also be challenging due to a large number of variables involved.

The difference between the principles of the classical MWC and KNF models reflects two philosophically different mechanisms by which conformational dynamics of the protein couple with binding events—conformational selection (MWC) or induced fit (KNF).[43] Both models describe the sigmoidal curves for positive cooperativity equally well. Over the years, MWC models have gained much more popularity in the literature.[44] At the same time, the extension, modification, and combination of both models have continued to be developed,[5,45–48] while new models such as probabilistic models have been proposed and tested against experimental data.[42,49–52]

12.3 Cooperativity of Conformational Interconversion in the Absence of Ligand

12.3.1 Coupled Conformational Exchange

In essence, cooperativity of ligand binding, which we have discussed so far, arises from coupling of conformational dynamics between different subunits in a complex. The strength of coupling can vary from 'completely coupled' so that the symmetry of the complex is always preserved regardless of the number of bound ligands (MWC model), to 'limited locally' in which case only the adjacent neighbors are allosterically connected (KNF model). From this perspective, the intrinsic coupling between the conformational interconversions of the subunits can be explored even in the absence of ligand. Take a symmetric square tetramer as an example, assume that each of its protomers exchanges between two conformations A and B (Figure 12.1D). Supposing that the protomers can somehow 'uncouple' from each other, *e.g.* *via* dissociation of the complex, the conformational equilibrium between the two states (A and B) is solely dependent on the intrinsic energy landscape of the protomer (Figure 12.1D, right). In the absence of cooperativity, the energy landscape of the complex is a linear combination of the states of each protomer (Figure 12.1E, with only a pair of protomers considered). However, due to inter-subunit interactions, certain conformational states might be stabilized or destabilized, deviating the energy landscapes from the case of independent exchange (Figure 12.1E). As a result, the conformational interconversion could be either positively or negatively coupled between subunits.

From a microscopic perspective, the complex comprises all combinatorial microstates at all times (Figure 12.1F). Cooperativity is manifested when the population distribution of the states deviates from that in an 'uncoupled' case. Assuming an extreme scenario of 100% positive cooperativity, only two states—all-A or all-B—would be observed (Figure 12.1F, a and f). It is also expected that, with positive cooperativity, microstate (c) is more populated than (d) due to concerted conformational interconversion between neighboring subunits (Figure 12.1F), and *vice versa*. Notably, if the conformational equilibrium between A and B can be biased by bound ligand, then cooperativity in ligand binding is expected as a consequence.

12.3.2 Strategies to Probe Conformational Cooperativity

As discussed above, cooperativity of conformational exchange gives rise to characteristic population distribution of the microstates in the complex. In reverse, the populations of all states, if determined, can be fit with models to extract information about cooperativity. A notable example is given later in the chapter (Section 12.5.2) which demonstrates the feasibility of this approach with a homo-oligomeric p97 complex.[52] The populations of all 14 microstates in the hexameric complex were obtained by cryo-EM and fitted with probabilistic models to inform on cooperativity. However, it is usually quite challenging to quantify the fractional populations of all microstates, especially in higher order oligomers.

An alternative strategy relies on perturbing the conformational equilibrium of some of the protomers in the complex and examining the effect of perturbation on the remaining protomers (Figure 12.1G) or the complex as a whole. In the absence of ligand, perturbation is typically achieved by introducing mutation(s) on residue(s) that critically impact the equilibrium of exchange between the states. In order to introduce mutations to specific subunit(s) in the complex, one method is to artificially concatenate multiple subunits in one polypeptide chain which is translated from a single gene product (Figure 12.1H).[53] In this case, mutation can be introduced by simple site-directed mutagenesis at the corresponding position of the targeted subunit in the DNA sequence. The concatenated subunits are connected 'head-to-tail' by flexible linkers, which are designed in such a way as to not interfere with the folding and assembling of the subunits into a fully functional enzyme (Figure 12.1H). This approach is well suitable for complexes with relatively small subunits and lower order multimers to limit the total length of the polypeptide chain and the encoding DNA.

A second way to introduce mutations to some but not all subunits of the complex is through reconstitution. Hetero-oligomeric complexes can be generated from two different types of subunits, with one of them bearing the mutation that perturbs the conformational equilibrium. When studied by NMR spectroscopy, which is discussed in the next section, strategic isotope labeling can be applied so that only one of the two types of subunits is 'NMR active,' serving as probes for structure and dynamics, while the other is 'NMR inactive' to which mutations are introduced. The inter-subunit allosteric effect is then read out from the 'NMR active' subunits when the conformational equilibrium of the neighboring 'NMR inactive' subunits is perturbed by mutations. Notably, such hetero-oligomers are particularly interesting for enzymes that display a dominant-negative effect with pathological mutations. In heterozygous patients, both WT subunits and those bearing disease mutations are expressed in cells and may assemble into mixed complexes. For example, a few enzymes in the AAA+ family have been shown to naturally form hetero-oligomers consisting of WT and mutant proteins both *in vitro*[21,54–57] and *in vivo*.[58–60] Incorporation of subunits containing disease mutations could negatively impact the activity of the WT subunits, or even render the entire complex catalytically inactive.[61–63]

12.3.3 Reconstitution of Hetero-oligomers and Differential Isotope Labeling

Reconstitution of a hypothetical trimeric complex for NMR studies is given as an example (Figure 12.2A). Two types of subunits (Type I and Type II) are used which differ both in the isotope labeling scheme and in the conformational equilibrium of interest. For example, Type I subunit can be a WT and 'NMR active' (*e.g.* Ile-δ1 $^{13}CH_3$ labeled) subunit and Type II can be an NMR inactive (*e.g.* perdeuterated) subunit bearing a mutation that perturbs its conformational equilibrium. Differential isotope labeling allows us to distinguish the two types of subunits in the NMR spectra and prevent spectral overcrowding by multiple sets of peaks due to conformational heterogeneity. Expressed and purified separately, the two types of subunits are mixed at a specific ratio followed by a stochastic reconstitution process. A common reconstitution method for high-affinity oligomers involves unfolding the mixture of the homo-oligomers in 6 M GdnHCl followed by refolding,[21,35,36,42] while reconstitution may also be achieved by changing buffer conditions, such as salt concentration, pH, and temperature, to reversibly dissociate and re-assemble the oligomers from different types of subunits. For complexes that undergo subunit exchange, heterogeneous

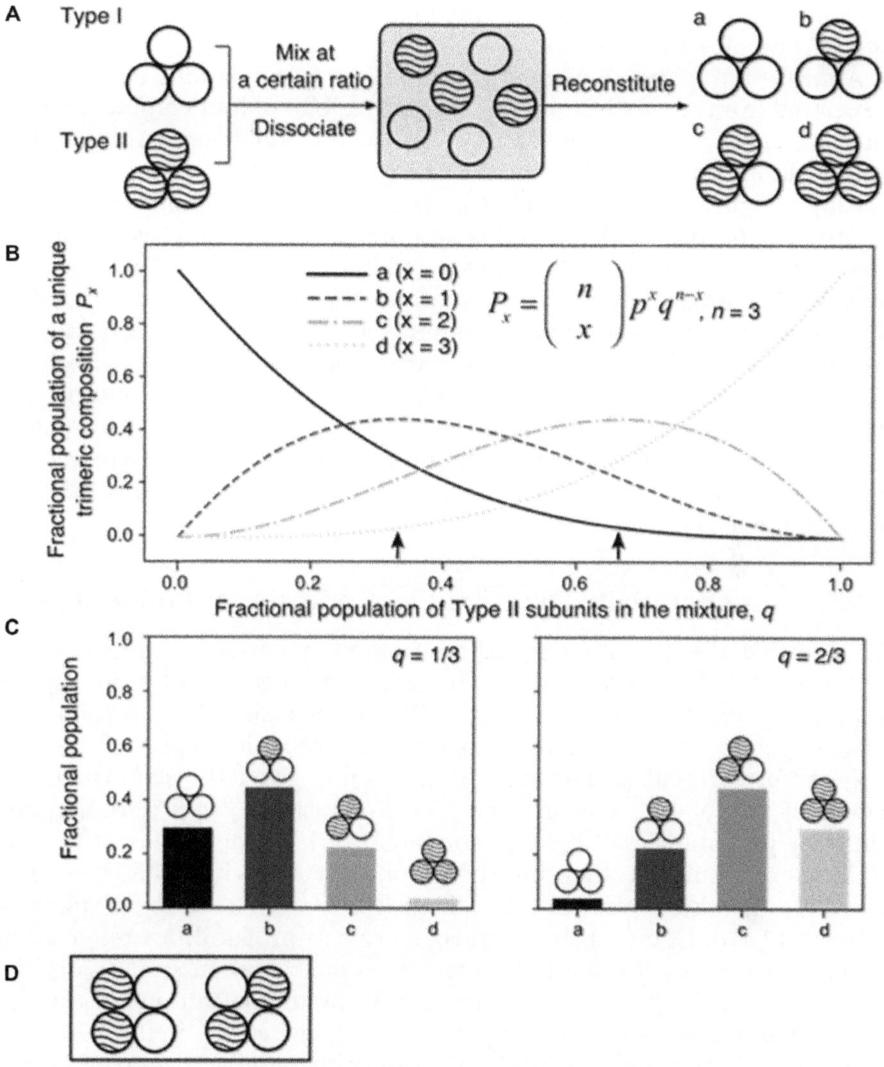

Figure 12.2 (A) Reconstitution of a trimeric complex from two types of subunits, generating four unique compositions (a–d). (B) Populations of the species (a–d) as a function of the fraction of Type II subunits during reconstitution. The fractional population of each species (a–d), $P_{\times}$, is calculated by the general binomial distribution equation, shown in the insert, in which n is the total number of subunits in the oligomer ($n=3$ for a trimer), x is the number of Type II subunits in the species ($x \in \{0,1,2,3\}$), and p and q are the fractions of Type I and II subunits in the entire mixture, respectively ($p+q=1$). With the mixing ratios of Type I and II subunits being 2 : 1 ($q=1/3$) and 1 : 2 ($q=2/3$) (indicated by the arrows in B), the fractional populations of species (a–d) are plotted in the left and right panels in C, respectively. (D) Two different configurations for the same subunit composition in a tetramer with two of each Type I and Type II subunits.

complexes were found by simply mixing the two types of homo-oligomers under physiological conditions.[21,64]

A mixture of hetero-oligomers with different compositions are inevitably generated from reconstitution. For example, a total of four compositions are produced for a trimer (Figure 12.2A, a–d). The fractional populations of these compositions are determined by the ratio of the two types of subunits upon mixing (Figure 12.2, B and C), following a binormal distribution (Figure 12.2B, insert). Notably, this assumes that the two types of subunits have equal propensities to be incorporated into the complex. A simple way to test this hypothesis is to include an affinity tag (*e.g.* His_6-tag) on one of the subunits and quantify the amount of retained complexes after reconstitution as a function of mixing ratios, which is then compared with what is expected from a binomial distribution.[42] For higher order oligomers, multiple configurations emerge with the same composition of subunits. A simple example (Figure 12.2D) shows two unique configurations for a tetramer with two Type I and Type II subunits each, which might need to be distinguished depending on the cooperativity model.

12.3.4 Cooperativity Described by Conditional Probabilities

To describe the cooperativity of conformational interconversion in the absence of ligand, probabilistic models have been shown to be conceptionally intuitive and mathematically approachable.[42,52] In such models, cooperativity is reported by conditional probabilities, which describe the conformational equilibrium of a subunit given the conformational states of its neighbors in the complex. For instance, in a tetramer, conditional probabilities $P(A|AAA)$ and $P(A|BBB)$ are defined as the probabilities of a subunit adopting the A conformation given the other three subunits in the complex are all in the A or B state, respectively (Figure 12.1G). Comparison between these two probabilities reflects how cooperative the conformational interconversion between subunits is. The framework of the probabilistic model, namely the definitions of the conditional probabilities underlying the model, depends on the system of interest and the proposed mechanism of cooperativity. Examples of such models will be provided and discussed in detail later in case studies 12.5.1 and 12.5.2.

12.4 Methyl-TROSY

12.4.1 Overview of Methyl-TROSY

Transverse relaxation optimized spectroscopy (TROSY)[27] was first demonstrated in ^1H-^{15}N spin systems. It takes advantage of the interference between the two major relaxation mechanisms—dipole–dipole interaction (DD) and chemical shift anisotropy (CSA)—of the coupled ^1H–^{15}N spin to obtain narrow resonance lines and improve spectral quality. Since CSA is dependent on the strength of the magnetic field squared, the optimal TROSY effect is achieved at a magnetic field of around 21.1 T (900 MHz proton frequency),[27] with the maximum peak height expected at 1.5 GHz proton frequency for ^{1}H-detected amide TROSY.[65] The TROSY effect is also exploited in aromatic ^{1}H–^{13}C[66] and ^{19}F–^{13}C systems.[67]

A particular type of TROSY experiment, termed methyl-TROSY,[28] takes advantage of the relaxation properties of a $^{13}CH_3$ methyl group. The symmetry of the methyl group leads to unique components of magnetization that have very slow relaxation rates. Such coherences can be carefully selected by design of the pulse sequences to obtain narrow linewidth[28] even for high-molecular-weight complexes up to 1 MDa.[29] The principles underlying the methyl-TROSY effect are well explained in several reviews.[11,68] It is worth noting that the slow relaxing transitions in the $^{13}CH_3$ system result from the interference between proton–proton and proton–carbon dipolar relaxation instead of cross-correlation between dipolar interactions and CSA relaxing mechanisms. In fact, CSAs in the methyl group are too small to interfere with DD significantly. Therefore, unlike the amide TROSY effect, the methyl-TROSY effect is rarely dependent on the magnetic field strength.[69]

12.4.2 Isotope Labeling of Methyl Groups

In order to obtain high-quality spectra of large biological molecules exploiting the methyl-TROSY effect, $^{13}CH_3$ methyl groups are introduced into proteins[70] or nucleic acids[71] to serve as probes of structure and dynamics in an otherwise highly deuterated background. Notably, it is usually crucial to achieve a high level of deuteration of the molecule except for the methyl sites of interest to obtain narrow lines for the methyl resonances. It has been shown by simulations that the external protons in the protein, resulting from incomplete deuteration, protonated precursors, and other $^{13}CH_3$-labeled methyl groups, can lead to significant increases in the relaxation rates of the slowly relaxing resonances in the methyl groups.[11] Therefore, decreasing the density of protons in the molecule, in particular in high-molecular-weight assemblies, is key to high-quality methyl spectra.

12.4.2.1 Methyl Labeling in E. coli

There are six types of methyl-group-containing amino acids—isoleucine, leucine, valine, methionine, alanine, and threonine—which can be labeled with $^{13}CH_3$ biosynthetically. To date, the most common strategy to prepare protein samples for methyl-TROSY spectroscopy involves overexpression in *E. coli* cells using minimal D_2O media supplemented by isotopically labeled amino acids or amino acid precursors. For example, the methyl group in methionine can be labeled by adding ε-^{13}C-methionine into the growth medium.[72] Yet, for most of the other methyl groups, addition of the labeled amino acids alone in the medium likely leads to isotope scrambling. This is due to the interconnected metabolic pathways that are involved in the biogenesis of the methyl-containing amino acids.[11] Therefore, it is usually necessary to supplement the medium with additional isotopically labeled metabolites to prevent undesired background labeling. For example, in order to label the methyl group in Ala, one of the cost-effective strategies involves supplementing the M9/D_2O medium with 2-[^{2}H], 3-[^{13}C] L-alanine together with perdeuterated α-ketoisovalerate, succinate, and isoleucine.[73] Alternatively, unique metabolic precursors can be used in which the specific $^{13}CH_3$-labeled methyl group in the precursor is transferred to the corresponding amino acid. For instance, labeling of the δ1-methyl group in Ile can be achieved by using isotopically labeled α-ketobutyrate,[74] while a common way to label the methyl groups in Leu and Val is to use α-ketoisovalerate as a precursor in which the isopropyl group is transferred to Leu and Val.[75] The two geminal methyl groups in α-ketoisovalerate, subsequently transferred to Leu and Val, can be both labeled with $^{13}CH_3$ or one with $^{13}CH_3$ and the other with $^{12}CD_3$. The latter labeling scheme bears two advantages in comparison: (i) no *J*-coupling between the two methyl carbons and (ii) smaller methyl relaxation rates due to the absence of protons in close vicinity. Notably, this approach using [$^{13}CH_3$, $^{12}CD_3$]-α-ketoisovalerate as precursor leads to labeling of both pro-(*S*) and pro-(*R*) methyls in Leu and Val with no stereospecificity, and only one of the two methyl groups is labeled in each Leu/Val residue. Stereospecific labeling of either the pro-(*S*) or the pro-(*R*) methyl group in Leu and Val can be achieved by the 2-acetolactate precursor either with the labeled methyl group attached to the C2 carbon or with the C4 methyl labeled with $^{13}CH_3$, respectively.[76] Stereospecific labeling effectively reduces the number of Leu and Val methyl resonances by half [with only pro-(*S*) or the pro-(*R*) methyls in the spectra], which is advantageous for large systems challenged by spectral overcrowding. A list of isotope labeling strategies for methyl groups has been thoroughly summarized by Schütz and Sprangers.[11]

In order to increase the overall cost efficiency of labeling and the number of probes per sample, it is common to label several types of methyl-containing amino acids at the same time. Many combinations out of the six types of amino acids can be achieved, including IM, ILV, ILVM, ILVA, ILVMT, and ILVMAT,[21,29,77–80] among which IM, ILV, and ILVM are perhaps the most commonly used. The combinations of the amino acid types to be labeled are usually decided based on a few factors including spectral information, the risk of overcrowding, limiting the total number of protons in the protein, cost, and the compatibility of labeling schemes. For example, α-ketoisovalerate can be readily combined with other precursors to label Leu and Val with non-stereospecific $^{13}CH_3/^{12}CD_3$ methyl groups. In contrast, labeling Ile(δ1) with α-ketobutyrate is not completely compatible with stereospecific labeling of the pro-(*R*)/(*S*) methyl in Leu and Val with 2- acetolactate, simultaneous usage of which results in inefficient labeling of LV methyl groups.[81] Significant improvement can be achieved, if such a combination of labeling scheme is desired, by a two-step approach of precursor addition, in which α-ketobutyrate is added to the growth medium 20 min after the addition of 2-acetolactate.[81] In addition, genetically modified *E. coli* strains in which genes that encode the specific enzymes involved in the biogenesis of the methyl-containing amino acids are deleted from the *E. coli* genome can be used to reduce scrambling.[80,82,83]

12.4.2.2 Methyl Labeling by Other Approaches

Although using *E. coli* as the expression host provides a relatively simple and economical way to produce methyl-labeled protein samples, it lacks the complex machineries that perform post-translational modifications for eukaryotic proteins. So far, the bottleneck for producing protein samples in eukaryotic cells for methyl-TROSY NMR studies is the requirement for high-level deuteration. Labeling of the Ile-δ1 methyl group has been achieved in yeast species *Pichia pastoris*[84,85] and *Kluyveromyces lactis*[86,87] with an incorporation level of 50–70%. However, methyl labeling in Leu and Val appears to be more challenging in yeast with a labeling efficiency of no more than a few percent. Labeling the methyl group in methionine has also been shown to be possible in inset cells with partial deuteration.[88–90] Although mammalian cells have little tolerance for high concentrations of D_2O, selectively deuterated and $^{13}CH_3$-methyl labeled leucine can be introduced to the growth medium in an otherwise protonated background to achieve local deuteration at the Leu residues,[91,92] which was shown to significantly improve the spectral quality compared to the case when the protonated counterparts were used.[91,92] Alternatively, a cell-free protein expression system[93] could be used, in particular for proteins that are toxic to host cells or that undergo rapid degradation *in vivo*. Deuteration in a cell-free system can be achieved by supplying deuterated amino acids and deuterated cell extract.[94–97] Notably, unlike *E. coli*-based expression systems, cell-free expression cannot intake specifically isotopically labeled precursors since it lacks the machinery for incorporating precursors into amino acids; therefore isotopically labeled amino acids are supplemented to the medium to achieve labeling on the desired sites.[93,98–100] Interestingly, by introducing hydrolyzed ILV-labeled inclusion body, produced from *E. coli*, into the cell-free medium, the same labeling pattern is transferred to the target protein, which provides an effective way to achieve a specific labeling scheme in a cell-free system.[101]

$^{13}CH_3$-labeled methyl groups can also be introduced to expressed and purified proteins *in vitro via* chemical or enzymatic reactions. One simple method is to attach $^{13}CH_3$ to the thiol group of a cysteine by reacting the protein with ^{13}C-methyl-methanethiosulfonate (^{13}C-MMTS).[102] The reaction results in formation of S-methylthiocysteine (MTC), which is structurally similar to methionine. This approach produces highly dynamic methyl probes that give intense peaks in the spectra, which is particularly advantageous for investigating high-molecular-weight molecular assemblies[35,102,103] and membrane proteins.[104–106] To achieve site-specific labeling, the target protein is usually genetically modified so that the sites of interest are mutated to Cys while the intrinsic Cys residues are mutated to other compatible amino acids. The targeted Cys residues should be solvent exposed if the reaction is to be performed in a non-denaturing condition. Deuteration of the molecule is typically needed in order to obtain optimal spectral quality; however, this method has also been successfully used in fully protonated proteins.[104,105]

In addition to cysteines, lysine side chains can be used to introduce labeled methyl groups as well. In fact, lysine methylation is a critical post-translational modification *in vivo* which regulates protein functions.[107] The amino group of lysine can accept up to three methyl groups, forming mono-, di-, or trimethyl lysine. *In vitro* reductive methylation reaction can achieve near complete dimethylation of the amino groups under non-denaturing conditions,[108,109] and $^{13}CH_3$ methyl groups can be introduced if ^{13}C-labeled formaldehyde is used as the methyl donor in this reaction.[110] By judiciously choosing the stoichiometry between aldehyde and the reactive amines, mono-methylation can also be achieved.[111] A major advantage of this method is that the target protein does not need to be isotopically labeled,[110–116] while, on the other hand, this approach bears a potential risk of structural modification of the protein due to lysine methylation, in particular for lysine residues buried in the core of the structure. By optimizing the amount of formaldehyde, it is possible to exclude the buried lysine residues from the reaction.[117] Alternatively, enzymatic reaction can be utilized to specifically modify lysine to the mono-, or di- and tri-methylated form using either WT or mutated lysine methyltransferase.[111]

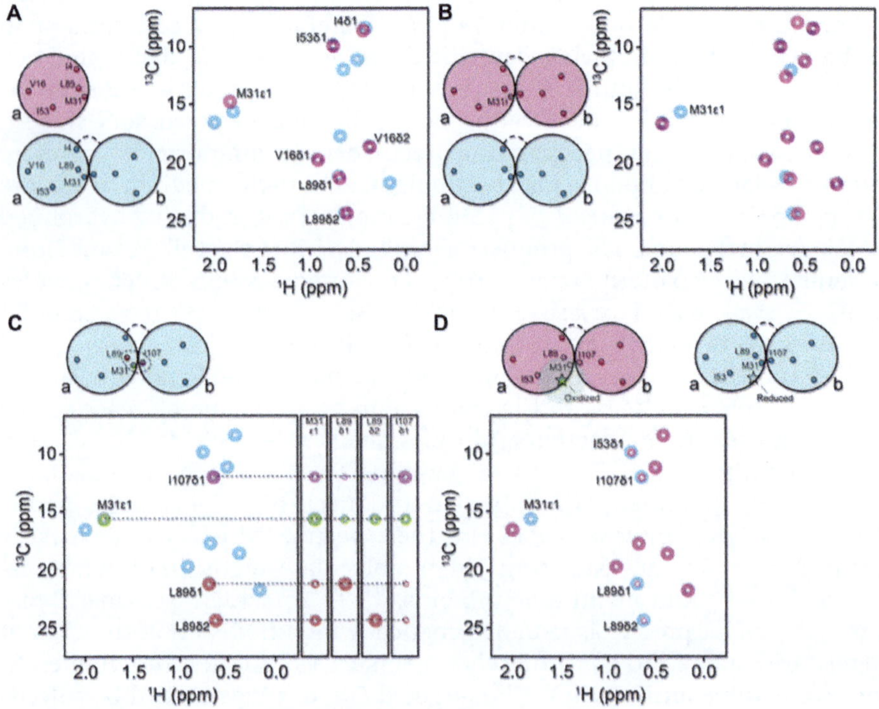

12.4.3 Assignment of Methyl Resonances

Conventionally, the assignment of side-chain methyl groups is achieved by first obtaining backbone assignment followed by correlating the methyl resonances to the assigned backbone resonances *via* TOCSY schemes.[118] However, for proteins or molecular complexes over 50 kDa, backbone assignment becomes very challenging. Therefore, a few strategies have been developed to obtain methyl assignment particularly for high-molecular-weight complexes, which we summarize here.

12.4.3.1 "Divide and Conquer"

The "divide-and-conquer" approach involves dissecting the target multi-domain protein (or multimeric complex) into individual domains (or subunits) so that the smaller fragments can be assigned by the conventional triple-resonance approach.[118] The assignments of the methyl resonances on the smaller fragments are then transferred to the full-length protein or intact complex (Figure 12.3A) by comparing chemical shifts and side-chain nuclear Overhauser effect (NOE) patterns. This approach requires the individual

Figure 12.3 Strategies to assign methyl resonances in large molecular assemblies. (A) Schematics of a 'divide- and-conquer' strategy. Assignment of methyl resonances on an isolated domain (or subunit) (pink) can be obtained *via* the traditional triple-resonance approach and then transferred to the full-length protein (or the complex) (cyan). Note that for some resonances chemical shift perturbations are expected between the smaller fragment (or subunit) and the full-length protein (or complex), such as M31 located at the inter-domain (or inter-subunit) interface. (B) Schematics of assigning methyl resonances by site-directed mutagenesis. The spectrum of an M31I mutant (pink) is compared with that of the WT protein (cyan), and M31 is assigned in the WT spectrum (cyan) due to the absence of the resonance in the mutant spectrum (pink). Note that an additional resonance (I31) appears in the Ile region in the mutant spectrum (pink) and some resonances in the vicinity of the mutation site show chemical shift perturbations. (C) Schematics of an NOE-based methyl assignment strategy. The measured distances (or connectivity) between the methyl resonances from the NOESY experiments are compared with those obtained from a high-resolution structure to solve for unambiguous methyl assignment either by automated assignment programs or manually. The strips from a C-C-H NOESY experiment show NOE cross-peaks between methyl groups in M31, L89, and I107. (D) Schematics of a PRE-based strategy to assign methyl resonances. A spin label (star) is attached to the protein in its oxidized (green, left) and reduced (transparent, right) states, respectively, and the spectra of the protein in both states are collected and compared. When the spin label is oxidized (green star, left), the closer to the spin label the methyl groups are located, the weaker their intensities are (*e.g.* M31, L89, I53, and I107), while the reduced spectrum (cyan) serves as a control in which the intensities of the resonances are not affected by the presence of the spin label. Note that the spectra in the schematics represent a protein that is labeled with [U-^{2}H, Ile δ1-$^{13}CH_3$, Met ε1-$^{13}CH_3$, Leu/Val δ/γ-($^{13}CH_3$,$^{12}CD_3$)].

domain (or subunit) to be prepared in a stable form amenable for triple-resonance experiments. Mutations on the domain (or subunit) interface may be needed in order to stabilize the fragment in a monomeric form.[29] Chemical shift perturbations between the fragment and the intact protein (or complex) are expected; yet it has been shown that these changes are largely limited to the domain (or subunit) interface. This approach has been successfully applied in both homo- and hetero-oligomeric assemblies[29,119] as well as in large multi-domain proteins and complexes.[22,72,120–122]

12.4.3.2 Site-directed Mutagenesis

Mutagenesis is perhaps the most straight-forward approach to assign methyl resonances in large protein complexes. Here a methyl-containing residue is mutated to a different type of amino acid, and the spectrum of the mutant is compared with that of the wild type to identify the resonance(s) corresponding to the mutated residue (Figure 12.3B). The mutation should largely preserve the property of the original amino acid so as to minimize the perturbation to the overall fold and the local structure of the protein, *e.g.* methionine to leucine and isoleucine to valine, and *vice versa*. The amino acid type of a methyl resonance in the spectrum can sometimes be speculated based on its chemical shift values and the labeling scheme of the sample,[11] which is useful to assist mutation-based assignment. In practice, chemical shift perturbations can occur to resonances in the vicinity of the mutation, making it challenging to assign solely based on inspection of the spectra. In such cases, combined analysis of several different mutants could help distinguish the elimination of resonance from peak shift. In addition, a methyl-TROSY-based NOE experiment[123–125] is highly recommended to aid assignment if a high-resolution structure is available. Combining the mutation-based approach with NOESY analysis also effectively reduces the number of mutations needed for assignment.[36] Although in general deemed labor-intensive and costly, the mutagenesis approach has been shown to be quite effective in obtaining assignments in a number of large biomolecular assemblies,[35,36,77,120,122,126–128] in particular for systems that lack structural information.

12.4.3.3 Structure-based Automatic Assignment

The availability of high-resolution structures greatly aids methyl resonance assignment for large molecular assemblies. Over the past decade, several structure-based automatic methyl assignment programs have been developed. In a nutshell, these programs search for assignments that give the highest agreement between the experimental structural constraints of the methyl resonances and the high-resolution structure. Experimental constraints are typically obtained from nuclear Overhauser effect (NOE) and paramagnetic-based experiments.

12.4.3.3.1 NOE-based Assignment. 3D or 4D HMQC-NOESY-HMQC experiments can be used to obtain NOE connectivity between methyl groups[123–125] (Figure 12.3C). For large protein complexes, 4D HMQC-NOESY-HMQC is advantageous for acquiring unambiguous connectivity between the methyl resonances, in which case non-uniform sampling is often used in conjunction to reduce acquisition time. The methyl NOE data can then be inputted into the automated assignment algorithms to obtain optimal assignment solutions.[129] Due to the high dimensionality of the search space entailed by the task, different computational strategies have been exploited in order to reduce the computational time. MAP-XSII[130] and FLAMEnGO2.0[131] use a Monte Carlo (MC) swapping routine to search for the optimal solution evaluated by a global score. MAGMA[132] and MAGIC[133] make use of graph algorithms which employ exhaustive sampling to match the NOE network from a NOESY experiment with that derived from the high-resolution structure. Methyl-FLYA[134] is conceptually similar to MAGMA and MAGIC and yet takes advantage of an evolutionary algorithm[135] to correlate the two NOE networks. These programs yield different accuracies and total numbers of successfully assigned methyl groups depending on the system at hand, among which MAGMA shows the highest accuracy and methyFLYA delivers the largest number of correct assignments considering all the benchtop examples.[129] Interestingly, taking the intersection of any two aforementioned methods (except for that of FLAMEnGO2.0 and MAP-XSII) leads to 100% accurate assignments.[129]

These assignment programs in general require an unassigned NOE peak list (or an extracted NOE contact list for MAGMA) and a PDB format of the protein structure with protons added. Most of the programs also require methyl chemical shifts input with indicated amino acid types. Depending on the labeling scheme, it might be possible to determine the amino acid type of a methyl resonance by its chemical shift values. For example, Ile residues can be easily distinguished due to their high field chemical shifts on the carbon dimension. In other cases, it requires the individual amino acid type to be specifically labeled in order to obtain the identities of the methyl resonances (*e.g.* Val *versus* Leu or Ala). The strategies for single amino acid type labeling are summarized in the review by Schütz and Sprangers.[11] MAGMA and MethylFLYA provide the option of ambiguously assigning Leu and Val residue types[129] which are difficult to distinguish by inspection of chemical shift values, and thus potentially reduce the cost of individual labeling of either of the residue types. For most of the programs, in particular for those that rely on exhaustive graph search algorithms (*i.e.* MAGMA and MAGIC), it is beneficial to identify the two geminal methyl groups from the same Leu or Val residue, which leads to a significant reduction of the computational time. In large proteins, the linkage between the geminal methyl pair can be obtained by simultaneous labeling of both methyl groups with $^{13}CH_3$ in Leu and Val and recording either a NOESY experiment with a short mixing time[29] or a 3D-HMBC-HMQC experiment that connects the protons of one methyl group with the carbon of the other methyl *via* a $^3J_{CH}$ coupling transfer.[136]

12.4.3.3.2 Paramagnetic-based Assignment. The assignment of methyl resonances in large proteins can also be obtained based on measurements of paramagnetic parameters, *i.e.* paramagnetic relaxation enhancement (PRE) and pseudo-contact shift (PCS), given that high-resolution structures are available. A paramagnetic tag is introduced onto a specific site of the protein, and its paragenetic effect on the methyl groups is measured. An automated algorithm then searches for assignments that give agreement between the experimentally determined paramagnetic parameters and those predicted from the structure. Unlike the NOE-based approach that relies on clusters of methyl groups in close vicinity (<5–10 Å), methyl groups distant from the paramagnetic center (*e.g.* ~25 Å for the nitroxide radical) can be probed by this method due to the large gyromagnetic ratio of the unpaired electron in the paramagnetic moiety. Two classes of paramagnetic species have been employed in NMR studies: the ones with high anisotropy of magnetic susceptibility, such as lanthanide metal center, which induce large PCSs, and those whose magnetic susceptibility tensors are mostly isotropic, such as nitroxide radical, which enable PRE measurement (Figure 12.3D).

Paramagnetic tags can be introduced into the protein by one of the following means: (i) disulfide linkage to a Cys residue[137,138] or simultaneously to two proximal Cys residues to reduce the mobility of the tag;[138,139] and (ii) peptide motif engineered into the protein sequence that binds paramagnetic metals.[140–142] The sample with the paramagnetic tag is then compared with a diamagnetic reference sample to obtain PCS and/or PRE values. These values are used as input into automatic assignment algorithms, such as PRE-ASSIGN[138] (PRE) and PARAssign[138] (PCS), to predict assignment based on the protein structure. In order to increase the accuracy of the assignments, it is common to obtain multiple datasets by attaching the paramagnetic tag at a few different sites.[138]

12.5 Application of Methyl-TROSY in the Investigation of Cooperativity in Large Protein Complexes

In the following, we provide a few examples which have utilized methyl-TROSY to study cooperativity in large homo-oligomeric complexes. The goal here is to show the versatility of this approach in different biomolecular systems and discuss the rationale and development for case-dependent cooperativity models.

12.5.1 Cooperative Gating of Archaeal 20S Proteasome

The proteasome is a supramolecular complex that plays a critical role in cellular homeostasis by degrading proteins that are misfolded, damaged, or no longer needed for function.[143] The proteasome 20S core particle (CP) exists ubiquitously in eukaryotic cells, as well as in most archaea and some bacteria.[144,145] A simple type of CP from *Thermoplasma acidophilum* is a D7 symmetric barrel-shaped complex consisting of four stacked heptameric rings (α7-β7-β7-α7) (Figure 12.4A), with the catalytic sites sequestered in the interior chamber of the complex.[146] To access the proteolytic sites, substrates enter the CP chamber *via* a 13-Å-diameter pore at each end of the barrel. These pores are gated by unstructured N-termini of the α-subunits, each of which adopts one of the two distinct conformations—it either stays on top of the α-ring ('out' conformation) (Figure 12.4B, green) or penetrates the pore and resides inside the chamber ('in' conformation) so as to occlude the pore and prevent substrate entry (Figure 12.4B, orange).[147] In what follows, we refer to these disordered termini as proteasome 'gates'. Important structural and dynamic properties of the gates have been unveiled by methyl TROSY NMR studies. For instance, it was established that on average two out of the seven gates adopt the 'in' conformation (Figure 12.4C), and the exchange between the 'in' and 'out' conformers occurs on the seconds timescale.[147] In addition, motions on picosecond-to-nanosecond as well as millisecond timescales have also been characterized for the gating residues.[147,148]

In order to probe the cooperativity between the 'in'/'out' conformational exchange of the gates, a differential labeling strategy, described in Section 12.3.3, was employed, whereby hetero-heptameric α-rings were reconstituted with two types of α subunits—WT subunits with $^{13}CH_3$-labeled methyl groups (NMR active) and mutant subunits with gates removed (gateless) and deuterated (NMR inactive)[42] (Figure 12.4D). With a small fraction (<5%) of WT subunits in the mixture, the reconstituted rings are predominantly gateless (Figure 12.4D, species a) or contain only one gate per ring (Figure 12.4D, species b). This allows the intrinsic conformational equilibrium of an individual gate to be probed in the absence of any inter-gate interaction. As the percentage of WT subunits increases during reconstitution, complexes with more than one gate per ring (Figure 12.4D, species c–t) become more populated, in which potential inter-gate interactions may occur and lead to cooperative conformational interconversion of the gates. In this regard, the overall population distribution of the in/out gates in the mixture is examined, from which cooperative conformational exchange can be probed.

Depending on the mechanism of cooperativity, two different models were constructed with the assumption that cooperativity arises from the interaction between either the 'in-gates' (Model 1, Figure 12.4E) or the 'out-gates' (Model 2, Figure 12.4F).[42] In these models, conditional probabilities are used to describe the conformational equilibrium of a gate given a certain neighboring environment. By comparing the probabilities in different neighboring environments, cooperativity in gating is then manifested. For example, in Model 1 P_1 and P_2 are defined as the probabilities of a gate adopting the 'in' conformation given that it is the first or second gate going 'in', respectively (Figure 12.4E). Based on this definition, $P_2 > P_1$ translates into a higher probability for a second gate to enter the lumen than the first one, implying positive cooperativity to adopt the 'in' conformation. This scenario could easily arise from potential interactions between the 'in-gates'. In contrast $P_2 < P_1$ describes a reduced probability for a second gate to go 'in' compared with the first one—a case of negative cooperativity—which could originate from repulsive forces between the 'in- gates' or diminished interaction of the second gate with the interior of the chamber due to partial occupancy of the binding sites by the first 'in-gate'. $P_2 = P_1$ simply implies no cooperativity. Conversely, in Model 2, it is assumed that cooperativity originates from potential interactions between the neighboring 'out-gates'. Therefore, three probabilities P_0', P_1', and P_2' were defined as the conditional probabilities of a gate adopting the 'in' conformation when it has 0, 1, or 2 neighbors that are 'out', respectively (Figure 12.4F). Under the

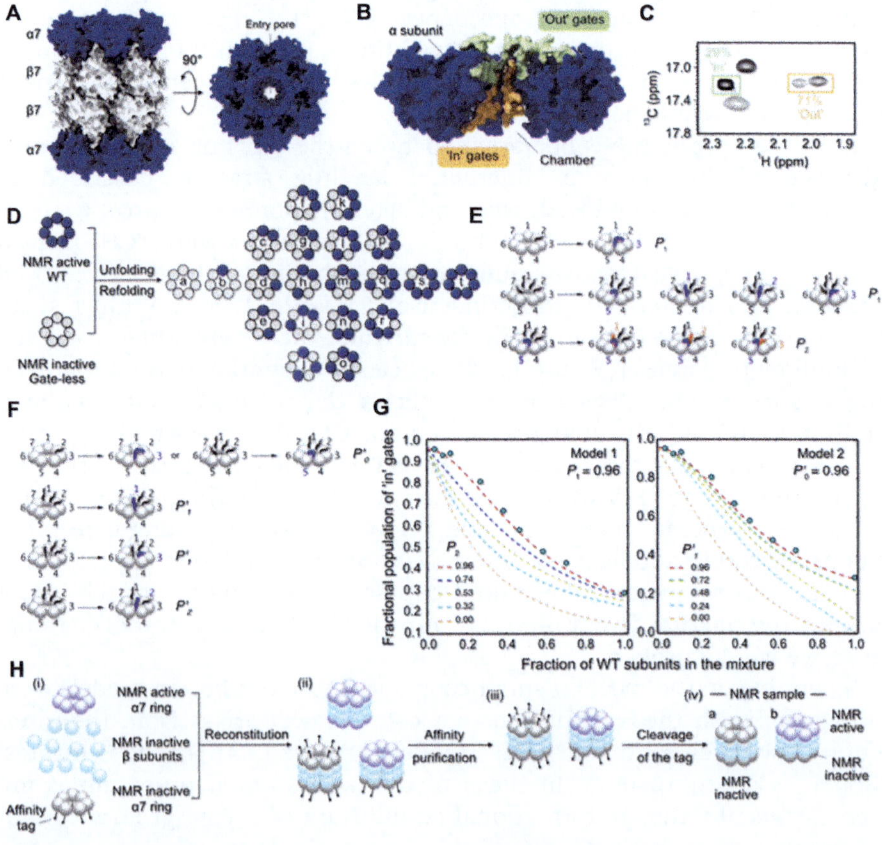

assumption of positive cooperativity whereby the 'out' conformation of a gate is stabilized by its 'out' neighbors, the probability for a gate to adopt the 'in' conformation given both of its neighbors are 'out' (P_2') is expected to be smaller than that when the gate has only one (P_1') or no (P_0') 'out' neighbors, *i.e.* $P_2' < P_1' < P_0'$. *Vice versa*, $P_2' > P_1' > P_0'$ implies negative cooperativity, while $P_2' = P_1' = P_0'$ implies independent conformational exchange for the gates. Notably, in both models, due to the limited size of the entry pore (13-Å-diameter, Figure 12.4A), there is an additional parameter constraining the maximum number of 'in-gates' per ring, which was set to either two or three in each model.

A series of complexes were reconstituted from the two types of α subunits—WT, NMR active subunits and gateless, NMR inactive subunits—at various mixing ratios, and the fractional population of 'in'/'out' gates was measured at each ratio, generating a 'titration' curve as shown in Figure 4G.[42] Note that at each 'titration' point, an ensemble of unique species was created (Figure 12.4D, species a–t). The fractional population of each species is described by the following equation,

$$F_y = n_y p^{x_y} q^{7-x_y} \tag{12.2}$$

Figure 12.4 Probing cooperative gating in the 20S proteasome by methyl-TROSY NMR. (A) Side (left) and top (right) views of the *T. acidophilum* 20S proteasome core particle (PDB ID 1PMA[146]). (B) Enlarged side view of an α-ring highlighting the N-terminal gates adopting either an 'in' (orange) or an 'out' (green) conformation. The front two α-subunits are removed to show the interior of the ring. (C) The fractional populations of the 'in'/'out' gates can be quantified by the intensities of the methyl resonances on the M-1 residue (located on the gate) in its corresponding 'in' (orange box) and 'out' (green box) conformations. (D) Reconstitution of the α-ring with the WT, NMR active subunits (blue circles) and the gateless, NMR inactive subunits (gray circles) leads to an ensemble of unique species a–t. (E–F) Two models constructed to describe the cooperativity of the 'in'/'out' conformational interconversion between the gates on the same α-ring. *P* is defined as the probability of a gate adopting the 'in' conformation under various given conditions. In Model 1 (panel E), P_1 (or P_2) is the conditional probability given the gate of interest is the first (or the second) 'in-gate' in the ring, while in Model 2 (panel F), P_0', P_1', and P_2' refer to the probabilities assuming the gate of interest has 0, 1, or 2 neighboring gates that are in the 'out' conformation, respectively. (G) Fitting of the 'titration' data (cyan dots) with both models reveals $P_1 = P_2$ for Model 1 and $P_0' = P_1' = P_2'$ for Model 2, suggesting a lack of gating cooperativity. Simulations of cases with $P_1 > P_2$ (Model 1, left) and $P_0' > P_1'$ (Model 2, right) are shown by the dashed lines. (H) Differential labeling of the two α-rings in the 20S CP facilitates investigation of long-range allostery and cooperativity across the entire CP. The schematics describe the process of preparing 20S CP with asymmetrically labeled α-rings for methyl-TROSY study.[42,151] Panels (B-G): Reproduced from ref. 42 with permission from PNAS, Copyright 2017. Panel (H): Reproduced from ref. 151 with permission from PNAS, Copyright 2020.

in which F_y is the fractional population of the species y, y∈(a–t), n_y is the degeneracy of the configuration due to symmetry [$n_y = 1$, y∈(a,t) and $n_y = 7$, y∈(b–s)], p and q are the fractional populations of WT and gateless subunits during reconstitution, respectively, and x_y is the number of WT subunits in species y. For each unique species, the fractional population of the 'in-gates' can be described by the probabilities defined in the models. Therefore, the overall population of the 'in-gates' in the sample is expressed through the probabilities and the mixing ratio of the subunits. By this means, the 'titration' curve can be fit to obtain the probabilities which provide insights into cooperativity. Fitting of the data showed that the conditional probabilities obtained for both models are all the same, namely $P_1 = P_2$ for Model 1 and $P_0' = P_1' = P_2'$ for Model 2 (Figure 12.4G), indicating a lack of cooperativity, with a maximum of two 'in-gates' at any time.[42] Interestingly, mutations that reduce the size of the gate (Y8G, D9G) or modify the property of the interior of the chamber (F91R) led to changes in the maximum number of 'in-gates' per ring and the intrinsic conformational equilibrium of an isolated gate, respectively. And yet for both mutants the data consistently showed no cooperativity between the gates.[42] These results implied that the interaction between the gate and the interior of the chamber dominates the 'in'/'out' conformational equilibrium, and the size of the entry pore limits the maximum number of 'in-gates' per ring. On the other hand, because of the lack of cooperativity, interactions between the gates on the same ring are likely negligible, if not absent, in the naked 20S CP from *T. acidophilum*.

The inter-gate interaction, although not manifested in the 20S CP alone, was found to play an essential role in gate-opening triggered by the 11S regulatory particle (RP).[42] The 11S RP enhances peptide hydrolysis by the proteasome partly *via* stabilizing the proteasome gates in the 'out' conformation to grant substrate access to the catalytic sites.[149,150] Following a similar strategy, hetero-oligomeric α-rings were generated from (i) WT and NMR active subunits, and (ii) NMR inactive subunits with mutations on the gates to abolish potential inter-gate communication. It was shown that the 11S RP fails to perturb the conformational equilibrium of the WT gates in this reconstituted hetero-oligomeric complex, emphasizing the importance of inter-gate cooperation in the gate-opening mechanism regulated by the 11S RP.[42]

The 20S CP (Figure 12.4A) is a highly allosteric molecular machine in which the effect of activator binding at the α-rings has been shown to communicate to distal active sites in the β subunits 75 Å away at the center of the complex.[20] In order to explore whether this allosteric effect extends throughout the whole CP and connects the two α-rings at the ends of the barrel, Rennella *et al.* investigated the cooperativity between the gating properties of the two α-rings in a complex.[151] The study took advantage of the fact that the *T. acidophilum* 20S CP can be reconstituted with purified α-rings and β-subunits. In this case, two types of α-rings—one NMR active and the other NMR inactive and containing affinity tags—were mixed during reconstitution (Figure 12.4H, i), generating a total of three types of CPs

(Figure 12.4H, ii). Two of these three types were selected by further affinity purification (Figure 12.4H, iii), among which only one gives NMR signals (Figure 12.4H, iv). With the two α-rings differentially labeled, mutations or binding of the 11S RP were then introduced to the NMR inactive α-ring to perturb its gating equilibrium while the allosteric effect of the perturbation was monitored on the other α-ring at the remote end of the CP by NMR. Notably, a K66A mutation was introduced to the protomers in the NMR active α-ring in order to diminish the binding of the 11S RP so that the long-range allosteric effect due to 11S binding to the distal end could be evaluated. The result showed that gating of the two α-rings are independent of each other, and the allosteric network does not extend all the way through the D7 axis of the CP. Uncoupling of the two half-proteasomes (one α7-β7 unit) could be functionally important in hybrid proteasome complexes in which the α-rings are bound with different RPs or are necessary to ensure a closed chamber when the CP is asymmetrically bound with only one RP at one of the two ends.

12.5.2 Cooperative Domain Dynamics in p97

p97/VCP, also known as Cdc48 in yeast and plants, is a homo-hexameric complex involved in a number of cellular processes including proteasomal and lysosomal degradation, cell cycle regulation, Golgi membrane fusion, and intracellular signaling.[152–154] Due to its important role in protein homeostasis, p97 has also been identified as a potential drug target for cancer therapeutics.[155] Each subunit of p97 consists of an N-terminal domain (NTD) and two ATPase domains, which organize into two stacked rings, named D1 and D2. The NTDs are located on the periphery of the D1 ring (Figure 12.5A). p97 undergoes conformational changes upon nucleotide hydrolysis, and perhaps the most significant change involves a large-scale domain rearrangement of the NTDs—when D1 is loaded with ATP, the NTDs are positioned above the D1 ring, referred to as an 'up' conformation (Figure 12.5A, left), while upon ATP hydrolysis the NTDs switch to a co-planar position with respect to the D1, referred to as a 'down' conformation (Figure 12.5A, right).[156] Conformational dynamics of the NTDs not only impact the ATP hydrolysis rate of p97[157] but also regulate the interaction between the enzyme and its various adaptor proteins,[77,158–161] which is crucial for its function. The importance of the NTD 'up'/'down' equilibrium is underscored by the occurrence of a series of disease mutations in p97 which are implicated in neurodegenerative disorders,[162] including an autosomal dominant multisystem disorder termed IBMPFD.[163] The missense mutations in IBMPFD patients are mostly located in the NTD–D1 interface and were found to disrupt the regulation of the NTD 'up'/'down' equilibrium.[77,164]

Using methyl-TROSY NMR, Schütz and Kay characterized the conformational dynamics of NTD in WT p97 as wells as in various IBMPFD disease mutants.[77] The 'up' and 'down' conformations of NTD lead to distinct chemical shifts for many residues located at the NTD–D1 interface and on the inter-domain linker. The disease mutations perturb the NTD conformational equilibrium mostly in the ADP-bound state. Interestingly, in various mutants, the methyl chemical shifts of key residues that report on

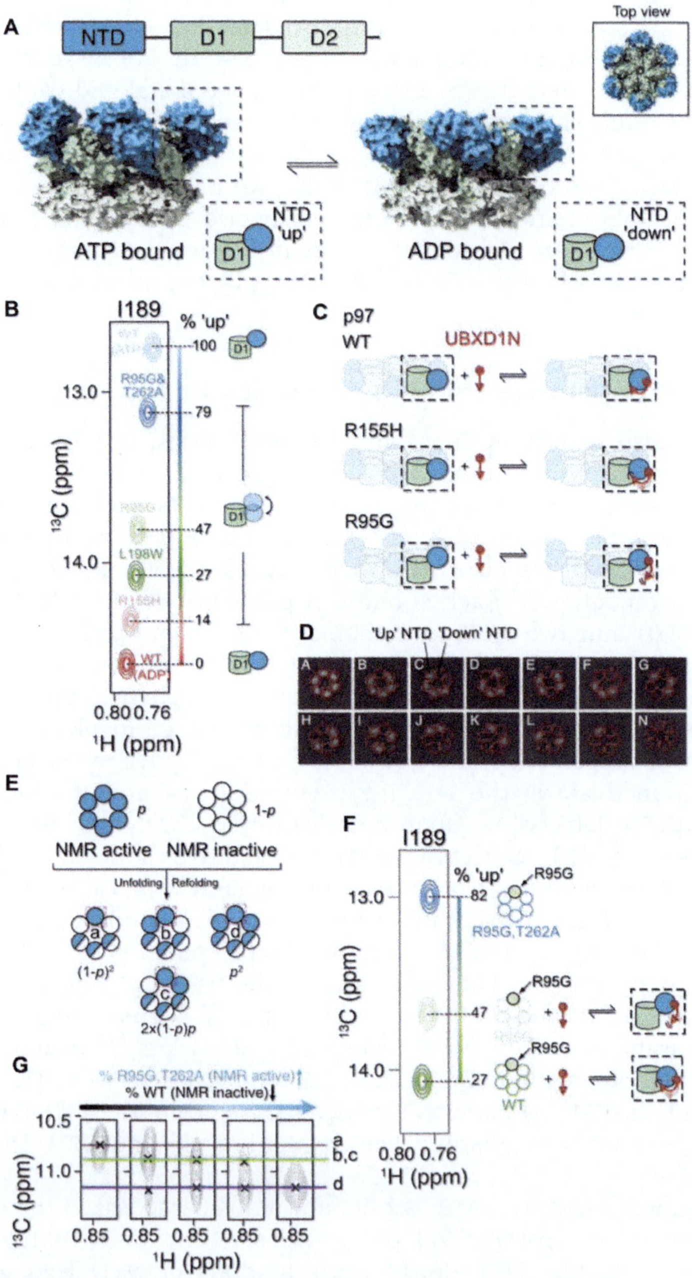

the NTD 'up'/'down' equilibrium fall on a linear range between the ADP-bound (100% 'down' conformation) and the ATP-bound (100% 'up' conformation) states (Figure 12.5B), implying that the NTD undergoes fast

Figure 12.5 Cooperative domain dynamics in p97. (A) Domain architecture of a p97 monomer (top). Top and side views of the p97 hexamer bound with ATPγS (PDB ID 5FTN[156]) and ADP (PDB ID 5FTK[156]). NTDs (blue) undergo large-scale domain motion depending on the nucleotide state of D1. The cartoons illustrate the 'up' and 'down' conformations of an NTD focusing on one NTD-D1 pair for clarity (in dashed-line boxes). (B) Disease mutations (*e.g.* R155H, L198W, R95G, and double-mutant R95G,T262A, as noted) perturb the conformational equilibrium of the NTD in the ADP-bound state, shifting the NTD from the 'down' conformation (WT-ADP) to various populations of the 'up' conformation. The population of the 'up' conformation can be read out from chemical shifts of certain methyl probes, such as Ile189. In this study, a truncated construct that includes the NTD and D1 (residue 1–480) of p97 was used, and the sample was uniformly labeled with [U-^{2}H, Ile δ1, Met ε1, Leu/Val δ1/γ1(*proR*)-$^{13}CH_3$]. (C) The N-terminal domain of an adaptor protein UBXD1 (UBXD1N, red) interacts with WT p97 (top) *via* a two-pronged mechanism (circle and triangle). As a consequence of a shift in the NTD 'up'/'down' equilibrium due to disease mutations, binding of UBXD1N is impaired, with one of the binding sites (H1/H2 motif, triangle) disengaged from p97 at a more (R95G, bottom) or less (R155H, middle) extend. (D) Cryo-EM classification of 14 unique species of a R95G mutant of p97-ADP with each NTD adopts either an 'up' or 'down' conformation. z-slices of 3D reconstructed maps (A–N) show the densities of the 'up'-NTDs, while the 'down'-NTDs are reflected by the empty red circles. (E) *In-vitro* reconstitution of hetero-hexameric p97 complexes with NMR active and inactive subunits *via* an unfolding-refolding approach. Considering only the allosteric effect of the nearest neighbors, an NMR active subunit (in red dashed-line boxes) has four distinct neighboring environments (a–d). Distant neighbors are denoted by half-blue/white circles as their allosteric effect is not considered in this case. With the fractional population of the NMR active subunits in the mixture defined as p, the populations of cases (a–d) are $(1-p)^2$, $(1-p)p$, $(1-p)p$, and p^2, respectively. (F) The NTD 'up'/'down' equilibrium of the R95G mutant is affected by its neighboring subunits, with more (less) 'down' NTD population with WT (R95G,T262A mutant) subunits at the adjacent positions than that in the homo-hexameric R95G complex. The allosteric effect of the WT neighbors partially restores the two-pronged binding of UBXD1N (red) to the R95G mutant (in dashed-line boxes). (G) In a 'titration' experiment, spectra were recorded on reconstituted p97 complexes consisting of the NMR active, mutant (R95G,T262A) subunits and the NMR inactive, WT subunits. With an increasing population of the mutant subunits in the mixture, species (a), (b)/(c), and (d), illustrated in panel E, are distinguished by the chemical shifts of the methyl group in Ile 274.[21,77] Panels (B-C): Reproduced from ref. 77, https://doi.org/10.7554/eLife.20143, under the terms of the CC BY 4.0 licence, https://creativecommons.org/licenses/by/4.0/. Panel (D): Adapted from ref. 52 with permission from John Wiley & Sons, Copyright © 2020 Wiley-VCH GmbH. Panels (F-G): Reproduced from ref. 21 with permission from PNAS, Copyright 2019.

exchange between these two conformations (>15 000 s^{-1}) in the mutants. In turn, the population distribution of the 'up'/'down' states can be read out from the chemical shifts of certain methyl probes, such as Ile189 (Figure 12.5B). It was revealed that in the ADP-bound state, the disease mutations shift the NTD from the 'down' conformation in WT p97 to various extents of the 'up' conformation depending on the specific mutation (Figure 12.5B). Intriguingly, the extent of disruption to the NTD conformational equilibrium, *i.e.* the population of the 'up' conformation due to the mutation, is correlated with disease severity.[77] As a consequence, the mutations also impair the two-pronged binding of an important adaptor protein, UBXD1 (Figure 12.5C), which recruits p97 into endosomal degradation pathways and is implicated in IBMPFD pathology.[77]

Structural studies of substrate-bound p97[165] as well as its yeast and archaeal homologs Cdc48[166,167] and VAT[168] by cryo-electron microscopy (cryo-EM) have revealed highly coordinated sequential ATP hydrolysis and subunit movement in the complex, which is proposed to be critical for substrate translocation. Yet, it is less clear whether the 'up'/'down' conformational interconversion of the six NTDs is cooperative. Focusing on the NTDs, a hexameric p97 can have a total of up to 14 unique conformers with different combinations of the NTDs in the complex adopting either the 'up' or 'down' conformation. In a recent cryo-EM investigation of an R95G p97 mutant, all 14 configurations of the hexamer were classified with their respective populations quantified (Figure 12.5D).[52] These populations, as mentioned in Section 12.3.2, report on the cooperativity of the NTD conformational interconversion, and thus can be fitted to a probabilistic model to obtain such information. In the case of the R95G homo-hexamers, the population distribution of the 14 conformers is consistent with a random binomial distribution, indicating no observable cooperativity between the NTD dynamics in the mutant complex. Notably, this approach is suitable to cases in which the overall populations of the 'up'/'down' states are comparable (the R95G mutant has ~50% 'up' conformation for the NTDs) so that the population distribution is not overly skewed for some of the conformers, which otherwise challenges classification and population quantification by cryo-EM.

IBMPFD is an autosomal dominant disease with the heterozygous patients possessing one copy each of WT and disease mutant p97 alleles. Formation of heterogeneous p97 complexes consisting of both WT and mutant subunits have been shown *in vitro*[21] as well as *in vivo*.[58,59,169] In this regard, it is of interest to investigate the NTD 'up'/'down' equilibrium in the context of hetero-hexamers by methyl-TROSY NMR.[21] To generate hetero-hexameric p97, NMR active (U-^{2}H, Ile δ1, Met ε1-$^{13}CH_3$-labeled) and inactive (U-^{2}H labeled) subunits are mixed at a specific ratio and reconstituted *via* an unfolding/refolding approach (Figure 12.5E). The two types of subunits also differ in the conformational equilibrium of the NTD through introduction of disease mutation(s). With only the nearest neighbors taken into account, for any NMR active subunit there are a total of four types of neighboring environments, as illustrated in Figure 12.5E (a–d). By controlling the mixing ratio during reconstitution, the majority of the NMR active subunits have at least one NMR inactive subunit in the adjacent positions (Figure 12.5E, b–d). It was revealed then that depending on the neighboring subunit(s), the NTD of the R95G mutant displays distinct populations between the 'up'/'down' conformations (Figure 12.5F, left). With the WT subunits in the adjacent positions, in which the NTDs are 'locked' in the 'down' position, the R95G mutant exhibits a lower percentage (27%) of 'up' NTDs, compared with that in the homo-hexameric R95G ring (47%), while with a double mutant R95G, T262A—an even more dysregulated mutant than R95G—as neighbors, a higher percentage of 'up' NTDs (82%) is observed for R95G. This apparent cooperative behavior highlights the effect of inter-subunit allostery on the NTD dynamics.

In a 'titration' experiment where the mixing ratio of the two types of subunits varies, the fractional populations of the NMR active subunits in different neighboring environments (Figure 12.5E, a–d) change accordingly. It is possible to distinguish between cases (a–d; Figure 12.5E) if each of the two neighboring subunits has a distinct and accumulated effect on the conformational equilibrium of the subunit in the middle. As in p97, cases (a), (b–c), and (d) lead to different 'up' populations (Figure 12.5G) when NMR active R95G,T262A subunits are mixed with NMR inactive WT subunits. By comparing the chemical shifts of the methyl probes in the NMR active subunit situated in different neighboring environments, a potential inter-subunit allosteric pathway was identified, which provides a structural basis for the cooperativity between the NTD dynamics.[21] As a result of cooperative 'up'/'down' NTD equilibrium, binding of the UBXD1 adaptor is impacted in the hetero-hexamers. While the R95G mutation impairs the two-pronged binding of UBXD1N in homo-hexameric p97, this binding can be partially restored by placing the R95G mutant subunit amongst the WT subunits (Figure 12.5F, right).[21]

12.5.3 A Highly Cooperative N-terminal Conformational Switch in SaClpP

ClpP is a barrel-shaped proteolytic complex which degrades misfolded or aggregated proteins in many prokaryotes and also in chloroplasts and mitochondria. It plays an essential role in protein homeostasis in bacteria and hence has become an emerging drug target for a new class of antibiotics.[170] ClpP consists of two stacked homo-heptameric rings with each subunit composed of an N-terminal, head, and handle region. A methyl-TROSY NMR study of ClpP from *Staphylococcus aureus* (SaClpP) has shown the N-terminal region, predominantly consisting of a β-hairpin and a flexible loop connecting the two β-strands, is structurally heterogeneous, leading to peak multiplicity in the NMR spectra for residues throughout the entire structure (Figure 12.6A, multi-colored single contour). A hydrophobic cluster at the interface of adjacent protomers was discovered to be critical in controlling the conformation of the complex. It was observed that a single mutation in this region, such as a V7A mutation, results in (i) disappearance of the N-terminal resonances due to unfolding of the N-terminal β-hairpin, (ii) chemical shift perturbations throughout the entire complex, and (iii) collapsing of multiplicity into a single set of peaks for most of the residues (Figure 12.6A, magenta multi-contour). Interestingly, the V7A mutation, whose location is removed from the active sites, completely abolishes the catalytic activity of the enzyme. The structure of the mutant was then determined using cryo-EM, revealing a large conformational rearrangement for the complex (Figure 12.6B). The tetradecameric SaClpP switches from a D7 symmetric organization to an asymmetric split-ring conformation upon the V7A mutation (Figure 12.6B). Discovery of this conformational switch highlights the structural plasticity of the SaClpP complex and emphasizes how the function of the enzyme is allosterically controlled by mutations *via* conformational rearrangement of the entire complex.

The extensive allostery within the complex is also manifested in highly cooperative conformational conversion and catalysis between protomers. To this end, hetero-oligomeric SaClpP complexes were reconstituted with WT and V7A mutant protomers at various mixing ratios and subjected to proteolytic assays and NMR measurements. The catalytic activities of the reconstituted complexes are consistent with the hypothesis that incorporation

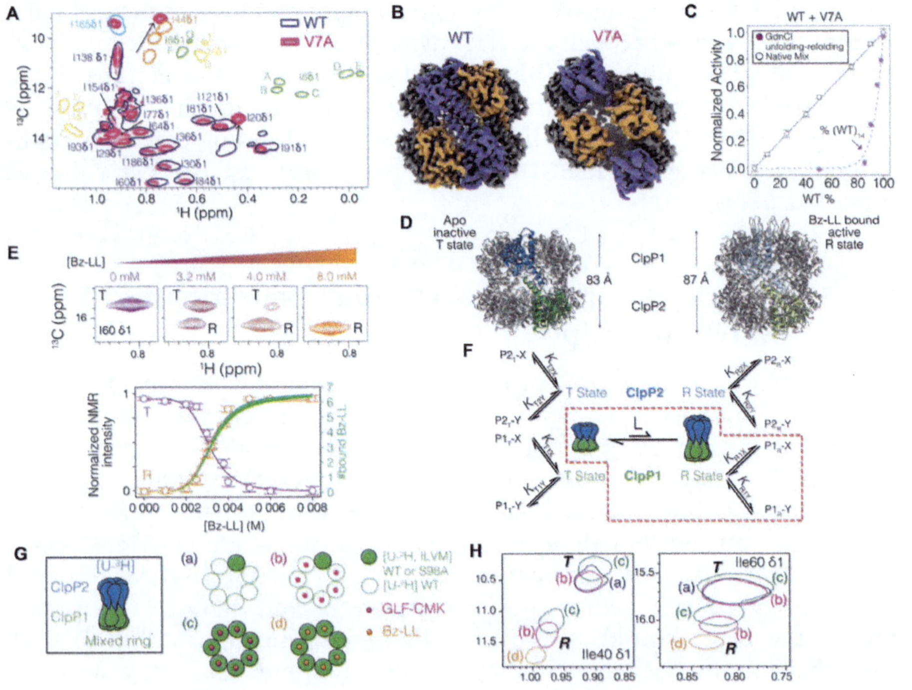

of one V7A subunit into the tetradecamer renders the entire complex catalytically inactive (Figure 12.6C). The loss of activity is attributed to the formation of the asymmetric spilt-ring conformation triggered by the V7A mutation on a single protomer in the complex, indicated by NMR analysis. In the split-ring conformation, concomitant increased disorder in the N-terminal region leads to the disappearance of the N-terminal resonances. The researchers then reconstituted mixed complexes with NMR active WT protomers and NMR inactive V7A protomers and subjected them to intensity analysis. It was observed that the intensities of the N-terminal resonances are proportional to WT subunits in complexes consisting of only WT protomers. In other words, incorporation of V7A promoters into the same complexes with the WT protomers 'silences' the WT N-terminal signal most probably *via* switching the complex to the 'split-ring' conformation. it

Figure 12.6 Highly cooperative ClpP degradation machines. (A) ^{1}H-^{13}C HMQC spectra of WT SaClpP (single contour) shows peak multiplicity for many residues, *e.g.* I14δ1(yellow single contours, A–E) and I8δ1 (green single contours, A–G). A V7A mutation (magenta, multiple contours) causes disappearance of the N-terminal resonances (*e.g.* I4δ1 and I8δ1), chemical shift perturbations (*e.g.* I20δ1 and I44δ1), and collapsing of peak multiplicity (*e.g.* I91δ1). (B) Comparison of the cryo-EM maps of the D7 symmetric conformation of the WT SaClpP and a 'split-ring' conformation of the V7A mutant. (C) Peptide hydrolysis of the reconstituted complexes consisting of WT and V7A protomers (purple) indicates that one V7A subunit in the tetradecamer is sufficient to abolish the catalytic activity of the entire complex. (D) Cryo-EM structures of mtClpP1P2 reveal an inactive 'compact' conformation in the apo state (PDB ID 6VGK)[36] and an active 'extended' conformation with an activator Bz-LL bound to the active sites (PDB ID 5DZK).[195] (E) Titration of the Bz-LL activator into [ILVM-$^{13}CH_3$]-labeled mtClpP1P2 leads to a highly cooperative transition of the complex between two distinct states—the T (left in panel D) and the R (right in panel D) state. (F) An extended MWC model describes simultaneous binding of the Bz-LL activator (X) and the substrate (Y) to the T/R states of the P1/P2 rings. Each binding event is characterized by a microscopic association constant *K*. A simplified model considering ligand binding only to the R state of the P1 ring is shown in the red dashed-line box and used for a global fit of the NMR data and functional assays. (G) Reconstituted mtClpP1P2 complexes consisting of [U-^{2}H] P2 ring and various types of P1 rings (a–d) are generated to probe inter-subunit allostery. The NMR active (or inactive) subunits are indicated by filled (or open) green circles. Hybrid P1 ring were produced with 95% [U-^{2}H] WT and 5% [ILVM-$^{13}CH_3$] S98A subunits (a–b). GLF-CMK (pink dots) was then used to covalently modify the WT subunits either in the hybrid (b) or homogeneous WT (c) P1 ring. WT P1 ring was also partially loaded with the Bz-LL activator (orange dots) (d). (H) Methyl-TROSY spectra of the complexes in (G) indicates inter-subunit allostery and cooperative activation of the mtClpP1P2 complex. The single contours are color-coded accordingly to panel G. Panels (A–C): Reproduced from ref. 35 with permission from PNAS, Copyright 2018. Panels (D–H): Reproduced from ref. 36 with permission from PNAS, Copyright 2020.

is noteworthy that this highly cooperative behavior is specific to mutations in the 'hotspots' which control the conformational switch. In contrast, the detrimental effect of an active-site mutation S98A is limited only to the mutated protomer with no allosteric effect on the neighboring subunits.

12.5.4 Cooperative Activation of mtClpP1P2

The ClpP protease from *Mycobacterium tuberculosis* consists of two heptameric rings, P1 and P2, encoded by two separate *clpP* genes. The degradation complex, referred to as mtClpP1P2 from here on, plays an essential role in maintaining proteostasis in the *M. tuberculosis* pathogen, and hence has gained great interest as a novel target for antibiotics.[171–173] It has been shown that the catalytic function of mtClpP1P2 requires the presence of small-molecule activators, such as benzoyl-leucyl-leucine (Bz-LL),[173] while the mechanism of activation was elusive. Vahidi *et al.* carried out a combined cryo-EM and methyl-TROSY NMR study which provided the structural basis for the mechanism of mtClpP1P2 activation.[36] The cryo-EM structure of the apo mtClpP1P2 reveals disruptions to an important β-sheet structure in the handle region, which leads to improper positioning of the catalytic triad and hence lack of activity in the apo complex. In contrast, addition of the Bz-LL activator triggers the formation of this handle β-sheet and the engagement of the oligomeric sensors, and at the same time converts the complex from a 'compact' inactive conformation in the apo state to an 'extended' active conformation (Figure 12.6D).[36] In accord with the cryo-EM structures, titration of Bz-LL into [U-^{2}H, Ileδ1-$^{13}CH_3$, Met ε1-$^{13}CH_3$, Leu/Val δ/γ-($^{13}CH_3$,$^{12}CD_3$)] (from here on referred to as [ILVM-$^{13}CH_3$]) labeled mtClpP1P2 leads to a conformational conversion of the complex in the slow exchange timescale (Figure 12.6E, top), observed by methyl-TROSY NMR, which corresponds to the transition of the complex from an inactive to an active state (Figure 12.6D). Interestingly, plotting the NMR intensities of the two conformations as a function of Bz-LL concentration reveals a characteristic sigmoidal binding curve with Hill coefficients of 7.4 for residues in mtClpP1(Figure 12.6E, bottom), indicating a highly cooperative transition.

To understand the activation mechanism and the cooperativity in mtClpP1P2, an impressive extended MWC model was constructed to describe the competitive binding of the Bz-LL activator and the substrate during hydrolysis (Figure 12.6F). Here all the protomers in the P1 or P2 ring are assumed to be microscopically equivalent and exist in one of the two states—the low-activity state (T state) or the high-activity state (R state), as implied in the classical MWC model. The Bz-LL activator (X) and the substrate (Y) bind to either state (T or R state) of the protomers in each ring (P1 or P2) with different microscopic affinities (Figure 12.6F). For instance, K_{R1X} is the microscopic association constant for the binding of Bz-LL (X) to the R state of the protomers in the P1 ring. Two additional parameters are defined in the model—an equilibrium constant L (=[T]/[R]) describing the ratio of the populations of the T and R states of the enzyme in the absence of ligands, and k_{cat} which correlates the concentration of bound substrates in the R state to the observed rate of substate hydrolysis. Using this model, a global fit was carried out using (i) the NMR titration data, which directly reports on the populations of the states (R and T), and (ii) the functional assays which measure the catalytic rates as a function of the substrate and activator concentrations. Notably, a simplified version of this model which only considers binding of the substrate and the activator to the R state of the P1 ring (Figure 12.6F, red dashed-line box) gave robust and impressively good fits to both datasets. The result of the fitting gives remarkable insights into the mechanism of the enzyme. For example, the fitted value of the equilibrium constant L ($\sim 2 \times 10$) indicates that in the absence of the Bz-LL activator, the enzyme is predominantly (>99.99%) in the T state (low activity), while binding of Bz-LL shifts the equilibrium to the active R state resulting in the sigmoidal curves observed in the NMR titration and the activity assays. It is also reflected in the fitting that the maximum activity for the P1 ring is achieved when approximately four Bz-LL molecules are bound to the ring, switching the complex to an active state, with the remaining two to three unoccupied sites carrying out substrate degradation.

In order to further evaluate the cooperativity and allostery in the P1 ring, an active site inhibitor Z-Gly-Leu- Phe-chloromethyl ketone (GLF-CMK) was used (Figure 12.6G), which covalently attaches to the active site serine residue and induces the 'extended' active conformation of the complex. The maximum hydrolysis activity was observed when around half of the active sites in P1 are modified by GLF-CMK, indicating that modification of only part of the ring leads to allosteric activation of the entire complex. To provide more insights into the inter-protomer communication, reconstituted samples were prepared with $^{13}CH_3$ methyl-labeled protomers with the methyl groups reporting on the conformational states of the complex. The P1 rings were prepared with 95% of NMR-inactive WT and 5% of ILVM-$^{13}CH_3$-labeled S98A (catalytically inactive) subunits *via* an unfolding–refolding strategy as described before (Figure 12.6G, a) followed by covalent modification of GLF-CMK (Figure 12.6G, b, pink dots). Note that only the WT subunits were covalently modified, while the S98A subunits do not react with GLF-CMK due to the mutation of the active site serine (Figure 12.6G, b). The NMR spectrum of ILVM-$^{13}CH_3$-labeled S98A in the reconstituted ring shows a striking difference before and after GLF-CMK modification of the neighboring WT subunits. In the absence of GLF-CMK, the S98A subunits show one set of peaks corresponding to the T state of the complex (Figure 12.6H, a, purple contour), while upon GLF-CMK modification of the neighbors, two distinct sets of peaks, with the positions of one set similar to those in the T state and the other to the R state, were observed (Figure 12.6H, b, magenta contours). Two similar sets of peaks were also observed when a homogeneous WT P1 ring is modified by GLF- CMK on all sites (Figure 12.6G, c, and 12.6H, c, green contours). Moreover, incomplete binding of the Bz-LL activator to the P1 rings (Figure 12.6G, d) results in only one set of resonances in the R state (Figure 12.6H, d), implying a complete conversion of the entire ring. These results clearly indicate that mtClpP1P2 is a highly allosteric and cooperative degradation machine in which activation of the complex is achieved by partial occupancy of the active sites.

12.6 Conclusions and Outlook

In this chapter, we have demonstrated the unique role of methyl-TROSY NMR in probing allosteric cooperativity in large molecular complexes, in particular, in homo-oligomeric molecular machines. During the revision of this chapter, another elegant work was published which utilized methyl-TROSY NMR in combination with the differential labeling strategy to characterize the cooperativity in the homo-trimeric mitochondrial protease HtrA2, revealing crucial details of the activation mechanism of HtrA2.[174] With the rapid progress of the cryo-EM technique, an increasing number of high-resolution structures of large molecular complexes has been uncovered, and the importance of coordinated motions between subunits in such complexes has been increasingly emphasized. Methyl-TROSY NMR spectroscopy is situated in a unique position to investigate allostery and cooperativity in high-molecular-weight complexes. It allows characterization of both the conformational equilibrium and dynamics of different subunits in the complex, providing a quantitative description of the free energy landscape of the system, while it can also shed light into the mechanism of cooperativity and allostery through identification of allosteric pathways and networks. With the continuous development of methyl-TROSY NMR methodology, including novel isotope-labeling strategies,[71,81,82,84,175–180] an ever-expanding pulse sequence toolbox tailored for methyl groups,[181–187] and automation of methyl resonance assignment,[130,132–134,188] in combination with the progress of NMR spectroscopy in general, such as high magnetic fields, state-of-the-art hardware, optimized non-uniform sampling strategies, and spectral reconstitution algorithms,[189–191] methyl-TROSY NMR will continue to play an indispensable role in structure biology of large molecular complexes. In addition, a new approach which integrates methyl-TROSY and cryo-EM techniques to investigate supramolecular complexes has emerged in recent years,[103,192–194] including a few of the examples discussed in this chapter.[21,35,36,52] The combined approach has opened a new venue to probe the mechanism of these complexes in great detail from both structural and dynamic perspectives. We believe methyl-TROSY NMR will continue to bring exciting possibilities to the field of structural biology in the near future.

Acknowledgements

The author would like to thank Dr. Lewis Kay for the valuable comments and suggestions on the manuscript. The author especially would like to thank Dr. Lewis Kay and Dr. Ayyalusamy Ramamoorthy for their continuous support and mentorship. The author apologizes to authors of studies that have not been covered and cited. This work has been supported by the Natural Sciences and Engineering Research Council of Canada Discovery Grant RGPIN-2020-05066.

References

1. A. Whitty, *Nat. Chem. Biol.*, 2008, **4**, 435.
2. D. E. Koshland and K. Hamadani, *J. Biol. Chem.*, 2002, **277**, 46841.
3. C. A. Hunter and H. L. Anderson, *Angew. Chem., Int. Ed.*, 2009, **48**, 7488.
4. C. Bohr, K. Hasselbalch and A. Krogh, *Skand. Arch. Physiol.*, 1904, **16**, 401.
5. Q. Cui and M. Karplus, *Protein Sci.*, 2008, **17**, 1295.
6. M. T. J. Smith and J. L. Rubinstein, *Science*, 2014, **345**, 617.
7. J. Ognjenović, R. Grisshammer and S. Subramaniam, *Annu. Rev. Biomed. Eng.*, 2019, **21**, 395.
8. K. Murata and M. Wolf, *Biochim. Biophys. Acta – Gen. Subj.*, 2018, **1862**, 324.
9. Y. Cheng, *Science*, 2018, **361**, 876 LP.
10. R. Rosenzweig and L. E. Kay, *Annu. Rev. Biochem.*, 2014, **83**, 291.
11. S. Schütz and R. Sprangers, *Prog. Nucl. Magn. Reson. Spectrosc.*, 2020, **116**, 56.
12. Z. Boswell and M. P. Latham, *Biochemistry*, 2019, **58**, 144.
13. B. T. Falk, P. J. Sapienza and A. L. Lee, *Proc. Natl. Acad. Sci. U. S. A.*, 2016, **113**, 9533.
14. K. Henzler-Wildman and D. Kern, *Nature*, 2007, **450**, 964.
15. T. Xie, T. Saleh, P. Rossi and C. G. Kalodimos, *Science*, 2020, **370**, eabc2754.
16. R. M. Culik, A. Sekhar, J. Nagesh, H. Deol, J. A. O. Rumfeldt, E. M. Meiering and L. E. Kay, *Proc. Natl. Acad. Sci.*, 2018, **115**, E2546.
17. D. D. Boehr, D. McElheny, H. J. Dyson and P. E. Wrightt, *Science*, 2006, **313**, 1638.
18. A. Sekhar, J. A. O. Rumfeldt, H. R. Broom, C. M. Doyle, R. E. Sobering, E. M. Meiering and L. E. Kay, *Proc. Natl. Acad. Sci. U. S. A.*, 2016, **113**, E6939.
19. Y. Toyama, K. Kontani, T. Katada and I. Shimada, *Sci. Adv.*, 2019, **5**, eaav894.
20. A. M. Ruschak and L. E. Kay, *Proc. Natl. Acad. Sci. U. S. A.*, 2012, **109**, E3454.
21. R. Huang, Z. A. Ripstein, J. L. Rubinstein and L. E. Kay, *Proc. Natl. Acad. Sci. U. S. A.*, 2019, **116**, 158.
22. G. E. Karagöz, A. M. S. Duarte, H. Ippel, C. Uetrecht, T. Sinnige, M. Van Rosmalen, J. Hausmann, A. J. R. Heck, R. Boelens and S. G. D. Rüdiger, *Proc. Natl. Acad. Sci. U. S. A.*, 2011, **108**, 580.

23. R. Selvaratnam, S. Chowdhury, B. VanSchouwen and G. Melacini, *Proc. Natl. Acad. Sci. U. S. A.*, 2011, **108**, 6133.
24. A. G. Palmer and H. Koss, *Methods Enzymol.*, 2019, **615**, 177.
25. N. Popovych, S. Sun, R. H. Ebright and C. G. Kalodimos, *Nat. Struct. Mol. Biol.*, 2006, **13**, 831.
26. S. R. Tzeng and C. G. Kalodimos, *Nature*, 2009, **462**, 368.
27. K. Pervushin, R. Riek, G. Wider and K. Wuthrich, *Proc. Natl. Acad. Sci.*, 1997, **94**, 12366.
28. V. Tugarinov, P. M. Hwang, J. E. Ollerenshaw and L. E. Kay, *J. Am. Chem. Soc.*, 2003, **125**, 10420.
29. R. Sprangers and L. E. Kay, *Nature*, 2007, **445**, 618.
30. P. Rossi, Y. R. Monneau, Y. Xia, Y. Ishida and C. G. Kalodimos, *Methods Enzymol.*, 2019, **614**, 107.
31. R. Rosenzweig and L. E. Kay, *J. Am. Chem. Soc.*, 2016, **138**, 1466.
32. Y. Jiang and C. G. Kalodimos, *J. Mol. Biol.*, 2017, **429**, 2667.
33. A. V. Hill, *Biochem. J.*, 1913, **7**, 471.
34. C. Puchades, C. R. Sandate and G. C. Lander, *Nat. Rev. Mol. Cell Biol.*, 2020, **21**, 43.
35. S. Vahidi, Z. A. Ripstein, M. Bonomi, T. Yuwen, M. F. Mabanglo, J. B. Juravskyd, K. Rizzolob, A. Velyvis, W. A. Houry, M. Vendruscolo, J. L. Rubinstein and L. E. Kay, *Proc. Natl. Acad. Sci. U. S. A.*, 2018, **115**, E6447.
36. S. Vahidi, Z. A. Ripstein, J. B. Juravsky, E. Rennella, A. L. Goldberg, A. K. Mittermaier, J. L. Rubinstein and L. E. Kay, *Proc. Natl. Acad. Sci. U. S. A.*, 2020, **117**, 5895.
37. J. Ma and M. Karplus, *Proc. Natl. Acad. Sci. U. S. A.*, 1998, **95**, 8502.
38. J. Monod, J. Wyman and J. P. Changeux, *J. Mol. Biol.*, 1965, **12**, 88.
39. L. Pauling, *Proc. Natl. Acad. Sci. U. S. A.*, 1935, **21**, 186.
40. D. E. Koshland, G. Némethy and D. Filmer, *Biochemistry*, 1966, **5**, 365.
41. H. G. Garcia, J. Kondev, N. Orme, J. A. Theriot and R. Phillips, *Methods Enzymol.*, 2011, **492**, 27.
42. R. Huang, F. Pérez and L. E. Kay, *Proc. Natl. Acad. Sci.*, 2017, **114**, E9846.
43. V. J. Hilser, J. O. Wrabl and H. N. Motlagh, *Annu. Rev. Biophys.*, 2012, **41**, 585.
44. A. Cornish-Bowden, *FEBS J.*, 2014, **281**, 621.
45. J. P. Changeux, *Philos. Trans. R. Soc. B Biol. Sci.*, 2018, **373**, 20170174.
46. H. N. Motlagh, J. O. Wrabl, J. Li and V. J. Hilser, *Nature*, 2014, **508**, 331.
47. T. A. J. Duke, N. Le Novère and D. Bray, *J. Mol. Biol.*, 2001, **308**, 541.
48. P. Csermely, R. Palotai and R. Nussinov, *Trends Biochem. Sci.*, 2010, **35**, 539.
49. X. Fei, T. A. Bell, S. Jenni, B. M. Stinson, T. A. Baker, S. C. Harrison and R. T. Sauer, *eLife*, 2020, **9**, e52774.
50. R. Gruber and A. Horovitz, *Philos. Trans. R. Soc. B Biol. Sci.*, 2018, **373**, 20170176.
51. A. Dyachenko, R. Gruber, L. Shimon, A. Horovitz and M. Sharon, *Proc. Natl. Acad. Sci. U. S. A.*, 2013, **110**, 7235.

52. R. Huang, Z. A. Ripstein, J. L. Rubinstein and L. E. Kay, *Angew. Chem., Int. Ed.*, 2020, **59**, 22423.
53. A. Martin, T. A. Baker and R. T. Sauer, *Nature*, 2005, **437**, 1115.
54. T. Eckert, S. Link, D. Tuong-Van Le, J. P. Sobczak, A. Gieseke, K. Richter and G. Woehlke, *J. Biol. Chem.*, 2012, **287**, 26278.
55. J. R. Hoskins, S. M. Doyle and S. Wickner, *Proc. Natl. Acad. Sci.*, 2009, **106**, 22233.
56. M. Stotz, O. Mueller-Cajar, S. Ciniawsky, P. Wendler, F. U. Hartl, A. Bracher and M. Hayer-Hartl, *Nat. Struct. Mol. Biol.*, 2011, **18**, 1366.
57. G. L. Hersch, R. E. Burton, D. N. Bolon, T. A. Baker and R. T. Sauer, *Cell*, 2005, **121**, 1017.
58. K. Arhzaouy, K. H. Strucksberg, S. M. Tung, K. Tangavelou, M. Stumpf, J. Faix, R. Schröder, C. S. Clemen and L. Eichinger, *PLoS One*, 2012, **7**, e46879.
59. C. C. Weihl, S. Dalal, A. Pestronk and P. I. Hanson, *Hum. Mol. Genet.*, 2006, **15**, 189.
60. Y. Matsushima, K. Takahashi, S. Yue, Y. Fujiyoshi, H. Yoshioka, M. Aihara, D. Setoyama, T. Uchiumi, S. Fukuchi and D. Kang, *Commun. Biol.*, 2021, **4**, 974.
61. D. T. Van Le, T. Eckert and G. Woehlke, *PLoS One*, 2013, **8**, e67815.
62. T. Eckert, S. Link, D. T. Le, J. Sobczak, A. Gieseke, K. Richter and G. Woehlke, *J. Biol. Chem.*, 2012, **287**, 26278.
63. J. R. Hoskins, S. M. Doyle and S. Wickner, *Proc. Natl. Acad. Sci.*, 2009, **106**, 22233.
64. N. D. Werbeck, S. Schlee and J. Reinstein, *J. Mol. Biol.*, 2008, **378**, 178.
65. K. Takeuchi, H. Arthanari and G. Wagner, *J. Biomol. NMR*, 2016, **66**, 221.
66. K. Pervushin, R. Riek, G. Wider and K. Wüthrich, *J. Am. Chem. Soc.*, 1998, **6**, 6394.
67. A. Boeszoermenyi, S. Chhabra, A. Dubey, D. L. Radeva, N. T. Burdzhiev, C. D. Chanev, O. I. Petrov, V. M. Gelev, M. Zhang, C. Anklin, H. Kovacs, G. Wagner, I. Kuprov, K. Takeuchi and H. Arthanari, *Nat. Methods*, 2019, **16**, 333.
68. J. E. Ollerenshaw, V. Tugarinov and L. E. Kay, *Magn. Reson. Chem.*, 2003, **41**, 843.
69. N. Bolik-Coulon, S. F. Cousin, P. Kadeřávek, J. N. Dumez and F. Ferrage, *J. Chem. Phys.*, 2019, **150**, 1.
70. V. Tugarinov, V. Kanelis and L. E. Kay, *Nat. Protoc.*, 2006, **1**, 749.
71. G. Abramov, A. Velyvis, E. Rennella, L. E. Wong and L. E. Kay, *Proc. Natl. Acad. Sci. U. S. A.*, 2020, **117**, 12836.
72. I. Gelis, A. M. J. J. Bonvin, D. Keramisanou, M. Koukaki, G. Gouridis, S. Karamanou, A. Economou and C. G. Kalodimos, *Cell*, 2007, **131**, 756.
73. I. Ayala, R. Sounier, N. Usé, P. Gans and J. Boisbouvier, *J. Biomol. NMR*, 2009, **43**, 111.
74. K. H. Gardner and L. E. Kay, *J. Am. Chem. Soc.*, 1997, **119**, 7599.
75. N. K. Goto, K. H. Gardner, G. A. Mueller, R. C. Willis and L. E. Kay, *J. Biomol. NMR*, 1999, **13**, 369.

76. P. Gans, O. Hamelin, R. Sounier, I. Ayala, M. A. Durá, C. D. Amero, M. Noirclerc-Savoye, B. Franzetti, M. J. Plevin and J. Boisbouvier, *Angew. Chem., Int. Ed.*, 2010, **49**, 1958.
77. A. K. Schuetz and L. E. Kay, *eLife*, 2016, **5**, e20143.
78. A. Proudfoot, A. O. Frank, F. Ruggiu, M. Mamo and A. Lingel, *J. Biomol. NMR*, 2016, **65**, 15.
79. R. Godoy-Ruiz, C. Guo and V. Tugarinov, *J. Am. Chem. Soc.*, 2010, **132**, 18340.
80. T. Saio, X. Guan, P. Rossi, A. Economou and C. G. Kalodimos, *Science*, 2014, **344**, 1250494.
81. R. Kerfah, M. J. Plevin, O. Pessey, O. Hamelin, P. Gans and J. Boisbouvier, *J. Biomol. NMR*, 2015, **61**, 73.
82. Y. R. Monneau, Y. Ishida, P. Rossi, T. Saio, S.-R. Tzeng, M. Inouye and C. G. Kalodimos, *J. Biomol. NMR*, 2016, **65**, 99.
83. Y. Miyanoiri, Y. Ishida, M. Takeda, T. Terauchi, M. Inouye and M. Kainosho, *J. Biomol. NMR*, 2016, **65**, 109.
84. L. Clark, J. A. Zahm, R. Ali, M. Kukula, L. Bian, S. M. Patrie, K. H. Gardner, M. K. Rosen and D. M. Rosenbaum, *J. Biomol. NMR*, 2015, **62**, 239.
85. L. D. Clark, I. Dikiy, K. Chapman, K. E. J. Rödström, J. Aramini, M. V LeVine, G. Khelashvili, S. G. F. Rasmussen, K. H. Gardner and D. M. Rosenbaum, *eLife*, 2017, **6**, e28505.
86. T. Sugiki, I. Shimada and H. Takahashi, *J. Biomol. NMR*, 2008, **42**, 159.
87. M. Miyazawa-Onami, K. Takeuchi, T. Takano, T. Sugiki, I. Shimada and H. Takahashi, *J. Biomol. NMR*, 2013, **57**, 297.
88. Y. Kofuku, T. Ueda, J. Okude, Y. Shiraishi, K. Kondo, T. Mizumura, S. Suzuki and I. Shimada, *Angew. Chem., Int. Ed.*, 2014, **53**, 13376.
89. R. Nygaard, Y. Zou, R. O. Dror, T. J. Mildorf, D. H. Arlow, A. Manglik, A. C. Pan, C. W. Liu, J. J. Fung, M. P. Bokoch, F. S. Thian, T. S. Kobilka, D. E. Shaw, L. Mueller, R. S. Prosser and B. K. Kobilka, *Cell*, 2013, **152**, 532.
90. Y. Kofuku, T. Ueda, J. Okude, Y. Shiraishi, K. Kondo, M. Maeda, H. Tsujishita and I. Shimada, *Nat. Commun.*, 2012, **3**, 1045.
91. A. Dubey, N. Stoyanov, T. Viennet, S. Chhabra, S. Elter, J. Borggräfe, A. Viegas, R. P. Nowak, N. Burdzhiev, O. Petrov, E. S. Fischer, M. Etzkorn, V. Gelev and H. Arthanari, *Angew. Chem., Int. Ed.*, 2021, **60**, 13783.
92. S. Yanaka, H. Yagi, R. Yogo, M. Yagi-Utsumi and K. Kato, *J. Biomol. NMR*, 2018, **71**, 193.
93. T. Kigawa, Y. Muto and S. Yokoyama, *J. Biomol. NMR*, 1995, **6**, 129.
94. Y. Kofuku, T. Ueda, J. Okude, Y. Shiraishi, K. Kondo, M. Maeda, H. Tsujishita and I. Shimada, *Nat. Commun.*, 2012, **3**, 1045.
95. A. Koglin, C. Klammt, N. Trbovic, D. Schwarz, B. Schneider, B. Schäfer, F. Löhr, F. Bernhard and V. Dötsch, *Magn. Reson. Chem.*, 2006, **44**, S17.

96. T. Etezady-Esfarjani, S. Hiller, C. Villalba and K. Wüthrich, *J. Biomol. NMR*, 2007, **39**, 229.
97. M. Etzkorn, T. Raschle, F. Hagn, V. Gelev, A. J. Rice, T. Walz and G. Wagner, *Structure*, 2013, **21**, 394.
98. T. Torizawa, M. Shimizu, M. Taoka, H. Miyano and M. Kainosho, *J. Biomol. NMR*, 2004, **30**, 311.
99. D. A. Vinarov, B. L. Lytle, F. C. Peterson, E. M. Tyler, B. F. Volkman and J. L. Markley, *Nat. Methods*, 2004, **1**, 149.
100. M. Kainosho, T. Torizawa, Y. Iwashita, T. Terauchi, A. Mei Ono and P. Güntert, *Nature*, 2006, **440**, 52.
101. R. Linser, V. Gelev, F. Hagn, H. Arthanari, S. G. Hyberts and G. Wagner, *J. Am. Chem. Soc.*, 2014, **136**, 11308.
102. T. L. Religa, A. M. Ruschak, R. Rosenzweig and L. E. Kay, *J. Am. Chem. Soc.*, 2011, **133**, 9063.
103. R. Huang, Z. A. Ripstein, R. Augustyniak, M. Lazniewski, K. Ginalski, L. E. Kay and J. L. Rubinstein, *Proc. Natl. Acad. Sci. U. S. A.*, 2016, **113**, E4190.
104. S. J. Abraham, R. C. Cheng, T. A. Chew, C. M. Khantwal, C. W. Liu, S. Gong, R. K. Nakamoto and M. Maduke, *J. Biomol. NMR*, 2015, **61**, 209.
105. A. R. Galiakhmetov, E. A. Kovrigina, C. Xia, J. J. P. Kim and E. L. Kovrigin, *J. Biomol. NMR*, 2018, **70**, 21.
106. I. Goba, D. Goricanec, D. Schum, M. Hillenbrand, A. Plückthun and F. Hagn, *ChemBioChem*, 2021, **22**, 139.
107. D. Han, M. Huang, T. Wang, Z. Li, Y. Chen, C. Liu, Z. Lei and X. Chu, *Cell Death Dis.*, 2019, **10**, 290.
108. G. E. Means and R. E. Feeney, *Biochemistry*, 1968, **7**, 2192.
109. G. E. Means, *J. Protein Chem.*, 1984, **3**, 121.
110. S. J. Abraham, S. Hoheisel and V. Gaponenko, *J. Biomol. NMR*, 2008, **42**, 143.
111. S. T. Larda, M. P. Bokoch, F. Evanics and R. S. Prosser, *J. Biomol. NMR*, 2012, **54**, 199.
112. M. P. Bokoch, Y. Zou, S. G. F. Rasmussen, C. W. Liu, R. Nygaard, D. M. Rosenbaum, J. J. Fung, H. J. Choi, F. S. Thian, T. S. Kobilka, J. D. Puglisi, W. I. Weis, L. Pardo, R. S. Prosser, L. Mueller and B. K. Kobilka, *Nature*, 2010, **463**, 108.
113. K. Möbius, K. Nordsieck, A. Pichert, S. A. Samsonov, L. Thomas, J. Schiller, S. Kalkhof, M. Teresa Pisabarro, A. G. Beck-Sickinger and D. Huster, *Glycobiology*, 2013, **23**, 1260.
114. R. Sounier, C. Mas, J. Steyaert, T. Laeremans, A. Manglik, W. Huang, B. K. Kobilka, H. Déméné and S. Granier, *Nature*, 2015, **524**, 375.
115. G. R. Moore, M. C. Cox, D. Crowe, M. J. Osborne, F. I. Rosell, J. Bujons, P. D. Barker, M. R. Mauk and A. G. Mauk, *Biochem. J.*, 1998, **332**, 439.
116. S. J. Abraham, T. Kobayashi, R. J. Solaro and V. Gaponenko, *J. Biomol. NMR*, 2009, **43**, 239.
117. Y. Hattori, K. Furuita, I. Ohki, T. Ikegami, H. Fukada, M. Shirakawa, T. Fujiwara and C. Kojima, *J. Biomol. NMR*, 2013, **55**, 19.

118. M. Sattler, J. Schleucher and C. Griesinger, *Prog. Nucl. Magn. Reson. Spectrosc.*, 1999, **34**, 93.
119. M. J. C. Audin, G. Dorn, S. A. Fromm, K. Reiss, S. Schütz, M. K. Vorländer and R. Sprangers, *Angew. Chem., Int. Ed.*, 2013, **52**, 8312.
120. M. A. Cvetkovic, J. P. Wurm, M. J. Audin, S. Schütz and R. Sprangers, *Nat. Chem. Biol.*, 2017, **13**, 522.
121. R. Rosenzweig, P. Farber, A. Velyvis, E. Rennella, M. P. Latham and L. E. Kay, *Proc. Natl. Acad. Sci.*, 2015, **112**, E6872.
122. O. Faust, M. Abayev-Avraham, A. S. Wentink, M. Maurer, N. B. Nillegoda, N. London, B. Bukau and R. Rosenzweig, *Nature*, 2020, **587**, 489.
123. C. Zwahlen, K. H. Gardner, S. P. Sarma, D. A. Horita, R. A. Byrd and L. E. Kay, *J. Am. Chem. Soc.*, 1998, **120**, 7617.
124. G. W. Vuister, G. M. Clore, A. M. Gronenborn, R. Powers, D. S. Garrett, R. Tschudin and A. Bax, *J. Magn. Reson., Ser. B*, 1993, **101**, 210.
125. V. Tugarinov, L. E. Kay, I. Ibraghimov and V. Y. Orekhov, *J. Am. Chem. Soc.*, 2005, **127**, 2767.
126. C. Amero, M. Asunción Durá, M. Noirclerc-Savoye, A. Perollier, B. Gallet, M. J. Plevin, T. Vernet, B. Franzetti and J. Boisbouvier, *J. Biomol. NMR*, 2011, **50**, 229.
127. A. Neu, U. Neu, A.-L. Fuchs, B. Schlager and R. Sprangers, *Nat. Chem. Biol.*, 2015, **11**, 697.
128. Y. Toyama, R. W. Harkness, T. Y. T. Lee, J. T. Maynes and L. E. Kay, *Proc. Natl. Acad. Sci. U. S. A.*, 2021, **118**, e2025022118.
129. I. Pritišanac, T. R. Alderson and P. Güntert, *Prog. Nucl. Magn. Reson. Spectrosc.*, 2020, **118–119**, 54.
130. Y. Xu and S. Matthews, *J. Biomol. NMR*, 2013, **55**, 179.
131. F.-A. Chao, L. Shi, L. R. Masterson and G. Veglia, *J. Magn. Reson.*, 2012, **214**, 103.
132. I. Pritišanac, M. T. Degiacomi, T. R. Alderson, M. G. Carneiro, E. AB, G. Siegal and A. J. Baldwin, *J. Am. Chem. Soc.*, 2017, **138**, 9523.
133. Y. R. Monneau, P. Rossi, A. Bhaumik, C. Huang, Y. Jiang, T. Saleh, T. Xie, Q. Xing and C. G. Kalodimos, *J. Biomol. NMR*, 2017, **69**, 215.
134. I. Pritišanac, J. M. Würz, T. R. Alderson and P. Güntert, *Nat. Commun.*, 2019, **10**, 1.
135. E. Schmidt and P. Güntert, *J. Am. Chem. Soc.*, 2012, **134**, 12817.
136. L. Siemons, H. W. Mackenzie, V. K. Shukla and D. F. Hansen, *J. Biomol. NMR*, 2019, **73**, 749.
137. Y. Z. Pan, B. Quade, K. D. Brewer, M. Szabo, J. D. Swarbrick, B. Graham and J. Rizo, *J. Biomol. NMR*, 2016, **66**, 281.
138. V. Venditti, N. L. Fawzi and G. M. Clore, *J. Biomol. NMR*, 2011, **51**, 319.
139. T. Saio, K. Ogura, M. Yokochi, Y. Kobashigawa and F. Inagaki, *J. Biomol. NMR*, 2009, **44**, 157.
140. A. Bahramzadeh, H. Jiang, T. Huber and G. Otting, *Angew. Chem., Int. Ed.*, 2018, **57**, 6226.

141. A. W. Barb and G. P. Subedi, *J. Biomol. NMR*, 2016, **64**, 75.
142. K. Barthelmes, A. M. Reynolds, E. Peisach, H. R. A. Jonker, N. J. DeNunzio, K. N. Allen, B. Imperiali and H. Schwalbe, *J. Am. Chem. Soc.*, 2011, **133**, 808.
143. Y. Mao, *Annu. Rev. Biochem.*, 2019, **87**, 697.
144. C. Volker and A. N. Lupas, in *The Proteasome—Ubiquitin Protein Degradation Pathway*, ed. P. Zwickl and W. Baumeister, Springer, Berlin, Heidelberg, 2002, pp. 1.
145. J. B. Jastrab and K. H. Darwin, *Annu. Rev. Microbiol.*, 2015, **69**, 109.
146. J. Löwe, D. Stock, B. Jap, P. Zwickl, W. Baumeister and R. Huber, *Science*, 1995, **268**, 533.
147. T. L. Religa, R. Spranger and L. E. Kay, *Science*, 2010, **328**, 98.
148. A. J. Baldwin, T. L. Religa, D. F. Hansen, G. Bouvignies and L. E. Kay, *J. Am. Chem. Soc.*, 2010, **132**, 10992.
149. B. M. Stadtmueller and C. P. Hill, *Mol. Cell*, 2011, **41**, 8.
150. A. Förster, E. I. Masters, F. G. Whitby, H. Robinson, C. P. Hill and N. York, *Mol. Cell*, 2005, **18**, 589.
151. E. Rennella, R. Huang, Z. Yu and L. E. Kay, *Proc. Natl. Acad. Sci. U. S. A.*, 2020, **117**, 5298.
152. H. Meyer, M. Bug and S. Bremer, *Nat. Cell Biol.*, 2012, **14**, 117.
153. Y. Ye, W. K. Tang, T. Zhang and D. Xia, *Front. Mol. Biosci.*, 2017, **4**, 39.
154. H. Meyer and C. C. Weihl, *J. Cell Sci.*, 2014, **127**, 3877.
155. D. J. Anderson, R. Le Moigne, S. Djakovic, B. Kumar, J. Rice, S. Wong, J. Wang, B. Yao, E. Valle, S. Kiss von Soly, A. Madriaga, F. Soriano, M.-K. Menon, Z. Y. Wu, M. Kampmann, Y. Chen, J. S. Weissman, B. T. Aftab, F. M. Yakes, L. Shawver, H.-J. Zhou, D. Wustrow and M. Rolfe, *Cancer Cell*, 2015, **28**, 653.
156. S. Banerjee, A. Bartesaghi, A. Merk, P. Rao, S. L. Bulfer, Y. Yan, N. Green, B. Mroczkowski, R. J. Neitz, P. Wipf, V. Falconieri, R. J. Deshaies, J. L. S. Milne, D. Huryn, M. Arkin and S. Subramaniam, *Science*, 2016, **351**, 871.
157. H. Niwa, C. A. Ewens, C. Tsang, H. O. Yeung, X. Zhang and P. S. Freemont, *J. Biol. Chem.*, 2012, **287**, 8561.
158. N. O. Bodnar and T. A. Rapoport, *Cell*, 2017, **169**, 722.
159. S. L. Bulfer, T. F. Chou and M. R. Arkin, *ACS Chem. Biol.*, 2016, **11**, 2112.
160. M. V Rao, D. R. Williams, S. Cocklin and P. J. Loll, *J. Biol. Chem.*, 2017, **292**, 18392.
161. A. E. Conicella, R. Huang, Z. A. Ripstein, A. Nguyen, E. Wang, T. Löhr, P. Schuck, M. Vendruscolo, J. L. Rubinstein and L. E. Kay, *Proc. Natl. Acad. Sci. U. S. A.*, 2020, **117**, 26226.
162. W. K. Tang and D. Xia, *Front. Mol. Biosci.*, 2016, **3**, 79.
163. G. D. J. Watts, J. Wymer, M. J. Kovach, S. G. Mehta, S. Mumm, D. Darvish, A. Pestronk, M. P. Whyte and V. E. Kimonis, *Nat. Genet.*, 2004, **36**, 377.
164. A. K. Schütz, E. Rennella and L. E. Kay, *Proc. Natl. Acad. Sci.*, 2017, **114**, E6822.

165. M. Pan, Y. Yu, H. Ai, Q. Zheng, Y. Xie, L. Liu and M. Zhao, bioRxiv, 2021, 2021.02.08.430295.
166. E. Twomey, Z. Ji, T. E. Wales, N. O. Bodnar, S. B. Ficarro, J. A. Marto, J. R. Engen and T. A. Rapoport, *Science*, 2019, **365**, eaax1033.
167. I. Cooney, H. Han, M. G. Stewart, R. H. Carson, D. T. Hansen, J. H. Iwasa, J. C. Price, C. P. Hill and P. S. Shen, *Science*, 2019, **48**, 502.
168. Z. A. Ripstein, R. Huang, R. Augustyniak, L. E. Kay and J. L. Rubinstein, *eLife*, 2017, **6**, e25754.
169. S. Dalal, M. Rosser, D. M. Cyr and P. I. Hanson, *Mol. Biol. Cell*, 2004, **15**, 638.
170. B. P. Conlon, E. S. Nakayasu, L. E. Fleck, M. D. Lafleur, V. M. Isabella, K. Coleman, S. N. Leonard, R. D. Smith, J. N. Adkins and K. Lewis, *Nature*, 2013, **503**, 365.
171. G. Wei, K. Jin-Yong, J. R. Anderson, A. Tatos, H. Seungpyo, J. Ying-Yu, K. Olga, K. Jong-Woo, L. In-Ae, L. Sun-Young, J. B. McAlpine, M. Surafel, S. Suhair, W. Yuehong, Y. Seung-Hwan, Y. Tae-Mi, A. L. Goldberg, G. F. Pauli, S. Joo-Won, S. G. Franzblau and C. Sanghyun, *Antimicrob. Agents Chemother.*, 2015, **59**, 880.
172. E. Gavrish, C. S. Sit, S. Cao, O. Kandror, A. Spoering, A. Peoples, L. Ling, A. Fetterman, D. Hughes, Bissell, H. Torrey, T. Akopian, A. Mueller, S. Epstein, A. Goldberg, J. Clardy and K. Lewis, *Chem. Biol.*, 2014, **21**, 509.
173. M. P. Choules, N. M. Wolf, H. Lee, J. R. Anderson, E. M. Grzelak, Y. Wang, R. Ma, W. Gao, J. B. McAlpine, Y. Jin, J. Cheng, H. Lee, J. Suh, N. M. Duc, S. Paik, J. H. Choe, E. Jo, C. L. Chang, J. S. Lee, B. U. Jaki, G. F. Pauli, S. G. Franzblau and S. Cho, *Antimicrob. Agents Chemother.*, 2019, **63**, e02204.
174. Y. Toyama, R. W. Harkness and L. E. Kay, *Proc. Natl. Acad. Sci. U. S. A.*, 2021, **118**, e2111257118.
175. M. Cai, Y. Huang, R. Yang, R. Craigie and G. M. Clore, *J. Biomol. NMR*, 2016, **66**, 85.
176. S. B. Azatian, N. Kaur and M. P. Latham, *J. Biomol. NMR*, 2019, **73**, 11.
177. E. S. O'Brien, D. W. Lin, B. Fuglestad, M. A. Stetz, T. Gosse, C. Tommos and A. J. Wand, *J. Biomol. NMR*, 2018, **71**, 263.
178. J. Klopp, A. Winterhalter, R. Gébleux, D. Scherer-Becker, C. Ostermeier and A. D. Gossert, *J. Biomol. NMR*, 2018, **71**, 247.
179. R. Suzuki, M. Sakakura, M. Mori, M. Fujii, S. Akashi and H. Takahashi, *J. Biomol. NMR*, 2018, **71**, 213.
180. Y. Kofuku, T. Yokomizo, S. Imai, Y. Shiraishi, M. Natsume, H. Itoh, M. Inoue, K. Nakata, S. Igarashi, H. Yamaguchi, T. Mizukoshi, E. Suzuki, T. Ueda and I. Shimada, *J. Biomol. NMR*, 2018, **71**, 185.
181. T. Yuwen, R. Huang, P. Vallurupalli and L. E. Kay, *Angew. Chem., Int. Ed. Engl.*, 2019, **58**, 6250.
182. T. Yuwen, A. Sekhar, A. J. Baldwin, P. Vallurupalli and L. E. Kay, *Angew. Chem., Int. Ed.*, 2018, **57**, 16777.

183. A. B. Gopalan, T. Yuwen, L. E. Kay and P. Vallurupalli, *J. Biomol. NMR*, 2018, **72**, 79.
184. S. Kumar, S. R. Akabayov, N. Kessler, L. S. Cohen, J. Solanki, F. Naider, L. E. Kay and J. Anglister, *J. Biomol. NMR*, 2020, **74**, 681.
185. R. Huang, J. P. Brady, A. Sekhar, T. Yuwen and L. E. Kay, *J. Biomol. NMR*, 2017, **68**, 249.
186. T. Yuwen, R. Huang and L. E. Kay, *J. Biomol. NMR*, 2017, **68**, 215.
187. T. Yuwen, P. Vallurupalli and L. E. Kay, *Angew. Chem., Int. Ed.*, 2016, **55**, 11490.
188. F.-A. Chao, J. Kim, Y. Xia, M. Milligan, N. Rowe and G. Veglia, *J. Magn. Reson.*, 2014, **245**, 17.
189. S. G. Hyberts, A. G. Milbradt, A. B. Wagner, H. Arthanari and G. Wagner, *J. Biomol. NMR*, 2012, **52**, 315.
190. J. Ying, F. Delaglio, D. A. Torchia and A. Bax, *J. Biomol. NMR*, 2017, **68**, 101.
191. D. F. Hansen, *J. Biomol. NMR*, 2019, **73**, 577.
192. D. F. Gauto, L. F. Estrozi, C. D. Schwieters, G. Effantin, P. Macek, R. Sounier, A. C. Sivertsen, E. Schmidt, R. Kerfah, G. Mas, J. P. Colletier, P. Güntert, A. Favier, G. Schoehn, P. Schanda and J. Boisbouvier, *Nat. Commun.*, 2019, **10**, 1.
193. Z. A. Ripstein, S. Vahidi, J. L. Rubinstein and L. E. Kay, *J. Am. Chem. Soc.*, 2020, **142**, 20519.
194. J. Kitevski-Leblanc, A. Fradet-Turcotte, P. Kukic, M. D. Wilson, G. Portella, T. Yuwen, S. Panier, S. Duan, M. D. Canny, H. Van Ingen, C. H. Arrowsmith, J. L. Rubinstein, M. Vendruscolo, D. Durocher and L. E. Kay, *eLife*, 2017, **6**, 1.
195. M. Li, O. Kandror, T. Akopian, P. Dharkar, A. Wlodawer, M. R. Maurizi and A. L. Goldberg, *J. Biol. Chem.*, 2016, **291**, 7465.

CHAPTER 13

Characterizing Conformational Diversity of G Protein-coupled Receptors by Solution NMR Spectroscopy

FENG-JIE WU,[a,b] FABIAN BUMBAK,[a,b] MOHAMMAD HOSSEIN TANIPOUR,[a,b,c] KAZEM ASADOLLAHI,[a,b,c] TASNEEM M. VAID,[a,b,c] ASHISH SETHI,[a,c] DANIEL J. SCOTT[a,b] AND PAUL R. GOOLEY*[a,c]

[a] Department of Biochemistry and Pharmacology, University of Melbourne, Parkville 3052VIC, Australia; [b] The Florey Institute of Neuroscience and Mental Health, University of Melbourne, Parkville 3052VIC, Australia; [c] Bio21 Institute of Molecular Sciences and Biotechnology, University of Melbourne, Parkville 3052VIC, Australia

*Email: prg@unimelb.edu.au

13.1 Introduction

G protein-coupled receptors (GPCRs) are the largest receptor family in eukaryotes, comprising 800 members. GPCRs are integral plasma membrane proteins sensing an extensive array of extracellular stimuli, including hormones, neurotransmitters, proteins, ions, photons, odorants, tastants, and vitamins to initiate a variety of signalling cascades in the cell. Their location in the plasma membrane and essential role in almost all physiological processes makes them ideal drug targets. However, targeting a specific

New Developments in NMR No. 26
NMR Spectroscopy for Probing Functional Dynamics at Biological Interfaces
Edited by Anirban Bhunia, Hanudatta S. Atreya and Neeraj Sinha

Published by the Royal Society of Chemistry, www.rsc.org

receptor and/or signalling pathway is challenging since all GPCRs have evolved from a common ancestor (*i.e.* they share the same structural fold) and a single receptor can activate multiple pathways. In the last decade significant structural details of GPCRs, isolated or complexed with ligands and effector proteins, have emerged. These structures provide the essential groundwork that has been taken up by NMR to further our understanding of GPCR function.

13.1.1 Overview of GPCR Structure

GPCRs are typically activated by the extracellular binding of an agonist ligand. These ligands can be classified based on their efficacy, where efficacy is defined as the extent to which a ligand can activate a specific signalling pathway. Full agonists maximally activate the receptor, while partial agonists only partially activate receptors, even when fully saturated. In the absence of an agonist, GPCRs have various levels of basal activity.[1–3] Inverse agonists reduce or ablate this basal activity, whereas neutral antagonists occupy the ligand-binding pocket, and compete with agonists, but do not suppress the basal activity. Agonist binding typically leads to GPCR-driven intracellular signalling, mostly through coupling to G-protein heterotrimers or arrestins (Figure 13.1). Heterotrimeric G proteins comprise three subunits: Gα, Gβ and Gγ. In humans, G proteins are classified into four major families, based on Gα isotypes: G_s, $G_{i/o}$, $G_{q/11}$ and G_{12}.[4] G-protein activation triggers exchange of GDP with GTP in the Gα subunit and the subsequent dissociation of the heterotrimer, with Gα and Gβγ binding effector proteins that activate independent signalling cascades. Effectors targeted by Gα-GTP are Gα subtype dependent.[5] $G\alpha_s$-GTP stimulates adenylyl cyclase1–9 (AC1–9) increasing cyclic adenosine monophosphate (cAMP) production, whereas $G\alpha_i$-GTP inhibits AC5 and AC6 decreasing cAMP levels. cAMP mainly regulates the activation and further signalling of protein kinase A (PKA). The $G\alpha_{q/11}$ subfamily mainly activates phospholipase Cβ isoforms (PLCβ1–4) which hydrolyse phosphatidylinositol biphosphate (PIP_2) to inositol triphosphate (IP_3) and diacylglycerol (DAG), resulting in intracellular Ca^{2+} mobilization and protein kinase C (PKC) activation, respectively.[6] Ca^{2+} and PKC participate in diverse downstream signalling to evoke various cellular events. Other effectors of $G\alpha_{q/11}$-GTP include Rho Guanine Nucleotide Exchange Factor (p63RhoGEF), which stimulates the Rho family protein RhoA, responsible for cell morphological change, cell migration, and proliferation.[7–9] Gβγ, after dissociating from Gα-GTP, can up-regulate AC (2, 4, 7) and PLCβ, and down-regulate AC1. Gβγ also modulates potassium channels, voltage-dependent Ca^{2+} channels, phosphoinositide 3 kinase and mitogen-activated protein kinases.[10,11] Further, Gβγ recruits the G protein-coupled receptor kinase (GRK) to the cell membrane to enable receptor phosphorylation, internalization, and signalling *via* arrestins.

There are four isoforms of arrestin (arrestin 1–4), of which two (arrestin 1 and 4) are restricted to the visual system. Arrestin 2 and 3 (also known as

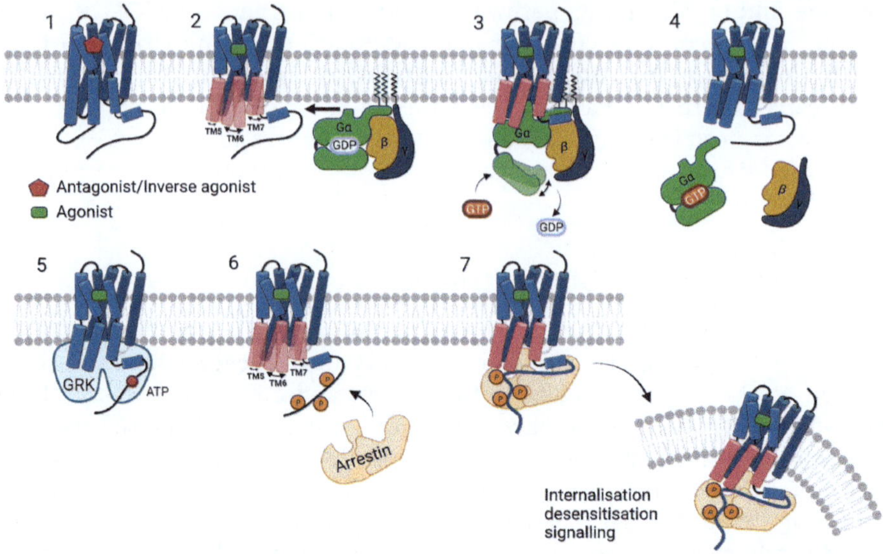

Figure 13.1 GPCR signalling through interaction with G protein and GRK/arrestin. (1) Inverse agonist blocks GPCR signalling, whereas antagonists block other ligands and maintain basal activity. (2) Agonist stabilizes pre-active GPCR conformations, enabling coupling of a GDP-bound heterotrimeric G protein with the GPCR. (3) Association of a G-protein heterotrimer with the GPCR stimulates GDP-to-GTP exchange in the Gα subunit. (4) The GTP-bound Gα and Gβγ subunits dissociate from the GPCR to activate independent signalling cascades. (5) Agonist-activated GPCR is phosphorylated by GPCR kinase (GRK). (6) Arrestin recognizes the phosphorylated GPCR. (7) Arrestin couples to the GPCR and induces G protein-independent signalling and internalization of the GPCR. Figure created with BioRender.com.

β-arrestin-1 and 2, respectively) couple to non-visual GPCRs and initiate receptor internalization and desensitization. Additionally, activated arrestins have their own range of signalling outputs distinct from G-protein signalling.[12] The coupling of arrestins requires activated GPCRs to be phosphorylated by G protein-coupled receptor kinases (GRKs) whereby specific phosphorylation "barcodes" regulate both arrestin/GPCR interaction and arrestin signalling output.[13]

All GPCRs share the same architecture of a transmembrane domain (7TM) comprising seven α-helices (TM1–7) connected by three extracellular loops (ECL1–3) and three intracellular loops (ICL1–3) (Figure 13.2). Studies of GPCRs over the last two decades have largely focused on solving structures in a bid to understand the molecular basis of GPCR function. The first GPCR crystal structure, of inactive bovine rhodopsin, was determined in 2000.[14] Starting 2007, the extensive work on β_2-adrenergic receptor (β_2-AR) structures established crystallization methodologies[15–17] that have since been widely applied to many other GPCRs.[18–22] Most recently, due to the advances

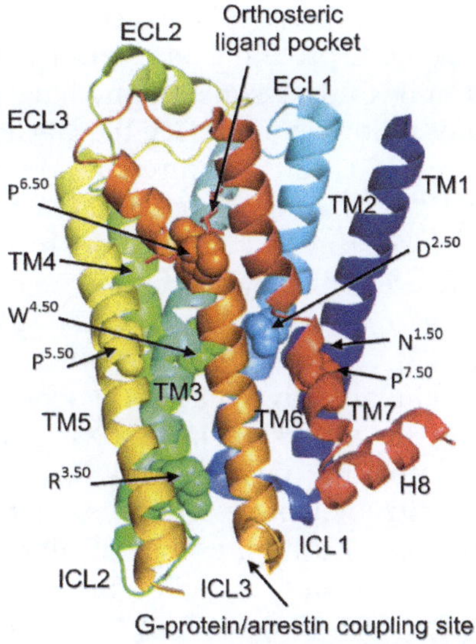

Figure 13.2 GPCRs share a common transmembrane domain fold. GPCRs are characterized by a transmembrane domain (7TM) comprising seven helices (TM1–7). The example of inactive β_2-AR (PDB ID: 3NY8) is drawn in rainbow colours ranging from blue (N-terminal, extracellular) to red (C-terminal, intracellular). An eighth helix (H8) at the C-terminal end runs parallel to the cytoplasmic side of the membrane. The helices are connected by three intracellular loops (ICL1–3) and three extracellular loops (ECL1–3). The orthosteric ligand (inverse agonist, ICI 118,551) is shown as red sticks, located in the orthosteric binding site. ICL3 was not resolved in the crystal structure. Ballesteros–Weinstein numbering[57] is used, where the most conserved residue of each helix (shown as spheres) is assigned the number 50, and other residues within the helix are numbered relative to this position. A particular residue is then assigned with a superscript x.y, where x represents the TM(1–7) and y the position in the helix, for example $W^{6.48}$ is a tryptophan in TM6, two residues towards the N-terminus of the helix from $P^{6.50}$, the most conserved residue in TM6.

in cryogenic electron microscopy (cryo-EM), several GPCR and G protein/arrestin complex structures have been determined at near-atomic resolution.[23–26]

Initially, GPCR structures were limited to antagonist-bound inactive states.[27] Antagonists can largely reduce the conformational heterogeneity of GPCRs and thus improve receptor stability, which is essential for crystallization. This is evident from β_2-AR crystal structures solved in the presence of multiple antagonists and inverse agonists.[28] The cytoplasmic sides of these inactive β_2-AR structures display a similar inactive conformation. Structures of solely agonist-bound GPCRs have rarely been reported, usually requiring thermostabilization,[21,29–38] likely due to the more dynamic nature of wild-type GPCRs in the presence of agonists.

The first report of an active-state GPCR structure was of opsin bound to a C-terminal fragment of the G protein transducin ($G\alpha_tCT$).[39] Technical developments in cryo-EM and X-ray crystallography have led to an increasing number of new structures being reported (see the GPCR database at https://gpcrdb.org/). Active states of GPCRs require the presence of active-state-stabilizing nanobodies,[19,20,40–43] engineered mini-G proteins,[44–46] heterotrimeric G proteins[23] or arrestins.[24–26] Compared to inactive GPCR structures, active-state structures display common rearrangements at the intracellular side: an outward tilt of TM6, rotations of TM3 and TM5, and an inward displacement of TM7 (Figure 13.3). Together, these conformational changes create an intracellular cavity to accommodate the insertion of the Gα C-terminus. Despite the successes in resolving structures of GPCRs coupled to different G proteins (G_s, $G_{i/o}$ and $G_{q/11}$) using cryo-EM, the question as to how G-protein sub-type selectivity is determined remains to be answered. To date, structures in the presence of different G proteins have only been solved for three GPCRs, namely glucagon receptor (GCGR),[47] cholecystokinin 1 receptor (CCK1R or CCK_AR)[48,49] and neurokinin-1 receptor (NK1R).[50] The GCGR structures determined in complex with G_s and G_{i1} display equivalent receptor backbone conformations reminiscent of other GPCR/G_s complexes, with G_{i1} binding to the same pocket albeit with less interactions.[47] Sulfated cholecystokinin 8 (CCK-8)-bound CCK1R structures determined in the presence of G_s and an engineered G_q variant also feature similar receptor backbone conformations, but in this case the G-protein binding cavity is shallower and the C-terminal α5-helix of G_s unfolds to fit the cavity.[48] In another study, CCK-8-bound CCK1R active structures were solved in complex with three different G proteins (G_s, G_i and G_q).[49] Again, the

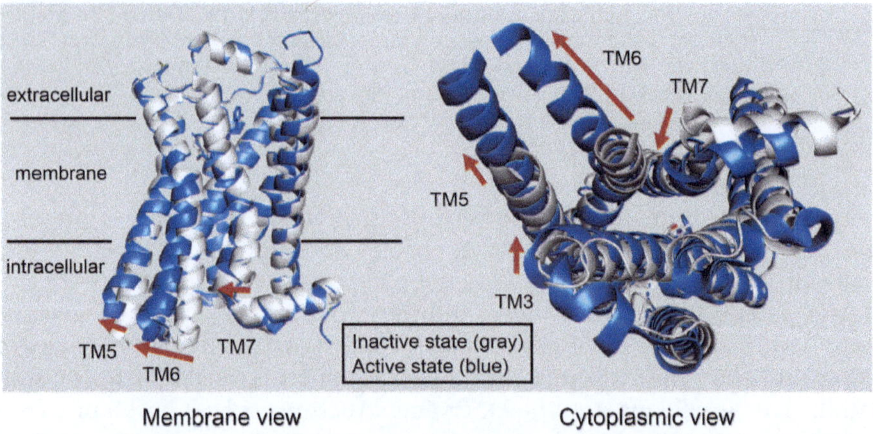

Figure 13.3 Conformational changes distinguish inactive and active GPCR states. Superposition of inactive (grey, PDB ID: 3NY8) and active (blue, PDB ID: 3SN6) β_2-AR shown from the membrane and cytoplasmic perspective. Significant conformational changes of TMs are highlighted with red arrows.

CCK1R conformations were almost identical but the G_q, G_s and G_i proteins adopted different orientations with varying sizes of the receptor–G protein interface. Besides differences between the C-terminal ('wavy hook') residues of the Gα α5-helices, the study revealed an unconventional ICL/G_q interaction that was not observed in the G_s- and $G_{i/o}$-coupled CCK1R structures, hinting at a possible role of ICL3 for Gα subtype coupling specificity. For NK1R interactions between the N-terminal region of the peptide ligand, substance P (SP) and NK1R were found to be required for balanced signalling *via* both the G_q and G_s signalling pathways, while loss of these interactions led to G_q-selective signalling, but cryo-EM structures of SP/NK1R complex with engineered G_s and G_q proteins were practically identical.[50] Interestingly, even a structure of NK1R–G_q activated with N-terminally truncated, and hence G_q-biased SP, revealed a similar conformation compared to full-length SP-bound NK1R, but with notably less electron density for the truncated peptide. It was thus suggested that the loss of extracellular contacts increases the conformational flexibility of the buried region of SP which destabilizes G_s but not G_q interaction. Generally, structures of GPCRs in the presence of the cognate $G_{i/o}$ and $G_{q/11}$ feature a less tilted TM6 and hence narrower intracellular cavity compared to GPCRs preferentially coupling to G_s.

Coupling of GPCRs to arrestins in addition to G proteins adds another level of complexity. Depending on the ligand, signalling can be biased in favour of either arrestin or G protein. Furthermore, phosphorylation of the GPCR C-terminus or/and ICL3 by G protein-coupled receptor kinases (GRKs) is a perquisite for arrestin coupling. So far, structures of only four GPCR-arrestin complexes have been solved: rhodopsin,[51] neurotensin receptor 1 (NTS_1),[26,52] β_1-AR[25] and M2 muscarinic receptor (M_2R).[24] While the arrestin-stabilized conformations of rhodopsin, NTS_1 or M_2R were similar to their corresponding G_i-coupled conformations, TM6 of arrestin-coupled β_1-AR was tilted less compared to the G_s-coupled state.[25]

13.1.2 Structural Rearrangement of GPCRs upon Activation

Agonist binding to the orthosteric pocket of the GPCR initiates conformational changes in the receptor to allosterically open the G-protein binding site on the intracellular side. Class A GPCRs have four highly conserved structural motifs in common, namely the CWxP, PIF, NPxxY and E/DRY motifs (Figure 13.4A). Significant structural rearrangements of these motifs have been observed by comparing inactive and active GPCR structures (Figure 13.4B–E). These motifs are critical GPCR microswitches that mediate allosteric signal transmission across the plasma membrane.[13,53–56]

The CWxP motif is located on TM6 near the ligand-binding pocket (Figure 13.4B). $W^{6.48}$ of the CWxP motif (superscript refers to Balesteros-Weinstein numbering[57]), also called the "toggle switch", connects the ligand-binding site with the PIF motif (also known as the "core triad") located in the transmission region or connector region of GPCRs[19,58–60]

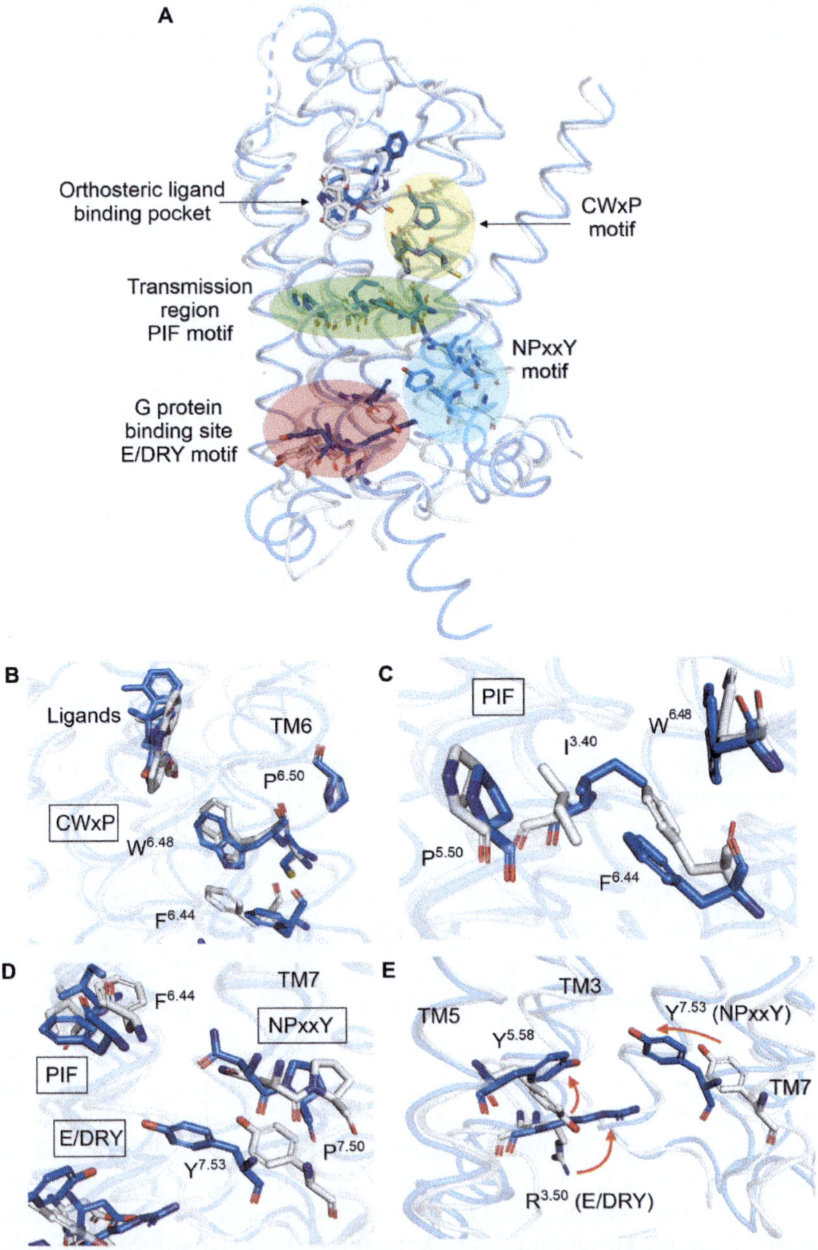

Figure 13.4 Conserved structural motifs of class A GPCRs. (A) Overview of several conserved microswitches or motifs in β_2-AR. (B–E) Conformational changes and connections of the CWxP, PIF, NPxxY and E/DRY motifs in the inactive (grey, PDB ID: 3NY8) and active (blue, PDB ID: 3SN6) structures of β_2-AR. Ballesteros–Weinstein numbering[57] is used.

(Figure 13.4C). The three PIF residues $P^{5.50}$, $I/V^{3.40}$ and $F^{6.44}$ display the same structural arrangement in most GPCR structures.[19] Relocation of $F^{6.44}$ in active-state structures is thought to be critical for the outward tilt of TM6 to accommodate transducer proteins. The NPxxY motif is formed by five consecutive residues at the C-terminal end of TM7 (Figure 13.4D). In high-resolution GPCR structures, $N^{7.49}$ of the NPxxY motif was found to reach across and contact residues in the transmission region through a water-mediated hydrogen bond.[61] The E/DRY motif comprises three conserved residues on TM3 near the G-protein binding site. Comparing inactive- to active-state receptor structures, $R^{3.50}$ of the E/DRY motif, $Y^{7.53}$ of the NPxxY motif and $Y^{5.57}$ undergo large conformational rearrangements (Figure 13.4E). In active-state GPCR structures, these three residues adopt near-identical positions connecting the NPxxY and E/DRY motifs through water-mediated hydrogen bonds.[62,63] To elucidate how these conformational changes are linked to signal transduction and ligand efficacy, complementary tools that provide direct access to conformational dynamics are needed. NMR spectroscopy is uniquely capable of providing atomic-level information on dynamic processes covering picosecond to second timescales.[64,65]

13.2 Preparing GPCRs for NMR Spectroscopy

13.2.1 Expression Systems

Due to the relative insensitivity of NMR experiments, milligram amounts of purified GPCR are generally required. Furthermore, NMR investigations usually require isotopic labelling (^{13}C, ^{15}N or ^{19}F) and incorporation of the GPCR into a suitable membrane mimetic system (Table 13.1). Only a few wild-type GPCRs (β_2-AR, A_{2A}AR, μOR and M_2R) are sufficiently stable in detergent micelles and can be highly expressed, making them suitable for NMR studies. For GPCRs that are unstable in detergent micelles, mutations can be introduced to increase thermostability, longevity and expression yields. For example, alanine-scanning of turkey β_1-AR was employed in a bid to increase thermostability.[66] The resulting mutant was reported to be expressed at high levels in insect cells and to be stable in detergents such as DM (decyl-maltoside), DDM (dodecyl-maltoside) and LMNG (lauryl maltose neopentyl glycol) thus proving useful for NMR studies.[67–72] Other GPCRs, including α_{1A}-AR, α_{1B}-AR and NTS_1, have been stabilized through directed evolution where large libraries of mutant proteins are randomly generated and undergo rounds of selection on the basis of GPCR expression levels in *Escherichia coli*, detergent stability and ligand-binding competency.[36,73–78]

As GPCRs are integral membrane systems they present challenges for expressing sufficient quantities of correctly folded protein for biophysical analyses. Most GPCRs prepared for solution NMR (Table 13.1) have been produced using the insect cell-baculovirus expression system (BVES). BVES enables post-translational modifications, such as glycosylation, acylation

Table 13.1 List of labelled NMR Studies on Non-vision Complete GPCRs.

GPCR	Expression system	Labelling strategy[a]	Environment[b]	Refs
β_2-AR	Sf9	(CM) ε-N[$^{13}CH_3$]$_2$-Lys	DDM	90,91
β_2-AR	Sf9	(CM) ^{19}F-TET	DDM/CHS	92–94
β_2-AR	expresSF+	(IL) $^{13}C^{\varepsilon}H_3$-Met	DDM	95
β_2-AR	Sf9	(CM) ^{19}F-BTFA	DDM and LMNG	96
β_2-AR	Sf9	(IL) $^{13}C^{\varepsilon}H_3$-Met	DDM	97
β_2-AR	expresSF+	(IL) ^{2}H-9AA, αβγ-^{2}H-, $^{13}C^{\varepsilon}H_3$-Met	Nanodisc	98,99
β_2-AR	expresSF+	(IL) C-term-[u-^{2}H,^{13}C,^{15}N] C-term-[u-^{2}H, [^{13}C,^{1}H]-Thr$^{\gamma2}$,-Ile$^{\delta1}$]	Nanodisc	96
β_2-AR	expresSF+	(IL) [2,3,3-^{2}H,^{15}N]-Leu	LMNG	100
β_2-AR	Sf9	(CM) ^{19}F-BTFA	DDM and LMNG	101
β_2-AR	Sf9	(CM) ^{19}F-BTFMA	LMNG and nanodiscs	102,103
β_1-AR	Sf9	(IL) ^{15}N-Val	DM	67–69
β_1-AR	Sf9/Sf21	(IL) $^{13}C^{\varepsilon}H_3$-Met	DM, LMNG and salipro nanoparticle	70,72,104
β_1-AR	Sf9	(CM) ^{19}F-TET	LMNG	71
β_1-AR	*E. coli*	(IL) u-^{2}H/^{15}N	DM	105
A_{2A}AR	*P. pastoris*	(CM) ^{19}F-BTFMA	LMNG/CHS	80,84–86
A_{2A}AR	*P. pastoris*	(IL) [u-^{2}H, $^{13}C^{\delta1}H_3$-Ile]	DDM	83
A_{2A}AR	*P. pastoris*	(IL) u-^{15}N, ~70% ^{2}H	LMNG/CHS	81,106
A_{2A}AR	*P. pastoris*	(IL) [u-^{2}H, α,β,β-^{2}H, $^{13}C^{\varepsilon}H_3$-Met]	Nanodisc	107
μOR	expresSF+	(IL) ^{2}H-8AA, αβ-^{2}H-, $^{13}C^{\varepsilon}H_3$-Met	Nanodisc	108
μOR	Sf9	(CM) ε-N[$^{13}CH_3$]$_2$-Lys	LMNG/CHS	109
M_2R	Sf9	(IL) $^{13}C^{\varepsilon}H_3$-Met	LMNG/CHS	110
α_{1A}-AR	*E. coli*	(IL) $^{13}C^{\varepsilon}H_3$-Met	DDM	111
α_{1B}-AR	*E. coli*	(IL) u-^{2}H/^{15}N/^{13}C and ILV	LMNG and nanodisc	112
NTS_1	*E. coli*	(IL) $^{13}C^{\varepsilon}H_3$-Met	DDM	113
NTS_1	*E. coli*	(CM) $^{13}CH_3$-MMTS	DDM	114
$CXCR_1$	*E. coli*	(IL) u-^{13}C/^{15}N and ^{13}C/^{15}N-Phe	DMPC proteoliposome	115
BLT_2 CB1	*E. coli* Sf9	(IL) [u-^{2}H,$^{13}C^{\delta1}H_3$-Ile]/ [u-^{2}H,$^{13}C^{\varepsilon}H_3$-Met](GI) ^{19}F-mtfF	NanodiscLMNG/ CHS	116,117

[a]CM, chemical modification; IL, isotopic labelling; GI, genetic incorporation; TET, 2,2,2-trifluoroethanethiol; BTFA, 3-bromo-1,1,1-trifluoroacetone; BTFMA, 2-bromo-4-trifluoromethylacetanilide; MMTS, S-methyl-^{13}C-methanethiosulfonate. ^{19}F-mtfF, 3′-trifluoromenthyl-phenylalanine.

[b]DDM, dodecyl-maltoside; LMNG, lauryl maltose neopentyl glycol; CHS, cholesterol hemisuccinate.

and disulfide bond formation, and the folding mechanisms of eukaryotic proteins.[79] Yeast systems, particularly *Pichia pastoris*, have similar advantages to BVES and have been used to express several GPCRs, such as

$A_{2A}AR$[80–86] and the histamine H_1 receptor.[87] Less popular expression systems for GPCRs are *E. coli* and mammalian cells. Cell-free expression systems have the potential for cost-effective isotope-labelling, but this approach has been challenging to produce good yields of correctly folded and functional protein.[88,89]

13.2.2 Isotope-labelling and Assignment Strategies

GPCRs for solution NMR studies are either isotopically labelled or chemically modified post-expression using NMR-active tags (Table 13.1). Among isotopic labelling strategies, ^{13}C-methyl methionine ($^{13}C^{\varepsilon}H_3$-Met) is most commonly used for GPCRs. This is attributed to the intense single peak for the methyl group located at the end of a long side chain, free of coupling to other side chain ^{1}H resonances; $^{13}C^{\varepsilon}H_3$-Met resonances are found near 17 ppm (^{13}C) and 2 ppm (^{1}H) generally resolved from detergent signals; the rarity of methionine in proteins making resonances relatively easy to assign by site-directed mutagenesis; and its propensity to be located in protein interfaces making it sensitive to conformational change. These features make methionine a desirable probe that can be applied in non-deuterated systems up to a size of about 100 kDa. $^{13}C^{\varepsilon}H_3$-Met has also been used as a probe in the nanodisc system where the apparent size requires the receptor to be either fully deuterated in *E. coli* or yeast expression systems,[107,116] or incorporated together with other deuterated amino acids in the insect expression system.[98,108] Although methionines often reside in the hydrophobic transmembrane regions of GPCRs, they are not part of any of the important microswitches (such as CWxP, PIF, E/DRY and NPxxY; Figure 13.4) and thus only indirectly report on conformational changes. $^{13}C^{\delta}H_3$-isoleucine labelling of $A_{2A}AR$ has been achieved in *P. pastoris* using methyl-^{13}C α-ketobutyric acid in a completely deuterated background.[83] While the $^{13}C^{\delta}H_3$-isoleucine is expected to have a well-resolved ^{13}C signal (13.6 ppm), its ^{1}H (0.8 ppm) overlaps with residual detergent signals, thus requiring deuterated detergent.

^{15}N-Val isotopic backbone labelling was successfully applied to stabilized turkey β_1-AR in a non-deuterated environment.[67–69] Here the receptor was reconstituted in DM micelles, a relatively harsh environment not ideal for naturally unstable GPCRs, and the data were collected at temperatures as high as 304 K. For other receptors that have been ^{15}N-labelled, deuteration appears to be required. Uniform ^{15}N backbone labelling and partial deuteration have been achieved for $A_{2A}AR$ in *P. pastoris*[81,82,106] and for β_1-AR and C-C chemokine receptor type 5 (CCR5) in insect cells.[118,119] Due to insufficient back-exchange of u-^{15}N, 70% ^{2}H $A_{2A}AR$ amide protons only residues from solvent-exposed regions gave rise to sufficiently intense signals, while amides that are buried in the transmembrane region were not well observed. However, additional signals, possibly from the 7TM were detected for u-^{15}N, ~60% ^{2}H β_1-AR using ^{15}N/^{2}H yeastolate in insect cells.[118,119] Imai and colleagues employed ^{15}N-labelled leucine in a non-deuterated background in β_2-AR but only observed broad signals.[100] However, labelling with

^{15}N-leucine enzymatically deuterated at the α and β positions ([2,3,3-^{2}H,^{15}N]-Leu) resulted in high-quality spectra. To assign resonances in these ^{15}N-labelled receptors, triple resonance experiments such as HNCO and HNCA have had very limited success.[67] Instead, residue-specific assignment has been achieved by site-directed mutagenesis.

To study the effect of phosphorylation of the predominantly disordered C-terminus of β_2-AR, Shiraishi and colleagues combined amino acid-selective and segment-selective isotope labelling approaches.[99] The 63-residue C-terminal region of β_2-AR (residues 349–413) was expressed in *E. coli* labelled with either [u-^{2}H,^{13}C,^{15}N] or [u-^{2}H, [^{13}C,^{1}H]-Thr$^{\gamma2}$,-Ile$^{\delta1}$] and then attached to the unlabelled, insect-cell-derived TM region of β_2-AR (1–348) using intein-mediated protein *trans*-splicing (PTS).

^{19}F-NMR has become increasingly popular for studying GPCR conformational dynamics.[120] Typically, a CF_3-probe is introduced by covalently modifying cysteine residues with 2,2,2-trifluoroethanethiol (TET), 3-bromo-1,1,1-trifluoroacetone (BTFA) or 2-bromo-4-trifluoromethylacetanilide (BTFMA). Cysteines utilized in this approach are either native or introduced and located at the cytoplasmic tips of TM5, TM6 and TM7 that are expected to undergo large conformational changes. ^{19}F is a convenient probe as it is the most naturally abundant fluorine isotope and because fluorine itself is absent from proteins. The downside of ^{19}F chemical-labelling, as with any labelling strategy that requires chemical modification of proteins, is that solvent inaccessible positions within the 7TM core of the GPCR cannot be labelled. Recently, ^{19}F-labelled unnatural amino acids, such as 3′-trifluromethyl-phenylalanine (tmfF), have been genetically incorporated into the cannabinoid receptor 1 (CB1).[117] To incorporate this label required the development of the orthogonal tRNA/aminoacyl-tRNA synthase pair to incorporate tmfF using the TAG amber codon expressing in Sf9 cells. This method has significant promise in incorporating ^{19}F-labels in poorly solvent accessible positions within a GPCR.

13.2.3 Membrane Mimetic Systems

Detergent micelles are the most common mimetic that have been utilized for membrane protein characterization.[121,122] Especially, DM, DDM and LMNG have been used to efficiently extract GPCRs from cell membranes. DM has a shorter aliphatic chain making it a harsh detergent suitable for GPCRs that are more stable. DDM and LMNG are more frequently used for unstable and detergent-sensitive GPCRs. LMNG comprises two hydrophobic chains, thus resembling the lipid architecture of the membrane.[123] LMNG has a lower critical micelle concentration (CMC) compared to DDM (10 μM *vs.* 165 μM respectively),[124,125] which translates to a lower off-rate for LMNG monomers compared to DDM.[96] A recent molecular dynamics (MD) study found that LMNG enhances the density of aliphatic chains around the hydrophobic regions of A_{2A}AR and that more hydrogen bonds were formed between head groups.[25] They also showed that the two polar groups of LMNG form persistent bifurcated hydrogen bonds between TM helices of A_{2A}AR, thus reducing the flexibility of intracellular regions. It was also found that NMR experiments using LMNG reconstituted A_{2A}AR can be conducted at higher temperatures compared to DDM reconstituted A_{2A}AR and that the significantly lower off-rate of LMNG allows the detection of distinct functional conformations.[96] However, the reliability of structural and dynamic studies in a micelle environment has been questioned due to various disadvantages, such as micelle formation might adversely affect membrane protein function by removing annular lipids, interruption of natural protein folding, not being a proper geometrical mimetic of the membrane bilayer because of the presence of strong curvature at the micelle surface, and denaturing detergent-sensitive associated proteins such as G proteins.[126–128]

Over the past few decades various alternatives to detergent micelles, have been introduced, including bicelles,[129] liposomes,[130] SMALPs,[131] amphipols,[132] peptergents,[133] lipopeptides[134] and β-strand peptides.[135] Currently, bilayer systems including nanodiscs,[136] beltides,[137] saposin nanoparticles[138] and peptidiscs[139] offer the best approximation of a native-like membrane environment for GPCR structural and dynamic studies. All bilayer systems share a similar overall architecture comprising lipid-interactive protein belts that wrap around a phospholipid bilayer with the embedded membrane protein.

The protein belts in nanodiscs, referred to as membrane scaffold proteins (MSPs), are engineered variants of apolipoprotein A-1 (Apo-A1) available in different sizes[140] and as circularized versions.[141,142] Peptidiscs use a reverse sequence of small peptides derived from Apo-A1, termed nanoscaffold protein (NSP_r).[139] In saposin (salipro) nanoparticles, the lysosomal lipid-interactive protein, saposin, is used as a protein belt.[136,138,139] A variety of natural and synthetic lipids have been utilized in the reconstitution of these bilayer systems, however neutral lipids like 1-palmitoyl-2-oleoyl-*sn*-glycero-3-phosphocholine (POPC), 1-palmitoyl-2-oleoyl-*sn*-glycero-3-phosphoglycerol (POPG), 1,2-dimyristoyl-*sn*-glycero-3-phosphocholine (DMPC) and 1,2-dihexadecanoyl-*sn*-glycero-3-phosphocholine (DPPC) are preferred unless the membrane protein requires a negatively charged environment which can be provided by adding lipids like 1-palmitoyl-2-oleoyl-*sn*-glycero-3-phospho-L-serine (POPS).[104,143,144]

Nanodiscs were utilized in a number of solution[98,99,107,145] and solid-state[145,146] NMR studies of GPCRs, and in one solution NMR study, salipro nanoparticles were used.[104] Phospholipid bilayer systems have rightfully attracted attention as a membrane mimetic, due to improved protein longevity, the potential to work at higher temperatures, a more physiologically relevant environment, and the capability to stabilize ternary GPCR-G protein complexes.[147] A disadvantage is the relatively larger size of these systems, compared to detergent micelles, resulting in faster relaxation of the NMR signal. Therefore, for wider and more general NMR applications challenges remain, including low efficiency of receptor incorporation into nanodiscs, applicability to a wider range of GPCRs and the need for deuteration of the complex for ^{1}H-detected experiments.[98,148]

13.3 Receptor-observed NMR Spectroscopy

Crystal and cryo-EM structures of GPCRs mainly describe two stable states: the antagonist-bound inactive state and the active state comprising a ternary GPCR/agonist/transducer complex. Despite this structural evidence, mechanisms such as partial agonism and biased agonism remain largely obscure. To characterize such dynamic processes of GPCR activation, techniques like NMR are well suited. Over the past decade, NMR studies on several diffusible ligand-activated GPCRs have given insights into how a ligand modulates the conformational landscape, namely β_2-AR,[90–98,101–103] β_1-AR,[67–71,86,104,106] A_{2A}AR,[80–86,106,107] μOR,[108,109] M_2R,[110] α_{1A}-AR,[111] NTS_1[113,114] and leukotriene B4 receptor (BLT_2)[116] (Table 13.1).

In these studies, GPCRs were either labelled with isotopes such as ^{13}C-methyl,[83,90,95,97–99,104,108–111,113,116] ^{15}N-valine,[67–69] ^{15}N-tryptophan and ^{15}N-glycine[81,82,106] or chemically modified with ^{13}C-methyl[90,91,109] and ^{19}F probes.[71,80,84–86,92–94,96,101–103,149] While ^{19}F data are typically collected as one-dimensional experiments, ^{13}C and ^{15}N are collected in two-dimensional data sets: ^{13}C, ^{1}H HMQC (heteronuclear multiple quantum correlation),[150] ^{13}C,^{1}H SOFAST (Selective Optimized-Flip-Angle Short-Transient) HMQC[151] and ^{15}N, ^{1}H TROSY (transverse optimized spectroscopy).[152] Interpretation of conformational change can be obtained from monitoring chemical shift and/or peak intensity changes of the resolved NMR signals. For example, the chemical shift of side chain ^{13}C resonances may reflect on the side chain dihedral angles of residues;[153,154] the ^{1}H and ^{19}F chemical can be affected by the local environment, including solvent exposure and the ring current effects from neighbouring aromatic residues.[95,155] Moreover, utilizing relaxation dispersion CPMG,[71,84,101] saturation transfer[71,80,93,101] and EXSY,[80,93] the exchange rates and heterogeneity of the conformational populations making up the GPCR conformational equilibria can be determined. Thus, the signalling processes related to the dynamic equilibria of conformational ensembles can be observed by NMR in solution. Examples of these studies follow.

13.3.1 β_2-AR

β_2-AR is one of the most extensively studied GPCRs by any technique including NMR. β_2-AR was either isotopically labelled with $^{13}C^{\varepsilon}H_3$-Met[95,97,98] or chemically modified to introduce ^{13}C-methyl and ^{19}F probes.[90–94,96,101–103] $^{13}C^{\varepsilon}H_3$-Met-labelled β_2-AR resonances were observed using either SOFAST-HMQC[95,98] or HSQC[97] pulse sequences. In these studies, β_2-AR was found to exist in two slow-exchanging inactive states, based on the $^{13}C^{\varepsilon}H_3$ group of Met82$^{2.53}$ displaying two resonances in the apo- and antagonist-bound states.[95,97] Met82$^{2.53}$ is located near the "toggle switch" residue Trp$^{6.48}$ of the CWxP motif and the PIF motif in the transmission region (Figure 13.4). Since Met82$^{2.53}$ is distant (7–9 Å) from the aromatic component of the inverse agonist carazolol, the chemical shifts of the Met82$^{2.53}$ methyl likely reflect conformational changes in the transmission region rather than a direct interaction with the ligand. A pre-active conformation (active-like state or active intermediate state) was observed in the presence of agonists with varying efficacy. All transmembrane probes (Met82$^{2.53}$, Met215$^{5.54}$ and Met279$^{6.41}$) displayed ligand-efficacy correlated chemical shift changes[95] and the peak intensities of Met215$^{5.54}$ and Met279$^{6.41}$ decreased with the addition of agonist, suggesting that the exchange rate of the intracellular domain shifted to the intermediate timescale.[95,97] The exchange rate (k_{ex}) between the inactive and pre-active states was found to be 2500–4000 s^{-1} in DDM micelles, whereas in nanodiscs this exchange rate appeared to be slower (250–500 s^{-1}).[98] NMR analysis of a fully active, G_s protein-coupled β_2-AR conformation has only been achieved in the presence of the full agonist (BI-167107) and a G-protein mimetic nanobody (Nb80).[97] In this complex, the peak intensities of Met215$^{5.54}$ and Met279$^{6.41}$ were recovered along with large chemical shift changes, which indicates that the fully active state is characterized by a distinct and stable conformation.

To study the extent of lateral movement of intracellular regions of β_2-AR on extracellular ligand interaction, Liu and colleagues[92] attached TET CF_3 probes to Cys265$^{6.27}$ and Cys327$^{7.54}$ located at the intracellular tips of TM6 and TM7, respectively. Even in the apo-state ^{19}F-NMR signals were indicative of both probes adopting two pre-existing conformations, representing inactive and active states. The conformational equilibrium of ^{19}F-labelled Cys265$^{6.27}$ correlated with the pharmacological efficacy of agonists, indicating that G-protein signalling capacity depends on the extent of displacement of TM6. In contrast, β-arrestin-biased ligands predominantly affected the conformation of TM7 rather than TM6.

Manglik and colleagues[101] labelled Cys265$^{6.27}$ of β_2-AR with BTFA and deconvoluted ^{19}F spectra to fit four different conformations: two inactive conformations with a fast exchange rate of $k_{ex} > 2000\ s^{-1}$, one pre-active state and one active state. Apo-state and inverse-agonist bound β_2-AR were found to exist in a conformational equilibrium between the two inactive states. In the presence of a full agonist the pre-active state was observed to be the dominant conformation, with the two inactive states being less populated. This influence of ligand efficacy on the conformational equilibrium was also verified using electron paramagnetic resonance spectroscopy (double electron–electron resonance, or DEER) of nitroxide spin-labelled β_2-AR.[101] The fully active state was only dominantly populated in the presence of both a full agonist and a G-protein mimetic nanobody. This state is distinct from the pre-active state, which is consistent with $^{13}C^{\varepsilon}H_3$-Met-labelled β_2-AR NMR experiments.[97]

Backbone dynamics of β_2-AR reconstituted in LMNG micelles were investigated using an amino acid-specific [2,3,3-^{2}H,^{15}N]-Leu labelling approach.[100] Out of the 28 well-dispersed leucines, four showed chemical shift changes >0.4 ppm between the inverse and full agonist bound state. Three out of those four leucine residues (L212$^{5.51}$, L284$^{6.46}$ and L287$^{6.49}$) clustered around the PIF motif, while L287 corresponds to the "x" position in the

CWxP motif, suggesting that this region is most affected by ligand binding. Using an MTSL label attached to I334 on helix-8, the authors showed that paramagnetic resonance enhancements (PREs) were in good agreement for inverse agonist-bound receptor compared with calculated PREs of the corresponding, inverse agonist-bound crystal structure. By combining MTSL labels at different positions with a simulated annealing approach, it was further shown that even the full agonist bound receptor is more likely to resemble crystal structures with a closed than those with a more open intracellular cavity. However, most of the NMR peaks from the intracellular cavity region were broadened in the presence of the full agonist suggesting a dynamic ensemble.

An elegant segmental labelling approach was used to study the effect of phosphorylation by GRK2 on the β_2-AR C-terminus to elucidate conformational changes enabling β-arrestin binding.[99] The 63-residue [u-^{2}H,^{13}C,^{15}N]-labelled C-terminal fragment was fused to unlabelled β_2-AR *via* PTS prior to incorporation of the segmentally labelled receptor into nanodiscs. Assignment of the C-terminal amide groups was achieved using TROSY-HNCA and TROSY-HN(CO)CA spectra for both unphosphorylated and phosphorylated β_2-AR. ^{1}H,^{15}N-HSQC spectra suggested that both peptides remained flexible without specific secondary structure. Since TM-proximal residues G353–T360 were not observed in the ^{1}H,^{15}N-HSQC spectrum of the phosphorylated peptide, the authors turned to site-specific $^{13}CH_3$ labelling of threonine γ2 and isoleucine δ1 methyl groups [u-^{2}H, [^{13}C,^{1}H]-Thr$^{\gamma2}$,-Ile$^{\delta1}$]. Using the $^{13}C^{\gamma2}H_3$ group of T360 it was found that the TM-proximal region of the C-terminal peptide of β_2-AR is not fully disordered when phosphorylated and likely to adopt multiple conformations that exchange on the order of microseconds to milliseconds. Using further evidence from cross-saturation experiments and perturbations observed for $^{13}C^{\varepsilon}H_3$-labelled methionine resonances in the β_2-AR TM domain, the authors suggest that the TM-proximal region of the phosphorylated C-terminal peptide electrostatically interacts with the TM domain, leading to a conformational change that results in a narrowing of the cytoplasmic cavity favourable for β-arrestin binding (Figure 13.5).

Other NMR studies of β_2-AR utilized $^{13}CH_3$ or $^{13}CH_2D$ groups introduced through reductive methylation on lysine residues focused on positions at the extracellular surface and cytoplasmic side.[90,91] A methylated lysine in ICL2 showed that the conformational dynamics of this loop plays a vital role in G_s and G_{i1} protein coupling selectivity.

13.3.2 β_1-AR

Solution NMR studies of β_1-AR have focused on thermostabilized versions of turkey β_1-AR, isotopically labelled with ^{15}N-Val[67–69] and $^{13}C^{\varepsilon}H_3$-Met,[70] or chemically modified with ^{19}F-TET.[71] Spectra of valine ^{1}H–^{15}N resonances provided a global view of the response of β_1-AR dynamics to bound ligands of varying efficacy.[67] Of particular importance was the signal of Val226$^{5.57}$, a residue located at the intracellular end of TM5. For samples of β_1-AR bound to different ligands, the chemical shifts of the $^{1}H,^{15}N$ signal of Val226$^{5.57}$ fell on a "straight line" and the positions correlated with the biological efficacies of the corresponding ligands. The authors suggested that this reflects tilting of TM5 towards the active conformation.[67,69]

Three major exchanging conformations were observed for β_1-AR: an inactive conformation, a pre-active conformation and an active conformation.[69] The exchange rate between the inactive and pre-active conformations was found to be on the microsecond scale, while the exchange rate between the pre-active and active conformations was slower than ~5 ms. The slow exchange between pre-active and active conformations is dependent on two conserved tyrosine residues (Y$^{5.58}$, Y$^{7.53}$) that are essential for stabilizing the active state. High pressure has been applied to ^{15}N-Valine-labelled β_1-AR to observe effects on the populations of the inactive, pre-active and active states.[68] Under high pressure, most of the resonances of full agonist-bound β_1-AR shifted towards positions observed for the ternary complexed active state and are consistent with a reduced volume of the orthosteric binding pocket. Such a transition was not observed for antagonist-bound β_1-AR (Figure 13.5).

Exchange rates were also determined for CF_3-labelled A282$^{6.27}$C and C344$^{7.54}$, which are conformational sensors located on the cytoplasmic side of TM6 and within the NPxxY motif at TM7, respectively.[71] Two rapidly exchanging ($k_{ex} = 1500$–$3000\ s^{-1}$) inactive states were determined, suggesting a similar range as reported for β_2-AR.[101] The (invisible) pre-active state was identified using ^{19}F saturation transfer experiments and the exchange rate between inactive states and the pre-active state was at the slow sub-second timescale ($k_{ex} = 3$–$7\ s^{-1}$), which differs from the exchange rate determined for ^{15}N-Val-labelled β_1-AR (microsecond scale).[67] Addition of nanobody to the agonist-bound β_1-AR resulted in an additional active state, whose population correlated with agonist efficacy.

For $^{13}C^{\varepsilon}H_3$-Met-labelled β_1-AR, three methionines, Met90$^{2.53}$, Met223$^{5.54}$ and Met296$^{6.41}$, located in the same positions as Met82$^{2.53}$, Met215$^{5.54}$ and Met279$^{6.41}$ in β_2-AR, displayed significant chemical shift changes in response to ligand binding.[70] Like Met82$^{2.53}$ in β_2-AR, Met90$^{2.53}$ was observed to sample two conformations in the apo state. The two transmembrane methionine probes, Met223$^{5.54}$ and Met296$^{6.41}$, displayed ligand-efficacy correlated linear chemical shift changes, consistent with observations for Met215$^{5.54}$ and Met279$^{6.41}$ in β_2-AR. This similarity suggests a shared activation mechanism between the two β-AR subtypes. Peak broadening was observed for Met223$^{5.54}$ and Met296$^{6.41}$ in the presence of agonists, indicating that the exchange rate slowed to the intermediate (μs-to-ms) timescale. The addition of a G-protein mimetic nanobody increased the peak intensities of Met223$^{5.54}$ and Met296$^{6.41}$, suggesting reduced mobility of these probes. Furthermore, $^{13}C^{\varepsilon}H_3$-Met β_1-AR has been incorporated into a lipidic environment, using salipro nanoparticles,[104] showing no noticeable differences in chemical shifts when compared to spectra of LMNG reconstituted β_1-AR. However, significant peak broadening was observed in the salipro nanoparticle system, which is likely due to faster relaxation resulting from the larger overall particle size.

13.3.3 $A_{2A}AR$

^{19}F NMR has been extensively applied to BTFMA-labelled $A_{2A}AR$ at position $V229^{6.31}C$, which senses the conformational change of the intracellular tip of TM6.[84] An ensemble of two inactive and two active states in equilibrium was observed. The two inactive states were preferentially populated in the presence of an inverse agonist and showed millisecond exchange, consistent with β_2-AR and β_1-AR NMR studies. The two active states were stabilized by a partial and a full agonist, respectively, consistent with the notion of a conformational selection mechanism. It was proposed that partial agonism was achieved by conformational selection of a "less open" active state that compromises G-protein coupling. Utilizing the same ^{19}F approach, the

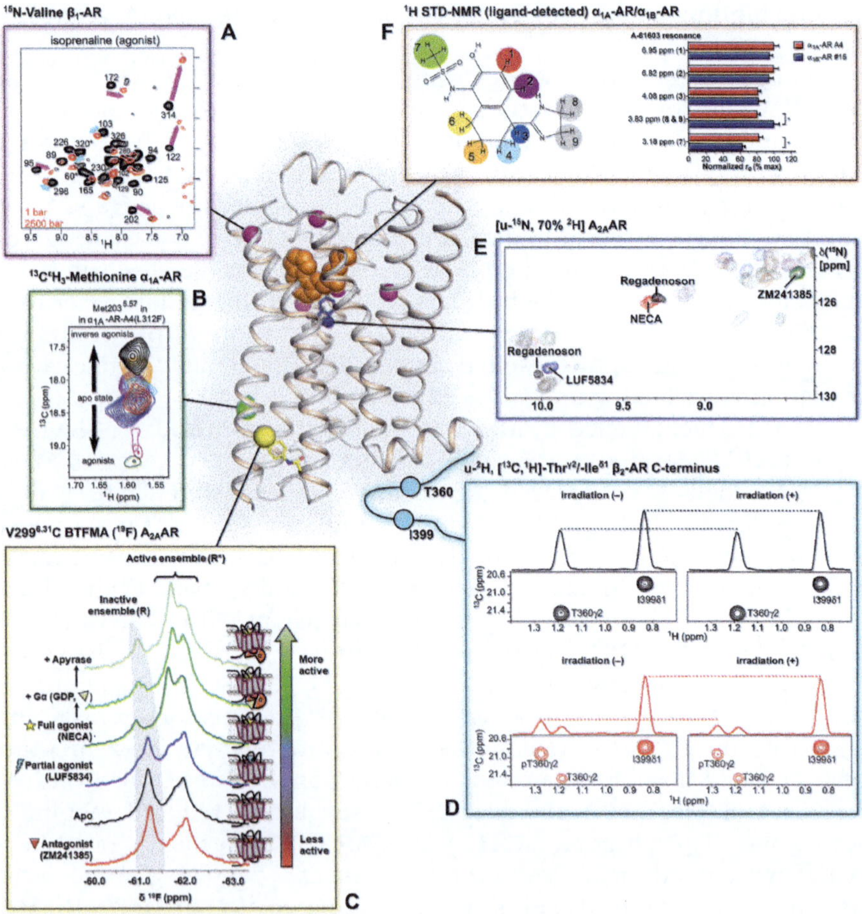

mechanism of negative allosteric modulation by Na^+ and positive allosteric modulation by divalent cations (Ca^{2+} and Mg^{2+}) was investigated.[85] Titration of Na^+ to apo $A_{2A}AR$ gradually shifted the equilibrium towards the

Figure 13.5 Experimental approaches for monitoring conformational changes in GPCRs by solution NMR spectroscopy. Examples of observing ligand-dependent conformational changes by solution NMR applied to a variety of class A GPCRs are shown. Locations of residues in these receptors used to investigate ligand-dependent conformational changes are mapped onto the modified structure of inactive β_2-AR (PDB ID: 2RH1) bound to the inverse agonist carazolol. Different experimental approaches include: (A) ^{15}N-Valine labelling, high-pressure and ^{15}N-TROSY to monitor the conformational change from pre-active to active state of agonist-bound β_1-AR incorporated into DM micelles. These results suggest that when GPCRs bind agonists in the G-protein-associated complex the orthosteric binding site is reduced in volume (positions indicated to shift in the spectrum are highlighted as magenta spheres in the structure). Reproduced from ref. 68 with permission from American Chemical Society, Copyright 2019. (B) $^{13}C^{\varepsilon}H_3$-Methionine labelling and ^{13}C SOFAST-HMQC were applied to α_{1A}-AR in DDM micelles. The residue $M203^{5.57}$ shows efficacy-dependent chemical shift changes (the position corresponding to the $^{13}C^{\varepsilon}H_3$ group of $M203^{5.57}$ is highlighted as a green sphere in the structure). Reproduced from ref. 111, https://doi.org/10.1074/jbc.RA120.012842 under the terms of the CC BY 4.0 license, https://creativecommons.org/licenses/by/4.0/. (C) ^{19}F NMR applied to BTFMA-labelled (at $V229^{6.31}C$) $A_{2A}AR$ in nanodiscs. These experiments monitored population changes in the presence of ligands of different efficacy, Gα and GDP (the position corresponding to the CF_3-moiety of BTFMA-labelled $V229^{6.31}C$ is highlighted as a yellow sphere in the structure). Reproduced from ref. 78 with permission from Elsevier, Copyright 2021. (D) Site-specific u-^2H, [^{13}C,^1H]-Thr$^{\gamma2}$,-Ile$^{\delta1}$ labelling of the C-terminal tail of β_2-AR in nanodiscs. Panels show the results of cross-saturation experiments for unphosphorylated (upper panel) and phosphorylated (lower panel) β_2-AR and consequently indicate that the C-terminal tail associates with the TMD when phosphorylated (cyan discs drawn onto the structure representative of the residue positions probed in the spectra). Reproduced from ref. 99, https://doi.org/10.1038/s41467-017-02632-8, under the terms of the CC BY 4.0 license, http://creativecommons.org/licenses/by/4.0/. (E) [u-^{15}N, 70% ^2H] labelling of $A_{2A}AR$, in LMNG micelles, and ^{15}N-TROSY were used to reveal conformational differences in partial agonist and full agonist bound receptor. In the spectrum the indole resonance of $W^{6.48}$ of the CWxP motif (shown as a blue sphere in the structure) was monitored. ZM241385 is an antagonist, NECA is a full agonist, whereas regadenoson and LUF5834 are partial agonists. Reproduced from ref. 106 with permission from Elsevier, Copyright 2021. (F) STD-NMR experiments reveal subtype-specific interactions for α_{1A}-AR and α_{1B}-AR. Unlabelled receptor was incorporated into DDM micelles and STD build-up experiments were acquired in the presence of A-61603, an α_{1A}-AR selective agonist. Reproduced from ref. 78 with permission from American Chemical Society, Copyright 2018.

inactive and the partial-agonist-stabilized active state. The addition of Ca^{2+} or Mg^{2+}, however, shifted the equilibrium toward the full-agonist-stabilized active state.

Using TET attached to A289$^{7.54}$C (a residue sequentially next to the tyrosine of the NPxxY motif; Figure 13.4), Susac and colleagues[80] found that $A_{2A}AR$ activation features aspects of both induced-fit and conformational selection mechanisms. In this study, the apo and inverse-agonist bound $A_{2A}AR$ simultaneously populated two states. In contrast to the BTFMA study,[84,85] agonist binding resulted in two entirely new agonist-stabilized conformers rather than shifting the equilibrium between two pre-existing states.

$A_{2A}AR$ has been subject to other labelling schemes such as u-^{15}N[81,82,106] and $^{13}C^{\delta}H_3$-Ile,[83] both in a deuterated background. For [u-^{15}N, 70% ^{2}H]-labelled $A_{2A}AR$,[81] the resolved ^{1}H,^{15}N resonances from six tryptophan indoles and eight glycine backbone groups provided probes for ligand-induced conformational changes throughout the receptor. This study highlighted the interplay between the "toggle switch" (CWxP) residue, Trp246$^{6.48}$ (Figure 13.4) and Asp52$^{2.50}$, which is important for coordinating the allosteric modulator sodium, and provided evidence that the response of Trp246$^{6.48}$ to ligands of various efficacy was still retained even when Asp52$^{2.50}$ was mutated to Asn to inactivate G-protein coupling. An additional study using [u-^{15}N, 70% ^{2}H]-labelled $A_{2A}AR$ showed that the resonance for Trp246$^{6.48}$ was distinctly different for partial and full agonists, indicating structural differences for these states that have not been resolved by X-ray crystallography (Figure 13.5).[106] In the [u-^{2}H,^{12}C, $^{13}C^{\delta1}H_3$-Ile] $A_{2A}AR$ NMR study,[83] inverse agonist binding reduced the conformational entropy of the receptor, which is consistent with studies on other GPCRs. Interestingly, in the presence of both agonist and G_s heterotrimer, most of the peaks were severely broadened, which was speculated to be due to the larger size of the macromolecular complex rather than a change in dynamics.

The conformational dynamics of $A_{2A}AR$ have also been subject to NMR studies in nanodiscs utilizing [u-^{2}H, α,β,β-^{2}H,$^{13}C^{\varepsilon}H_3$-Met] in a fully deuterated background.[107] The authors demonstrated that agonist-bound $A_{2A}AR$ exists in equilibrium of multiple NPxxY states with exchange rates faster than the chemical shift difference. The addition of docosahexaenoic acid (DHA) to the nanodiscs shifted the conformational equilibrium toward the G protein-coupled fully active state.

13.3.4 μOR

Exposed lysine residues of μOR were modified by reductive methylation and subsequent introduction of ^{13}C-dimethylamines to generate ε-N[$^{13}CH_3$]$_2$-Lys probes.[109] It was found that either agonist or G-protein mimetic nanobody alone are incapable of completely stabilizing the fully active conformation of the G protein-coupling interface (TM5 and TM6), which is in line with NMR spectroscopic observations made for other class A GPCRs.

Okude and colleagues[108] incorporated $^{13}C^{\varepsilon}H_3$-Met-labelled μOR into nanodiscs and found that the conformational changes of TM3, TM5, TM6 and TM7 at the intracellular side are coupled, describing an equilibrium between multiple closed and open conformations. Evidence of biased agonism was established using the chemical shift of Met245$^{5.49}$, for which the addition of balanced and biased agonists resulted in distinct chemical shifts that correlated with the bias factor (the ratio of β-arrestin signalling efficacy to G-protein signalling efficacy).

13.3.5 Other GPCRs, M_2R, α_{1A}-AR, NTS_1 and BLT_2

Other GPCRs subject to conformational analysis by NMR were M_2R, α_{1A}-AR, NTS_1 and BLT_2. M_2R was selectively labelled with $^{13}C^{\varepsilon}H_3$-Met and reconstituted with LMNG and CHS (cholesterol hemisuccinate).[110] The methionine resonances in M_2R frequently showed multiple peaks in equilibrium, thus representing a more complex conformational landscape compared to other GPCRs. Interestingly, agonists of M_2R induced two distinct chemical shifts for M202$^{5.54}$ (located in a region commonly reporting on ligand efficacy) assigned to two conformations: one conformation was observed in the presence of two super-agonists and a partial agonist, while the other conformation was stabilized by a full agonist and another partial agonist. The addition of an active-state stabilizing nanobody further stabilized the conformation observed for the super-agonist class of agonists, while the addition of the nanobody to the other agonist-bound M_2R shifted the resonances to completely new positions.

$^{13}C^{\varepsilon}H_3$-Met-labelled thermostabilized α_{1A}-AR was studied in DDM micelles.[111] In this study, three methionine residues, M292$^{6.55}$, M115$^{3.41}$ and M203$^{5.57}$, served as probes sensing the conformational changes of the ligand-binding pocket, the PIF motif and the E/DRY motif, respectively (Figure 13.4). Although the receptor was signalling incompetent, binding of ligands with varying efficacy modulated the chemical shifts of particularly M203$^{5.57}$ (a E/DRY motif sensor) in an efficacy- and not affinity-dependent manner (Figure 13.5). This efficacy-dependent equilibrium of active and inactive states resulted in a "straight line" terminated by chemical shifts induced by inverse agonists on one and full agonists on the other side. Following this analysis, back mutations were performed to restore signalling competency. Importantly, back-mutation of the stabilizing mutation N322$^{7.49}$K restored the NPxxY microswitch. The chemical shifts still correlated with efficacy, but M115$^{3.41}$ now adopted three states: an (inverse agonist) inactive, an intermediate (basal) and an active state that progressively shifted from partial to full agonist.

Conformational dynamics of bacterially expressed thermostabilized NTS_1[36] reconstituted in DDM micelles has been investigated by two different ^{13}C-labelling approaches, $^{13}C^{\varepsilon}H_3$-Met[113] and chemical labelling of cysteine residues by S-methyl-^{13}C-methanethiosulfonate (MMTS).[114] ^{1}H,^{13}C HMQC spectra of signalling incompetent variants, that were acquired in the presence of agonist and antagonist, show little change in $^{13}C,^{1}H$ chemical shift or peak intensity.[114] However, for signalling competent (L167$^{3.50}$R back-mutated) variants chemical shift perturbations and line-broadening effects consistent with different conformational states were observed. Although resonances of $^{13}C^{\varepsilon}H_3$-Met-labelled NTS_1 have not been assigned, distinct subpopulations for active and inactive states are observed for individual methionines.[113] $^{13}CH_3$-labelling of exposed cysteines, particularly Cys172$^{3.55}$ and the introduced mutation, V300$^{6.27}$C (the equivalent residue previously ^{19}F-labelled in β_2-AR[92,101] and β_1-AR[71]) show four distinct states, two corresponding to inactive and two active states. Chemical shift perturbation analysis indicated that although the antagonist stabilizes the inactive state of the receptor, agonist alone is not sufficient for full activation of the receptor, consistent with previous data on other GPCRs. The addition of a peptide derived from the C-terminal helix5 of $G\alpha_{i,1}$ pushes the receptor toward the most active conformation with line broadening consistent with millisecond–microsecond timescale motion. Similar to the α_{1A}-AR study,[111] a series of back-mutations combined with monitoring of active/inactive populations in $^{13}C,^{1}H$ HMQC spectra facilitated the rescue of functional variants that could still be expressed in bacteria.

BLT_2 was labelled with [u-$^{2}H,^{13}C^{\delta1}H_3$-Ile], [u-$^{2}H,^{13}C^{\varepsilon}H_3$-Met] and reconstituted into nanodiscs.[116] The methyl resonances of Ile229$^{6.40}$ and Met105$^{3.35}$ from BLT_2 were observed to populate four peaks in the unliganded BLT_2, hinting at a complex landscape that comprises at least four different conformations. Agonists and the cholesterol derivative CHS perturbed the observed chemical shifts, suggesting a conformational coupling between agonists and the membrane lipid composition, which is thought to play an important role in GPCR signalling.

13.3.6 Complexes of GPCRs and Heterotrimeric Gαβγ Proteins

Current NMR studies on GPCRs have revealed the dynamic plasticity required for signal transduction which is not observed in crystal and cryo-EM structures. These studies revealed that GPCRs exist in equilibria between multiple conformations that are simultaneously populated[156] (Figure 13.6). In the absence of a bound ligand, GPCRs are highly dynamic, sampling inactive and active conformational states, which provides a mechanistic basis for basal receptor activity. GPCR activation is thought to predominantly operate *via* a conformational selection mechanism, whereby agonist binding shifts the receptor conformational equilibrium towards active states, thus increasing the probability of G-protein or β-arrestin coupling; whereas the binding of inverse agonists or antagonists would enrich the populations of inactive conformations. MD simulations on β_2-AR[13,157] support the co-existence of multiple states where ligand binding shifts the equilibrium to certain states, increasing the likelihood of the receptor being inactivated or activated.

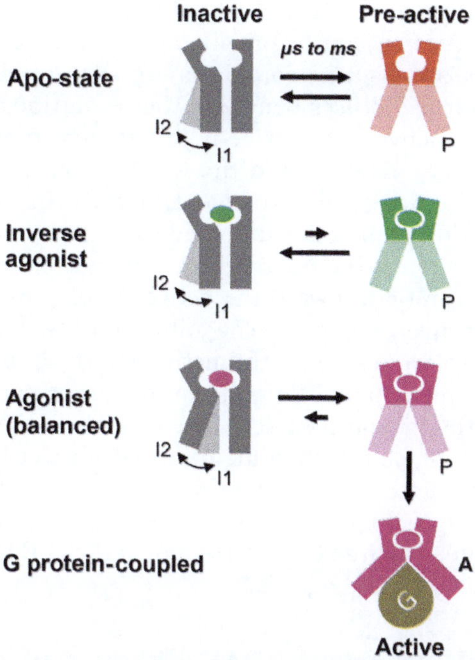

Figure 13.6 Dynamic model of GPCR activation.[108] GPCRs co-exist as an ensemble of open and closed states. Apo-state receptor (red) equally populates inactive (I) and pre-active (P) states. For inactive states, it has been proposed that they may represent an equilibrium between a closed (I1) and an open (I2) ionic lock.[101] In this case, inverse-agonist binding (green) would favour the closed ionic lock state. Agonist binding (magenta) shifts the equilibrium toward the pre-active state (P). G-protein binding induces the fully active state (A) of the receptor.

G protein-coupled active states of GPCRs in reported NMR studies were mainly achieved with the help of G-protein mimetics (nanobodies, miniG and G-protein C-terminal peptide). Coupling to heterotrimeric G proteins has recently been shown to be a more complex process. Using single-molecule fluorescence resonance energy transfer (smFRET) imaging it was shown that GDP-bound G_s protein couples to β_2-AR forming an intermediate state that is distinct from the nucleotide-free G_s protein stabilized active conformation.[158] To further understand the mechanism of activation, NMR observations of direct coupling of heterotrimeric $G_{\alpha\beta\gamma}$ proteins to GPCR will be required. However, this is experimentally difficult due to severe line-broadening of ^{1}H-observed signals in solution NMR because of the formation of slow tumbling macromolecular complexes. Solid-state NMR, in principle, may remove the limits in terms of protein size to explore intermediate states of GPCRs with GDP-bound G-protein coupling and G-protein selectivity, while G-protein/β-arrestin bias mechanisms could also be investigated. Nevertheless, nanodisc-incorporated A_{2A}AR was labelled with ^{19}F at V229$^{6.31}$C to investigate coupling to a G-protein heterotrimer composed of shortened-Gα, β1 and γ2 subunits (Figure 13.5).[86] This study provided a direct comparison to previous work in detergent,[84,85] revealing that the dynamics of $A_{2A}AR$ in nanodiscs are slower, while sharpened resonances observed in nanodiscs benefited the resolution of additional states (one inactive, three active). The probe is considered to measure the state of the ionic lock between E228$^{6.30}$ and the E/DRY motif, where locked is consistent with the inactive state and unlocked with the active states. In the absence of G protein, the inactive state had the same chemical shift for both apo- and inverse agonist-bound receptor. On addition of G protein, the resonances shifted upfield toward the active states (unlocked E/DRY), even in the presence of inverse agonist. The addition of GDP-bound G protein to the apo-receptor selected one of the active states, implying that it represented the pre-coupled state. The addition of full or partial agonists to the apo-receptor/G protein complex revealed that the other active states are signalling competent states that facilitate GDP release. Consistent with single-molecule fluorescence experiments,[158] the study further suggested that among the signalling competent states, one corresponds to a full agonist bound state, which releases GDP more efficiently than the other state which is favoured by partial agonists.

13.4 Ligand-observed NMR Studies of GPCRs

Although GPCR structural studies are mostly focused on the structure and dynamics of the 7TM bundle, investigating the ligand-binding mode and ligand conformation in the bound state is critical for the development of drugs with respect to affinity, potency, efficacy, and subtype specificity. NMR is a powerful technique for studying protein–ligand interactions and is particularly well-suited for characterizing interactions of weak-binding ligands. To date, several ligand-observed NMR experiments have been applied for studying GPCR–ligand interactions including transferred nuclear Overhauser effect spectroscopy (tr-NOE),[159,160] saturation-transfer difference (STD) NMR,[161] interligand Overhauser effect (ILOE)[162] and interligand NOEs for pharmacophore mapping (INPHARMA).[163] All these experiments are based on the same principle, the nuclear Overhauser effect (NOE), but have different applications and provide different information about ligand–GPCR interactions.[164,165]

13.4.1 Small-molecule Ligands

NMR has been utilized for fragment-based screening (FBS) using thermostabilized GPCRs.[166] Ligand-observed NMR methods, such as STD-NMR and target-immobilized NMR screening (TINS) have been applied to $A_{2A}AR$[167,168] and $β_1$-AR[169] where fragments were identified that inhibited the binding of tritium-labelled ligands with IC_{50} (the half-maximal inhibitory concentration) values ranging from 5 μM to 5 mM. STD-NMR has been applied to determine the bound conformations and binding epitopes of ligands on

CXCR4,[170] Human Sweet Receptor,[171] α_{1A}- and α_{1B}-AR[78] and others[172–175] (Figure 13.5).

An advantage these methods offer is their capability to monitor relatively weak interactions of low-molecular-weight compounds with GPCRs, making them suitable for fragment-based lead discovery approaches. The crucial stage in FBS is lead optimization, in which the structural information is needed to decide how to link, merge or grow the fragments.[176] Such information is provided by tr-NOESY, ILOE and INPHARMA experiments. Tr-NOESY experiments have led to the determination of the bound conformations of ligands at the leukotriene receptor BLT_2,[172] pituitary adenylyl cyclase-activating polypeptide receptor[177] and κ-opioid receptor (κOR)[178] and INPHARMA has been applied for the selection of docking models for human $A_{2A}AR$,[179] GPR40[180] and the α1-adrenreceptors (α_{1A}-AR and α_{1B}-AR).[181] As ILOE and INPHARMA are essentially identical experiments it is not easy to distinguish the two. ILOEs are interligand NOEs generated between two simultaneously bound ligands proximal to each other. INPHARMA are interligand NOEs between two ligands that sequentially bind to a similar site. In the latter case the NOEs are transferred from the protons of the first to the subsequently binding ligand *via* the receptor. A powerful way of distinguishing between ILOE and INPHARMA is to fully or partially deuterate the receptor which quenches the INPHARMA-generated NOEs.[181]

13.4.2 Peptide Ligands

Many endogenous peptides target GPCRs and most of them are of clinical importance. 118 GPCRs are activated by endogenous peptide ligands including all members of the secretin family (class B) and several class A GPCRs.[182] While small-molecule ligands often have a relatively rigid structure and bind to a specific binding pocket, peptide ligands, especially those that are intrinsically disordered (IDPs), are likely to follow an induced fit mechanism of binding.[183] It is proposed that peptide ligands search through different trajectories of binding, starting from the lipidic environment around the receptor, to pass an energy barrier required for adopting the final bound-state conformation.[184–187] Thus, this mechanism of interaction by peptide ligands with GPCRs paves a way for finding new allosteric binding sites through capturing encounter complexes and provides critical information about the recognition of ligands by GPCRs.

The earliest study of an interaction of a peptide ligand with its cognate GPCR by solution-state NMR used tr-NOE to determine the structure of the bound state of the α-helical pituitary adenylate cyclase activating polypeptide PACAP21, in complex with the PACAP receptor.[177] PACAP21 binds to the detergent micelle before it finds its path to the orthosteric binding pocket of the receptor. Such a "reduction in dimensionality" was found to be a critical step which assists diffusion of the peptide to the receptor after which only the N-terminal (residues 3–7 with residues 1–3 remaining extended) adopts a beta-coil structure which is critical for specific receptor binding and activation.

In contrast to ligands of class B GPCRs, the peptide ligands of class A receptors are mostly disordered in solution.[188–193] Dynorphin, an opioid peptide, is one of the few peptides in this class that has been investigated for conformational changes in the bound state by solution state NMR.[178] Like other opioid peptides, dynorphin consists of an N-terminal message sequence responsible for receptor activation and a C-terminal address sequence for receptor subtype recognition and binding. Using tr-NOE and $^{15}N\text{-}R_1$ and $^{15}N\text{-}R_2$ analysis, dynorphin was shown to adopt a helical conformation in the middle region (L5–R9) upon binding to κOR. Interestingly, the C- and N-termini of the peptide remained highly mobile even in the bound state, highlighting the importance of peptide dynamics in binding and activation of GPCRs. Furthermore, the binding to micelles is supposed to play a critical role in the binding and conformational change of dynorphin. Importantly, dynorphin binds strongly ($K_D = 200$ nM) to κOR, while a condition for tr-NOE experiments is a fast off-rate which is typically not observed for such affinities. However, as noted[178] fast off-rates under such conditions of strong binding are possible if the on-rates are faster than the diffusion limit ($k_{on} > 10^{-10}$ $M^{-1}\,s^{-1}$) which are potentially achievable through phenomena such as electrostatically driven interactions.[194]

A similar approach combined with molecular dynamics (MD) simulations was applied to investigate the structure and dynamics of ghrelin in the bound state.[195] It was observed that the N-terminal part of ghrelin forms a hydrophobic core with the C8 chain of an octanoyl moiety that enables efficient ligand binding to the receptor. Moreover, MD simulations suggested that the lipid environment plays a role in conformational dynamics and ligand binding of GPCRs, supporting the hypothesis of reduction in dimensionality for the interaction of peptide ligands with GPCRs.[177] However, the ghrelin structures derived from solid-state NMR propose a different structure for bound ghrelin including α-helical structures in the middle and at the C-terminus of the peptide.[196] These discrepancies perhaps point to the dynamic nature of peptide ligands even in the bound state.

MD and NMR studies have indicated that the C-terminus of neurotensin (NT) which is buried in the orthosteric binding pocket of NTS_1 is likely to remain dynamic.[197,198] MD simulations of the NT8-13/NTS_1 complex suggest that the side chain of Y11 toggles between $\chi 1$ gauche(+) and gauche(−) conformations and that this motion correlates with the rotameric state of $Trp^{6.48}$ of the conserved CWxP motif (Figure 13.4) in NTS_1.[198] A combination of MD and STD-NMR epitope mapping of the low-affinity NT analogue NT10–13 supports the presence of two different conformations for the side chain of Y11 of NT in the receptor bound state.[197]

Most peptide GPCR families, such as Neurotensin receptors, Kinin receptors and neuropeptide Y receptors, comprise several subtypes with different physiological activities. Ligand-detected NMR experiments can decipher the determinants of subtype selectivity at atomic resolution.[78,199]

Solid-state NMR studies complemented with MD simulation and docking experiments on bradykinin receptors (B1R and B2R) and their high-affinity ligands have revealed the complex network of ligand–receptor interactions deriving the subtype selectivity of bradykinin (BK) and desArg^{10}-Kallidin (DAKD).[199] Although BK and DAKD differ by only two residues, they are highly selective for either B2R or B1R, respectively. These studies illustrate that peptide GPCRs subtype selectivity is encoded by a combination of conformational changes, electrostatic and hydrophobic interactions.

13.5 Conclusions

In this chapter we have highlighted the progress and contribution of solution NMR spectroscopy to investigations of GPCRs. Development of overexpression methods, isotope labelling strategies and membrane mimetic systems have expanded the number of GPCR targets that can be studied and the experimental approaches that can be applied. NMR has revealed that the conformational landscape of GPCRs is complex and that conformations can reach beyond simple active and inactive state models. Similarly, NMR has enabled investigations of the ligand-binding mechanism and pathways to the orthosteric binding site of GPCRs, and the conformational diversity of the bound ligand. However, the biological and pharmaceutical importance of conformational dynamics in GPCR signalling remains to be fully explored but will undoubtably have a significant impact on the discovery of signalling-specific drugs.

References

1. R. J. Lefkowitz, S. Cotecchia, P. Samama and T. Costa, *Trends Pharmacol. Sci.*, 1993, **14**, 303–307.
2. C. A. Milano, P. C. Dolber, H. A. Rockman, R. A. Bond, M. E. Venable, L. F. Allen and R. J. Lefkowitz, *Proc. Natl. Acad. Sci. U. S. A.*, 1994, **91**, 10109–10113.
3. K. A. Berg and W. P. Clarke, *Int. J. Neuropsychopharmacol.*, 2018, **21**, 962–977.
4. V. Cabrera, T. Cabrera Vera, J. Vanhauwe, T. Thomas, M. Medkova, A. Preininger, M. Mazzoni and H. E. Hamm, *Endocr. Rev.*, 2003, **24**, 765–781.
5. K. Kristiansen, *Pharmacol. Ther.*, 2004, **103**, 21–80.
6. N. Mizuno and H. Itoh, *Neurosignals*, 2009, **17**, 42–54.
7. S. Lutz, A. Freichel Blomquist, Y. Yang, U. Rümenapp, K. Jakobs, M. Schmidt and T. Wieland, *J. Biol. Chem.*, 2005, **280**, 11134–11139.
8. S. Lutz, A. Shankaranarayanan, C. Coco, M. Ridilla, M. Nance, C. Vettel, D. Baltus, C. Evelyn, R. Neubig, T. Wieland and J. J. G. Tesmer, *Science*, 2007, **318**, 1923–1927.
9. A. Ridley, *Trends Cell Biol.*, 2006, **16**, 522–529.

10. S. M. Khan, R. Sleno, S. Gora, P. Zylbergold, J. P. Laverdure, J. C. Labbe, G. J. Miller and T. E. Hebert, *Pharmacol. Rev.*, 2013, **65**, 545–577.
11. A. V. Smrcka, *Cell. Mol. Life Sci.*, 2008, **65**, 2191–2214.
12. R. J. Lefkowitz, *Angew. Chem., Int. Ed. Engl.*, 2013, **52**, 6366–6378.
13. N. R. Latorraca, A. J. Venkatakrishnan and R. O. Dror, *Chem. Rev.*, 2017, **117**, 139–155.
14. K. Palczewski, T. Kumasaka, T. Hori, C. A. Behnke, H. Motoshima, B. A. Fox, I. Le Trong, D. C. Teller, T. Okada, R. E. Stenkamp, M. Yamamoto and M. Miyano, *Science*, 2000, **289**, 739–745.
15. S. G. F. Rasmussen, H.-J. Choi, D. Rosenbaum, T. Kobilka, F. Thian, P. Edwards, M. Burghammer, V. R. P. Ratnala, R. Sanishvili, R. Fischetti, G. F. X. Schertler, W. Weis and B. K. Kobilka, *Nature*, 2007, **450**, 383–387.
16. D. M. Rosenbaum, V. Cherezov, M. A. Hanson, S. G. Rasmussen, F. S. Thian, T. S. Kobilka, H. J. Choi, X. J. Yao, W. I. Weis, R. C. Stevens and B. K. Kobilka, *Science*, 2007, **318**, 1266–1273.
17. V. Cherezov, D. M. Rosenbaum, M. A. Hanson, S. G. Rasmussen, F. S. Thian, T. S. Kobilka, H. J. Choi, P. Kuhn, W. I. Weis, B. K. Kobilka and R. C. Stevens, *Science*, 2007, **318**, 1258–1265.
18. D. M. Thal, B. Sun, D. Feng, V. Nawaratne, K. Leach, C. C. Felder, M. G. Bures, D. A. Evans, W. I. Weis, P. Bachhawat, T. S. Kobilka, P. M. Sexton, B. K. Kobilka and A. Christopoulos, *Nature*, 2016, **531**, 335–340.
19. W. Huang, A. Manglik, A. J. Venkatakrishnan, T. Laeremans, E. N. Feinberg, A. L. Sanborn, H. E. Kato, K. E. Livingston, T. S. Thorsen, R. C. Kling, S. Granier, P. Gmeiner, S. M. Husbands, J. R. Traynor, W. I. Weis, J. Steyaert, R. O. Dror and B. K. Kobilka, *Nature*, 2015, **524**, 315–321.
20. A. C. Kruse, A. M. Ring, A. Manglik, J. Hu, K. Hu, K. Eitel, H. Hubner, E. Pardon, C. Valant, P. M. Sexton, A. Christopoulos, C. C. Felder, P. Gmeiner, J. Steyaert, W. I. Weis, K. C. Garcia, J. Wess and B. K. Kobilka, *Nature*, 2013, **504**, 101–106.
21. J. F. White, N. Noinaj, Y. Shibata, J. Love, B. Kloss, F. Xu, J. Gvozdenovic-Jeremic, P. Shah, J. Shiloach, C. G. Tate and R. Grisshammer, *Nature*, 2012, **490**, 508–513.
22. K. Haga, A. Kruse, H. Asada, T. Yurugi Kobayashi, M. Shiroishi, C. Zhang, W. Weis, T. Okada, B. Kobilka, T. Haga and T. Kobayashi, *Nature*, 2012, **482**, 547–551.
23. J. Garcia-Nafria and C. G. Tate, *Mol. Cell. Endocrinol.*, 2019, **488**, 1–13.
24. D. P. Staus, H. Hu, M. J. Robertson, A. L. W. Kleinhenz, L. M. Wingler, W. D. Capel, N. R. Latorraca, R. J. Lefkowitz and G. Skiniotis, *Nature*, 2020, **579**, 297–302.
25. Y. Lee, T. Warne, R. Nehmé, S. Pandey, H. Dwivedi-Agnihotri, M. Chaturvedi, P. C. Edwards, J. García-Nafría, A. G. W. Leslie, A. K. Shukla and C. G. Tate, *Nature*, 2020, **583**, 862–866.

26. W. Huang, M. Masureel, Q. Qu, J. Janetzko, A. Inoue, H. E. Kato, M. J. Robertson, K. C. Nguyen, J. S. Glenn, G. Skiniotis and B. K. Kobilka, *Nature*, 2020, **579**, 303–308.
27. G. Pándy-Szekeres, C. Munk, T. M. Tsonkov, S. Mordalski, K. Harpsøe, A. S. Hauser, A. J. Bojarski and D. E. Gloriam, *Nucleic Acids Res.*, 2017, **46**, D440–D446.
28. D. Wacker, G. Fenalti, M. A. Brown, V. Katritch, R. Abagyan, V. Cherezov and R. C. Stevens, *J. Am. Chem. Soc.*, 2010, **132**, 11443–11445.
29. A. Jazayeri, M. Rappas, A. J. H. Brown, J. Kean, J. C. Errey, N. J. Robertson, C. Fiez-Vandal, S. P. Andrews, M. Congreve, A. Bortolato, J. S. Mason, A. H. Baig, I. Teobald, A. S. Dore, M. Weir, R. M. Cooke and F. H. Marshall, *Nature*, 2017, **546**, 254–258.
30. T. Hua, K. Vemuri, S. P. Nikas, R. B. Laprairie, Y. Wu, L. Qu, M. Pu, A. Korde, S. Jiang, J. H. Ho, G. W. Han, K. Ding, X. Li, H. Liu, M. A. Hanson, S. Zhao, L. M. Bohn, A. Makriyannis, R. C. Stevens and Z. J. Liu, *Nature*, 2017, **547**, 468–471.
31. B. E. Krumm, J. F. White, P. Shah and R. Grisshammer, *Nat. Commun.*, 2015, **6**, 7895.
32. T. Warne, R. Moukhametzianov, J. G. Baker, R. Nehme, P. C. Edwards, A. G. Leslie, G. F. Schertler and C. G. Tate, *Nature*, 2011, **469**, 241–244.
33. G. Lebon, T. Warne, P. C. Edwards, K. Bennett, C. J. Langmead, A. G. Leslie and C. G. Tate, *Nature*, 2011, **474**, 521–525.
34. D. M. Rosenbaum, C. Zhang, J. A. Lyons, R. Holl, D. Aragao, D. H. Arlow, S. G. Rasmussen, H. J. Choi, B. T. Devree, R. K. Sunahara, P. S. Chae, S. H. Gellman, R. O. Dror, D. E. Shaw, W. I. Weis, M. Caffrey, P. Gmeiner and B. K. Kobilka, *Nature*, 2011, **469**, 236–240.
35. J. Park, P. Scheerer, K. Hofmann, H.-W. Choe and O. P. Ernst, *Nature*, 2008, **454**, 183–187.
36. P. Egloff, M. Hillenbrand, C. Klenk, A. Batyuk, P. Heine, S. Balada, K. M. Schlinkmann, D. J. Scott, M. Schutz and A. Pluckthun, *Proc. Natl. Acad. Sci. U. S. A.*, 2014, **111**, E655–662.
37. Y. Peng, J. D. McCorvy, K. Harpsoe, K. Lansu, S. Yuan, P. Popov, L. Qu, M. Pu, T. Che, L. F. Nikolajsen, X. P. Huang, Y. Wu, L. Shen, W. E. Bjorn-Yoshimoto, K. Ding, D. Wacker, G. W. Han, J. Cheng, V. Katritch, A. A. Jensen, M. A. Hanson, S. Zhao, D. E. Gloriam, B. L. Roth, R. C. Stevens and Z. J. Liu, *Cell*, 2018, **172**, 719–730, e714.
38. H. Zhang, G. W. Han, A. Batyuk, A. Ishchenko, K. L. White, N. Patel, A. Sadybekov, B. Zamlynny, M. T. Rudd, K. Hollenstein, A. Tolstikova, T. A. White, M. S. Hunter, U. Weierstall, W. Liu, K. Babaoglu, E. L. Moore, R. D. Katz, J. M. Shipman, M. Garcia-Calvo, S. Sharma, P. Sheth, S. M. Soisson, R. C. Stevens, V. Katritch and V. Cherezov, *Nature*, 2017, **544**, 327–332.
39. P. Scheerer, J. H. Park, P. W. Hildebrand, Y. J. Kim, N. Krauss, H. W. Choe, K. P. Hofmann and O. P. Ernst, *Nature*, 2008, **455**, 497–U430.

40. S. G. Rasmussen, H. J. Choi, J. J. Fung, E. Pardon, P. Casarosa, P. S. Chae, B. T. Devree, D. M. Rosenbaum, F. S. Thian, T. S. Kobilka, A. Schnapp, I. Konetzki, R. K. Sunahara, S. H. Gellman, A. Pautsch, J. Steyaert, W. I. Weis and B. K. Kobilka, *Nature*, 2011, **469**, 175–180.
41. T. Che, S. Majumdar, S. A. Zaidi, P. Ondachi, J. D. McCorvy, S. Wang, P. D. Mosier, R. Uprety, E. Vardy, B. E. Krumm, G. W. Han, M.-Y. Lee, E. Pardon, J. Steyaert, X.-P. Huang, R. T. Strachan, A. R. Tribo, G. W. Pasternak, F. I. Carroll, R. C. Stevens, V. Cherezov, V. Katritch, D. Wacker and B. L. Roth, *Cell*, 2018, **172**, 55–67, e15.
42. A. M. Ring, A. Manglik, A. C. Kruse, M. D. Enos, W. I. Weis, K. C. Garcia and B. K. Kobilka, *Nature*, 2013, **502**, 575–579.
43. M. Masureel, Y. Zou, L.-P. Picard, E. van der Westhuizen, J. P. Mahoney, J. P. G. L. M. Rodrigues, T. J. Mildorf, R. O. Dror, D. E. Shaw, M. Bouvier, E. Pardon, J. Steyaert, R. K. Sunahara, W. I. Weis, C. Zhang and B. K. Kobilka, *Nat. Chem. Biol.*, 2018, **14**, 1059–1066.
44. B. Carpenter, R. Nehme, T. Warne, A. G. Leslie and C. G. Tate, *Nature*, 2016, **536**, 104–107.
45. C. J. Tsai, F. Pamula, R. Nehme, J. Muhle, T. Weinert, T. Flock, P. Nogly, P. C. Edwards, B. Carpenter, T. Gruhl, P. Ma, X. Deupi, J. Standfuss, C. G. Tate and G. F. X. Schertler, *Sci. Adv.*, 2018, **4**, eaat7052.
46. J. Garcia-Nafria, R. Nehme, P. C. Edwards and C. G. Tate, *Nature*, 2018, **558**, 620–623.
47. A. Qiao, S. Han, X. Li, Z. Li, P. Zhao, A. Dai, R. Chang, L. Tai, Q. Tan, X. Chu, L. Ma, T. S. Thorsen, S. Reedtz-Runge, D. Yang, M.-W. Wang, P. M. Sexton, D. Wootten, F. Sun, Q. Zhao and B. Wu, *Science*, 2020, **367**, 1346–1352.
48. J. I. Mobbs, M. J. Belousoff, K. G. Harikumar, S. J. Piper, X. Xu, S. G. B. Furness, H. Venugopal, A. Christopoulos, R. Danev, D. Wootten, D. M. Thal, L. J. Miller and P. M. Sexton, *PLoS Biol.*, 2021, **19**, e3001295.
49. Q. Liu, D. Yang, Y. Zhuang, T. I. Croll, X. Cai, A. Dai, X. He, J. Duan, W. Yin, C. Ye, F. Zhou, B. Wu, Q. Zhao, H. E. Xu, M. W. Wang and Y. Jiang, *Nat. Chem. Biol.*, 2021, **17**, 1238–1244.
50. J. A. Harris, B. Faust, A. B. Gondin, M. A. Damgen, C. M. Suomivuori, N. A. Veldhuis, Y. Cheng, R. O. Dror, D. M. Thal and A. Manglik, *Nat. Chem. Biol.*, 2021, **18**, 109–115.
51. Y. Kang, X. E. Zhou, X. Gao, Y. He, W. Liu, A. Ishchenko, A. Barty, T. A. White, O. Yefanov, G. W. Han, Q. Xu, P. W. de Waal, J. Ke, M. H. Tan, C. Zhang, A. Moeller, G. M. West, B. D. Pascal, N. Van Eps, L. N. Caro, S. A. Vishnivetskiy, R. J. Lee, K. M. Suino-Powell, X. Gu, K. Pal, J. Ma, X. Zhi, S. Boutet, G. J. Williams, M. Messerschmidt, C. Gati, N. A. Zatsepin, D. Wang, D. James, S. Basu, S. Roy-Chowdhury, C. E. Conrad, J. Coe, H. Liu, S. Lisova, C. Kupitz, I. Grotjohann, R. Fromme, Y. Jiang, M. Tan, H. Yang, J. Li, M. Wang, Z. Zheng, D. Li, N. Howe, Y. Zhao, J. Standfuss, K. Diederichs, Y. Dong, C. S. Potter, B. Carragher, M. Caffrey, H. Jiang, H. N. Chapman, J. C. Spence, P. Fromme, U. Weierstall, O. P. Ernst, V. Katritch, V. V. Gurevich,

P. R. Griffin, W. L. Hubbell, R. C. Stevens, V. Cherezov, K. Melcher and H. E. Xu, *Nature*, 2015, **523**, 561–567.
52. W. Yin, Z. Li, M. Jin, Y. L. Yin, P. W. de Waal, K. Pal, Y. Yin, X. Gao, Y. He, J. Gao, X. Wang, Y. Zhang, H. Zhou, K. Melcher, Y. Jiang, Y. Cong, X. Edward Zhou, X. Yu and H. Eric Xu, *Cell Res.*, 2019, **29**, 971–983.
53. X. Deupi and J. Standfuss, *Curr. Opin. Struct. Biol.*, 2011, **21**, 541–551.
54. S. Filipek, *Curr. Opin. Struct. Biol.*, 2019, **55**, 114–120.
55. B. Trzaskowski, D. Latek, S. Yuan, U. Ghoshdastider, A. Debinski and S. Filipek, *Curr. Med. Chem.*, 2012, **19**, 1090–1109.
56. L. H. Zhao, S. Ma, I. Sutkeviciute, D. D. Shen, X. E. Zhou, P. W. de Waal, C. Y. Li, Y. Kang, L. J. Clark, F. G. Jean-Alphonse, A. D. White, D. Yang, A. Dai, X. Cai, J. Chen, C. Li, Y. Jiang, T. Watanabe, T. J. Gardella, K. Melcher, M. W. Wang, J. P. Vilardaga, H. E. Xu and Y. Zhang, *Science*, 2019, **364**, 148–153.
57. J. A. Ballesteros and H. Weinstein, *Methods Neurosci.*, 1995, **25**, 366–428.
58. D. M. Rosenbaum, S. G. Rasmussen and B. K. Kobilka, *Nature*, 2009, **459**, 356–363.
59. T. W. Schwartz, T. M. Frimurer, B. Holst, M. M. Rosenkilde and C. E. Elling, *Annu. Rev. Pharmacol. Toxicol.*, 2006, **46**, 481–519.
60. L. Shi, G. Liapakis, R. Xu, F. Guarnieri, J. A. Ballesteros and J. A. Javitch, *J. Biol. Chem.*, 2002, **277**, 40989–40996.
61. J. Schoppe, J. Ehrenmann, C. Klenk, P. Rucktooa, M. Schutz, A. S. Dore and A. Pluckthun, *Nat. Commun.*, 2019, **10**, 17.
62. A. Manglik and A. C. Kruse, *Biochemistry*, 2017, **56**, 5628–5634.
63. B. Carpenter and C. G. Tate, *Curr. Opin. Struct. Biol.*, 2017, **45**, 124–132.
64. M. Kovermann, P. Rogne and M. Wolf-Watz, *Q. Rev. Biophys.*, 2016, **49**, e6.
65. A. Sekhar and L. E. Kay, *Annu. Rev. Biophys.*, 2019, **48**, 297–319.
66. M. J. Serrano-Vega, F. Magnani, Y. Shibata and C. G. Tate, *Proc. Natl. Acad. Sci. U. S. A.*, 2008, **105**, 877–882.
67. S. Isogai, X. Deupi, C. Opitz, F. M. Heydenreich, C. J. Tsai, F. Brueckner, G. F. Schertler, D. B. Veprintsev and S. Grzesiek, *Nature*, 2016, **530**, 237–241.
68. L. A. Abiko, A. Grahl and S. Grzesiek, *J. Am. Chem. Soc.*, 2019, **141**, 16663–16670.
69. A. Grahl, L. A. Abiko, S. Isogai, T. Sharpe and S. Grzesiek, *Nat. Commun.*, 2020, **11**, 2216.
70. A. S. Solt, M. J. Bostock, B. Shrestha, P. Kumar, T. Warne, C. G. Tate and D. Nietlispach, *Nat. Commun.*, 2017, **8**, 1795.
71. J. N. Frei, R. W. Broadhurst, M. J. Bostock, A. Solt, A. J. Y. Jones, F. Gabriel, A. Tandale, B. Shrestha and D. Nietlispach, *Nat. Commun.*, 2020, **11**, 669.
72. P. Rossler, D. Mayer, C. J. Tsai, D. B. Veprintsev, G. F. X. Schertler and A. D. Gossert, *Molecules*, 2020, **25**, 5984.
73. I. Dodevski and A. Pluckthun, *J. Mol. Biol.*, 2011, **408**, 599–615.

74. C. A. Sarkar, I. Dodevski, M. Kenig, S. Dudli, A. Mohr, E. Hermans and A. Pluckthun, *Proc. Natl. Acad. Sci. U. S. A.*, 2008, **105**, 14808–14813.
75. M. Schutz, J. Schoppe, E. Sedlak, M. Hillenbrand, G. Nagy-Davidescu, J. Ehrenmann, C. Klenk, P. Egloff, L. Kummer and A. Pluckthun, *Sci. Rep.*, 2016, **6**, 21508.
76. D. J. Scott, L. Kummer, P. Egloff, R. A. Bathgate and A. Pluckthun, *Biochim. Biophys. Acta*, 2014, **1838**, 2817–2824.
77. D. J. Scott and A. Pluckthun, *J. Mol. Biol.*, 2013, **425**, 662–677.
78. K. J. Yong, T. M. Vaid, P. J. Shilling, F. J. Wu, L. M. Williams, M. Deluigi, A. Pluckthun, R. A. D. Bathgate, P. R. Gooley and D. J. Scott, *ACS Chem. Biol.*, 2018, **13**, 1090–1102.
79. J. Boivineau, M. Haffke and V. P. Jaakola, *Methods Mol. Biol.*, 2020, **2127**, 63–80.
80. L. Susac, M. T. Eddy, T. Didenko, R. C. Stevens and K. Wuthrich, *Proc. Natl. Acad. Sci. U. S. A.*, 2018, **115**, 12733–12738.
81. M. T. Eddy, M. Y. Lee, Z. G. Gao, K. L. White, T. Didenko, R. Horst, M. Audet, P. Stanczak, K. M. McClary, G. W. Han, K. A. Jacobson, R. C. Stevens and K. Wuthrich, *Cell*, 2018, **172**, 68–80 e12.
82. M. Eddy, Z.-G. Gao, P. Mannes, N. Patel, K. Jacobson, V. Katritch, R. Stevens and K. Wüthrich, *J. Am. Chem. Soc.*, 2018, **140**, 8228–8235.
83. L. D. Clark, I. Dikiy, K. Chapman, K. E. Rodstrom, J. Aramini, M. V. LeVine, G. Khelashvili, S. G. Rasmussen, K. H. Gardner and D. M. Rosenbaum, *eLife*, 2017, **6**, e28505.
84. L. Ye, N. Van Eps, M. Zimmer, O. P. Ernst and R. S. Prosser, *Nature*, 2016, **533**, 265–268.
85. L. Ye, C. Neale, A. Sljoka, B. Lyda, D. Pichugin, N. Tsuchimura, S. T. Larda, R. Pomes, A. E. Garcia, O. P. Ernst, R. K. Sunahara and R. S. Prosser, *Nat. Commun.*, 2018, **9**, 1372.
86. S. K. Huang, A. Pandey, D. P. Tran, N. L. Villanueva, A. Kitao, R. K. Sunahara, A. Sljoka and R. S. Prosser, *Cell*, 2021, **184**, 1884–1894 e1814.
87. T. Shimamura, M. Shiroishi, S. Weyand, H. Tsujimoto, G. Winter, V. Katritch, R. Abagyan, V. Cherezov, W. Liu, G. Han, T. Kobayashi, R. Stevens and S. Iwata, *Nature*, 2011, **475**, 65–70.
88. P. J. Shilling, F. Bumbak, D. J. Scott, R. A. D. Bathgate and P. R. Gooley, *Sci. Rep.*, 2017, **7**, 1094.
89. Y. Suzuki, T. Ogasawara, Y. Tanaka, H. Takeda, T. Sawasaki, M. Mogi, S. Liu and K. Maeyama, *Front. Pharmacol.*, 2018, **9**, 38.
90. M. P. Bokoch, Y. Zou, S. G. Rasmussen, C. W. Liu, R. Nygaard, D. M. Rosenbaum, J. J. Fung, H. J. Choi, F. S. Thian, T. S. Kobilka, J. D. Puglisi, W. I. Weis, L. Pardo, R. S. Prosser, L. Mueller and B. K. Kobilka, *Nature*, 2010, **463**, 108–112.
91. X. Ma, Y. Hu, H. Batebi, J. Heng, J. Xu, X. Liu, X. Niu, H. Li, P. W. Hildebrand, C. Jin and B. K. Kobilka, *Proc. Natl. Acad. Sci. U. S. A.*, 2020, **117**, 23096–23105.

92. J. J. Liu, R. Horst, V. Katritch, R. C. Stevens and K. Wuthrich, *Science*, 2012, **335**, 1106–1110.
93. R. Horst, J. Liu, R. Stevens and K. Wüthrich, *Angew. Chem.*, 2013, **125**, 10962–10965.
94. M. T. Eddy, T. Didenko, R. C. Stevens and K. Wuthrich, *Structure*, 2016, **24**, 2190–2197.
95. Y. Kofuku, T. Ueda, J. Okude, Y. Shiraishi, K. Kondo, M. Maeda, H. Tsujishita and I. Shimada, *Nat. Commun.*, 2012, **3**, 1045.
96. K. Chung, T. Kim, A. Manglik, R. Alvares, B. Kobilka and R. S. Prosser, *J. Biol. Chem.*, 2012, **287**, 36305–36311.
97. R. Nygaard, Y. Zou, R. O. Dror, T. J. Mildorf, D. H. Arlow, A. Manglik, A. C. Pan, C. W. Liu, J. J. Fung, M. P. Bokoch, F. S. Thian, T. S. Kobilka, D. E. Shaw, L. Mueller, R. S. Prosser and B. K. Kobilka, *Cell*, 2013, **152**, 532–542.
98. Y. Kofuku, T. Ueda, J. Okude, Y. Shiraishi, K. Kondo, T. Mizumura, S. Suzuki and I. Shimada, *Angew. Chem., Int. Ed. Engl.*, 2014, **53**, 13376–13379.
99. Y. Shiraishi, M. Natsume, Y. Kofuku, S. Imai, K. Nakata, T. Mizukoshi, T. Ueda, H. Iwai and I. Shimada, *Nat. Commun.*, 2018, **9**, 194.
100. S. Imai, T. Yokomizo, Y. Kofuku, Y. Shiraishi, T. Ueda and I. Shimada, *Nat. Chem. Biol.*, 2020, **16**, 430–439.
101. A. Manglik, T. H. Kim, M. Masureel, C. Altenbach, Z. Yang, D. Hilger, M. T. Lerch, T. S. Kobilka, F. S. Thian, W. L. Hubbell, R. S. Prosser and B. K. Kobilka, *Cell*, 2015, **161**, 1101–1111.
102. D. P. Staus, L. M. Wingler, D. Pichugin, R. S. Prosser and R. J. Lefkowitz, *J. Biol. Chem.*, 2019, **294**, 13218–13223.
103. D. Staus, R. Strachan, A. Manglik, B. Pani, A. Kahsai, T. Kim, L. Wingler, S. Ahn, A. Chatterjee, A. Masoudi, A. Kruse, E. Pardon, J. Steyaert, W. Weis, R. S. Prosser, B. Kobilka, T. Costa and R. J. Lefkowitz, *Nature*, 2016, **535**, 448–452.
104. C.-T. Chien, L. Helfinger, M. Bostock, A. Solt, Y. Tan and D. Nietlispach, *J. Am. Chem. Soc.*, 2017, **139**, 14829–14832.
105. L. A. Abiko, M. Rogowski, A. Gautier, G. Schertler and S. Grzesiek, *J. Biomol. NMR*, 2021, **75**, 25–38.
106. M. T. Eddy, B. T. Martin and K. Wuthrich, *Structure*, 2021, **29**, 170–176, e173.
107. T. Mizumura, K. Kondo, M. Kurita, Y. Kofuku, M. Natsume, S. Imai, Y. Shiraishi, T. Ueda and I. Shimada, *Sci. Adv.*, 2020, **6**, eaay8544.
108. J. Okude, T. Ueda, Y. Kofuku, M. Sato, N. Nobuyama, K. Kondo, Y. Shiraishi, T. Mizumura, K. Onishi, M. Natsume, M. Maeda, H. Tsujishita, T. Kuranaga, M. Inoue and I. Shimada, *Angew. Chem., Int. Ed. Engl.*, 2015, **54**, 15771–15776.
109. R. Sounier, C. Mas, J. Steyaert, T. Laeremans, A. Manglik, W. Huang, B. K. Kobilka, H. Demene and S. Granier, *Nature*, 2015, **524**, 375–378.
110. J. Xu, Y. Hu, J. Kaindl, P. Risel, H. Hubner, S. Maeda, X. Niu, H. Li, P. Gmeiner, C. Jin and B. K. Kobilka, *Mol. Cell*, 2019, **75**, 53–65, e57.

111. F. J. Wu, L. M. Williams, A. Abdul-Ridha, A. Gunatilaka, T. M. Vaid, M. Kocan, A. R. Whitehead, M. D. W. Griffin, R. A. D. Bathgate, D. J. Scott and P. R. Gooley, *J. Biol. Chem.*, 2020, **295**, 7404–7417.
112. M. Schuster, M. Deluigi, M. Pantic, S. Vacca, C. Baumann, D. J. Scott, A. Pluckthun and O. Zerbe, *Biochim. Biophys. Acta, Biomembr.*, 2020, **1862**, 183354.
113. F. Bumbak, A. C. Keen, N. J. Gunn, P. R. Gooley, R. A. D. Bathgate and D. J. Scott, *Biochim. Biophys. Acta*, 2018, **1860**, 1372–1383.
114. I. Goba, D. Goricanec, D. Schum, M. Hillenbrand, A. Pluckthun and F. Hagn, *ChemBioChem*, 2021, **22**, 139–146.
115. S. H. Park, B. B. Das, F. Casagrande, Y. Tian, H. J. Nothnagel, M. Chu, H. Kiefer, K. Maier, A. A. De Angelis, F. M. Marassi and S. J. Opella, *Nature*, 2012, **491**, 779–783.
116. M. Casiraghi, M. Damian, E. Lescop, E. Point, K. Moncoq, N. Morellet, D. Levy, J. Marie, E. Guittet, J. L. Baneres and L. J. Catoire, *J. Am. Chem. Soc.*, 2016, **138**, 11170–11175.
117. X. Wang, D. Liu, L. Shen, F. Li, Y. Li, L. Yang, T. Xu, H. Tao, D. Yao, L. Wu, K. Hirata, L. M. Bohn, A. Makriyannis, X. Liu, T. Hua, Z. J. Liu and J. Wang, *J. Am. Chem. Soc.*, 2021, **143**, 16320–16325.
118. B. Franke, C. Opitz, S. Isogai, A. Grahl, L. Delgado, A. D. Gossert and S. Grzesiek, *J. Biomol. NMR*, 2018, **71**, 173–184.
119. C. Opitz, S. Isogai and S. Grzesiek, *J. Biomol. NMR*, 2015, **62**, 373–385.
120. C. Di Pietrantonio, A. Pandey, J. Gould, A. Hasabnis and R. S. Prosser, *Methods Enzymol.*, 2019, **615**, 103–130.
121. M. le Maire, P. Champeil and J. V. Moller, *Biochim. Biophys. Acta*, 2000, **1508**, 86–111.
122. P. A. McDonnell and S. J. Opella, *J. Magn. Reson.*, 1993, **102**, 120–125.
123. P. S. Chae, S. G. Rasmussen, R. R. Rana, K. Gotfryd, R. Chandra, M. A. Goren, A. C. Kruse, S. Nurva, C. J. Loland, Y. Pierre, D. Drew, J. L. Popot, D. Picot, B. G. Fox, L. Guan, U. Gether, B. Byrne, B. Kobilka and S. H. Gellman, *Nat. Methods*, 2010, **7**, 1003–1008.
124. K. H. Cho, M. Husri, A. Amin, K. Gotfryd, H. J. Lee, J. Go, J. W. Kim, C. J. Loland, L. Guan, B. Byrne and P. S. Chae, *Analyst*, 2015, **140**, 3157–3163.
125. T. VanAken, S. Foxall-VanAken, S. Castleman and S. Ferguson-Miller, *Methods Enzymol.*, 1986, **125**, 27–35.
126. L. Frey, N. A. Lakomek, R. Riek and S. Bibow, *Angew. Chem., Int. Ed. Engl.*, 2017, **56**, 380–383.
127. M. B. Jones and J. C. Garrison, *Anal. Biochem.*, 1999, **268**, 126–133.
128. M. Orwick-Rydmark, T. Arnold and D. Linke, *Curr. Protoc. Protein Sci.*, 2016, **84**, 481–4835.
129. C. R. Sanders and R. S. Prosser, *Structure*, 1998, **6**, 1227–1234.
130. A. D. Bangham, M. M. Standish and J. C. Watkins, *J. Mol. Biol.*, 1965, **13**, 238–252.
131. T. J. Knowles, R. Finka, C. Smith, Y. P. Lin, T. Dafforn and M. Overduin, *J. Am. Chem. Soc.*, 2009, **131**, 7484–7485.

132. M. Zoonens, F. Giusti, F. Zito and J. L. Popot, *Biochemistry*, 2007, **46**, 10392–10404.
133. J. I. Yeh, S. Du, A. Tortajada, J. Paulo and S. Zhang, *Biochemistry*, 2005, **44**, 16912–16919.
134. C. L. McGregor, L. Chen, N. C. Pomroy, P. Hwang, S. Go, A. Chakrabartty and G. G. Prive, *Nat. Biotechnol.*, 2003, **21**, 171–176.
135. H. Tao, S. C. Lee, A. Moeller, R. S. Roy, F. Y. Siu, J. Zimmermann, R. C. Stevens, C. S. Potter, B. Carragher and Q. Zhang, *Nat. Methods*, 2013, **10**, 759–761.
136. T. H. Bayburt and S. G. Sligar, *Protein Sci.*, 2003, **12**, 2476–2481.
137. A. N. Larsen, K. K. Sorensen, N. T. Johansen, A. Martel, J. J. Kirkensgaard, K. J. Jensen, L. Arleth and S. R. Midtgaard, *Soft Matter*, 2016, **12**, 5937–5949.
138. J. Frauenfeld, R. Loving, J. P. Armache, A. F. Sonnen, F. Guettou, P. Moberg, L. Zhu, C. Jegerschold, A. Flayhan, J. A. Briggs, H. Garoff, C. Low, Y. Cheng and P. Nordlund, *Nat. Methods*, 2016, **13**, 345–351.
139. M. L. Carlson, J. W. Young, Z. Zhao, L. Fabre, D. Jun, J. Li, J. Li, H. S. Dhupar, I. Wason, A. T. Mills, J. T. Beatty, J. S. Klassen, I. Rouiller and F. Duong, *eLife*, 2018, **7**, e34085.
140. F. Hagn, M. L. Nasr and G. Wagner, *Nat. Protoc.*, 2018, **13**, 79–98.
141. J. Miehling, D. Goricanec and F. Hagn, *ChemBioChem*, 2018, **19**, 1927–1933.
142. M. L. Nasr, D. Baptista, M. Strauss, Z. J. Sun, S. Grigoriu, S. Huser, A. Pluckthun, F. Hagn, T. Walz, J. M. Hogle and G. Wagner, *Nat. Methods*, 2017, **14**, 49–52.
143. T. K. Ritchie, Y. V. Grinkova, T. H. Bayburt, I. G. Denisov, J. K. Zolnerciks, W. M. Atkins and S. G. Sligar, *Methods Enzymol.*, 2009, **464**, 211–231.
144. P. Xu, S. Huang, H. Zhang, C. Mao, X. E. Zhou, X. Cheng, I. A. Simon, D. D. Shen, H. Y. Yen, C. V. Robinson, K. Harpsoe, B. Svensson, J. Guo, H. Jiang, D. E. Gloriam, K. Melcher, Y. Jiang, Y. Zhang and H. E. Xu, *Nature*, 2021, **592**, 469–473.
145. I. G. Denisov and S. G. Sligar, *Chem. Rev.*, 2017, **117**, 4669–4713.
146. J. M. Boettcher, R. L. Davis-Harrison, M. C. Clay, A. J. Nieuwkoop, Y. Z. Ohkubo, E. Tajkhorshid, J. H. Morrissey and C. M. Rienstra, *Biochemistry*, 2011, **50**, 2264–2273.
147. M. Zhang, M. Gui, Z. F. Wang, C. Gorgulla, J. J. Yu, H. Wu, Z. J. Sun, C. Klenk, L. Merklinger, L. Morstein, F. Hagn, A. Pluckthun, A. Brown, M. L. Nasr and G. Wagner, *Nat. Struct. Mol. Biol.*, 2021, **28**, 258–267.
148. A. Viegas, T. Viennet and M. Etzkorn, *Biol. Chem.*, 2016, **397**, 1335–1354.
149. S. Chen, M. Lu, D. Liu, L. Yang, C. Yi, L. Ma, H. Zhang, Q. Liu, T. M. Frimurer, M.-W. Wang, T. W. Schwartz, R. C. Stevens, B. Wu, K. Wüthrich and Q. Zhao, *Nat. Commun.*, 2019, **10**, 638.
150. V. Tugarinov, P. M. Hwang, J. E. Ollerenshaw and L. E. Kay, *J. Am. Chem. Soc.*, 2003, **125**, 10420–10428.

151. P. Schanda, E. Kupce and B. Brutscher, *J. Biomol. NMR*, 2005, **33**, 199–211.
152. K. Pervushin, R. Riek, G. Wider and K. Wuthrich, *Proc. Natl. Acad. Sci. U. S. A.*, 1997, **94**, 12366–12371.
153. G. Butterfoss, E. DeRose, S. Gabel, L. Perera, J. Krahn, G. Mueller, X. Zheng and R. London, *J. Biomol. NMR*, 2010, **48**, 31–47.
154. R. London, B. Wingad and G. Mueller, *J. Am. Chem. Soc.*, 2008, **130**, 11097–11105.
155. D. Liu and K. Wuthrich, *J. Biomol. NMR*, 2016, **65**, 1–5.
156. I. Shimada, T. Ueda, Y. Kofuku, M. T. Eddy and K. Wuthrich, *Nat. Rev. Drug Discovery*, 2019, **18**, 59–82.
157. R. O. Dror, D. H. Arlow, P. Maragakis, T. J. Mildorf, A. C. Pan, H. Xu, D. W. Borhani and D. E. Shaw, *Proc. Natl. Acad. Sci. U. S. A.*, 2011, **108**, 18684–18689.
158. G. G. Gregorio, M. Masureel, D. Hilger, D. S. Terry, M. Juette, H. Zhao, Z. Zhou, J. M. Perez-Aguilar, M. Hauge, S. Mathiasen, J. A. Javitch, H. Weinstein, B. K. Kobilka and S. C. Blanchard, *Nature*, 2017, **547**, 68–73.
159. G. M. Clore and A. M. Gronenborn, *J. Magn. Reson.*, 1982, **48**, 402–417.
160. G. M. Clore and A. M. Gronenborn, *J. Magn. Reson.*, 1983, **53**, 423–442.
161. M. Mayer and B. Meyer, *J. Am. Chem. Soc.*, 2001, **123**, 6108–6117.
162. D. Li, E. F. DeRose and R. E. London, *J. Biomol. NMR*, 1999, **15**, 71–76.
163. J. Orts, C. Griesinger and T. Carlomagno, *J. Magn. Reson.*, 2009, **200**, 64–73.
164. W. Becker, K. C. Bhattiprolu, N. Gubensak and K. Zangger, *Chem. Phys. Chem.*, 2018, **19**, 895–906.
165. C. B. Post, *Curr. Opin. Struct. Biol.*, 2003, **13**, 581–588.
166. S. P. Andrews, G. A. Brown and J. A. Christopher, *ChemMedChem*, 2014, **9**, 256–275.
167. D. Chen, J. C. Errey, L. H. Heitman, F. H. Marshall, A. P. Ijzerman and G. Siegal, *ACS Chem. Biol.*, 2012, **7**, 2064–2073.
168. S. Igonet, C. Raingeval, E. Cecon, M. Pucic-Bakovic, G. Lauc, O. Cala, M. Baranowski, J. Perez, R. Jockers, I. Krimm and A. Jawhari, *Sci. Rep.*, 2018, **8**, 8142.
169. M. Congreve, R. L. Rich, D. G. Myszka, F. Figaroa, G. Siegal and F. H. Marshall, *Methods Enzymol.*, 2011, **493**, 115–136.
170. B. D. Cox, A. K. Mehta, J. O. DiRaddo, D. C. Liotta, L. J. Wilson and J. P. Snyder, *Biochem. Biophys. Res. Commun.*, 2015, **466**, 28–32.
171. F. M. Assadi-Porter, M. Tonelli, E. Maillet, K. Hallenga, O. Benard, M. Max and J. L. Markley, *J. Am. Chem. Soc.*, 2008, **130**, 7212–7213.
172. L. J. Catoire, M. Damian, F. Giusti, A. Martin, C. van Heijenoort, J. L. Popot, E. Guittet and J. L. Baneres, *J. Am. Chem. Soc.*, 2010, **132**, 9049–9057.
173. Y. Kofuku, C. Yoshiura, T. Ueda, H. Terasawa, T. Hirai, S. Tominaga, M. Hirose, Y. Maeda, H. Takahashi, Y. Terashima, K. Matsushima and I. Shimada, *J. Biol. Chem.*, 2009, **284**, 35240–35250.

174. C. Yoshiura, Y. Kofuku, T. Ueda, Y. Mase, M. Yokogawa, M. Osawa, Y. Terashima, K. Matsushima and I. Shimada, *J. Am. Chem. Soc.*, 2010, **132**, 6768–6777.
175. C. Yoshiura, T. Ueda, Y. Kofuku, M. Matsumoto, J. Okude, K. Kondo, Y. Shiraishi and I. Shimada, *J. Biomol. NMR*, 2015, **63**, 333–340.
176. P. J. Hajduk and J. Greer, *Nat. Rev. Drug Discovery*, 2007, **6**, 211–219.
177. H. Inooka, T. Ohtaki, O. Kitahara, T. Ikegami, S. Endo, C. Kitada, K. Ogi, H. Onda, M. Fujino and M. Shirakawa, *Nat. Struct. Biol.*, 2001, **8**, 161–165.
178. C. O'Connor, K. L. White, N. Doncescu, T. Didenko, B. L. Roth, G. Czaplicki, R. C. Stevens, K. Wuthrich and A. Milon, *Proc. Natl. Acad. Sci. U. S. A.*, 2015, **112**, 11852–11857.
179. K. Fredriksson, P. Lottmann, S. Hinz, I. Onila, A. Shymanets, C. Harteneck, C. E. Muller, C. Griesinger and T. E. Exner, *Angew. Chem., Int. Ed. Engl.*, 2017, **56**, 5750–5754.
180. S. Bartoschek, T. Klabunde, E. Defossa, V. Dietrich, S. Stengelin, C. Griesinger, T. Carlomagno, I. Focken and K. U. Wendt, *Angew. Chem., Int. Ed. Engl.*, 2010, **49**, 1426–1429.
181. T. M. Vaid, D. K. Chalmers, D. J. Scott and P. R. Gooley, *Chemistry*, 2020, **26**, 11796–11805.
182. S. R. Foster, A. S. Hauser, L. Vedel, R. T. Strachan, X. P. Huang, A. C. Gavin, S. D. Shah, A. P. Nayak, L. M. Haugaard-Kedstrom, R. B. Penn, B. L. Roth, H. Brauner-Osborne and D. E. Gloriam, *Cell*, 2019, **179**, 895–908 e821.
183. K. Sugase, H. J. Dyson and P. E. Wright, *Nature*, 2007, **447**, 1021–1025.
184. R. Schwyzer, *Biochemistry*, 1986, **25**, 4281–4286.
185. R. Schwyzer, *EMBO J.*, 1987, **6**, 2255–2259.
186. R. Schwyzer, *Biopolymers*, 1995, **37**, 5–16.
187. R. Schwyzer, *J. Mol. Recognit.*, 1995, **8**, 3–8.
188. G. Da Costa, A. Bondon, O. Delalande, L. Mouret and J. P. Monti, *J. Biomol. Struct. Dyn.*, 2013, **31**, 809–817.
189. A. Gayen, S. K. Goswami and C. Mukhopadhyay, *Biochim. Biophys. Acta*, 2011, **1808**, 127–139.
190. H. Heise, S. Luca, B. L. de Groot, H. Grubmuller and M. Baldus, *Biophys. J.*, 2005, **89**, 2113–2120.
191. A. Kukol, *Vitam. Horm.*, 2008, **77**, 1–12.
192. S. A. Monks, G. Karagianis, G. J. Howlett and R. S. Norton, *J. Biomol. NMR*, 1996, **8**, 379–390.
193. M. R. Tessmer and D. A. Kallick, *Biochemistry*, 1997, **36**, 1971–1981.
194. L. J. Catoire, M. Damian, M. Baaden, E. Guittet and J. L. Baneres, *J. Biomol. NMR*, 2011, **50**, 191–195.
195. G. Ferre, M. Louet, O. Saurel, B. Delort, G. Czaplicki, C. M'Kadmi, M. Damian, P. Renault, S. Cantel, L. Gavara, P. Demange, J. Marie, J. A. Fehrentz, N. Floquet, A. Milon and J. L. Baneres, *Proc. Natl. Acad. Sci. U. S. A.*, 2019, **116**, 17525–17530.

196. B. J. Bender, G. Vortmeier, S. Ernicke, M. Bosse, A. Kaiser, S. Els-Heindl, U. Krug, A. Beck-Sickinger, J. Meiler and D. Huster, *Structure*, 2019, **27**, 537–544 e534.
197. F. Bumbak, T. Thomas, B. J. Noonan-Williams, T. M. Vaid, F. Yan, A. R. Whitehead, S. Bruell, M. Kocan, X. Tan, M. A. Johnson, R. A. D. Bathgate, D. K. Chalmers, P. R. Gooley and D. J. Scott, *ACS Pharmacol. Transl. Sci.*, 2020, **3**, 690–705.
198. X. Cong, S. Fiorucci and J. Golebiowski, *J. Chem. Theory Comput.*, 2018, **14**, 4467–4473.
199. L. Joedicke, J. Mao, G. Kuenze, C. Reinhart, T. Kalavacherla, H. R. A. Jonker, C. Richter, H. Schwalbe, J. Meiler, J. Preu, H. Michel and C. Glaubitz, *Nat. Chem. Biol.*, 2018, **14**, 284–290.

CHAPTER 14

Characterising Intrinsically Disordered Proteins Using NMR Spectroscopy and MD Simulations

G. T. HELLER,[†] L. YU[†] AND D. F. HANSEN*

Department of Structural and Molecular Biology, Division of Biosciences, University College London, London WC1E 6BT, UK
*Email: d.hansen@ucl.ac.uk

14.1 Introduction

Intrinsically disordered proteins (IDPs) and proteins with intrinsically disordered regions (IDRs) form a class of dynamic biomolecules that do not adopt a single three-dimensional structure, but instead exist as ensembles of rapidly interconverting states.[1–3] The conformations of IDPs differ from those predicted by statistical random coils, as IDPs can adopt transiently populated secondary structures and long-range tertiary interactions crucial for their functions.[4] This high degree of structural plasticity provides various functional advantages, such as the ability to bind multiple interaction partners including DNA, RNA, and structured and unstructured proteins,[5] sensitivity to local environments, and accessibility to various post-translational modifications in signalling pathways. It is also emerging that disorder

[†]These authors contributed equally.

New Developments in NMR No. 26
NMR Spectroscopy for Probing Functional Dynamics at Biological Interfaces
Edited by Anirban Bhunia, Hanudatta S. Atreya and Neeraj Sinha

Published by the Royal Society of Chemistry, www.rsc.org

plays important roles in the formation of both dynamic, reversible assemblies through liquid–liquid phase separation and solid fibrillar structures.[1]

Bioinformatics predicts that approximately one-third of the human proteome is comprised of intrinsic disorder,[6] yet current understandings of disordered proteins lag behind those of folded proteins, whose tertiary structures can now be accurately predicted from sequence alone using advances in machine and deep learning.[7–9] Intrinsically disordered proteins and IDRs not only extend our understanding of structural biology but also push the limits of techniques able to characterise extreme conformational heterogeneity.

In this chapter, we focus on recent advances in two highly complementary techniques, namely nuclear magnetic resonance spectroscopy (NMR) and molecular dynamics simulations (MD), which offer unparalleled accuracy and precision, respectively, into the structures and functions of IDPs. As an experimental technique, NMR requires minimal perturbations to the native structural ensembles and measurements can even be performed within cellular environments.[10] After reviewing classical methods and their limitations for characterising highly dynamic systems, we highlight recent applications of NMR to characterise conformational exchange of IDPs, including Carr–Purcell–Meiboom–Gill (CPMG) relaxation dispersion and chemical exchange saturation transfer (CEST), which have revealed important mechanistic details on interactions of IDPs with structured binding partners. Subsequently we discuss MD, a highly complementary technique to NMR which, in principle, can resolve individual conformations within a heterogeneous structural ensemble, whereas NMR alone can only provide ensemble-averaged measurements, both over time and the many molecules in solution. We review recent advances and limitations in MD for IDPs and IDRs, with a special emphasis on enhanced sampling techniques that facilitate a full exploration of shallow free energy landscapes typically adopted by disordered systems. Finally, we conclude with a discussion of integrative approaches for combining the experimental accuracy from NMR with the precision offered by MD, to fully and accurately resolve structural ensembles of IDPs.

14.2 NMR Spectroscopy to Characterise IDPs

NMR spectroscopy is a powerful experimental technique that provides information with atomic resolution across various timescales ranging from picoseconds to minutes and hours. Recent advances in spectrometer hardware, including increasing static magnetic field strengths and cryogenically cooled probes as well as non-uniform sampling techniques[11–13] coupled with creative pulse sequence design[14] have established NMR as one of the most well-suited experimental techniques to probe IDPs directly in solution at physiologically relevant conditions, including temperature, pH, and salt concentrations.[15] Furthermore, NMR can be applied to study IDPs and IDRs in isolation, within large complexes, and in living cells.[14] A wealth of structural and dynamic insight can be obtained from a wide variety of NMR measurements, but here, we focus on solution-state NMR techniques that have been particularly informative for characterising protein disorder.

14.2.1 Chemical Shifts and the Fingerprints of IDPs

The most easily accessible NMR parameters, chemical shifts, are powerful reporters that provide ensemble-averaged residue-specific information. Whereas aliphatic ^{13}C chemical shifts predominantly report on local dihedral angles, carbonyl ^{13}CO reflect the primary structure of a protein as well as hydrogen-bond formation, and ^{1}H and ^{15}N backbone chemical shifts additionally report on long-range electrostatics and ring-currents generated by aromatic groups.

The two-dimensional backbone $^{1}H–^{15}N$ heteronuclear single quantum coherence (HSQC) experiment is a routine method that allows one to quickly determine whether a protein is folded, intrinsically disordered, or a molten globule; even in the absence of a sequence-specific chemical shift assignment. Relative to folded proteins, IDPs have favourable sharp peaks as a result of their rapid conformational dynamics.[14] Even large IDPs, such as the tau protein with over 400 residues,[16] give rise to sharp cross-peaks in $^{1}H-^{15}N$ HSQC spectra. Another common feature of IDPs is the narrow dispersion of $^{1}H_N$ chemical shift peaks, which results from conformational averaging of the structural ensemble.[15] An elegant demonstration of the use of $^{1}H–^{15}N$ HSQC spectra to distinguish between folded and disordered states is shown in Figure 14.1, where the intrinsically disordered eukaryotic translation initiation factor 4E-binding protein 2 (4E-BP2) folds upon phosphorylation.[17]

Although $^{1}H–^{15}N$ HSQC spectra can provide a quick assessment of whether a protein is folded or disordered, this method has limitations, particularly near physiological temperature and pH (pH ∼ 7.4, 37 °C). Whereas the backbone amides of a folded protein often form long-lived, stable intra-molecular hydrogen bonds, these are usually extremely transient in the solvent-exposed backbones of IDPs, leading to a fast exchange of amide protons with the bulk solvent, which causes signal loss and line broadening. An alternative to the $^{1}H–^{15}N$ HSQC experiment in these physiological conidtions is the complementary $^{13}CO–^{15}N$ (CON) experiment, which is insensitive to hydrogen exchange.[18] While the CON experiment has reduced sensitivity compared to the $^{1}H–^{15}N$ HSQC, the CON experiment offers better spectral resolution and the ability to detect proline residues, which are often abundant in disordered amino acid sequences.[3] One does need to bear in mind that folded and intrinsically disordered proteins are not as easily distinguished using CON spectra as $^{1}H–^{15}N$ HSQC spectra. Under certain experimental conditions additional sensitivity can be obtained by employing the band-selective excitation short-transient (BEST) or BEST-transverse

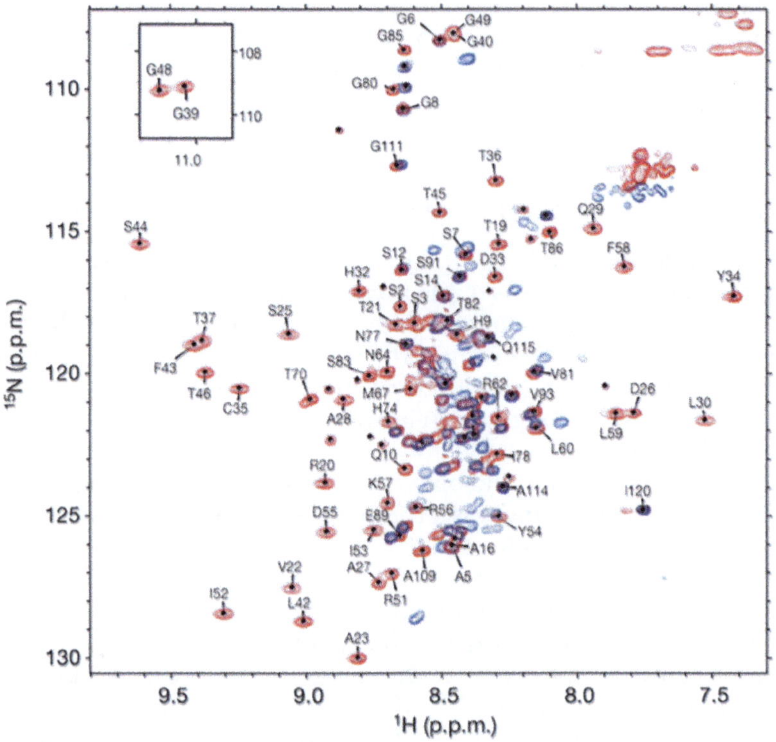

Figure 14.1 Two-dimensional ^{1}H–^{15}N HSQC spectra provide a simple method to determine whether a given protein is structured or disordered. Bah *et al.* used ^{1}H–^{15}N HSQC spectra to demonstrate the folding-upon phosphorylation of 4E-BP2. The disordered, unphosphorylated form of 4E-BP2 (blue) shows intense peaks with a narrow chemical shift dispersion in the ^{1}H dimension, suggesting that it is intrinsically disordered. In contrast, the phosphorylated form of 4E-BP2 (red) shows significant upfield (<8 ppm) and downfield (>9 ppm) chemical shifts in the ^{1}H dimension and peak broadening, suggesting partial folding upon phosphorylation. Reproduced from ref. 17 with permission from Springer Nature, Copyright 2014.

relaxation optimised spectroscopy (TROSY) techniques[19,20] and a combination of ^{1}H-excite and ^{13}C-detect methods.[14,21]

The initial step prior to fully investigating an IDP using NMR is to obtain a chemical shift assignment. A comprehensive guide and overview of the chemical shift assignment procedure for IDPs is already available[14] and this process will therefore not be discussed here. As soon as the chemical shift assignment is complete, residue-specific information about the conformational sampling of the IDP in question becomes readily available. Also, following a chemical shift assignment, a battery of experiments become accessible to probe the residual structure and dynamics of IDPs in a variety of conditions. Most of these methods use either the ^{1}H–^{15}N HSQC or the

$^{13}C–^{15}N$ CON fingerprint spectrum to encode parameters reporting on residual structure or dynamics of the IDP in question.

14.2.2 Information on Residual Structure

14.2.2.1 Secondary Chemical Shifts

For a ^{13}C, ^{15}N uniformly labelled protein, several chemical shifts, including those of $^{1}H_{\alpha}$, $^{1}H_{N}$, ^{15}N, $^{13}C'$, $^{13}C_{\alpha}$, and $^{13}C_{\beta}$, can be measured for each residue, except for prolines, which lack amide protons. These chemical shift values are sensitive to backbone conformations and are especially useful reporters of secondary structural propensities.[22,23] In particular, the 'secondary chemical shift', δ_{2nd}, which determines how much a nucleus deviates from a random coil environment, can be obtained by $\delta_{2nd} = \delta_{obs} - \delta_{rc}$, where the observed chemical shift, δ_{obs}, is subtracted by the random coil chemical shift value, δ_{rc}.[22,24] There are several reference tables of δ_{rc} values in the literature, accounting for neighbour effects, pH, and solvent.[14,25] Furthermore, chemical shifts can be implemented into algorithms such as the secondary structure propensity (SSP) programme, which reports on site-specific helical or sheet propensities of IDPs[26] (Figure 14.2). Even the chemical shifts of methyl-bearing side chains report on secondary structure.[27]

14.2.2.2 Scalar Couplings

Scalar couplings, particularly three-bond ^{3}J-couplings, provide valuable structural information on the dihedral angle formed by the three bonds through which the coupling is mediated. The ^{3}J-coupling between two atoms can in many cases be well approximated by the Karplus relationship:[28]

$$^{3}J(\theta) = A\cos^{2}(\theta + \Delta\theta) + B\cos(\theta + \Delta\theta) + C \qquad (14.1)$$

where A, B, C, and $\Delta\theta$ are empirically determined parameters and θ represents the dihedral angle. Commonly measured ^{3}J-couplings are those between $^{1}H_{\alpha}$(residue i) and ^{15}N(residue $i+1$), which provide information on the ψ dihedral angle, and ^{3}J-couplings between $^{1}H_{\alpha}$ (residue i) and $^{1}H_{N}$, (residue i), which report on the ϕ dihedral angles. Scalar couplings have also been used in combination with chemical shifts and nuclear Overhauser effects (NOEs) to demonstrate that two forms of the Alzheimer's-associated amyloid-β peptide, namely the 40- and 42-residue variants, have nearly identical experimental NMR observables in their monomeric forms, despite dramatically different aggregation propensities.[29] The measurement of accurate ^{3}J-couplings can be very time-consuming and following advancements in predicting secondary structure from sequence and chemical shifts, chemical shifts are now often used as the main source of experimental information on secondary structure.

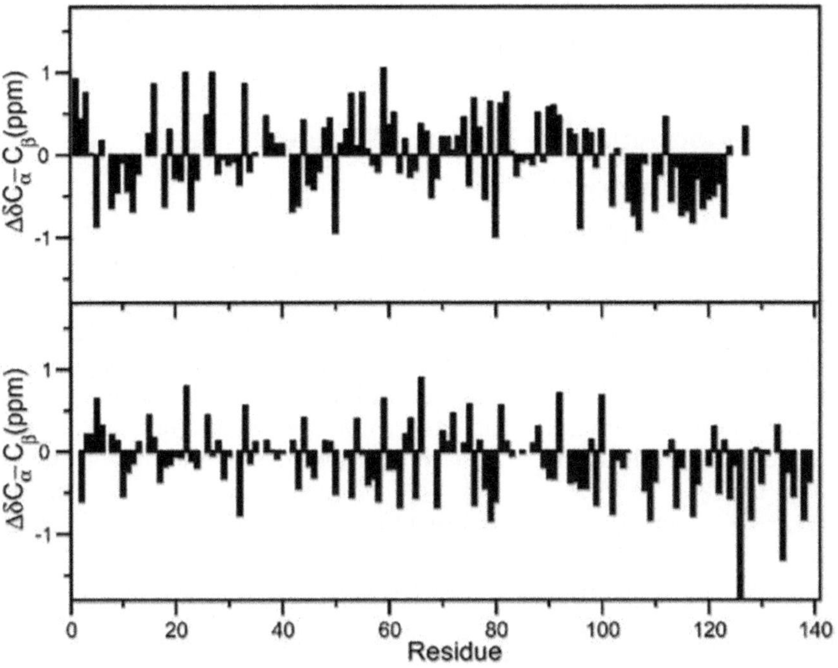

Figure 14.2 Accessing a secondary structure from ^{13}C chemical shifts. $\Delta\delta(^{13}C_\alpha) - \Delta\delta(^{13}C_\beta)$ secondary chemical shifts for γ-synuclein (top) and α-synuclein (bottom). Values less than and greater than zero indicate β-structure and α-structure propensities, respectively. Reproduced from ref. 26 with permission from John Wiley & Sons, Copyright © 2006 The Protein Society.

14.2.2.3 Dipolar Couplings

The magnetic moments of nuclear spins interact through space with each other at a strength that is proportional to the inverse-cube of distance between two nuclei, r^{-3}, *via* dipolar couplings. In isotropic solutions, where all orientations of the interaction vector are equally probable, this interaction averages to zero. However, it is possible to reintroduce a small fraction of the dipolar couplings, referred to as residual dipolar coupling (RDC), as this provides information on local and global structures as well as on dynamics.[30–33] In practice, this is done by creating an environment with preferred orientations, for example, taking advantage of the natural magnetic anisotropy of the molecule[34] or by explicitly placing the protein of interest in an anisotropic alignment medium, such as bicelles or Pf1 phage, which partly aligns the protein.[35] It is important that the alignment medium does not alter the conformational sampling of the IDP. If addition of the alignment medium does not lead to chemical shift changes it is generally assumed that the conformational sampling has not been altered significantly.

Residual dipolar couplings report on the orientation of the interaction vector in question. For example, for $^{1}H–^{15}N$ backbone RDCs, the value of the RDCs report on the orientation of the $^{1}H–^{15}N$ vector.[36] It is important to note that, because of degeneracy, an RDC cannot directly be transformed into an orientation and in theory five independent measurements need to be obtained.[37] One option is to measure the RDC using many different alignment media with different alignment properties. In practice this is often not possible, particularly for IDPs, and instead several RDCs are often measured in a couple of alignment media. Typical measured RDCs include $^{1}D_{NH}$ (backbone $^{1}H_{N}$ and ^{15}N), $^{1}D_{C\alpha H\alpha}$ (backbone $^{13}C_{\alpha}$ and $^{1}H_{\alpha}$), $^{2}D_{HCO}$ (backbone $^{1}H^{N}$ and CO), and $^{1}D_{C\alpha C\beta}$ ($^{13}C_{\alpha}$ and $^{13}C_{\beta}$).[36,38] For folded proteins, the preferred orientation of the protein in each alignment medium can often be described by a second alignment tensor. For flexible and disordered proteins, the alignment is often considered on a local scale, where a persistence length of *ca.* five residues is usually used. Previously RDC measurements have been employed to characterise the formation of transient helicies.[39]

14.2.3 Other NMR Observables

Structural and dynamic information can also be obtained from pulsed-field-gradient diffusion experiments. These provide the translational diffusion coefficient from which the hydrodynamic radius can be determined, reporting on the overall compactness of the protein of interest.[40] This observable is highly complementary to the radius of gyration determined by small-angle X-ray scattering (SAXS).[41] Both the hydrodynamic radius and the radius of gyration provide global insight into the conformations of IDPs.

Structural and dynamic information can also be obtained from various other NMR measurements including NOEs, which report on an internuclear distance up to about 6 Å and are useful for detecting stable secondary structures such as α-helices. A source of long-range structural information comes from paramagnetic relaxation enhancements (PREs). Typically, a paramagnetic spin-label, *e.g.* *S*-(1-oxyl-2,2,5,5-tetramethyl-2,5-dihydro-1*H*-pyrrol-3-yl)methyl methanesulfonothioate (MTSL), is covalently attached to a cysteine residue. The presence of the spin-label leads to a PRE, which is proportional to r^{-6}, and can provide long-range distance information up to about 15–25 Å.[42] However the attachment of a spin-label, that is often hydrophobic, may affect the conformational ensemble of the IDP of interest, which needs to be considered.[43]

14.2.4 Characterising Conformational and Chemical Exchange

Characterising the complex mechanisms by which IDPs and IDRs bind to their partners has been greatly facilitated by the development of techniques to characterise conformational and chemical exchange using NMR. NMR has contributed substantially to characterisations of IDP interactions, particularly for cases of folding-upon-binding.

Two NMR techniques have thus far dominated in the characterisation of non-covalent interactions formed by IDPs, that is, Carr–Purcell–Meiboom–Gill (CPMG) relaxation dispersion and the Chemical Exchange Saturation Transfer (CEST) technique. With both techniques, minor low-populated states can be characterised when these are in exchange with the major state and the population of the minor state exceeds a couple of per cent. The NMR signal corresponding to the minor state is most-often absent in NMR spectra. A common feature of both CPMG and CEST is that the minor state is characterised by observations made on the exchange-averaged state signals corresponding mainly to the major state.

CPMG and CEST experiments for characterising non-covalent interactions of IDPs are generally performed by adding a sub-stoichiometric amount of binding partner to the IDP. In such a case, and if there is an exchange between the free and bound states, then the free state will be a major state, while the bound state will be the low-populated minor state. For a simple two-site exchange-binding mechanism, with the binding partner at a constant concentration, the IDP in question can be modelled as being in dynamic exchange between the free and bound states, expressed as (Scheme 14.1):

$$\text{free} \underset{k_{bf}}{\overset{k_{fb}}{\rightleftharpoons}} \text{bound}$$

Scheme 14.1

where k_{fb} and k_{bf} describe the first-order rate constants for the interconversion between them. Specifically, $k_{fb} = [L]k_{on}$, where [L] is the concentration of the binding partner and k_{on} is the second-order on-rate; k_{bf} is the off-rate for the complex formed. Parameters that can be derived from CPMG and CEST measurements to characterise a low-populated bound state include: the overall exchange rate, $k_{ex} = k_{fb} + k_{bf}$, the chemical shift difference, $\Delta\omega = \omega_{bound} - \omega_{free}$, which provides information on the secondary structure, and the population of the minor state, p_b.[44] In some cases, one can also obtain $\Delta R_2 = R_{2,bound} - R_{2,free}$, which provides information on the correlation time of the minor state. The overall rate of exchange, k_{ex}, determines which technique, CPMG or CEST, is more useful to characterise the minor state.

Low populated states with lifetimes between *ca.* 0.5 and 100 ms are ideally suited for the CPMG and CEST techniques. Notably, a significant share of IDPs and IDRs form transient but specific complexes within this range of lifetimes.[45] Conformational exchange between free and bound states on millisecond timescales leads to moderate line broadening of the NMR signals corresponding to the free state (enhanced transverse relaxation rate R_2) which, although it hampers the data analysis in conventional NMR experiments, can be employed to provide local structural and dynamic knowledge about exchange processes that are otherwise elusive.[38,46]

14.2.4.1 CPMG Relaxation Dispersion

CPMG relaxation dispersion experiments are most sensitive to exchange processes with 200 $s^{-1} \leq k_{ex} \leq 2000$ s^{-1} *i.e.* on the millisecond timescale.[47] In the CPMG relaxation dispersion experiment the effective transverse relaxation rate, $R_{2,eff}$, is measured during a constant-time delay where a variable number of refocusing π pulses are applied. If the nucleus in question is only in one state, and there is no fluctuation of its chemical shift over time, then the $R_{2,eff}$ will be independent of the number of refocusing pulses applied. However, if the nucleus is exchanging between states on the millisecond timescale and its chemical shift becomes time-dependent then a relaxation dispersion profile is observed (Figure 14.3). Chemical exchange generally leads to a line broadening of NMR signals, where the purpose of the CPMG π pulses is to quench this line broadening by bringing the exchanging system towards the fast-exchange regime. Thus, at low CPMG frequencies substantial line broadenings are still present and high $R_{2,eff}$ values are obtained. On the contrary, at high CPMG pulsing frequencies the line broadenings are substantially quenched, leading to lower values of $R_{2,eff}$. Analysis of the CPMG relaxation dispersion profiles can provide the difference in chemical shift, $\Delta\omega$, between a free and bound conformation, as well as the exchange rate, k_{ex}, and the population of the bound species, p_b.

14.2.4.2 Chemical Exchange Saturation Transfer

For chemical exchange processes with $k_{ex} < 200$ s^{-1} the exchange contribution to $R_{2,eff}$ becomes relatively small, which means that the relaxation dispersions are too small to be analysed accurately by CPMG[48] experiments. For slower exchange rates, 20 $s^{-1} \leq k_{ex} \leq 200$ s^{-1}, CEST, a longitudinal

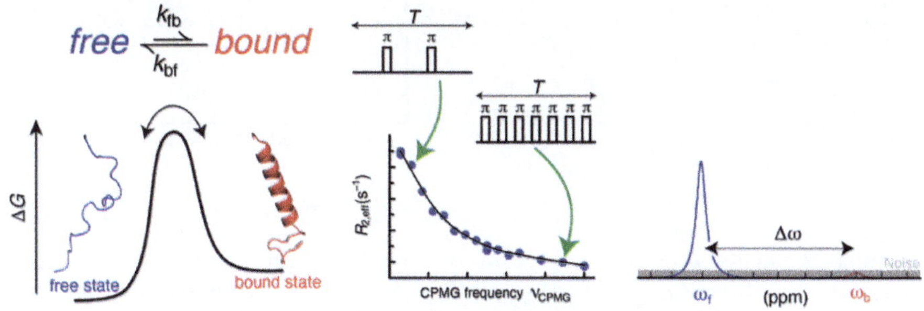

Figure 14.3 Schematic representation of the CPMG experiment. An IDP is in exchange between two states, namely a free state and a bound state. A variable number of pulses is applied during a constant time, T, which in turn modulate the relaxation rates of NMR probes on the molecule (*e.g.* ^{15}N). A conformational exchange will lead to a relaxation dispersion, that is, $R_{2,eff}$ decreases with increased CPMG frequency. Analysis of the CPMG relaxation dispersion profiles provide parameters such as $\Delta\omega$, k_{ex}, and p_b.

magnetisation-based experiment, can offer an attractive avenue for the characterisation of the exchanging system.[48] The basic crux of the CEST technique is that longitudinal magnetisation is saturated for a time T_{CEST} with a weak radiofrequency B_1 field. When the saturation frequency coincides with the frequency of the low-populated state, the saturation is transferred from the minor state to the major state, because each molecule is in constant exchange between the two states (Figure 14.4). As a result, the chemical shift of the minor state, which is otherwise invisible in NMR spectra, becomes visible in the CEST profile.[49] Analysis of the CEST profile, $I(\omega_{sat})/I_0$, also provides the rate of exchange, k_{ex}, as well as the population of the minor state, p_b.

CPMG dispersion profiles and CEST profiles can be analysed either independently or simultaneously, as they provide complementary information. Whereas a simple two-state binding interaction can often be characterised accurately using either a CPMG or CEST experiment, a simultaneous analysis of several types of experiments becomes necessary when the model of exchange is more complex. For example, for folding-upon-binding, *e.g.* ref. 50, several possible models for the exchange have been proposed as possible scenarios (Figure 14.5).[51] Generally, least-squares fitting is used and the mechanism is deduced from the model that best fits the available data. For more complex models and for three-site exchange, additional data are often

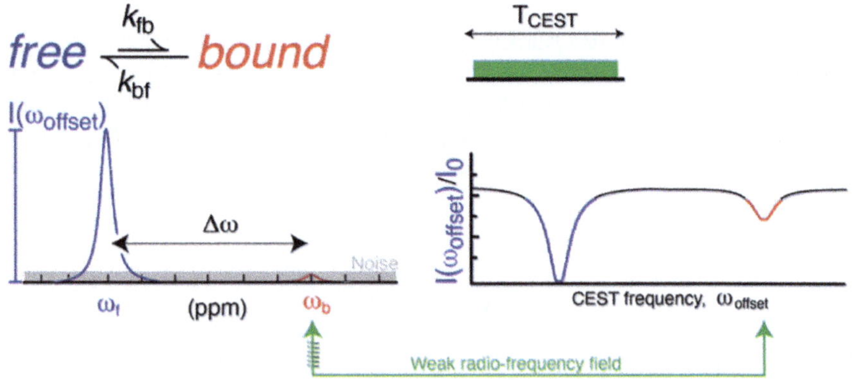

Figure 14.4 Schematic representation of the CEST experiment. A weak radiofrequency B_1 field is applied at a range of frequencies, ω_{offset}. In the scenario where a low-populated bound state is in dynamic exchange with the major free form, and where the signal from the bound state is broadened, a perturbation at the radio-frequency that corresponds to ω_b will result in a decreased intensity of the major state, as result of the exchange of 'saturation'. The intensity of the visible free state, I/I_0, where I_0 is the intensity without the weak radio-frequency pulse, is plotted as a function of the frequency of the applied B_1 field (right panel). When B_1 is applied at a frequency of ω_b a dip of I/I_0 can be observed on CEST intensity profile (green arrow).[49]

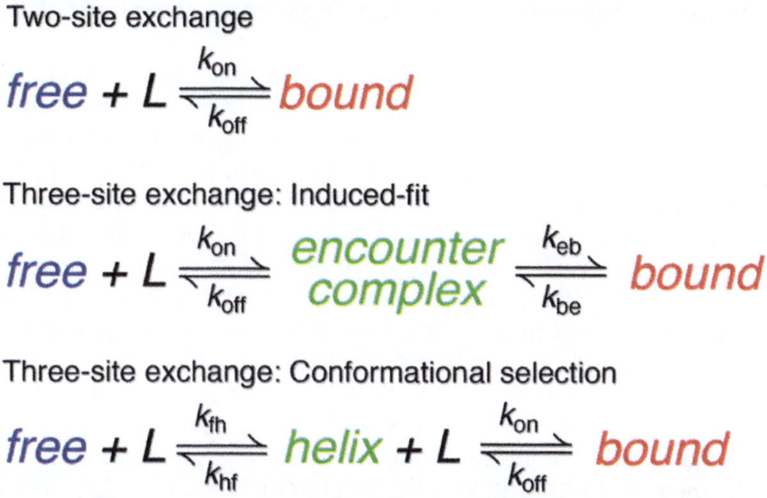

Figure 14.5 Exchange models that are typically used to analyse IDP folding-upon-binding to a ligand (L). The rate constants k_{off} and k_{on} represent the first-order off-rate and the second-order on-rate, respectively, k_{eb} and k_{be} are the rates describing the reaction between the encounter complex and the bound state, while k_{hf} and k_{fh} are the first-order rates describing the unimolecular reaction where the bound conformation of the IDP is formed.

required, which could consist of a simultaneous analysis of CPMG and CEST data or inclusion of the chemical exchange-induced shifts.[52]

Amongst other techniques, NMR chemical exchange methods have been used to characterise the coupled folding and binding of the disordered transcription factor c-Myc to the KIX domain of the transcriptional coactivator CREB-binding protein. In that case it was observed that c-Myb binds the KIX domain in a way that involves features of conformational selection (or binding of a pre-folded conformation in the free state) and induced fit (folding-upon-binding).[51] Overall, CPMG relaxation dispersion and CEST experiments have facilitated the mechanistic characterisation of complex IDP behaviour such as multi-step folding-upon-binding,[45,50,51,53] multivalent IDP interactions,[54] and IDP fuzzy complex formation, where the IDPs remain disordered in the bound state.[55]

14.2.5 Ensemble Description Methods without MD

NMR measurements provide site-specific conformational information rather than yielding atomic coordinates as is typical of X-ray crystallography and cryogenic electron microscopy (cryo-EM) data. The interpretation of NMR observables is therefore often complemented by computational modelling techniques including statistical coil generators and MD. As MD-based methods are the major focus of the following sections, here we initially focus on ensemble description methods in the absence of MD.

A commonly used modelling scheme in this context is to first generate an initial pool of physically possible conformers and then select a sub-ensemble that reproduces experimental data. Statistical coil generators produce many structures (often millions) by random assignment of ϕ/ψ backbone dihedral angles based on experimental statistical potentials. In contrast to all-atom MD, these approaches are computationally accessible approaches for the generation of large conformational ensembles. Examples include trajectory-directed ensemble sampling (TraDES),[56] Flexible-Meccano,[57] database Markov method,[58] and three-residue-based sampling.[59] It should be noted that it is also possible to use structures generated from MD as the initial structural pool. Once large conformational ensembles are produced, one can select for conformations which agree with experimental data. Two algorithms with similar workflows based on this conformational selection principle include ENSEMBLE[60] and ASTEROIDS,[61] which are introduced here as examples of ensemble description methods in the absence of MD.

In the case of ENSEMBLE, a conformer pool is first generated, for example using TraDES. Subsequently, a Monte Carlo-based approach weights each structure present in this initial pool. Then, the ENSEMBLE algorithm will attempt to identify the smallest set of conformations that is consistent with various types of experimental data including secondary chemical shifts, NOEs, and PREs. This process is repeated iteratively, adding new conformers each round to increase sampling.[62] Similarly, in the ASTEROIDS approach, a conformational pool is first generated, for example, by Flexible-Meccano. Then, ASTEROIDS employs a genetic algorithm approach to select sets of structures of a certain size from this large conformational pool that are best fit to experimental data. Self-avoiding random evolution proceeds by randomly generating new ensembles each round. Among the pool of new ensembles and previous generation, 100 best ensembles are selected to 'survive' to the next generation, and this process is iteratively repeated.[61] For example, Flexible-Meccano combined with ASTEROIDS was used to model the binding of the disordered domain of the Sendai virus nucleoprotein to the C-terminal domain of the phosphoprotein using chemical shifts as restraints. This model was combined with extensive $^{1}H_{N}$, ^{13}CO, and ^{15}N relaxation dispersion measurements to probe this complex binding interaction.[50]

In the next section, we introduce MD as a powerful tool to complement information from NMR experiments, underscoring challenges encountered with simulating IDPs. We highlight how NMR data have aided force field restraints and validation towards MD parameters with increased accuracy.

14.3 MD Simulations of IDPs

The NMR techniques discussed above offer unparalleled accuracy and resolution for experimentally characterising IDPs in solution, with minimal perturbations to their structural ensembles. Nevertheless, NMR data are average measurements, which report on the extreme conformational heterogeneity of IDPs, both temporally and conformationally, preventing a full atomic-resolution description of structural ensembles. For example, NMR parameters typically report on ensemble averages of 10^{16}–10^{17} dynamic molecules and provide limited information on the underlying structural distributions, especially in the case of extremely flexible biomolecules. MD simulations offer complementary insights and can, in theory, provide time-resolved descriptions of IDP motions, thereby providing structural, dynamical, and mechanistic information that is inherently more detailed than that afforded by experimental techniques, including NMR. Simulations of IDPs and IDRs can range dramatically in precision, from highly precise *ab initio* approaches to empirically derived all-atom force fields to coarse grained models. In this section, we outline the major challenges associated with characterising the ensembles of IDPs with a focus on all-atom explicit solvent MD, highlighting recent advances to overcome these bottlenecks.

14.3.1 Force Fields for IDPs: Bottlenecks and Progress

All-atom MD simulations employ empirical molecular mechanics force fields to describe inter-atomic interactions. Over the past two decades, as the biological importance of intrinsic disorder has emerged, it has become increasingly clear that several all-atom force fields, which had been optimised for folded proteins, were overestimating the secondary structural propensities and compactness of IDPs and IDRs.[63] The predictive power of all-atom force fields for disorder was therefore improved by optimising the backbone ϕ and ψ dihedral potentials yielding the OPLS-AA/L,[64] AMBER03,[65] CHARMM27,[66,67] AMBER99SB,[68] AMBER99SB*, AMBER03*,[69] AMBER03w[70] CHARMM22*,[71] CHARMM36,[72] AMBER14SB,[73] AMBER99IDPs,[74] AMBER14-IDPSFF,[75] CHARMM36m,[76] and CHARMM36IDPSFF[77] force fields. Similarly, some of these force fields were further improved by optimising side-chain torsion potentials.[78] Notably, these force fields were all validated by comparing back-calculated and experimentally measured NMR observables such as scalar ^3J-couplings and RDCs and some were additionally validated using SAXS data.

While these advancements in force fields improved local predictions of IDP behaviour, these force fields generally underestimate global compactness of disordered polypeptide chains, now understood to be a result of an underestimation of dispersion interactions between water and amino acids.[63,79,80] This has been addressed in a few ways including (*a*) rescaling the Lennard–Jones well-depth parameter between water oxygen and protein atoms (yielding AMBER03ws with the four-point TIP4P/2005s water model),[79] (*b*) introducing a new water model, called TIP4P-D, with larger C_6 dispersion parameters,[80] or (*c*) altering the Lennard–Jones well-depth parameters for water hydrogens.[76] The predictive power of these force fields has been validated by comparing back-calculated SAXS profiles to experimental measurements as well as using NMR measurements.[81]

One major drawback of modulating the interaction strengths between water and amino acids is the destabilisation of folded proteins, making it challenging to accurately characterise complex scenarios such as folding-upon-binding or partially folded systems. A comprehensive investigation of folded and disordered systems resulted in a force field that combines the TIP4P-D water model with a variant of the AMBER99SB-ILDN with optimised torsion potentials. This force field, called AMBER99SB-disp, is very promising for simulation of IDPs, folded proteins, and partially folded systems.[81,82]

While force fields are undoubtedly improving for modelling protein disorder, it is still essential to test diverse models to determine which most accurately describes both local features (consistent with NMR chemical shifts, ^3J-couplings, and residual dipolar couplings) and global features such as radii of gyration (consistent with SAXS intensity profiles and radii of hydration).[63] All force fields, no matter how accurate, are approximations and will not be able to fully and accurately predict the full behavioural properties of a given biomolecule. This is especially relevant for IDPs and IDRs, where minor force field inaccuracies propagated across extremely heterogeneous structural ensembles can lead to incorrect predictions[83–85] (Figure 14.6).

14.3.2 The Sampling Problem and Solutions for IDPs

All-atom explicit solvent simulations of IDPs are extremely computationally expensive. In a typical setup, one models forces between tens of thousands

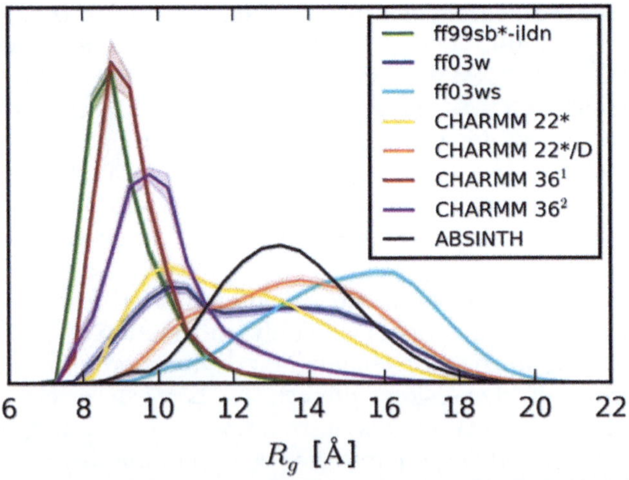

Figure 14.6 Histograms of radii of gyration, R_g, for the structural ensembles calculated using various force fields for a model IDP comprised of arginine and serine residues, called the 'RS' peptide. These histograms demonstrate that different force fields predict dramatically different behaviours for the same biomolecule. Reproduced from ref. 84 with permission from American Chemical Society, Copyright 2015.

of protein and solvent atoms or more, which are assessed every few femtoseconds of simulation time. Simulating one microsecond of protein dynamics corresponds to approximately one billion steps, generally requiring weeks or months on today's most powerful computers with hundreds or thousands of processors. The long timescales required to reliably sample the vast conformational heterogeneity of protein disorder remains a major challenge in the field.

Tremendous advances in biomolecular simulation employing graphics processing units (GPUs), purpose built hardware (MDGRAPE,[86] Anton,[87] Anton 2[88]), and massive parallelisation *via* grid computing, such as folding@home[89] and using cloud computing infrastructure,[90] have met the sampling problem head on with massive computational power.[91] However, even with these developments, many relevant functional motions of IDPs occur on timescales beyond those currently accessible to MD.

Enhanced sampling techniques (Figure 14.7) represent an alternative solution to the sampling problem that is computationally more efficient, allowing exploration of the configurational space while preserving the equilibrium distribution of conformers in the given statistical ensemble. For example, parallel tempering, also called replica exchange MD, is a popular enhanced sampling technique, in which one simulates the system at multiple temperatures, thus increasing the available conformational space available to the system. The configurations of neighbouring replicas are allowed to exchange periodically using the Metropolis criterion where the probability of exchange is proportional to $\exp(-\Delta\beta\Delta E)$, where $\beta = (k_B T)^{-1}$,

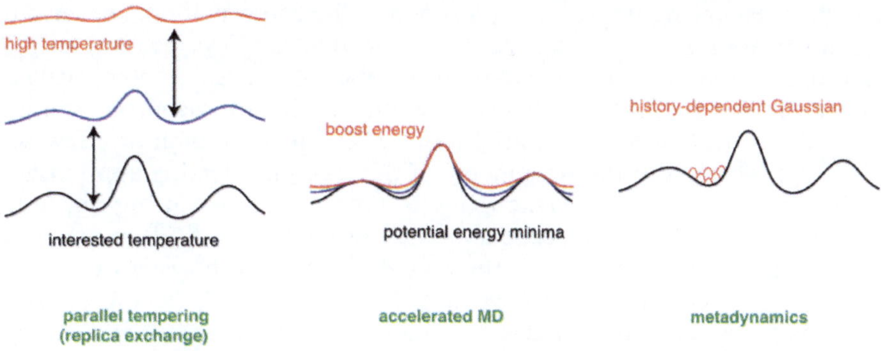

Figure 14.7 Representation of enhanced sampling approaches for MD. In the parallel tempering scheme, independent replicas of the system are simulated in varying conditions, such as temperature, and exchanges are attempted between the replicas at regular intervals with a probability given by the Metropolis criterion. In the accelerated MD scheme, the potential energy surface of the system is smoothed by adding a boost potential, allowing the system to overcome potential energy barriers. In the metadynamics scheme, Gaussian functions are added to certain collective variables to discourage the system from visiting conformational space that it has already explored. Adapted from ref. 139 with permission from Elsevier, Copyright 2016.

where k_B is the Boltzmann constant, T is the temperature, and ΔE is the potential energy difference between the two states.[92–95] One disadvantage of parallel tempering is that it depends on sufficient overlap between the energy distributions of the neighbouring replicas to ensure practical acceptance rates. As a result, more replicas with moderate temperature differences between them are required to simulate large systems, which can amount to prohibitively expensive computational costs for explicit solvent simulations. One solution is the use of Replica Exchange with Solute Tempering (REST),[96] in which the total interaction energy of the system is comprised of three components, namely the protein intramolecular energy (E_{pp}), the interaction energy between the protein and water (E_{pw}), and the interaction energy between water molecules (E_{ww}). Instead of simulating each replica at a different temperature, in REST, each replica is run on various altered potential energy surfaces which arise by scaling these three components. The energy of the first replica (known as 'replica 0') is not scaled, but increasing replicas have the E_{pp} and E_{pw} terms progressively decreased, with exchanges between replicas following Metropolis criteria. As the acceptance probability only depends on differences in the E_{pp} and E_{pw} terms, fewer replicas are required. Modifications to REST increase efficiency[97] and enhance the exploration of free-energy landscapes with high barriers.[98]

In the case of IDPs, which have similar energies in distinct conformational states, other enhanced sampling methods may better aid conformational sampling. In particular, using accelerated MD,[99] one 'boosts' the energy along internal degrees of freedom for the IDP of interest, such as ϕ and ψ dihedral angles, thus allowing the system to overcome potential energy barriers and allowing improved exploration of the conformational space. An alternative approach is metadynamics,[100] which employs coarse-grained quantities of the system, *e.g.* radius of gyration or helical content, called collective variables (CVs), to enhance sampling. In this approach, an external history-dependent biasing potential term, which is a function of a few selected CVs is added to the free energy of the system. Selecting appropriate CVs has traditionally been challenging, but recent progress using machine learning to select CVs promises to alleviate this challenge.[101–103] In practice, the biasing potential created by the CVs discourages the system from returning to configurations that have already been explored, forcing the system to visit new conformational space (Figure 14.7). CVs can be any quantity that can be calculated from the atomic positions throughout the simulation and should capture all the slow degrees of freedom of the system that cannot otherwise be sufficiently sampled. Several strategies have been proposed to simplify the choice of CV selection including combination of metadynamics with parallel tempering[104] or further enhance sampling by bias-exchange metadynamics.[105] More recently, an approach called parallel-bias metadynamics was proposed in which multiple low-dimensional bias potentials are simultaneously applied in parallel.[106,107]

Enhanced sampling methods are powerful tools to explore the conformational ensembles adopted by IDPs, but these approaches generally come at the cost of the dimension of time, which can provide useful information on the kinetic transitions of disordered systems. For example, using a combination of unbiased trajectories (315-μs of all-atom MD trajectories) with the CHARMM22* force field and deep learning approaches, it was possible to probe the structures and dynamics of the amyloid-β peptide, whose aggregation is associated with the onset of Alzheimer's disease. Here, the IDP was characterised by a Markov state model, which describes both populations of states and their rates of interconversion[108] (Figure 14.8).

14.3.3 Other MD Approaches for Characterising Protein Disorder

The computational costs of explicit solvent all-atom MD simulations of IDPs can be considerably reduced by directly estimating the solvation free energy of the system through the use of implicit solvent protein force fields, which allow one to reduce the system size by more than ten-fold.[109–112] Another approach for reducing the complexity of these systems is to group atoms together into larger particles called beads using coarse-graining approaches. This approach is extremely general and covers a wide range of resolutions, ranging from a few beads per amino acid to one bead per thousands of nucleic acid bases. In particular, course-graining approaches have been employed to probe a picomolar-affinity protein complex between two IDPs, namely the human proteins histone H1 and its nuclear chaperone prothymosin-α associate, in which both partners remain disordered in the bound state.[113] Furthermore, coarse-grained models have shown promising results for probing liquid–liquid phase separation involving intrinsic disorder.[114,115]

14.4 Integrative Modelling of IDPs by Combining NMR and MD

NMR experiments provide highly accurate measurements of protein disorder in physiological conditions; however, these data are sparse and report on both dynamic and conformationally averaged structures, and are thus degenerate since various structural ensembles may fit the same data. MD, on the other hand, in theory, has the potential to provide a full – often atomistic – picture of the structural ensembles of IDPs, but despite recent advances, still suffers from force field inaccuracies and sampling limitations. The ability to integrate these highly complementary approaches holds enormous promise for the accurate and precise structural characterisation of IDPs.

14.4.1 Comparing MD Ensembles with NMR Observables

To validate and restrain structural ensembles with experimental NMR data, one must have forward models to predict observables from structural

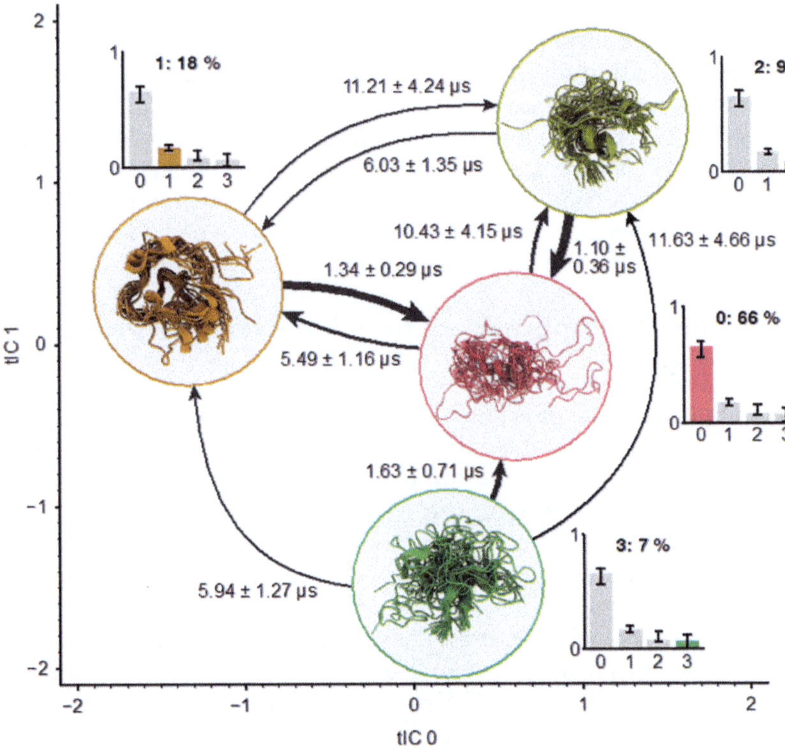

Figure 14.8 Four-state Markov state model describing structures and interconversion rates of the disordered amyloid-β peptide. Thicker arrows correspond to higher transition probabilities. The state location is projected onto the two slowest time-independent coordinates (tICs). Reproduced from ref. 108 with permission from Springer Nature, Copyright 2021.

ensembles. For example, chemical shifts can be predicted from structural ensembles using one of several algorithms including SPARTA+[116] and CamShift.[117] SPARTA+ is a neural network-based approach to predict chemical shifts for backbone and $^{13}C_{\beta}$ nuclei in proteins. Notably, it was trained on structured proteins that have been crystallised.[116] CamShift instead predicts chemical shifts using an empirically derived polynomial function, which uses interatomic distances to predict chemical shifts based on the Re-referenced protein chemical shift DataBase (RefDB).[117] A key advantage of CamShift is that the predictions are differentiable with respect to the atomic coordinates, which makes it possible to use the predictions along with experimental chemical shifts as restraints. Scalar ^{3}J-couplings, which often report on backbone dihedral angles, can be easily calculated from structural ensembles using eqn (14.1) and compared with experimentally determined values. RDCs can be predicted from structural ensembles using the Prediction of ALignmEnt from Structure (PALES) method.[36] When the alignment of the IDP is purely steric, a local alignment tensor can be predicted from the shape of the IDP using a simple steric obstruction model.[36,118] More recently, an alternative tensor-free approach, called the ϑ method, has been proposed to refine structural ensembles using data from RDC measurements.[119] This method is based on the observation that the RDC between two nuclei depends on the angle ϑ between the internuclear vector and the external magnetic field.[119] Additionally, distances from NOEs and PREs can be employed as structural restraints as the interaction scales as r^{-6}. Furthermore, radii of hydration can be estimated from calculated radii of gyration from MD simulations,[120] which can be compared to diffusion measurements. It is considerably more challenging to quantitatively compare nuclear-spin relaxation experiments, for example ^{15}N R_1 and R_2, to MD simulations due to the many structural and dynamic processes that contribute to the relaxation process.[121] Nevertheless, some progress has been made in this area for structured proteins by employing Markov state models that characterise the dynamic interconversion between metastable states.[122]

14.4.2 Maximum Entropy *versus* Maximum Parsimony Methods

Over the past decade, tremendous progress has been made towards methods that combine experimental data with computational approaches. In general, these approaches fall into two categories, namely those inspired by either the maximum entropy principle or the maximum parsimony principle. Maximum entropy methods typically involve the generation of a large structural ensemble to match experimental data, which can be done 'on-the-fly' by restraining a simulation or by reweighting a simulation after it has completed.[123] This perturbation aims to match the ensemble with the experimental data while remaining close to the *a priori* ensemble. Alternatively, the maximum parsimony approach aims to identify the minimal number of conformations required to match experimental data.

In our view, the choice between maximum entropy and maximum parsimony methods depends on the conformational entropy and dynamics of the system, as well as on the resolution limits of experimental data and the available computational resources. For example, either maximum parsimony or maximum entropy methods can be useful for structured proteins with few states that can be well-characterised experimentally. In the context of probing the conformations of IDPs, however, maximum entropy methods are generally superior over maximum parsimony methods, as current experiments cannot fully resolve the vast conformational heterogeneity of these dynamic biomolecules. Therefore combining experimental data with *prior* information, *e.g.* from force fields, allows us to accurately and precisely understand the structures of IDPs. To illustrate this, consider the simple example shown in Figure 14.9: suppose an NMR measurement, such as a chemical shift is observed for an atom A within an IDP (black dashed lines in

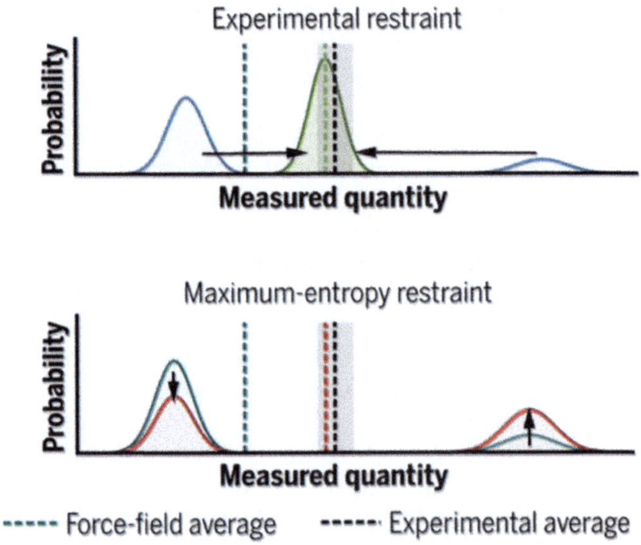

Figure 14.9 Comparison of maximum parsimony and maximum entropy restraints. In this simplified schematic, black dashed lines represent an NMR observable (averaged out over many conformations), blue curves represent distributions calculated from MD simulations, and blue dashed lines represent experimental observables calculated from MD ensembles. Maximum parsimony methods restrain the MD ensemble to values close to the experimental observable (green curves). In maximum entropy methods, the distributions are minimally perturbed so that the predicted observable matches the experimental one. Reproduced from ref. 91 with permission from The American Association for the Advancement of Science, Copyright 2018.

Figure 14.9). This observable represents the average environment of atom *A* for all conformations that are in rapid exchange and could be due to many potential distributions. MD simulations use prior information to predict these distributions (blue curves, Figure 14.9). To compare the distributions from MD simulations to the experimental observable, we calculate the ensemble-averaged observable (blue dashed lines). Maximum parsimony methods enforce close agreement between these observables, resulting in a new distribution which is close to the ensemble average from the experiment (green curve, Figure 14.9). This procedure, however, assumes the experimental observable was indeed produced by conformations near the ensemble average, which, in the case of the extreme conformational heterogeneity of IDPs, is often not the case. In contrast, maximum entropy methods make no assumptions about the conformational entropy of the system. Instead, the distributions predicted by MD are minimally perturbed (red curve) so that their predicted observables match experimental observables (red dashed line).[91,124]

14.4.3 Accounting for Errors with Bayesian Inference

While maximum entropy methods are powerful approaches for the structural characterisation of IDPs, they do not inherently account for sources of error, both from experimental measurements themselves and the forward models used to predict observables from the structural ensembles. This is particularly important in the context of weighting averaged experimental data from multiple sources, such as different kinds of NMR techniques. Bayesian approaches, such as the inferential structure determination method,[125] are well-suited to integrate multiple kinds of experimental data, while accounting for variable and potentially unknown errors.

Recent progress in combining maximum entropy approaches with Bayesian methods holds great promise for the precise, accurate, and rigorous determination of structural ensembles of IDPs. There are several Bayesian *a posteriori* reweighting approaches for structural ensemble determination including the 'Bayesian ensemble refinement' method[126] and the 'Bayesian/maximum entropy (BME)' approach.[127] The BME approach was used to characterise the binding of chaperones to integral membrane proteins with hydrophilic disordered domains by combining all-atom MD guided by SAXS data, and these were validated against experimental NMR parameters. The simulations demonstrated the interplay of hydrophobic and hydrophilic interactions underpinning chaperone function.[128]

There are only a handful of methods that combine maximum entropy and Bayesian approaches to directly incorporate experimental data as structural restraints on-the-fly. These include the 'Bayesian ensemble refinement' method[126] and metainference.[129] Using these approaches, one simulates several replicas of an IDP in parallel, combining both prior information (from the force field) and an energy term that relates the experimental data (measured as an average of many conformations within a sample) to average predicted values of the observables calculated over the replicas. The effect of the restraints on the simulation can vary, is sensitive to the unknown noise level in the data, and scales with the number of replicas in the presence of experimental noise. Metainference has been combined with metadynamics and parallel bias for enhanced sampling[130] and is available in the open-source PLUMED library.[131]

Metainference metadynamics has been used to combine all-atom MD with chemical shift restraints to characterise small-molecule binding to IDPs. In particular, the interaction between a small, drug-like molecule called 10058-F4 and a peptide from the oncogenic transcription factor, c-myc, demonstrated that the binding interaction is extremely dynamic, the disordered peptide remains disordered in the bound form, and that there is some degree of sequence specificity in the binding.[132] Furthermore, this approach was also used to characterise small-molecule binding to the monomeric form of the amyloid-β. The interaction was again observed to be extremely dynamic, and the conformational entropy of amyloid-β was observed to increase in the presence of the compound, thereby contributing favourably to the binding free energy *via* a mechanism called entropic expansion.[133,134]

14.5 Conclusion and Future Perspectives

Tremendous recent progress in both the NMR and MD fields is allowing scientists to meet the enormous experimental and computational challenge of describing the heterogeneous structures, functions, and behaviours of IDPs and IDRs. Here, we have outlined key approaches in both fields and the exciting new ways in which these approaches can be integrated. NMR and MD approaches can be further complemented with other biophysical techniques, including SAXS,[135] single-molecule spectroscopy, circular dichroism, fluorescence resonance energy transfer (FRET),[135,136] Fourier transform infrared spectroscopy, electron paramagnetic resonance spectroscopy, and fluorescence correlation spectroscopy (FCS) to further probe these complex biomolecules and their interactions.[137]

Recent progress in the NMR and MD fields including ultra-high field NMR magnets (1.2 GHz), development of ^{13}C-detection methods, improved force fields, increased computational power, advanced integrative techniques, and the sharing of structural ensembles [for example on the Protein Ensemble Database (https://proteinensemble.org/)][138] make the present an incredibly exciting time to study protein disorder.

Acknowledgements

The authors would like to acknowledge Thomas Löhr for helpful discussions. This work was supported by Schmidt Science Fellows, the Wellcome Trust (101569/z/13/z), and the Biotechnology and Biological Sciences Research Council UK (BBSRC, BB/T011831/1).

References

1. V. Csizmok, A. V. Follis, R. W. Kriwacki and J. D. Forman-Kay, *Chem. Rev.*, 2016, **116**, 6424–6462.
2. C. J. Oldfield and A. K. Dunker, *Annu. Rev. Biochem.*, 2014, **83**, 553–584.
3. P. Tompa, *Trends Biochem. Sci.*, 2002, **27**, 527–533.
4. M. Fuxreiter, I. Simon, P. Friedrich and P. Tompa, *J. Mol. Biol.*, 2004, **338**, 1015–1026.
5. P. Tompa, E. Schad, A. Tantos and L. Kalmar, *Curr. Opin. Struct. Biol.*, 2015, **35**, 49–59.
6. G. T. Heller, P. Sormanni and M. Vendruscolo, *Trends Biochem. Sci.*, 2015, **40**, 491–496.
7. J. Jumper, R. Evans, A. Pritzel, T. Green, M. Figurnov, K. Tunyasuvunakool, O. Ronneberger, R. Bates, A. Zidek and A. Bridgland, *Fourteenth Critical Assessment of Techniques for Protein Structure Prediction (Abstract Book)*, 2020, 22, 24.

8. C. Norn, B. I. Wicky, D. Juergens, S. Liu, D. Kim, D. Tischer, B. Koepnick, I. Anishchenko, D. Baker and S. Ovchinnikov, *Proc. Natl. Acad. Sci. U. S. A.*, 2021, **118**, e2017228118.
9. A. W. Senior, R. Evans, J. Jumper, J. Kirkpatrick, L. Sifre, T. Green, C. Qin, A. Žídek, A. W. Nelson and A. Bridgland, *Nature*, 2020, **577**, 706–710.
10. F.-X. Theillet, A. Binolfi, B. Bekei, A. Martorana, H. M. Rose, M. Stuiver, S. Verzini, D. Lorenz, M. Van Rossum and D. Goldfarb, *Nature*, 2016, **530**, 45–50.
11. S. G. Hyberts, A. G. Milbradt, A. B. Wagner, H. Arthanari and G. Wagner, *J. Biomol. NMR*, 2012, **52**, 315–327.
12. J. Ying, F. Delaglio, D. A. Torchia and A. Bax, *J. Biomol. NMR*, 2017, **68**, 101–118.
13. G. Karunanithy and D. F. Hansen, *J. Biomol. NMR*, 2021, **75**, 179–191.
14. B. Brutscher, I. C. Felli, S. Gil-Caballero, T. Hošek, R. Kümmerle, A. Piai, R. Pierattelli and Z. Sólyom, *Intrinsically Disordered Proteins Studied by NMR Spectroscopy*, 2015, 49–122.
15. H. J. Dyson and P. E. Wright, *Chem. Rev.*, 2004, **104**, 3607–3622.
16. S. Ambadipudi, J. G. Reddy, J. Biernat, E. Mandelkow and M. Zweckstetter, *Chem. Sci.*, 2019, **10**, 6503–6507.
17. A. Bah, R. M. Vernon, Z. Siddiqui, M. Krzeminski, R. Muhandiram, C. Zhao, N. Sonenberg, L. E. Kay and J. D. Forman-Kay, *Nature*, 2015, **519**, 106–109.
18. W. Bermel, I. Bertini, I. C. Felli, R. Kümmerle and R. Pierattelli, *J. Magn. Reson.*, 2006, **178**, 56–64.
19. E. Lescop, P. Schanda and B. Brutscher, *J. Magn. Reson.*, 2007, **187**, 163–169.
20. Z. Solyom, M. Schwarten, L. Geist, R. Konrat, D. Willbold and B. Brutscher, *J. Biomol. NMR*, 2013, **55**, 311–321.
21. G. Karunanithy, J. Reinstein and D. F. Hansen, *J. Phys. Chem. Lett.*, 2020, **11**, 5649–5654.
22. D. S. Wishart, B. D. Sykes and F. M. Richards, *J. Mol. Biol.*, 1991, **222**, 311–333.
23. S. Spera and A. Bax, *J. Am. Chem. Soc.*, 1991, **113**, 5490–5492.
24. M. Kjaergaard and F. M. Poulsen, *J. Biomol. NMR*, 2011, **50**, 157–165.
25. J. T. Nielsen and F. A. Mulder, *J. Biomol. NMR*, 2018, **70**, 141–165.
26. J. A. Marsh, V. K. Singh, Z. Jia and J. D. Forman-Kay, *Protein Sci.*, 2006, **15**, 2795–2804.
27. M. Kjaergaard, V. Iešmantavičius and F. M. Poulsen, *Protein Sci.*, 2011, **20**, 2023–2034.
28. M. Karplus, *J. Chem. Phys.*, 1959, **30**, 11–15.
29. J. Roche, Y. Shen, J. H. Lee, J. Ying and A. Bax, *Biochemistry*, 2016, **55**, 762–775.
30. O. F. Lange, N.-A. Lakomek, C. Farès, G. F. Schröder, K. F. Walter, S. Becker, J. Meiler, H. Grubmüller, C. Griesinger and B. L. De Groot, *Science*, 2008, **320**, 1471–1475.

31. G. Bouvignies, P. Bernado, S. Meier, K. Cho, S. Grzesiek, R. Brüschweiler and M. Blackledge, *Proc. Natl. Acad. Sci. U. S. A.*, 2005, **102**, 13885–13890.
32. M. R. Jensen, P. R. Markwick, S. Meier, C. Griesinger, M. Zweckstetter, S. Grzesiek, P. Bernadó and M. Blackledge, *Structure*, 2009, **17**, 1169–1185.
33. N. Tjandra and A. Bax, *Science*, 1997, **278**, 1111–1114.
34. J. Tolman, J. Flanagan, M. A. Kennedy and J. Prestegard, *Proc. Natl. Acad. Sci. U. S. A.*, 1995, **92**, 9279–9283.
35. A. Bax and A. Grishaev, *Curr. Opin. Struct. Biol.*, 2005, **15**, 563–570.
36. M. Zweckstetter and A. Bax, *J. Am. Chem. Soc.*, 2000, **122**, 3791–3792.
37. P. R. Markwick, G. Bouvignies, L. Salmon, J. A. McCammon, M. Nilges and M. Blackledge, *J. Am. Chem. Soc.*, 2009, **131**, 16968–16975.
38. I. R. Kleckner and M. P. Foster, *Biochim. Biophy. Acta*, 2011, **1814**, 942–968.
39. M. R. Jensen, K. Houben, E. Lescop, L. Blanchard, R. W. Ruigrok and M. Blackledge, *J. Am. Chem. Soc.*, 2008, **130**, 8055–8061.
40. D. K. Wilkins, S. B. Grimshaw, V. Receveur, C. M. Dobson, J. A. Jones and L. J. Smith, *Biochemistry*, 1999, **38**, 16424–16431.
41. W.-Y. Choy, F. A. Mulder, K. A. Crowhurst, D. Muhandiram, I. S. Millett, S. Doniach, J. D. Forman-Kay and L. E. Kay, *J. Mol. Biol.*, 2002, **316**, 101–112.
42. R. Konrat, *J. Magn. Reson.*, 2014, **241**, 74–85.
43. S. Sasmal, J. Lincoff and T. Head-Gordon, *Biophys. J.*, 2017, **113**, 1002–1011.
44. A. J. Baldwin and L. E. Kay, *Nat. Chem. Biol.*, 2009, **5**, 808–814.
45. C. Charlier, G. Bouvignies, P. Pelupessy, A. Walrant, R. Marquant, M. Kozlov, P. De Ioannes, N. Bolik-Coulon, S. Sagan and P. Cortes, *J. Am. Chem. Soc.*, 2017, **139**, 12219–12227.
46. D. M. Korzhnev, P. Neudecker, A. Mittermaier, V. Y. Orekhov and L. E. Kay, *J. Am. Chem. Soc.*, 2005, **127**, 15602–15611.
47. D. F. Hansen, P. Vallurupalli, P. Lundström, P. Neudecker and L. E. Kay, *J. Am. Chem. Soc.*, 2008, **130**, 2667–2675.
48. P. Vallurupalli, G. Bouvignies and L. E. Kay, *J. Am. Chem. Soc.*, 2012, **134**, 8148–8161.
49. P. Vallurupalli, A. Sekhar, T. Yuwen and L. E. Kay, *J. Biomol. NMR*, 2017, **67**, 243–271.
50. R. Schneider, D. Maurin, G. Communie, J. Kragelj, D. F. Hansen, R. W. Ruigrok, M. R. Jensen and M. Blackledge, *J. Am. Chem. Soc.*, 2015, **137**, 1220–1229.
51. M. Arai, K. Sugase, H. J. Dyson and P. E. Wright, *Proc. Natl. Acad. Sci. U. S. A.*, 2015, **112**, 9614–9619.
52. P. Vallurupalli, G. Bouvignies and L. E. Kay, *J. Phys. Chem. B*, 2011, **115**, 14891–14900.
53. K. Sugase, H. J. Dyson and P. E. Wright, *Nature*, 2007, **447**, 1021–1025.

54. J. Kragelj, A. Palencia, M. H. Nanao, D. Maurin, G. Bouvignies, M. Blackledge and M. R. Jensen, *Proc. Natl. Acad. Sci. U. S. A.*, 2015, **112**, 3409–3414.
55. E. Delaforge, J. Kragelj, L. Tengo, A. S. Palencia, S. Milles, G. Bouvignies, N. Salvi, M. Blackledge and M. R. Jensen, *J. Am. Chem. Soc.*, 2018, **140**, 1148–1158.
56. H. J. Feldman and C. W. Hogue, *Proteins: Struct., Funct., Bioinf.*, 2002, **46**, 8–23.
57. V. Ozenne, F. Bauer, L. Salmon, J.-R. Huang, M. R. Jensen, S. Segard, P. Bernadó, C. Charavay and M. Blackledge, *Bioinformatics*, 2012, **28**, 1463–1470.
58. R. I. Cukier, *J. Phys. Chem. B*, 2018, **122**, 9087–9101.
59. A. Estaña, N. Sibille, E. Delaforge, M. Vaisset, J. Cortés and P. Bernadó, *Structure*, 2019, **27**, 381–391, e382.
60. J. A. Marsh, C. Neale, F. E. Jack, W.-Y. Choy, A. Y. Lee, K. A. Crowhurst and J. D. Forman-Kay, *J. Mol. Biol.*, 2007, **367**, 1494–1510.
61. G. Nodet, L. Salmon, V. Ozenne, S. Meier, M. R. Jensen and M. Blackledge, *J. Am. Chem. Soc.*, 2009, **131**, 17908–17918.
62. T. Mittag and J. D. Forman-Kay, *Curr. Opin. Struct. Biol.*, 2007, **17**, 3–14.
63. M. Chan-Yao-Chong, D. Durand and T. Ha-Duong, *J. Chem. Inf. Model.*, 2019, **59**, 1743–1758.
64. G. A. Kaminski, R. A. Friesner, J. Tirado-Rives and W. L. Jorgensen, *J. Phys. Chem. B*, 2001, **105**, 6474–6487.
65. Y. Duan, C. Wu, S. Chowdhury, M. C. Lee, G. Xiong, W. Zhang, R. Yang, P. Cieplak, R. Luo and T. Lee, *J. Comput. Chem.*, 2003, **24**, 1999–2012.
66. A. D. MacKerell Jr, M. Feig and C. L. Brooks, *J. Am. Chem. Soc.*, 2004, **126**, 698–699.
67. A. D. Mackerell Jr, M. Feig and C. L. Brooks III, *J. Comput. Chem.*, 2004, **25**, 1400–1415.
68. V. Hornak, R. Abel, A. Okur, B. Strockbine, A. Roitberg and C. Simmerling, *Proteins: Struct., Funct., Bioinf.*, 2006, **65**, 712–725.
69. R. B. Best and G. Hummer, *J. Phys. Chem. B*, 2009, **113**, 9004–9015.
70. R. B. Best and J. Mittal, *J. Phys. Chem. B*, 2010, **114**, 14916–14923.
71. S. Piana, K. Lindorff-Larsen and D. E. Shaw, *Biophys. J.*, 2011, **100**, L47–L49.
72. J. Huang and A. D. MacKerell Jr, *J. Comput. Chem.*, 2013, **34**, 2135–2145.
73. J. A. Maier, C. Martinez, K. Kasavajhala, L. Wickstrom, K. E. Hauser and C. Simmerling, *J. Chem. Theory Comput.*, 2015, **11**, 3696–3713.
74. W. Wang, W. Ye, C. Jiang, R. Luo and H. F. Chen, *Chem. Biol. Drug Des.*, 2014, **84**, 253–269.
75. D. Song, R. Luo and H.-F. Chen, *J. Chem. Inf. Model.*, 2017, **57**, 1166–1178.
76. J. Huang, S. Rauscher, G. Nawrocki, T. Ran, M. Feig, B. L. De Groot, H. Grubmüller and A. D. MacKerell, *Nat. Methods*, 2017, **14**, 71–73.
77. H. Liu, D. Song, H. Lu, R. Luo and H. F. Chen, Wiley Online Library, 2018.

78. K. Lindorff-Larsen, S. Piana, K. Palmo, P. Maragakis, J. L. Klepeis, R. O. Dror and D. E. Shaw, *Proteins*, 2010, **78**, 1950–1958.
79. R. B. Best, W. Zheng and J. Mittal, *J. Chem. Theory Comput.*, 2014, **10**, 5113–5124.
80. S. Piana, A. G. Donchev, P. Robustelli and D. E. Shaw, *J. Phys. Chem. B*, 2015, **119**, 5113–5123.
81. A. Kuzmanic, R. B. Pritchard, D. F. Hansen and F. L. Gervasio, *J. Phys. Chem. Lett.*, 2019, **10**, 1928–1934.
82. P. Robustelli, S. Piana and D. E. Shaw, *Proc. Natl. Acad. Sci. U. S. A.*, 2018, **115**(7), E1346–E1355.
83. F. Palazzesi, M. K. Prakash, M. Bonomi and A. Barducci, *J. Chem. Theory Comput.*, 2015, **11**, 2–7.
84. S. Rauscher, V. Gapsys, M. J. Gajda, M. Zweckstetter, B. L. de Groot and H. Grubmuller, *J. Chem. Theory Comput.*, 2015, **11**, 5513–5524.
85. J. Henriques, C. Cragnell and M. Skepo, *J. Chem. Theory Comput.*, 2015, **11**, 3420–3431.
86. I. Ohmura, G. Morimoto, Y. Ohno, A. Hasegawa and M. Taiji, *Philos. Trans. R. Soc. A*, 2014, **372**, 20130387.
87. D. E. Shaw, P. Maragakis, K. Lindorff-Larsen, S. Piana, R. O. Dror, M. P. Eastwood, J. A. Bank, J. M. Jumper, J. K. Salmon, Y. Shan and W. Wriggers, *Science*, 2010, **330**, 341–346.
88. D. E. Shaw, J. Grossman, J. A. Bank, B. Batson, J. A. Butts, J. C. Chao, M. M. Deneroff, R. O. Dror, A. Even and C. H. Fenton, *SC'14: Proceedings of the International Conference for High Performance Computing, Networking, Storage and Analysis*, 2014.
89. M. Shirts and V. S. Pande, *Science*, 2000, **290**, 1903–1904.
90. K. J. Kohlhoff, in *Biomolecular Simulations*, Springer, 2019, pp. 291–309.
91. S. Bottaro and K. Lindorff-Larsen, *Science*, 2018, **361**, 355–360.
92. R. Zhou, B. J. Berne and R. Germain, *Proc. Natl. Acad. Sci. U. S. A.*, 2001, **98**, 14931–14936.
93. G. l. H. Zerze, C. M. Miller, D. Granata and J. Mittal, *J. Chem. Theory Comput.*, 2015, **11**, 2776–2782.
94. Y. Sugita and Y. Okamoto, *Chem. Phys. Lett.*, 1999, **314**, 141–151.
95. U. H. E. Hansmann, *Chem. Phys. Lett.*, 1997, **281**, 140–150.
96. P. Liu, B. Kim, R. A. Friesner and B. Berne, *Proc. Natl. Acad. Sci. U. S. A.*, 2005, **102**, 13749–13754.
97. L. Wang, R. A. Friesner and B. Berne, *J. Phys. Chem. B*, 2011, **115**, 9431–9438.
98. R. Appadurai, J. Nagesh and A. Srivastava, *Nat. Commun.*, 2021, **12**, 1–11.
99. D. Hamelberg, J. Mongan and J. A. McCammon, *J. Chem. Phys.*, 2004, **120**, 11919–11929.
100. A. Laio and M. Parrinello, *Proc. Natl. Acad. Sci. U. S. A.*, 2002, **99**, 12562–12566.
101. E. Chiavazzo, R. Covino, R. R. Coifman, C. W. Gear, A. S. Georgiou, G. Hummer and I. G. Kevrekidis, *Proc. Natl. Acad. Sci. U. S. A.*, 2017, **114**, E5494–E5503.

102. L. Bonati, V. Rizzi and M. Parrinello, *J. Phys. Chem. Lett.*, 2020, **11**, 2998–3004.
103. P. Gkeka, G. Stoltz, A. Barati Farimani, Z. Belkacemi, M. Ceriotti, J. D. Chodera, A. R. Dinner, A. L. Ferguson, J.-B. Maillet and H. Minoux, *J. Chem. Theory Comput.*, 2020, **16**, 4757–4775.
104. G. Bussi, F. L. Gervasio, A. Laio and M. Parrinello, *J. Am. Chem. Soc.*, 2006, **128**, 13435–13441.
105. S. Piana and A. Laio, *J. Phys. Chem. B*, 2007, **111**, 4553–4559.
106. J. Pfaendtner and M. Bonomi, *J. Chem. Theory Comput.*, 2015, **11**, 5062–5067.
107. A. Prakash, C. D. Fu, M. Bonomi and J. Pfaendtner, *J. Chem. Theory Comput.*, 2018, **14**, 4985–4990.
108. T. Löhr, K. Kohlhoff, G. T. Heller, C. Camilloni and M. Vendruscolo, *Nat. Comput. Sci.*, 2021, **1**, 71–78.
109. J.-M. Choi and R. V. Pappu, *J. Chem. Theory Comput.*, 2019, **15**, 1367–1382.
110. J. Chen, W. Im and C. L. Brooks, *J. Am. Chem. Soc.*, 2006, **128**, 3728–3736.
111. K. H. Lee and J. Chen, *J. Comput. Chem.*, 2017, **38**, 1332–1341.
112. T. Löhr, A. Jussupow and C. Camilloni, *J. Chem. Phys.*, 2017, **146**, 165102.
113. A. Borgia, M. B. Borgia, K. Bugge, V. M. Kissling, P. O. Heidarsson, C. B. Fernandes, A. Sottini, A. Soranno, K. J. Buholzer and D. Nettels, *Nature*, 2018, **555**, 61–66.
114. G. Krainer, T. J. Welsh, J. A. Joseph, J. R. Espinosa, S. Wittmann, E. de Csilléry, A. Sridhar, Z. Toprakcioglu, G. Gudiškytė and M. A. Czekalska, *Nat. Commun.*, 2021, **12**, 1–14.
115. G. Tesei, T. K. Schulze, R. Crehuet and K. Lindorff-Larsen, *Proc. Natl. Acad. Sci.*, 2021, e2111696118.
116. Y. Shen and A. Bax, *J. Biomol. NMR*, 2010, **48**, 13–22.
117. K. J. Kohlhoff, P. Robustelli, A. Cavalli, X. Salvatella and M. Vendruscolo, *J. Am. Chem. Soc.*, 2009, **131**, 13894–13895.
118. M. Zweckstetter, *Nat. Protoc.*, 2008, **3**, 679.
119. C. Camilloni and M. Vendruscolo, *J. Phys. Chem. B*, 2014, **119**, 653–661.
120. M. Nygaard, B. B. Kragelund, E. Papaleo and K. Lindorff-Larsen, *Biophys. J.*, 2017, **113**, 550–557.
121. F. Kümmerer, S. Orioli, D. Harding-Larsen, F. Hoffmann, Y. Gavrilov, K. Teilum and K. Lindorff-Larsen, *J. Chem. Theory Comput.*, 2021, **17**, 5262–5275.
122. S. Olsson and F. Noé, *J. Am. Chem. Soc.*, 2017, **139**, 200–210.
123. R. Rangan, M. Bonomi, G. T. Heller, A. Cesari, G. Bussi and M. Vendruscolo, *J. Chem. Theory Comput.*, 2018, **14**, 6632–6641.
124. M. Bonomi, G. T. Heller, C. Camilloni and M. Vendruscolo, *Curr. Opin. Struct. Biol.*, 2017, **42**, 106–116.
125. W. Rieping, M. Habeck and M. Nilges, *Science*, 2005, **309**, 303–306.
126. G. Hummer and J. Kofinger, *J. Chem. Phys.*, 2015, **143**, 243150.

127. S. Bottaro, T. Bengtsen and K. Lindorff-Larsen, in *Structural Bioinformatics*, Springer, 2020, pp. 219–240.
128. I. Sučec, Y. Wang, O. Dakhlaoui, K. Weinhäupl, T. Jores, D. Costa, A. Hessel, M. Brennich, D. Rapaport and K. Lindorff-Larsen, *Sci. Adv.*, 2020, **6**, eabd0263.
129. M. Bonomi, C. Camilloni, A. Cavalli and M. Vendruscolo, *Sci. Adv.*, 2016, **2**, e1501177.
130. M. Bonomi, C. Camilloni and M. Vendruscolo, *Sci. Rep.*, 2016, **6**, 31232.
131. The PLUMED Consortium, *Nat. Methods*, 2019, **16**, 670–673.
132. G. T. Heller, F. A. Aprile, M. Bonomi, C. Camilloni, A. De Simone and M. Vendruscolo, *J. Mol. Biol.*, 2017, **429**, 2772–2779.
133. G. T. Heller, M. Bonomi and M. Vendruscolo, *J. Mol. Biol.*, 2018, 2288–2292.
134. G. T. Heller, F. A. Aprile, T. C. Michaels, R. Limbocker, M. Perni, F. S. Ruggeri, B. Mannini, T. Löhr, M. Bonomi and C. Camilloni, *Sci. Adv.*, 2020, **6**, eabb5924.
135. K. A. Merchant, R. B. Best, J. M. Louis, I. V. Gopich and W. A. Eaton, *Proc. Natl. Acad. Sci. U.S.A.*, 2007, **104**, 1528–1533.
136. G. Fuertes, N. Banterle, K. M. Ruff, A. Chowdhury, D. Mercadante, C. Koehler, M. Kachala, G. E. Girona, S. Milles and A. Mishra, *Proc. Natl. Acad. Sci. U.S.A.*, 2017, **114**, E6342–E6351.
137. G. T. Heller, F. A. Aprile and M. Vendruscolo, *Cell. Mol. Life Sci.*, 2017, **74**, 3225–3243.
138. M. Varadi, S. Kosol, P. Lebrun, E. Valentini, M. Blackledge, A. K. Dunker, I. C. Felli, J. D. Forman-Kay, R. W. Kriwacki, R. Pierattelli, J. Sussman, D. I. Svergun, V. N. Uversky, M. Vendruscolo, D. Wishart, P. E. Wright and P. Tompa, *Nucleic Acids Res.*, 2014, **42**, D326–D335.
139. J. R. Allison, *Curr. Opin. Struct. Biol.*, 2017, **43**, 79–87.

CHAPTER 15

Structure- and Dynamics-guided Drug Development Using NMR and its Application to Diverse Pharmaceutical Modalities

KOH TAKEUCHI

Graduate School of Pharmaceutical Sciences, The University of Tokyo, 7-3-1 Hongo, Bunkyo, Tokyo 113-0033
Email: koh-takeuchi@mol.f.u-tolyo.ac.jp

15.1 Introduction

The unique strength of solution-state nuclear magnetic resonance (NMR) is to contribute not only to determine high-resolution structures of macromolecules but also to provide quantitative atomic-resolution information on functional interaction and dynamics of the systems of interest. The quantitative dynamics information obtained by NMR includes the population of each conformation in equilibrium, the exchange rates between the different conformational states, and the structural characteristics of each state.[1–23] These can be utilized in characterizing protein conformational equilibrium in both unligated and ligated conditions and conformational changes upon the binding to ligands to exert certain functions.[24–27] It also allows analyzing transiently formed complexes, often associated with multiple unstable bound states.[28–30] NMR also tells us the positions of transiently formed structural elements in intrinsically disordered regions and intrinsically

New Developments in NMR No. 26
NMR Spectroscopy for Probing Functional Dynamics at Biological Interfaces
Edited by Anirban Bhunia, Hanudatta S. Atreya and Neeraj Sinha

Published by the Royal Society of Chemistry, www.rsc.org

disordered proteins (IDP),[31–34] which serve as general target-binding sites and/or multimerization sites.[35–38] These include the characterization of fuzzy complexes representing a continuum from the fully ordered states to the fully disordered ensembles.[39] The tautomeric states of histidine and other functional moieties in a macromolecule can be directly and separately monitored by solution-state NMR,[40–44] which can be reflected in the efficiency of enzymatic reactions[45,46] and sequence-dependent misincorporation in replication and translation,[44] *etc.* Furthermore, NMR can also describe the behavior of macromolecules in a dynamic system, such as membrane-less compartments like liquid–liquid phase separations.[47–54] Thus, NMR provides a unique avenue toward such characterization that is inaccessible to other techniques, such as X-ray crystallography and cryo-electron microscopy.

The ability of NMR to analyze macromolecule dynamics was already acknowledged when it started to be applied to macromolecules. For example, NMR characterized the aromatic ring flips in bovine pancreatic trypsin inhibitor protein families in the 1970s.[55,56] Such a ring flip is independent of the protein unfolding process and is permitted with the conformational breathing of the protein. In combination with stable-isotope labeling techniques,[57–68] NMR can analyze protein structure and dynamics with atomic resolution without introducing any chemical modifications that could spoil their functionally important dynamic properties.

Taking advantage of these unique characteristics, the solution-state NMR technique has unveiled the functional dynamics of many critical biological systems, including enzymes,[3,25,27,69–74] membrane proteins,[72,75] signaling molecules,[76] transcription factors,[27,71,73] molecular chaperones,[77] nucleic acids,[78] and others.[3,25,69,70] It is also important to point out that NMR can obtain various independent dynamics parameters that characterize internal motions in multiple time scales in a macromolecule, presenting a considerably detailed description of their function (Figure 15.1). Quantitative dynamics measurements by NMR also allow estimating the difference in the conformational entropy upon binding to the ligand,[18,79–85] which are somewhat overlooked under the hood of the changes of the desolvation entropy. In addition, NMR can access the lowly populated high-energy protein conformations in an equilibrium that is not readily accessible or invisible for other structural methods.[1,43,86–88] These features of NMR are indispensable to expand static structure-guided drug development (SGDD) into *dynamic* SGDD, which actively utilizes the dynamic aspect of macromolecule–ligand interactions in the rational design of functional molecules.[89–94]

The development of new functional molecules (modalities) that act on living organisms will broaden the scope of biological research and substantially impact human life. Especially in a clinical area, new pharmaceutical modalities would cure diseases that could not be treated before. High-molecular-weight (HMW) biomedicines, such as monoclonal antibodies and medium-size molecules, are currently drawing much attention in this sense. These modalities are beneficial because they can express bioactivities that small molecules cannot realize. Well-known examples would be the human

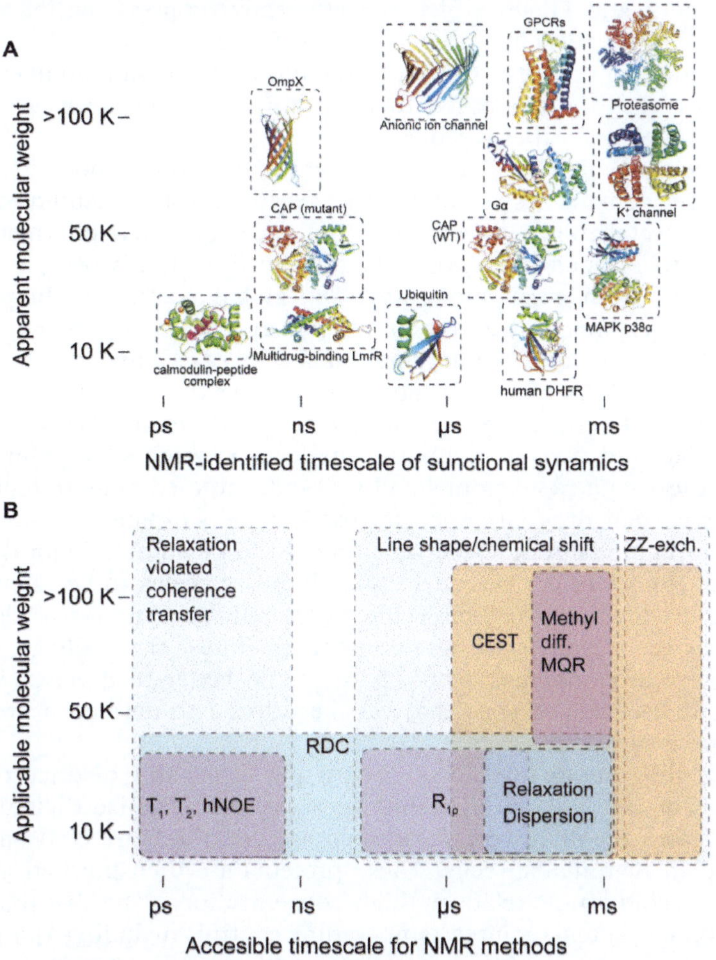

Figure 15.1 NMR-identified functional dynamics of proteins. (A) The timescale of NMR-identified functional dynamics and the apparent molecular weights of systems. For membrane proteins, the apparent molecular weight includes those of micelle and nanodiscs. This figure highlights the timescale shown to be associated with the function of the proteins. It should be noted that the proteins might harbor functional dynamics with different timescales that are not depicted in this figure. Reproduced from ref. 237, https://doi.org/10.3390/ijms21051829, under the terms of the CC BY 4.0 license, https://creativecommons.org/licenses/by/4.0/. (B) The timescale of the dynamics and molecular weight range that is accessible for each NMR experiment. Reproduced from ref. 2 with permission from Elsevier, Copyright 2019.

anti-programmed cell death protein 1 (PD1) monoclonal antibodies, which can inhibit the protein–protein interaction (PPI) between PD1 and PD-L1 to cure certain types of cancer.[95] Examples of the medium-size molecules are cyclosporine A and FK506, which can efficiently suppress the intracellular interaction between calcineurin and nuclear factor of activated T-cells to inhibit the immune response. Currently, there is an increasing number of oral drugs that exceed the rule of five by Lipinski,[96] which has been used as the practical guideline for these medications.

As small molecule drugs are much cheaper than other modalities, their development remains essential for the worldwide sustainability of good health and well-being. However, with the success of new pharmaceutical modalities mentioned above, the global trend in drug discovery is rapidly moving away from small molecules. The global market for biopharmaceuticals, which was only about $67 billion in 2002, had grown eight-fold to $530 billion in 2012, while small-molecule sales almost halved from $415 billion to $218 billion. In addition, among the top 20 drugs in the 2016 sales rankings, there are eight drugs with molecular weights exceeding 500, and two of them have molecular weights exceeding 1000, clearly indicating a trend from small molecules to medium-size molecules. Thus, the current drug-discovery process is rapidly expanding its scope to include novel modalities, which were thought to be unfeasible for drugs 20 years ago. Not all of the pharmaceutical industry seems to be catching up with this trend, and small-molecule drugs still account for a high percentage of the products in development pipelines, especially in smaller pharmaceutical companies. Thus, it might be better to develop versatile and easy-to-use technologies that can be utilized to develop a new drug with these new modalities.

Solving the unique problems originating from the distinct physical properties of new modalities would be a way to maximize their use. For example, most biopharmaceuticals, such as antibodies, are HMW proteins expressed in mammalian cells. These proteins are then purified and formulated in solution at relatively high concentrations. The distribution of these products often requires temperature control, including refrigerated transportation using cold-chain technology. On the other hand, the higher-order structure (HOS) of biomedicine can be modified or distorted relatively easily by chemical degradations, conversions to oxidized forms, and/or partial denaturation that can cause irreversible aggregations.[97,98] These structural deteriorations of biopharmaceuticals would not only reduce their activity but also cause side effects. Therefore, technology that enables a non-invasive evaluation of the HOS of HMW biomedicine in solution is wanted. As for medium-size molecules, their structures are not large enough to retain a defined hydrophobic core. Therefore, their structural flexibilities are generally more substantial than those of a small molecule as well as biopharmaceuticals. Rational optimization of medium-size molecules with a larger number of conformational degrees of freedom is an issue that needs to be addressed. In addition, while the ability of the medium-size molecules to permeate the cell is an important aspect, no technologies exist to predict and improve this desired feature.

15.2 NMR Techniques to Promote the Use of New Pharmaceutical Modalities

The unique characteristics of NMR to unveil the structure and dynamics of molecules of interest would contribute to solving the problems associated with these new modalities. In particular, we believe that characteristics of NMR to perform a variety of non-destructive analyses, such as atomic-level structural analysis, kinetic analysis, and interaction analysis, under various solution conditions would make the technique the most promising approach among other structural methods. NMR is characterized by the ability to conduct both structure and dynamics analysis in solution. Using NMR, the structural information of each component in a structurally heterogeneous mixture can be obtained quantitatively and separately without any preceding separation and purification. In addition, unlike other structural strategies, NMR can handle a substantial versatility of sample conditions. The solvent is not necessarily limited to an aqueous solution, and organic solvents and solutions with a high concentration of co-solute can be used. It is also possible to vary the temperature and pressure to assess the thermodynamic characteristics of proteins and protein–ligand complexes.[99–101] Solution NMR can also access the interface between the solution and solid phases, such as fibrils, nanoparticles, and bead-attached proteins.[102–105] Furthermore, structure and dynamics information of the molecules of interest in living cells can be obtained by using in-cell or *in vivo* NMR strategies.[106–110]

On the other hand, the solution-state NMR has some limitations. The most considerable and fundamental limitation of NMR is its intrinsic insensitivity compared to other spectroscopic techniques. This limitation has been alleviated partially by high external magnetic field strengths, cryogenic probe technology, and stable isotopic labeling strategies. In addition, when solution-state NMR is utilized in HMW systems, the method suffers from further attenuation of sensitivity and loss of resolution. Toward this end, the developments of new measurement techniques, including our own, have elevated the applicable molecular weight and widened the scopes of solution-state NMR applications. In this chapter, we review the development of new NMR technologies to promote new pharmaceutical modalities. We discuss NMR methods that non-invasively perform HOS analysis of biopharmaceuticals using direct detection of nuclei with lower gyromagnetic ratios (γ), such as ^{13}C and ^{15}N. We also overview the NMR analysis that enables structural optimization of ligands, including medium-size molecules, utilizing quantitative conformational dynamics information. In addition, the versatility of NMR in sample conditions allows us to conduct structural comparisons in different solution conditions, including those that mimic membrane environments, to understand the cell permeabilization activity of medium-size molecules.

15.3 NMR Characterization of HMW Biopharmaceuticals

15.3.1 Limitation of Conventional and 1D NMR in the HOS Analyses of HMW Biopharmaceuticals

As mentioned earlier, most biopharmaceuticals are HMW proteins such as antibodies, and their three-dimensional structures can change depending on various environmental factors such as expression, purification, and solution conditions. Therefore, to evaluate the HOS of biopharmaceuticals, it is essential to have a technique for a non-invasive direct evaluation of the three-dimensional structures under their formulation conditions. The NMR method is one of the most suitable techniques to meet this requirement, as it can perform a non-invasive HOS analysis at the atomic level in various solution conditions. However, when applying conventional multidimensional NMR techniques to biopharmaceuticals with a molecular weight exceeding 100 kDa, including antibodies, we encounter a problem of deuteration.

Current primary multidimensional NMR techniques suitable for HMW proteins, such as $^{1}H^{15}N$ transverse relaxation optimized spectroscopy (TROSY)[111] and $^{1}H^{13}C$ methyl heteronuclear multiple quantum coherence (HMQC),[57,112] require perdeuteration for highly sensitive detection. On the other hand, antibodies and other biopharmaceuticals are primarily expressed in mammalian cells that are not amenable for perdeuteration. Although heterologous expression in bacteria and yeast cells can achieve high-level perdeuteration, monoclonal antibodies (mAbs) obtained using this approach do not retain mammalian-type glycosylation.[67,113,114] Since the compositions of sugar chains can modulate the stability, bioactivity, and immunogenicity of the biopharmaceuticals,[115,116] the heterologous expression in bacteria or yeast cells would not be appropriate for the HOS characterization of biopharmaceuticals. It should also be noted that there are a couple of reports showing that selective deuteration near detecting methyl moieties can improve the sensitivity of $^{1}H^{13}C$ methyl HMQC;[117–119] however, these strategies require deuterated amino acids that are often too expensive to be used casually at the production level. Thus, the current major multidimensional NMR technique for HMW proteins cannot be readily applicable to biopharmaceuticals.

Utilization of ^{1}H-detected or ^{13}C-detected one-dimensional (1D) NMR can partially solve such a problem, as these experiments are often sufficiently sensitive even without deuteration.[120–125] However, a site-specific detection of resonances is not straightforward with 1D NMR experiments. Additional tricks such as specific labeling[123–125] and extensive post-acquisition processing would be required to obtain site-specific information.[123–130] In contrast, multidimensional NMR spectra can offer a comprehensive fingerprint of a protein, in which most of the sites in the protein are individually resolved in the spectrum. The general strategy to obtain the two-dimensional (2D) NMR spectra of mAb is the fragmentation of intact IgG into Fab and Fc portions, each with one-third of the molecular weight of intact IgG. Indeed, an almost complete assignment is reported for the Fc portion.[131] Although the fragmentation of IgG by selective protease digestions is well established, it is not always straightforward to find such a condition in other proteins of interest.

15.3.2 Variable-temperature NMR Experiments

Another strategy to obtain the 2D NMR spectra of mAb is a high-temperature measurement. Higher temperature accelerates the molecular tumbling as the viscosity of water at 50 °C is ~27% and ~57% lower than those at 4 °C and 25 °C, respectively. With the implementation of high-sensitive cryogenic probe technology[132] and the utilization of rapid acquisition methods,[133–135] high-temperature recording at natural isotopic abundance is possible for both ^{13}C and ^{15}N.[136] It is reported that more than 95% of IgG ^{1}H^{13}C methyl resonances can be observed in a spectrum acquired at 50 °C with <300 μM natural isotopic abundance samples.[136] As for 2D ^{1}H^{15}N correlations, a band-selective optimized flip angle short transient (SOFAST)-HMQC spectrum of the National Institute of Standards and Technology (NIST) mAb was obtained with a total experimental duration of approximately 68 h at 900 MHz and 50 °C.[137] Such spectra have been shown to distinguish the different mAbs,[137] the effect of additives,[138] and chemical modifications.[139,140] However, measuring at a higher temperature would possibly induce the deterioration of the biologics, such as structural and chemical modifications and aggregation that might not exist at lower temperatures.

15.3.3 Application of ^{15}N-direct Observation to HOS Analyses of HMW Biopharmaceuticals

The author and colleagues have recently developed the ^{15}N-direct detected cross-correlated relaxation-enhanced polarization transfer (CRINEPT) method,[141] enabling NMR observation of HMW proteins under non-deuterated conditions. In this method, the authors selected ^{15}N as the detecting nuclei instead of commonly used ^{1}H. Although ^{15}N is one of the most insensitive nuclei with low γ, the feature also makes the nuclei yield narrower NMR resonances and are less susceptive to higher molecular weight conditions.[142] In addition, due to its low γ, the ^{15}N tends to be isolated, and its transverse relaxation rates are less sensitive to the presence of remote protons. Thus, ^{15}N-direct detection does not require perdeuteration that is not feasible for mammalian cells. Furthermore, due to its low natural abundance (0.1%), a ^{15}N direct detection provides background-free NMR spectra. These features are essential for the structure and dynamic characterization of HMW biopharmaceuticals in formulated conditions that often contain high concentrations of additives. It should also be noted that the ^{15}N transverse relaxation rate is insensitive to high salt concentrations.[143]

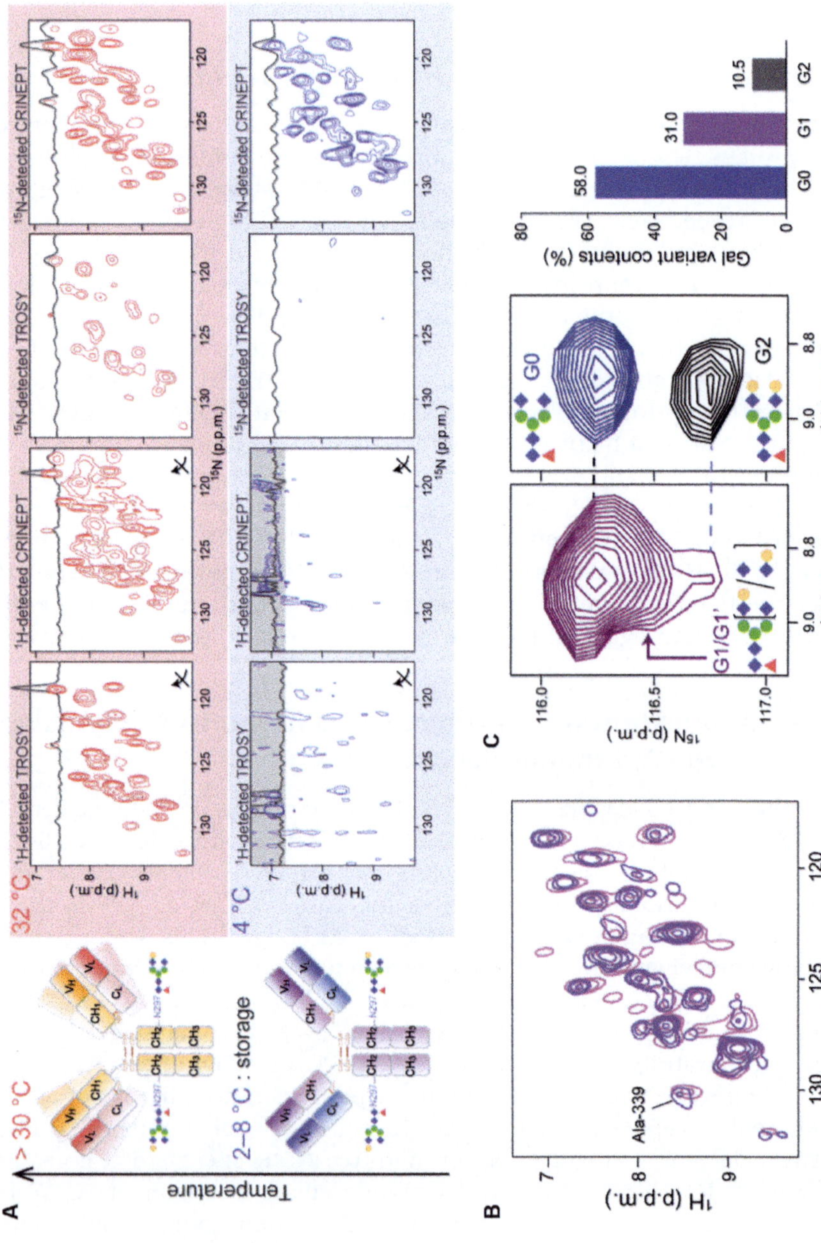
A
Temperature
> 30 °C
2–8 °C : storage
32 °C
4 °C
^{1}H-detected TROSY
^{1}H-detected CRINEPT
^{15}N-detected TROSY
^{15}N-detected CRINEPT
^{1}H (p.p.m.)
^{15}N (p.p.m.)
B
Ala-339
C
G1/G1'
G0
G2
Gal variant contents (%)
58.0
31.0
10.5

The ^{15}N-detected CRINEPT was applied to an intact mAb at a wide range of temperatures (4–32 °C) and formulation conditions for subcutaneous injection (SC) and intravenous injection (IV). A large fraction of ^{1}H–^{15}N correlation signals of mAb was successfully observed under formulated storage conditions (4 °C), where the conventional ^{1}H-detected experiments did not work (Figure 15.2A). For a 150 kDa protein at 4 °C, the sensitivity of the ^{15}N-detected CRINEPT was expected to be >20-fold greater than that of the previously reported insensitive nuclei enhanced by a polarization transfer (INEPT)-based experiment.[142,143] The principles of the coherence transfer in the ^{15}N-detected CRINEPT are the same as in the ^{1}H-detected CRINEPT;[144] however, several tricks have been introduced for high-sensitivity detection. The ^{1}H-chemical shift evolution and the ^{1}H to ^{15}N coherence transfer periods are concatenated at the beginning of the pulse program. In the acquisition period, the ^{15}N-direct detection of antiphase coherence is carried out without refocusing $^{1}H^{15}N$ coupling. In addition, ^{15}N-detected CRINEPT spectra are free from artificial resonances from additives in the formulation conditions, such as amino acids, surfactants, and sugars. It should be noted that the spectral patterns of mAb obtained by ^{15}N-direct detected experiment in different formulation conditions for SC and IV were not identical. This observation indicates that the conformation of mAb is slightly different in these formulations. Especially, Ala-339 in the domain linker showed different chemical shifts, indicating that the linker region might be sensitive to the difference in formulations (Figure 15.2B).[141]

In addition, using the ^{15}N-detected CRINEPT spectra, the heterogeneity of galactosyl modifications at the tip of the N-linked sugar chain was quantitatively characterized in the intact mAb (Figure 15.2C).[141] The Lys-248 resonance observed in ^{15}N-detected CRINEPT spectra of intact mAb consists of multiple components dispersed in the ^{15}N dimension. Since the heterogeneities of the Lys-248 resonance are resolved by enzymatic degalactosylation and galactosylation of the sugar chain, this reflects the polymorphic glycosylation states in

Figure 15.2 Application ^{15}N-detected CRINEPT experiments to mAb. (A) The spectra were recorded at (top) 32 °C and (bottom) 4 °C for an intact [^{15}N-Ala] IgG. For each condition, the ^{1}H-detected TROSY (most left), ^{1}H-detected CRINEPT (middle left), ^{15}N-detected TROSY (middle right), and ^{15}N-detected CRINEPT were shown. For ^{1}H-detected spectra, the ^{1}H and ^{15}N axes of the spectra are transposed as indicated by arrows. (B) Overlaid ^{15}N-detected CRINEPT spectra of [^{15}N-Ala] mAb in *SC* (pink) and *IV* (blue) formulations. The assignments for Ala-339 are based on the chemical shifts of the Ala resonances of the Fc fragment.[131] (C) The heterogeneity of the galactosylation states was analyzed by the ^{15}N-detected CRINEPT experiment. (Left) The ^{1}H–^{15}N resonance of Lys-248 in the ^{15}N-detected CRINEPT spectrum of the intact mAb is shown along with those from the ΔGal (G0, blue) and Gal (G2, black) samples, which are enzymatically galactosidated and galactosylated, respectively. (Right) Ratios of different galactosylation states, estimated from the deconvolution of the Lys-248 resonance in the intact mAb. Reproduced from ref. 238, https://doi.org/10.3390/ijms22010046, under the terms of the CC BY 4.0 license, https://creativecommons.org/licenses/by/4.0/.

the intact mAb. Indeed, Lys-248 is located close to the glycochain termini, where heterogeneity in galactosylation states would be observed. The galactosylation states quantified by NMR agreed with those estimated by liquid chromatography–mass spectrometry. Therefore, the ^{15}N-detected CRINEPT is suitable for a non-invasive HOS evaluation of the biologics. This advantage of the ^{15}N-detected CRINEPT will enable the broader application of solution-state NMR for large eukaryotic proteins, such as membrane proteins and hetero-oligomeric protein complexes, which are typically expressed only in eukaryotic cells. Given the importance of the dynamic properties of these proteins for their function, the ^{15}N-detected CRINEPT provides the appropriate means to examine the structure and dynamics of these biologically and pharmacologically important systems using solution-state NMR for an advanced structure-guided drug development that includes conformational dynamics information.

15.3.4 ^{19}F–^{13}C TROSY Measurement Technique

Although the ^{15}N-detected CRINEPT experiment enables us to observe the mainchain amide resonances of HMW proteins without deuteration, a detection of aromatic sidechain resonances is still challenging, despite the importance of aromatic residues at the interaction sites. The sensitivity of the aromatic sidechain signals often decreases more rapidly with molecular weight than mainchain ^{1}H^{15}N, due to larger dipole–dipole (DD) interaction and chemical shift anisotropy (CSA) for the aromatic ^{1}H–^{13}C moiety. In addition, one-bond carbon–carbon couplings ($^{1}J_{CC}$) complicate signal separation and/or require a constant time scheme to resolve the problem in exchange for sensitivity. The TROSY strategy, which reduces the transverse relaxation utilizing an interplay between DD and CSA interactions, partially resolved the problem;[145,146] however, the TROSY effect on aromatic ^{1}H–^{13}C moiety is not substantial due to the mismatch of the absolute DD and CSA values, and principal tensors. This limits the application of conventional methods up to ~30 kDa unless a more complicated labeling strategy such as alternate ^{12}C–^{13}C labeling is used.[147]

The author and colleagues recently developed the ^{19}F–^{13}C TROSY method to improve the sensitivity of the NMR signals of aromatic side chains.[148] Utilization of the strong TROSY interference between fluorine (^{19}F) and carbon (^{13}C) nuclei in aromatic moiety for high-sensitivity and high-resolution NMR detection has not gathered attention in the past.[148] ^{19}F is one of the most sensitive nuclei due to its sizeable gyromagnetic ratio ($\gamma = 0.941$ relative to ^{1}H) and its 100% natural abundance. In addition, ^{19}F chemical shifts are extremely sensitive to the local chemical environment,[149] resulting in an extensive chemical shift range of ~200 ppm in biological molecules.[150] Taking advantage of the high sensitivity of ^{19}F chemical shifts to the local chemical environment, site-specific ^{19}F labeling has been successfully utilized to examine the functional dynamics of membrane proteins and HMW systems, including GPCRs.[151–156] Site-specific ^{19}F labeling has been mainly introduced by the chemical modification of cysteine using trifluoromethyl (CF_3) probes; however, a multidimensional experiment of ^{19}F nuclei is not pursued due to their fast transverse relaxation rates originating from large CSA.

We have recently found that the efficient mutual interference between DD and CSA in ^{19}F–^{13}C systems in aromatic amino acids, which makes TROSY resonances of ^{19}F-attached ^{13}C nuclei ($^{13}C_F$) significantly narrower than those of the corresponding ^{1}H–^{13}C pairs.[148] Although the angle between the DD and CSA principal tensor is not that different between ^{19}F–^{13}C and ^{1}H–^{13}C pairs, the mismatch of the absolute DD and CSA values is largely resolved in the ^{19}F–^{13}C pair due to smaller DD and larger CSA. Thus, the TROSY effect occurs more efficiently in the ^{19}F–^{13}C system. Due to this preferable effect, the molecular weight-dependent enhancement of ^{13}C transverse relaxation is largely reduced for aromatic ^{19}F–^{13}C systems. In addition, the larger $^{1}J_{FC}$ coupling (240 Hz) than those of $^{1}J_{HC}$ coupling (160 Hz) allows efficient coherence transfer *via* the scalar coupling.

While only a smaller number of signals than expected were observed in the conventional ^{1}H–^{13}C TROSY spectra of a [Tyr-ε ^{1}H^{13}C] 42 kDa maltose-binding protein (MBP), the introduction of a ^{19}F–^{13}C pair into the same positions yielded the expected number of signals with higher sensitivity.[148] Thus, the ^{19}F–^{13}C TROSY method has superior sensitivity compared to the current ^{1}H–^{13}C TROSY method. Indeed, for the $^{13}C_F$ coherence, our estimates indicate that the $^{13}C_F$ TROSY component relaxes ∼8 times slower than the $^{13}C_H$ TROSY component incorporated in a 42 kDa ($\tau_c = 25$ ns, at 25 °C) system at 600 MHz (^{1}H frequency). It should be noted that any coherence with transverse ^{19}F resonance experiences ∼20-times faster relaxation than that of ^{13}C. Thus, we design the pulse sequence with an out-and-stay coherence transfer instead of the conventional out-and-back experiment to minimize the period with transverse ^{19}F. Between two possible experiments with different starting nuclei, an experiment that starts with ^{13}C magnetization and ends on ^{19}F detection was more sensitive than the other for the MBP sample.

On the other hand, an experiment that starts with ^{19}F magnetization and ends on ^{13}C detection had a much better resolution. The ^{19}F–^{13}C TROSY experiment detected 15 intense and isolated ^{19}F–^{13}C pairs.[148] The sensitivity and resolution of the experiment demonstrate the advantage of the 2D ^{19}F–^{13}C TROSY experiments, as both the ^{19}F and ^{13}C 1D spectra recorded for the same sample ([Tyr-ε ^{19}F^{13}C] MBP; 500 μM) cannot resolve individual resonances. In addition, the aromatic ^{19}F–^{13}C signals were observed for a 180 kDa protein in the ^{19}F–^{13}C TROSY experiment at a concentration of 1 mM. Similar to the ^{15}N direct observation TROSY method, the ^{19}F–^{13}C TROSY does not require deuteration. Thus, the strategy can also be used to evaluate the HOS of HMW proteins such as antibodies. Although both the ^{15}N-direct detection and the ^{19}F–^{13}C TROSY method require stable isotope labeling, such labeling costs are relatively low. In addition, the high-resolution spectra obtained from these measurements can be "bridged" to an NMR analysis that does not require stable isotope labeling.

The ^{19}F–^{13}C TROSY method is also applicable to nucleic acids. On average, the TROSY component of $^{13}C_F$ relaxes four times slower than the equivalent decoupled resonance for a nucleobase harboring a $^{19}F-^{13}C$ pair. Indeed, a favorable TROSY effect was reported for DNA[148] and the 61 nt hHBV ϵ RNA.[157] Thus, the $^{19}F-^{13}C$ TROSY will facilitate NMR studies of large RNAs and RNA–protein complexes.

In addition, the new method can be applied to a wide range of biological analyses, such as the active site of HMW enzymes and the ligand-binding site of receptors. Furthermore, the strategy can be used in combination with in-cell NMR technology to analyze the structure and dynamics of HMW proteins and nucleic acids in living cells. Since ^{19}F nuclei do not exist in cells, the $^{19}F-^{13}C$ TROSY method would allow a selective observation of target proteins separately from other proteins in the cell, even at low concentrations.

15.4 Dynamic Structural Optimization of Ligands, Including Medium-size Molecules

Medium-size molecules are a modality that is expected to target intracellular PPIs, whose functions are difficult to control with small molecules and antibodies. However, in order to realize drug discovery of medium-size molecules, it remains necessary to establish a way to design molecules that specifically recognize such intracellular targets. In contrast, the degree of structural freedom in medium-size molecules is more prominent than in small molecules or biopharmaceuticals. Small-molecule drugs typically have only a few angular freedoms in the structure, and biopharmaceuticals have a hydrophobic core to stabilize the native conformation. On the other hand, medium-size molecules typically have a larger number of degrees of freedom and do not have a defined hydrophobic core to stabilize the conformation. Therefore, it is not easy to predict the bound structures of middle-size molecules using *in silico* methods. In addition, the interactions of medium-size molecules are often realized by several weak interactions scattered in the interface.

15.4.1 Utilization of Forbidden Coherence Transition (FCT) for Dynamic Ligand Optimization

The author and colleagues have developed a new NMR technique, FCT analysis, to evaluate the local mobility of bound ligands and their spatial complementarity at the interfaces.[158–160] This technique utilizes an NMR phenomenon called relaxation violated coherence transfer,[161,162] which has been used to analyze the fast ps–ns dynamics of methyl groups in HMW proteins, such as 360 kDa half-proteasome complex,[163] ClpP protease,[164] and kinases.[165,166] The applications of the strategy are not limited to proteins but also nucleic acids and protein–nucleic acid complexes, such as nucleosome core particles.[167]

Although the technique has been mostly utilized to extract dynamics information, the analyses, in theory, can extract two relaxation parameters: the intramethyl $^{1}H-^{1}H$ dipolar cross-correlated relaxation rate, η, and the sum of the dipolar cross-relaxation rates with external protons, δ. The parameter η is proportional to the generalized order parameter, S^2_{axis}, which reports the amplitude of the ps–ns motion of the methyl threefold axis. The parameter δ is the cross-correlated relaxation rate derived from external protons and is proportional to the sum of the −6 power of the distance between the external protons and the methyl group. The contribution of external protons to δ is substantial when $r_{eff} = (\Sigma_{ext} r^{-6}_{ext,\ HH})^6 < 3.0$ Å,[163] while the effects are only marginal when r_{eff} is > 4.0 Å. Therefore, the FCT method provides the local surface complementarity information. Although these features have not been utilized extensively, we proposed that changing the proton density of the receptor by fractional deuteration would affect the δ_b values of the bound ligand without changing η values. Thus, the difference of the δ_b values ($\Delta\delta_b$) between the perdeuterated and fractionally deuterated conditions would reflect surface complementarity to the receptor.[168] In this case, $\Delta\delta_b$ is more substantial for methyl groups that are buried deeply in the receptor with optimal surface complementarity. Therefore, by observing the time-dependent development of FCT in NMR spectra using perdeuterated and partially deuterated receptors, it is possible to simultaneously analyze the mobility and spatial complementarity of the ligand methyl moiety in the complex. Since the FCT method is applicable to HMW proteins, it is attractive for structure-guided drug development in general by contributing rational design of better ligands with restricted local dynamics and improved surface complementarity in the receptor-bound state. It should be noted that the strategy was recently used to estimate the effect of local deuteration on the methyl TROSY effect.[119]

We also have shown that the FCT method could be extended to weak-affinity ligands that experience a fast equilibrium between free and bound states (Ex-FCT).[168] In the study, the local dynamics and the surface complementarity of weak-affinity ligands in the receptor-bound state can be obtained in the same scheme as FCT. In the fast exchange situation, observed η and δ are the population averages of the values in the free and bound states. Nevertheless, since the magnitudes of the η and δ values are proportional to the molecular weight of the system, the contribution of the bound-state coherences to the Ex-FCT profile is more substantial than those of the free state. Therefore, even in ligand excess conditions, the Ex-FCT profile would effectively reflect the methyl dynamics in the bound state if the apparent molecular weight of ligands changes substantially upon binding. Practically, from two separate experiments with a distinct ratio of ligand/receptor, η and δ for each methyl in the bound state can be quantitatively determined.[168]

The FCT analysis of an inhibitor peptide in complex with MAP kinase p38 revealed that Val-8 γ, which is more exposed to the solvent, was less dynamic as compared to Val-7 γ and Ile-9 δ1, that are located in the binding pocket of p38α. The higher S^2 values for Val-8 γ seemed to originate from the limited rotameric states of the methyl groups due to proximal water. The results also indicate that the FCT method is sensitive enough to detect the mobile part of the interface. In addition, the dynamics of the methyl groups and the degrees of surface exposure are not correlated. Indeed, it is reported that some methyl groups are highly dynamic in ligand–receptor interfaces in other systems, as well.[169] The surface complementarity around each methyl group deduced from the experiment is consistent with the X-ray structure of the peptide–p38α complex. When Val-8 was replaced by slightly larger isoleucine, the binding constant was improved several-fold, and the thermodynamic parameters of binding became more enthalpy-dependent, indicating an improvement in the thermodynamic properties of the interaction. Thus, dynamic optimization of ligand that considers the local dynamics of the bound state is essential for designing medium-size molecules with a wide interaction surface.

The FCT analysis was also extended to CF_3 groups.[170] It has been shown that approximately 25% of therapeutic drugs on the market contain at least one fluorine atom, and the CF_3 group is one of the major fluorine-containing substituents. The dynamics property of the CF_3 group of AST-487 in the p38α-bound state was analyzed with this strategy. The order parameter was estimated to be 0.64 for the system, indicating that the CF_3 group and its directly attached phenyl ring of AST-487 retain substantial mobility in the p38α-bound state. This complements the lack of NOE-based information about the intermolecular interaction for the trifluoromethyl-phenyl group.

15.4.2 Consideration of Multiple Binding Modes in Ligand Optimization

Although current SGDD mainly relies on static structures, estimating the enthalpic contributions remains challenging. The enthalpic contribution in the bound state can be obtained from quantum-mechanical calculations such as the fragment molecular orbital method,[171–173] using a high-resolution structure of the complex. However, one should be careful that structural variation in the bound state would largely affect the calculation. Although the conformational dynamics of bound ligands have rarely been investigated, Lee *et al.* reported that the inhibitor of LpxC, a novel antibiotic target, accesses alternative, minor population states in solution, besides the major conformation observed in crystal structures.[174] A novel inhibitor that incorporates both the major and minor interaction sites increased the inhibition constants of the ligand.

In the context of protein–ligand interactions, the formation of interaction may not be an isolated event. For example, forming a hydrogen bond requires breaking hydrogen bonds with water present in the unligated state. In addition, affinity or free energy change upon the binding is also affected by the entropic change of the system. Molecular dynamics simulations might estimate both enthalpic and entropic terms; however, it still harbors errors on the order of ~1 kcal mol^{-1},[175–177] corresponding to ~10-times difference in K_d values. Therefore, a strategy to extract quantitative conformational dynamics in the protein–ligand complex is of importance in ligand optimization.

Since entropy is a measure of disorder in the system, it is connected to the change in the number of independent states.[178] In this sense, the changes in the fast dynamics in protein can be connected to the conformational entropy of the system. It is shown that changes in the fast dynamics of methyl-bearing amino acids can be used to empirically estimate the changes in conformational entropy in protein interaction systems.[79,80,84,179–181]

Although the first dynamics of the protein–ligand complex contribute to the entropy of the system, a ligand can have slower conformational dynamics, which might change upon binding to a receptor. A ^{13}C relaxation dispersion (RD) experiment can identify slower ligand flexibility using natural isotopic abundance samples.[182] The strategy was applied to three structurally similar Pin1 ligands with distinct flexibility. One of the ligands is a substrate of Pin1 with the FFpSPR sequence, which isomerizes between the *cis-trans* conformations at the pSer-Pro (pSP) peptide bond. The other two inhibitors retain alkene isosteres instead of the core pSP motif, thereby locking the conformational equilibrium. The authors successfully identified the site-specific ^{13}C RD in the pSP motif of the substrate peptide, while neither inhibitors exhibited any RD profile.[182,183] Thus, the NMR strategy can precisely identify the position and time scales of motion in ligands in complex with the receptor.

It should be noted that binding kinetics in the millisecond range often leads to an exchange broadening and contribute to the RD profile. Thus, it is often difficult to distinguish an exchange between free/bound states from an exchange between multiple binding modes in the bound form without having additional experiments. A simple strategy to distinguish these two phenomena is to carry out experiments with increasing concentration. For the compound that does not have multiple conformations in the bound form, the broadening was solely attributed to the free/bound interaction kinetics. In such a case, the line broadening can be overcome with a high molar excess of the ligand. On the other hand, if the line broadening is due to an exchange between multiple bound conformations, the broadening could not be resolved by an excess ligand. This was clearly shown by analyzing two homologous small-molecule ligands of Bcl-xL with mid-micromolar dissociation constants.[184]

As we discussed here, utilizing conformational dynamics information of ligands in receptor-bound states is quite important to understand the thermodynamics and kinetics of ligands and improve the ligands' efficacy. To experimentally quantify and evaluate the dynamics–activity and dynamics–function relationships of ligands, a suite of NMR experiments discussed here would contribute to defining the ligand dynamics in a wide range of time scales.

15.5 Controlling the Dynamics of a Target Protein to Facilitate SGDD

The contribution of NMR in SGDD is evident from its utilization in the development of breakthrough medications. A hallmark example is Venetoclax, ABT-199, a therapeutic for chronic lymphocytic leukemia. ABT-199 and its analogs inhibit the activity of anti-apoptotic proteins, such as B-cell lymphoma-2 (Bcl-2), by orthosterically inhibiting the PPIs with their Bcl-2 homology domains and the helical interaction segment in the partners. The early development of the Bcl-2 inhibitor was achieved by using NMR followed by further optimization to improve the selectivity that leads to the development of ABT-199.[185]

As exemplified above, PPIs contain many attractive drug targets due to their biological importance. However, in general, the development of PPI inhibitors is not easy as the target sites are too flat and broad to be interfered with by small molecule ligands.[186,187] It has been increasingly recognized that successful PPI inhibitors utilize the structural flexibility of proteins. In particular, they target "cryptic" binding pockets that are not apparent in the absence of a binding ligand but are evident upon the binding of the ligands.[188–190] These hidden binding sites provide additional interaction surfaces to the ligand, thereby enhancing their affinity and specificity. However, most of the cryptic sites have been discovered serendipitously after solving the structures of the protein–ligand complexes. Therefore, a means of discovering and utilizing cryptic sites in advance would be beneficial to the development of PPI inhibitors in general.

For this purpose, we recently developed a new NMR strategy to detect low-populated cryptic sites without ligand and utilized them in an early stage of drug development.[191] Using the temperature-dependent ^{13}C chemical shift changes (CSCs) of methyl groups and ^{13}C methyl RD experiments, the authors showed that Bcl-xL, a target of PPI inhibitors, samples cryptic site *open* conformation even in the absence of ligand (Figure 15.3A). The authors then focused on the allosteric regions that showed temperature-dependent CSCs but are located outside the cryptic pocket. Guided by the CSCs upon the introduction of mutations, the authors found an allosteric mutant that stabilizes the cryptic site *open* conformation (Figure 15.3B). Binding experiments using isothermal titration calorimetry confirmed that the cryptic-*site-open* mutant showed an increased affinity for the BAK peptide, an endogenous ligand known to bind to the cryptic site.

Furthermore, the cryptic site *open* mutant protein showed elevated efficiency in obtaining hit peptides using a phage display library screening (Figure 15.3C), and a hit peptide is confirmed to bind to the cryptic site of wild-type Bcl-xL protein. This observation indicates that NMR-based control of structural equilibrium at the cryptic site can improve the druggability of the target proteins. Since half of the proteins in the human proteome have a cryptic site,[188] the strategy for identifying and stabilizing cryptic sites will promote the development of PPI inhibitors.

15.6 Kinetics Parameter of Interaction by NMR Dynamics Analysis

The importance of the resident time in ligand optimization is now widely recognized in drug discovery,[192,193] as it has substantial impacts on their pharmacological efficacy.[194,195] Quantifying dynamics in the μs and ms range in a

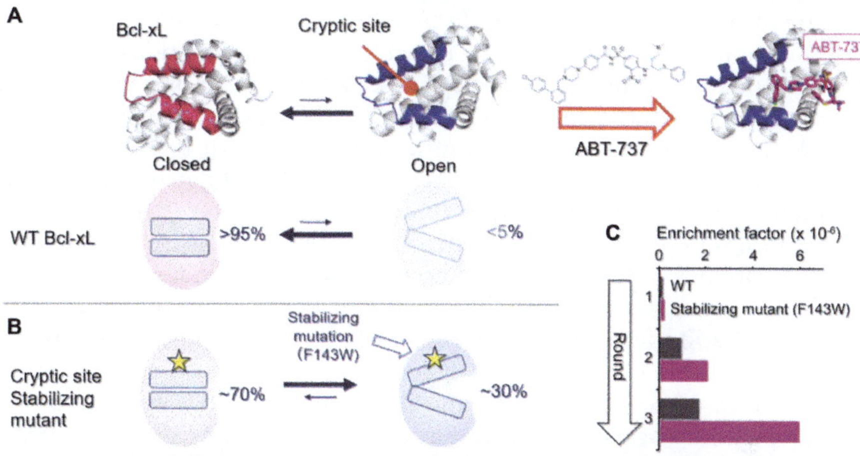

Figure 15.3 The NMR strategy to target cryptic sites. (A) Schematic view of the ABT-737 binding to the cryptic site in Bcl-xL. NMR identified a low-populated cryptic site *open* conformation (middle) in the unligated Bcl-xL. ABT-737 presumably binds to the cryptic site *open* conformation in a conformational selection manner. (B) Cryptic site stabilizing mutation, F143W, allosterically increased the population of the cryptic site *open* conformation in the unligated state. (C) Efficient screening of the Bcl-xL binding peptides by the F143W mutant. Comparison of the enrichment factors in each round of phage display screening. The ratio of the hit clone against the whole clone *cfu* was shown for each panning round.

free/bound exchange system can be characterized by using $R_{1\rho}$ and R_2 relaxation dispersion (RD) experiments,[196–198] respectively. Slower sub-second dynamics can be handled by the ZZ-exchange experiments.[199–202] These experiments provide essential insight into the ligand interaction kinetics.[203] Both RD experiments can define k_{on} and k_{off} rates from the fitting of dispersion curves measured at different concentration ratios between ligand and proteins.[203,204] The kinetics parameter provides complementary information in understanding the protein–ligand interaction crucial for drug development.[205,206] For small-molecule ligands, label-free ^{1}H $R_{1\rho}$ and R_2 relaxation dispersion (RD) experiments were reported to define the resident time of the ligands.[205,207]

15.7 Elucidation of Cell Permeabilization Activity of Medium-size Molecules by NMR

As mentioned earlier, medium-size molecules, especially those that are cyclic, are emerging as new drug-discovery modalities. One of the most attractive targets of the medium-size cyclic molecules is *intracellular* PPIs. It is shown that some of the cyclic peptides can enter mammalian cells to inhibit intracellular PPIs.[208–213] However, most cyclic peptides are not cell-permeable. Thus, clarifying the factors that define their cell permeabilization activity is wanted to achieve a rational design of the cell-permeable peptides.

Historically, both solution- and solid-state NMR has been used to characterize the structures of membrane permeabilizing antimicrobial peptides in membrane environments to unveil structure–function relationships.[214–218] Conformational dynamics of cyclic peptides have also been implicated as another critical factor for passive cell permeability. A classic example is cyclosporine A, a cyclic undecapeptide immunosuppressant with seven methylated backbone amides.[219] Molecular dynamics simulations validated against NMR experiments revealed that cell permeability of the cyclosporine A and its analogs is attributed to a dynamic conformational interplay between aqueous and membrane-preferred conformations, as well as their exchange rates.[220–222] Similar mechanisms have been described for other cyclic peptides.[223,224] However, it is still unclear whether this mechanism of cell permeability can be extended to other cyclic peptides in general.[225]

Our recent study addressed the question. In the study, we focused on cyclorasin, a group of cyclic peptides known to exhibit anticancer activity.[226] Cyclorasins are undecapeptide cyclic peptides that show inhibitory activity against Ras–Raf interaction *in vitro*.[227] While cyclorasins share a cell permeability peptide (CPP)-like motif that contributes to cell membrane permeability, the activity of these peptides in a cell is substantially different. For example, cyclorasin 9A5 inhibits the Ras–Raf protein interaction in both a cell-free assay and H358 lung cancer cells with IC_{50} values of 120 nM and 3.1 μM, respectively.[227] Cyclorasin 9A54, which shows a more potent PPI inhibition activity ($IC_{50} = 18$ nM) in the cell-free assay, was less active in the cellular assay ($IC_{50} = 7.2$ μM). There are three amino acid substitutions between cyclorasins 9A5 and 9A54; Thr-2, dAla-3, and Fpa-9 in cyclorasin 9A5 are substituted by Tle-2, dVal-3, and F_2pa-9 in cyclorasin 9A54, respectively (where dAla, dVal, Tle, Fpa, and F_2pa are D-alanine, D-valine, L-tertiallyleucine, L-4-fluorophenylalanine, and L-3,4-difluorophenylalanine). Except for the fluorination at position 9, all substitutions are outside of the CPP-like motif, and these substitutions make cyclorasin 9A54 more lipophilic than the cyclorasin 9A5; therefore, loss of the cell permeabilization activity is somewhat unexpected.

In order to clarify the differences in the cell permeabilization activity of cyclorasins 9A5 and 9A54, the structures of these peptides in aqueous solution and dimethyl sulfoxide (DMSO), a solvent to mimic the surface of the cell membrane, were compared by NMR. We found that the distribution of aromatic and basic residues in the structures of cyclorasins 9A5 and 9A54 do not differ significantly in water; however, the structure of cyclorasin 9A5 changed significantly in DMSO (Figure 15.4A). The cyclorasin 9A5 forms an amphiphilic structure with aromatic residues clustered on one side and multiple Arg residues arranged around the aromatic residues in DMSO, which is known to be advantageous for intracellular migration *via* endocytosis. In contrast, cyclorasin 9A54 cannot form the amphiphilic structure in DMSO. At the sites with distinct sequences between cyclorasins 9A5 and 9A54, 9A5 forms a tighter turn structure with Thr-2 and dAla-3 at the tip, which allows 9A5 to form the compact amphipathic structure. On the other hand, residues 2 and 3 of

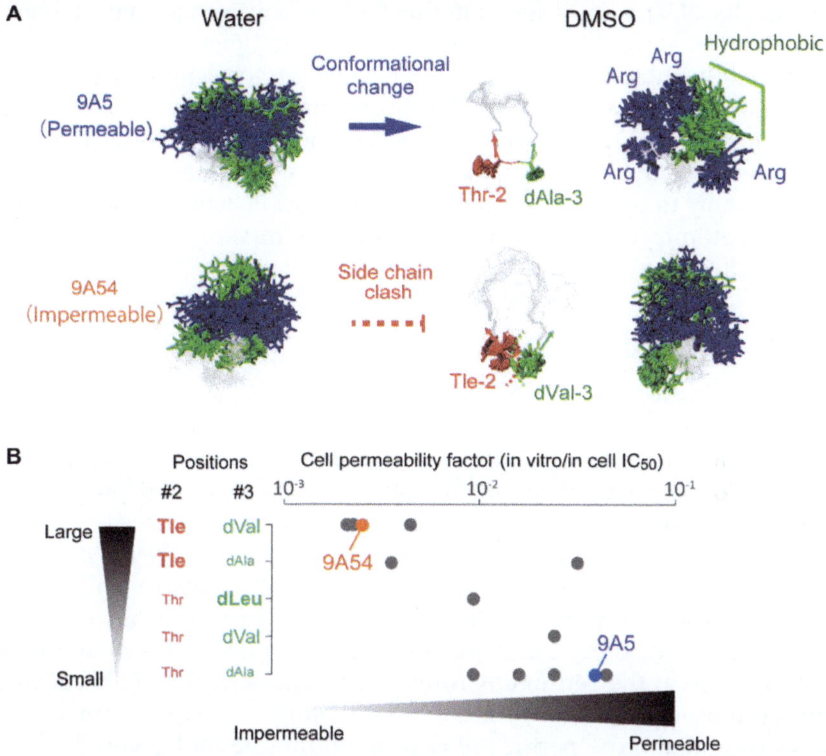

Figure 15.4 Elucidation of cell permeabilization activity of cyclorasins by NMR. (A) Structures of cyclorasins 9A5 and 9A54 in water and DMSO. Cyclorasin 9A5 (upper panel) changes the conformation in DMSO to adopt the amphiphilic structure, in which a cluster of the hydrophobic residues (green) is surrounded by Arg residues. In contrast, cyclorasin 9A54 (lower panel) cannot undergo sufficient structural change to form the amphiphilic structure in DMSO. (B) Anti-correlation between the side chain bulkiness and the cell permeabilization activities of 14 cyclorasin analogs. Some of the peptide data are overlapping with each other. The cell permeability factors are calculated from the ratio between the *in vitro* and *in vivo* IC_{50} values.[227] Adapted from ref. 226 with permission from John Wiley & Sons, Copyright © 2021 Wiley-VCH GmbH.

cyclorasin 9A54 are bulkier Tle and dVal. Due to the steric hindrance between the sidechains, cyclorasin 9A54 cannot take the turn structure similar to cyclorasin 9A5 and thus cannot form the amphipathic structure.

For the other 14 cyclorasin peptides, the bulkiness of the side chains at the second and third positions and cell permeabilization activity were inversely correlated (Figure 15.4B). Furthermore, the additional structural analysis showed that cyclorasins 9A12 and 12A, which show cell permeabilization activity, can undergo structural transformation and form an amphipathic structure similar to 9A5 in DMSO. In contrast, cyclorasin 9A44d, which shows no cell permeabilization activity, cannot take the same structure.[226]

These results clearly indicate that the flexible conformational change in response to the solvent environment is an essential factor determining the cell permeabilization activity of the cyclic peptides. In cyclorasin groups, the steric hindrance between the side chains of consecutive amino acid residues defines the ability of the peptide to show cell permeabilization activity. The result also shows the importance of performing conformational analyses in various solvents to predict their cell permeabilization activity, and the ability to perform atomic-level conformational analysis in various solvents makes NMR essential for achieving this goal.

15.8 Conclusion

As described above, the author and others have developed NMR strategies to promote the use of new drug-discovery modalities. By utilizing this fundamental technique, more applicational research can be conducted. For example, NMR can be used to test the structural identity and properties of biosimilars compared to original biopharmaceuticals produced in different protocols.[228] As shown in evaluating the heterogeneity of glycan structures, NMR can quantitatively distinguish different components in a mixture.[141] The feature would also enable the structural characterization of the biopharmaceuticals in the various formulation and physiological conditions, including those in the serum environment.[229] The versatility of NMR to the sample conditions also contributes to establishing a strategy to theoretically improve the membrane permeability of medium-size molecules.[226,230] The feature would further enable the characterization and development of bioactive ligands in cellular conditions. NMR will analyze if the dynamic behaviors of protein–ligand complexes in cellular conditions are distinct from diluted solvent conditions and how much the interaction of a ligand would be specific to the target in a crowded cellular environment. Ultimately, NMR would enable in-cell SGDD, in which the whole target-based process from hit identification to lead optimization process can be conducted by an NMR platform. Many of the important physiological functions of proteins involve dynamic conformational changes. In addition, as was the case for the cryptic sites, physiologically important conformations might only be evident in conformational equilibrium in a physiological solution.[191] There is no doubt that an integrated approach involving computational strategy with experimental strategies, such as X-ray crystal structure and cryo-electron microscopy, will visualize the dynamics of proteins and protein–ligand complexes.[231,232] We believe that NMR would be a key technology that grips the casting vote in integrated structural biology by providing critical conformational dynamics information. The target of structure and dynamics analysis of NMR is not limited to proteins. Other macromolecules and structural elements, such as RNA and sugar chains, also retain functional dynamics, which have been analyzed by NMR.[78,233–236] Toward this end, NMR would continue to contribute to the development of yet unknown modalities in the future.

Abbreviations

NMR	Nuclear magnetic resonance
SGDD	Structure-guided drug development
PD1	Programmed cell death protein 1
PPI	Protein–protein interaction
HOS	High-order structure
HMW	High-molecular-weight
TROSY	Transverse relaxation optimized spectroscopy
HMQC	Heteronuclear multiple quantum coherence
mAbs	Monoclonal antibodies
1D	One-dimensional
2D	Two-dimensional
SOFAST	Band-selective optimized flip angle short transient
CRINEPT	Cross-correlated relaxation-enhanced polarization transfer
SC	Subcutaneous injection
IV	Intravenous injection
INEPT	Insensitive nuclei enhanced by polarization transfer
DD	Dipole–dipole
CSA	Chemical shift anisotropy
^{1}J	One-bond couplings
FCT	Forbidden coherence transition
Bcl-2	B-cell lymphoma-2
CSCs	Chemical shift changes
AMPs	Antimicrobial peptides
CPP	Cell permeability peptide
DMSO	Dimethyl sulfoxide

Acknowledgements

This work was supported by grants from the Japan Society for the Promotion of Science, KAKENHI Grant Numbers 18K19415, 20K21494, 20H03378, and 20H04722. This chapter is dedicated to the memory of a recently deceased sincere friend and an editor of this book, Prof. Hanudatta S. Atreya.

References

1. T. R. Alderson and L. E. Kay, Unveiling invisible protein states with NMR spectroscopy, *Curr. Opin. Struct. Biol.*, 2020, **60**, 39–49.
2. H. Arthanari, K. Takeuchi, A. Dubey and G. Wagner, Emerging solution NMR methods to illuminate the structural and dynamic properties of proteins, *Curr. Opin. Struct. Biol.*, 2019, **58**, 294–304.
3. A. Bax and G. M. Clore, Protein NMR: Boundless opportunities, *J. Magn. Reson.*, 2019, **306**, 187–191.
4. M. J. Bostock, A. S. Solt and D. Nietlispach, The role of NMR spectroscopy in mapping the conformational landscape of GPCRs, *Curr. Opin. Struct. Biol.*, 2019, **57**, 145–156.

5. T. Ikeya, D. Ban, D. Lee, Y. Ito, K. Kato and C. Griesinger, Solution NMR views of dynamical ordering of biomacromolecules, *Biochim. Biophys. Acta, Gen. Subj.*, 2018, **1862**(2), 287–306.
6. Y. Jiang and C. G. Kalodimos, NMR Studies of Large Proteins, *J. Mol. Biol.*, 2017, **429**(17), 2667–2676.
7. D. Ban, C. A. Smith, B. L. de Groot, C. Griesinger and D. Lee, Recent advances in measuring the kinetics of biomolecules by NMR relaxation dispersion spectroscopy, *Arch. Biochem. Biophys.*, 2017, **628**, 81–91.
8. M. Kovermann, P. Rogne and M. Wolf-Watz, Protein dynamics and function from solution state NMR spectroscopy, *Q. Rev. Biophys.*, 2016, **49**, e6.
9. G. P. Lisi and J. P. Loria, Using NMR spectroscopy to elucidate the role of molecular motions in enzyme function, *Prog. Nucl. Magn. Reson. Spectrosc.*, 2016, **92–93**, 1–17.
10. K. Oxenoid and J. J. Chou, A functional NMR for membrane proteins: dynamics, ligand binding, and allosteric modulation, *Protein Sci.*, 2016, **25**(5), 959–973.
11. A. G. Palmer, A dynamic look backward and forward, *J. Magn. Reson.*, 2016, **266**, 73–80.
12. L. E. Kay, New Views of Functionally Dynamic Proteins by Solution NMR Spectroscopy, *J. Mol. Biol.*, 2016, **428**(2 Pt A), 323–331.
13. P. J. Farber and A. Mittermaier, Relaxation dispersion NMR spectroscopy for the study of protein allostery, *Biophys. Rev.*, 2015, **7**(2), 191–200.
14. B. M. Burmann and S. Hiller, Chaperones and chaperone-substrate complexes: Dynamic playgrounds for NMR spectroscopists, *Prog. Nucl. Magn. Reson. Spectrosc.*, 2015, **86–87**, 41–64.
15. A. G. Palmer, Enzyme dynamics from NMR spectroscopy, *Acc. Chem. Res.*, 2015, **48**(2), 457–465.
16. C. Göbl, T. Madl, B. Simon and M. Sattler, NMR approaches for structural analysis of multidomain proteins and complexes in solution, *Prog. Nucl. Magn. Reson. Spectrosc.*, 2014, **80**, 26–63.
17. M. Osawa, K. Takeuchi, T. Ueda, N. Nishida and I. Shimada, Functional dynamics of proteins revealed by solution NMR, *Curr. Opin. Struct. Biol.*, 2012, **22**(5), 660–669.
18. M. Akke, Conformational dynamics and thermodynamics of protein-ligand binding studied by NMR relaxation, *Biochem. Soc. Trans.*, 2012, **40**(2), 419–423.
19. J. L. Kitevski-LeBlanc and R. S. Prosser, Current applications of 19F NMR to studies of protein structure and dynamics, *Prog. Nucl. Magn. Reson. Spectrosc.*, 2012, **62**, 1–33.
20. K. Chen and N. Tjandra, The use of residual dipolar coupling in studying proteins by NMR, *Top. Curr. Chem.*, 2012, **326**, 47–67.
21. R. Ishima, Recent developments in (15)N NMR relaxation studies that probe protein backbone dynamics, *Top. Curr. Chem.*, 2012, **326**, 99–122.
22. L. E. Kay, NMR studies of protein structure and dynamics, *J. Magn. Reson.*, 2005, **173**(2), 193–207.

23. D. D. Boehr, H. J. Dyson and P. E. Wright, An NMR perspective on enzyme dynamics, *Chem. Rev.*, 2006, **106**(8), 3055–3079.
24. T. Ueda, Y. Kofuku, J. Okude, S. Imai, Y. Shiraishi and I. Shimada, Function-related conformational dynamics of G protein-coupled receptors revealed by NMR, *Biophys. Rev.*, 2019, **11**(3), 409–418.
25. D. D. Boehr, R. Nussinov and P. E. Wright, The role of dynamic conformational ensembles in biomolecular recognition, *Nat. Chem. Biol.*, 2009, **5**(11), 789–796.
26. E. Delaforge, S. Milles, J. R. Huang, D. Bouvier, M. R. Jensen, M. Sattler, D. J. Hart and M. Blackledge, Investigating the Role of Large-Scale Domain Dynamics in Protein-Protein Interactions, *Front. Mol. Biosci.*, 2016, **3**, 54.
27. K. Takeuchi, M. Imai and I. Shimada, Conformational equilibrium defines the variable induction of the multidrug-binding transcriptional repressor QacR, *Proc. Natl. Acad. Sci. U. S. A.*, 2019, **116**(40), 19963–19972.
28. G. M. Clore, C. Tang and J. Iwahara, Elucidating transient macromolecular interactions using paramagnetic relaxation enhancement, *Curr. Opin. Struct. Biol.*, 2007, **17**(5), 603–616.
29. Z. Liu, Z. Gong, X. Dong and C. Tang, Transient protein-protein interactions visualized by solution NMR, *Biochim. Biophys. Acta*, 2016, **1864**(1), 115–122.
30. A. M. Bonvin, R. Boelens and R. Kaptein, NMR analysis of protein interactions, *Curr. Opin. Chem. Biol.*, 2005, **9**(5), 501–508.
31. H. J. Dyson and P. E. Wright, Perspective: the essential role of NMR in the discovery and characterization of intrinsically disordered proteins, *J. Biomol. NMR*, 2019, **73**(12), 651–659.
32. N. Salvi, A. Abyzov and M. Blackledge, Atomic resolution conformational dynamics of intrinsically disordered proteins from NMR spin relaxation, *Prog. Nucl. Magn. Reson. Spectrosc.*, 2017, **102–103**, 43–60.
33. D. Kurzbach, G. Kontaxis, N. Coudevylle and R. Konrat, NMR Spectroscopic Studies of the Conformational Ensembles of Intrinsically Disordered Proteins, *Adv. Exp. Med. Biol.*, 2015, **870**, 149–185.
34. M. R. Jensen, R. W. Ruigrok and M. Blackledge, Describing intrinsically disordered proteins at atomic resolution by NMR, *Curr. Opin. Struct. Biol.*, 2013, **23**(3), 426–435.
35. W. M. Borcherds; and G. W. Daughdrill, Chapter Six – Using NMR Chemical Shifts to Determine Residue-Specific Secondary Structure Populations for Intrinsically Disordered Proteins, in *Methods in Enzymology*, ed. E. Rhoades, Academic Press, 2018, vol. 611, pp. 101–136.
36. M. Fuxreiter, I. Simon, P. Friedrich and P. Tompa, Preformed Structural Elements Feature in Partner Recognition by Intrinsically Unstructured Proteins, *J. Mol. Biol.*, 2004, **338**(5), 1015–1026.
37. H. J. Dyson and P. E. Wright, Coupling of folding and binding for unstructured proteins, *Curr. Opin. Struct. Biol.*, 2002, **12**(1), 54–60.
38. S. A. Kotler, V. Tugarinov, T. Schmidt, A. Ceccon, D. S. Libich, R. Ghirlando, C. D. Schwieters and G. M. Clore, Probing initial transient oligomerization events facilitating Huntingtin fibril nucleation at

atomic resolution by relaxation-based NMR, *Proc. Natl. Acad. Sci.*, 2019, **116**(9), 3562.
39. M. Fuxreiter; and P. Tompa, Fuzzy Complexes: A More Stochastic View of Protein Function, in *Fuzziness: Structural Disorder in Protein Complexes*, ed. M. Fuxreiter and P. Tompa, Springer US, New York, NY, 2012; pp. 1–14.
40. U. Weininger, Chapter Three - Optimal Isotope Labeling of Aromatic Amino Acid Side Chains for NMR Studies of Protein Dynamics, in *Methods in Enzymology*, ed. A. J. Wand, Academic Press, 2019, vol. 614, pp. 67–86.
41. N. Shimba, Z. Serber, R. Ledwidge, S. M. Miller, C. S. Craik and V. Dötsch, Quantitative identification of the protonation state of histidines in vitro and in vivo, *Biochemistry*, 2003, **42**(30), 9227–9234.
42. J. G. Pelton, D. A. Torchia, N. D. Meadow and S. Roseman, Tautomeric states of the active-site histidines of phosphorylated and unphosphorylated IIIGlc, a signal-transducing protein from Escherichia coli, using two-dimensional heteronuclear NMR techniques, *Protein Sci.*, 1993, **2**(4), 543–558.
43. A. L. Hansen and L. E. Kay, Measurement of histidine pKaa values and tautomer populations in invisible protein states, *Proc. Natl. Acad. Sci.*, 2014, **111**(17), E1705.
44. I. J. Kimsey, E. S. Szymanski, W. J. Zahurancik, A. Shakya, Y. Xue, C. C. Chu, B. Sathyamoorthy, Z. Suo and H. M. Al-Hashimi, Dynamic basis for dG•dT misincorporation via tautomerization and ionization, *Nature*, 2018, **554**(7691), 195–201.
45. U. Weininger, K. Modig, A. J. Geitner, P. A. Schmidpeter, J. R. Koch and M. Akke, Dynamics of Aromatic Side Chains in the Active Site of FKBP12, *Biochemistry*, 2017, **56**(1), 334–343.
46. D. H. Meadows, J. L. Markley, J. S. Cohen and O. Jardetzky, Nuclear magnetic resonance studies of the structure and binding sites of enzymes. I. Histidine residues, *Proc. Natl. Acad. Sci.*, 1967, **58**(4), 1307.
47. L. Emmanouilidis, L. Esteban-Hofer, F. F. Damberger, T. de Vries, C. K. X. Nguyen, L. F. Ibáñez, S. Mergenthal, E. Klotzsch, M. Yulikov, G. Jeschke and F. H. Allain, NMR and EPR reveal a compaction of the RNA-binding protein FUS upon droplet formation, *Nat. Chem. Biol.*, 2021, **17**(5), 608–614.
48. A. E. Conicella, G. L. Dignon, G. H. Zerze, H. B. Schmidt, A. M. D'Ordine, Y. C. Kim, R. Rohatgi, Y. M. Ayala, J. Mittal and N. L. Fawzi, TDP-43 α-helical structure tunes liquid-liquid phase separation and function, *Proc. Natl. Acad. Sci. U. S. A.*, 2020, **117**(11), 5883–5894.
49. J. P. Brady, P. J. Farber, A. Sekhar, Y.-H. Lin, R. Huang, A. Bah, T. J. Nott, H. S. Chan, A. J. Baldwin, J. D. Forman-Kay and L. E. Kay, Structural and hydrodynamic properties of an intrinsically disordered region of a germ cell-specific protein on phase separation, *Proc. Natl. Acad. Sci. U. S. A.*, 2017, **114**(39), E8194.
50. K. A. Burke, A. M. Janke, C. L. Rhine and N. L. Fawzi, Residue-by-Residue View of In Vitro FUS Granules that Bind the C-Terminal Domain of RNA Polymerase II, *Mol. Cell*, 2015, **60**(2), 231–241.

51. V. H. Ryan, G. L. Dignon, G. H. Zerze, C. V. Chabata, R. Silva, A. E. Conicella, J. Amaya, K. A. Burke, J. Mittal and N. L. Fawzi, Mechanistic View of hnRNPA2 Low-Complexity Domain Structure, Interactions, and Phase Separation Altered by Mutation and Arginine Methylation, *Mol. Cell*, 2018, **69**(3), 465–479, e7.
52. T. Yuwen, J. P. Brady and L. E. Kay, Probing Conformational Exchange in Weakly Interacting, Slowly Exchanging Protein Systems via Off-Resonance R(1ρ) Experiments: Application to Studies of Protein Phase Separation, *J. Am. Chem. Soc.*, 2018, **140**(6), 2115–2126.
53. D. T. Murray, M. Kato, Y. Lin, K. R. Thurber, I. Hung, S. L. McKnight and R. Tycko, Structure of FUS Protein Fibrils and Its Relevance to Self-Assembly and Phase Separation of Low-Complexity Domains, *Cell*, 2017, **171**(3), 615–627, e16.
54. A. Siegert, M. Rankovic, F. Favretto, T. Ukmar-Godec, T. Strohäker, S. Becker and M. Zweckstetter, Interplay between tau and α-synuclein liquid-liquid phase separation, *Protein Sci.*, 2021, **30**(7), 1326–1336.
55. K. Wuthrich and G. Wagner, NMR investigations of the dynamics of the aromatic amino acid residues in the basic pancreatic trypsin inhibitor, *FEBS Lett.*, 1975, **50**(2), 265–268.
56. G. Wagner, A. DeMarco and K. Wuthrich, Dynamics of the aromatic amino acid residues in the globular conformation of the basic pancreatic trypsin inhibitor (BPTI). I. 1H NMR studies, *Biophys. Struct. Mech.*, 1976, **2**(2), 139–158.
57. S. Schütz and R. Sprangers, Methyl TROSY spectroscopy: A versatile NMR approach to study challenging biological systems, *Prog. Nucl. Magn. Reson. Spectrosc.*, 2020, **116**, 56–84.
58. H. Zhang and H. van Ingen, Isotope-labeling strategies for solution NMR studies of macromolecular assemblies, *Curr. Opin. Struct. Biol.*, 2016, **38**, 75–82.
59. R. Kerfah, M. J. Plevin, R. Sounier, P. Gans and J. Boisbouvier, Methyl-specific isotopic labeling: a molecular tool box for solution NMR studies of large proteins, *Curr. Opin. Struct. Biol.*, 2015, **32**, 113–122.
60. A. D. Gossert and W. Jahnke, Isotope labeling in insect cells, *Adv. Exp. Med. Biol.*, 2012, **992**, 179–196.
61. S. Yanaka, H. Yagi, R. Yogo, M. Yagi-Utsumi and K. Kato, Stable isotope labeling approaches for NMR characterization of glycoproteins using eukaryotic expression systems, *J. Biomol. NMR*, 2018, **71**(3), 193–202.
62. P. Lundström, A. Ahlner and A. T. Blissing, Isotope labeling methods for large systems, *Adv. Exp. Med. Biol.*, 2012, **992**, 3–15.
63. N. K. Goto and L. E. Kay, New developments in isotope labeling strategies for protein solution NMR spectroscopy, *Curr. Opin. Struct. Biol.*, 2000, **10**(5), 585–592.
64. T. Kigawa, T. Yabuki, Y. Yoshida, M. Tsutsui, Y. Ito, T. Shibata and S. Yokoyama, Cell-free production and stable-isotope labeling of milligram quantities of proteins, *FEBS Lett.*, 1999, **442**(1), 15–19.

65. T. Torizawa, M. Shimizu, M. Taoka, H. Miyano and M. Kainosho, Efficient production of isotopically labeled proteins by cell-free synthesis: a practical protocol, *J. Biomol. NMR*, 2004, **30**(3), 311–325.
66. M. Kainosho, T. Torizawa, Y. Iwashita and T. Terauchi, Mei Ono, A.; Güntert, P., Optimal isotope labelling for NMR protein structure determinations, *Nature*, 2006, **440**(7080), 52–57.
67. K. Kato, Y. Yamaguchi and Y. Arata, Stable-isotope-assisted NMR approaches to glycoproteins using immunoglobulin G as a model system, *Prog. Nucl. Magn. Reson. Spectrosc.*, 2010, **56**(4), 346–359.
68. C. Prasanna, A. Dubey and H. S. Atreya, Amino acid selective unlabeling in protein NMR spectroscopy, *Methods Enzymol.*, 2015, **565**, 167–189.
69. T. R. Alderson and L. E. Kay, NMR spectroscopy captures the essential role of dynamics in regulating biomolecular function, *Cell*, 2021, **184**(3), 577–595.
70. K. Henzler-Wildman and D. Kern, Dynamic personalities of proteins, *Nature*, 2007, **450**(7172), 964–972.
71. K. Takeuchi, Y. Tokunaga, M. Imai, H. Takahashi and I. Shimada, Dynamic multidrug recognition by multidrug transcriptional repressor LmrR, *Sci. Rep.*, 2014, 4.
72. N. Nishida, M. Osawa, K. Takeuchi, S. Imai, P. Stampoulis, Y. Kofuku, T. Ueda and I. Shimada, Functional dynamics of cell surface membrane proteins, *J. Magn. Reson.*, 2014, **241**, 86–96.
73. K. Takeuchi, M. Imai and I. Shimada, Dynamic equilibrium on DNA defines transcriptional regulation of a multidrug binding transcriptional repressor, LmrR, *Sci. Rep.*, 2017, **7**(1), 267.
74. Y. Tokunaga, K. Takeuchi, H. Takahashi and I. Shimada, Allosteric enhancement of MAP kinase p38α's activity and substrate selectivity by docking interactions, *Nat. Struct. Mol. Biol.*, 2014, **21**(8), 704–711.
75. I. Shimada, T. Ueda, Y. Kofuku, M. T. Eddy and K. Wüthrich, GPCR drug discovery: integrating solution NMR data with crystal and cryo-EM structures, *Nat. Rev. Drug Discovery*, 2018, **18**, 59.
76. D. A. Torchia, NMR studies of dynamic biomolecular conformational ensembles, *Prog. Nucl. Magn. Reson. Spectrosc.*, 2015, **84–85**, 14–32.
77. C. Huang and C. G. Kalodimos, Structures of Large Protein Complexes Determined by Nuclear Magnetic Resonance Spectroscopy, *Annu. Rev. Biophys.*, 2017, **46**, 317–336.
78. B. Liu, H. Shi and H. M. Al-Hashimi, Developments in solution-state NMR yield broader and deeper views of the dynamic ensembles of nucleic acids, *Curr. Opin. Struct. Biol.*, 2021, **70**, 16–25.
79. A. J. Wand and K. A. Sharp, Measuring Entropy in Molecular Recognition by Proteins, *Annu. Rev. Biophys.*, 2018, **47**, 41–61.
80. M. S. Marlow, J. Dogan, K. K. Frederick, K. G. Valentine and A. J. Wand, The role of conformational entropy in molecular recognition by calmodulin, *Nat. Chem. Biol.*, 2010, **6**(5), 352–358.
81. C. Diehl, O. Engström, T. Delaine, M. Håkansson, S. Genheden, K. Modig, H. Leffler, U. Ryde, U. J. Nilsson and M. Akke, Protein

flexibility and conformational entropy in ligand design targeting the carbohydrate recognition domain of galectin-3, *J. Am. Chem. Soc.*, 2010, **132**(41), 14577–14589.

82. C. G. Kalodimos, NMR reveals novel mechanisms of protein activity regulation, *Protein Sci.*, 2011, **20**(5), 773–782.
83. A. J. Wand, The dark energy of proteins comes to light: conformational entropy and its role in protein function revealed by NMR relaxation, *Curr. Opin. Struct. Biol.*, 2012, **23**(1), 75–81.
84. V. Kasinath, K. A. Sharp and A. J. Wand, Microscopic insights into the NMR relaxation-based protein conformational entropy meter, *J. Am. Chem. Soc.*, 2013, **135**(40), 15092–15100.
85. K. Takeuchi, Y. Tokunaga, M. Imai, H. Takahashi and I. Shimada, Dynamic multidrug recognition by multidrug transcriptional repressor LmrR, *Sci. Rep.*, 2014, **4**, 6922.
86. P. Vallurupalli, D. F. Hansen and L. E. Kay, Structures of invisible, excited protein states by relaxation dispersion NMR spectroscopy, *Proc. Natl. Acad. Sci. U. S. A.*, 2008, **105**(33), 11766–11771.
87. A. J. Baldwin and L. E. Kay, NMR spectroscopy brings invisible protein states into focus, *Nat. Chem. Biol.*, 2009, **5**(11), 808–814.
88. N. J. Anthis and G. M. Clore, Visualizing transient dark states by NMR spectroscopy, *Q. Rev. Biophys.*, 2015, **48**(1), 35–116.
89. A. D. Gossert and W. Jahnke, NMR in drug discovery: A practical guide to identification and validation of ligands interacting with biological macromolecules, *Prog. Nucl. Magn. Reson. Spectrosc.*, 2016, **97**, 82–125.
90. A. Dubey, K. Takeuchi, M. Reibarkh and H. Arthanari, The role of NMR in leveraging dynamics and entropy in drug design, *J. Biomol. NMR*, 2020, **74**(10), 479–498.
91. M. J. Harner, A. O. Frank and S. W. Fesik, Fragment-based drug discovery using NMR spectroscopy, *J. Biomol. NMR*, 2013, **56**(2), 65–75.
92. T. Carlomagno, Ligand-target interactions: what can we learn from NMR?, *Annu. Rev. Biophys. Biomol. Struct.*, 2005, **34**, 245–266.
93. W. Jahnke, M. J. J. Blommers, C. Fernández, C. Zwingelstein and R. Amstutz, Strategies for the NMR-Based 	 Identification and Optimization of Allosteric Protein Kinase Inhibitors, *ChemBioChem*, 2005, **6**(9), 1607–1610.
94. M. Pellecchia, I. Bertini, D. Cowburn, C. Dalvit, E. Giralt, W. Jahnke, T. L. James, S. W. Homans, H. Kessler, C. Luchinat, B. Meyer, H. Oschkinat, J. Peng, H. Schwalbe and G. Siegal, Perspectives on NMR in drug discovery: a technique comes of age, *Nat. Rev. Drug Discovery*, 2008, **7**(9), 738–745.
95. J. Kline and T. F. Gajewski, Clinical development of mAbs to block the PD1 pathway as an immunotherapy for cancer, *Curr. Opin. Investig. Drugs*, 2010, **11**(12), 1354–1359.
96. C. A. Lipinski, F. Lombardo, B. W. Dominy and P. J. Feeney, Experimental and computational approaches to estimate solubility and

permeability in drug discovery and development settings, *Adv. Drug Delivery Rev.*, 2001, **46**(1–3), 3–26.
97. A. Beck, E. Wagner-Rousset, D. Ayoub, A. Van Dorsselaer and S. Sanglier-Cianférani, Characterization of Therapeutic Antibodies and Related Products, *Anal. Chem.*, 2013, **85**(2), 715–736.
98. Y. Le Basle, P. Chennell, N. Tokhadze, A. Astier and V. Sautou, Physicochemical Stability of Monoclonal Antibodies: A Review, *J. Pharm. Sci.*, 2020, **109**(1), 169–190.
99. R. Kitahara, High-Pressure NMR Spectroscopy Reveals Functional Substates of Ubiquitin and Ubiquitin-Like Proteins, *Subcell. Biochem.*, 2015, **72**, 199–214.
100. T. R. Alderson, C. Charlier, D. A. Torchia, P. Anfinrud and A. Bax, Monitoring Hydrogen Exchange During Protein Folding by Fast Pressure Jump NMR Spectroscopy, *J. Am. Chem. Soc.*, 2017, **139**(32), 11036–11039.
101. M. Xue, T. Wakamoto, C. Kejlberg, Y. Yoshimura, T. A. Nielsen, M. W. Risør, K. W. Sanggaard, R. Kitahara and F. A. A. Mulder, How internal cavities destabilize a protein, *Proc. Natl. Acad. Sci.*, 2019, **116**(42), 21031.
102. N. L. Fawzi, J. Ying, D. A. Torchia and G. M. Clore, Probing exchange kinetics and atomic resolution dynamics in high-molecular-weight complexes using dark-state exchange saturation transfer NMR spectroscopy, *Nat. Protoc.*, 2012, **7**(8), 1523–1533.
103. M. Yokogawa, K. Takeuchi and I. Shimada, Bead-linked proteoliposomes: a reconstitution method for nmr analyses of membrane protein-ligand interactions, *J. Am. Chem. Soc.*, 2005, **127**(34), 12021–12027.
104. N. Nishida, H. Sumikawa, M. Sakakura, N. Shimba, H. Takahashi, H. Terasawa, E. Suzuki and I. Shimada, Collagen-binding mode of vWF-A3 domain determined by a transferred cross-saturation experiment, *Nat. Struct. Biol.*, 2003, **10**(1), 53–58.
105. I. Pal, V. P. Brahmkhatri, S. Bera, D. Bhattacharyya, Y. Quirishi, A. Bhunia and H. S. Atreya, Enhanced stability and activity of an antimicrobial peptide in conjugation with silver nanoparticle, *J. Colloid Interface Sci.*, 2016, **483**, 385–393.
106. Y. Ito and P. Selenko, Cellular structural biology, *Curr. Opin. Struct. Biol.*, 2010, **20**(5), 640–648.
107. G. Siegal and P. Selenko, Cells, drugs and NMR, *J. Magn. Reson.*, 2019, **306**, 202–212.
108. N. Nishida, Y. Ito and I. Shimada, In situ structural biology using in-cell NMR, *Biochim. Biophys. Acta, Gen. Subj.*, 2020, **1864**(2), 129364.
109. E. Luchinat and L. Banci, In-cell NMR: a topical review, *IUCrJ*, 2017, **4**(Pt 2), 108–118.
110. R. Hänsel, L. M. Luh, I. Corbeski, L. Trantirek and V. Dötsch, In-cell NMR and EPR spectroscopy of biomacromolecules, *Angew. Chem., Int. Ed. Engl.*, 2014, **53**(39), 10300–10314.
111. K. Pervushin, R. Riek, G. Wider and K. Wuthrich, Attenuated T2 relaxation by mutual cancellation of dipole-dipole coupling and chemical

shift anisotropy indicates an avenue to NMR structures of very large biological macromolecules in solution, *Proc. Natl. Acad. Sci. U. S. A.*, 1997, **94**(23), 12366–12371.
112. V. Tugarinov, P. M. Hwang, J. E. Ollerenshaw and L. E. Kay, Cross-correlated relaxation enhanced 1H[bond]13C NMR spectroscopy of methyl groups in very high molecular weight proteins and protein complexes, *J. Am. Chem. Soc.*, 2003, **125**(34), 10420–10428.
113. W. D. Morgan, A. Kragt and J. Feeney, Expression of deuterium-isotope-labelled protein in the yeast pichia pastoris for NMR studies, *J. Biomol. NMR*, 2000, **17**(4), 337–347.
114. M. Miyazawa-Onami, K. Takeuchi, T. Takano, T. Sugiki, I. Shimada and H. Takahashi, Perdeuteration and methyl-selective (1)H, (13)C-labeling by using a Kluyveromyces lactis expression system, *J. Biomol. NMR*, 2013, **57**(3), 297–304.
115. L. Zhang, S. Luo and B. Zhang, Glycan analysis of therapeutic glycoproteins, *MAbs*, 2016, **8**(2), 205–215.
116. R. Jefferis, Glycosylation of recombinant antibody therapeutics, *Biotechnol. Prog.*, 2005, **21**(1), 11–16.
117. Y. Kofuku, T. Ueda, J. Okude, Y. Shiraishi, K. Kondo, T. Mizumura, S. Suzuki and I. Shimada, Functional dynamics of deuterated β2 -adrenergic receptor in lipid bilayers revealed by NMR spectroscopy, *Angew. Chem., Int. Ed. Engl.*, 2014, **53**(49), 13376–13379.
118. Y. Kofuku, T. Yokomizo, S. Imai, Y. Shiraishi, M. Natsume, H. Itoh, M. Inoue, K. Nakata, S. Igarashi, H. Yamaguchi, T. Mizukoshi, E.-I. Suzuki, T. Ueda and I. Shimada, Deuteration and selective labeling of alanine methyl groups of β2-adrenergic receptor expressed in a baculovirus-insect cell expression system, *J. Biomol. NMR*, 2018, **71**(3), 185–192.
119. A. Dubey, N. Stoyanov, T. Viennet, S. Chhabra, S. Elter, J. Borggräfe, A. Viegas, R. P. Nowak, N. Burdziev, O. Petrov, E. S. Fischer, M. Etzkorn, V. Gelev and H. Arthanari, Local deuteration enables NMR observation of methyl groups in proteins from eukaryotic and cell-free expression systems, *Angew. Chem., Int. Ed.*, 2021, **60**(25), 13783–13787.
120. P. Kheddo, M. J. Cliff, S. Uddin, C. F. van der Walle and A. P. Golovanov, Characterizing monoclonal antibody formulations in arginine glutamate solutions using 1H NMR spectroscopy, *mAbs*, 2016, **8**(7), 1245–1258.
121. P. Kheddo, J. E. Bramham, R. J. Dearman, S. Uddin, C. F. van der Walle and A. P. Golovanov, Investigating Liquid–Liquid Phase Separation of a Monoclonal Antibody Using Solution-State NMR Spectroscopy: Effect of Arg·Glu and Arg·HCl, *Mol. Pharmaceutics*, 2017, **14**(8), 2852–2860.
122. K. Chen, D. S. Long, S. C. Lute, M. J. Levy, K. A. Brorson and D. A. Keire, Simple NMR methods for evaluating higher order structures of monoclonal antibody therapeutics with quinary structure, *J. Pharm. Biomed. Anal.*, 2016, **128**, 398–407.
123. K. Kato, C. Matsunaga, Y. Nishimura, M. Waelchli, M. Kainosho and Y. Arata, Application of 13C nuclear magnetic resonance spectroscopy

to molecular structural analyses of antibody molecules, *J. Biochem.*, 1989, **105**(6), 867–869.

124. K. Kato, C. Matsunaga, T. Igarashi, H. Kim, A. Odaka, I. Shimada and Y. Arata, Complete assignment of the methionyl carbonyl carbon resonances in switch variant anti-dansyl antibodies labeled with [1-13C]methionine, *Biochemistry*, 1991, **30**(1), 270–278.
125. M. Kainosho and T. Tsuji, Assignment of the three methionyl carbonyl carbon resonances in Streptomyces subtilisin inhibitor by a carbon-13 and nitrogen-15 double-labeling technique. A new strategy for structural studies of proteins in solution, *Biochemistry*, 1982, **21**(24), 6273–6279.
126. L. Poppe, J. B. Jordan, K. Lawson, M. Jerums, I. Apostol and P. D. Schnier, Profiling formulated monoclonal antibodies by (1)H NMR spectroscopy, *Anal. Chem.*, 2013, **85**(20), 9623–9629.
127. J. Franks, J. N. Glushka, M. T. Jones, D. H. Live, Q. Zou and J. H. Prestegard, Spin Diffusion Editing for Structural Fingerprints of Therapeutic Antibodies, *Anal. Chem.*, 2016, **88**(2), 1320–1327.
128. K. Kato, C. Matsunaga, A. Odaka, S. Yamato, W. Takaha, I. Shimada and Y. Arata, Carbon-13 NMR study of switch variant anti-dansyl antibodies: antigen binding and domain-domain interactions, *Biochemistry*, 1991, **30**(26), 6604–6610.
129. C. Matsunaga, K. Kato and Y. Arata, A13C NMR study of the hinge region of a mouse monoclonal antibody, *J. Biomol. NMR*, 1991, **1**(4), 379–390.
130. Y. Arata; K. Kato; H. Takahashi; and I. Shimada, [15] Nuclear magnetic resonance study of antibodies: A multinuclear approach, in *Methods in Enzymology*, Academic Press, 1994, vol. 239, pp. 440–464.
131. H. Yagi, Y. Zhang, M. Yagi-Utsumi, T. Yamaguchi, S. Iida, Y. Yamaguchi and K. Kato, Backbone (1)H, (13)C, and (15)N resonance assignments of the Fc fragment of human immunoglobulin G glycoprotein, *Biomol. NMR Assign.*, 2015, **9**(2), 257–260.
132. H. Kovacs, D. Moskau and M. Spraul, Cryogenically cooled probes–a leap in NMR technology, *Prog. Nucl. Magn. Reson. Spectrosc.*, 2005, **46**(2–3), 131–155.
133. P. Schanda, E. Kupce and B. Brutscher, SOFAST-HMQC experiments for recording two-dimensional heteronuclear correlation spectra of proteins within a few seconds, *J. Biomol. NMR*, 2005, **33**(4), 199–211.
134. M. W. Maciejewski, M. Mobli, A. D. Schuyler, A. S. Stern and J. C. Hoch, Data sampling in multidimensional NMR: fundamentals and strategies, *Top. Curr. Chem.*, 2012, **316**, 49–77.
135. S. G. Hyberts, H. Arthanari and G. Wagner, Applications of non-uniform sampling and processing, *Top. Curr. Chem.*, 2012, **316**, 125–148.
136. L. W. Arbogast, R. G. Brinson and J. P. Marino, Mapping Monoclonal Antibody Structure by 2D 13C NMR at Natural Abundance, *Anal. Chem.*, 2015, **87**(7), 3556–3561.
137. L. W. Arbogast, R. G. Brinson, T. Formolo, J. T. Hoopes and J. P. Marino, 2D 1HN, 15N Correlated NMR Methods at Natural Abundance for Obtaining Structural Maps and Statistical Comparability of Monoclonal Antibodies, *Pharm. Res.*, 2016, **33**(2), 462–475.

138. S. M. Singh, S. Bandi, D. N. M. Jones and K. M. G. Mallela, Effect of Polysorbate 20 and Polysorbate 80 on the Higher-Order Structure of a Monoclonal Antibody and Its Fab and Fc Fragments Probed Using 2D Nuclear Magnetic Resonance Spectroscopy, *J. Pharm. Sci.*, 2017, **106**(12), 3486–3498.
139. S. Majumder, A. Saati, S. Philip, L. L. Liu, E. Stephens and J. C. Rouse, Alphonse Ignatius, A., Utility of High Resolution NMR Methods to Probe the Impact of Chemical Modifications on Higher Order Structure of Monoclonal Antibodies in Relation to Antigen Binding, *Pharm. Res.*, 2019, **36**(9), 130.
140. A. Hinterholzer, V. Stanojlovic, C. Cabrele and M. Schubert, Unambiguous Identification of Pyroglutamate in Full-Length Biopharmaceutical Monoclonal Antibodies by NMR Spectroscopy, *Anal. Chem.*, 2019, **91**(22), 14299–14305.
141. Y. Tokunaga, K. Takeuchi, J. Okude, K. Ori, T. Torizawa and I. Shimada, Structural Fingerprints of an Intact Monoclonal Antibody Acquired under Formulated Storage Conditions via 15N Direct Detection Nuclear Magnetic Resonance, *J. Med. Chem.*, 2020, **63**(10), 5360–5366.
142. K. Takeuchi, H. Arthanari, I. Shimada and G. Wagner, Nitrogen detected TROSY at high field yields high resolution and sensitivity for protein NMR, *J. Biomol. NMR*, 2016, **64**, 143–151.
143. K. Takeuchi, H. Arthanari, M. Imai, G. Wagner and I. Shimada, Nitrogen-detected TROSY yields comparable sensitivity to proton-detected TROSY for non-deuterated, large proteins under physiological salt conditions, *J. Biomol. NMR*, 2016, **64**(2), 143–151.
144. R. Riek, G. Wider, K. Pervushin and K. Wuthrich, Polarization transfer by cross-correlated relaxation in solution NMR with very large molecules, *Proc. Natl. Acad. Sci. U. S. A.*, 1999, **96**(9), 4918–4923.
145. K. Pervushin, R. Riek, G. Wider and K. Wüthrich, Transverse Relaxation-Optimized Spectroscopy (TROSY) for NMR Studies of Aromatic Spin Systems in 13C-Labeled Proteins, *J. Am. Chem. Soc.*, 1998, **120**(25), 6394–6400.
146. A. Meissner and O. W. Sørensen, Optimization of Three-Dimensional TROSY-Type HCCH NMR Correlation of Aromatic 1H–13C Groups in Proteins, *J. Magn. Reson*, 1999, **139**(2), 447–450.
147. A. G. Milbradt, H. Arthanari, K. Takeuchi, A. Boeszoermenyi, F. Hagn and G. Wagner, Increased resolution of aromatic cross peaks using alternate C-13 labeling and TROSY, *J. Biomol. NMR*, 2015, **62**(3), 291–301.
148. A. Boeszoermenyi, S. Chhabra, A. Dubey, D. L. Radeva, N. T. Burdzhiev, C. D. Chanev, O. I. Petrov, V. M. Gelev, M. Zhang, C. Anklin, H. Kovacs, G. Wagner, I. Kuprov, K. Takeuchi and H. Arthanari, Aromatic 19F-13C TROSY: a background-free approach to probe biomolecular structure, function, and dynamics, *Nat. Methods*, 2019, **16**(4), 333–340.
149. R. Campos-Olivas, R. Aziz, G. L. Helms, J. N. Evans and A. M. Gronenborn, Placement of 19F into the center of GB1: effects on structure and stability, *FEBS Lett.*, 2002, **517**(1–3), 55–60.

150. J. T. Gerig, Fluorine NMR of proteins, *Prog. Nucl. Magn. Reson. Spectrosc.*, 1994, **26**, 293–370.
151. A. Manglik, Kim, H. Tae, M. Masureel, C. Altenbach, Z. Yang, D. Hilger, Lerch, T. Michael, Kobilka, S. Tong, Thian, S. Foon, Hubbell, L. Wayne, R. S. Prosser, Kobilka and K. Brian, Structural Insights into the Dynamic Process of β2-Adrenergic Receptor Signaling, *Cell*, 2015, **161**(5), 1101–1111.
152. M. T. Eddy, T. Didenko, R. C. Stevens and K. Wüthrich, β2-Adrenergic Receptor Conformational Response to Fusion Protein in the Third Intracellular Loop, *Structure*, 2016, **24**(12), 2190–2197.
153. L. Ye, N. Van Eps, M. Zimmer, O. P. Ernst and R. Scott Prosser, Activation of the A2A adenosine G-protein-coupled receptor by conformational selection, *Nature*, 2016, **533**, 265.
154. L. Ye, C. Neale, A. Sljoka, B. Lyda, D. Pichugin, N. Tsuchimura, S. T. Larda, R. Pomes, A. E. Garcia, O. P. Ernst, R. K. Sunahara and R. S. Prosser, Mechanistic insights into allosteric regulation of the A2A adenosine G protein-coupled receptor by physiological cations, *Nat. Commun.*, 2018, **9**(1), 1372.
155. J. J. Liu, R. Horst, V. Katritch, R. C. Stevens and K. Wüthrich, Biased Signaling Pathways in β22-Adrenergic Receptor Characterized by 19F-NMR, *Science*, 2012, **335**(6072), 1106.
156. F. Yang, X. Yu, C. Liu, C.-X. Qu, Z. Gong, H.-D. Liu, F.-H. Li, H.-M. Wang, D.-F. He, F. Yi, C. Song, C.-L. Tian, K.-H. Xiao, J.-Y. Wang and J.-P. Sun, Phospho-selective mechanisms of arrestin conformations and functions revealed by unnatural amino acid incorporation and 19F-NMR, *Nat.* Commun., 2015, **6**, 8202.
157. F. Nußbaumer, R. Plangger, M. Roeck and C. Kreutz, Aromatic 19F–13C TROSY—[19F, 13C]-Pyrimidine Labeling for NMR Spectroscopy of RNA, *Angew. Chem., Int. Ed.*, 2020, **59**(39), 17062–17069.
158. Y. Mizukoshi, K. Takeuchi, M. Arutaki, Y. Tokunaga, T. Takizawa, H. Hanzawa and I. Shimada, Improvement of Ligand Affinity and Thermodynamic Properties by NMR-Based Evaluation of Local Dynamics and Surface Complementarity in the Receptor-Bound State, *Angew. Chem., Int. Ed. Engl.*, 2016, **55**(47), 14606–14609.
159. Y. Tokunaga, K. Takeuchi and I. Shimada, Forbidden Coherence Transfer of 19F Nuclei to Quantitatively Measure the Dynamics of a CF3-Containing Ligand in Receptor-Bound States, *Molecules*, 2017, **22**(9), 1492.
160. H. Sun, L. E. Kay and V. Tugarinov, An Optimized Relaxation-Based Coherence Transfer NMR Experiment for the Measurement of Side-Chain Order in Methyl-Protonated, Highly Deuterated Proteins. *The, J. Phys. Chem. B*, 2011, **115**(49), 14878–14884.
161. V. Tugarinov, R. Sprangers and L. E. Kay, Probing side-chain dynamics in the proteasome by relaxation violated coherence transfer NMR spectroscopy, *J. Am. Chem. Soc.*, 2007, **129**(6), 1743–1750.
162. L. E. Kay and J. H. Prestegard, Methyl group dynamics from relaxation of double quantum filtered NMR signals. Application to deoxycholate, *J. Am. Chem. Soc.*, 1987, **109**(13), 3829–3835.

163. V. Tugarinov, R. Sprangers and L. E. Kay, Probing Side-Chain Dynamics in the Proteasome by Relaxation Violated Coherence Transfer NMR Spectroscopy, *J. Am. Chem. Soc.*, 2007, **129**(6), 1743–1750.
164. T. L. Religa, A. M. Ruschak, R. Rosenzweig and L. E. Kay, Site-Directed Methyl Group Labeling as an NMR Probe of Structure and Dynamics in Supramolecular Protein Systems: Applications to the Proteasome and to the ClpP Protease, *J. Am. Chem. Soc.*, 2011, **133**(23), 9063–9068.
165. Y. Wang, M. VS, J. Kim, G. Li, L. G. Ahuja, P. Aoto, S. S. Taylor and G. Veglia, Globally correlated conformational entropy underlies positive and negative cooperativity in a kinase's enzymatic cycle, *Nat. Commun.*, 2019, **10**(1), 799.
166. W. M. Marsiglia, J. Katigbak, S. Zheng, M. Mohammadi, Y. Zhang and N. J. Traaseth, A Conserved Allosteric Pathway in Tyrosine Kinase Regulation, *Structure*, 2019, **27**(8), 1308–1315, e3.
167. G. Abramov, A. Velyvis, E. Rennella, L. E. Wong and L. E. Kay, A methyl-TROSY approach for NMR studies of high-molecular-weight DNA with application to the nucleosome core particle, *Proc. Natl. Acad. Sci.*, 2020, **117**(23), 12836.
168. Y. Mizukoshi, K. Takeuchi, M. Arutaki, Y. Tokunaga, T. Takizawa, H. Hanzawa and I. Shimada, Improvement of Ligand Affinity and Thermodynamic Properties by NMR-Based Evaluation of Local Dynamics and Surface Complementarity in the Receptor-Bound State, *Angew. Chem., Int. Ed.*, 2016, **55**(47), 14606–14609.
169. M. S. Marlow, J. Dogan, K. K. Frederick, K. G. Valentine and A. J. Wand, The role of conformational entropy in molecular recognition by calmodulin, *Nat. Chem. Biol.*, 2010, **6**(5), 352–358.
170. Y. Tokunaga, K. Takeuchi and I. Shimada, Forbidden Coherence Transfer of F-19 Nuclei to Quantitatively Measure the Dynamics of a CF3-Containing Ligand in Receptor-Bound States, *Molecules*, 2017, **22**(9), 1492.
171. D. G. Fedorov, T. Nagata and K. Kitaura, Exploring chemistry with the fragment molecular orbital method, *Phys. Chem. Chem. Phys.*, 2012, **14**(21), 7562–7577.
172. D. G. Fedorov; and K. Kitaura, *The Fragment Molecular Orbital Method: Practical Applications to Large Molecular Systems*. CRC Press, Boca Raton, FL, 2009.
173. K. Kitaura, E. Ikeo, T. Asada, T. Nakano and M. Uebayasi, Fragment molecular orbital method: an approximate computational method for large molecules, *Chem. Phys. Lett.*, 1999, **313**(3), 701–706.
174. C.-J. Lee, X. Liang, Q. Wu, J. Najeeb, J. Zhao, R. Gopalaswamy, M. Titecat, F. Sebbane, N. Lemaitre, E. J. Toone and P. Zhou, Drug design from the cryptic inhibitor envelope, *Nat. Commun.*, 2016, **7**(1), 10638.
175. M. Aldeghi, V. Gapsys and B. L. de Groot, Accurate Estimation of Ligand Binding Affinity Changes upon Protein Mutation, *ACS Cent. Sci.*, 2018, **4**(12), 1708–1718.
176. H. Öztürk, A. Özgür and E. Ozkirimli, DeepDTA: deep drug–target binding affinity prediction, *Bioinformatics*, 2018, **34**(17), i821–i829.

177. M. Thafar, A. B. Raies, S. Albaradei, M. Essack and V. B. Bajic, Comparison Study of Computational Prediction Tools for Drug-Target Binding Affinities, *Front. Chem.*, 2019, **7**, 782.
178. K. A. Dill; S. Bromberg, LLC, C. R. C. P., *Molecular driving forces: statistical thermodynamics in biology, chemistry, physics, and nanoscience*. 2011.
179. J. A. Caro, K. W. Harpole, V. Kasinath, J. Lim, J. Granja, K. G. Valentine, K. A. Sharp and A. J. Wand, Entropy in molecular recognition by proteins, *Proc. Natl. Acad. Sci. U. S. A.*, 2017, **114**(25), 6563–6568.
180. K. K. Frederick, M. S. Marlow, K. G. Valentine and A. J. Wand, Conformational entropy in molecular recognition by proteins, *Nature*, 2007, **448**(7151), 325–329.
181. S.-R. Tzeng and C. G. Kalodimos, Protein activity regulation by conformational entropy, *Nature*, 2012, **488**(7410), 236–240.
182. A. T. Namanja, X. J. Wang, B. Xu, A. Y. Mercedes-Camacho, B. D. Wilson, K. A. Wilson, F. A. Etzkorn and J. W. Peng, Toward flexibility-activity relationships by NMR spectroscopy: dynamics of Pin1 ligands, *J. Am. Chem. Soc.*, 2010, **132**(16), 5607–5609.
183. J. S. Zintsmaster, B. D. Wilson and J. W. Peng, Dynamics of ligand binding from 13C NMR relaxation dispersion at natural abundance, *J. Am. Chem. Soc.*, 2008, **130**(43), 14060–14061.
184. M. Reibarkh, T. J. Malia and G. Wagner, NMR distinction of single- and multiple-mode binding of small-molecule protein ligands, *J. Am. Chem. Soc.*, 2006, **128**(7), 2160–2161.
185. A. M. Petros, J. R. Huth, T. Oost, C.-M. Park, H. Ding, X. Wang, H. Zhang, P. Nimmer, R. Mendoza, C. Sun, J. Mack, K. Walter, S. Dorwin, E. Gramling, U. Ladror, S. H. Rosenberg, S. W. Elmore, S. W. Fesik and P. J. Hajduk, Discovery of a potent and selective Bcl-2 inhibitor using SAR by NMR, *Bioorg. Med. Chem. Lett.*, 2010, **20**(22), 6587–6591.
186. E. Valkov, T. Sharpe, M. Marsh, S. Greive and M. Hyvonen, Targeting protein-protein interactions and fragment-based drug discovery, *Top. Curr. Chem.*, 2012, **317**, 145–179.
187. M. R. Arkin, Y. Tang and J. A. Wells, Small-Molecule Inhibitors of Protein-Protein Interactions: Progressing toward the Reality, *Chem. Biol.*, 2014, **21**(9), 1102–1114.
188. P. Cimermancic, P. Weinkam, T. J. Rettenmaier, L. Bichmann, D. A. Keedy, R. A. Woldeyes, D. Schneidman-Duhovny, O. N. Demerdash, J. C. Mitchell, J. A. Wells, J. S. Fraser and A. Sali, CryptoSite: Expanding the Druggable Proteome by Characterization and Prediction of Cryptic Binding Sites, *J. Mol. Biol.*, 2016, **428**(4), 709–719.
189. J. A. Wells and C. L. McClendon, Reaching for high-hanging fruit in drug discovery at protein–protein interfaces, *Nature*, 2007, **450**(7172), 1001–1009.
190. A. Kuzmanic, G. R. Bowman, J. Juarez-Jimenez, J. Michel and F. L. Gervasio, Investigating Cryptic Binding Sites by Molecular Dynamics Simulations, *Acc. Chem. Res.*, 2020, **53**(3), 654–661.

191. Y. Mizukoshi, K. Takeuchi, Y. Tokunaga, H. Matsuo, M. Imai, M. Fujisaki, H. Kamoshida, T. Takizawa, H. Hanzawa and I. Shimada, Targeting the cryptic sites: NMR-based strategy to improve protein druggability by controlling the conformational equilibrium, *Sci. Adv.*, 2020, **6**(40), eabd0480.
192. R. A. Copeland, D. L. Pompliano and T. D. Meek, Drug–target residence time and its implications for lead optimization, *Nat. Rev. Drug Discovery*, 2006, **5**(9), 730–739.
193. J. Gabrielsson, H. Dolgos, P. G. Gillberg, U. Bredberg, B. Benthem and G. Duker, Early integration of pharmacokinetic and dynamic reasoning is essential for optimal development of lead compounds: strategic considerations, *Drug Discovery Today*, 2009, **14**(7–8), 358–372.
194. B. Disse, G. A. Speck, K. L. Rominger, T. J. Witek and R. Hammer, Tiotropium (Spiriva): mechanistical considerations and clinical profile in obstructive lung disease, *Life Sci.*, 1999, **64**(6–7), 457–464.
195. J. A. van Noord, T. A. Bantje, M. E. Eland, L. Korducki and P. J. Cornelissen, A randomised controlled comparison of tiotropium nd ipratropium in the treatment of chronic obstructive pulmonary disease. The Dutch Tiotropium Study Group, *Thorax*, 2000, **55**(4), 289–294.
196. J. P. Loria, M. Rance and A. G. Palmer, A Relaxation-Compensated Carr – Purcell – Meiboom – Gill Sequence for Characterizing Chemical Exchange by NMR Spectroscopy, *J. Am. Chem. Soc.*, 1999, **121**(10), 2331–2332.
197. D. F. Hansen, P. Vallurupalli and L. E. Kay, An Improved 15N Relaxation Dispersion Experiment for the Measurement of Millisecond Time-Scale Dynamics in Proteins. *The, J. Phys. Chem. B*, 2008, **112**(19), 5898–5904.
198. F. Massi, E. Johnson, C. Wang, M. Rance and A. G. Palmer, NMR R1ρ Rotating-Frame Relaxation with Weak Radio Frequency Fields, *J. Am. Chem. Soc.*, 2004, **126**(7), 2247–2256.
199. N. A. Farrow, O. Zhang, J. D. Forman-Kay and L. E. Kay, A heteronuclear correlation experiment for simultaneous determination of 15N longitudinal decay and chemical exchange rates of systems in slow equilibrium, *J. Biomol. NMR*, 1994, **4**(5), 727–734.
200. K. Sugase, Elucidating slow binding kinetics of a protein without observable bound resonances by longitudinal relaxation NMR spectroscopy, *J. Biomol. NMR*, 2011, **50**(3), 219–227.
201. Y. Li and A. G. Palmer, TROSY-selected ZZ-exchange experiment for characterizing slow chemical exchange in large proteins, *J. Biomol. NMR*, 2009, **45**(4), 357–360.
202. M. P. Latham, G. R. Zimmermann and A. Pardi, NMR Chemical Exchange as a Probe for Ligand-Binding Kinetics in a Theophylline-Binding RNA Aptamer, *J. Am. Chem. Soc.*, 2009, **131**(14), 5052–5053.
203. A. Furukawa, T. Konuma, S. Yanaka and K. Sugase, Quantitative analysis of protein–ligand interactions by NMR, *Prog. Nucl. Magn. Reson. Spectrosc.*, 2016, **96**, 47–57.

204. K. Sugase, H. Dyson and P. Wright, Mechanism of coupled folding and binding of an intrinsically disordered protein, *Nature*, 2007, **447**(7147), 1021–1025.
205. P. Trigo-Mourino, C. Griesinger and D. Lee, Label-free NMR-based dissociation kinetics determination, *J. Biomol. NMR*, 2017, **69**(4), 229–235.
206. W. Keighley *The need for high throughput kinetics early in the drug discovery process*. www.ddw-online.com/media/32/3832/the-need-for-high-throughput-kinetics.pdf.
207. T. Moschen, S. Grutsch, M. A. Juen, C. H. Wunderlich, C. Kreutz and M. Tollinger, Measurement of Ligand–Target Residence Times by 1H Relaxation Dispersion NMR Spectroscopy, *J. Med. Chem.*, 2016, **59**(23), 10788–10793.
208. C. N. Birts, S. K. Nijjar, C. A. Mardle, F. Hoakwie, P. J. Duriez, J. P. Blaydes and A. Tavassoli, A cyclic peptide inhibitor of C-terminal binding protein dimerization links metabolism with mitotic fidelity in breast cancer cells, *Chem. Sci.*, 2013, **4**(8), 3046–3057.
209. Y. S. Chang, B. Graves, V. Guerlavais, C. Tovar, K. Packman, K. H. To, K. A. Olson, K. Kesavan, P. Gangurde, A. Mukherjee, T. Baker, K. Darlak, C. Elkin, Z. Filipovic, F. Z. Qureshi, H. Cai, P. Berry, E. Feyfant, X. E. Shi, J. Horstick, D. A. Annis, A. M. Manning, N. Fotouhi, H. Nash, L. T. Vassilev and T. K. Sawyer, Stapled alpha-helical peptide drug development: a potent dual inhibitor of MDM2 and MDMX for p53-dependent cancer therapy, *Proc. Natl. Acad. Sci. U. S. A.*, 2013, **110**(36), E3445–E3454.
210. C. L. Ahlbach, K. W. Lexa, A. T. Bockus, V. Chen, P. Crews, M. P. Jacobson and R. S. Lokey, Beyond cyclosporine A: conformation-dependent passive membrane permeabilities of cyclic peptide natural products, *Future Med. Chem.*, 2015, **7**(16), 2121–2130.
211. P. G. Dougherty, J. Wen, X. Pan, A. Koley, J.-G. Ren, A. Sahni, R. Basu, H. Salim, G. Appiah Kubi, Z. Qian and D. Pei, Enhancing the Cell Permeability of Stapled Peptides with a Cyclic Cell-Penetrating Peptide, *J. Med. Chem.*, 2019, **62**(22), 10098–10107.
212. P. G. Dougherty, A. Sahni and D. Pei, Understanding Cell Penetration of Cyclic Peptides, *Chem. Rev.*, 2019, **119**(17), 10241–10287.
213. K. Ito, T. Passioura and H. Suga, Technologies for the synthesis of mRNA-encoding libraries and discovery of bioactive natural product-inspired non-traditional macrocyclic peptides, *Molecules*, 2013, **18**(3), 3502–3528.
214. F. Porcelli, A. Ramamoorthy, G. Barany and G. Veglia, On the role of NMR spectroscopy for characterization of antimicrobial peptides, *Methods Mol. Biol.*, 2013, **1063**, 159–180.
215. K. Takeuchi, H. Takahashi, M. Sugai, H. Iwai, T. Kohno, K. Sekimizu, S. Natori and I. Shimada, Channel-forming membrane permeabilization by an antibacterial protein, sapecin: determination of membrane-buried and oligomerization surfaces by NMR, *J. Biol. Chem.*, 2004, **279**(6), 4981–4987.

216. T. Nakamura, H. Takahashi, K. Takeuchi, T. Kohno, K. Wakamatsu and I. Shimada, Direct determination of a membrane-peptide interface using the nuclear magnetic resonance cross-saturation method, *Biophys. J.*, 2005, **89**(6), 4051–4055.
217. K. Wakamatsu, A. Takeda, T. Tachi and K. Matsuzaki, Dimer structure of magainin 2 bound to phospholipid vesicles, *Biopolymers*, 2002, **64**(6), 314–327.
218. A. Naito, N. Matsumori and A. Ramamoorthy, Dynamic membrane interactions of antibacterial and antifungal biomolecules, and amyloid peptides, revealed by solid-state NMR spectroscopy, *Biochim. Biophys. Acta, Gen. Subj.*, 2018, **1862**(2), 307–323.
219. S. L. Schreiber and G. R. Crabtree, The mechanism of action of cyclosporin A and FK506, *Immunol. Today*, 1992, **13**(4), 136–142.
220. J. Witek, M. Muhlbauer, B. G. Keller, M. Blatter, A. Meissner, T. Wagner and S. Riniker, Interconversion Rates between Conformational States as Rationale for the Membrane Permeability of Cyclosporines, *Chem. Phys. Chem.*, 2017, **18**(23), 3309–3314.
221. J. Witek, B. G. Keller, M. Blatter, A. Meissner, T. Wagner and S. Riniker, Kinetic Models of Cyclosporin A in Polar and Apolar Environments Reveal Multiple Congruent Conformational States, *J. Chem. Inf. Model.*, 2016, **56**(8), 1547–1562.
222. C. K. Wang, J. E. Swedberg, P. J. Harvey, Q. Kaas and D. J. Craik, Conformational Flexibility Is a Determinant of Permeability for Cyclosporin, *J. Phys. Chem. B*, 2018, **122**(8), 2261–2276.
223. C. L. Ahlbach, K. W. Lexa, A. T. Bockus, V. Chen, P. Crews, M. P. Jacobson and R. S. Lokey, Beyond cyclosporine A: conformation-dependent passive membrane permeabilities of cyclic peptide natural products, *Future Med. Chem.*, 2015, **7**(16), 2121–2130.
224. S. Ono, M. R. Naylor, C. E. Townsend, C. Okumura, O. Okada and R. S. Lokey, Conformation and Permeability: Cyclic Hexapeptide Diastereomers, *J. Chem. Inf. Model.*, 2019, **59**(6), 2952–2963.
225. A. Furukawa, J. Schwochert, C. R. Pye, D. Asano, Q. D. Edmondson, A. C. Turmon, V. G. Klein, S. Ono, O. Okada and R. S. Lokey, Drug-Like Properties in Macrocycles above MW 1000: Backbone Rigidity versus Side-Chain Lipophilicity, *Angew. Chem., Int. Ed.*, 2020, **59**(48), 21571–21577.
226. K. Takeuchi, I. Misaki, Y. Tokunaga, M. Fujisaki, H. Kamoshida, T. Takizawa, H. Hanzawa and I. Shimada, Conformational Plasticity of Cyclic Ras-Inhibitor Peptides Defines Cell Permeabilization Activity, *Angew. Chem., Int. Ed. Engl.*, 2021, **60**(12), 6567–6572.
227. P. Upadhyaya, Z. Qian, N. G. Selner, S. R. Clippinger, Z. Wu, R. Briesewitz and D. Pei, Inhibition of Ras Signaling by Blocking Ras–Effector Interactions with Cyclic Peptides, *Angew. Chem., Int. Ed.*, 2015, **54**(26), 7602–7606.
228. B. Japelj, G. Ilc, J. Marušič, J. Senčar, D. Kuzman and J. Plavec, Biosimilar structural comparability assessment by NMR: from small proteins to monoclonal antibodies, *Sci. Rep.*, 2016, **6**(1), 32201.

229. S. Yanaka, T. Yamazaki, R. Yogo, M. Noda, S. Uchiyama, H. Yagi and K. Kato, NMR Detection of Semi-Specific Antibody Interactions in Serum Environments, *Molecules*, 2017, **22**(10), 1619.
230. K. Moritsugu, K. Takeuchi, N. Kamiya, J. Higo, I. Yasumatsu, Y. Fukunishi and I. Fukuda, Flexibility and Cell Permeability of Cyclic Ras-Inhibitor Peptides Revealed by the Coupled Nosé-Hoover Equation, *J. Chem. Inf. Model*, 2021, **61**(4), 1921–1930.
231. D. F. Gauto, L. F. Estrozi, C. D. Schwieters, G. Effantin, P. Macek, R. Sounier, A. C. Sivertsen, E. Schmidt, R. Kerfah, G. Mas, J. P. Colletier, P. Güntert, A. Favier, G. Schoehn, P. Schanda and J. Boisbouvier, Integrated NMR and cryo-EM atomic-resolution structure determination of a half-megadalton enzyme complex, *Nat. Commun.*, 2019, **10**(1), 2697.
232. F. Delhommel, F. Gabel and M. Sattler, Current approaches for integrating solution NMR spectroscopy and small-angle scattering to study the structure and dynamics of biomolecular complexes, *J. Mol. Biol.*, 2020, **432**(9), 2890–2912.
233. A. Kotar, H. N. Foley, K. M. Baughman and S. C. Keane, Advanced approaches for elucidating structures of large RNAs using NMR spectroscopy and complementary methods, *Methods*, 2020, **183**, 93–107.
234. J. L. Markley, Current Solution NMR Techniques for Structure-Function Studies of Proteins and RNA Molecules, *Adv. Exp. Med. Biol.*, 2018, **1105**, 43–58.
235. J. R. Bothe, E. N. Nikolova, C. D. Eichhorn, J. Chugh, A. L. Hansen and H. M. Al-Hashimi, Characterizing RNA dynamics at atomic resolution using solution-state NMR spectroscopy, *Nat. Methods*, 2011, **8**(11), 919–931.
236. K. Kato and T. Yamaguchi, Paramagnetic NMR probes for characterization of the dynamic conformations and interactions of oligosaccharides, *Glycoconj. J.*, 2015, **32**(7), 505–513.
237. Y. Tokunaga, T. Viennet, H. Arthanari and K. Takeuchi, Spotlight on the Ballet of Proteins: The Structural Dynamic Properties of Proteins Illuminated by Solution NMR, *Int. J. Mol. Sci.*, 2020, **21**(5), 1829.
238. Y. Tokunaga and K. Takeuchi, Role of NMR in High Ordered Structure Characterization of Monoclonal Antibodies, *Int. J. Mol. Sci.*, 2020, **22**(1), 46.

CHAPTER 16

Locating Hydrogen Atoms Using Fast-MAS Solid-state NMR and microED

YUSUKE NISHIYAMA[a,b]

[a] JEOL RESONANCE Inc., Akishima, Tokyo 196-8558, Japan; [b] RIKEN-JEOL Collaboration Center, Yokohama, Kanagawa 230-0045, Japan
Email: yunishiy@jeol.co.jp

16.1 Introduction

Hydrogen atoms play crucial roles in chemistry, biology, material sciences, *etc.* through hydrogen-bonding, proton/hydrogen exchange, proton conductivity, to name a few. However, it is not a straightforward task to determine the position of H atoms using diffraction-based methods. Single-crystal X-ray diffraction (SCXRD), as a gold standard of structural solutions, locates hydrogens poorly due to the low density of the electron cloud around the hydrogen atoms. Three dimensional (3D) electron diffraction (microED), in principle, provides a better accuracy compared to SCXRD if dynamic scattering is considered in the data analysis,[1] however, microED still sometimes fails to locate hydrogens. Neutron diffraction (ND) is a suitable method to observe hydrogen atoms. However, ND requires a large facility which is not easily accessible for most researchers, large single crystals, and sometimes deuteration to suppress the background signals. On the other hand, ^{1}H, which is the major isotope of hydrogen, is the most sensitive nucleus for NMR observation and possesses a suitable spin number of 1/2 with a very high natural abundance of >99%. Thus, NMR can be the most

New Developments in NMR No. 26
NMR Spectroscopy for Probing Functional Dynamics at Biological Interfaces
Edited by Anirban Bhunia, Hanudatta S. Atreya and Neeraj Sinha

Published by the Royal Society of Chemistry, www.rsc.org

suitable method to provide complementary information on hydrogens to the diffraction methods. In fact, it is widely accepted that a combination of solid-state NMR and diffraction method, dubbed *NMR crystallography*, is useful to obtain fruitful information on crystalline structures, especially the positions of hydrogen.[2–5] While NMR crystallography can be carried out solely using experimental NMR data together with crystal structure prediction calculations,[6] major NMR crystallography approaches utilize a combination of solid-state NMR and diffraction methods. This combined approach can be classified into two different categories to locate the hydrogen atoms. (1) Distance measurements between hydrogen and neighboring nuclei: in this approach, determination of dipolar couplings is crucial owing to its inverse cubic dependence to internuclear distances. (2) Chemical shift calculation: experimentally available isotropic chemical shifts are used to validate the candidate structures together with first-principles quantum computations which predict chemical shifts. In this evaluation, not only ^{1}H chemical shifts but also the chemical shifts of the other nuclei including ^{13}C, ^{15}N, *etc.* are involved. Quadrupolar couplings can also be used. In either case, crystalline structures of non-hydrogen atoms are solved by the diffraction methods, while hydrogen is located and/or validated based on solid-state NMR involving ^{1}H. However, solid-state NMR measurements including the ^{1}H nucleus are not straightforward. Intense ^{1}H–^{1}H dipolar couplings together with a high abundance of ^{1}H inevitably introduce complex strong ^{1}H–^{1}H coupling networks, resulting in large residual dipolar couplings to the higher order, although they are averaged out to the first order. Fortunately, recent advances in MAS technology allow sample spinning faster than 100 kHz even for commercially available systems,[7–27] enabling easy handling of the ^{1}H nucleus in rigid solids.

As an example of the former approach (1), in a system with N…H…O interactions, dipolar coupling between nitrogen and proton is utilized to determine the distances between N and H, while distances between N and O are obtained from the diffraction-based structure. Finally, the hydrogen position is precisely determined by combining these pieces of information. In order to determine the size of dipolar couplings under MAS, special techniques should be used, since MAS introduces time-dependence to the dipolar interactions so that their time averages go to zero. Dipolar interactions can be recovered by interfering with the averaging process under MAS, by applying a radio-frequency (rf) field synchronous to the MAS modulation frequency, adjusting MAS frequency to a specific internal NMR interaction like differences of isotropic chemical shift, *etc.* It should be noted that the recoupling has to be achieved at the first-order average Hamiltonian. Otherwise, quantitative distance information is not available. This is because the higher-order average Hamiltonian includes various pieces of information other than a single dipolar coupling, making the distance information from higher-order recoupling qualitative.

In the latter approach, either SCXRD or powder XRD-based structures are used for structure solution in most cases. Hydrogen positions are optimized using quantum computation and validated by comparing calculated NMR chemical shift with the experimental one. ^{1}H isotropic chemical shifts, which are easily available for fast-MAS ^{1}H NMR, are sensitive to molecular packings and play a crucial role in chemical shift validation. Of course, ^{13}C and ^{15}N isotropic chemical shift measurements are trivial. However, the high quality of crystals for SCXRD or isomorphic microcrystalline samples for powder XRD for crystal structure solutions are not always readily available. A notable example can be found in pharmaceuticals. Active pharmaceutical ingredients (API) in pharmaceutical tables often form nano- to micro-sized single crystals which are too small for SCXRD. In addition, the presence of excipient hampers the application of powder XRD. Quite recently, the emerging method of microED has been introduced to solve the crystalline structure from nano- to micro-sized single crystals, potentially solving problems.[28–31] Indeed, this method is useful for NMR crystallography.

In this chapter, we focus on the NMR methodology for quantitative distance measurements using the first-order hetero/homonuclear dipolar recouplings under fast-MAS and also briefly introduce the emerging method of microED in NMR crystallography, mainly focusing on the recently developed methods in my laboratory.

16.2 Heteronuclear Dipolar Interactions

Two different approaches to quantitatively determine heteronuclear dipolar coupling are widely used. In one approach, the initial magnetization starts from the observation channel. Subsequently, time evolution of the magnetization is observed under the influence of heteronuclear interactions. These are known as separated-local-field (SLF) experiments.[32] Notably, REDOR, DIPSHIFT, and related methods are widely used for the observation of heteronuclear dipolar intractions[33] (Figures 16.1 and 16.2, respectively). In this approach, the abundance of non-observation nuclei should be large enough, since the effect of heteronuclear dipolar couplings appears only if the non-observation nuclei are NMR active. Thus, ^{1}H, a nucleus with high natural abundance, or an isotopically enriched nucleus is typically assigned to the non-observation nuclei. Otherwise, distance information is hardly obtained. For example, if ^{13}C is used at natural abundance (1.1%) as a non-observation nucleus, 99% of the observed nuclei do not experience the heteronuclear dipolar interactions and only 1.1% of those exhibit the dipolar coupling oscillation. As these signals appear overlapped with each other, the dipolar coupling oscillation appears to be only 1.1% of the total signal intensity at maximum, making the extraction of distance information challenging. In the other approach, the initial magnetization starts from the non-observation channel and transferred magnetization to the observation channel is monitored as a build-up curve. Cross-polarization (CP) and TEDOR techniques can be categorized into this approach.[15,34,35] In this approach, as the magnetization transfer occurs only between NMR active

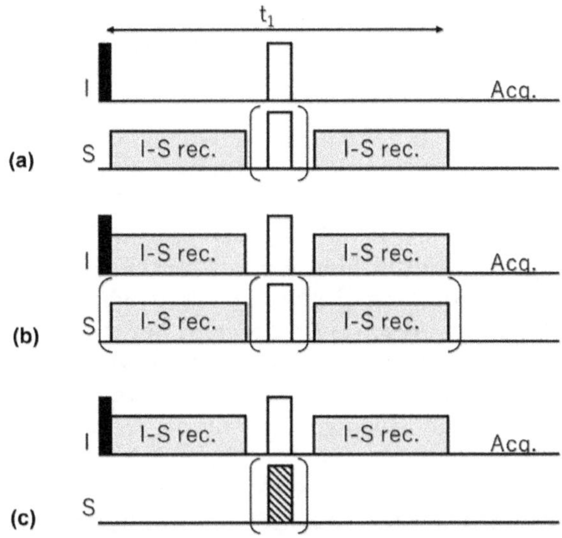

Figure 16.1 Implementation of REDOR experiments. While *I* spins undergo spin-echo in all the sequences, *I–S* heteronuclear recoupling is achieved by irradiating on the (a) *S* channel, (b) *I* and *S* channels simultaneously, and (c) *I* channel. The heteronuclear recoupling sequence which is responsible for the *I–S* distance measurements is shaded. Black and white rectangles represent π/2 and π pulses, respectively. While the rectangular with stripe is typically a π pulse for spin-1/2 nuclei, that can be π, π/2, mixing, or saturation pulses for quadrupolar nuclei.

nuclei, the observed signal is always associated with the NMR active non-observation nuclei. Thus, the abundance of the non-observation nuclei is not an issue. The low natural abundance could still be an issue because of the low sensitivity.

16.2.1 REDOR, DIPSHIFT and Related Methods: Working Principle

REDOR and DIPSHIFT experiments observe time evolution of the magnetization under the influence of heteronuclear dipolar interactions. Hereafter, we refer to the nucleus whose magnetization experiences the heteronuclear dipolar evolution as *I*, and that coupled to *I* as *S* (see Figure 16.1 for REDOR and Figure 16.2 for DIPSHIFT); *I* spin corresponds to observation nuclei and *S* to non-observation. As mentioned above, the abundance of *S* spin should be large enough. It should be noted that additional magnetization transfer may result in observation of other than *I* spin, thus we try to avoid using the terminology of observation and non-observation channels as much as possible. While these sequences are invented and developed independently, they share the same principal mechanism.[36] The experiments are essentially spin echo experiments on the *I* channel during which *I* magnetization

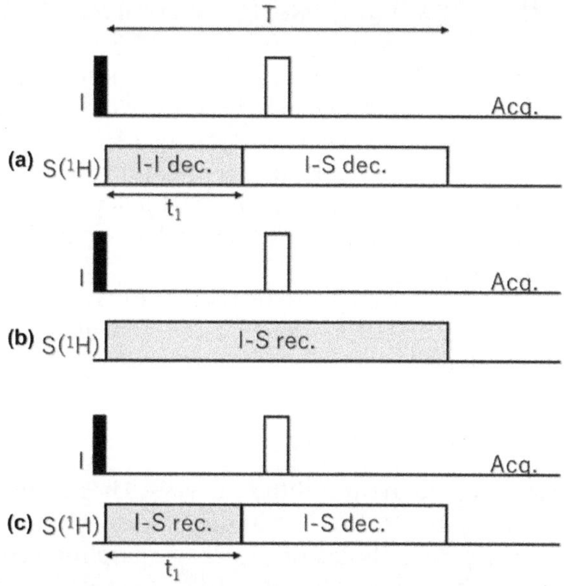

Figure 16.2 Implementation of DIPSHIFT experiments. While the total echo time T is kept constant, heteronuclear dipolar interaction is reintroduced by (a) turn on dipolar coupling in a fraction of the MAS cycle time, (b) varying scaling factor, and (c) turn on dipolar coupling with a multiple integer of the MAS cycle time. The heteronuclear recoupling sequences which are responsible for the I–S distance measurements are shaded. Black and white rectangles represent $\pi/2$ and π pulses, respectively.

evolves under I–S heteronuclear dipolar interactions. In REDOR experiments, two independent experiments were performed with and without the influence of heteronuclear dipolar interactions at various recoupling times, giving dephased (S') and reference (S_0) NMR signals. The heteronuclear dipolar interaction can be controlled either by switching on/off the recoupling sequences or applying/removing the refocusing π pulse on the S channel. The effect of T_2' relaxation during spin echo can be eliminated by evaluating $(S_0–S')/S_0$, which gives the fraction curve of REDOR experiments. On the other hand, DIPSHIFT experiments are implemented in a constant-time manner. While total echo duration is fixed, the effective heteronuclear dipolar coupling is varied. There are several different implementations. In the original implementation, the heteronuclear dipolar coupling is turned on during a fraction of MAS rotor cycles (Figure 16.2a). Since MAS averaging is completed after a full rotor cycle, this approach leads to imperfect averaging, resulting in partial reintroduction of heteronuclear dipolar interactions. The other implementation uses variable scaling factors or effective dipolar couplings. This can be achieved by changing the pulse timings (Figure 16.2b) or by varying the number of repetitions of the recoupling sequence which is synchronous to the MAS cycle time (Figure 16.2c). Since the DIPSHIFT curve is observed with the constant time implementation, the effect of T_2' decay is eliminated. In both REDOR and DIPSHIFT experiments, the heteronuclear dipolar coupling can be obtained by fitting the REDOR fraction curve and DIPSHIFT curve, respectively. In the ideal case, *i.e.* perfect spin echo, ideal pulses, absence of interfering molecular motion, *etc.*, the analytical curve can be used with only a single fitting parameter of heteronuclear dipolar coupling. Otherwise, numerical calculation including various parameters may be used. Nevertheless, the size of heteronuclear dipolar couplings, and thus internuclear distances, can be extracted from the REDOR and DIPSHIFT experiments.

In the following, first we discuss the feature of REDOR experiments, and subsequently introduce DIPSHIFT based on the considerations in REDOR owing to its close similarity.

16.2.1.1 REDOR: Implementation

REDOR can be implemented in three different ways. The recoupling sequence can be applied on S (Figure 16.1a), *I*/*S* simultaneously (Figure 16.1b), and *I* (Figure 16.1c) channels. The choice of dipolar recoupling sequence is also important as it affects the range of distance to be measured and the effect of CSA. When the recoupling sequence is applied only for either the *I* or *S* channel, it inevitably introduces CSA on the irradiated nuclei (*S* in Figure 16.1a and *I* in Figure 16.1c), as CSA and heteronuclear dipolar couplings have the same symmetry with respect to rf and sample rotation. On the other hand, simultaneous irradiation on *I* and *S* allows exclusive recoupling of heteronuclear dipolar coupling, suppressing CSA of both channels, if the sequences are properly chosen. The form of the recoupled spin part is also an important factor. If the interactions are reintroduced so that they don't commute each other, the effect of the weaker dipolar interaction is suppressed by the stronger interaction, *i.e.* it is truncated. However, at the same time, this reduces the effect of long-distance coupling, making the observation of short contact easy and precise.[37] In addition to dipolar truncation between heteronuclear couplings, CSA interaction may prevent the observation of dipolar coupling due to the truncation effect.[38] On the other hand, if the dipolar couplings are recoupled in a way that the recoupled interactions commute each other, no dipolar truncation occurs and each heteronuclear dipolar coupling doesn't interfere. This allows simultaneous observation of the long distances in the presence of a short-distance pair. Moreover, the recoupled *S* spin CSA interaction in Figure 16.1a doesn't play any role in the *I* spin evolution, and the recoupled *I* spin CSA evolution is refocused by the central π pulse on the *I* channel in Figure 16.1c. Nevertheless, these recoupled CSAs ideally don't play anything in all the cases, however, they behave differently in the presence of MAS fluctuation.[39] The sequence in Figure 16.1a and b is immune to CSA from the beginning, and thus to MAS fluctuation. On the other hand, the sequence in Figure 16.1c relies on refocusing by the π pulse to suppress recoupled CSA. Thus, MAS fluctuation may hamper this refocusing process, resulting in additional signal decay and noise.[40–42]

16.2.1.2 REDOR: ^{1}H Channel Assignments; *I* or *S*?

When the ^{1}H–X distance is measured, ^{1}H can be assigned to either the *I* or *S* channel, resulting in ^{1}H and X observation, respectively, in the absence of additional magnetization transfers. While ^{1}H observation is beneficial in terms of the sensitivity enhancement owing to its high gyromagnetic ratio, abundance, and thus thermal polarization, the spectral resolution is somewhat limited due to its rather broad ^{1}H linewidth together with its limited chemical shift range. ^{1}H observation is mandatory if the X nuclei are quadrupolar nuclei or spin-1/2 with large anisotropies like paramagnetic interactions or chemical shift anisotropies of heavy spin-1/2 nuclei, which are not able to be observed directly. On the other hand, X observation typically provides higher resolution than ^{1}H observation, especially when X is spin-1/2 nuclei. However, it suffers from lower sensitivity than the ^{1}H detection. Thus, the choice of nuclei for detection depends on the compromise between sensitivity and resolution. It should be noted that the magnetization after the REDOR sequence may be transferred to the other nuclei for observation,[43] especially in fully labeled protein samples. In such a case, careful consideration is further required for the choice of sequences.

In addition to heteronuclear dipolar coupling and CSA interactions, recoupling pulses may introduce homonuclear dipolar coupling. This is especially problematic when recoupling pulses are applied to ^{1}H. A traditional REDOR sequence, which consists of a train of rotor synchronous π pulses, suffers from ^{1}H–^{1}H dipolar coupling in a finite pulse regime. In this case, extensive deuteration is required to get rid of ^{1}H–^{1}H dipolar coupling. Instead, the other recoupling sequence which is designed to suppress ^{1}H–^{1}H couplings can be used.

16.2.1.3 REDOR: Application to Quadrupolar Nuclei

The REDOR approach may be applied to quadrupolar nuclei ($I > 1/2$) if the size of the quadrupolar couplings is small enough. However, the rf field strength is often not strong enough to cover the whole spectral range because of the large quadrupolar coupling. This makes π pulses on the quadrupolar nuclei inefficient. For half-integer quadrupolar nuclei, central transition (CT) selective pulses may be used.[44] The CT pulse should be carefully optimized to avoid any interference from outer energy levels, which results in enhanced REDOR dephasing but makes the fraction curve less quantitative.[45] Instead of inversion pulses, rotor-assisted spin state mixing in REAPDOR[46] and saturation in RESPDOR[47] can be applied to recover heteronuclear dipolar evolution. Nevertheless, in all these measurements involving quadrupolar nuclei, dipolar recoupling pulses should be applied only to the spin-1/2 nuclei and the number of pulses on quadrupolar nuclei should be minimized.

16.2.1.4 REDOR: ^{1}H–X Distance Measurements with Recoupling on the S Channel

Figure 16.1a depicts ^{1}H–X distance measurements with recoupling on the S channel. If $I = {}^1$H is chosen, the observed signal intensity is largely sacrificed by short ^{1}H T_2' decay, making the long-distance measurements difficult. In addition, spectral resolution is compromised due to relatively limited ^{1}H resolution. Thus, application to ^{1}H–X distance measurements in a fully protonated sample is unrealistic unless sufficiently fast MAS is applicable. In addition, the natural abundance of the S spin has to be large enough, otherwise isotopic labeling is required. ^{1}H T_2' can be elongated by using intensive deuteration combined with fast-MAS. Long-range ^{1}H–^{13}C distances are successfully measured in back-exchanged and fully ^{13}C,^{15}N-labeled protein to elucidate secondary/ternary structures.[48] In this example, the ^{1}H resolution is largely improved by deuteration and the additional ^{15}N dimension also helps to improve the spectral separation.

When $S = {}^1$H in Figure 16.1a is chosen, more than 99% of I spin signals experience the REDOR effect because of the high natural abundance of ^{1}H. This implementation is favored in the sense that the experiments typically benefit from the longer T_2' relaxation time of the I nuclei than ^{1}H. However, ^{1}H–^{1}H dipolar coupling can be problematic during recoupling sequences. The traditional REDOR π pulse train was used on deuterated samples[36,49,50] for dipolar recoupling measurements between covalently bonded ^{1}H–^{15}N and ^{1}H–^{13}C pairs. The deuteration reduces S(^{1}H) spin abundance, and thus deuteration has to be selectively applied to non-observed proton sites or additional ^{1}H–^{15}N and ^{1}H–^{13}C filters should be used to suppress signals from ^{2}H–^{15}N and ^{2}H–^{13}C pairs. As the size of dipolar couplings is too large, the REDOR fraction curve shows too steep build-up if regular REDOR timing is used. The scaling factor of the REDOR π pulse train can be adjusted by changing the pulse timings so that precise determination of large dipolar coupling is possible.[49,50] Instead of the π pulse train, I–S heteronuclear recoupling sequences can be designed with simultaneous decoupling of ^{1}H-^{1}H interactions in fully protonated systems using the symmetry-based framework.[44,51–54] SR4, which is a very robust heteronuclear dipolar recoupling sequence with ^{1}H–^{1}H decoupling, is introduced and applied to measure ^{1}H–^{17}O distances. Since the recoupled heteronuclear interactions by SR4 commute each other, multiple distances were monitored in the presence of stronger couplings. The selective π pulse on ^{1}H allows site-selective distance measurements. SR4 and $R12_3^5$ sequences are used to measure covalently bonded ^{1}H–^{13}C distances.[55]

16.2.1.5 REDOR: ^{1}H–X Distance Measurements with Recoupling on the I and S Channels Simultaneously

This implementation (Figure 16.1b) is widely used in ^{13}C–^{15}N distance measurements with REDOR π pulse trains to both channels because of its robustness. The other example can be found in the PARS experiment which is designed based on this approach to measure ^{1}H–X distances.[56,57] While a pair of symmetry-based sequences consisting of RN_0 and RN_π is applied on $S = {}^1$H channel, π pulses are applied on the *I* channel between each symmetry sequence. As each symmetry sequence recouples both ^{1}H CSA and heteronuclear dipolar interactions in a non-commuting form, ^{1}H CSA and dipolar interactions can interfere with each other. However, the phase-alternated pair together with the π pulses on the *S* channel result in selective dipolar recoupling with suppression of ^{1}H CSA, removing the effect of ^{1}H CSA from the *I* spin time evolution. Additionally, since no CSA of either *I* or *S* spins is recoupled, the sequence is expected to be robust with respect to MAS fluctuation. While the PARS experiment is a kind of DIPSHIFT experiment rather than REDOR, the idea can be applied in the REDOR framework. In addition, the combination of a dual symmetry-based sequence, in principle, can be used.[58]

16.2.1.6 REDOR: ^{1}H–X Distance Measurements with Recoupling on the I Channel

This implementation (Figure 16.1c) has its unique advantage that the number of pulses on the *S* channel is minimized since *I–S* dipolar coupling is reintroduced by irradiating only on the *I* channel. This allows for the *S* channel to be assigned to nuclei with large anisotropies like quadrupolar nuclei and heavy spin-1/2 nuclei bearing large CSA, however, the sequence is sensitive to MAS fluctuation. The sequence reintroduces *I* spin CSA, which is refocused at the end of the echo period, requiring very stable MAS, otherwise imperfect refocusing results in signal decay and t_1 noise. As the sequences are accompanied with an obvious disadvantage of sensitivity to MAS fluctuation, it is advisable to use the sequences in Figure 16.1a and b if *S* is a manageable nucleus. CW irradiation on the *I* channel to recouple heteronuclear dipolar coupling can partially alleviate the sensitivity to MAS fluctuation.[40] On the other hand, the implementation solely allows access to nuclei with large anisotropies which can even be invisible in the direct observation. These nuclei should be assigned to the *S* channel, and thus ^{1}H is assigned to the *I* channel. Note that the abundance of S spin should be large enough. While ^{1}H–S dipolar coupling can easily be recoupled regardless of the anisotropy of the S spin, the large anisotropy of the *S* spin poses the difficulty in implementing the inversion pulse on the *S* channel. When the perfect inversion of *S* nuclei is difficult, partial inversion was demonstrated by rf field irradiation with 1/3 of the cycle time of rotation in REAPDOR experiments.[46,59] Instead of an inversion pulse, a composite π/2 pulse is used to manage ^{2}H nuclei.[46] an innovative idea, RIDER, of spin-state mixing was demonstrated using longitudinal relaxation of quadrupolar nuclei with a framework of CODEX-type experiments.[60,61] The saturation pulse, which is typically easier than the inversion pulse, is used to measure distances between ^{13}C and ^{14}N.[47] Besides RIDER, rf pulses are used to mix the spin state, introducing the dependence of quadrupolar couplings on the REDOR effect. Thus, in principle, quadrupolar interactions have to be involved in the REDOR curve analysis. Fortunately, it has been shown that REAPDOR is not sensitive to the size of quadrupolar coupling, relative orientation of dipolar and quadrupolar interactions, *etc.*, enabling distance observation without knowing the quadrupolar parameters. Phase-modulated saturation pulses almost completely saturate the quadrupolar spin systems even though the rf field strength is several times smaller than the quadrupolar interactions.[62–64] In these experiments, the REDOR/REAPDOR/RESPDOR curve can be fit only varying the size of heteronuclear dipolar couplings (plus optionally the fraction curve intensity).

The distance measurement between ^{1}H and ^{14}N has been demonstrated using a combination of RESPDOR sequence and phase-modulated (PM) saturation pulses at the fast MAS conditions using the sequence shown in Figure 16.1c.[42] An application of this method will be shown later (see Section 16.2.1.7) The SR4 sequence is applied to decouple ^{1}H–^{1}H interactions. Since SR4 recouples dipolar interactions so that they commute each other, multiple ^{1}H–^{14}N distances are observed from a single set of the experiment. Although ^{14}N is a $S=1$ nucleus with quadrupolar interactions typically in 1–6 MHz, it is shown that phase-modulated pulses can saturate the spin system well at a reasonable rf field strength even at the fast-MAS conditions. The high abundance of ^{14}N (99.6%) enables most of NH protons to experience the RESPDOR effect. This results in very rapid observation of the PM-S-RESPDOR fraction curve, typically several tens of minutes to several hours. More importantly, the PM-S-RESPDOR fraction curve is fitted with the analytical expression with the one variable parameter, that is ^{1}H–^{14}N heteronuclear dipolar couplings, without any vertical scaling. This means the ^{1}H–^{14}N distances can be measured even if the fraction curves do not reach their maxima, allowing long distance measurements. In addition, the method can be combinedly with selective excitation schemes to improve the throughput.[65]

^{1}H–^{14}N distances are readily obtained by ^{1}H–^{14}N PM-S-RESPDOR experiments, however, there is ambiguity if multiple ^{14}N sites exist because of the broadband nature of the phase-modulated saturation pulses; it is not clear which nitrogen is responsible for RESPDOR dephasing since all nitrogens are simultaneously saturated. This problem can be overcome with use of ^{14}N overtone irradiation (OT) in which the rf field frequency of twice the ^{14}N Larmor frequency is applied.[66] As the nutation frequency of ^{14}N OT is small due to its second-order nature, ^{14}N OT pulses enable the selective observation of a specific ^{14}N site because of its narrow bandwidth.[67–70] ^{14}N OT pulses can be combined in the REDOR framework toward selective ^{1}H–^{14}N distance measurements, even in the presence of multiple ^{14}N sites. Although phase-modulated rf irradiation enables complete and broadband saturation of all the ^{14}N energy levels when the ^{14}N Larmor frequency is utilized, ^{14}N OT irradiation introduces exchanges exclusively between $m=1$ and $m=-1$ states (Figure 16.3). As the ^{14}N OT pulse makes nutation frequencies

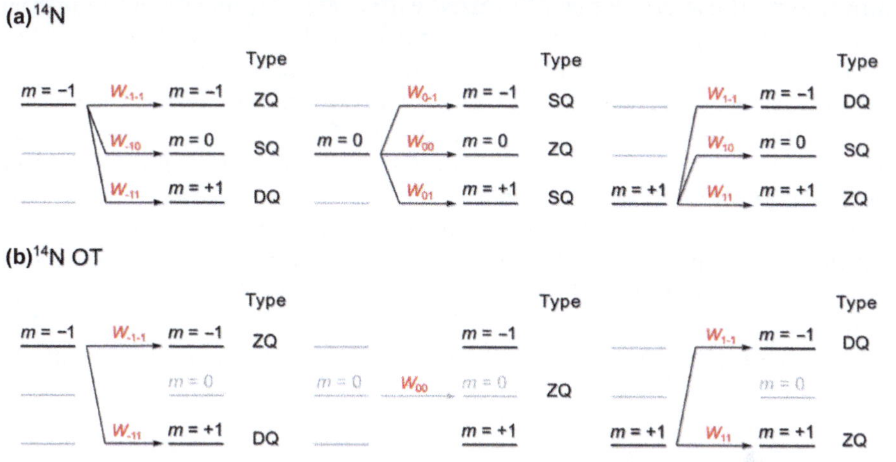

Figure 16.3 Transitions induced by (a) the ^{14}N fundamental and (b) ^{14}N overtone frequencies. While phase-modulated ^{14}N saturation pulses introduce all the possible energy mixing with the ^{14}N fundamental frequency, ^{14}N overtone irradiation exchanges between $m = -1$ and $+1$ exclusively, keeping $m = 0$ untouched. Reproduced from ref. 66, https://doi.org/10.3389/fmolb.2021.645347, under the terms of the CC BY 4.0 license, https://creativecommons.org/licenses/by/4.0/.

orientation-dependent, ^{14}N OT-REDOR includes both REDOR (complete inversion) and RESPDOR (complete saturation) regimes. The ratio between REDOR and RESPDOR can be quantified by introducing the *f* factor which describes the magnetization exchange between $m = \pm 1$ states of a certain ^{14}N site. In the experiments, the *f* factor varies between 0 to 1 depending on the rf field strength, length of ^{14}N OT pulse, size of quadrupolar coupling, to name a few. However, *f* can be experimentally determined, once oscillation appears in the ^{14}N OT-REDOR curve by analytical fitting, as the *f* factor defines the plateau of the fraction curve. The *f* factor thus determined can be used to observe longer distances between ^{1}H and ^{14}N which matches to the ^{14}N OT resonance. For example, ^{1}H–^{14}**NH** (Figure 16.4c and d) and ^{1}H–^{14}**NH**$_3^+$ (Figure 16.4e and f) in L-alanyl-L-alanine are selectively determined by separately irradiating ^{14}**NH** and ^{14}**NH**$_3^+$, respectively. In this example, the *f* factor for each of the ^{14}**NH** and ^{14}**NH**$_3^+$ sites is first determined by a covalently bonded ^{1}H–^{14}N pair (Figure 16.4c and e). The *f* factor thus determined is applied to non-bonded ^{1}H–^{14}N for long-distance measurements (Figure 16.4d).

16.2.1.7 *Application of ^{1}H-^{14}N PM-S-RESPDOR*

^{1}H–^{14}N PM-S-RESDOR is applied to distinguish salt and co-crystal in pharmaceutical samples as quantitative distance measurements between hydrogen and nitrogen give the clear answer.[71] In pharmaceutical applications, the crystal of API may exhibit poor solubility or stability. To improve or overcome these problems, API mixed with inert coformer is widely used to form multi-component crystals. The most notable approach is mixing a basic API with an acidic coformer to form salts. This typically happens if the pK_a difference is larger than 3. The acidic proton is transferred to basic nitrogen, forming intermolecular ionic interactions (Figure 16.5b). On the other hand, if the pK_a difference is smaller than 3, it could form intermolecular hydrogen bonding, resulting in a new class of molecular crystals, *i.e.* co-crystal (Figure 16.5a). The situation in between is also observed as a continuum (Figure 16.5c). Although it is crucial to identify salt or co-crystal from the view point of intellectual properties, direct evidence of salt and co-crystal formation cannot be obtained from the widely used SCXRD method because the hydrogen position in SCXRD is obscured due to its low electron

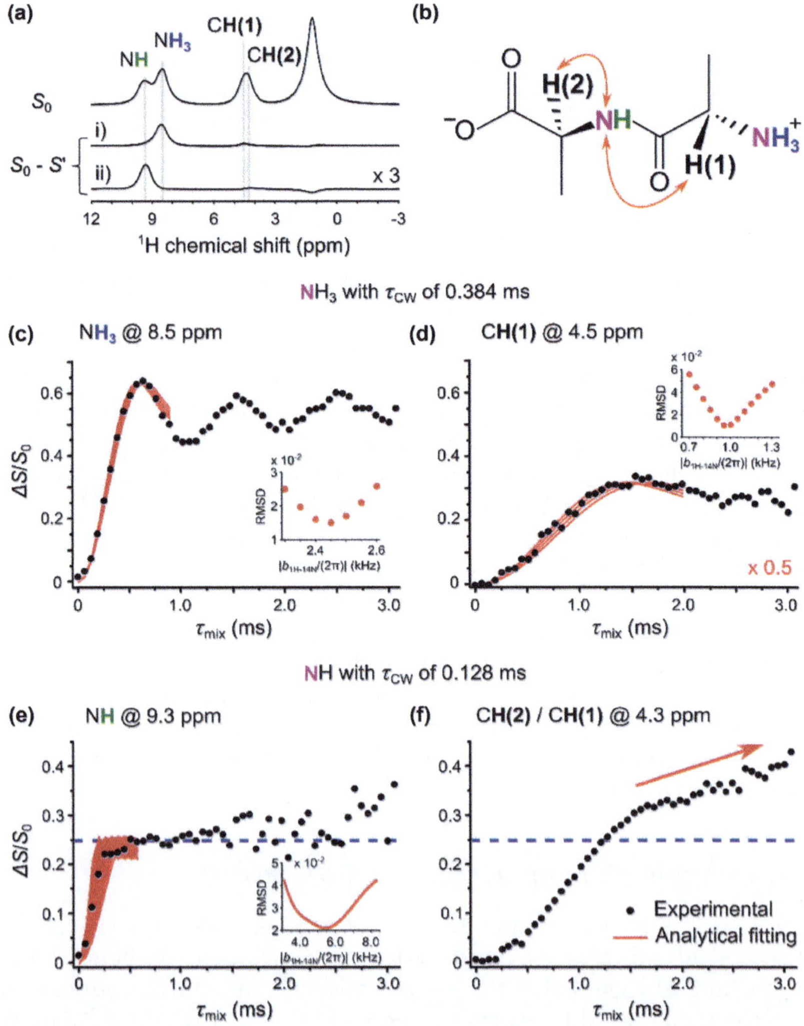

density; typically salt and co-crystals are identified indirectly from the distances between N and O. On the other hand, solid-state NMR gives direct evidence on it by measuring proton–nitrogen distances. Two components of ethionamide–salicylic acid crystals were prepared by two different recrystallizations.[71] SCXRD structures evidence the presence of intermolecular contact between nitrogen and oxygen. In order to identify salt or co-crystal, ^{1}H–^{14}N PM-S-RESPDOR measurements were performed. In one crystal, a very short internuclear distance of 1.07 Å is observed, revealing salt formation (Figure 16.6a). Very interestingly, although the pK_a difference is exactly the same, the other crystal exhibits a much longer distance of 1.50 Å, evidencing the co-crystal formation (Figure 16.6b). This result clearly indicates the importance of experimental identification of salt/co-crystals using solid-state NMR, as the pK_a difference may not be a reliable criterion.

16.2.1.8 Application of ^{1}H–^{63}Cu PM-S-RESPDOR

The PM-S-RESPDOR experiment was also used to elucidate the glassy metal-organic framework (MOF) structures. MOFs are self-assembled organic–inorganic coordinated polymers formed from organic linkers and metal ions. As MOFs exhibit various exciting characteristics including gas absorption, separation, ion conductor, proton conductor, *etc.*, they have been extensively investigated.[73,74] Atomic-level structures of MOFs are of importance since they deeply relate to their characteristics. Fortunately, most MOFs typically exhibit high crystallinity, and the crystalline structure of MOFs is readily available using SCXRD, provided large high-quality single crystals are prepared. Recently, a new class of MOF, glassy MOF, has been emerging, which is potentially useful for practical applications due to its excellent

Figure 16.4 (a) The 1D ^{1}H spin echo and the two difference spectra where (i) NH_3^+ and (ii) NH sites are saturated/inversed during ^{1}H-^{14}N OT-REDOR experiments. (b) The NH, NH_3^+, CH(1), and CH(2) sites, assigned in (a), are located in the molecular structure of Ala-Ala. The two arrows show the proximity of both CH(2)/CH(1) to the ^{14}NH site. (c–f) The experimental ^{1}H–^{14}N OT-REDOR fraction curves (black circles) fitted by the universal curves (red lines) for NH_3^+ at 8.5 ppm in (c) and CH(1) at 4.5 ppm in (d) when $^{14}NH_3^+$ OT was saturated/inverted by the CW pulse with τ_{CW} of 0.384 ms and NH at 9.3 ppm in (e) when ^{14}NH OT was saturated/inverted by the CW pulse with τ_{CW} of 0.128 ms. The fitting parameter *f* for universal curves is (c, d) 0.94 and (e) 0.37. For (d), the universal curves are halved, highlighted by a factor of 0.5 in red. For (e), the maximum signal fraction $\Delta S/S_0$ of 0.25 for NH is presented by the horizontal dashed line. For (f), the signal fraction $\Delta S/S_0$ of CH(2)/CH(1) at 4.3 ppm with ^{14}NH OT irradiation is over this line and is still increasing with τ_{mix} (highlighted by the arrow). The RMSD analyses (inset) were calculated for the best-fitting ^{1}H–^{14}N dipolar couplings. Reproduced from ref. 66, https://doi.org/10.3389/fmolb.2021.645347, under the terms of the CC BY 4.0 license, https://creativecommons.org/licenses/by/4.0/.

Figure 16.5 Schematic presentation of (a) a co-crystal, (b) a salt, and (c) a continuum (where the H-atom position is shared between the two non-H atoms) in a typical O...H...N interaction. Reproduced from ref.72, https://doi.org/10.1107/S205225251700687X, under the terms of the CC BY 2.0 license, https://creativecommons.org/licenses/by/2.0/uk/legalcode.

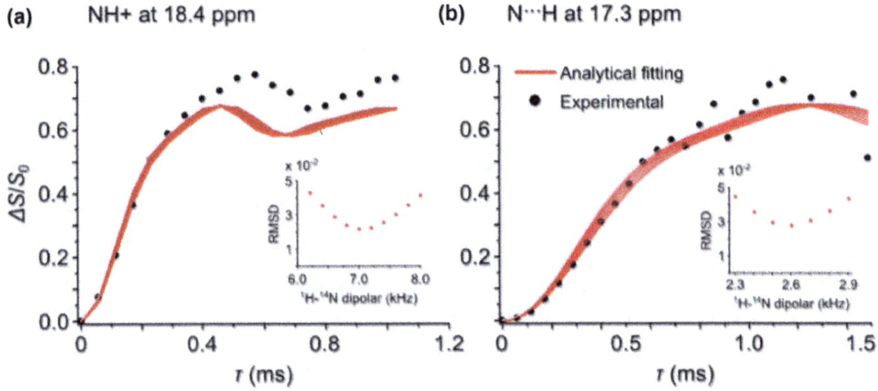

Figure 16.6 ^1H–^{14}N PM-S-RESPDOR fraction curve of (a) salt ETH^+SAL^- and (b) co-crystal ETH.SAL fitted with an analytical curve while varying only the size of the heteronuclear dipolar couplings. Insets show the RMSD of each fraction curve. Reproduced from ref. 71 with permission from American Chemical Society, Copyright 2019.

formability, high transparency, and the absence of inter-crystalline gaps.[75] Despite the importance of their structures, the lack of long-range periodicity poses the difficulty to structural elucidation of glassy MOFs. Indeed, solid-state NMR can be a good candidate to monitor a local structure of amorphous/glassy materials. ^{13}C CPMAS, ^1H–^{13}C HETCOR, *etc.* have been widely used for qualitative structure validations. Here, we demonstrated semi-quantitative measurements of ^1H–^{63}Cu distances toward structural

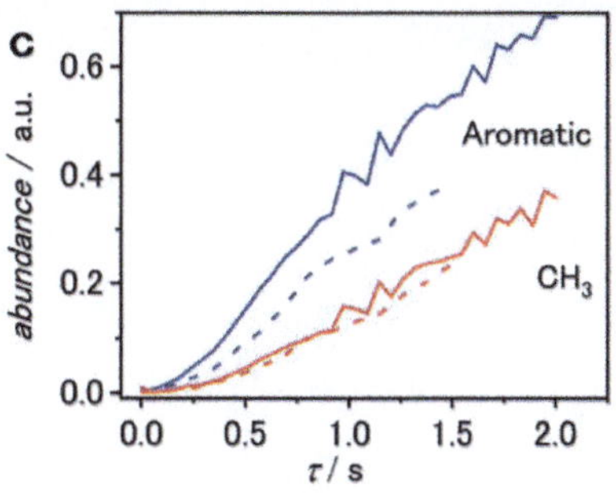

Figure 16.7 ^{1}H–^{63}Cu PM-S-RESPDOR fraction curves of crystalline (solid line) and glassy (dotted line) $[Cu_2(SCN)_3(C_2bpy)]$ MOF for aromatic (blue) and CH_3 protons (red). Reproduced from ref. 76 with permission from the Royal Society of Chemistry.

elucidation of a glassy MOF.[76] MOF with a combination of Cu^{+} and 4,4-bi-pyridin-1-ium derivatives has been designed based on an anti-crystal engineering approach, exhibiting crystal melt and subsequent glass formation. ^{1}H–^{63}Cu PM-S-RESPDOR is observed for crystal and glassy MOFs (Figure 16.7). Although the presence of multiple cuppers hampers the quantitative distance measurements, and semi-quantitative distance information is obtained. While the distance between CH_3 proton to ^{63}Cu remained almost the same after vitrification (glass formation), a clear increase in internuclear distances is observed between the aromatic proton to ^{63}Cu. This indicates the dissociation of π–π stacking, while keeping the local proximity between CH_3 proton and ^{63}Cu metal ions.

16.2.1.9 *Application of RESPDOR Filter*

RESPDOR together with ^{1}H–^{1}H spin diffusion is used to observe NMR signals selectively from API in pharmaceutical tablets.[77] While pharmaceutical formulations are composed of complex mixtures of APIs and excipients, APIs often exclusively include special elements, which do not appear in the excipient. Typical examples can be found in fluorine, chlorine, nitrogen, *etc.*, depending on the API. Thus, the RESPDOR fraction spectrum of these elements provides selective observation of ^{1}H in API. As the RESPDOR effect appears very locally (<3 Å), only a part of ^{1}H in API is polarized by RESPDOR filtering. The subsequent ^{1}H–^{1}H diffusion can be utilized to uniformly and exclusively polarize ^{1}Hs among the API crystals, as inter-crystal spin diffusion is unlikely to happen.

As an example, a pharmaceutical tablet which includes L-cysteine as an API is chosen. In this tablet, nitrogen is chosen as the nucleus for selective detection of API signals. Although intense signals appear from excipients (Figure 16.8a), they are largely suppressed in the ^{1}H–^{14}N PM-S-RESPDOR filtered spectrum, which exhibits the ^{1}H NMR signal next to nitrogen. ^{1}H–^{1}H spin diffusion using RFDR uniformly polarizes ^{1}Hs among the API crystals.

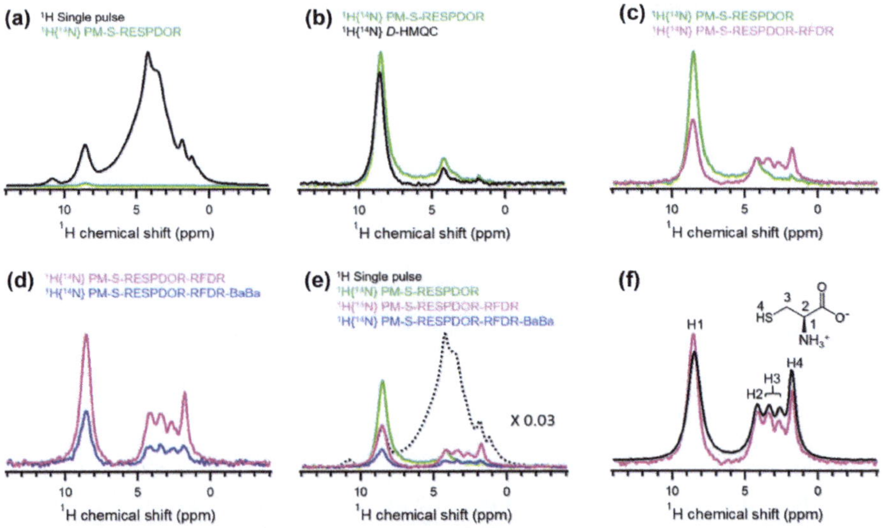

Figure 16.8 ^{1}H NMR spectra of a drug formulation which include L-cysteine as an API. Details are shown in the figure and main text. Reproduced from ref. 77 with permission from Elsevier, Copyright 2020.

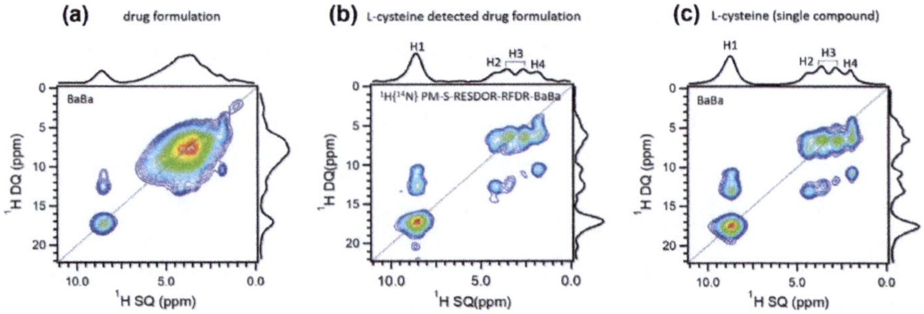

Figure 16.9 ^{1}H DQ/SQ 2D correlation spectra of (a, b) drug formulation and (c) pristine L-cysteine. A ^{1}H–^{14}N PM-S-RESPDOR filter combined with RFDR spin diffusion is applied prior to DQ/SQ observation in (b) for selective detection of L-cysteine (API). Reproduced from ref. 77 with permission from Elsevier, Copyright 2020.

In fact, the NMR spectrum thus obtained agrees well with the ^{1}H NMR spectrum of the pristine API (Figure 16.8c and f). ^{1}H double quantum (DQ)/single quantum (SQ) correlation spectra were observed (Figure 16.8d and 16.9). Although the ^{1}H DQ/SQ spectrum of the tablet is dominated by intense signals originating from the excipients (Figure 16.9a), the signals from the API are exclusively observed, using PM-S-RESPDOR filter with RFDR mixing. The spectrum thus obtained agrees well with that of the pristine API (Figure 16.9b and c). In this example, the PM-S-RESPDOR filtering process yields 7.1% of the original signal and finally 1.3% of the signal survives after

PM-S-RESPDOR/RFDR/BaBa filters. This is still a large enough polarization for ^{1}H observation. Indeed, while the API loading is only 10.6% in this case, the limit of detection is expected to be below 5%. This method can be used for ^{19}F and ^{35}Cl, depending on the API.

16.2.2 DIPSHIFT

DIPSHIFT experiments have been widely used to probe the local molecular dynamics in polymers and biomolecules by monitoring the heteronuclear dipolar couplings at slow to moderate MAS rates.[32,78,79] Assuming the CH/NH bond length, the heteronuclear dipolar couplings between C/N and ^{1}H, is known, local dynamics can be elucidated from the motionally averaged dipolar couplings. In the absence of dynamics, the DIPSHIFT method can also be used to determine the bond length and locate the hydrogen accurately.[80–82] Although DIPSHIFT experiments can in principle be implemented in the same way as REDOR, the practical implementations always follow the way used in REDOR in Figure 16.1a with $S = {}^{1}$H (Figure 16.2). Here, we limit our discussion to the DIPSHIFT implementation shown in Figure 16.2. As no CSA is reintroduced on the I channel, the sequence is robust with respect to MAS fluctuation. In addition, most of the I spin neighboring the hydrogen experiences the ^{1}H–I DIPSHIFT effect because of the high abundance of ^{1}H. Thus, the abundance of the I spin is not a problem except for the sensitivity. When the sensitivity is the limiting factor, DNP can be applied to overcome the problem.[80]

The heart of DIPSHIFT experiments is to determine the heteronuclear dipolar coupling by varying the effect of the ^{1}H–I heteronuclear coupling with suppressing of ^{1}H–^{1}H dipolar couplings. Traditionally, dipolar coupling is reintroduced by applying ^{1}H–^{1}H decoupling pulses in the fraction of the MAS cycle time during which heteronuclear dipolar coupling is activated, preventing the MAS averaging of the heteronuclear couplings (Figure 16.2a).[78,83] The rest of the echo time is filled with ^{1}H heteronuclear decoupling, which removes the effect of heteronuclear couplings as well as ^{1}H–^{1}H homonuclear couplings. The isotropic chemical shift evolution of the I spin is refocused by the π pulse. As it is implemented using the constant time manner, the effect of T_2' of the I spin is effectively suppressed. While the DIPSHIFT intensity drops in the intermediate t_1 evolution due to recoupled heteronuclear dipolar coupling, it returns to the original intensity when t_1 reaches the MAS cycle time in the absence of molecular dynamics. It should be noted that the refocusing may not be perfect in the presence of intermediate molecular motion, providing additional insight into the dynamics. Nevertheless, the size of the heteronuclear dipolar coupling can be obtained from the DIPSHIFT curve. As dipolar evolution is observed only in one rotor cycle, the MAS frequency should intentionally be set low as 2–4 kHz in order to probe weak dipolar interactions by increasing the evolution time. This limitation is partially alleviated by monitoring DIPSHIFT dephasing for longer than one rotor cycle.[84] However, a fast-MAS frequency

>60 kHz results in even sub-cycle time of ^1H–^1H decoupling sequences comparable to the MAS cycle time, complicating the DIPSHIFT implementation. In this regime, different DIPSHIFT implementations are applicable as shown in Figure 16.2b and c. The total echo time is arbitrarily set to integer multiples of the MAS cycle time. While the original approach relies on incomplete MAS averaging to reintroduce heteronuclear dipolar interactions, active recoupling sequences synchronized to MAS are used. The recoupling sequence can span the whole (Figure 16.2b) or a part of the spin echo period (Figure 16.2c). In Figure 16.2b, the dipolar scaling factor can be varied by adjusting the pulse positions in the REDOR recoupling sequence.[36] Although deuteration is required due to a lack of ^1H–^1H decoupling capability, modified REDOR recoupling block is successfully applied at a MAS rate of 60 kHz. The total echo time can be arbitrarily set according to the size of the heteronuclear dipolar interaction of interest. Alternatively, the number of recoupling sequences which is synchronous to MAS can be varied as shown Figure 16.2c. The ^1H heteronuclear decoupling is also applied during the rest of the spin echo duration, nulling the heteronuclear dipolar evolution. This approach results in the same effect as the variable scaling factor method. A similar approach is also found in CSA recoupling sequences.[85] In most implementations, the recoupling block is designed based on the symmetry principle with suppression of the ^1H–^1H dipolar interaction.[37,44,86–88] This enables distance measurements of fully protonated samples. Notably, only two (excitation and refocusing) pulses are applied on the I channel, and the sequence is suitable for quadrupolar nuclei also. The distances between ^1H and ^{17}O are thus successfully determined.[86,87] In principle, the sequences used in ^1H CSA recoupling can be used to recouple ^1H–X dipolar interaction with suppression of ^1H–^1H dipolar couplings at the fast-MAS rate.[26,89–95] In fact, the identical recoupling sequence is used both for DIPSHIFT and ^1H CSA measurements.[37,89] The application of the DIPSHIFT approach at fast-MAS is emerging and rapid growth is expected, since most of the theoretical background is well established.

16.2.3 Heteronuclear Dipolar Coupling Measurements Using Magnetization Transfer

The size of the heteronuclear dipolar coupling can be measured by monitoring the rate of magnetization transfer or build-up (Figure 16.10), while the methods listed above, including REDOR and DIPSHIFT, utilize the evolution of the magnetization under the influence of heteronuclear dipolar couplings. Hereafter, the source spin is referred to as the S spin and observation spin as the I spin to which the S magnetization is transferred. The differences from REDOR/DIPSHIFT experiments are found in (1) the effect of abundance of the S spin, (2) T_2' relaxation of I spin, and (3) t_1-noise. (1) In REDOR/DIPSHIFT and related experiments, the abundance of the S spin has to be large enough to minimize the interference of the I spin signals

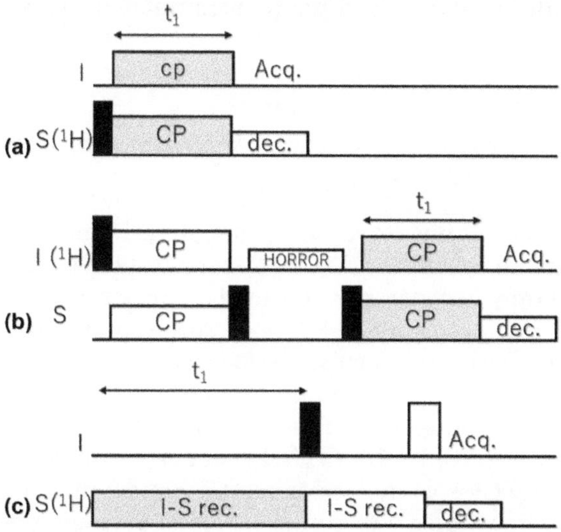

Figure 16.10 Sequences used to determine ^{1}H–X heteronuclear dipolar couplings by monitoring the build-up curve as a function of t_1. (a) Dipolar-mediated ^{1}H to *I* magnetization transfer is observed under fast-MAS (CP-VC). (b) ^{1}H-detected CP-VC (invCP-VC). (c) PREST-II sequence. The heteronuclear recoupling sequences which are responsible for the *I–S* distance measurements are shaded. Black and white rectangles represent π/2 and π pulses, respectively.

which are not coupled to the *S* spin. Thus, nuclei at high natural abundance or isotopically enriched nuclei should be assigned to the *S* spin. On the other hand, as all the transferred *I* spin magnetization obviously originates from the *S* magnetization in the current approach, there is no interference from isolated or uncoupled *I* spin. This allows observation of the *I* spin build-up regardless of the abundance of the *S* spin. This feature could be useful for ^{1}H detection of rare *S* spin, *e.g.* ^{1}H(*I*)–^{13}C(*S*) distance measurement at natural abundance, with suppressing of the uncoupled *I* spin (^{1}H) signals. This experiment is typically accompanied by preparation of the *S* polarization with another CP transfer from ^{1}H to the *S* spin prior to build-up observation (see Figure 16.10b and Section 16.2.3.1). (2) The effect of the T_2' relaxation of the I spin during the spin echo is minimized by calculating the fraction curves in REDOR and constant time implementation in DIPSHIFT. However, the absolute sensitivity suffers from T_2' of the *I* spin. This could be a limiting factor for the observation of weak dipolar couplings which require long-term echo. On the other hand, as magnetization transfer is typically observed as a build-up curve, the curve suffers from relaxation which is different from T_2' in REDOR/DIPSHIFT. Depending on the relative size of these two different relaxations, either the REDOR/DIPSHIFT or magnetization transfer method is preferable. (3) While REDOR experiments observe the difference in NMR signals, the build-up curve is directly accumulated. This results in less t_1-noise, as is often observed in CP-based experiments.

16.2.3.1 Cross-polarization-based Distance Measurements Between ^1H and Spin-1/2

A commonly used method to transfer magnetization between ^{1}H and X nuclei is cross-polarization (CP).[34] The vast majority of CP experiments have been performed only between spin-1/2 nuclei, since it is technically difficult for quadrupolar nuclei to spin-lock the magnetization under MAS.[96] $S = {}^1$H is typically chosen for CP between spin-1/2 nuclei, as this benefits from large thermal polarizations of ^{1}H and a rapid repetition rate owing to short T_1 of ^{1}H, resulting in sensitivity enhancements per unit of time. Subsequently non-^{1}H I spin is observed. The rate of magnetization transfer can be observed by collecting CP spectra by varying the contact time, giving a build-up curve. Since CP recouples heteronuclear dipolar interactions in a non-commuting way, the recoupled Hamiltonian experiences dipolar truncation. Thus, CP-based methods are suitable to observe the distances between the closest spin pairs. In the isolated two-spin system, heteronuclear dipolar coupling can easily be extracted from the dipolar oscillation.[97] An example can be found in a perdeuterated and back-exchanged protein sample to observe ^{1}H–^{15}N dipolar couplings.[98] However, the transient oscillation, and thus the heteronuclear dipolar coupling constant, is rarely obtained in fully protonated samples at moderate MAS rates. This is because there are still dominant residual ^{1}H–^{1}H interactions, although the ^{1}H–^{1}H dipolar coupling is suppressed to the first order under the CP conditions. These residual interactions can be suppressed by applying ^{1}H–^{1}H decoupling combined with CP. the most notable example is LG-CP and its variant under MAS.[99–104] These experiments are typically performed under slow to moderate MAS as the LG conditions depend on the MAS rate and thus scaling factor, introducing additional complexity to spectral analysis. These difficulties can be avoided by fast-MAS. Fast-MAS suppresses residual ^{1}H–^{1}H dipolar terms during CP without applying any tricks, since these effects come from higher-order terms. CP with variable contact time (CP-VC) can be used to extract dipolar transient oscillation.[105–109] $S = {}^1$H are chosen similarly to the usual CP experiments (Figure 16.10a) and transient oscillation of I signal intensities is analyzed after applying Fourier transformation (FT). Very interestingly, the FT powder pattern is too sensitive to rf inhomogeneity, and only singularities are experimentally observed.[105] As the position of singularity is robust with respect to experimental imperfection, accurate heteronuclear dipolar couplings are determined. No special curve fitting is required to extract the distance information. As CP-VC is observed on non-^{1}H I spin, the sensitivity is partially sacrificed but highly resolved peaks of the I spins help to selectively observe each ^{1}H–I distance. As CP-VC suffers from dipolar truncation, the shortest, in most cases covalently bonded, ^{1}H–I coupling can be extracted from each CP-VC build-up curve without interference from long-range couplings. This is particularly useful when the *I* spins exist at multiple sites.

The low sensitivity can be alleviated by the ^{1}H detection. The sequence consists of the first CP to prepare the initial polarization of the non-^{1}H *S* spin and the second CP to transfer the *S* magnetization to $I = {}^{1}$H nuclei for CP-VC detection (Figure 16.10b: invCP-VC).[106] This approach is essentially the same as the widely used ^{1}H-detected 2D heteronuclear correlation experiment at fast-MAS to improve the sensitivity[110–117] without the t_1 evolution time. This is especially useful if *S* nuclei are low-gamma nuclei such as ^{15}N, despite the inferior ^{1}H resolution to ^{15}N. Examples are shown in Sections 16.2.3.2 and 16.4.

16.2.3.2 Application of invCP-VC to Distinguish Salt and Co-crystals

The invCP-VC method is applied to determine ^{1}H–^{15}N distances in a prototypical multi-component system to distinguish salt and co-crystals.[72] This is a ^{15}N counterpart of the research shown in Section 16.2.1.7. In those systems, there is only one ^{15}N nucleus per molecule involved. Thus, a lack of the ^{15}N resolution is not a big issue. The practical issue for the pharmaceutical application is the long ^{1}H T_1 relaxation time. To solve this problem, ^{1}H–^{1}H spin diffusion is enhanced by RFDR irradiation during repetition delay to efficiently reduce the repetition time.[9] Finally, ^{1}H–^{15}N invCP-VC spectra were observed for each system (Figure 16.11). The 2D invCP-VC spectra clearly exhibit dipolar splitting along the vertical axis at the NH proton chemical shift. Theoretically, the powder pattern appears in the dipolar dimension, however, the sensitivity to rf inhomogeneity makes the most of the signals at the background level except for the signals at the singularities. As a result, ^{1}H–^{15}N dipolar couplings, and thus internuclear distances, can be readily obtained from the splitting between two peaks in the dipolar dimension, as shown in Figure 16.11. The ^{1}H–^{15}N distances undoubtfully assign salt/co-crystal/continuum for each crystal. Unfortunately, the experimental time (3–5 days for each sample) for invCP-VC is significantly longer than those of ^{1}H–^{14}N PM-S-RESPDOR as mentioned before. However, invCP-VC can be a good alternative method when (1) t_1-noise hampers RESPDOR observation, (2) molecular motion is present, and (3) multiple nitrogen atoms are involved. As discussed above, invCP-VC is robust with respect to MAS fluctuation compared to RESPDOR, alleviating t_1-noise problems. It is known that the ^{14}N experiment is sensitive to molecular motion, which interferes with quadrupolar interactions.[118] Although rapid molecular motion merely results in scaling the dipolar coupling in most cases, it can interfere with ^{14}N observation because of large quadrupolar couplings. Even ~100 ns motion, which is comparable to the inverse of quadrupolar coupling, hampers ^{14}N observation. On the other hand, such

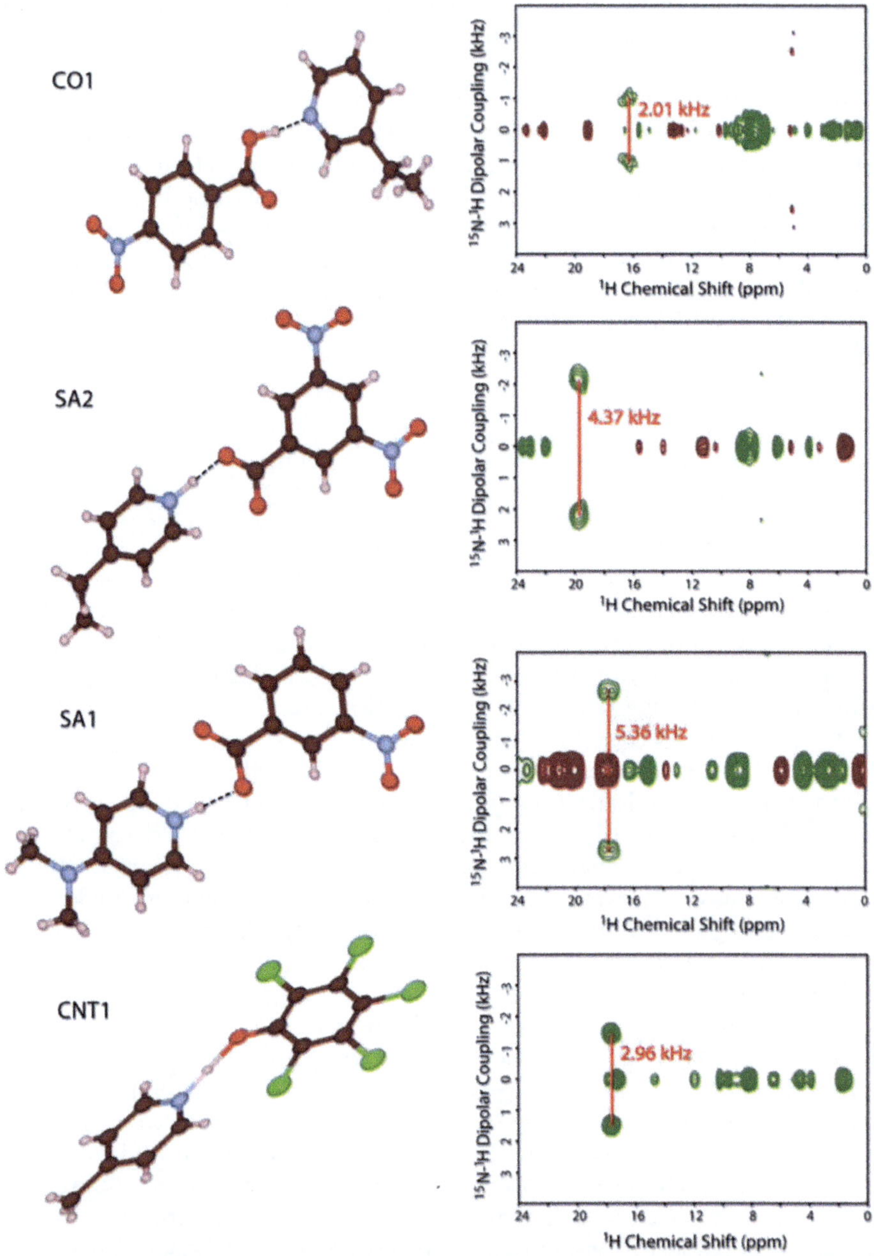

Figure 16.11 Crystalline structures of prototypical multi-component systems determined by SCXRD (left) and their invCP-VC spectra (right). The dipolar couplings in ^1H–^{15}N in N. . .H. . .O systems are obtained from splitting of invCP-VC spectra. The coupling constants are directly converted into ^1H–^{15}N internuclear distances: 1.62 Å (CO1), 1.25 Å (SA2), 1.17 Å (SA1), and 1.43 Å (CNT1), corresponding to co-crystal, continuum, salt, and co-crystal, respectively. Reproduced from ref.72, https://doi.org/10.1107/S205225251700687X, under the terms of the CC BY 2.0 license, https://creativecommons.org/licenses/by/2.0/uk/legalcode.

a rapid molecular motion doesn't interfere with invCP-VC observation except for dipolar scalings. When multiple ^{14}N atoms are present in the vicinity of the observed ^{1}H nucleus, the ^{1}H fraction intensities are affected by multiple ^{1}H–^{14}N coupling, complicating the data analysis (an example can be found in Section 16.4.). This can be avoided by ^{1}H–^{14}N OT-REDOR and invCP-VC. In addition, the ^{15}N-based method potentially gives much higher resolution than ^{1}H–^{14}N RESPDOR if an additional ^{15}N dimension is added using ^{1}H-detected 3D invCP-VC.

16.2.3.3 Distance Measurements Between ^{1}H and Quadrupolar Nuclei

The magnetization transfer from/to quadrupolar nuclei is not straightforward compared to the spin-1/2 systems. As mentioned above, CP involving quadrupolar nuclei is very inefficient due to poor spin lock capability under MAS and is rarely used for sensitivity enhancement, but rather applied for spectral editing. In general, the number of pulses on quadrupolar nuclei should be minimized for efficient magnetization transfer. The PREST sequence, which is based on a gamma-encoded heteronuclear dipolar recoupling sequence, is successfully applied to magnetization transfer from ^{1}H to quadrupolar nuclei. In the PREST sequence, $S = {}^{1}$H as a source nucleus and non-^{1}H I spin is observed (Figure 16.10c). While ^{1}H–^{1}H dipolar couplings are suppressed by properly designed recoupling sequence, ^{1}H–I dipolar coupling is reintroduced in the gamma-encoded manner.[119] As PREST requires only one excitation pulse on the I channel for magnetization transfer with an optional π pulse to refocus time evolution before detection, the sequence provides high transfer efficiency even for quadrupolar nuclei. It is possible to extract the size of heteronuclear dipolar couplings from the rate of the PREST transfer as demonstrated by ^{1}H–^{15}N distance measurements,[80] however, its application to quadrupolar nuclei remains a challenge. It is possible to transfer magnetization using the INEPT mechanism[120] with heteronuclear dipolar recoupling, namely D-RINPET.[121,122] Unlike PREST, where recoupled dipolar Hamiltonians do not commute each other, the recoupled Hamiltonians can commute each other in D-RINEPT if the recoupling sequence is properly chosen. This, in principle, allows long-distance measurements. However, it is still practically a challenge to extract the size of dipolar couplings including quadrupolar nuclei. In addition, as the number of pulses on the I channel in D-RINEPT is higher than those of PREST, this sacrifices sensitivities in D-RINEPT.

16.3 Homonuclear Dipolar Coupling

^{1}H–^{1}H homonuclear dipolar couplings are an ideal source of structural information, especially for ^{1}H positions. (1) ^{1}H possesses the highest gyromagnetic ratio among the stable isotopes. Thus, ^{1}H–^{1}H homonuclear dipolar couplings, which are proportional to the square of the gyromagnetic ratio, exhibit the largest values at a given internuclear distance. This is a suitable property to measure long internuclear distances. It is important to determine the long distance, since long-range correlation provides structurally relevant information, compared to short-range distance which is readily estimated from chemical structure and typical bond-length. (2) The abundance of ^{1}H is greater than 99%. Along with the ubiquitous nature of ^{1}H, proximate ^{1}H–^{1}H pairs can be easily found in many systems. This, in principle, brings distances between various moieties without isotopic labeling. In addition, the sensitivity of ^{1}H NMR is generally good owing to the high gyromagnetic ratio and abundance. (3) Homonuclear couplings give a larger amount of distance information than heteronuclear couplings with a given number of spins. For example, two different N spin systems are assumed for comparison between homonuclear and heteronuclear couplings; the system consisted of one S spin and N–1 I spin for heteronuclear couplings, and a system of N I spin for homonuclear couplings. Although the number of heteronuclear couplings is $N-1$, that of homonuclear couplings is $N(N-1)/2$, giving additional detailed structural information. Despite the favorable feature of ^{1}H–^{1}H homonuclear couplings, however, their practical use is not straightforward because of the complex nature of homonuclear couplings. Here, we first discuss the methods to extract homonuclear couplings not only for ^{1}Hs but also the other nuclei, typically ^{13}C in fully ^{13}C-labeled systems, then focus on ^{1}H–^{1}H distance measurements.

The distance between like nuclei can be easily obtained from the build-up curve in the case of two isolated spin systems. These spin systems can be found in selectively labeled two spin pairs,[123–126] and naturally occurring NMR active spin pairs in rare nuclei.[127] These experiments include homonuclear dipolar recoupling sequences which are similar to heteronuclear dipolar recouplings. Typically, rotor synchronous rf irradiation reintroduces homonuclear dipolar interactions in ZQ, SQ, and DQ forms, while the internal Hamiltonian may be used for recoupling as well. These recoupling schemes should be first-order recouplings, otherwise the build-up curve is determined by several different interactions other than homonuclear coupling between the spin pair of interest, making the quantitative distance estimation difficult.

However, application of these approach to abundant nuclei, for example ^{1}H, uniformly labeled ^{13}C *etc.*, is not straightforward. This is due to the non-commute nature of homonuclear dipolar Hamiltonians. While a broadband dipolar recoupling sequence introduces all the interactions, these terms do not commute each other regardless of the form (ZQ, SQ, and DQ). This results in well-known dipolar truncation phenomena:[128] the spin dynamics in two dipolar couplings sharing one nucleus is dominated by the stronger dipolar coupling. This obscures the dipolar oscillation from the weaker dipolar coupling. Generally, the build-up rate for the weaker dipolar coupling is more rapid than that expected from the internuclear distance, and the transfer efficiency is suppressed compared to those in isolated two-spin systems. This makes it difficult to determine weak dipolar couplings in the presence of stronger couplings. This is particularly problematic for observation of structurally informative weaker couplings. Full spin dynamics calculation including various ^{1}H–^{1}H couplings (and ^{1}H–^{13}C couplings) reproduces the build-up curve,[129,130] although quantitative ^{1}H–^{1}H distance measurements are still challenging. Extensive research works have been conducted for uniformly ^{13}C-labeled systems to avoid dipolar truncation. A number of second-order recoupling sequences have been developed to alleviate the dipolar truncation effects in ^{13}C–^{13}C correlation experiments.[131–133] The reversed MIRROR sequence is applied to the back-exchanged fully ^{15}N-labeled protein to accelerate ^{1}H–^{1}H correlations.[134] Selective ^{1}H–^{1}H recoupling mediated by second-order Hamiltonian is proposed using spin-lock with weak rf field strength (BASS-SD).[135] Although longer correlations are observed with the above sequences due to suppression of the dipolar truncation effect, the distance information is qualitative in nature of the second-order recouplings. ^{1}H–^{1}H spin diffusion can also be used to probe the proximity among ^{1}Hs at moderate MAS rates. The data can be analyzed using the simple diffusion equation or quantum spin dynamics calculations, providing fruitful information to validate the model structures.[3,136–139] Although ^{1}H–^{1}H spin diffusion is largely suppressed at the fast-MAS regime, fpRFDR can recover the ZQ dipolar Hamiltonian to promote ^{1}H–^{1}H spin diffusion.[22,140,141] Nevertheless, it remains a big challenge to extract explicit distance information between ^{1}Hs using second- and higher-order processes.

Selective recoupling of a specific dipolar coupled pair provides a straightforward solution toward the dipolar truncation issue. Selective ^{13}C–^{13}C recoupling can be achieved by the rotational resonance (RR) experiment and related experiments.[142–146] RR phenomena exclusively occur when the difference of isotropic chemical shifts matches the spinning frequency. This reintroduces homonuclear coupling between the selected pair, keeping the other coupling decoupled, thus avoiding dipolar truncation. However, application of RR to ^{1}H recoupling is challenging due to the limited peak dispersion of ^{1}H (~20 ppm) and rather broad lineshape.

Dipolar truncation doesn't occur if interactions commute each other. The effective homonuclear dipolar coupling can be prepared in a commutable way, by introducing isotropic chemical shift terms. This procedure resembles heteronuclear dipolar couplings. Consider the heteronuclear dipolar couplings in the rotating frame defined by the Zeeman interactions of I and S spins. While the planar parts of the dipolar Hamiltonians $(I_+S_- + I_-S_+)$ are truncated in the rotating frame by the large differences in Larmor frequencies between I and S spins, the Ising operator (IzSz) is kept unchanged. The resulting heteronuclear dipolar couplings (IzSz) commute each other. Similarly, introduction of isotropic chemical shift together with scaled dipolar coupling results in a commutable homonuclear dipolar

Hamiltonian. This can be achieved by simply adjusting the spinning axis slightly off from the magic angle.[147–149] ^{13}C–^{13}C dipolar couplings were successfully determined using fully ^{13}C-labeled samples. Although spectral resolution is compromised in the off-angle experiments, it is recovered by using a switching-angle spinning system.[148] The same concept can be applied using sophisticated pulse sequences which introduce homonuclear ZQ Hamiltonian and isotropic chemical shifts under MAS. ZQ-SHESHORE consists of ZQ recoupling block and free precession to prepare Ising operators.[150,151] TDR recouples ZQ Hamiltonian and isotropic/anisotropic chemical shift.[152] TOFU and FOLD are designed to reintroduce ZQ and isotropic chemical shift interactions.[153–155] Selectivities in these experiments are ensured by either selective pulse or 2D expansion, giving selective ^{13}C–^{13}C distance measurements in fully ^{13}C-labeled samples.

It is interesting that simultaneous introduction of isotropic chemical shifts and DQ Hamiltonian also enables selective recoupling of DQ homonuclear interactions even if broadband DQ recoupling sequence is applied. The homonuclear DQ Hamiltonian can be written in the interaction frame defined by isotropic chemical shift interactions as

$$H = \omega_{DQ} I_1^+ I_2^+ \exp(i\Sigma t) + \omega_{DQ}^* I_1^- I_2^- \exp(-i\Sigma t),$$

where ω_{DQ} represents the DQ dipolar coupling and $\Sigma = \delta_1 + \delta_2$ stands for the summation of recoupled isotropic chemical shifts with respect to the rf offset. The isotropic chemical shift poses time dependence on the DQ Hamiltonian, or in other words the DQ Hamiltonian is truncated, unless $\Sigma = 0$ (or $\delta_1 = -\delta_2$) is satisfied. This allows selective DQ recoupling for spin pairs satisfying $\Sigma = 0$. Selective recoupling of ^{13}C–^{13}C couplings using this mechanism has been demonstrated using C^zN,[156] SDR,[157,158] DQ-SEASHORE,[159] and COSMIC.[160]

Simultaneous recoupling of DQ and isotropic chemical shifts can successfully be used to selectively recouple ^1H–^1H dipolar couplings using SERP and SPR.[161–165] Notably, SERP provides quantitative ^1H–^1H distance measurements in fully protonated (or natural abundance) samples. The SERP sequence is essentially the same as the PMLG unit with z-rotation but with different phase ramps of φ (Figure 16.12). While φ is typically set to 208° for PMLG ^1H–^1H *decoupling*, it is set to either 480° or 680° in the SERP *recoupling*. In addition, PMLG block is applied synchronously to sample spinning. This implementation can be used to recouple homonuclear dipolar interactions, although PMLG is originally introduced to decouple them. The sequence can be categorized into the CN symmetry with a symmetry element of P_0 and used previously for ^1H–^1H recoupling.[166] The SERP implementation has $C2_2^1$ symmetry, which recouples isotropic/anisotropic chemical shift, heteronuclear dipolar coupling, and ZQ/DQ homonuclear dipolar couplings. It is shown that $\varphi = 480°$ or 680° conditions at an rf field strength 1.58 times the MAS frequency provide zero scaling factors for anisotropic chemical shift, heteronuclear dipolar coupling, and ZQ homonuclear dipolar couplings simultaneously,

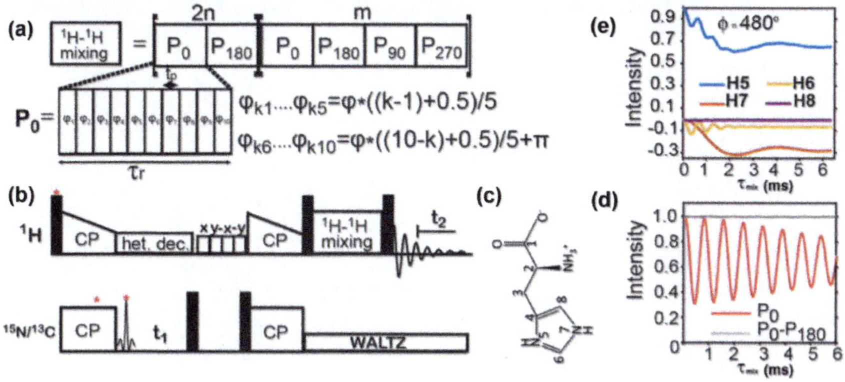

Figure 16.12 (a) Schematic representation of the SERP DQ recoupling block. (b) Pulse sequence to obtain the SERP build-up curve from NH/CH to the other ^{1}Hs. (c) Representation of the four-spin system consisting of H5, H6, H7, and H8 taken from the imidazole ring of L-histidine·HCl·H_2O. (d) Behavior of Iz magnetization under the SERP sequence. (e) Magnetization transfer from H5 to H7. While the source magnetization is placed on H5, SERP offset is set in the middle of H5 and H7 resonances. This ensures the selective magnetization transfer from H5 to H7 in the presence of strong H5–H6, H7–H6, and H7–H8 homonuclear couplings. Reproduced from ref. 161 with permission from American Chemical Society, Copyright 2018.

resulting in isotropic chemical shift and DQ homonuclear recoupling.[164] This ensures the selective recoupling of the DQ Hamiltonian when the SERP sequence is applied at the center of two ^{1}H peaks of interest. Four-spin numerical simulation is shown in Figure 16.12e, where source magnetization is located at H5 and build-up curves of the other ^{1}Hs during SERP are plotted. The resonance offset is set to the middle of H5 and H7 to ensure selective recoupling between H5 and H7. It is shown that despite the presence of strong dipolar coupling between H5 and H6, the H5 magnetization is selectively transferred to H7, suppressing the dipolar truncation effect. The negative magnetization transfer comes from the DQ recoupling nature. Quantitative distance measurements can be obtained by fitting the build-up curve with the numerical simulation.

The ^{1}H–^{1}H distances determined by SERP systematically exhibit shorter distances than the estimated distances from XRD structures (Figure 16.13).[164] This is because of the presence of the third spin or imperfect suppression of the other dipolar coupling to ^{1}H spins possessing close chemical shifts. Thus, better accuracy is obtained at a higher magnetic field due to larger ^{1}H chemical shift separations. The accuracy of the measured distances can be improved by incorporating the third spin into the numerical simulation or reducing the effect of the third spin by partial deuteration.[164] The method to determine quantitative ^{1}H–^{1}H distances is still at a very early stage, however, rapid progress is expected due to methodological and instrumental development, especially with fast-MAS.

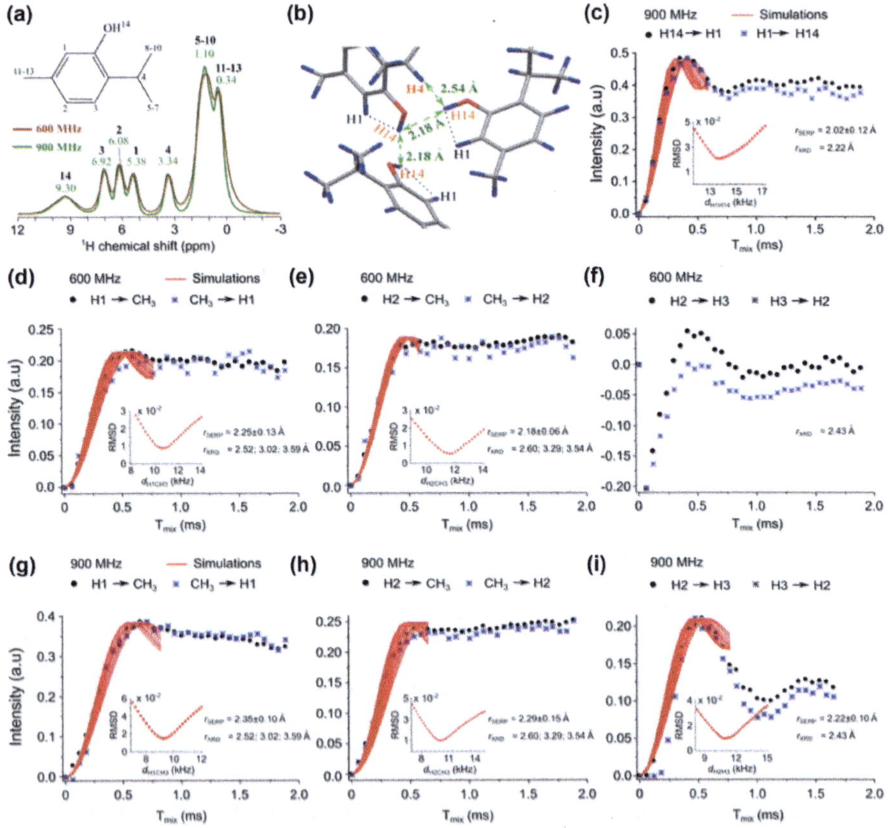

Figure 16.13 (a) Molecular structure and ^{1}H NMR spectra of thymol. (b) Intra- and intermolecular proximities between H4 and H14. (c)–(i) SERP build-up curves and numerical fittings. RMSD between experimental and numerical fittings are also plotted. While (d–f) were observed at 600 MHz (14.1 T), (c, g–i) were at 900 MHz (21.1 T). Reproduced from ref. 164 with permission from The American Institute of Physics, Copyright 2020.

16.4 Electron and NMR Crystallography

It has been demonstrated that microED, that is 3D ED, is capable of elucidating the crystalline structure from very tiny crystals of less than 1 μm. Even from a tiny crystal which is invisible with optical microscopes, microED successfully solves the crystalline structure.[167] This is due to the much stronger interactions of electrons to matter than those of X-ray. In addition, unlike X-ray, an electron beam can be focused using electron lens, enabling electron microscopy. In fact, one can easily switch between a transmission microscope (TEM) image and ED pattern. Since nanobeam diffraction and selected area ED allows selective observation of a particular single crystal, it is possible to determine the structure from mixed micro-crystals by choosing appropriate single crystals.[29–31] The experimental setup of microED is rather straightforward. A set of ED patterns is observed under the continuous rotation of the sample around the axis perpendicular to the electron beam. A very low dose rate on the order of 0.01 el$Å^{-2}$ s^{-1} is used to minimize dose damage together with a highly sensitive camera. The crystalline structure can be solved from the set of diffraction patterns thus obtained.

The basic principle of the structure solution using microED shares the same theoretical background as SCXRD. The behavior can be easily understood in the reciprocal space rather than the real space. The 2D diffraction pattern essentially represents a 2D slice on the Ewald sphere in the reciprocal space. Since the orientation of the Ewald sphere is determined by the angle of incidence with respect to the crystal, the whole of the 3D reciprocal space can be swept by changing the angle of incidence or orientation of the crystal with respect to the beam axis. The 3D reciprocal space thus reconstructed can be converted into the 3D real space by inverse Fourier transformation (FT). This resembles FIDs (real space) and spectra (reciprocal space) for NMR measurements. For example, a periodic spikelet due to rotor-echoes resembles the periodic crystalline structure. The FT of rotor-echo spikelet results in periodic spinning sidebands in the spectrum, where the separation of the spinning sideband is determined by the MAS rate, that is the inverse of the spacing between each rotor-echo. Similarly, the FT of the periodic structure gives periodic spots in the reciprocal space, whose period is determined by the inverse of the lattice parameters. The shape of each rotor-echo spikelet represents the envelope of SSBs. Similarly, the inner structure of the lattice is reflected by the distribution of intensity of diffraction spots. Thus, accurate observation of diffraction intensity is important for structural solutions. The decay of FID affects the shape of each SSB. In the diffraction experiments, the size of the crystal (or area where the beam is irradiated) affects the shape of each diffraction spot (excitation error). Nevertheless, the real space can easily be reconstructed from 3D reciprocal space using inverse FT. Similarly to SCXRD, phase information is lost in ED as the diffraction intensity is proportional to the square of the structure factor. Fortunately, a well-established method for SCXRD can be used for phasing the microED pattern, using molecular replacement, direct method, *etc.*

There are several stumbling blocks to experimentally measure ED patterns for structural solutions. These are well summarized in the literature,[168] and here we only briefly discuss the key aspects. (1) Dynamic scattering; it is often observed that the diffracted beams experience multiple diffractions in the crystal owing to the very strong interaction of electrons. This results in redistribution of diffraction intensities, hampering the kinetic approach used in the SCXRD analysis. (2) Excitation error, or finite size of the crystal size/thickness; in order to evaluate the intensity of the diffraction spot correctly, it is required to integrate the volume of the finite size of the diffraction spot. This is achieved by sweeping the crystalline orientations during observation. (3) The finite rotation angle of TEM; this limits the total sweep range of the crystalline orientation, resulting in a missing wedge. This problem can be avoided by merging a set of data from several different crystal orientations. (4) Dose damage; application to organic molecules requires low-dose irradiation since they are sensitive to dose damage. microED has been developed to address these issues using the following revolutionary approaches. (A) The collection of non-oriented patterns to observe 3D reciprocal space.[169] While ED is traditionally measured at oriented zone axis patterns, non-oriented patterns minimize the diffracted beams and thus the effect of dynamic scattering. Lacking the crystalline alignment procedure in microED automatically satisfies this condition. (B) Continuous sample rotation mimicking SCXRD.[28] Observation of the diffraction spot under continuous rotation results in accumulating the volume intensity of each diffraction spot. The volume integration can be done by using software dedicated for SCXRD analysis, avoiding problems related to excitation error. This simple strategy allows microED observation of virtually all the TEM as no additional or special equipment is used besides precise setting of the eucentric height. (C) Highly sensitive detectors allow observations at very low-dose conditions. The cryo-holder which maintains the sample at a temperature of around 100 K helps to reduce dose damage also. A combination of the above approaches enables microED to be feasible even for beam-sensitive organic crystals.

De novo structure determination of cimetidine form-B has been demonstrated using solid-state NMR and microED. Although cimetidine is a well-known pharmaceutical molecule, the crystalline structure of form-B is still unsolved. This is because form-B develops needle-shaped crystals which are not easily grown to SCXRD quality of crystals. Moreover, form-B often associates with contamination of form-C, hampering the application of powder XRD for structural solutions. As the crystal size of cimetidine form-B is large enough for microED, it is a straightforward task to solve the structure. However, the initial structure solved by microED includes two major errors. (1) Hydrogen is invisible and (2) atoms with similar atomic numbers are assigned incorrectly. Although the latter problem results in almost random assignments of carbon, nitrogen, and oxygen in organic molecules, most of these assignments can be easily corrected, given the chemical structure. However, the former is still problematic in understanding the hydrogen bonding structures. Solid-state NMR provides complementary insights to microED-based structures, since solid-state NMR directly observes ^{1}H, ^{13}C, ^{14}N, and ^{15}N, which are the nuclei of interest in microED. The $^{1}H/^{15}N$ 2D correlation spectrum exhibits three groups of covalently bonded N–H peaks (N1/N8, N3/C9, and N4/N10) and each group is resolved into two peaks from form-B together with a peak from contamination of form-C (Figure 16.14). The signals from form-C can be easily assigned by comparing spectra of different lots. The splitting of each peak indicates that there are two magnetically inequivalent molecules in a single unit cell. This agrees well with the structure candidates based on

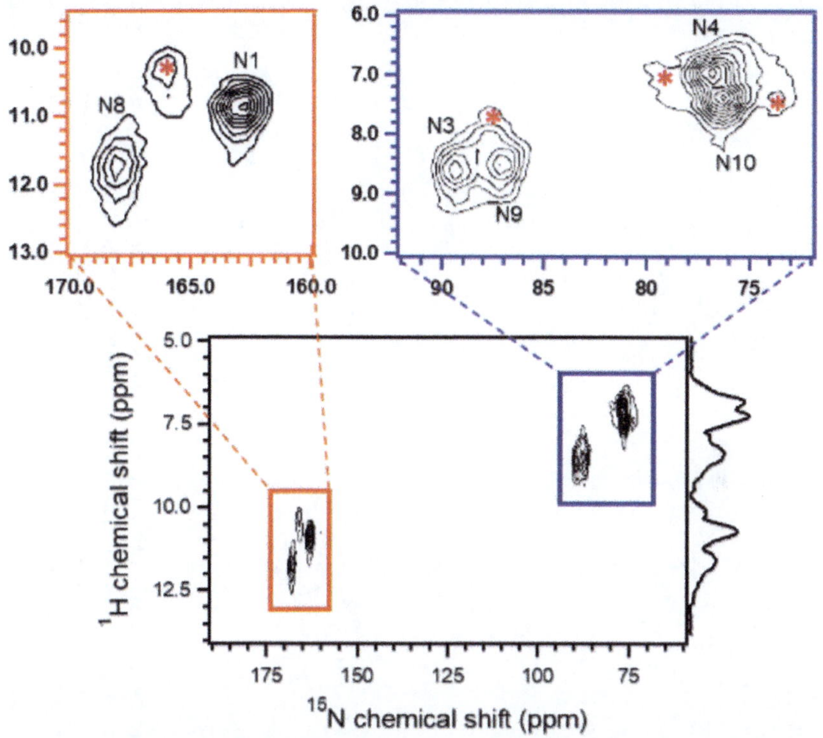

Figure 16.14 $^1H/^{15}N$ 2D correlation spectra of cimetidine form-B. Resonances from the covalently bonded N–H pairs only appear. Asterisks indicate the signal from the contaminated form-C. Reproduced from ref. 31, https://doi.org/10.1038/s41467-019-11469-2, under the terms of the CC BY 4.0 license, http://creativecommons.org/licenses/by/4.0/.

the initial structure solved by microED (Figure 16.15); the candidates include two molecules, namely extended (I) and bended (II) molecules. In addition, the presence of three N-H groups indicates that only one nitrogen on the imidazole ring is protonated, keeping the other nitrogen unprotonated. This results in four different candidate structures which differ in protonated states of the imidazole rings (Figure 16.15). As the difference of R factors is very minor, microED data alone do not indicate which is the correct structure. Each structure was optimized and isotropic ^1H, ^{13}C, and ^{15}N chemical shifts were calculated using the GIPAW method. The RMSD of experimental and calculated chemical shift is shown in Table 16.1. The best matching and energy minima are found in CB2. Thus, it is concluded that CB2 is the correct structure of form-B. The C–H and N–H distances are also observed using the invCP-VC methods and agree well with that obtained for the CB2 crystalline structure. N–H distances also can be obtained using RESPDOR experiments, however, invCP-VC is preferable to avoid interference from non-bonded ^{14}N, if sensitivity is high enough.

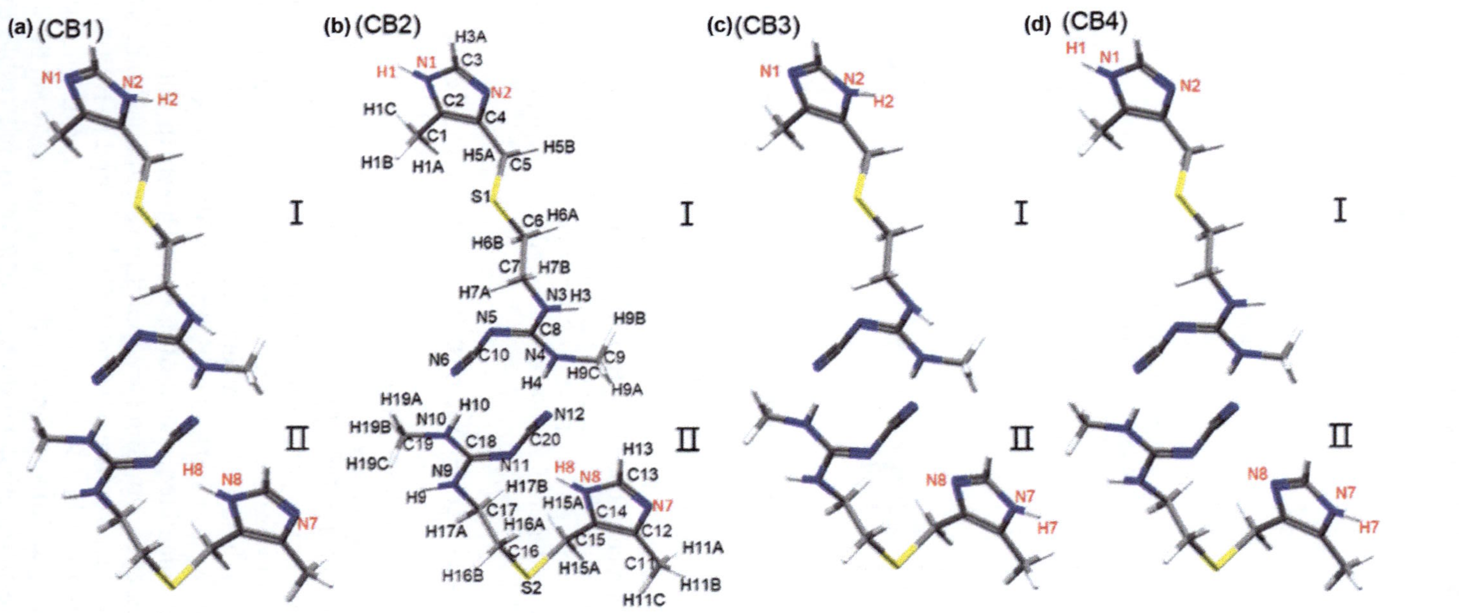

Figure 16.15 Four candidate structures of cimetidine form-B solved by microED. Although the initial structure includes mis-assignment of carbon, nitrogen, and oxygen, they are corrected based on the chemical structure. Protons are also attached. It is not clear which nitrogen is protonated on the imidazole ring (N1 or N2/N7 or N8), resulting in four candidates. Reproduced from ref. 31, https://doi.org/10.1038/s41467-019-11469-2, under the terms of the CC BY 4.0 license, http://creativecommons.org/licenses/by/4.0/.

Table 16.1 RMSD of calculated and experimental chemical shifts of cimetidine form-B for each of the candidate structures shown in Figure 16.15. Reproduced from ref. 31, https://doi.org/10.1038/s41467-019-11469-2, under the terms of the CC BY 4.0 license, http://creativecommons.org/licenses/by/4.0/.

	^{1}H (ppm)	^{13}C (ppm)	^{15}N (ppm)	Energy (Ry)
CB1	1.23	4.50	12.77	− 2192.289
CB2	0.73	3.35	3.96	− 2192.543
CB3	1.61	6.28	17.04	− 2192.069
CB4	1.21	5.68	13.31	− 2192.308

16.5 Summary

This chapter summarizes the methods to locate the H-atoms accurately using solid-state NMR and microED. Solid-state NMR can be used to extract internuclear distances involving H-atoms through heteronuclear ^{1}H–X and homonuclear ^{1}H–^{1}H dipolar couplings. The heteronuclear couplings between ^{1}H–X are measured by observing the time-evolution of transversal magnetization under the influence of heteronuclear couplings (REDOR, DIPSHIFT, and related methods) and the build-up curve with dipolar-mediated magnetization transfer (CP-VC, PREST, *etc.*). The choice of the method depends on (1) abundance of X nuclei, (2) size of anisotropic interactions of X nuclei (quadrupolar, CSA, *etc.*), (3) relaxation of X and ^{1}H during the sequence, (4) size of ^{1}H–X coupling, and (5) t_1-noise, *etc.* Identification of salt and co-crystal using ^{1}H–^{14}N/^{15}N distances in pharmaceutical applications and probing the local structure in glassy MOFs using ^{1}H–^{63}Cu distances are demonstrated. These methods provide not only ^{1}H–X distances but also NMR signals of the vicinity of the X nuclei; selective observation of the API signal from a pharmaceutical formulation is shown. While methods to determine distances between ^{1}H and spin-1/2 X nucleus are well established, large anisotropy in the quadrupolar nuclei and heavy spin-1/2 nuclei pose challenges. PM-S-RESPDOR partially addresses this issue. However, a method which is applicable to rare X nuclei with large anisotropies still needs to be developed. Although homonuclear ^{1}H–^{1}H distances are potentially useful, the experimental methods are still at a very early stage due to the complex nature of the ^{1}H–^{1}H spin network, strong ^{1}H–^{1}H couplings, limited ^{1}H chemical shift range, and poor ^{1}H resolution. Selective recoupling of a particular ^{1}H–^{1}H pair to the first order is achieved by simultaneous introduction of isotropic chemical shifts and ZQ/DQ homonuclear interactions. Notably, quantitative distance measurements are possible by SERP, which recouples CS and DQ simultaneously. However, there still remains the problem of selectivity; the effect of the third ^{1}H spin in spectral/spatial vicinity of the selected ^{1}H–^{1}H pair is not negligible, resulting in underestimation of the ^{1}H–^{1}H distances. Finally, the combination of microED and solid-state NMR in the NMR crystallography framework is demonstrated. microED is a potentially useful method for structural solution of nano- to microcrystals, however, the structures solved by microED have issues in regard to poor/invisible H-position and mis-assignment of carbon/nitrogen/oxygen atoms. While the crystalline structures involving non-H atoms are readily obtained from microED, missing H atoms are accurately placed by solid-state NMR. Solid-state NMR is also useful to correct the assignment through ^{1}H, ^{13}C, ^{14}N, and ^{15}N observation. The method does not need large and high-quality single crystals and paves a way for structural elucidation of various systems where large crystals are not available.

Acknowledgements

This work was supported by the JSPS KAKENHI (Grant No. 20K05483).

References

1. L. Palatinus, P. Brázda, P. Boullay, O. Perez, M. Klementová, S. Petit, V. Eigner, M. Zaarour and S. Mintova, Hydrogen Positions in Single Nanocrystals Revealed by Electron Diffraction, *Science*, 2017, **355**(6321), 166–169.
2. C. Martineau, J. Senker and F. Taulelle, *Chapter One - NMR Crystallography*, ed. Webb, G. A. B. T.-A. R. on N. M. R. S., Academic Press, 2014, vol. 82, pp. 1–57.
3. D. L. Bryce, NMR Crystallography: Structure and Properties of Materials from Solid-State Nuclear Magnetic Resonance Observables, *IUCrJ*, 2017, **4**, 350–359.
4. S. E. Ashbrook and D. McKay, Combining Solid-State NMR Spectroscopy with First-Principles Calculations-a Guide to NMR Crystallography, *Chem. Commun.*, 2016, **52**(45), 7186–7204.
5. P. Hodgkinson, NMR Crystallography of Molecular Organics, *Prog. Nucl. Magn. Reson. Spectrosc.*, 2020, **118–119**, 10–53.
6. E. Salager, R. S. Stein, C. J. Pickard, B. Elena and L. Emsley, Powder NMR Crystallography of Thymol, *Phys. Chem. Chem. Phys.*, 2009, **11**(15), 2610–2621.
7. A. Samoson and H-Mas, *J. Magn. Reson.*, 2019, **306**, 167–172.
8. T. Kobayashi, K. Mao, P. Paluch, A. Nowak-Krõl, J. Sniechowska, Y. Nishiyama, D. T. Gryko, M. J. Potrzebowski and M. Pruski, Study of Intermolecular Interactions in the Corrole Matrix by Solid-State NMR under 100 KHz MAS and Theoretical Calculations, *Angew. Chem., Int. Ed.*, 2013, **52**(52), 14108–14111.
9. Y. Q. Ye, M. Malon, C. Martineau, F. Taulelle and Y. Nishiyama, Rapid Measurement of Multidimensional 1H Solid-State NMR Spectra at Ultra-Fast MAS Frequencies, *J. Magn. Reson.*, 2014, **239**, 75–80.
10. V. Agarwal, S. Penzel, K. Szekely, R. Cadalbert, E. Testori, A. Oss, J. Past, A. Samoson, M. Ernst, A. Böckmann and B. H. Meier, DeNovo 3D Structure Determination from Sub-Milligram Protein Samples by

Solid-State 100kHz MAS NMR Spectroscopy, *Angew. Chem., Int. Ed.*, 2014, **53**(45), 12253–12256.

11. L. B. Andreas, K. Jaudzems, J. Stanek, D. Lalli, A. Bertarelloa, T. Marchand, Le, D. C. De Paepe, S. Kotelovica, I. Akopjana, B. Knott, S. Wegner, F. Engelke, A. Lesage, L. Emsley, K. Tars, T. Herrmann and G. Pintacuda, Structure of Fully Protonated Proteins by Proton-Detected Magic- Angle Spinning NMR, *Proc. Natl. Acad. Sci. U. S. A.*, 2016, **113**(33), 9187–9192.
12. J. P. Demers, V. Chevelkov and A. Lange, Progress in Correlation Spectroscopy at Ultra-Fast Magic-Angle Spinning: Basic Building Blocks and Complex Experiments for the Study of Protein Structure and Dynamics, *Solid State Nucl. Magn. Reson.*, 2011, **40**(3), 101–113.
13. Y. Su, L. Andreas and R. G. Griffin, Magic Angle Spinning NMR of Proteins: High-Frequency Dynamic Nuclear Polarization and ^{1}H Detection, *Annu. Rev. Biochem.*, 2015, **84**, 465–497.
14. S. P. Brown, Applications of High-Resolution ^{1}H Solid-State NMR, *Solid State Nucl. Magn. Reson.*, 2012, **41**, 1–27.
15. A. Lesage, Recent Advances in Solid-State NMR Spectroscopy of Spin $I = 1/2$ Nuclei, *Phys. Chem. Chem. Phys.*, 2009, **11**(32), 6876.
16. M. Deschamps, *Ultrafast Magic Angle Spinning Nuclear Magnetic Resonance*, 1st ed., 2014, Elsevier Ltd., vol. 81. 10.1016/B978-0-12-800185-1.00003-6.
17. M. H. Levitt, D. P. Raleigh, F. Creuzet and R. G. Griffin, Theory and Simulations of Homonuclear Spin Pair Systems in Rotating Solids, *J. Chem. Phys.*, 1990, **92**(11), 6347–6364.
18. A. Samoson, T. Tuherm and Z. Gan, High-Field High-Speed MAS Resolution Enhancement in ^{1}H NMR Spectroscopy of Solids, *Solid State Nucl. Magn. Reson.*, 2001, **20**(3–4), 130–136.
19. Y. Nishiyama, *Solid-State NMR under Ultrafast MAS Rate of 40-120 KHz*, 2017, DOI: 10.1007/978-981-10-5966-7_6.
20. A.-C. Pöppler, J.-P. Demers, M. Malon, A. P. Singh, H. W. Roesky, Y. Nishiyama and A. Lange, Ultrafast Magic-Angle Spinning: Benefits for the Acquisition of Ultrawide-Line NMR Spectra of Heavy Spin- 1/2 Nuclei, *ChemPhysChem*, 2016, **17**(6), 812–816.
21. Y. Nishiyama, N. T. Duong and Y.-L. Hong, *Chapter 21: Very Fast MAS NMR >60 KHz for Structural Elucidation*, 2019, DOI: 10.1039/9781788016483-00506.
22. J. Struppe, C. M. Quinn, S. Sarkar, A. M. Gronenborn and T. Polenova, Ultrafast ^{1}H MAS NMR Crystallography for Natural Abundance Pharmaceutical Compounds, *Mol. Pharm.*, 2020, **17**(2), 674–682.
23. T. Schubeis, T. Le Marchand, L. B. Andreas and G. Pintacuda, ^{1}H Magic-Angle Spinning NMR Evolves as a Powerful New Tool for Membrane Proteins, *J. Magn. Reson.*, 2018, **287**, 140–152.
24. V. E. Zorin, S. P. Brown and P. Hodgkinson, Origins of Linewidth in H^{1} Magic-Angle Spinning NMR, *J. Chem. Phys.*, 2006, **125**(14), 144508.

25. Y. Nishiyama, Fast Magic-Angle Sample Spinning Solid-State NMR at 60–100 KHz for Natural Abundance Samples, *Solid State Nucl. Magn. Reson.*, 2016, **78**, 24–36.
26. J. Struppe, C. M. Quinn, M. Lu, M. Wang, G. Hou, X. Lu, J. Kraus, L. B. Andreas, J. Stanek, D. Lalli, A. Lesage, G. Pintacuda, W. Maas, A. M. Gronenborn and T. Polenova, Expanding the Horizons for Structural Analysis of Fully Protonated Protein Assemblies by NMR Spectroscopy at MAS Frequencies above 100 KHz, *Solid State Nucl. Magn. Reson.*, 2017, **87**, 117–125.
27. Y. Nishiyama, Y. Endo, T. Nemoto, H. Utsumi, K. Yamauchi, K. Hioka and T. Asakura, Very Fast Magic Angle Spinning ^{1}H-^{14}N 2D Solid-State NMR: Sub-Micro-Liter Sample Data Collection in a Few Minutes, *J. Magn. Reson.*, 2011, **208**(1), 44–48.
28. I. Nederlof, E. Van Genderen, Y. W. Li and J. P. Abrahams, A Medipix Quantum Area Detector Allows Rotation Electron Diffraction Data Collection from Submicrometre Three-Dimensional Protein Crystals, *Acta Crystallogr. Sect. D Biol. Crystallogr*, 2013, **69**(7), 1223–1230.
29. C. G. Jones, M. W. Martynowycz, J. Hattne, T. J. Fulton, B. M. Stoltz, J. A. Rodriguez, H. M. Nelson and T. Gonen, The CryoEM Method MicroED as a Powerful Tool for Small Molecule Structure Determination, *ACS Cent. Sci.*, 2018, **4**(11), 1587–1592.
30. T. Gruene, J. T. C. Wennmacher, C. Zaubitzer, J. J. Holstein, J. Heidler, A. Fecteau-Lefebvre, S. De Carlo, E. Müller, K. N. Goldie, I. Regeni, T. Li, G. Santiso-Quinones, G. Steinfeld, S. Handschin, E. van Genderen, J. A. van Bokhoven, G. H. Clever and R. Pantelic, Rapid Structure Determination of Microcrystalline Molecular Compounds Using Electron Diffraction, *Angew. Chem., Int. Ed.*, 2018, **57**(50), 16313–16317.
31. C. Guzmán-Afonso, Y. L. Hong, H. Colaux, H. Iijima, A. Saitow, T. Fukumura, Y. Aoyama, S. Motoki, T. Oikawa, T. Yamazaki, K. Yonekura and Y. Nishiyama, Understanding Hydrogen-Bonding Structures of Molecular Crystals via Electron and NMR Nanocrystallography, *Nat. Commun.*, 2019, **10**(1), 1–10.
32. M. G. Munowitz, R. G. Griffin, G. Bodenhausen and T. H. Huang, Two-Dimensional Rotational Spin-Echo Nuclear Magnetic Resonance in Solids: Correlation of Chemical Shift and Dipolar Interactions, *J. Am. Chem. Soc.*, 1981, **103**(10), 2529–2533.
33. T. Gullion and J. Schaefer, Rotational-Echo Double-Resonance NMR, *J. Magn. Reson*, 1989, **81**, 196–200.
34. A. Pines, M. G. Gibby and J. S. Waugh, Proton-Enhanced Nuclear Induction Spectroscopy. a Method for High Resolution Nmr of Dilute Spins in Solids, *J. Chem. Phys.*, 1972, **56**(4), 1776–1777.
35. A. W. Hing, S. Vega and J. Schaefer, *Transferred-Echo Double-Resonance NMR*, 1992, **209**, 205–209.
36. M. G. Jain, G. Rajalakshmi, V. Agarwal, P. K. Madhu and K. R. Mote, On the Direct Relation between REDOR and DIPSHIFT Experiments in Solid-State NMR, *J. Magn. Reson.*, 2019, **308**, 106563.

37. G. Hou, I. J. L. Byeon, J. Ahn, A. M. Gronenborn and T. Polenova, ^{1}H-^{13}C/^{1}H-^{15}N Heteronuclear Dipolar Recoupling by R-Symmetry Sequences under Fast Magic Angle Spinning for Dynamics Analysis of Biological and Organic Solids, *J. Am. Chem. Soc.*, 2011, **133**(46), 18646–18655.
38. G. Hou, S. Paramasivam, S. Yan, T. Polenova and A. J. Vega, Multidimensional Magic Angle Spinning NMR Spectroscopy for Site-Resolved Measurement of Proton Chemical Shift Anisotropy in Biological Solids, *J. Am. Chem. Soc.*, 2013, **135**(4), 1358–1368.
39. Y. Nishiyama, V. Agarwal and R. Zhang, T1-Noise Suppression by γ-Free Recoupling Sequences in Solid-State NMR for Structural Characterization of Fully Protonated Molecules at Fast MAS, *J. Phys. Chem. C*, 2020, **124**(48), 26332–26343.
40. Z. Gan, Rotary Resonance Echo Double Resonance for Measuring Heteronuclear Dipolar Coupling under MAS, *J. Magn. Reson.*, 2006, **183**(2), 235–241.
41. L. Chopin, R. Rosanske and T. Gullion, Simple Improvements in Spinning-Speed Control for MAS N M R Experiments, *J. Magn. Reson. – Ser. A*, 1996, **122**(2), 237–239.
42. N. T. Duong, F. Rossi, M. Makrinich, A. Goldbourt, M. R. Chierotti, R. Gobetto and Y. Nishiyama, Accurate ^{1}H-^{14}N Distance Measurements by Phase-Modulated RESPDOR at Ultra-Fast MAS, *J. Magn. Reson.*, 2019, **308**(xxxx), 106559.
43. S. L. Carnahan, B. J. Lampkin, P. Naik, M. P. Hanrahan, I. I. Slowing, B. Vanveller, G. Wu, A. J. Rossini and O.-H. Probing, Bonding through Proton Detected ^{1}H-^{17}O Double Resonance Solid-State NMR Spectroscopy, *J. Am. Chem. Soc.*, 2019, **141**(1), 441–450.
44. A. Brinkmann and A. P. M. Kentgens, Proton-Selective ^{17}O-^{1}H Distance Measurements in Fast Magic-Angle-Spinning Solid-State NMR Spectroscopy for the Determination of Hydrogen Bond Lengths, *J. Am. Chem. Soc.*, 2006, **128**(46), 14758–14759.
45. L. Chopin, S. Vega and T. Gullion, A MAS NMR Method for Measuring ^{13}C-^{17}O Distances, *J. Am. Chem. Soc.*, 1998, **120**(18), 4406–4409.
46. T. Gullion, Measurement of Dipolar Interactions between Spin- 1 2 and Quadrupolar Nuclei by Rotational-Echo, Adiabatic-Passage, Double-Resonance NMR, *Chem. Phys. Lett.*, 1995, **246**(3), 325–330.
47. Z. Gan, Measuring Multiple Carbon-Nitrogen Distances in Natural Abundant Solids Using R-RESPDOR NMR, *Chem. Commun.*, 2006, **No. 45**, 4712–4714.
48. M. Ghosh and C. M. Rienstra, ^{1}H-Detected REDOR with Fast Magic-Angle Spinning of a Deuterated Protein, *J. Phys. Chem. B*, 2017, **121**(36), 8503–8511.
49. P. Schanda, B. H. Meier and M. Ernst, Quantitative Analysis of Protein Backbone Dynamics in Microcrystalline Ubiquitin by Solid-State NMR Spectroscopy, *J. Am. Chem. Soc.*, 2010, **132**(45), 15957–15967.
50. M. G. Jain, K. R. Mote, J. Hellwagner, G. Rajalakshmi, M. Ernst, P. K. Madhu and V. Agarwal, Measuring Strong One-Bond Dipolar

Couplings Using REDOR in Magic-Angle Spinning Solid-State NMR, *J. Chem. Phys.*, 2019, **150**(13), 134201.
51. M. H. Levitt, Symmetry-Based Pulse Sequences in Magic-Angle Spinning Solid-State NMR, *Encycl. Magn. Reson.*, 2007, **9**, 165–196.
52. J. C. C. Chan, C-REDOR: Rotational Echo Double Resonance under Very Fast Magic-Angle Spinning, *Chem. Phys. Lett.*, 2001, **335**(3–4), 289–297.
53. J. C. C. Chan and H. Eckert, C-Rotational Echo Double Resonance: Heteronuclear Dipolar Recoupling with Homonuclear Dipolar Decoupling, *J. Chem. Phys.*, 2001, **115**(13), 6095–6105.
54. F. C. Chou, S. J. Huang and J. C. C. Chan, Heteronuclear Dipolar Recoupling in Multiple-Spin System under Fast Magic-Angle Spinning, *J. Magn. Reson.*, 2009, **197**(1), 96–99.
55. L. Chen, Q. Wang, B. Hu, O. Lafon, J. Trébosc, F. Deng and J. P. Amoureux, Measurement of Hetero-Nuclear Distances Using a Symmetry-Based Pulse Sequence in Solid-State NMR, *Phys. Chem. Chem. Phys.*, 2010, **12**(32), 9395–9405.
56. G. Hou, X. Lu, A. J. Vega and T. Polenova, Accurate Measurement of Heteronuclear Dipolar Couplings by Phase-Alternating R-Symmetry (PARS) Sequences in Magic Angle Spinning NMR Spectroscopy, *J. Chem. Phys.*, 2014, **141**(10), 104202.
57. X. Lu, H. Zhang, M. Lu, A. J. Vega, G. Hou and T. Polenova, Improving Dipolar Recoupling for Site-Specific Structural and Dynamics Studies in Biosolids NMR: Windowed RN-Symmetry Sequences, *Phys. Chem. Chem. Phys.*, 2016, **18**(5), 4035–4044.
58. A. Brinkmann and M. H. Levitt, Symmetry Principles in the Nuclear Magnetic Resonance of Spinning Solids: Heteronuclear Recoupling by Generalized Hartmann-Hahn Sequences, *J. Chem. Phys.*, 2001, **115**(1), 357–384.
59. T. Gullion and A. J. Vega, Measuring Heteronuclear Dipolar Couplings for $I = 1/2$, $S > 1/2$ Spin Pairs by REDOR and REAPDOR NMR, *Prog. Nucl. Magn. Reson. Spectrosc.*, 2005, **47**(3–4), 123–136.
60. K. Saalwächter and K. Schmidt-Rohr, Relaxation-Induced Dipolar Exchange with Recoupling - An MAS NMR Method for Determining Heteronuclear Distances without Irradiating the Second Spin, *J. Magn. Reson.*, 2000, **145**(2), 161–172.
61. T. Gullion, Measuring 13 C- 2 D Dipolar Couplings with a Universal REDOR Dephasing Curve, *J. Magn. Reson.*, 2000, **146**(1), 220–222.
62. M. Makrinich, E. Nimerovsky and A. Goldbourt, Pushing the Limit of NMR-Based Distance Measurements – Retrieving Dipolar Couplings to Spins with Extensively Large Quadrupolar Frequencies, *Solid State Nucl. Magn. Reson.*, 2018, **92**(April), 19–24.
63. E. Nimerovsky, R. Gupta, J. Yehl, M. Li, T. Polenova and A. Goldbourt, Phase-Modulated LA-REDOR: A Robust, Accurate and Efficient Solid-State NMR Technique for Distance Measurements between a Spin-1/2 and a Quadrupole Spin, *J. Magn. Reson.*, 2014, **244**, 107–113.

64. E. Nimerovsky, M. Makrinich and A. Goldbourt, Analysis of Large-Anisotropy-Spin Recoupling Pulses for Distance Measurement under Magic-Angle Spinning NMR Shows the Superiority and Robustness of a Phase Modulated Saturation Pulse, *J. Chem. Phys.*, 2017, **146**(12), 124202.
65. A. V. Wijesekara, A. Venkatesh, B. J. Lampkin, B. VanVeller, J. W. Lubach, K. Nagapudi, I. Hung, P. L. Gor'kov, Z. Gan and A. J. Rossini, Fast Acquisition of Proton-Detected HETCOR Solid-State NMR Spectra of Quadrupolar Nuclei and Rapid Measurement of NH Bond Lengths by Frequency Selective HMQC and RESPDOR Pulse Sequences, *Chem. - Eur. J*, 2020, **26**(35), 7881–7888.
66. N. T. Duong, Z. Gan and Y. Nishiyama, Selective ^{1}H-^{14}N Distance Measurements by ^{14}N Overtone Solid-State NMR Spectroscopy at Fast MAS, *Front. Mol. Biosci.*, 2021, **8**, 71.
67. Z. Gan, I. Hung, Y. Nishiyama, J.-P. Amoureux, O. Lafon, H. Nagashima, J. Trébosc and B. Hu, ^{14}N Overtone Nuclear Magnetic Resonance of Rotating Solids, *J. Chem. Phys.*, 2018, **149**(6), 064201.
68. Y. Nishiyama, M. Malon, Z. Gan, Y. Endo and T. Nemoto, Proton-Nitrogen-14 Overtone Two-Dimensional Correlation NMR Spectroscopy of Solid-Sample at Very Fast Magic Angle Sample Spinning, *J. Magn. Reson.*, 2013, **230**, 160–164.
69. L. A. O'Dell, R. He and J. Pandohee, Identifying H-N Proximities in Solid-State NMR Using ^{14}N Overtone Irradiation under Fast MAS, *CrystEngComm*, 2013, **15**(43), 8657–8667.
70. M. K. Pandey and Y. Nishiyama, A One-Dimensional Solid-State NMR Approach for ^{14}NH/^{14}NH Overtone Correlation through ^{1}H/^{1}H Mixing under Fast MAS, *Phys. Chem. Chem. Phys.*, 2018, **20**(40), 25849–25853.
71. D. Bernasconi, S. Bordignon, F. Rossi, E. Priola, C. Nervi, R. Gobetto, D. Voinovich, D. Hasa, N. T. Duong, Y. Nishiyama and M. R. Chierotti, Selective Synthesis of a Salt and a Cocrystal of the Ethionamide-Salicylic Acid System, *Cryst. Growth Des.*, 2020, **20**(2), 906–915.
72. L. Rajput, M. Banik, J. R. Yarava, S. Joseph, M. K. Pandey, Y. Nishiyama and G. R. Desiraju, Exploring the Salt-Cocrystal Continuum with Solid-State NMR Using Natural-Abundance Samples: Implications for Crystal Engineering, *IUCrJ*, 2017, **4**, 466–475.
73. O. M. Yaghi, M. O'Keeffe, N. W. Ockwig, H. K. Chae, M. Eddaoudi and J. Kim, Reticular Synthesis and the Design of New Materials, *Nature*, 2003, **423**(6941), 705–714.
74. S. Kitagawa, R. Kitaura and S. I. Noro, Functional Porous Coordination Polymers, *Angew. Chem., Int. Ed.*, 2004, **43**(18), 2334–2375.
75. T. D. Bennett and S. Horike, Liquid, Glass and Amorphous Solid States of Coordination Polymers and Metal–Organic Frameworks, *Nat. Rev. Mater.*, 2018, **3**(11), 431–440.
76. S. S. Nagarkar, H. Kurasho, N. T. Duong, Y. Nishiyama, S. Kitagawa and S. Horike, Crystal Melting and Glass Formation in Copper Thiocyanate

Based Coordination Polymers, *Chem. Commun.*, 2019, **55**(38), 5455–5458.
77. Y. L. Hong, G. N. Manjunatha Reddy and Y. Nishiyama, Selective Detection of Active Pharmaceutical Ingredients in Tablet Formulations Using Solid-State NMR Spectroscopy, *Solid State Nucl. Magn. Reson.*, 2020, **106**(December 2019), 101651.
78. E. R. DeAzevedo, K. Saalwachter, O. Pascui, A. A. De Souza, T. J. Bonagamba and D. Reichert, Intermediate Motions as Studied by Solid-State Separated Local Field NMR Experiments, *J. Chem. Phys.*, 2008, **128**(10), 104505.
79. R. K. Hester, J. L. Ackerman, B. L. Neff and J. S. Waugh, Separated Local Field Spectra in NMR: Determination of Structure of Solids, *Phys. Rev. Lett.*, 1976, **36**(18), 1081–1083.
80. L. Zhao, M. P. Hanrahan, P. Chakravarty, A. G. Dipasquale, L. E. Sirois, K. Nagapudi, J. W. Lubach and A. J. Rossini, Characterization of Pharmaceutical Cocrystals and Salts by Dynamic Nuclear Polarization-Enhanced Solid-State NMR Spectroscopy, *Cryst. Growth Des.*, 2018, **18**(4), 2588–2601.
81. S. Li and M. Hong, Protonation, Tautomerization, and Rotameric Structure of Histidine: A Comprehensive Study by Magic-Angle-Spinning Solid-State NMR, *J. Am. Chem. Soc.*, 2011, **133**(5), 1534–1544.
82. M. Hohwy, C. P. Jaroniec, B. Reif, C. M. Rienstra and R. G. Griffin, Local Structure and Relaxation in Solid-State NMR: Accurate Measurement of Amide N-H Bond Lengths and H-N-H Bond Angles [2], *J. Am. Chem. Soc.*, 2000, **122**(13), 3218–3219.
83. M. Munowitz, W. P. Aue and R. G. Griffin, Two-Dimensional Separation of Dipolar and Scaled Isotropic Chemical Shift Interactions in Magic Angle NMR Spectra, *J. Chem. Phys.*, 1982, **77**(4), 1686–1689.
84. M. Hong, J. D. Gross, C. M. Rienstra, R. G. Griffin, K. K. Kumashiro and K. Schmidt-Rohr, Coupling Amplification in 2D MAS NMR and Its Application to Torsion Angle Determination in Peptides, *J. Magn. Reson.*, 1997, **129**(1), 85–92.
85. K. J. Fritzsching, E. G. Keeler, C. He and A. E. McDermott, Scaled Recoupling of Chemical Shift Anisotropies at High Magnetic Fields under MAS with Interspersed C -Elements, *J. Chem. Phys.*, 2020, **153**(10), 104201.
86. A. Brinkmann and A. P. M. Kentgens, Sensitivity Enhancement and Heteronuclear Distance Measurements in Biological ^{17}O Solid-State NMR, *J. Phys. Chem. B*, 2006, **110**(32), 16089–16101.
87. J. D. van Beek, R. Dupree and M. H. Levitt, Symmetry-Based Recoupling of ^{17}O-^{1}H Spin Pairs in Magic-Angle Spinning NMR, *J. Magn. Reson.*, 2006, **179**(1), 38–48.
88. X. Zhao, M. Edén and M. H. Levitt, Recoupling of Heteronuclear Dipolar Interactions in Solid-State NMR Using Symmetry-Based Pulse Sequences, *Chem. Phys. Lett.*, 2001, **342**(3–4), 353–361.
89. H. K. Miah, D. A. Bennett, D. Iuga and J. J. Titman, Measuring Proton Shift Tensors with Ultrafast MAS NMR, *J. Magn. Reson.*, 2013, **235**, 1–5.

90. M. K. Pandey, M. Malon, A. Ramamoorthy and Y. Nishiyama, Composite-180° Pulse-Based Symmetry Sequences to Recouple Proton Chemical Shift Anisotropy Tensors under Ultrafast MAS Solid-State NMR Spectroscopy, *J. Magn. Reson.*, 2015, **250**, 45–54.
91. M. K. Pandey and Y. Nishiyama, Determination of NH Proton Chemical Shift Anisotropy with ^{14}N-^{1}H Heteronuclear Decoupling Using Ultrafast Magic Angle Spinning Solid-State NMR, *J. Magn. Reson.*, 2015, **261**, 133–140.
92. D. H. Brouwer and J. A. Ripmeester, Symmetry-Based Recoupling of Proton Chemical Shift Anisotropies in Ultrahigh-Field Solid-State NMR, *J. Magn. Reson.*, 2007, **185**(1), 173–178.
93. L. Duma, D. Abergel, P. Tekely and G. Bodenhausen, Proton Chemical Shift Anisotropy Measurements of Hydrogen-Bonded Functional Groups by Fast Magic-Angle Spinning Solid-State NMR Spectroscopy, *Chem. Commun.*, 2008, **3**(20), 2361–2363.
94. L. Liang, G. Hou and X. Bao, Measurement of Proton Chemical Shift Anisotropy in Solid-State NMR Spectroscopy, *Solid State Nucl. Magn. Reson.*, 2018, **93**(April), 16–28.
95. G. Hou, R. Gupta, T. Polenova and A. J. Vega, A Magic-Angle-Spinning NMR Spectroscopy Method for the Site-Specific Measurement of Proton Chemical-Shift Anisotropy in Biological and Organic Solids, *Isr. J. Chem.*, 2014, **54**(1–2), 171–183.
96. A. J. Vega, MAS NMR Spin Locking of Half-Integer Quadrupolar Nuclei, *J. Magn. Reson.*, 1992, **96**(1), 50–68.
97. L. Müller, A. Kumar, T. Baumann and R. R. Ernst, Transient Oscillations in NMR Cross-Polarization Experiments in Solids, *Phys. Rev. Lett.*, 1974, **32**(25), 1402–1406.
98. V. Chevelkov, U. Fink and B. Reif, Accurate Determination of Order Parameters From ^{1}H,^{15}N Dipolar Couplings in MAS Solid-State NMR Experiments, *J. Am. Chem. Soc.*, 2009, **131**(39), 14018–14022.
99. B. J. Van Rossum, C. P. De Groot, V. Ladizhansky, S. Vega and H. J. M. De Groot, A Method for Measuring Heteronuclear (^{1}H-^{13}C) Distances in High Speed MAS NMR, *J. Am. Chem. Soc.*, 2000, **122**(14), 3465–3472.
100. J. L. Lorieau and A. E. McDermott, Conformational Flexibility of a Microcrystalline Globular Protein: Order Parameters by Solid-State NMR Spectroscopy, *J. Am. Chem. Soc.*, 2006, **128**(35), 11506–11512.
101. S. V. Dvinskikh, H. Zimmermann, A. Maliniak and D. Sandström, Heteronuclear Dipolar Recoupling in Solid-State Nuclear Magnetic Resonance by Amplitude-, Phase-, and Frequency-Modulated Lee-Goldburg Cross-Polarization, *J. Chem. Phys.*, 2005, **122**(4), 044512.
102. S. V. Dvinskikh and D. Sandström, Frequency Offset Refocused PISEMA-Type Sequences, *J. Magn. Reson.*, 2005, **175**(1), 163–169.
103. V. Ladizhansky and S. Vega, Polarization Transfer Dynamics in Lee-Goldburg Cross Polarization Nuclear Magnetic Resonance Experiments on Rotating Solids, *J. Chem. Phys.*, 2000, **112**(16), 7158–7168.

104. A. Venkatesh, I. Hung, K. C. Boteju, A. D. Sadow, P. L. Gor'kov, Z. Gan and A. J. Rossini, Suppressing ^{1}H Spin Diffusion in Fast MAS Proton Detected Heteronuclear Correlation Solid-State NMR Experiments, *Solid State Nucl. Magn. Reson.*, 2020, **105**(August 2019), 101636.
105. P. Paluch, J. Trébosc, Y. Nishiyama, M. J. Potrzebowski, M. Malon and J. P. Amoureux, Theoretical Study of CP-VC: A Simple, Robust and Accurate MAS NMR Method for Analysis of Dipolar C-H Interactions under Rotation Speeds Faster than ca. 60 KHz, *J. Magn. Reson.*, 2015, **252**, 67–77.
106. Y. Nishiyama, M. Malon, M. J. Potrzebowski, P. Paluch and J. P. Amoureux, Accurate NMR Determination of C-H or N-H Distances for Unlabeled Molecules, *Solid State Nucl. Magn. Reson*, 2016, **73**, 15–21.
107. R. Zhang, J. Damron, T. Vosegaard and A. Ramamoorthy, A Cross-Polarization Based Rotating-Frame Separated-Local-Field NMR Experiment under Ultrafast MAS Conditions, *J. Magn. Reson.*, 2015, **250**, 37–44.
108. K. R. Thurber and R. Tycko, Measurement of Sample Temperatures under Magic-Angle Spinning from the Chemical Shift and Spin-Lattice Relaxation Rate of 79Br in KBr Powder, *J. Magn. Reson.*, 2009, **196**(1), 84–87.
109. P. Paluch, T. Pawlak, J.-P. Amoureux and M. J. Potrzebowski, Simple and Accurate Determination of X–H Distances under Ultra-Fast MAS NMR, *J. Magn. Reson.*, 2013, **233**, 56–63.
110. Y. Ishii and R. Tycko, Sensitivity Enhancement in Solid State ^{15}N NMR by Indirect Detection with High-Speed Magic Angle Spinning, *J. Magn. Reson.*, 2000, **142**(1), 199–204.
111. Y. Ishii, J. P. Yesinowski and R. Tycko, Sensitivity Enhancement in Solid-State ^{13}C NMR of Synthetic Polymers and Biopolymers by ^{1}H NMR Detection with High-Speed Magic Angle Spinning, *J. Am. Chem. Soc.*, 2001, **123**(12), 2921–2922.
112. E. K. Paulson, C. R. Morcombe, V. Gaponenko, B. Dancheck, R. A. Byrd and K. W. Zilm, Sensitive High Resolution Inverse Detection NMR Spectroscopy of Proteins in the Solid State, *J. Am. Chem. Soc.*, 2003, **125**(51), 15831–15836.
113. B. Reif and R. G. Griffin, ^{1}H Detected ^{1}H,^{15}N Correlation Spectroscopy in Rotating Solids, *J. Magn. Reson.*, 2003, **160**(1), 78–83.
114. D. H. Zhou and C. M. Rienstra, Rapid Analysis of Organic Compounds by Proton-Detected Heteronuclear Correlation NMR Spectroscopy with 40 KHz Magic-Angle Spinning, *Angew. Chem., Int. Ed.*, 2008, **47**(38), 7328–7331.
115. D. H. Zhou, G. Shah, M. Cormos, C. Mullen, D. Sandoz, M. Rienstra, C. Proton-Detected Solid-State and N. M. R. Spectroscopy, of Fully Protonated Proteins at 40 KHz Magic-Angle Spinning, *J. Am. Chem. Soc.*, 2007, **129**, 11791–11801.
116. J. W. Wiench, C. E. Bronnimann, V. S. Y. Lin and M. Pruski, Chemical Shift Correlation NMR Spectroscopy with Indirect Detection in Fast

Rotating Solids: Studies of Organically Functionalized Mesoporous Silicas, *J. Am. Chem. Soc.*, 2007, **129**(40), 12076–12077.

117. S. M. Althaus, K. Mao, J. A. Stringer, T. Kobayashi and M. Pruski, Indirectly Detected Heteronuclear Correlation Solid-State NMR Spectroscopy of Naturally Abundant 15N Nuclei, *Solid State Nucl. Magn. Reson.*, 2014, **57–58**, 17–21.
118. S. Cavadini, A. Abraham, S. Ulzega and G. Bodenhausen, Evidence for Dynamics on a 100 Ns Time Scale from Single- and Double-Quantum Nitrogen-14 NMR in Solid Peptides, *J. Am. Chem. Soc.*, 2008, **130**(33), 10850–10851.
119. X. Zhao, W. Hoffbauer, J. Schmedt Auf Der Günne and M. H. Levitt, Heteronuclear Polarization Transfer by Symmetry-Based Recoupling Sequences in Solid-State NMR, *Solid State Nucl. Magn. Reson.*, 2004, **26**(2), 57–64.
120. J. P. Amoureux, J. Trebosc, J. Wiench and M. Pruski, HMQC and Refocused-INEPT Experiments Involving Half-Integer Quadrupolar Nuclei in Solids, *J. Magn. Reson.*, 2007, **184**(1), 1–14.
121. A. Venkatesh, M. P. Hanrahan and A. J. Rossini, Proton Detection of MAS Solid-State NMR Spectra of Half-Integer Quadrupolar Nuclei, *Solid State Nucl. Magn. Reson.*, 2017, **84**(March), 171–181.
122. J. Trebosc, B. Hu, J. P. Amoureux and Z. Gan, Through-Space R3-HETCOR Experiments between Spin-1/2 and Half-Integer Quadrupolar Nuclei in Solid-State NMR, *J. Magn. Reson.*, 2007, **186**(2), 220–227.
123. M. Carravetta, M. Eden, O. G. Johannessen, H. Luthman, P. J. E. Verdegem, J. Lugtenburg, A. Sebald and M. H. Levitt, Estimation of Carbon - Carbon Bond Lengths and Medium-Range Internuclear Distances by Solid-State Nuclear Magnetic Resonance, *J. Am. Chem. Soc.*, 2001, **123**(43), 10628–10638.
124. M. Fritz, J. Kraus, C. M. Quinn, G. P. A. Yap, J. Struppe, I. V. Sergeyev, A. M. Gronenborn and T. Polenova, Measurement of Accurate Interfluorine Distances in Crystalline Organic Solis: A High-Frequency Magic Angle Spinning NMR Approach, *J. Phys. Chem. B*, 2019, **123**(50), 10680–10690.
125. R. Zhang, Y. Nishiyama and A. Ramamoorthy, Proton-Detected 3D $^{1}H/^{13}C/\ ^{1}H$ Correlation Experiment for Structural Analysis in Rigid Solids under Ultrafast-MAS above 60 KHz, *J. Chem. Phys.*, 2015, **143**(16), 164201.
126. R. Zhang and A. Ramamoorthy, Selective Excitation Enables Assignment of Proton Resonances and ^{1}H-^{1}H Distance Measurement in Ultrafast Magic Angle Spinning Solid State NMR Spectroscopy, *J. Chem. Phys.*, 2015, **143**(3), 034201.
127. D. H. Brouwer, P. E. Kristiansen, C. A. Fyfe and M. H. Levitt, Symmetry-Based ^{29}Si Dipolar Recoupling Magic Angle Spinning NMR Spectroscopy: A New Method for Investigating Three-Dimensional Structures of Zeolite Frameworks, *J. Am. Chem. Soc.*, 2005, **127**(2), 542–543.

128. M. J. Bayro, M. Huber, R. Ramachandran, T. C. Davenport, B. H. Meier, M. Ernst and R. G. Griffin, Dipolar Truncation in Magic-Angle Spinning NMR Recoupling Experiments, *J. Chem. Phys.*, 2009, **130**(11), 114506.
129. Y. Matsuki, H. Akutsu and T. Fujiwara, Precision ^{1}H-^{1}H Distance Measurement via 13C NMR Signals: Utilization of ^{1}H-^{1}H Double-Quantum Dipolar Interactions Recoupled under Magic Angle Spinning Conditions, *Magn. Reson. Chem.*, 2004, **42**(2), 291–300.
130. J. P. Bradley, C. Tripon, C. Filip and S. P. Brown, Determining Relative Proton–Proton Proximities from the Build-up of Two-Dimensional Correlation Peaks in ^{1}H Double-Quantum MAS NMR: Insight from Multi-Spin Density-Matrix Simulations, *Phys. Chem. Chem. Phys.*, 2009, **11**(32), 6941.
131. K. Takegoshi, S. Nakamura and T. Terao, ^{13}C–^{1}H Dipolar-Assisted Rotational Resonance in Magic-Angle Spinning NMR, *Chem. Phys. Lett.*, 2001, **344**(5–6), 631–637.
132. K. Takegoshi, S. Nakamura and T. Terao, ^{13}C-^{1}H Dipolar-Driven ^{13}C-^{13}C Recoupling without ^{13}C Rf Irradiation in Nuclear Magnetic Resonance of Rotating Solids, *J. Chem. Phys.*, 2003, **118**(5), 2325–2341.
133. C. R. Morcombe, V. Gaponenko, R. A. Byrd and K. W. Zilm, Diluting Abundant Spins by Isotope Edited Radio Frequency Field Assisted Diffusion, *J. Am. Chem. Soc.*, 2004, **126**(23), 7196–7197.
134. J. J. Wittmann, V. Agarwal, J. Hellwagner, A. Lends, R. Cadalbert, B. H. Meier and M. Ernst, Accelerating Proton Spin Diffusion in Perdeuterated Proteins at 100 KHz MAS, *J. Biomol. NMR*, 2016, **66**(4), 233–242.
135. M. G. Jain, D. Lalli, J. Stanek, C. Gowda, S. Prakash, T. S. Schwarzer, T. Schubeis, K. Castiglione, L. B. Andreas, P. K. Madhu, G. Pintacuda and V. Agarwal, Selective ^{1}H-^{1}H Distance Restraints in Fully Protonated Proteins by Very Fast Magic-Angle Spinning Solid-State NMR, *J. Phys. Chem. Lett.*, 2017, **8**(11), 2399–2405.
136. J. N. Dumez, M. C. Butler, E. Salager, B. Elena-Herrmann and L. Emsley, Ab Initio Simulation of Proton Spin Diffusion, *Phys. Chem. Chem. Phys.*, 2010, **12**(32), 9172–9175.
137. J. N. Dumez, M. E. Halse, M. C. Butler and L. Emsley, A First-Principles Description of Proton-Driven Spin Diffusion, *Phys. Chem. Chem. Phys.*, 2012, **14**(1), 86–89.
138. M. E. Halse, J. N. Dumez and L. Emsley, Quasi-Equilibria in Reduced Liouville Spaces, *J. Chem. Phys.*, 2012, **136**(22), 224511.
139. M. Baias, C. M. Widdifield, J. N. Dumez, H. P. G. Thompson, T. G. Cooper, E. Salager, S. Bassil, R. S. Stein, A. Lesage, G. M. Day and L. Emsley, Powder Crystallography of Pharmaceutical Materials by Combined Crystal Structure Prediction and Solid-State ^{1}H NMR Spectroscopy, *Phys. Chem. Chem. Phys.*, 2013, **15**(21), 8069–8080.
140. Y. Nishiyama, R. Zhang and A. Ramamoorthy, Finite-Pulse Radio Frequency Driven Recoupling with Phase Cycling for 2D^{1}H/^{1}H Correlation at Ultrafast MAS Frequencies, *J. Magn. Reson*, 2014, **243**, 25–32.

141. R. Zhang, Y. Nishiyama, P. Sun and A. Ramamoorthy, Phase Cycling Schemes for Finite-Pulse-RFDR MAS Solid State NMR Experiments, *J. Magn. Reson*, 2015, **252**, 55–66.
142. D. P. Raleigh, M. H. Levitt and R. G. Griffin, Rotational Resonance in Solid State NMR, *Chem. Phys. Lett.*, 1988, **146**(1), 71–76.
143. P. T. F. Williamson, A. Verhoeven, M. Ernst and B. H. Meier, Determination of Internuclear Distances in Uniformly Labeled Molecules by Rotational-Resonance Solid-State NMR, *J. Am. Chem. Soc.*, 2003, **125**(9), 2718–2722.
144. A. T. Petkova and R. Tycko, Rotational Resonance in Uniformly ^{13}C-Labeled Solids: Effects on High-Resolution Magic-Angle Spinning NMR Spectra and Applications in Structural Studies of Biomolecular Systems, *J. Magn. Reson.*, 2004, **168**(1), 137–146.
145. K. Nomura, K. Takegoshi, T. Terao, K. Uchida and M. Kainosho, Determination of the Complete Structure of a Uniformly Labeled Molecule by Rotational Resonance Solid-State NMR in the Tilted Rotating Frame, *J. Am. Chem. Soc.*, 1999, **121**(16), 4064–4065.
146. R. Ramachandran, V. Ladizhansky, V. S. Bajaj and R. G. Griffin, ^{13}C-^{13}C Rotational Resonance Width Distance Measurements in Uniformly ^{13}C-Labeled Peptides, *J. Am. Chem. Soc.*, 2003, **125**(50), 15623–15629.
147. G. Pileio, Y. Guo, T. N. Pham, J. M. Griffin, M. H. Levitt and S. P. Brown, Residual Dipolar Couplings by Off-Magic-Angle Spinning in Solid-State Nuclear Magnetic Resonance Spectroscopy, *J. Am. Chem. Soc.*, 2007, **129**(36), 10972–10973.
148. J. Becker-Baldus, T. F. Kemp, J. Past, A. Reinhold, A. Samoson and S. P. Brown, Longer-Range Distances by Spinning-Angle-Encoding Solid-State NMR Spectroscopy, *Phys. Chem. Chem. Phys.*, 2011, **13**(10), 4514–4518.
149. G. Pileio, S. Mamone, G. Mollica, I. M. Montesinos, A. Gansmüller, M. Carravetta, S. P. Brown and M. H. Levitt, Estimation of Internuclear Couplings in the Solid-State NMR of Multiple-Spin Systems. Selective Spin Echoes and off-Magic-Angle Sample Spinning, *Chem. Phys. Lett.*, 2008, **456**(1–3), 116–121.
150. T. Kobayashi, Z. Wang and M. Pruski, Homonuclear Dipolar Recoupling of Arbitrary Pairs in Multi-Spin Systems under Magic Angle Spinning: A Double-Frequency-Selective ZQ-SEASHORE Experiment, *Solid State Nucl. Magn. Reson.*, 2019, **101**, 76–81.
151. K. N. Hu and R. Tycko, Zero-Quantum Frequency-Selective Recoupling of Homonuclear Dipole-Dipole Interactions in Solid State Nuclear Magnetic Resonance, *J. Chem. Phys.*, 2009, **131**(4), 045101.
152. I. Marin-Montesinos, G. Mollica, M. Carravetta, A. Gansmüller, G. Pileio, M. Bechmann, A. Sebald and M. H. Levitt, Truncated Dipolar Recoupling in Solid-State Nuclear Magnetic Resonance, *Chem. Phys. Lett.*, 2006, **432**(4–6), 572–578.

153. L. A. Straasø, M. Bjerring, N. Khaneja and N. C. Nielsen, Multiple-Oscillating-Field Techniques for Accurate Distance Measurements by Solid-State NMR, *J. Chem. Phys.*, 2009, **130**(22), 225103.
154. N. Khaneja and N. C. Nielsen, Triple Oscillating Field Technique for Accurate Distance Measurements by Solid-State NMR, *J. Chem. Phys.*, 2008, **128**(1), 015103.
155. L. A. Straasø, J. T. Nielsen, M. Bjerring, N. Khaneja and N. C. Nielsen, Accurate Measurements of ^{13}C-^{13}C Distances in Uniformly ^{13}C-Labeled Proteins Using Multi-Dimensional Four-Oscillating Field Solid-State NMR Spectroscopy, *J. Chem. Phys.*, 2014, **141**(11), 114201.
156. Y. Matsuki, H. Akutsu and T. Fujiwara, Band-Selective Recoupling of Homonuclear Double-Quantum Dipolar Interaction with a Generalized Composite 0° Pulse: Application to ^{13}C Aliphatic Region-Selective Magnetization Transfer in Solids, *J. Magn. Reson.*, 2003, **162**(1), 54–66.
157. R. Tycko, Stochastic Dipolar Recoupling in Nuclear Magnetic Resonance of Solids, *Phys. Rev. Lett.*, 2007, **99**(18), 1–4.
158. R. Tycko *Theory of Stochastic Dipolar Recoupling in Solid-State Nuclear Magnetic Resonance* †, 2008, pp. 6114–6121.
159. A. K. Paravastu and R. Tycko, Frequency-Selective Homonuclear Dipolar Recoupling in Solid State NMR, *J. Chem. Phys.*, 2006, **124**(19), 194303.
160. G. De Paëpe, J. R. Lewandowski and R. G. Griffin, Spin Dynamics in the Modulation Frame: Application to Homonuclear Recoupling in Magic Angle Spinning Solid-State NMR, *J. Chem. Phys.*, 2008, **128**(12), 124503.
161. N. T. Duong, S. Raran-Kurussi, Y. Nishiyama and V. Agarwal, Quantitative ^{1}H-^{1}H Distances in Protonated Solids by Frequency-Selective Recoupling at Fast Magic Angle Spinning NMR, *J. Phys. Chem. Lett.*, 2018, **9**(20), 5948–5954.
162. Z. Zhang, A. Oss, M. Org, A. Samoson, M. Li, H. Tan, Y. Su and J. Yang, Selectively Enhanced ^{1}H-^{1}H Correlations in Proton-Detected Solid-State NMR under Ultrafast MAS Conditions, *J. Phys. Chem. Lett.*, 2020, **11**(19), 8077–8083.
163. Z. Zhang, A. Oss, M. L. Org, A. Samoson, M. Li, H. Tan, Y. Su and J. Yang, Erratum: Selectively Enhanced ^{1}H-^{1}H Correlations in Proton-Detected Solid-State NMR under Ultrafast MAS Conditions (Journal of Physical Chemistry Letters (2020) 11: 19 (8077 – 8083) DOI: 10.1021/Acs.Jpclett.0c02412), *J. Phys. Chem. Lett*, 2020, **11**(23), 10157.
164. L. R. Potnuru, N. T. Duong, S. Ahlawat, S. Raran-Kurussi, M. Ernst, Y. Nishiyama and V. Agarwal, Accuracy of ^{1}H-^{1}H Distances Measured Using Frequency Selective Recoupling and Fast Magic-Angle Spinning, *J. Chem. Phys.*, 2020, **153**(8), 084202.
165. N. T. Duong, S. Raran-Kurussi, Y. Nishiyama and V. Agarwal, Can Proton-Proton Recoupling in Fully Protonated Solids Provide Quantitative, Selective and Efficient Polarization Transfer?, *J. Magn. Reson.*, 2020, **317**, 106777.

166. G. Mollica, P. K. Madhu, F. Ziarelli, A. Thévand, P. Thureau and S. Viel, Towards Measurement of Homonuclear Dipolar Couplings in ^{1}H Solid-State NMR: Recoupling with a Rotor-Synchronized Decoupling Scheme, *Phys. Chem. Chem. Phys.*, 2012, **14**(13), 4359–4364.
167. J. A. Rodriguez, M. I. Ivanova, M. R. Sawaya, D. Cascio, F. E. Reyes, D. Shi, S. Sangwan, E. L. Guenther, L. M. Johnson, M. Zhang, L. Jiang, M. A. Arbing, B. L. Nannenga, J. Hattne, J. Whitelegge, A. S. Brewster, M. Messerschmidt, S. Boutet, N. K. Sauter, T. Gonen and D. S. Eisenberg, Structure of the Toxic Core of α-Synuclein from Invisible Crystals, *Nature*, 2015, **525**(7570), 486–490.
168. M. Gemmi and A. E. Lanza, 3D Electron Diffraction Techniques, *Acta Crystallogr. Sect. B Struct. Sci. Cryst. Eng. Mater.*, 2019, **75**, 495–504.
169. U. Kolb, T. Gorelik, C. Kübel, M. T. Otten and D. Hubert, Towards Automated Diffraction Tomography: Part I-Data Acquisition, *Ultramicroscopy*, 2007, **107**(6–7), 507–513.

CHAPTER 17

Membranes, Minerals and Magnets: Application of NMR Spectroscopy to Biological Interfaces

J. E. KENT,[†] K. SHIN[†] AND F. M. MARASSI*

Sanford Burnham Prebys Medical Discovery Institute, 10901 North Torrey Pines Road, La Jolla, CA 92037, USA
*Email: fmarassi@sbp.edu

17.1 Introduction

Nature has evolved sophisticated mechanisms of producing biomolecular structures that compartmentalize various biochemical and cellular environments. Biological membranes, for example, provide the amphiphilic architecture responsible for cells,[1] the building-block of life, and biomineral structures like bone are required for mechanical movement in vertebrates.[2,3] Within these structures, complex chemical reactions can be sorted and synchronized, while their interfaces also host many processes for additional levels of regulation.[4–8]

Processes that occur at interfacial surfaces, such as signal transduction cascades or endocrine signaling, play critical roles in mediating everything from homeostasis to development.[9] The associated biomolecular interactions

[†]These authors contributed equally.

New Developments in NMR No. 26
NMR Spectroscopy for Probing Functional Dynamics at Biological Interfaces
Edited by Anirban Bhunia, Hanudatta S. Atreya and Neeraj Sinha

Published by the Royal Society of Chemistry, www.rsc.org

are highly regulated and involve many types of biomolecules. These interactions are also highly dynamic to respond to changing external stimuli. Understanding the structural basis of these biological processes has broad implications for our fundamental understanding of biology as well as biomedical and therapeutic applications. Structural biology plays a central role in this regard. Protein structures are essential for understanding their functions and for guiding drug design,[10–12] and elucidating the molecular mechanisms of biomineral formation is key for advancing the fields of orthopedics, urology, dentistry, and other industrial implications in electronics.[13–15]

For comprehensive biophysical characterization of the players involved, it is important to consider methods of sample preparation to best mimic the native environment and techniques that can characterize both molecular conformation and dynamics. Nuclear magnetic resonance (NMR) spectroscopy is a uniquely versatile method suitable for a wide range of sample types that can achieve both criteria.[16–18] Different sample environments can span significantly different molecular sizes ranging from angstroms to centimeters, with varying physical properties and complexities. Here, we describe recent NMR studies on two key biological structures that are critical for mammalian life: membranes and biominerals. Recent advances in the field provide exciting new opportunities to gain insights about interfacial interactions. Solution and solid-state NMR experiments can be combined to examine complex multicomponent assemblies across the wide range of dynamics time and distance scales encountered between the atomic and macromolecular assembly levels. Together, they extend the range of molecular size, dynamics and sample types that can be examined, and the emerging field of *in situ* NMR presents new exciting opportunities for examining the structure–activity relationship (SAR) of proteins in native isolated cell membranes and whole cells, in future studies.

17.2 Membranes and Membrane Proteins

The assembly of amphiphilic lipid molecules to form biological membranes is critical for all cellular life and was a key step in evolution.[1,19] Eukaryotic cells contain numerous membranes, including: cell membranes that define the bounds of the cytoplasm; the nuclear membrane that protects the genetic contents;[20] and the various membranes of intracellular organelles.[21] Cells have developed many types of molecular machineries to relay essential signals and transport molecules across membranes.[22] Examples range from small peptide hormones, to integral membrane receptors and ion channels.[23,24] These membrane proteins all contain extended stretches of hydrophobic residues that predispose their membrane localization but have diverse structures and functions.[25] Moreover, membrane properties such as lipid bilayer thickness, curvature, and mosaicity constrain membrane proteins' conformational freedom and regulate their activities.[5–8]

Membrane surfaces are also important sites of biological interactions.[26–29] They can confine proteins to a 2D diffusion space and increase their local concentrations to improve the probability of functional ligand encounters.[30] Interactions with membranes, and the diverse assortment of membrane components (*e.g.*, lipids, sterols), can also alter each protein conformation and dynamics to regulate biological function.[4,30–35]

Membrane proteins represent a substantial portion of the genomes of many organisms (~30% of both the *Homo sapiens* and *Escherichia coli* genomes).[36,37] Because they regulate many fundamental biological processes, they are highly attractive targets for structural characterization and drug development.[38] However, the requirement of aptly recapitulating the hydrophobic lipid bilayer with the surrounding aqueous conditions *in vitro* can be challenging.[39]

The toolbox of the modern structural biologist has undergone major developments in recent years and is now adequately equipped to rise to these stringent requirements. Furthermore, the diverse choice of techniques has also driven the development of membrane mimetics that can be employed with varying levels of success (Figure 17.1). Detergent micelles, lipid bilayers, and native cellular preparations have each demonstrated their suitability for structural studies of membrane proteins (*vide infra*). X-ray crystallography and cryogenic electron microscopy (cryo-EM) have had many successes when applied to membrane proteins,[40–43] but the types of membrane mimetics suitable for these techniques, and the observations that can be made, are limited.[44,45] NMR spectroscopy, on the other hand, is adequately equipped to give both atomic-resolution structural data and dynamics information for membrane proteins across the wide range of membrane mimetics, and NMR is the premier technology for completely characterizing the SAR of these types of systems.[46] Although the focus of this section is on the relative strengths and weaknesses of multiple membrane mimetics and their application to NMR spectroscopy, practical considerations regarding sample preparation may be found elsewhere in recent reviews,[47–49] and published protocols.[50–54]

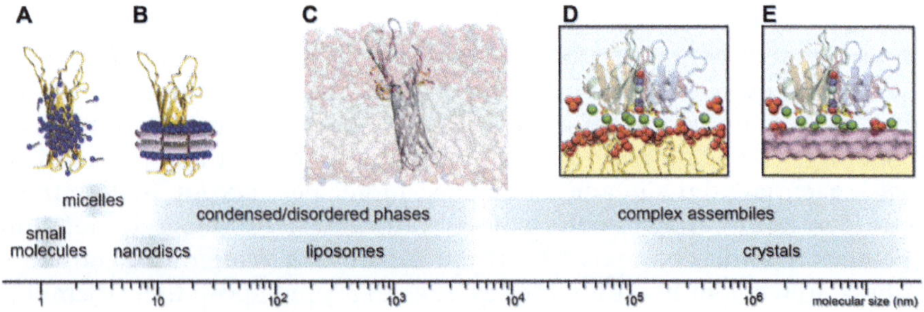

Figure 17.1 Examples of membrane mimetics suitable for biomolecular NMR studies. Micelles (A); nanodiscs (B); whole bacterial cell envelope preparations (C); protein (cartoon) and lipid (yellow) assemblies (D); and protein (cartoon), lipid (yellow) and biomineral (pink) assemblies (E).

17.2.1 Detergent Micelles

Micelles assemble from multiple amphiphilic molecules that contain a hydrophobic aliphatic moiety and a hydrophilic headgroup.[55] By contrast to the phospholipids that they mimic, detergent molecules are soluble and compatible with solution NMR methods.[56,57] Above their critical micelle concentration, detergents assemble into aggregates with the aliphatic chains forming a hydrophobic core and the hydrophilic headgroups facing the aqueous environment.[58] Like lipids, detergents can have electro-neutral, charged or glycosylated headgroups that represent natural post-synthetic modifications.[59]

17.2.1.1 Protein–Headgroup Interactions

Lipid headgroups at membrane surfaces interact with a wide range of proteins.[30] Due to their high curvature, chemical homogeneity and dynamics, micelles are highly simplified models for characterizing the interactions of proteins with membrane surfaces. Despite these drawbacks, micelles facilitate studies of highly disordered proteins, such as peptide hormones, which present highly overlapped NMR signals.[60] They yield high-resolution solution NMR spectra which are essential for resonance assignments and residue-specific analysis. Solution NMR experiments like diffusion-ordered spectroscopy (DOSY) are very well-suited for studying peptide hormone interactions with micelles,[61,62] and can be used to determine the translational diffusion coefficients of small peptides as they associate and dissociate with the relatively larger micelles. These studies can be used to quantitate micelle–protein interactions and evaluate the binding preferences for different headgroups, to gain insights that augment the information obtained from isothermal titration calorimetry, where protein–micelle interactions are complicated by the inherent concentration-dependent properties of micelles.[63]

Solution NMR has been used to study many peptide hormones in micelles.[4] For example, the 17-residue isoform of apelin, an important peptide hormone for cardiovascular homeostasis, has been shown to undergo β-turn structuring upon association with sodium dodecyl sulfate (SDS) micelles (Figure 17.2).[64,65] and solution NMR has been used to characterize the preferential headgroup interactions of apelin and apela isoforms, both of which bind to the same G-protein coupled receptor (GPCR). All apelin isoforms preferentially interact with anionic headgroup micelles,[66] but DOSY experiments showed that the 32-residue apela isoform has highly diminished preference, with near equivalent binding for zwitterionic and anionic micelles.[67] Alterations of protein secondary structure in response to these micelle interactions are also detected through chemical shift analysis. While apela adopts α-helical properties upon micelle association, apelin remains a largely random coil with some indications of potential β-turns.

Measurements of NMR transverse relaxation times (T_2) can provide further insights about the interactions of proteins with polar lipid headgroups,

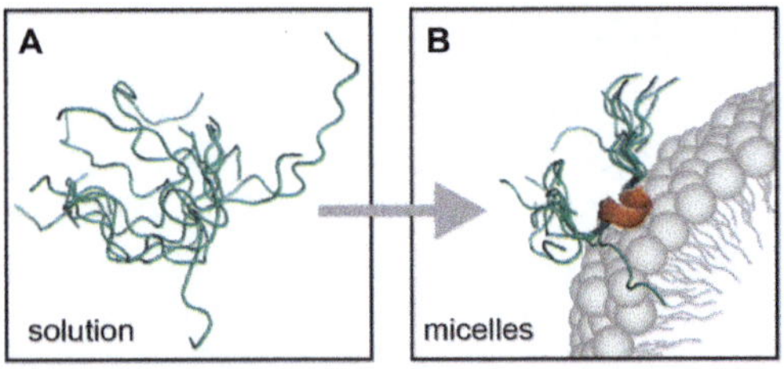

Figure 17.2 Induced structuring of the peptide apelin-17 upon micelle surface adsorption, as described.[4]

where micelle association results in shorter T_2 and broader line widths.[68] Solution NMR is also useful for characterizing the structural basis for protein–lipid recognition. In a recent study, $^1H/^{15}N$ HSQC chemical shifts were used to map the binding of phosphatidylinositol phosphate (PIP) lipids by the pleckstrin homology (PH) domain of the protein PLEKHA7.[69] PLEKHA7 is involved in intracellular signaling and cell–cell adhesion,[70] and the interactions of its PH domain with PIP lipids are central to this function. Chemical shift mapping experiments revealed three PIP binding sites and provided insights into the mechanism of PLEKHA7 membrane association.[69]

17.2.1.2 Membrane-embedded Proteins

When membrane proteins are reconstituted into detergent micelles, the hydrophobic transmembrane amino acids become encased within the micellar hydrophobic core whilst extramembrane regions localize near the hydrophilic micellar headgroups or remain fully exposed to the aqueous milieu. Detergent micelles offer some advantages over other membrane mimetics. The relatively small hydrodynamic radii (<4 nm) of these detergent–protein complexes yield high-resolution multidimensional solution NMR spectra that enable the characterization of membrane proteins.[54] The chemistry of available micellar detergent molecules and their successful application to membrane protein studies by solution NMR has been reviewed in depth.[71] Acquisition of high-resolution multidimensional NMR spectra is critical for structure determination and care must be taken to screen for suitable detergents. Although this can be labor-intensive and potentially cost-prohibitive, many detergents have been developed to improve the quality of solution NMR data.[72,73]

The outer membrane protein Ail from the deadly microbe *Yersinia pestis* has been characterized in multiple sample types by various NMR techniques and serves as a good example of the versatility of modern NMR technology.[74–80] Ail is essential for the pathogenicity of *Y. pestis*, the etiologic agent of plague, and interacts with both bacterial components and proteins from the human immune regulatory responses.[81–85] In this case, detergent selection was optimized to obtain the narrowest linewidths and most homogeneous line shapes, with the broadest chemical shift distribution. Samples of Ail in *n*-decylphosphocholine (DePC) micelles are well-suited for multidimensional solution NMR experiments and enabled the protein's eight-stranded β-barrel structure to be determined.[78] Notably, solution NMR revealed signals from extracellular loop sites that are too dynamic to be observed by X-ray crystallography.[86] These are important functional sites and their atomic-level detection paves the way for SAR studies.

NMR structures of many membrane proteins have been determined in micelles (Figure 17.3).[87–93] The β-barrel structures of outer-membrane proteins represent excellent examples of the power of NMR.[94–97] In these cases, backbone structure determination is facilitated by the ability to measure ^1H–^1H NOE contacts between hydrogen-bonded backbone sites on neighboring strands. The use of an implicit solvation potential during structure calculation further facilitates structure determination, and yields structures with higher precision, accuracy and conformational quality.[78,98–100] Recent examples of structures determined by solution NMR in detergent micelles have had great success in tackling complex protein classes, including seven-pass transmembrane proteins like the phototaxis receptor sensory rhodopsin II (pSRII) from *Natronomonas pharaonis* with a backbone RMSD as low as 0.48 Å.[90]

In addition to structure determination, solution NMR is uniquely well-suited for characterizing protein–ligand and protein–protein interactions.[101,102] Notable examples include the interactions between pSRII and its ligand all-*trans*-retinal (isomerized to 13-*cis*-retinal) in detergent micelles.[90] While micelles can permit observation of these interactions, reported cases of success are rather sparse. Instead, there are more instances of aberrant detergent binding and altered conformations, or protein unfolding in the presence of detergent molecules and their dynamic aggregates.[16] In the case of Ail, for example, although micelle reconstitution provides high-resolution spectra, it produces samples incapable of binding to its known ligands (Figure 17.4).[75]

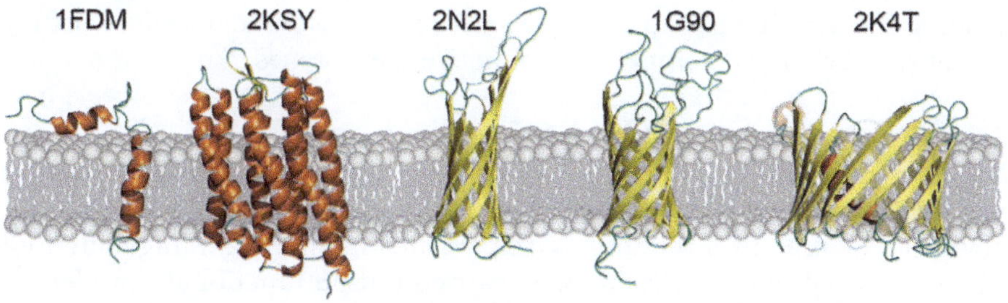

Figure 17.3 Examples of membrane protein structures determined by solution NMR in detergent micelles. Colors reflect secondary structure: coil (green), helix (orange), and strand (yellow). 1FDM, fd major coat protein; 2KSY, pSRII; 2N2L, Ail; 1G90, OmpA; 2K4T, VDAC-1. The proteins are inserted in a model lipid bilayer for visualization.

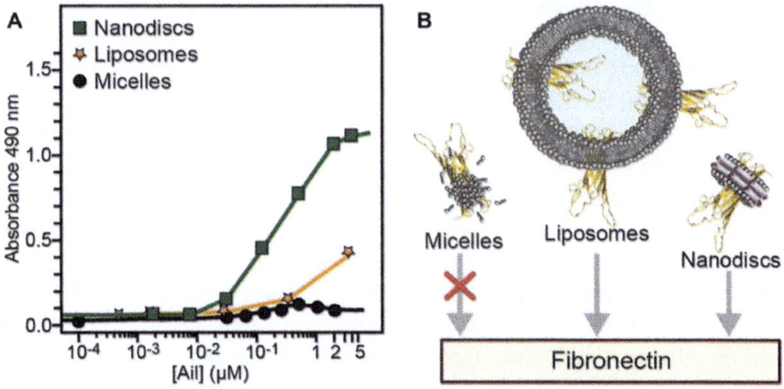

Figure 17.4 Influence of the membrane environment on fibronectin-binding properties of Ail.[75] Enzyme-linked immunosorbent assay data for Ail binding to fibronectin. Ail reconstituted into micelles (black circles), liposomes (orange stars), and nanodiscs (green squares).

17.2.1.3 Mixed Micelles and Isotropic Bicelles

A major difference between micelles and physiological membranes is their high curvature and lack of a proper lipid bilayer.[1,103,104] Detergent micelles are also highly homogeneous in composition. Mixed micelles including bicelles (bilayered micelles) can circumvent these problems. These structures can incorporate a wider range of native phospholipids.[105] For example, detergent micelles mixed with lipopolysaccharide (LPS) have been used in solution NMR studies to map the interactions of LPS with OprH, an outer-membrane protein from *Pseudomonas aeruginosa* which confers antibiotic resistance.[106] LPS-dependent chemical shift perturbations and signal intensity changes were observed in both backbone and side-chain resonances, LPS binding affinities in the range of ~200 μM could be determined by monitoring spectral changes, and a reduction in extracellular loop dynamics could be characterized by monitoring broadening of ^{13}C signals from Arg and Lys sidechains. Nevertheless, mixed micelles do not fully replicate the native membrane properties including the fact that LPS can only be added up ~9% of the micelle, woefully short of the ~50% LPS composition of native membranes.[107]

17.2.2 Detergent-free Lipid Bilayers

17.2.2.1 Liposomes and Planar Bilayers

Knowledge of protein structures is important for understanding function, and hence, structural studies are best carried out on functional samples.[108] While micelles can be useful membrane mimetics for NMR, they do not adequately recapitulate the physiological membrane properties and can distort structure and alter activity.

Detergent-free samples do not have these drawbacks. Phospholipids spontaneously assemble into spherical bilayer vesicles or liposomes whose curvature, diameter (30 nm to 10 μm), bilayer thickness, and bilayer organization (unilamellar or multilamellar) can be controlled by the lipid composition and the preparation protocol.[109,110] When layered on polar surfaces such as glass, liposomes spread to adopt a planar bilayer structure that can be useful for some types of NMR experiments (see below). The characteristics of liposomes and planar bilayers are similar, in many aspects, to native biological membranes and can support both the structures and biological functions of membrane proteins. In the case of *Y. pestis* Ail, protein reconstitution in detergent-free liposomes yielded an active sample that binds the known human ligands (Figure 17.4).[75] Of note, SAR studies with liposomes may be limited by the closed geometry and challenges associated with controlling inside–outside protein orientation of these spherical lipid assemblies.[75,111] Liposomes and planar bilayers are too large for solution NMR studies but are highly well-suited for solid-state NMR of membrane proteins.

17.2.2.2 Oriented Sample (OS) Solid-state NMR

Solid-state NMR can measure anisotropic spin interactions.[112] OS solid-state NMR experiments take full advantage of this property and provide uniquely useful information about molecular structure.[113–115] Specifically, OS solid-state NMR leverages the membrane-specific orientation of proteins within a lipid bilayer.[116] The samples can be aligned by adsorption on glass (planar bilayer) or magnetically when placed within the field of the NMR magnet (bicelles). Their preparation requires careful adjustment of protein and lipid compositions and concentrations, but once uniaxial alignment is achieved, the resulting high-resolution NMR spectra have the advantage that they directly reflect the structures and membrane-inserted orientations of membrane proteins.[117]

Membrane orientation is an important functional property of proteins and transmembrane helix orientation, dynamics and interactions are critical for regulating biological function. The function of many classes of α-helical transmembrane proteins, such as GPCRs, depends on changes in transmembrane helix orientation.[121] Notably, the intrinsic properties of biological membranes such as curvature and thickness restrict the possible dynamic interplay for the membrane proteins localized within it,[122] making it important to perform experiments in near-native detergent-free samples,[123] as demonstrated for the solid-state NMR structure of the chemokine receptor CXCR1 in liposomes.[124]

The first NMR structure of a transmembrane peptide, gramicidin, was determined in planar lipid bilayers using OS solid-state NMR.[125] In the case of the 50-residue fd bacteriophage pVIII coat protein, $^1H/^{15}N$ separated local field (SLF) spectra, obtained with Polarization Inversion Spin Exchange at the Magic Angle (PISEMA), yielded the three-dimensional structure of the membrane-embedded protein and provided the means for simultaneous resonance assignment and structure refinement[114] based on the special properties of OS NMR spectra.[126–128]

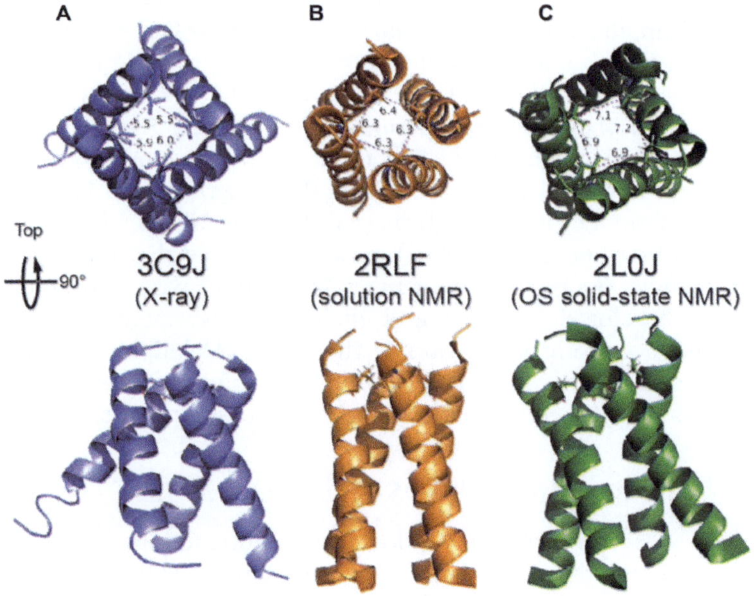

Figure 17.5 Influence of the membrane mimetic on M2 proton channel structure from influenza A. The structure of M2 determined by X-ray crystallography[118] (*n*-octyl-β-D-glucopyranoside micelles; PDB: 3C9J – amantidine molecule removed for simplicity), solution NMR[119] (DHPC micelles; PDB: 2RLF), and OS solid-state NMR[120] [1,2-dioleoyl-*sn-glycero-3*-phosphocholine/1,2-dioleoyl-*sn-glycero-3*-phosphoethanolamine (DOPC/DOPE) bilayer] shown from the top and side views. The structures demonstrate the distruption of α-helix packing in the crystal and in micelles, which alters the size of the channel. Distances between Val27 Cβ atoms in each of these calculated structure are shown in Ångstrom (magenta dashed lines).

Orientation restraints derived from OS NMR spectra are critical for structure determination and refinement. OS NMR studies of the M2 proton channel from influenza A yielded insights into the importance of the quaternary structure for its function.[120] Measurement of $^{1}H/^{15}N$ orientation restraints for M2 in planar bilayers enabled the M2 channel structure to be determined with atomic resolution. It is notable that the structure and rimantadine ligand-binding site of this functional sample are dramatically different from those determined in either detergent micelles by solution NMR or crystalline samples by X-ray diffraction (Figure 17.5).[118,119]

17.2.2.3 Magic Angle Spinning (MAS) Solid-state NMR

Narrow line widths in solution NMR stem from the averaging of the spatial components of the anisotropic interactions as proteins reorient freely and rapidly in the field of the NMR magnet.[112] In solid-state NMR, by contrast, the spin interactions are averaged to their isotropic value by spinning the sample at the magic angle (54.74°) relative to the external magnetic field at a rate greater than the anisotropic interaction frequency.[129–131] This affords significant increases in spectral resolution and the observation of only the isotropic component of the spin interaction tensors. MAS solid-state NMR spectroscopy is well-suited for studying the structure and dynamics of membrane proteins in liposomes,[132–138] due in part to the relative simplicity of sample preparation compared to oriented samples, and the broad collection of experiments that have been developed. Furthermore, since both solution NMR and solid-state MAS NMR provide isotropic resonances, data comparisons are possible across the two techniques. There are, however, challenges when studying membrane proteins under these conditions: firstly, the high relative centrifugal force (~930 000×*g* for a 3.2 mm MAS rotor spinning at 12 000 Hz) from spinning the sample at high frequency may adversely affect the protein. Finally, while the loss of anisotropic terms removes the possibility of directly extracting orientation restraints, recoupling techniques[139,140] enable a method for structure determination of membrane proteins in liposomes.[124,141–143]

A recent MAS NMR study focused on the transmembrane domain of the envelope protein E (ETM) from SARS-CoV-2 in ER-Golgi-like lipid bilayers.[144] Structure determination was based on ^{1}H–^{1}H distances from NHHC experiments and ^{19}F–^{13}C distances from $^{13}C/^{19}F$ rotation-echo double-resonance (REDOR) experiments.[145–147] Although prior work had determined the structure of ETM from SARS-CoV-1 in micelles,[148] the lowest energy conformation of the pentameric assembly calculated in ER-Golgi-like liposomes differed significantly. Major differences between the two structural ensembles included the lack of handedness and considerably tighter packing of the helices in the lipid bilayer. Moreover, a recent NMR structural study of a longer E protein segment in micelles retains activity and reveals a different ligand-binding site.[149]

In addition to structure calculation, the study of membrane proteins by MAS solid-state NMR can provide mechanistic insights about key interactions that occur within the membrane, with other membrane/protein components, or with soluble proteins. Biological membranes such as mammalian plasma membranes and Gram-negative bacterial outer membranes exhibit a high degree of asymmetry with molecules such as phosphatidylserine existing almost exclusively in the cytoplasmic facing leaflet and LPS molecules only in the outer leaflet.[107,150] Assessing the influence of these components and reconstituting the asymmetry of the native membrane is an important task that NMR is uniquely equipped to tackle.

While the complete reconstitution of native asymmetry for NMR-mediated characterization remains elusive, progress toward this goal has been made. Incorporation of LPS into liposomes was demonstrated to study its effect on Ail.[76] Through the addition of an 8% molar ratio of Kdo_2-Lipid A (KLA) during the liposome preparation, the impact of the components of the bacterial asymmetric membrane on the conformational dynamics of Ail could be visualized, to understand how the interplay between these molecules affects the virulence associated with *Y. pestis*.[81] Few assignable chemical shift perturbations (CSPs) were measured as a result of the newly established membrane composition, but an observable increase in resonances spanning 45–48 ppm of the ^{13}C dimension indicated more glycine residues being observed within the dynamics regime of the experiment. The global influence of this molecule, which is important to the unique nature of bacterial membranes, could be determined through the increased signal intensity in cross-polarization (CP) experiments, indicating a reduction in molecular dynamics.[151]

17.2.2.4 Lipid Nanodiscs and Macrodiscs

Nanodiscs are small (>6 nm in diameter) membrane mimetics produced by encasing a lipid bilayer with a scaffolding molecule, which can be either truncated constructs of human apolipoprotein-A1 (ApoA1) or styrene–maleic acid polymers (SMA).[152,153] Nanodiscs are detergent-free and can support functional samples of membrane proteins.[154] Unlike spherical liposomes, their open geometry facilitates quantitative functional assays.[155] Nanodisc size can be controlled through the use of different scaffolds,[156] and nanodiscs can be suitable for both solution and solid-state NMR studies.[74,157] Small nanodiscs (~6–8 nm in diameter) have been used to study multiple classes of proteins, including single-pass transmembrane proteins like Bcl-xL[158] and CD4,[159] multi-pass membrane proteins,[160] and β-barrels.[75,78,161,162]

Although proteins in nanodiscs can produce high-resolution solution NMR spectra, the large complex size can result in short T_2 times, severely limiting or outright prohibiting structural characterization.[163] The relationship between magnetic field strength and T_2 relaxation suggests that one may simply utilize higher field magnets to overcome the enhanced relaxation of these larger molecular assemblies,[68] but the reduction in linewidths at higher fields alone is insufficient to overcome the short T_2 times and securing a higher field magnet on a whim is seldom possible.

To circumvent this problem, several approaches are available. The simplest is ^{2}H incorporation to reduce the effect of spin diffusion for more favorable relaxation properties.[164] Transverse relaxation optimized spectroscopy (TROSY)-adapted pulse programs are also useful as they are designed to aid in the study of larger macromolecules and complexes.[165] With this strategy, ~58% of the backbone resonances of Ail were successfully assigned though corresponding predominantly to residues within the transmembrane regions of the protein.[76] Using the methyl-TROSY approach for site specifically labeled proteins is another notable strategy,[166,167] as demonstrated by recent studies characterizing archaea bacteriorhodopsin in lipid nanodiscs.[168] If a key-residue is known, selective labeling or incorporation of other NMR active nuclei circumvents the need for assignment—for example using ^{19}F-detected solution NMR spectroscopy, the allosteric modulation of the adenosine A2A receptor (V229C mutant conjugated with ^{19}F trifluoromethyl phenylalanine) could be observed.[169,170]

In cases where the same protein fold is retained in detergent micelles and nanodiscs, direct transfer of assignments can be performed. In other instances where direct transfer is ambiguous, the incorporation of detergent molecules into nanodiscs can transition the protein from nanodiscs to the micellar state. Provided that the detergent–lipid exchange is in the NMR fast regime, gradual changes in cross-peak chemical shifts can be observed with detergent titration for subsequent assignment.[171] Using such an approach is mainly applicable for 2D NMR experiments such as ^{1}H–^{15}N HSQC as in the case of the 283-residue β-barrel VDAC-1.[172]

Once the bottleneck of resonance assignment has been overcome, protein conformations can be studied in a near-native environment, whilst also benefitting from the high resolution associated with solution NMR. In the case of Ail, the structural ensemble calculated with a combination of dihedral angles, long range ^{1}H–^{1}H contacts, and hydrogen bond restraints in nanodiscs is similar to the eight-stranded β-barrel structure calculated in micelles.[76] Like other β-barrels, Ail presents an ideal scenario for NMR, as the tertiary structure is underpinned by the hydrogen bonding network that exists across each strand between backbone amide protons and carbonyl oxygens, only the HN resonances of residues which constitute the transmembrane region of the β-barrel are required for structure calculation.[173,174] In the case of membrane proteins with multiple transmembrane α-helices (*e.g.*, GPCRs) the amide to carbonyl hydrogen bonding network is important for defining the secondary but not tertiary structure. In this case, side-chain assignments are needed. While this can be challenging to obtain due to significant signal overlap, it is required for measuring meaningful long-range interhelical contacts for structure calculation.[175]

Nanodiscs can be prepared with many types of phospholipids. In the case of Ail, this feature was capitalized on by incorporating LPS from native *Y. pestis* LPS into nanodiscs for solution NMR studies.[80] Solution NMR yielded informative insights about the Ail–LPS interaction (Figure 17.6). Arg and Lys residues localizing to the base of the extracellular loops were found to constitute the LPS binding sites, as identified by both combined HN/N CSPs and reduced proton/deuterium (H/D) exchange for the Nε and Nη moieties of the sidechains.

Nanodiscs can also be used for studying protein to membrane surface interactions. By preparing nanodiscs with varying lipid composition, it is possible to characterize the lipid-headgroup selectivity and specificity involved in binding of surface interacting proteins. For example, the selective binding of the PH domain of PLEKHA7 with PIP lipids could be studied by incorporating PIPs in nanodiscs for solution NMR studies.[69] The PH domain presents a membrane exchange rate in the ms time scale when associated with nanodiscs at 15 °C, resulting in the loss of visible solution NMR resonances. However, elevating the experimental temperature to 45 °C increases the exchange dynamics and renders the system amenable for residue-specific characterization (Figure 17.6).

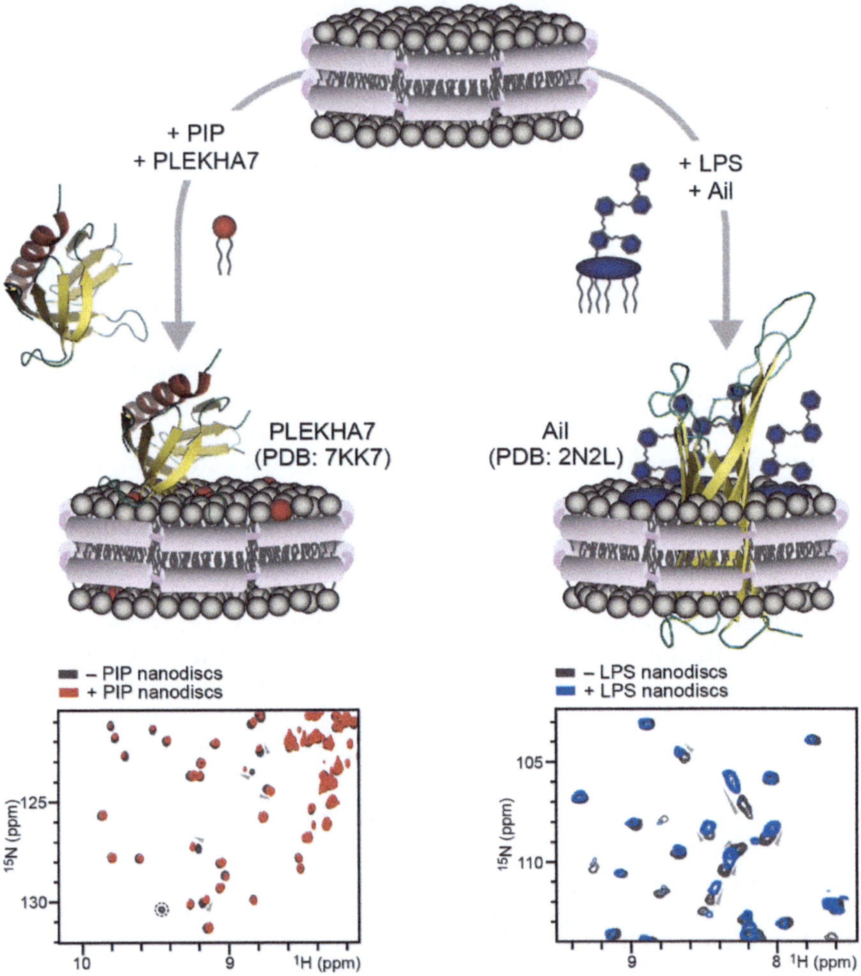

Figure 17.6 Use of nanodiscs (ND) to examine the effects of membrane components on protein structure and function. Different lipids, PIP (red) and LPS (blue), can be incorporated into nanodiscs to characterize their effects on PLEKHA7,[69] bound to the membrane surface, and *Y. pestis* Ail,[80] inserted in the membrane. Comparisons of $^{1}H/^{15}N$ HSQC NMR spectra yield CSPs (black arrows) and signal loss (black dashed circles) that reflect protein–lipid interactions.

Nanodiscs can also be studied by solid-state NMR,[74,157,176] enabling investigation across different dynamics ranges. Sedimentation of Ail nanodiscs produced MAS data with resolution comparable to Ail reconstituted into liposomes.[74] Interestingly, the sedimented nanodiscs could be resuspended for solution NMR spectroscopy, demonstrating that the centrifugal g forces associated with MAS had no deleterious influence on the conformation of the protein, in agreement with studies focusing on soluble proteins (Figure 17.7).[177]

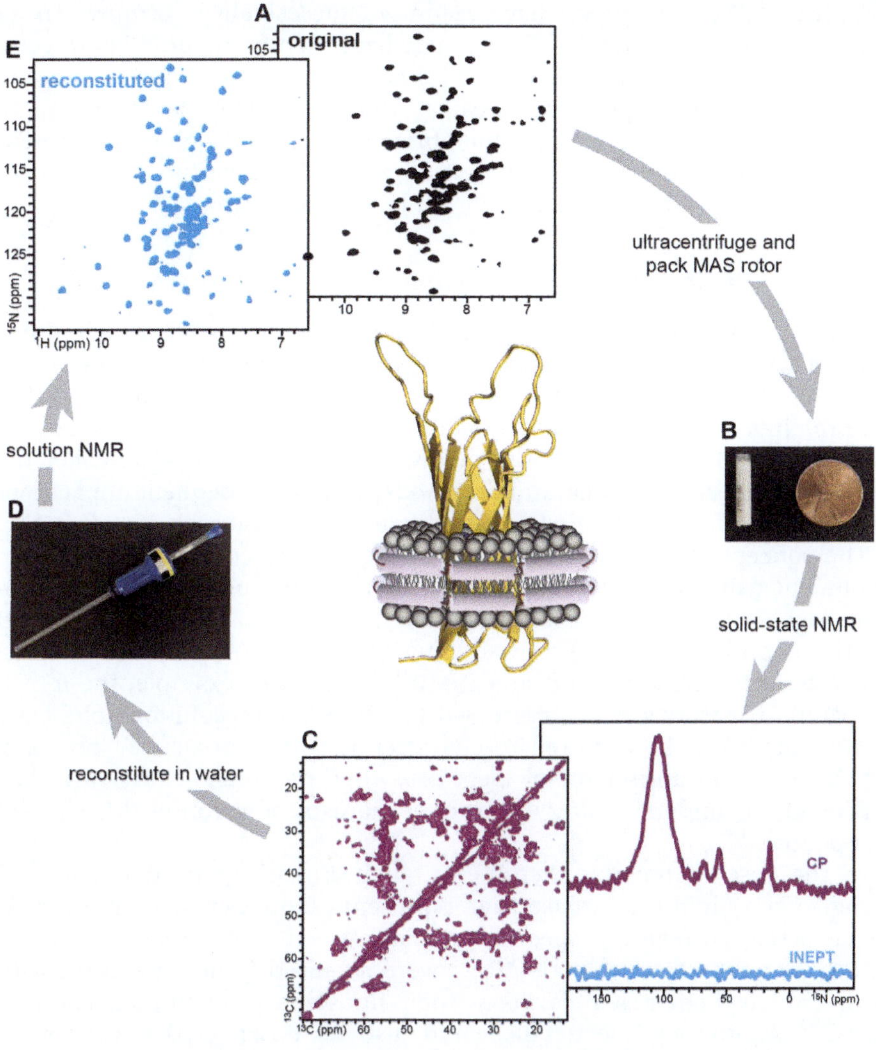

Figure 17.7 Reversibility of nanodisc sedimentation for multiple types of NMR spectroscopy. Schematic shows the experimental outline.[74] (A) Solution NMR $^{1}H/^{15}N$ HSQC spectrum of Ail embedded in nanodiscs prior to sedimentation; (B) ultracentrifugation and packing of MAS rotor; (C) MAS solid-state NMR (PDSD and ^{15}N direct CP) of sedimented Ail nanodiscs; (D) reconstitution with buffer for solution NMR acquisition; (E) solution NMR spectrum, acquired after reconstitution with additional buffer.

SMA copolymers offer an alternative means of encasing lipid nanodiscs and macrodiscs (>30 nm). Notably, SMA can extract bilayer discs from either liposomes or native membranes. The advantage of SMA is the ease of preparation relative to ApoA1-derived nanodiscs, which must be assembled with great control of the molar ratios of each component within the complex.[178] SMA macrodiscs have been used successfully to acquire $^{1}H/^{15}N$ SLF experiments for proteins from single-pass membrane proteins to complex multipass/β-barrel proteins.[79,179] SMA discs represent an exciting alternative to other aligned samples, as they also facilitate extraction of membrane proteins from native membranes,[180] and may allow for protein orientation to be determined *ex vivo* all whilst increasing the amount of detectable protein relative to mechanically aligned samples.

17.2.2.5 *In situ NMR Spectroscopy*

The need for studying proteins within an ever more native environment has been raised consistently over numerous works focusing on membrane proteins (*vide supra*). Central to the idea of increasing the sample nativity is that the protein's environment is essential for function.[181] This section will detail the initial work published in the field of *in situ* and in-cell NMR, key developments in sample preparation methods, and the subsequent impact that this has had on the field of structural biology for membrane proteins.

The concept of studying components of cells in a completely native environment using solid-state NMR spectroscopy is not new, with work spanning back ~40 years. Early examples include characterization of penicillin G on the peptidoglycan (PG) cross-linking within *Aerococcus viridans* using whole bacterial cells by solid-state NMR.[182] Since this example, the *in situ* preparation methods have progressed to where high-resolution solid-state NMR spectra can be acquired to characterize surface-associated proteins. In the case of antimicrobial peptides, this progression in NMR has furthered our understanding into the mechanism of action of this class of compounds.[183]

In the case of membrane proteins, both *E. coli* lipoprotein Lpp and *P. aeruginosa* Lipid A deacylase PagL represent examples where multiple 2D ^{13}C-detected correlation spectra (dipolar-assisted rotational resonance, DARR; and SPECIFIC-$^{15}N/^{13}C\alpha$)[184,185] were acquired, yielding spectra with astonishingly high resemblance to their liposome reconstituted counterparts.[186] Additional fine details could also be observed that cannot be studied using reconstituted preparations. In the case of Lpp, the observation of lysine Cβ/Cδ/Cγ resonances localizing to the C-terminus of the protein indicated a restricted motion of this usually flexible sidechain, providing evidence for cross-linking to the PG layer—a known function of this protein.

Following initial detailing of the application of high-resolution *in situ* NMR spectroscopy to membrane proteins, the classes of membrane proteins studied using such techniques have expanded rapidly. High-resolution data for multiple classes of membrane proteins, including ion channels[187,188] and insertases[189,190] have been reported. Over the course of these studies, the sample preparation strategies and NMR experiments have evolved.

To prepare *in situ* samples amenable for high-resolution MAS solid-state NMR, several requirements must be met: (1) the level of expression of the protein must be sufficiently high for signal detection; (2) the conformation occupied by the protein must be homogeneous and native; and (3) isotopic enrichment must be obtained predominantly for the protein of interest. Conditions (1) and (2) depend heavily on the protein identity and associated expression profile and are optimized on a case-by-case basis, a common challenge for any structural biology technique.[47] Condition (3), on the other hand, is a challenge specific to NMR and recent studies have focused on this problem. For example, rifampicin, an *E. coli* RNA polymerase inhibitor, can facilitate selective isotope incorporation into the protein of interest when using a T7-based expression system.[191–193] Approaches amenable to both bacterial and mammalian expression systems are also available relying on selective degradation of non-target proteins using the MazF protease.[188,194]

Development of ultrafast MAS (>111 kHz) for ^{1}H-detected NMR spectroscopy presents an exciting progression in the field, unlocking the potential for highly sensitive experiments which can detect long range ^{1}H–^{1}H contacts in native environments.[195] Work dedicated to the successful incorporation of ^{2}H nuclei for ^{1}H-detected experiments to be applied to these samples is also underway to achieve narrower linewidths,[196] for cases when ultrafast MAS is not possible.[197] When studying proteins that are refolded from inclusion bodies, the cells are cultured in ^2H_2O-based media with D-glucose as the carbon source and ^{2}H to ^{1}H exchange of the amide protons occurring during refolding. This approach is not amenable for *in situ* NMR, as the deuterated amide protons from any inaccessible transmembrane domains cannot be exchanged back to ^{1}H prior to data collection—preventing any signal from being observed. Methodologies like inverse fractional deuteration have been successfully applied to native preparations of the KcsA ion channel,[187] which allows for the observation of ^{1}H resonances corresponding to protons that are inaccessible to H/D exchange. With this labeling scheme, linewidths comparable to deuterated samples were attained for residues within the TM domain of KcsA.

An attractive aspect of studying proteins in their native environment is the ability to correlate the protein conformational interplay that occurs natively to the biological function that is often determined using *in vivo* assays. For example, the study of pathogenic bacteria relies heavily on cell-based assays to identify how specific virulence genes contribute to the invasive nature of these bacteria (*e.g.*, cell adhesion, cell invasion, serum resistance). Therefore, leveraging the ability of NMR spectroscopy to provide atomic-level details of complex multicomponent systems in functionally relevant environments, such as heterogeneous preparations of bacterial outer-membrane proteins, provides unique means to determine the mechanism of bacterial virulence. ^{1}H-detected (CP-HSQC)[198] and ^{13}C-detected (proton driven spin diffusion, PDSD and SPECIFIC-^{15}N/^{13}Cα)[199] correlation spectra of Ail expressed in the outer membrane of *E. coli* after rifampicin treatment reflect the specific incorporation of ^{13}C and ^{15}N nuclei into the protein of interest, however resonances attributed to non-protein components of the membrane (phospholipids and LPS) were also observed in the ^{13}C PDSD experiment.[77] 2D experiments enabled the transfer of resonance assignments performed on reconstituted liposome samples for both ^{13}C- and ^{1}H-detected data sets. Analysis of the Cα and Cβ chemical shifts of well-resolved peaks confirmed the retention of the β-stranded nature of Ail in the Gram-negative outer membrane. CSPs and enhanced relative signal intensities localizing to the outer leaflet facing residues of the transmembrane barrel, confirming the sensitivity of the protein to the asymmetry of the membrane environment (Figure 17.8). More interestingly, the response of Ail to the complex

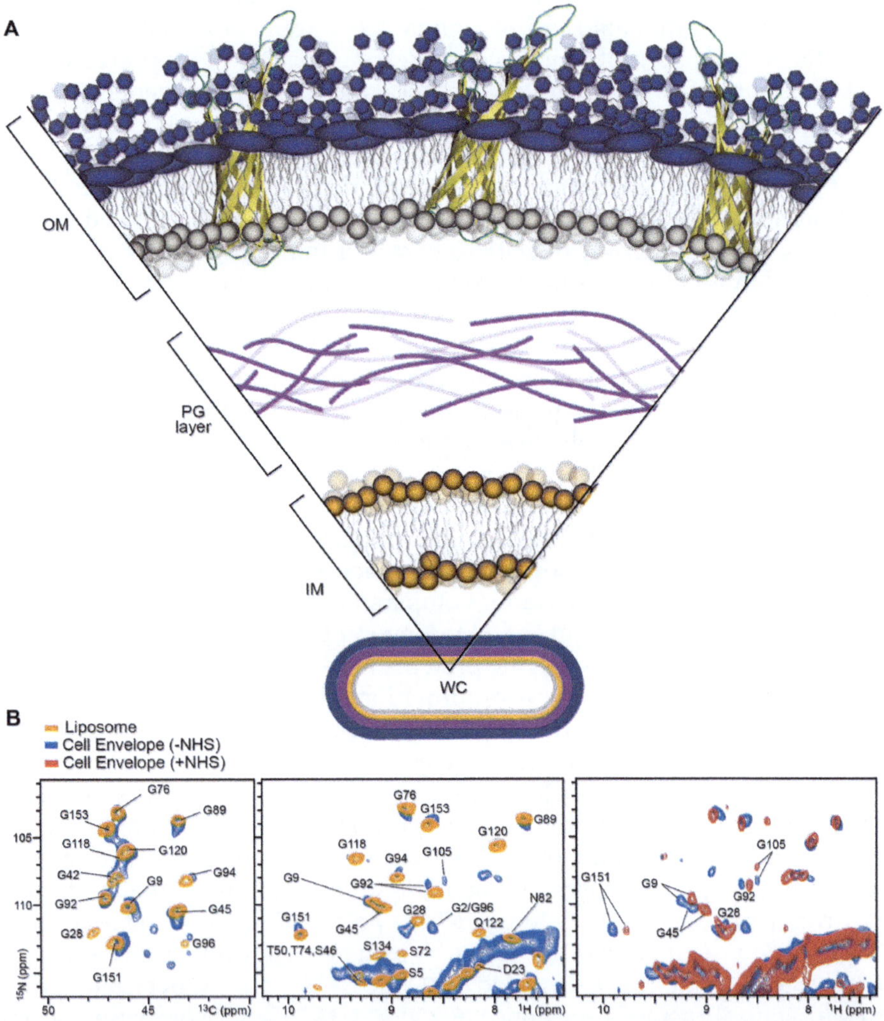

Figure 17.8 Solid-state NMR of Ail *in situ*. (A) Diagram of *E. coli* whole cell (WC).[77] The expanded cutout depicts the cell envelope, with Ail (yellow/green) embedded into the asymmetric outer membrane (OM), the peptidoglycan (PG) layer (magenta), and the inner membrane (IM). (B) MAS ^{1}H- and ^{13}C-detected solid-state NMR spectra acquired for liposomes (orange), whole cell envelopes (blue), and cell envelopes after incubation in normal human serum (NHS; red).

biological interface between *Y. pestis* and host blood was recapitulated inside of an MAS rotor. Incubation of *in situ* Ail samples with normal human serum yielded measurable CSPs throughout the protein (Figure 17.8). This work focusing on Ail demonstrates the power of NMR spectroscopy to provide insights into the alteration of protein dynamics localized to complex interfaces unlike other techniques.

The *in situ* samples described here all relied on the existence of reference data sets and prior NMR investigations using optimized reconstituted preparations of the protein of interest. One of the main reasons behind this is the difficulty associated with performing *de novo* assignments on these near-native preparations due to the relative insensitivity of these types of samples. As a result of this, the current state of the field limits the candidates which can be studied *in situ* to those already well characterized. Developments in the field of sensitivity-enhanced NMR spectroscopy using dynamic nuclear polarization (DNP) present a possible solution to this, as sensitivity enhancements on the order of 20–55-fold have been reported for membrane proteins,[200] with a theoretical limit of 658 (for ^{1}H).[201] This approach has the potential to significantly reduce the required level of protein expression, and potentially facilitate the study of transiently expressed proteins in cellular preparations. No examples of this approach being used to perform *de novo* assignments for *in situ* preparations have been published yet, but the authors await such strides and expect an increase in the utilization of these preparations for studying proteins with atomic resolution to occur shortly afterwards due to the fast turnaround time and lack of sample preparation required as well as the wealth of data within these samples.

17.3 Biominerals

Inorganic mineral phases play central roles in biology.[2,3] Through the process of biomineralization, living organisms convert ions to inorganic materials to engineer these hard mineral structures. This is perhaps best exemplified by calcium and phosphate ions, which mineralize to form hydroxyapatite (HAP), the crystalline constituent of bones, enamels, and other extracellular mineral deposits. HAP provides surfaces to recruit different molecules, and when mineralized into ordered structures, can provide physical barriers and binding surfaces to encapsulate, compartmentalize, and localize molecular and cellular materials.

Given the important roles associated with calcium ions, ranging from cellular signaling to bone development, many molecules have evolved to associate with this ion for regulation and homeostasis.[202] Proteins are no exception, and they typically mediate calcium ligation through negatively charged Asp and Glu residues. Carboxylate-rich matrix proteins, such as osteocalcin, osteopontin and collagen, mediate HAP crystal growth in bone and enamel and have been shown to bind ionic calcium and HAP through charged side chains.[203–205]

Interestingly, differences in structural properties of these proteins result in distinct mechanisms of biomineral nucleation and growth. For example, disordered dynamic structures at the mineral surface can block multiple crystal sites by steric hindrance.[206] Additionally, disordered proteins can even adopt new conformations upon interaction with biomineral surfaces.[207] Folded protein structures, on the other hand, can provide distinct binding sites, and recruit regulatory biomolecules.[208]

Studying these proteins in native conditions is critical for understanding their functions in health and disease. While X-ray crystallography and cryo-EM are not readily applicable to these systems,[209] solid-state NMR, alone or in combination with solution NMR, provides exciting avenues for the atomic-level characterization of these fundamental assemblies.

17.3.1 Protein–Metal Ion Interactions in Solution

Recent work on the mechanism of mineral ligation by the 459-residue human blood protein vitronectin (Vn), sheds light on its role in the progression of age-related macular degeneration (AMD).[210–213] Interest in the role of Vn in biomineralization is a classic tale of challenging novel functional insights gained from the structure of a protein. Structure determination and NMR characterization of its hemopexin (HX) domain (Vn-HX) identified a novel ion and biomineral binding motif.[214,215] Vn-HX encompasses most of the C-terminal 325-residue HX domain minus regions with predicted disorder, and forms a four-bladed β/α-propeller structure (Figure 17.9). Each blade contains three β-strands (β1–β3) and a short α-helix, and the four blades assemble to form a channel that coordinates a metal–chloride–metal ion triplet stabilized by the β1 backbone carbonyls and amides. NMR studies revealed that four Asp residues at the top of the channel come together to form a calcium-binding motif.

Leveraging the relatively small molecular size of Vn-HX (~23 kDa), the calcium ion interaction could be characterized by solution NMR.[215] Spectral analysis demonstrates that calcium ion binding occurs under slow-exchange NMR timescale,[171] as observed by the growth of a second set of cross-peaks with calcium titration, but binding resulted in negligible structural alterations. Rather, chemical shift perturbations were primarily associated with the residues located at the top of the channel and provided the first insight into how Vn can bind metal ions and HAP.[215]

Addition of calcium ions to identify biomineral binding sites was also utilized in NMR studies of other HAP binding proteins. For example, osteopontin is a 44-kDa intrinsically disordered protein that binds and inhibits HAP formation and growth.[216–218] It is highly negatively charged and can be phosphorylated to provide additional negative electric potential.[219] In this case, calcium binding is in the fast-exchange NMR time regime, and residue-specific calcium binding could be monitored by measuring the magnitude of chemical shift perturbation during titration.[206] This approach identified binding sites that correlate with HAP binding affinity. By contrast, solution

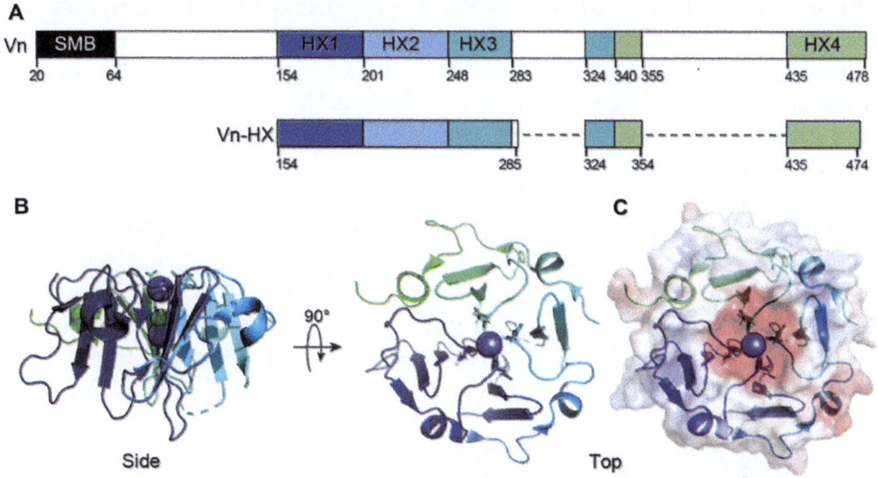

Figure 17.9 Structure and ion coordination of the Vn hemopexin (HX) domain (Vn-HX).[214] (A) Vn domain organization. Vn contains two structured domains: somatomedin B (SMB) and HX. The HX domain contains four repeating sequence motifs (HX1–4). Regions of predicted disorder are shown in white. (B) Orthogonal views of Vn-HX (PDB: 6O5E). The fold is a four-bladed β/α-propeller with two metal ions (blue spheres) and one chloride ion (green sphere) coordinated by backbone atoms in the propeller shaft. (C) Top view with surface colored by electrostatic potential. Four carboxylate groups from Asp residues (D162, C207, D255, D347) form a chiral arrangement that concentrates the negative charge at the top of the protein. Cartoon representations of the protein (A–C) are colored from N-terminus (blue) to C-terminus (green). Surface representation (C) is colored by electrostatic potential from -5 kT/e (red) to +5 kT/e (blue).

NMR studies of the 22-kDa HAP-binding protein amelogenin could not identify calcium-binding residues or provide insight into the mechanism of biomineral formation,[220] due to the protein's propensity to aggregate into nanospheres.[220–223]

17.3.2 Protein–Hydroxyapatite Interactions

Solid-state NMR is the ideal technique for studying on protein–HAP samples where the protein is effectively immobilized by its interactions with the mineral surface. This work is facilitated by methods that enable *in vitro* HAP formation and the commercial availability of HAP.[224–227] In the case of Vn-HX, sedimentation with crystalline HAP nanospheres yields very high-quality solid-state NMR spectra that can be interpreted to map HAP binding.[215] CP-mediated 2D NCA and HN correlation experiments can be compared directly with their solution NMR HNCA and HSQC counterparts to obtain a picture of both calcium ion and HAP binding.[228] Despite the broader line widths associated with solid-state NMR and the widely different physical states of the samples, the solution and solid-state spectra overlap significantly, demonstrating that Vn-HX maintains its overall structure in solution and upon sedimentation. Signals from four Ala residues near the binding site contribute significant insights,[215] since the distinct $^{15}N/^{13}C$ chemical shifts from these sites undergo major perturbations upon binding either ionic calcium or HAP (Figure 17.10).

In some cases, protein recruitment onto biomineral surfaces can result in substantial conformational changes and requires different NMR methods for characterization. For example, the 43-residue protein statherin interacts with HAP *via* a negatively charged N-terminal motif.[229,230] While statherin is largely disordered in solution, it adopts α-helical propensity when bound to HAP. Solid-state MAS NMR experiments, dipolar recoupling with a windowless sequence (DRAWS)[231] and REDOR,[145] were highly useful for determining torsion angles and ^{13}C–$^{15}N/^{19}F$ internuclear distances between residues that allowed structure determination.[229,232] The enamel-forming protein amelogenin is also disordered in solution but becomes structured when bound to HAP.[207,233,234] Solid-state NMR spectroscopy revealed that the N- and C-termini of amelogenin adopt a β-stranded fold when bound to HAP.[233,234]

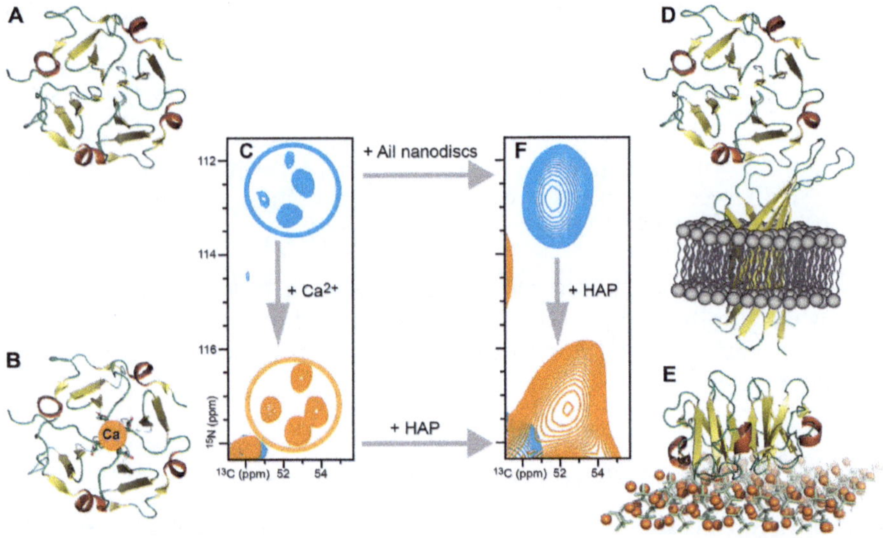

Figure 17.10 NMR characterization of the interaction of Vn-HX with calcium, HAP, and Ail nanodiscs.[215] Four $^{15}N/^{13}C\alpha$ resonances from Ala (A163, A208, A256, A348) are sensitive to calcium coordination. (A–C) Effect of calcium binding: calcium free protein (A); calcium-bound protein (B); selected region of solution NMR $^{1}H/^{15}N/^{13}C\alpha$ spectrum of Vn-HX in calcium-free (blue) or calcium-bound (orange) states (C). (D–F) Effect of cosedimentation with Ail nanodiscs or HAP: protein bound to Ail nanodiscs (D); protein bound to HAP surface (E); selected region of solid-state NMR $^{15}N/^{13}C\alpha$ spectrum of Vn-HX bound to Ail-nanodiscs without calcium (blue), or bound to HAP (orange) (F).

Statherin is a known inhibitor of mineral nucleation and growth in saliva while amelogenin is the dominant protein during enamel formation[209,235] and this functional knowledge guided NMR sample preparation. In the case of statherin, pre-formed HAP was utilized,[229,236] while amelogenin samples were prepared by supplementing HAP production conditions with amelogenin.[207,233] This will understandably result in different localization of the protein other than the biomineral surface but given the function of amelogenin to form enamels and statherin to inhibit mineral growth post-enamel formation, the two contrasting methods likely provide more physiologically and functionally relevant environments for the two proteins.

Although HAP-associated proteins usually require solid-state NMR for characterization, solution NMR may offer some insights. Osteopontin–HAP interactions, for example, have been studied in solution using samples of methylcellulose hydrogel[206] to obtain a viscous environment that maintains HAP in suspension, while allowing the protein to reorient freely. Using this approach, binding with HAP for osteopontin could be analyzed and compared to calcium ion-binding results.[206] Solution NMR Carr–Purcell–Meiboom–Gill (CPMG)[237,238] relaxation experiments and saturation transfer difference (STD)[239] experiments can also be useful for identifying HAP-bound protein.[206]

17.3.3 Moving Toward Ordered Biomineral Structures

Bone and enamel are highly ordered structures that vary significantly in size and shape. They are enriched in collagen and other proteins. In addition, biominerals associated with lipids are a hallmark of disorders like AMD, Alzheimer's disease, and atherosclerosis.[213] Like the biological membranes described in the previous sections, obtaining conditions that mimic the native environment or working *in situ* with native samples is possible and has been pursued for biominerals. Intact bone can be obtained for different mammalian models and mechanically or chemically modified to fit solid-state MAS NMR rotors for characterization.[240–245] The samples can retain their native complexity, including 3D structure, collagen protein matrix, different cell types, and complex extracellular matrix, while ultra-high-speed MAS with ^{1}H-detection and DNP have improved spectral resolution and signal intensity to better enable characterization of different molecules even at natural abundance.[246–249] Innovative approaches for sample preparation further enable studies with improved NMR signals, as exemplified by the "heavy mouse" model in which isotopically enriched *ex vivo* samples can be acquired by feeding the rodents a diet supplemented with ^{15}N- and ^{13}C-labeled amino acids.[250,251] In summary, the field of NMR has made many strides, making it significantly easier to characterize biomineralization in native conditions, and the probability of complete and comprehensive characterization continues to improve with each advancement.

17.4 Concluding Remarks

The essential chemistry which occurs at all biological interfaces and underpins life as we know it is complex and often difficult to study, particularly when considering the highly heterogeneous nature which defines these niches. The work reviewed in this chapter demonstrates that NMR spectroscopy is highly versatile and uniquely suited for structural studies of many types of biologically important systems. Exciting developments in NMR spectroscopy and sample preparation methods that include native samples are providing avenues to study ever more complicated biological phenomena with greater sensitivity and higher resolution than previously possible. Though the field of native NMR spectroscopy is relatively young, the progress being made toward studying native heterogeneous samples with atomic resolution presents a bright future for complete and comprehensive characterization of the dynamic molecular interplay that occurs on the complex biological interfaces.

Abbreviations

Ail	Attachment invasion locus
AMD	age-related macular degeneration
CP	cross-polarization
DARR	dipolar-assisted rotational resonance
DePC	*n*-decylphosphocholine
DHPC	dihexanoyl phosphatidylcholine
DNP	dynamic nuclear polarization
DMPC	1,2-dimyristoyl-*sn-glycero*-3-phosphocholine
DMPG	1,2-dimyristoyl-*sn-glycero*-3-phosphoryl-3′-rac-glycerol
DOPC	1,2-dioleoyl-*sn-glycero*-3-phosphocholine
DOPE	1,2-dioleoyl-*sn-glycero*-3-phosphoethanolamine
DOSY	diffusion-ordered spectroscopy
DRAWS	dipolar recoupling with a windowless sequence
ETM	transmembrane domain of envelope protein E
GPCR	G-protein coupled receptor
HAP	hydroxyapatite
H/D exchange	proton/deuterium exchange
HSQC	heteronuclear single quantum coherence
INEPT	insensitive nuclei enhanced by polarization transfer
KLA	Kdo_2-Lipid A
LPS	lipopolysaccharide
MAS	magic angle spinning
NMR	nuclear magnetic resonance
OS	oriented sample
PH	pleckstrin homology
PIP	phosphatidylinositol phosphate

PLEKHA7	pleckstrin homology domain-containing family A member 7
PISEMA	polarization inversion spin exchange at the magic angle
pSRII	phototaxis receptor sensory rhodopsin II
REDOR	rotation-echo double-resonance
RMSD	root-mean-square deviation
SAR	structure–activity relationship
SDS	sodium dodecyl sulfate
T_2	transverse relaxation time
Vn	vitronectin
VDAC-1	voltage-dependent anion channel-1
Vn-HX	vitronectin hemopexin domain construct

References

1. T. Harayama and H. Riezman, *Nat. Rev. Mol. Cell Biol.*, 2018, **19**, 281–296.
2. S. Weiner, *Rev. Mineral. Geochem.*, 2003, **54**, 1–29.
3. H. A. Lowenstam and S. Weiner, *On Biomineralization*, Oxford University Press, New York, 1989.
4. D. N. Langelaan and J. K. Rainey, *Biochem. Cell Biol.*, 2010, **88**, 203–210.
5. E. N. Frigini, E. E. Barrera, S. Pantano and R. D. Porasso, *Biochim. Biophys. Acta, Biomembr.*, 2020, **1862**, 183094.
6. B. Bozic, S. L. Das and S. Svetina, *Soft Matter*, 2015, **11**, 2479–2487.
7. K. M. Callahan, B. Mondou, L. Sasseville, J. L. Schwartz and N. D'Avanzo, *Channels*, 2019, **13**, 424–439.
8. K. Yoshida, S. Nagatoishi, D. Kuroda, N. Suzuki, T. Murata and K. Tsumoto, *Biochemistry*, 2019, **58**, 504–508.
9. S. Li, *Int. J. Biol. Sci.*, 2005, **1**, 152.
10. D. S. Goodsell and S. K. Burley, *Oncogene*, 2020, **39**, 6623–6632.
11. J. D. Westbrook and S. K. Burley, *Structure*, 2019, **27**, 211–217.
12. J. D. Westbrook, R. Soskind, B. P. Hudson and S. K. Burley, *Drug Discovery Today*, 2020, **25**, 837–850.
13. F. J. Schoen, H. Harasaki, K. M. Kim, H. C. Anderson and R. J. Levy, *J. Biomed. Mater. Res.*, 1988, **22**, 11–36.
14. F. L. Coe, J. H. Parks and J. R. Asplin, *N. Engl. J. Med.*, 1992, **327**, 1141–1152.
15. D. Losic, J. G. Mitchell and N. H. Voelcker, *Adv. Mater.*, 2009, **21**, 2947–2958.
16. H. X. Zhou and T. A. Cross, *Annu. Rev. Biophys.*, 2013, **42**, 361–392.
17. L. Banci, I. Bertini, C. Luchinat and M. Mori, *Prog. Nucl. Magn. Reson. Spectrosc.*, 2010, **56**, 247–266.
18. J. Keeler, *Understanding NMR Spectroscopy*, John Wiley and Sons, Chichester, 2010.
19. G. Griffiths, *Nat. Rev. Mol. Cell Biol.*, 2007, **8**, 1018–1024.
20. S. Lim, R. J. Quinton and N. J. Ganem, *Mol. Biol. Cell*, 2016, **27**, 3210–3213.

21. Y. Ravichandran, B. Goud and J. B. Manneville, *Curr. Opin. Cell Biol.*, 2020, **62**, 104–113.
22. J. B. Dacks and M. C. Field, *J. Cell Sci.*, 2007, **120**, 2977–2985.
23. B. I. Posner and S. A. Laporte, in *Neuroendocrinology: The Normal Neuroendocrine System*, 2010, DOI: 10.1016/s0079-6123(08)81001-1, pp. 1–16.
24. W. A. Catterall, *Trends Neurosci.*, 1993, **16**, 500–506.
25. M. J. McKay, F. Afrose, R. E. Koeppe, 2nd and D. V. Greathouse, *Biochim. Biophys. Acta, Biomembr.*, 2018, **1860**, 2108–2117.
26. W. D. Heo, T. Inoue, W. S. Park, M. L. Kim, B. O. Park, T. J. Wandless and T. Meyer, *Science*, 2006, **314**, 1458–1461.
27. H. Gerlach, V. Laumann, S. Martens, C. F. Becker, R. S. Goody and M. Geyer, *Nat. Chem. Biol.*, 2010, **6**, 46–53.
28. A. Nishikimi, H. Fukuhara, W. Su, T. Hongu, S. Takasuga, H. Mihara, Q. Cao, F. Sanematsu, M. Kanai, H. Hasegawa, Y. Tanaka, M. Shibasaki, Y. Kanaho, T. Sasaki, M. A. Frohman and Y. Fukui, *Science*, 2009, **324**, 384–387.
29. T. I. Strochlic, J. Viaud, U. E. Rennefahrt, T. Anastassiadis and J. R. Peterson, *Mol. Cell*, 2010, **40**, 493–500.
30. D. F. Sargent and R. Schwyzer, *Proc. Natl. Acad. Sci. U. S. A.*, 1986, **83**, 5774–5778.
31. G. Nawrocki, W. Im, Y. Sugita and M. Feig, *Proc. Natl. Acad. Sci. U. S. A.*, 2019, **116**, 24562–24567.
32. M. Bachar and O. M. Becker, *Biophys. J.*, 2000, **78**, 1359–1375.
33. C. C. Lee, Y. Sun and H. W. Huang, *Biophys. J.*, 2010, **98**, 2236–2245.
34. N. J. Hu, J. Bradshaw, H. Lauter, J. Buckingham, E. Solito and A. Hofmann, *Biophys. J.*, 2008, **94**, 1773–1781.
35. J. Coutant, P. A. Curmi, F. Toma and J. P. Monti, *Biochemistry*, 2007, **46**, 5656–5663.
36. T. Kalwarczyk, M. Tabaka and R. Holyst, *Bioinformatics*, 2012, **28**, 2971–2978.
37. M. S. Almen, K. J. Nordstrom, R. Fredriksson and H. B. Schioth, *BMC Biol.*, 2009, **7**, 50.
38. A. B. van Spriel, G. van den Bogaart and A. Cambi, *Front. Physiol.*, 2015, **6**, 172.
39. J. Birch, H. Cheruvara, N. Gamage, P. J. Harrison, R. Lithgo and A. Quigley, *Biology*, 2020 , 9, 401.
40. E. Eren, M. Murphy, J. Goguen and B. van den Berg, *Structure*, 2010, **18**, 809–818.
41. S. W. Cowan, T. Schirmer, G. Rummel, M. Steiert, R. Ghosh, R. A. Pauptit, J. N. Jansonius and J. P. Rosenbusch, *Nature*, 1992, **358**, 727–733.
42. J. Hu, L. J. Worrall, M. Vuckovic, C. Hong, W. Deng, C. E. Atkinson, B. Brett Finlay, Z. Yu and N. C. J. Strynadka, *Nat. Microbiol.*, 2019, **4**, 2010–2019.
43. J. Zhao, S. Benlekbir and J. L. Rubinstein, *Nature*, 2015, **521**, 241–245.

44. E. P. Carpenter, K. Beis, A. D. Cameron and S. Iwata, *Curr. Opin. Struct. Biol.*, 2008, **18**, 581–586.
45. D. Lyumkis, *J. Biol. Chem.*, 2019, **294**, 5181–5197.
46. B. Liang and L. K. Tamm, *Nat. Struct. Mol. Biol.*, 2016, **23**, 468–474.
47. A. Pandey, K. Shin, R. E. Patterson, X. Q. Liu and J. K. Rainey, *Biochem. Cell Biol.*, 2016, **94**, 507–527.
48. G. I. Danmaliki and P. M. Hwang, *Biochim. Biophys. Acta, Biomembr.*, 2020, **1862**, 183356.
49. S. Majeed, A. B. Ahmad, U. Sehar and E. R. Georgieva, *Membranes*, 2021, 11, 685.
50. G. D. Henry and B. D. Sykes, in *Methods in Enzymology*, Academic Press, 1994, vol. 239, pp. 515–535.
51. N. Das, D. T. Murray and T. A. Cross, *Nat. Protoc.*, 2013, **8**, 2256–2270.
52. L. A. Baker, G. E. Folkers, T. Sinnige, K. Houben, M. Kaplan, E. A. W. van der Cruijsen and M. Baldus, in *Methods in Enzymology*, ed. and A. K. Shukla, Academic Press, 2015, vol. 557, pp. 307–328.
53. S. C. Lee, T. J. Knowles, V. L. G. Postis, M. Jamshad, R. A. Parslow, Y.-P. Lin, A. Goldman, P. Sridhar, M. Overduin, S. P. Muench and T. R. Dafforn, *Nat. Protoc.*, 2016, **11**, 1149–1162.
54. F. Hagn, M. L. Nasr and G. Wagner, *Nat. Protoc.*, 2018, **13**, 79–98.
55. R. C. Oliver, J. Lipfert, D. A. Fox, R. H. Lo, S. Doniach and L. Columbus, *PLoS One*, 2013, **8**, e62488.
56. C. R. Sanders and F. Sonnichsen, *Magn. Reson. Chem.*, 2006, **44**(Spec No), S24–40.
57. J. T. Buboltz and G. W. Feigenson, *Langmuir*, 2005, **21**, 6296–6301.
58. H. Wennerström, *Phys. Rep.*, 1979, **52**, 1–86.
59. A. M. Seddon, P. Curnow and P. J. Booth, *Biochim. Biophys. Acta*, 2004, **1666**, 105–117.
60. B. Brutscher, I. C. Felli, S. Gil-Caballero, T. Hošek, R. Kümmerle, A. Piai, R. Pierattelli and Z. Sólyom, in *Intrinsically Disordered Proteins Studied by NMR Spectroscopy*, ed. I. C. Felli and R. Pierattelli, Springer International Publishing, Cham, 2015, pp. 49–122.
61. K. F. Morris and C. S. Johnson, *J. Am. Chem. Soc.*, 2002, **115**, 4291–4299.
62. D. H. Wu, A. D. Chen and C. S. Johnson, *J. Magn. Reson., Ser. A*, 1995, **115**, 260–264.
63. Z. Yang and C. G. Brouillette, *Methods Enzymol.*, 2016, **567**, 319–358.
64. D. N. Langelaan, E. M. Bebbington, T. Reddy and J. K. Rainey, *Biochemistry*, 2009, **48**, 537–548.
65. D. N. Langelaan and J. K. Rainey, *J. Phys. Chem. B*, 2009, **113**, 10465–10471.
66. K. Shin, M. Sarker, S. K. Huang and J. K. Rainey, *Sci. Rep.*, 2017, **7**, 15433.
67. S. K. Huang, K. Shin, M. Sarker and J. K. Rainey, *Biochim. Biophys. Acta, Biomembr.*, 2017, **1859**, 767–778.
68. G. S. Rule and T. K. Hitchens, in *Fundamentals of Protein NMR Spectroscopy*, Springer Netherlands, Dordrecht, 2006, ch. 19, pp. 431–474.

69. A. E. Aleshin, Y. Yao, A. Iftikhar, A. A. Bobkov, J. Yu, G. Cadwell, M. G. Klein, C. Dong, L. A. Bankston, R. C. Liddington, W. Im, G. Powis and F. M. Marassi, *Structure*, 2021, **29**(9), 1029–1039.e3.
70. W. Meng, Y. Mushika, T. Ichii and M. Takeichi, *Cell*, 2008, **135**, 948–959.
71. B. C. Choy, R. J. Cater, F. Mancia and E. E. Pryor, Jr., *Biochim. Biophys. Acta, Biomembr.*, 2021, **1863**, 183533.
72. P. M. Hwang and L. E. Kay, in *Methods in Enzymology*, Academic Press, 2005, vol. 394, pp. 335–350.
73. A. Yee, X. Chang, A. Pineda-Lucena, B. Wu, A. Semesi, B. Le, T. Ramelot, G. M. Lee, S. Bhattacharyya, P. Gutierrez, A. Denisov, C. H. Lee, J. R. Cort, G. Kozlov, J. Liao, G. Finak, L. Chen, D. Wishart, W. Lee, L. P. McIntosh, K. Gehring, M. A. Kennedy, A. M. Edwards and C. H. Arrowsmith, *Proc. Natl. Acad. Sci. U. S. A.*, 2002, **99**, 1825–1830.
74. Y. Ding, L. M. Fujimoto, Y. Yao and F. M. Marassi, *J. Biomol. NMR*, 2015, **61**, 275–286.
75. Y. Ding, L. M. Fujimoto, Y. Yao, G. V. Plano and F. M. Marassi, *Biochim. Biophys. Acta*, 2015, **1848**, 712–720.
76. S. K. Dutta, Y. Yao and F. M. Marassi, *J. Phys. Chem. B*, 2017, **121**, 7561–7570.
77. J. E. Kent, L. M. Fujimoto, K. Shin, C. Singh, Y. Yao, S. H. Park, S. J. Opella, G. V. Plano and F. M. Marassi, *Biophys. J.*, 2021, **120**, 453–462.
78. F. M. Marassi, Y. Ding, C. D. Schwieters, Y. Tian and Y. Yao, *J. Biomol. NMR*, 2015, **63**, 59–65.
79. S. H. Park, J. Wu, Y. Yao, C. Singh, Y. Tian, F. M. Marassi and S. J. Opella, *Biochim. Biophys. Acta, Biomembr.*, 2020, **1862**, 183333.
80. C. Singh, H. Lee, Y. Tian, S. Schesser Bartra, S. Hower, L. M. Fujimoto, Y. Yao, S. A. Ivanov, R. Z. Shaikhutdinova, A. P. Anisimov, G. V. Plano, W. Im and F. M. Marassi, *Mol. Microbiol.*, 2020, **114**, 510–520.
81. A. M. Kolodziejek, D. R. Schnider, H. N. Rohde, A. J. Wojtowicz, G. A. Bohach, S. A. Minnich and C. J. Hovde, *Infect. Immun.*, 2010, **78**, 5233–5243.
82. E. S. Krukonis and J. J. Thomson, *FEBS Lett.*, 2020, **594**, 2598–2620.
83. D. K. Ho, M. Skurnik, A. M. Blom and S. Meri, *Eur. J. Immunol.*, 2014, **44**, 742–751.
84. S. S. Bartra, Y. Ding, L. Miya Fujimoto, J. G. Ring, V. Jain, S. Ram, F. M. Marassi and G. V. Plano, *Microbiology*, 2015, **161**, 2174–2183.
85. J. J. Thomson, S. C. Plecha and E. S. Krukonis, *Mol. Microbiol.*, 2019, **111**, 82–95.
86. S. Yamashita, P. Lukacik, T. J. Barnard, N. Noinaj, S. Felek, T. M. Tsang, E. S. Krukonis, B. J. Hinnebusch and S. K. Buchanan, *Structure*, 2011, **19**, 1672–1682.
87. F. C. Almeida and S. J. Opella, *J. Mol. Biol.*, 1997, **270**, 481–495.
88. P. Teriete, C. M. Franzin, J. Choi and F. M. Marassi, *Biochemistry*, 2007, **46**, 6774–6783.

89. Q. Li, Y. L. Wong and C. Kang, *Biochim. Biophys. Acta*, 2014, **1838**, 1313–1321.
90. A. Gautier, H. R. Mott, M. J. Bostock, J. P. Kirkpatrick and D. Nietlispach, *Nat. Struct. Mol. Biol.*, 2010, **17**, 768–774.
91. S. C. Howell, M. F. Mesleh and S. J. Opella, *Biochemistry*, 2005, **44**, 5196–5206.
92. D. D. Mowrey, T. Cui, Y. Jia, D. Ma, A. M. Makhov, P. Zhang, P. Tang and Y. Xu, *Structure*, 2013, **21**, 1897–1904.
93. M. Bayrhuber, I. Maslennikov, W. Kwiatkowski, A. Sobol, C. Wierschem, C. Eichmann, L. Frey and R. Riek, *Biochemistry*, 2019, **58**, 4017–4027.
94. A. Arora, F. Abildgaard, J. H. Bushweller and L. K. Tamm, *Nat. Struct. Biol.*, 2001, **8**, 334–338.
95. C. Fernandez, K. Adeishvili and K. Wuthrich, *Proc. Natl. Acad. Sci. U. S. A.*, 2001, **98**, 2358–2363.
96. B. Liang and L. K. Tamm, *Proc. Natl. Acad. Sci. U. S. A.*, 2007, **104**, 16140–16145.
97. S. Hiller, R. G. Garces, T. J. Malia, V. Y. Orekhov, M. Colombini and G. Wagner, *Science*, 2008, **321**, 1206–1210.
98. Y. Tian, C. D. Schwieters, S. J. Opella and F. M. Marassi, *J. Biomol. NMR*, 2017, **67**, 35–49.
99. Y. Tian, C. D. Schwieters, S. J. Opella and F. M. Marassi, *Biophys. J.*, 2015, **109**, 574–585.
100. Y. Tian, C. D. Schwieters, S. J. Opella and F. M. Marassi, *J. Magn. Reson.*, 2014, **243**, 54–64.
101. G. Gao, J. G. Williams and S. L. Campbell, in *Protein-Protein Interactions*, ed. H. Fu, Methods in Molecular Biology, Humana Press, 2004, vol. 261.
102. S. Maity, R. K. Gundampati and T. K. Suresh Kumar, *Nat. Prod. Commun.*, 2019, **14**, 1934578X1984929.
103. T. L. Lin, S. H. Chen, N. E. Gabriel and M. F. Roberts, *J. Am. Chem. Soc.*, 2002, **108**, 3499–3507.
104. A. Tonnesen, S. M. Christensen, V. Tkach and D. Stamou, *Biophys. J.*, 2014, **106**, 201–209.
105. T. A. Harroun, M. Koslowsky, M. P. Nieh, C. F. de Lannoy, V. A. Raghunathan and J. Katsaras, *Langmuir*, 2005, **21**, 5356–5361.
106. T. C. Edrington, E. Kintz, J. B. Goldberg and L. K. Tamm, *J. Biol. Chem.*, 2011, **286**, 39211–39223.
107. Y. Kamio and H. Nikaido, *Biochemistry*, 1976, **15**, 2561–2570.
108. J. C. Whisstock and A. M. Lesk, *Q. Rev. Biophys.*, 2003, **36**, 307–340.
109. V. Nele, M. N. Holme, U. Kauscher, M. R. Thomas, J. J. Doutch and M. M. Stevens, *Langmuir*, 2019, **35**, 6064–6074.
110. A. A. Jovanović, B. D. Balanč, A. Ota, P. Ahlin Grabnar, V. B. Djordjević, K. P. Šavikin, B. M. Bugarski, V. A. Nedović and N. Poklar Ulrih, *Eur. J. Lipid Sci. Technol.*, 2018, **120**, 1800039.

111. A. Verchère, I. Broutin and M. Picard, in *Membrane Protein Structure and Function Characterization: Methods and Protocols*, ed. J.-J. Lacapere, Springer, New York, New York, NY, 2017, pp. 259–282.
112. S. K. Hansen, K. Bertelsen, B. Paaske, N. C. Nielsen and T. Vosegaard, *Prog. Nucl. Magn. Reson. Spectrosc.*, 2015, **88–89**, 48–85.
113. F. M. Marassi, A. Ramamoorthy and S. J. Opella, *Proc. Natl. Acad. Sci. U. S. A.*, 1997, **94**, 8551-8556.
114. F. M. Marassi and S. J. Opella, *Protein Sci.*, 2003, **12**, 403–411.
115. T. A. Cross, V. Ekanayake, J. Paulino and A. Wright, *J. Magn. Reson.*, 2014, **239**, 100–109.
116. T. A. Cross and S. J. Opella, *Curr. Opin. Struct. Biol.*, 1994, **4**, 574–581.
117. F. M. Marassi, C. Ma, H. Gratkowski, S. K. Straus, K. Strebel, M. Oblatt-Montal, M. Montal and S. J. Opella, *Proc. Natl. Acad. Sci.*, 1999, **96**, 14336.
118. A. L. Stouffer, R. Acharya, D. Salom, A. S. Levine, L. Di Costanzo, C. S. Soto, V. Tereshko, V. Nanda, S. Stayrook and W. F. DeGrado, *Nature*, 2008, **451**, 596–599.
119. J. R. Schnell and J. J. Chou, *Nature*, 2008, **451**, 591–595.
120. M. Sharma, M. Yi, H. Dong, H. Qin, E. Peterson, D. D. Busath, H. X. Zhou and T. A. Cross, *Science*, 2010, **330**, 509–512.
121. D. Hilger, M. Masureel and B. K. Kobilka, *Nat. Struct. Mol. Biol.*, 2018, **25**, 4–12.
122. M. R. de Planque and J. A. Killian, *Mol. Membr. Biol.*, 2003, **20**, 271–284.
123. S. Y. Liao, M. Lee and M. Hong, *J. Struct. Biol.*, 2019, **206**, 20–28.
124. S. H. Park, B. B. Das, F. Casagrande, Y. Tian, H. J. Nothnagel, M. Chu, H. Kiefer, K. Maier, A. A. De Angelis, F. M. Marassi and S. J. Opella, *Nature*, 2012, **491**, 779–783.
125. R. R. Ketchem, W. Hu and T. A. Cross, *Science*, 1993, **261**, 1457–1460.
126. J. Wang, J. Denny, C. Tian, S. Kim, Y. Mo, F. Kovacs, Z. Song, K. Nishimura, Z. Gan, R. Fu, J. R. Quine and T. A. Cross, *J. Magn. Reson.*, 2000, **144**, 162–167.
127. F. M. Marassi and S. J. Opella, *J. Magn. Reson.*, 2000, **144**, 150–155.
128. F. M. Marassi, *Biophys. J.*, 2001, **80**, 994–1003.
129. I. J. Lowe, *Phys. Rev. Lett.*, 1959, **2**, 285–287.
130. E. R. Andrew, A. Bradbury and R. G. Eades, *Nature*, 1958, **182**, 1659.
131. H. Kessemeier and R. E. Norberg, *Phys. Rev.*, 1967, **155**, 321–337.
132. A. McDermott, *Annu. Rev. Biophys.*, 2009, **38**, 385–403.
133. P. J. Judge and A. Watts, *Curr. Opin. Chem. Biol.*, 2011, **15**, 690–695.
134. W. T. Franks, A. H. Linden, B. Kunert, B. J. van Rossum and H. Oschkinat, *Eur. J. Cell Biol.*, 2012, **91**, 340–348.
135. M. Hong, Y. Zhang and F. Hu, *Annu. Rev. Phys. Chem.*, 2012, **63**, 1–24.
136. B. Reif, *J. Magn. Reson.*, 2012, **216**, 1–12.
137. L. Shi and V. Ladizhansky, *Methods Mol. Biol.*, 2012, **895**, 153–165.
138. J. Goncalves, M. Eilers, K. South, C. A. Opefi, P. Laissue, P. J. Reeves and S. O. Smith, *Methods Enzymol.*, 2013, **522**, 365–389.
139. R. G. Griffin, *Nat. Struct. Biol.*, 1998, **5**, 508–512.

140. R. Tycko, *Annu. Rev. Phys. Chem.*, 2011, **62**, 279–299.
141. B. B. Das, H. J. Nothnagel, G. J. Lu, W. S. Son, Y. Tian, F. M. Marassi and S. J. Opella, *J. Am. Chem. Soc.*, 2012, **134**, 2047–2056.
142. F. Tian, Z. Song and T. A. Cross, *J. Magn. Reson.*, 1998, **135**, 227–231.
143. F. M. Marassi, B. B. Das, G. J. Lu, H. J. Nothnagel, S. H. Park, W. S. Son, Y. Tian and S. J. Opella, *Methods*, 2011, **55**, 363–369.
144. V. S. Mandala, M. J. McKay, A. A. Shcherbakov, A. J. Dregni, A. Kolocouris and M. Hong, *Nat. Struct. Mol. Biol.*, 2020, **27**, 1202–1208.
145. T. Gullion and J. Schaefer, in *Advances in Magnetic and Optical Resonance*, ed. W. S. Warren, Academic Press, 1989, vol. 13, pp. 57–83.
146. A. Lange, S. Luca and M. Baldus, *J. Am. Chem. Soc.*, 2002, **124**, 9704–9705.
147. A. A. Shcherbakov and M. Hong, *J. Biomol. NMR*, 2018, **71**, 31–43.
148. W. Surya, Y. Li and J. Torres, *Biochim. Biophys. Acta, Biomembr.*, 2018, **1860**, 1309–1317.
149. S. H. Park, H. Siddiqi, D. V. Castro, A. A. De Angelis, A. L. Oom, C. A. Stoneham, M. K. Lewinski, A. E. Clark, B. A. Croker, A. F. Carlin, J. Guatelli and S. J. Opella, *PLoS Pathog.*, 2021, **17**, e1009519.
150. M. S. Bretscher, *Nat. New Biol.*, 1972, **236**, 11–12.
151. J. P. Amoureux and M. Pruski, *Mol. Phys.*, 2002, **100**, 1595–1613.
152. T. H. Bayburt, Y. V. Grinkova and S. G. Sligar, *Nano Lett.*, 2002, **2**, 853–856.
153. S. Scheidelaar, M. C. Koorengevel, J. D. Pardo, J. D. Meeldijk, E. Breukink and J. A. Killian, *Biophys. J.*, 2015, **108**, 279–290.
154. T. Raschle, S. Hiller, T. Y. Yu, A. J. Rice, T. Walz and G. Wagner, *J. Am. Chem. Soc.*, 2009, **131**, 17777–17779.
155. A. Nath, Y. V. Grinkova, S. G. Sligar and W. M. Atkins, *J. Biol. Chem.*, 2007, **282**, 28309–28320.
156. I. G. Denisov, Y. V. Grinkova, A. A. Lazarides and S. G. Sligar, *J. Am. Chem. Soc.*, 2004, **126**, 3477–3487.
157. A. Z. Kijac, Y. Li, S. G. Sligar and C. M. Rienstra, *Biochemistry*, 2007, **46**, 13696–13703.
158. P. Ryzhov, Y. Tian, Y. Yao, A. A. Bobkov, W. Im and F. M. Marassi, *Biophys. J.*, 2020, **119**, 1324–1334.
159. J. M. Gluck, B. W. Koenig and D. Willbold, *Anal. Biochem.*, 2011, **408**, 46–52.
160. M. Etzkorn, T. Raschle, F. Hagn, V. Gelev, A. J. Rice, T. Walz and G. Wagner, *Structure*, 2013, **21**, 394–401.
161. F. Hagn and G. Wagner, *J. Biomol. NMR*, 2015, **61**, 249–260.
162. T. Schubeis, T. Le Marchand, C. Daday, W. Kopec, K. Tekwani Movellan, J. Stanek, T. S. Schwarzer, K. Castiglione, B. L. de Groot, G. Pintacuda and L. B. Andreas, *Proc. Natl. Acad. Sci. U. S. A.*, 2020, **117**, 21014–21021.
163. R. Puthenveetil, K. Nguyen and O. Vinogradova, *Nanotechnol. Rev.*, 2017, **6**, 111–126.
164. M. Sattler and S. W. Fesik, *Structure*, 1996, **4**, 1245–1249.

165. K. Pervushin, R. Riek, G. Wider and K. Wüthrich, *Proc. Natl. Acad. Sci. U. S. A.*, 1997, **94**, 12366–12371.
166. V. Tugarinov and L. E. Kay, *J. Mol. Biol.*, 2003, **327**, 1121–1133.
167. J. E. Ollerenshaw, V. Tugarinov and L. E. Kay, *Magn. Reson. Chem.*, 2003, **41**, 843–852.
168. L. Kooijman, P. Ansorge, M. Schuster, C. Baumann, F. Lohr, S. Jurt, P. Guntert and O. Zerbe, *J. Biomol. NMR*, 2020, **74**, 45–60.
169. L. Ye, N. Van Eps, M. Zimmer, O. P. Ernst and R. S. Prosser, *Nature*, 2016, **533**, 265–268.
170. L. Ye, C. Neale, A. Sljoka, B. Lyda, D. Pichugin, N. Tsuchimura, S. T. Larda, R. Pomes, A. E. Garcia, O. P. Ernst, R. K. Sunahara and R. S. Prosser, *Nat. Commun.*, 2018, **9**, 1372.
171. M. P. Williamson, *Prog. Nucl. Magn. Reson. Spectrosc.*, 2013, **73**, 1–16.
172. S. Bibow, R. Bohm, S. M. Modaresi and S. Hiller, *Anal. Chem.*, 2020, **92**, 7786–7793.
173. M. G. Rossmann and P. Argos, *Annu. Rev. Biochem.*, 1981, **50**, 497–532.
174. V. Yeh, A. Goode and B. B. Bonev, *Biology*, 2020, **9**, 396.
175. A. Gautier, *Biochim. Biophys. Acta*, 2014, **1837**, 578–588.
176. N. A. Lakomek, L. Frey, S. Bibow, A. Bockmann, R. Riek and B. H. Meier, *J. Phys. Chem. B*, 2017, **121**, 7671–7680.
177. I. Bertini, C. Luchinat, G. Parigi, E. Ravera, B. Reif and P. Turano, *Proc. Natl. Acad. Sci. U. S. A.*, 2011, **108**, 10396–10399.
178. J. M. Dorr, S. Scheidelaar, M. C. Koorengevel, J. J. Dominguez, M. Schafer, C. A. van Walree and J. A. Killian, *Eur. Biophys. J.*, 2016, **45**, 3–21.
179. J. Radoicic, S. H. Park and S. J. Opella, *Biophys. J.*, 2018, **115**, 22–25.
180. M. C. Fiori, W. Zheng, E. Kamilar, G. Simiyu, G. A. Altenberg and H. Liang, *Sci. Rep.*, 2020, **10**, 9940.
181. U. Coskun and K. Simons, *Structure*, 2011, **19**, 1543–1548.
182. G. S. Jacob, J. Schaefer and G. E. Wilson, Jr., *J. Biol. Chem.*, 1983, **258**, 10824–10826.
183. J. Medeiros-Silva, S. Jekhmane, A. L. Paioni, K. Gawarecka, M. Baldus, E. Swiezewska, E. Breukink and M. Weingarth, *Nat. Commun.*, 2018, **9**, 3963.
184. K. Takegoshi, S. Nakamura and T. Terao, *Chem. Phys. Lett.*, 2001, **344**, 631–637.
185. M. Baldus, A. T. Petkova, J. Herzfeld and R. G. Griffin, *Mol. Phys.*, 1998, **95**, 1197–1207.
186. M. Renault, R. Tommassen-van Boxtel, M. P. Bos, J. A. Post, J. Tommassen and M. Baldus, *Proc. Natl. Acad. Sci. U. S. A.*, 2012, **109**, 4863–4868.
187. J. Medeiros-Silva, D. Mance, M. Daniels, S. Jekhmane, K. Houben, M. Baldus and M. Weingarth, *Angew. Chem., Int. Ed. Engl.*, 2016, **55**, 13606–13610.
188. L. Mao, K. Inoue, Y. Tao, G. T. Montelione, A. E. McDermott and M. Inouye, *J. Biomol. NMR*, 2011, **49**, 131–137.

189. C. Pinto, D. Mance, M. Julien, M. Daniels, M. Weingarth and M. Baldus, *J. Struct. Biol.*, 2019, **206**, 1–11.
190. L. A. Baker, T. Sinnige, P. Schellenberger, J. de Keyzer, C. A. Siebert, A. J. M. Driessen, M. Baldus and K. Grunewald, *Structure*, 2018, **26**, 161–170e163.
191. G. Hartmann, K. O. Honikel, F. Knusel and J. Nuesch, *Biochim. Biophys. Acta*, 1967, **145**, 843–844.
192. L. A. Baker, M. Daniels, E. A. van der Cruijsen, G. E. Folkers and M. Baldus, *J. Biomol. NMR*, 2015, **62**, 199–208.
193. F. C. Almeida, G. C. Amorim, V. H. Moreau, V. O. Sousa, A. T. Creazola, T. A. Americo, A. P. Pais, A. Leite, L. E. Netto, R. J. Giordano and A. P. Valente, *J. Magn. Reson.*, 2001, **148**, 142–146.
194. L. Mao, S. T. Vaiphei, T. Shimazu, W. M. Schneider, Y. Tang, R. Mani, M. J. Roth, G. T. Montelione and M. Inouye, *J. Struct. Funct. Genomics*, 2010, **11**, 81–84.
195. T. Schubeis, T. Le Marchand, L. B. Andreas and G. Pintacuda, *J. Magn. Reson.*, 2018, **287**, 140–152.
196. D. Cala-De Paepe, J. Stanek, K. Jaudzems, K. Tars, L. B. Andreas and G. Pintacuda, *Solid State Nucl. Magn. Reson.*, 2017, **87**, 126–136.
197. L. B. Andreas, K. Jaudzems, J. Stanek, D. Lalli, A. Bertarello, T. Le Marchand, D. Cala-De Paepe, S. Kotelovica, I. Akopjana, B. Knott, S. Wegner, F. Engelke, A. Lesage, L. Emsley, K. Tars, T. Herrmann and G. Pintacuda, *Proc. Natl. Acad. Sci. U. S. A.*, 2016, **113**, 9187–9192.
198. E. Barbet-Massin, A. J. Pell, J. S. Retel, L. B. Andreas, K. Jaudzems, W. T. Franks, A. J. Nieuwkoop, M. Hiller, V. Higman, P. Guerry, A. Bertarello, M. J. Knight, M. Felletti, T. Le Marchand, S. Kotelovica, I. Akopjana, K. Tars, M. Stoppini, V. Bellotti, M. Bolognesi, S. Ricagno, J. J. Chou, R. G. Griffin, H. Oschkinat, A. Lesage, L. Emsley, T. Herrmann and G. Pintacuda, *J. Am. Chem. Soc.*, 2014, **136**, 12489–12497.
199. N. Bloembergen, *Physica*, 1949, **15**, 386–426.
200. K. M. Visscher, J. Medeiros-Silva, D. Mance, J. Rodrigues, M. Daniels, A. Bonvin, M. Baldus and M. Weingarth, *Angew. Chem., Int. Ed. Engl.*, 2017, **56**, 13222–13227.
201. G. Pinter and H. Schwalbe, *Angew. Chem., Int. Ed. Engl.*, 2017, **56**, 8332–8334.
202. E. Carafoli and J. Krebs, *J. Biol. Chem.*, 2016, **291**, 20849–20857.
203. A. L. Boskey and R. Roy, *Chem. Rev.*, 2008, **108**, 4716–4733.
204. Q. Q. Hoang, F. Sicheri, A. J. Howard and D. S. Yang, *Nature*, 2003, **425**, 977–980.
205. L. A. Scudeller, S. Srinivasan, A. M. Rossi, P. S. Stayton, G. P. Drobny and D. G. Castner, *Biointerphases*, 2017, **12**, 02D411.
206. J. Holzinger, H. Kotisch, K. W. Richter and R. Konrat, *ChemBioChem*, 2021, **22**, 2300–2305.
207. R. J. Arachchige, S. D. Burton, J. X. Lu, B. Ginovska, L. K. Harding, M. E. Taylor, J. Tao, A. Dohnalkova, B. J. Tarasevich, G. W. Buchko and W. J. Shaw, *Biophys. J.*, 2018, **115**, 1666–1672.

208. J. R. Long, W. J. Shaw, P. S. Stayton and G. P. Drobny, *Biochemistry*, 2001, **40**, 15451–15455.
209. W. J. Shaw, B. J. Tarasevich, G. W. Buchko, R. M. J. Arachchige and S. D. Burton, *J. Struct. Biol.*, 2020, **212**, 107630.
210. R. F. Mullins, S. R. Russell, D. H. Anderson and G. S. Hageman, *FASEB J.*, 2000, **14**, 835–846.
211. T. M. Shin, J. M. Isas, C. L. Hsieh, R. Kayed, C. G. Glabe, R. Langen and J. Chen, *Mol. Neurodegener.*, 2008, **3**, 16.
212. G. S. Hageman, R. F. Mullins, S. R. Russell, L. V. Johnson and D. H. Anderson, *FASEB J.*, 1999, **13**, 477–484.
213. R. B. Thompson, V. Reffatto, J. G. Bundy, E. Kortvely, J. M. Flinn, A. Lanzirotti, E. A. Jones, D. S. McPhail, S. Fearn, K. Boldt, M. Ueffing, S. G. Ratu, L. Pauleikhoff, A. C. Bird and I. Lengyel, *Proc. Natl. Acad. Sci. U. S. A.*, 2015, **112**, 1565–1570.
214. K. Shin, B. C. Lechtenberg, L. M. Fujimoto, Y. Yao, S. S. Bartra, G. V. Plano and F. M. Marassi, *Sci. Adv.*, 2019, **5**, eaax5068.
215. K. Shin, J. E. Kent, C. Singh, L. M. Fujimoto, J. Yu, Y. Tian, W. Im and F. M. Marassi, *Proc. Natl. Acad. Sci. U. S. A.*, 2020, **117**, 18504–18510.
216. F. P. Reinholt, K. Hultenby, A. Oldberg and D. Heinegard, *Proc. Natl. Acad. Sci. U. S. A.*, 1990, **87**, 4473–4475.
217. S. A. Steitz, M. Y. Speer, M. D. McKee, L. Liaw, M. Almeida, H. Yang and C. M. Giachelli, *Am. J. Pathol.*, 2002, **161**, 2035–2046.
218. H. A. Goldberg and G. K. Hunter, *Ann. N. Y. Acad. Sci.*, 1995, **760**, 305–308.
219. B. Christensen, M. S. Nielsen, K. F. Haselmann, T. E. Petersen and E. S. Sorensen, *Biochem. J.*, 2005, **390**, 285–292.
220. G. W. Buchko, B. J. Tarasevich, J. Bekhazi, M. L. Snead and W. J. Shaw, *Biochemistry*, 2008, **47**, 13215–13222.
221. J. Moradian-Oldak, J. P. Simmer, E. C. Lau, P. E. Sarte, H. C. Slavkin and A. G. Fincham, *Biopolymers*, 1994, **34**, 1339–1347.
222. J. Moradian-Oldak, W. Leung and A. G. Fincham, *J. Struct. Biol.*, 1998, **122**, 320–327.
223. F. B. Wiedemann-Bidlack, E. Beniash, Y. Yamakoshi, J. P. Simmer and H. C. Margolis, *J. Struct. Biol.*, 2007, **160**, 57–69.
224. A. L. Boskey and A. S. Posner, *J. Phys. Chem.*, 2002, **80**, 40–45.
225. G. K. Hunter and H. A. Goldberg, *Proc. Natl. Acad. Sci. U. S. A.*, 1993, **90**, 8562–8565.
226. M. Tavafoghi Jahromi and M. Cerruti, *Cryst. Growth Des.*, 2015, **15**, 1096–1104.
227. L. Muller and F. A. Muller, *Acta Biomater.*, 2006, **2**, 181–189.
228. J. Cavanagh, *Protein NMR spectroscopy*, Academic Press, Amsterdam, 2008.
229. G. Goobes, R. Goobes, O. Schueler-Furman, D. Baker, P. S. Stayton and G. P. Drobny, *Proc. Natl. Acad. Sci. U. S. A.*, 2006, **103**, 16083–16088.
230. V. Raghunathan, J. M. Gibson, G. Goobes, J. M. Popham, E. A. Louie, P. S. Stayton and G. P. Drobny, *J. Phys. Chem. B*, 2006, **110**, 9324–9332.

231. D. M. Gregory, D. J. Mitchell, J. A. Stringer, S. Kiihne, J. C. Shiels, J. Callahan, M. A. Mehta and G. P. Drobny, *Chem. Phys. Lett.*, 1995, **246**, 654–663.
232. W. J. Shaw, J. R. Long, J. L. Dindot, A. A. Campbell, P. S. Stayton and G. P. Drobny, *J. Am. Chem. Soc.*, 2000, **122**, 1709–1716.
233. J. X. Lu, Y. S. Xu, G. W. Buchko and W. J. Shaw, *J. Dent. Res.*, 2013, **92**, 1000–1004.
234. J. Zhang, J. Wang, C. Ma and J. Lu, *Int. J. Mol. Sci.*, 2020, **21**, 2946.
235. A. Roehrich and G. Drobny, *Acc. Chem. Res.*, 2013, **46**, 2136–2144.
236. R. Goobes, G. Goobes, W. J. Shaw, G. P. Drobny, C. T. Campbell and P. S. Stayton, *Biochemistry*, 2007, **46**, 4725–4733.
237. H. Y. Carr and E. M. Purcell, *Phys. Rev.*, 1954, **94**, 630–638.
238. S. Meiboom and D. Gill, *Rev. Sci. Instrum.*, 1958, **29**, 688–691.
239. M. Mayer and B. Meyer, *Angew. Chem., Int. Ed.*, 1999, **38**, 1784–1788.
240. O. Nikel, D. Laurencin, S. A. McCallum, C. M. Gundberg and D. Vashishth, *Langmuir*, 2013, **29**, 13873–13882.
241. R. K. Rai and N. Sinha, *J. Phys. Chem. C*, 2011, **115**, 14219–14227.
242. C. Singh, R. K. Rai, A. M. Kayastha and N. Sinha, *Magn. Reson. Chem.*, 2016, **54**, 132–135.
243. C. Singh and N. Sinha, *J. Phys. Chem. C*, 2016, **120**, 9393–9398.
244. N. Tiwari, R. Rai and N. Sinha, *Solid State Nucl. Magn. Reson.*, 2020, **107**, 101666.
245. N. Tiwari, S. Wi, F. Mentink-Vigier and N. Sinha, *J. Phys. Chem. B*, 2021, **125**, 4757–4766.
246. L. B. Andreas, T. Le Marchand, K. Jaudzems and G. Pintacuda, *J. Magn. Reson.*, 2015, **253**, 36–49.
247. T. Azais, S. Von Euw, W. Ajili, S. Auzoux-Bordenave, P. Bertani, D. Gajan, L. Emsley, N. Nassif and A. Lesage, *Solid State Nucl. Magn. Reson.*, 2019, **102**, 2–11.
248. D. Lee, C. Leroy, C. Crevant, L. Bonhomme-Coury, F. Babonneau, D. Laurencin, C. Bonhomme and G. De Paepe, *Nat. Commun.*, 2017, **8**, 14104.
249. C. Singh, R. K. Rai, F. Aussenac and N. Sinha, *J. Phys. Chem. Lett.*, 2014, **5**, 4044–4048.
250. W. Y. Chow, R. Rajan, K. H. Muller, D. G. Reid, J. N. Skepper, W. C. Wong, R. A. Brooks, M. Green, D. Bihan, R. W. Farndale, D. A. Slatter, C. M. Shanahan and M. J. Duer, *Science*, 2014, **344**, 742–746.
251. V. W. Wong, D. G. Reid, W. Y. Chow, R. Rajan, M. Green, R. A. Brooks and M. J. Duer, *J. Biomol. NMR*, 2015, **63**, 119–123.

CHAPTER 18

Solid-state NMR Shows That the Structure and Dynamics of Specific Residues in the Membrane Receptor CXCR1 Are Altered by Interactions with Specific Residues in Its Agonist IL-8

R. K. RAI,[a] A. A. DE ANGELIS,[b] S. H. PARK,[b] S. BERKAMP[c] AND S. J. OPELLA*[b]

[a] Indiana University School of Medicine, Indianapolis, IN 46202-3082, USA; [b] Department of Chemistry and Biochemistry, University of California, San Diego, La Jolla, CA 92093, USA; [c] Ernst Ruska-Centre for Microscopy and Spectroscopy with Electrons, ER-C-3, 52425 Jülich, Germany
*Email: sopella@ucsd.edu

18.1 Introduction

18.1.1 Structure, Dynamics, and Interactions of Membrane Proteins in Lipid Bilayers

Living systems, even when examined in their simplest forms, require a great deal of information to specify the structures and functions of constituent

New Developments in NMR No. 26
NMR Spectroscopy for Probing Functional Dynamics at Biological Interfaces
Edited by Anirban Bhunia, Hanudatta S. Atreya and Neeraj Sinha

Published by the Royal Society of Chemistry, www.rsc.org

macromolecules and assemblies, such as the proteins associated with the membrane bilayers of cells and enveloped virus particles. The requisite information is stored in long, linear macromolecules such as DNA and RNA, as monomers of four nucleotides arranged in triplet codons, and subsequently translated into the 20-amino acid linear primary structures of polypeptides. Only at the final stages of protein synthesis do the proteins fold into their characteristic three-dimensional structures and engage with and assemble into asymmetric chemically heterogeneous environments where they display the chemical activities essential for expressing their biological functions. The sophisticated chemical reactions are highly specific and efficient as a result of the confluence of the protein structure, dynamics, and influence of the dynamic chemical environment.

The structures of native folded proteins are not static entities and undergo global and local motions. Consequently, understanding proteins requires descriptions of their chemical composition, linear and three-dimensional arrangements of the amino acids in the polypeptide chain, and of the motions of the side chains and backbones of the amino acids. Generally, these descriptions start with and are organized by the tertiary structures of individual proteins.

The atomic-resolution three-dimensional structures of many globular proteins have been determined in crystals and in aqueous solution. Notably, ambitious efforts to characterize the structures of all the proteins in the human proteome are proceeding. The structures have the potential to provide the initial steps of an analytical definition of the molecular basis of life in wild-type, mutated, and diseased organisms. And of how the definition can be altered by chemical interactions with other macromolecules as well as small molecules, including those added externally as drugs. About one-third of all expressed proteins are associated with membranes and are key to many signaling events, including those resulting from drug interactions. However, existing experimental and theoretical approaches to describing protein structures are sub-optimal for membrane proteins because of the nature of the chemical and physical properties of these systems, which are strongly influenced by phospholipid bilayers.[1]

Larger membrane proteins present extremely complex situations because of the close connection among the multiple chemical components of protein-containing phospholipid bilayers. To date, our understanding of membrane proteins has been limited primarily by technical challenges, especially the availability of suitable samples for the experimental methods of structural biology, which include X-ray crystallography, cryo-electron microscopy, and NMR spectroscopy. Nonetheless, these techniques have all contributed to important and complementary structures of membrane proteins as the samples have been developed and improved over time. And they are all complemented by advances in calculation methods that incorporate experimental structural constraints. As the most powerful form of spectroscopy, NMR leads the way towards functional studies of membrane proteins. A classic overview of a membrane protein system is illustrated in Figure 18.1, where individual polypeptides are distributed throughout a liquid crystalline phospholipid bilayer.[2] Notably, the proteins are immobilized on the relevant NMR timescales by the phospholipid bilayer environment, and it is this characteristic that dictates the use of solid-state NMR methods to obtain high-resolution spectra where individual resonances are observable for each atomic site in the polypeptide. As illustrated in Figure 18.1, the spectroscopic definition of immobilization is fulfilled in the presence of limited lateral and rotational global motions of proteins in bilayer samples because of the absence of complete motional averaging.

As alluded to above, in assessing potential NMR approaches for studies of proteins, an obvious but crucial demarcation of their global dynamics needs to be established. Despite their names, "solution NMR" and "solid-state NMR" are not readily differentiated by the physical states of the samples. Although only solid-state NMR can be applied to single crystals or powders, both forms of NMR can utilize clear, non-viscous liquid samples. The differences lie in the overall rotational diffusion rates of the protein molecules compared to the spectroscopic "rulers" established by the frequency spans of the operative nuclear spin interactions. In most cases these are the chemical shift anisotropy, heteronuclear dipole–dipole couplings, and

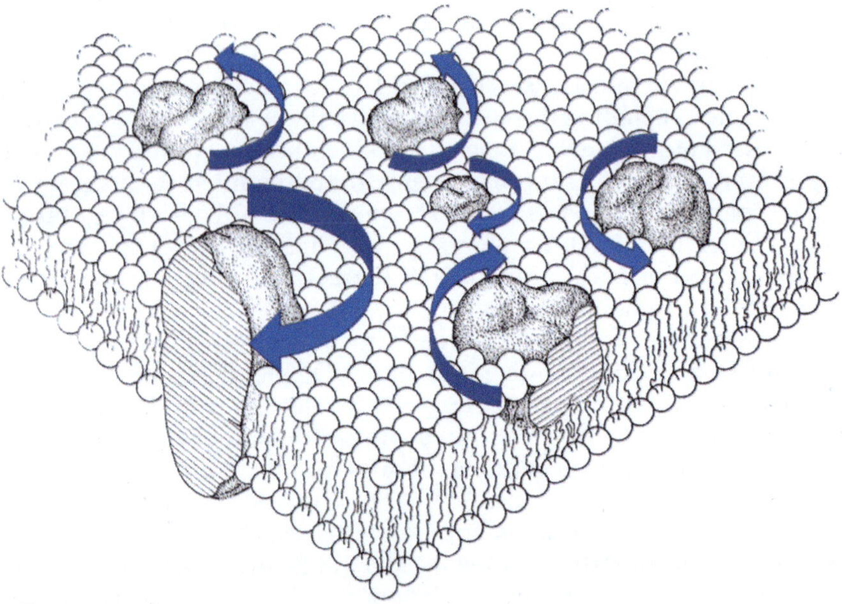

Figure 18.1 Cartoon representation of three-dimensional cross-sections of the "Fluid Mosaic Model" of membrane proteins that are completely or partially embedded in the matrix formed by phospholipid bilayers. Adapted from ref. 2 with permission from AAAS, Copyright 1972.

homonuclear dipole–dipole couplings. Relatively small soluble proteins have rotational correlation times on the nanosecond timescale. Since this is "fast" compared to the 10^4–10^6 Hz spans of the spin-interactions, the proteins respond like "liquids" in the NMR experiments: global motional averaging provides the mechanism for narrowing the resonance linewidths. In contrast, proteins in large complexes, such as membrane proteins in lipid bilayers, reorient slowly on this timescale and behave like "solids." The unaveraged spectral manifestations of the spin-interactions are anisotropic and provide detailed information about molecular structure and dynamics. However, high-resolution spectra are available only when molecular motion is replaced by radiofrequency irradiations or mechanical sample rotation as a line narrowing mechanism,[3–5] only then is it possible to observe the anisotropic properties of the spin interactions. This chapter emphasizes results from our own research, which has been highly focused on the development of NMR for application to membrane proteins in lipid environments that closely recapitulate the properties of biological membranes. The goal, as summarized here, is to exploit NMR methods to address the key challenges posed by larger membrane proteins, in general, and use them to advance our understanding of the structural basis of the functions of G Protein-Coupled Receptors (GPCRs), in particular. These membrane proteins are technically challenging because they contain seven trans-membrane helices and subtle changes in structure and dynamics are responsible for their transmitting signals across the cell membrane.

18.1.2 G Protein-coupled Receptors

GPCRs are the largest family of membrane proteins in the human genome. They are characterized by having seven transmembrane spanning helices separated by alternating intracellular and extracellular loop regions. They share a common signaling mechanism in which extracellular stimuli induce conformational changes in the receptor, which in turn activate intracellular heterotrimeric G proteins.[6] GPCRs regulate many cellular processes, such as cell differentiation, cell motility, and angiogenesis. Due to their connection to a myriad of signaling pathways, misregulation of GPCRs and their ligands stand at the basis of many diseases, such as cancer, asthma, and hypertension. Notably, approximately 35% of all pharmaceuticals target a GPCR.[7] Even though structures of around 100 different GPCRs have been determined,[8] many questions remain regarding how they interact with their ligands, what conformational changes take place upon receptor activation and how receptor dynamics influence function.

CXCR1 is a class A, rhodopsin-like GPCR, which belongs to the family of chemokine receptors. We determined the three-dimensional structure of human CXCR1 by rotationally aligned (RA) solid-state NMR spectroscopy[9] in phospholipid bilayers, without modification of its amino acid sequence and under physiological conditions, as shown in Figure 18.2B.[10] CXCR1 is

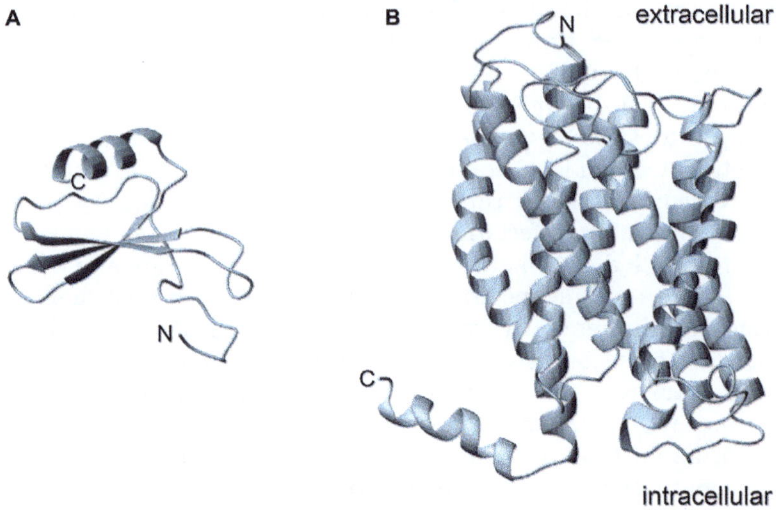

Figure 18.2 Structures of the chemokine IL-8 and its receptor CXCR1. (A) Structure of the monomeric chemokine IL-8(1–66) in aqueous solution (PDB: 5WDZ). (B) Structure of the chemokine receptor CXCR1(1–350) in phospholipid bilayers (PDB: 2LNL). Extracellular N-terminal and intracellular C-terminal residues of CXCR1 involved in ligand and G-protein interactions, respectively, are not included in the figure because they are mobile on the timescales of the NMR experiments used to determine the structure of the receptor. Adapted from ref. 18 with permission from Springer Nature, Copyright 2017.

activated by the chemokine interleukin-8 (IL-8, CXCL8), a major mediator of the inflammatory immune response. In case of infection, macrophages secrete IL-8, which then forms a chemotactic gradient. CXCR1 is expressed on the cell surface of several types of immune cells such as neutrophils and dendritic cells. IL-8 binds to the extracellular domains of CXCR1, which then activates the neutrophil and initiates its migration toward the site of inflammation, which ultimately leads to neutrophil degranulation.[11] IL-8 has a standard chemokine fold that consists of a disordered N-terminus, the N-loop, one turn of a 3_{10}-helix, three β-strands, and a C-terminal α-helix.[12–14] It has 72 residues in its most common form and forms a homodimer at micromolar concentrations. It can interact with CXCR1 as a monomer and as a dimer, but it is thought that the monomer binds to the N-terminal domain of CXCR1 with higher affinity than the dimer.[15–17] An obligatory form of monomeric IL-8 can be obtained by deleting the C-terminal residues 67–72 which contribute to the dimerization interface. The structure of monomeric interleukin-8, IL-8(1–66) from solution NMR is shown in Figure 18.2A.[18] This construct has been shown to bind CXCR1 with the same affinity as wild-type IL-8, and the same potency in neutrophil chemotaxis and Ca^{2+} mobilization assays. The existing model for CXCR1–IL-8 interaction involves two binding sites on the receptor; residues in both the flexible N-terminal domain (Binding Site-I) as well as the second and third extracellular loops (Binding Site-II) of CXCR1 have been shown to interact with IL-8. Based on biophysical and mutational studies, several residues in IL-8 have been found to be important for the interaction with Binding Site-I in CXCR1: mainly residues in the N-loop, in the third beta strand, and a small part of the alpha-helix.

NMR, molecular dynamics, and mutational studies have started to characterize the binding pattern between IL-8 and the N-terminal domain of CXCR1. To this end several different constructs of isolated Binding Site-I have been described in the literature; synthetic peptides with sequences corresponding to portions of the N-terminal domain of CXCR1 interspaced with hexanoic acid, peptides composed of residues 9–29 and ND-CXCR1 (1–38), a soluble peptide containing the first 38 residues of CXCR1 developed by our laboratory. The results of these interaction studies largely overlap; residues 9–29 are most important for the interaction with IL-8, especially residues D14, G19, P21, P22, D24, E25, D26, and Y27. Interestingly, the N-terminal domain has the highest sequence diversity among all GPCRs and even among chemokine receptors. The N-terminal domain does not only vary in length, but it can undergo post-translational modifications, and can be linked to the extracellular loops through a disulfide bridge. It is thought that this variety is the main mechanism through which specific chemokines are selected by chemokine receptors. The N-terminal domain is even linked to receptor internalization rates. When the N-terminal domain of CXCR2 was placed on CXCR1, the entire receptor internalized faster and matched the internalization rate of endogenous CXCR2. To better understand chemokine binding to the N-terminal domain of CXCR1 and the interplay between the N-terminal domain and the remainder of the receptor, several constructs of CXCR1 have been developed in our laboratory, including N-terminal truncated CXCR1 [NT-CXCR1(39–350)], C-terminal truncated CXCR1 [CT-CXCR1(1–319)], both N- and C-terminal double truncated CXCR1 [DT-CXCR1(23–319)] in addition to the wild-type receptor. Additionally, 1TM-CXCR1(1–72) was developed, which contains the flexible N-terminus, the first transmembrane helix, and the first intracellular loop of CXCR1. These constructs greatly facilitate the experimental studies described here.

Solid-state NMR spectroscopy has become a powerful tool for characterizing protein–ligand complexes; there are several examples where solid-state NMR has been used to describe the interaction of ligands bound to GPCRs.[19] As an example of NMR studies of interaction between a membrane protein and its biological ligand, we present results on the atomic-level characterization of the interaction between CXCR1 and monomeric IL-8, focused on Binding Site-I. We describe the interaction site of unlabeled IL-8(1–66) on uniformly ^{13}C,^{15}N-labeled 1TM-CXCR1(1–72) using experiments performed in a near-native environment of hydrated phospholipid bilayers and at physiological temperatures.

18.2 Dynamics of Membrane Proteins

18.2.1 Dynamics by Solid-state NMR

The native environment for a membrane protein is a phospholipid bilayer. It is their association with phospholipid bilayers that distinguishes them from soluble proteins. This not only means that they are influenced by a distinctive, highly asymmetric chemical environment but also, more importantly for NMR, that the polypeptides are immobilized on the timescales of the chemical shift and dipolar coupling spin interactions. It is this immobilization that links membrane proteins in phospholipid bilayers and solid-state NMR and makes solid-state NMR methods essential for obtaining high-resolution spectra. Approaches have been developed for both unoriented and oriented samples, however, they all rest on the foundation of the most fundamental aspects of solid-state NMR, and the chemical shift and homo- and hetero-nuclear dipole–dipole interactions.

In unoriented, immobile samples, the chemical shift anisotropy and heteronuclear dipole–dipole interactions yield powder pattern resonances with frequency breadths of 10^4–10^5 Hz, which is the critical factor for characterizing the dynamics of membrane proteins and defining the spectroscopic requirements for obtaining high-resolution spectra. Significantly, in modern NMR experiments, the radiofrequency irradiations can have magnitudes $>10^5$ Hz, the rates of magic angle sample spinning (MAS) are typically $>10^5$ Hz, and the receiver bandwidths can be $>10^6$ Hz. All these factors, which required many years of development, combine to make high-resolution solid-state NMR spectroscopy feasible and are essential for applications of NMR to membrane proteins in phospholipid bilayers.

Historically, high-resolution solid-state NMR spectroscopy and NMR of membrane proteins were developed in parallel, starting at essentially the same time, since the instrumentation and methods of solid-state NMR suitable for single crystals and powders of organic solids were also suitable for proteins and phospholipids in membrane bilayers. The initial papers describing the application of high-resolution double-resonance solid-state NMR to membranes were published within a year of those introducing the principles and methods of proton-enhanced nuclear induction spectroscopy.[20] Significantly, the origins of high-resolution solid-state NMR are the same as for the field of NMR spectroscopy itself.

Solid-state NMR is unparalleled as a method for elucidating molecular dynamics since it is possible to analyze motionally averaged powder patterns quite directly. Selectivity for powder patterns can be obtained in various solid-state NMR experiments based on several criteria, including the basic mechanism for generating magnetization through cross-polarization. The presence of heteronuclear dipolar couplings is necessary to obtain thermodynamic equilibrium and its presence or absence provides a method for discriminating between mobile and immobile sites on the 10 kHz timescale.

Figure 18.3A shows the proton decoupled ^{15}N NMR spectrum of fd coat protein in dimyristoyl phosphatidylcholine bilayers.[21] Panel B is a calculated chemical shift anisotropy (CSA) powder pattern based on the magnitudes of the principal values of the amide chemical shift tensor observed in peptides and proteins. The shape and frequency breadth of the experimental line shape reflect the motions affecting the labeled nitrogen sites. Large-amplitude motions that occur more often than the frequency breadth of the static powder pattern have the effect of averaging or reducing the observed breadth of the powder pattern. Also in some instances, the line shape may be altered characteristically by low-frequency or small-amplitude motions. The rapidly moving N-terminal and lysine sidechain amino groups contribute to the upfield narrow signals. The amide nitrogens of fd coat protein in bilayers give a resonance line shape with two distinct components. Most of the backbone sites of the protein are immobile on the 10 kHz timescale of the ^{15}N chemical shift anisotropy, but the narrow component superimposed on the broad powder pattern indicates that some of the backbone sites are undergoing large-amplitude, effectively isotropic motions on this timescale. Figure 18.3 clearly demonstrates that the

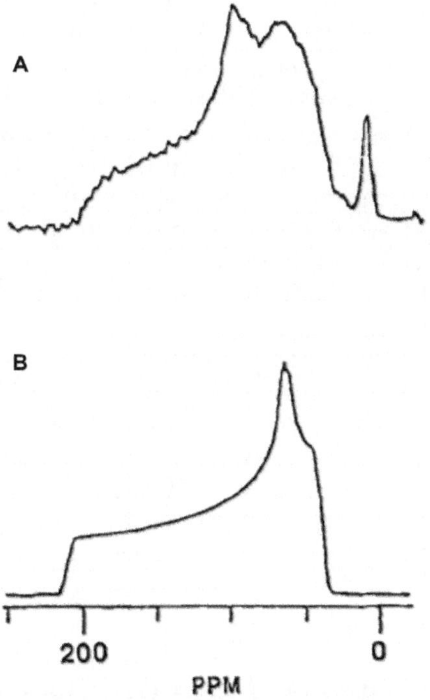

Figure 18.3 (A) Experimental solid-state ^{15}N NMR spectrum of the membrane-bound form of uniformly ^{15}N-labeled fd coat protein in phospholipid bilayers. (B) Calculated ^{15}N amide chemical shift anisotropy powder pattern. Reproduced from ref. 21 with permission from John Wiley and Sons, Copyright © 1988 Alan R. Liss, Inc.

membrane-bound form of fd coat protein has two classes of backbone amide dynamics.

Both monotopic and polytopic membrane proteins can undergo global rotational diffusion about the bilayer normal with diffusion coefficients (Dr) between 10^3 and 10^4 s^{-1} in reconstituted phospholipid vesicles and cell membranes. Therefore, all the signals reflect the angles of the principal axes of their dipole–dipole and chemical shift tensors to the axis defined by the bilayer normal. The frequency spans and signs of the axially symmetric powder patterns resulting from motional averaging about a common axis provide sufficient structural restraints for calculating the three-dimensional structure of a membrane protein in phospholipid bilayers.

Figure 18.4 illustrates that the seven transmembrane helical protein bacteriorhodopsin undergoes rapid rotational diffusion about the bilayer normal at temperatures above the gel-to-liquid crystalline phase transition temperature of the phospholipid bilayers.[22] The spectrum in Figure 18.4B is of bacteriorhodopsin at 3 °C, well below the DMPC phase transition temperature. The chemical shift powder pattern from the carbonyl carbons has the full width and asymmetry expected in a rigid lattice. In contrast, the spectrum of the same sample obtained at 30 °C (see Figure 18.4C and D), above the phase transition temperature, is much narrower and axially symmetric, as a result of motional narrowing from rotational diffusion about the bilayer normal. These spectra show that most of the labeled carbonyl groups are aligned in the protein with the C=O bond close to parallel to the bilayer normal, which is consistent with the arrangement of helices in the static structures. The rate of rotational diffusion about the bilayer normal can be switched between frequencies that are "slow" or "fast" on relevant NMR timescales that are defined by the frequency breadth of the static powder patterns by changing the temperature. Of particular importance to structure determination is the finding that under most conditions even quite large membrane proteins undergo fast rotational diffusion about the bilayer normal.[23–26] Residues that undergo large-amplitude internal backbone dynamics often separate protein domains; and can be identified by NMR experiments combined with protein engineering that produces truncated constructs. For example, proteolytic digestion of coat protein samples in bilayers with chymotrypsin results in cleavage of only the solvent-exposed regions of the protein and the removal of the residues in the N- and C-terminal portions of the protein. This removes the narrow spectral component from the backbone amide resonance intensity.[21]

18.2.2 Use of Membrane Protein Dynamics in Structure Determination

A universal aim of NMR spectroscopy, regardless of the state of matter, is to obtain "high-resolution" spectra. There are multiple aspects to the definition of high resolution in NMR, and one of the most important is that each

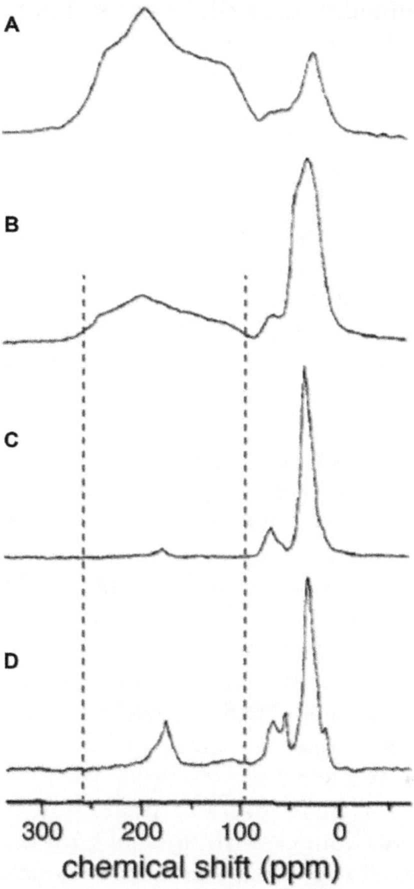

Figure 18.4 ^{13}C NMR powder spectra of ^{13}C′-leucine-labeled bacteriorhodopsin (bR) samples. The carbonyl carbon resonances are centered at 175 ppm, and the aliphatic resonances are in the region 20–80 ppm. (A) Hydrated native purple membrane at 3 °C. (B) Reconstituted bR-containing DMPC bilayer vesicles at 3 °C. (C) Same as in (B) except at 30 °C. (D) Same as in (C) except the spectrum was obtained with a Hahn echo sequence. The vertical dashed lines mark the discontinuities of the rigid lattice carbonyl carbon chemical shift powder pattern. Adapted from ref. 22 with permission from American Chemical Society, Copyright 1985.

resonance is associated with a single chemical site. It is essential to know which site this is. It is equally important that one spin interaction at a time is analyzed. The three basic approaches to obtaining high resolution—motional averaging, spin dilution, and spin manipulation—are used in solid-state NMR studies of membrane proteins, and this accounts for the relatively long development period from the first double-resonance spectra of a liposome sample to the complete structures of membrane proteins that are now being determined. Recent advances are built on the fundamentals of NMR. Solid-state NMR as applied to membrane proteins has advanced to the point that fully justifies the use of the term "high-resolution" spectroscopy.[27,28]

The three-dimensional structure of a protein can be visualized based on intramolecular distances or angles. In general, the use of distances is more familiar because of the long history of single crystal X-ray diffraction and the measurement of specific short-range distances in nuclear Overhauser enhancements (NOEs) solution NMR experiments. Notably, Ramachandran plots that represent the backbone angles of peptide bonds are fully validated descriptions of the backbone architecture of proteins. This background provides the foundation for three distinct approaches to protein structure determination by solid-state NMR spectroscopy. The first is the commonly employed magic angle spinning (MAS) solid-state NMR method applicable to unoriented polycrystalline or gel samples where internuclear distances are measured among the resolved isotropic resonances observable because of the extensive averaging of the anisotropic components by rapid mechanical sample rotation at 54.7° (the magic angle) relative to the direction of the static magnetic field. The second method, static solid-state NMR, utilizes stationary uniaxially aligned samples. The unaveraged anisotropic nuclear spin interactions are represented in the spectra in the form of angle-dependent frequencies of individual resonance and splittings due to dipole–dipole couplings. These measurements provide orientational information about bonds and functional groups relative to the axis of molecular alignment, a common reference direction.

The third method is derived from the second and it too is based on the measurement of orientational constraints but differs in using rotational alignment by MAS rather than magnetic or mechanical alignment. In these samples, the individual protein molecules undergo fast rotational diffusion about a defined axis, providing the reference for the angle measurements. With only single or few sites isotopically labeled, the static powder spectra consist of rotationally averaged powder patterns whose shapes and frequency spans provide the angle measurements relative to the common axis of molecular rotation equivalent to those obtained from uniaxially aligned samples relative to the common axis of sample alignment. However, with many signals, as is typically the case for uniformly isotopically labeled samples, the overlap of the powder patterns allows no or very limited possibilities for resolution of signals from individual sites. Here the concepts of MAS and sample alignment converge. The magic angle sample spinning resolves the overlapping resonances, and the underlying rotationally averaged powder patterns can be recovered through application of appropriate recoupling pulse sequences. The frequencies of resonances measured from aligned samples and the principal values of rotationally averaged powder patterns can be interpreted analogously with a common definition of the alignment axis. The ability to acquire powder patterns associated with isotropic chemical shifts lies at the heart of rotationally aligned (RA) solid-state

NMR.[29–31] This makes it possible to observe the effects of rotational diffusion on the powder patterns, which yields the angle between the spin interaction tensor and the direction of rotation. In membrane proteins this direction is generally the bilayer normal, so providing equivalence between the angles measured with an aligned sample in the form of single lines and the parallel edge of the motionally averaged powder pattern. Rotationally aligned solid-state NMR is precisely tailored for the unique properties of membrane proteins in liquid crystalline phospholipid bilayers. It combines features of magic angle spinning and oriented-sample solid-state NMR to resolve and assign resonances associated with each amino acid residue, measure site-specific orientation restraints relative to the bilayer, and calculate the three-dimensional structure of the protein and its orientation within the membrane bilayer. RA solid-state NMR differs from previously used oriented sample (OS) solid-state NMR methods in that it relies on the inherent rotational diffusion of membrane proteins in phospholipid bilayers to provide orientation-dependent motional averaging of heteronuclear dipolar coupling and chemical shift anisotropy powder patterns relative to the bilayer normal, rather than the orientation-dependent frequencies of single-line resonances observed in OS NMR of stationary, uniaxially aligned samples. It differs from more standard MAS solid-state NMR approaches that typically rely on short-range distance and angle measurements within the polypeptide chain rather than to an externally defined axis, in this case the bilayer normal.

The rotational diffusion of membrane proteins in phospholipid bilayers was shown originally to occur by optical methods. Notably, the same phenomenon was observed in the averaging of line shapes by NMR. McLaughlin and coworkers in 1975 used ^{31}P NMR of the phospholipids to show that they undergo fast rotational motion by two pertinent spectroscopic effects. One was the averaging of the powder pattern line shape and the other was the observation of single-line spectra when the lipid bilayers were mechanically aligned on glass plates over a wide range of angles (not just with the normal parallel to the field). Ten years later, in a similar, complementary experiment, Griffin and coworkers demonstrated that the protein bacteriorhodopsin undergoes rotational diffusion in phosphatidyl choline bilayers because of the averaging of the ^{13}C′ powder pattern.[22] Contemporaneously, Cornell and coworkers showed that ^{13}C′-labeled gramicidin undergoes the same type of motional averaging in phospholipid bilayers.[32,33] In a highly definitive experiment, Cross and coworkers utilized gramicidin labeled at a single site to show with a two-dimensional separated local field (SLF) experiment that the motionally averaged chemical shift and dipolar coupling principal elements were collinear while in a static sample they differ in alignment by about 17°.[34] In an experiment very similar to the original experiment of McLaughlin and coworkers, we were able to show single-line resonances from tilted coil samples over a wide range of angles, again something that could only be observed in the presence of rotational diffusion.[35] The spectral manifestations of rotational averaging provide the experimental foundation for RA solid-state NMR.

In the structure determination of a membrane protein by RA solid-state NMR, the essential first step is to verify that the protein undergoes rapid rotational diffusion about the bilayer normal and that the rotation can be switched between the slow and fast limits by changing the temperature. The powder pattern line shape is effective in demonstrating that a helical membrane protein is undergoing fast rotational diffusion, since the static powder pattern is highly asymmetric with a large frequency span. For transmembrane helices a large fraction of the backbone carbonyl bonds lies approximately parallel to the bilayer normal (*i.e.*, the axis of motional averaging); thus, when the protein undergoes fast rotational diffusion, the $^{13}C'$ powder pattern becomes axially symmetric and is significantly narrowed. A full two-dimensional $^{1}H-^{15}N$ dipolar coupling/^{13}C chemical shift SLF spectrum contains resonances from the protein's $^{13}C\alpha$ sites. Although there is considerable spectral overlap, an individual resonance can be identified. The rotationally averaged $^{1}H-^{15}N$ dipolar coupling and ^{15}N CSA powder patterns associated with each isotropic resonance are characterized with three-dimensional MAS solid-state NMR experiments. The data from other residues are combined and used to calculate the three-dimensional structure of the protein.[36] The frequencies are measured from spectra, such as those shown in Figure 18.5. The experimentally measured, motionally averaged $^{1}H-^{13}C$ and $^{1}H-^{15}N$ heteronuclear dipolar couplings are compared to calculated static powder patterns on the right side of the figure. On the left side,

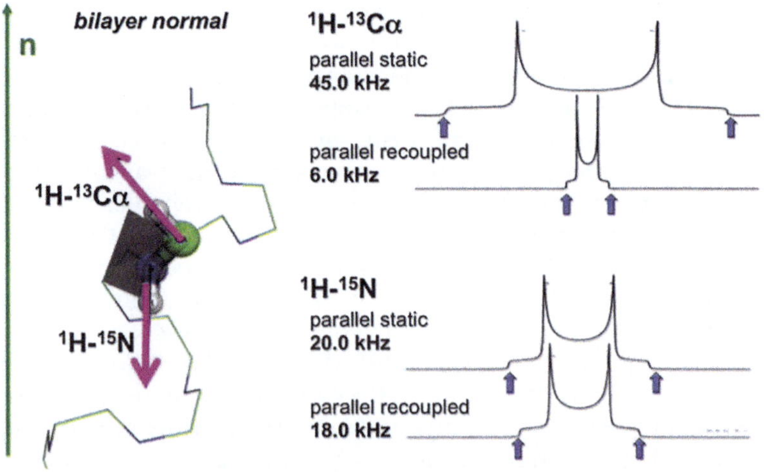

Figure 18.5 Orientation of the peptide plane associated with residue L31 of uniformly $^{13}C/^{15}N$-labeled MerFt in DMPC proteoliposomes at 25 °C as determined from experimental data obtained from rotationally aligned samples. The $^{1}H-^{15}N$ and $^{1}H-^{13}C_{\alpha}$ dipolar coupling restraints and ^{15}N chemical shift anisotropy restraints are sufficient to determine the orientation of the associated peptide plane relative to the axis of alignment. A complete set of peptide plane orientations yields the three-dimensional structure of the protein. Adapted from ref. 37 with permission from Elsevier, Copyright 2015.

the resulting angular vectors derived from the experimental data are superimposed on the peptide plane of Leu 31 of uniformly $^{13}C/^{15}N$-labeled MerFt in DMPC proteoliposomes at 25 °C.

18.2.3 Dynamics of CXCR1

Local and global motions are essential properties of proteins, since they strongly influence their structures, interactions, and functions. In general, the dynamics of membrane proteins[31,38–43] are not as well characterized as those for water-soluble globular proteins,[44] largely due to difficulties encountered in the expression, purification, and refolding of membrane proteins into uniform, active samples, as well as the limitations of current experimental methods and instrumentation. However, NMR spectroscopy is sensitive to a wide range of amplitudes and frequencies of motions, and it is a fundamentally sound approach to characterizing protein dynamics in almost any situation, including protein-containing micelles, bicelles,[45,46] and bilayers. In order to characterize motions on different timescales, we combined experiments that apply solution NMR to protein-containing isotropic bicelle samples and both oriented sample (OS) and RA magic angle spinning (MAS) solid-state NMR to aligned and unoriented phospholipid bilayer samples. This approach characterizes the local backbone and global motions of the 350-residue G protein-coupled receptor (GPCR) CXCR1. The results obtained on CXCR1 in phospholipid bilayers complement the spectroscopic studies of this protein in its native, functional environment of planar phospholipid bilayers under physiological conditions.[25] Although the methods for NMR studies of membrane protein dynamics are still being developed, it is well established that the chemically, physically, and dynamically asymmetric liquid crystalline bilayer environment in which they reside has a profound effect on their structures and dynamics. This is reflected in the two major classes of motions found in the initial studies of CXCR1:[1] there are local rapid backbone motions of residues near the N- and C- termini of the protein in all lipid environments examined; and[2] the protein undergoes rapid global rotational diffusion about the normal of liquid crystalline phospholipid bilayers. The motions of the flexible extracellular N-terminus are of direct relevance to the structure and function of CXCR1. The chemokine IL-8, the natural ligand of CXCR1, has been shown to bind to residues near the N-terminus of the receptor.[15,47] These protein–protein interactions are key to the physiological response to inflammatory stimuli. CXCR1 mediates cell migration and activation of polymorphonuclear neutrophils (PMNs)[48,49] and IL-8-induced chemotaxis that requires global motions of the receptor to direct the PMN to sites of inflammation where IL-8 is secreted by macrophages, lymphocytes, and epithelial and endothelial cells. Clearly, the interactions of specific amino acid residues of IL-8 with those of CXCR1 are directly relevant to the mechanisms of actions of its biological functions, and the motions of the N-terminal residues of CXCR1 undoubtedly affect these interactions. Thus, NMR is capable of providing an atomic resolution view of how the inflammatory response is mediated by membrane proteins.

The experimental studies are correlated with the primary structure CXCR1 in Figure 18.6C, which shows the amino acid sequence of full-length human CXCR1 arranged schematically so that it suggests the locations of the seven trans-membrane helices, interhelical loops, and terminal regions that constitute the main elements of the protein.[50] The blue (N-terminus) and red (C-terminus) colors designate the residues found to undergo rapid local backbone motions by NMR spectroscopy; only terminal residues are observed to undergo motions on the relevant NMR timescales (10^4 Hz).

The dynamics of CXCR1 consist of fast local backbone motions of the residues near the N- and C-termini, but not for the residues in the trans-membrane helices and structured interhelical loops, and rapid global rotational diffusion of the protein about the bilayer normal in liquid crystalline but not in gel phase phospholipid bilayers. The fast rotational diffusion of CXCR1 has dramatic effects on the shape and breadth of the ^{13}C CSA powder pattern from a carbonyl group in a peptide bond, as illustrated in Figure 18.7 with simulated spectra. Notably, a large difference is observed in the carbonyl ^{13}C chemical shift powder pattern between the high-temperature spectra (30 °C) and the low-temperature spectra (5 °C), which corresponds

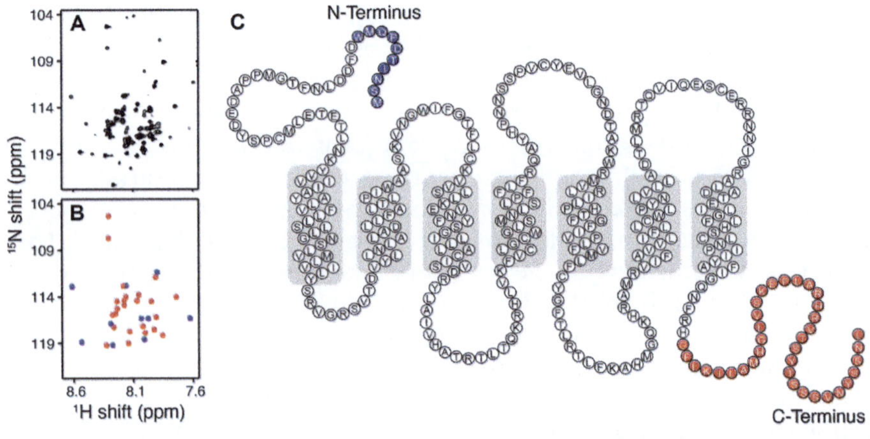

Figure 18.6 (A) Experimental solution ^{1}H/^{15}N HSQC NMR spectrum of uniformly ^{15}N-labeled CXCR1 in $q = 0.1$ isotropic bicelles at 50 °C. (B) Colored "dot" representations of the experimental solution NMR spectra that facilitate comparisons among the various polypeptide constructs of CXCR1. Blue dots correspond to resonances assigned to mobile residues near the N-terminus of the full-length protein. Red dots correspond to resonances assigned to mobile residues near the C-terminus of the full-length protein. (C) Representation of the linear amino acid sequence of CXCR1 arranged to highlight the transmembrane helices, interhelical loops, and terminal regions predicted by hydropathy analysis. The N- and C-terminal residues found to be mobile by NMR experiments are colored in blue and red. Reproduced from ref. 75 with permission from American Chemical Society, Copyright 2011.

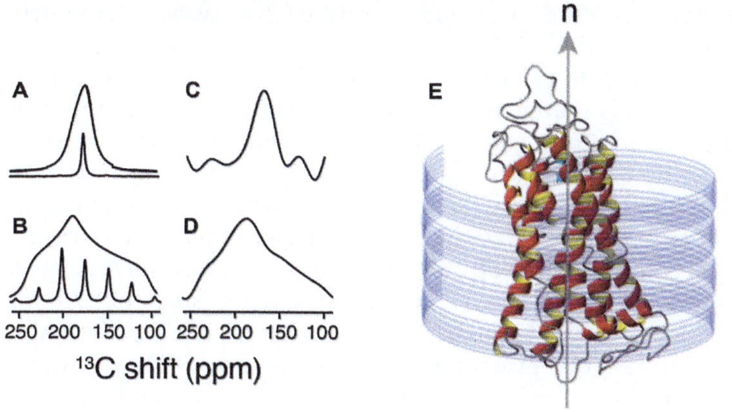

Figure 18.7 The effects of rotational diffusion on the chemical shift anisotropy powder pattern from the solid-state ^{13}C NMR spectra of randomly 35% ^{13}C-labeled, 100% ^{15}N-labeled CXCR1 in unoriented DMPC/POPC proteoliposomes. (A and C) Experiments performed at 30 °C, where CXCR1 undergoes fast rotational diffusion about the bilayer normal. (B and D) Experiments performed at 5 °C where CXCR1 is immobile. (A and B) Simulated MAS solid-state NMR spectra with 200 Hz and 3 kHz, respectively, of added line broadening are superimposed on the experimental spectra. (C and D) Experimental MAS solid-state ^{13}C NMR spectra of carbonyl powder patterns obtained at 30 °C and at 5 °C with the two-dimensional SUPER CSA recoupling experiment, at a MAS spinning rate of 5 kHz. (E) Schematic representation of CXCR1 undergoing rapid rotational diffusion about the lipid bilayer normal (*n*). Reproduced from ref. 75 with permission from American Chemical Society, Copyright 2011.

to the phospholipids being in the liquid crystalline phase at 30 °C and in the gel phase at 5 °C. The dramatic differences between the stationary sample and MAS spectra (Figure 18.7A and B) make it very easy to differentiate between immobile proteins and those undergoing fast rotational diffusion in lipid bilayers.

The static powder pattern for a carbonyl group is highly asymmetric, with a large frequency span of approximately 150 ppm, as observed in spectra from samples of a polycrystalline peptide at room temperature or a membrane protein in lipid bilayers at very low temperatures. When a labeled carbonyl group in a peptide bond in the backbone of an alpha-helix aligned with its long axis approximately parallel to the bilayer normal (*e.g.*, a transmembrane helix) and undergoing rotational diffusion about the bilayer normal along with the rest of the protein, the resulting powder pattern is substantially reduced in breadth and is axially symmetric, as is the case for all rotationally averaged powder patterns. The sign of the powder pattern and the frequency span between its parallel and perpendicular edges are determined by the angle between the principal axis of the chemical shift tensor and the axis of rotational diffusion. Thus, individual powder patterns can be used to define the orientation of each peptide plane relative to the bilayer normal, and each other.

The data in Figure 18.7B show that in a slow-spinning MAS experiment the powder pattern is mapped out by the intensities of the spinning side bands surrounding the center band at the isotropic chemical shift frequency.[44] The family of side bands is separated by the spinning frequency and spans the breadth of the entire static powder pattern, which is much greater than the relatively slow spinning rate used in the experiment. In contrast, for a protein undergoing rotational diffusion about the bilayer normal, no sidebands are observed because of the limited frequency span of the rotationally averaged powder pattern (Figure 18.7C). In practice, a trans-membrane helix in a uniaxially aligned protein undergoing rotational diffusion yields a single relatively narrow resonance. When magic angle spinning is applied, an even narrower signal is observed. The large differences observed between the MAS spectra in Figure 18.7A and B readily differentiate between proteins that are immobile, *e.g.*, at low temperatures, or are undergoing fast rotational diffusion in phospholipid bilayers at temperatures above their gel-to-liquid crystalline phase transition.

18.3 IL-8 Interactions with CXCR1

As an example of the biological insights that can be provided by high-resolution NMR spectroscopy, we summarize a study of the interactions between the water-soluble chemokine IL-8 and its membrane-bound receptor CXCR1. This starts with structural information about both proteins as affected by their local and global dynamics.

18.3.1 Solid-state NMR Experiments

The solid-state NMR experiments were performed on a Bruker AVANCE-III HD 750 MHz spectrometer with a 4 mm H/C/N triple-resonance Bruker E-free MAS probe (Bruker, https://www.bruker.com). The spinning rate was maintained at 11.11 kHz $\pm$ 2 Hz. The temperature controller was set to 7 °C; taking into account frictional heating reflected in the ^{1}H chemical shift of the water resonance, the actual sample temperature was estimated to be 10 $\pm$ 2 °C. The applied radio frequency (rf) fields corresponded to $\pi/2$ pulse lengths of 2.5 μs, 3.2 μs, and 5.1 μs for ^{1}H, ^{13}C, and ^{15}N, respectively.

Spectral measurements obtained with direct ^{13}C detection used 100 kHz rf irradiation for ^{1}H decoupling with swept frequency two-pulse phase modulation (SW-TPPM).[51,52] Cross-polarization (CP)[4] from ^{1}H to ^{15}N was optimized by applying amplitude-modulated rf irradiation corresponding to a 100%–70% ramp on the ^{1}H channel; the modulation affected the ~50 kHz rf irradiation on the ^{1}H channel and the ~27 kHz rf irradiation on the ^{15}N channel during the 1 ms contact time. Two-dimensional homonuclear ^{13}C–^{13}C correlation spectra were acquired using proton-driven spin diffusion

(PDSD) with 10 ms, 50 ms, and 100 ms mixing times.[53,54] Double cross-polarization from ^{15}N amide to ^{13}Cα sites was performed with spectrally induced filtering in combination with cross-polarization (SPECIFIC-CP).[55] Adiabatic tangential pulses were applied on the ^{13}C channel. For two-dimensional N–C correlation spectra, rf irradiation of ~27 kHz was applied on the ^{15}N channel and ~16 kHz on the ^{13}C channel for band-selective polarization transfer and continuous wave. Continuous wave (CW) 100 kHz ^{1}H irradiation was applied for decoupling during the 4.5 ms contact time required for NCA and NCO magnetization transfers. Two-dimensional ^{15}N-detected SLF experiments were carried out using a $R18^{7}{}_{1}$ symmetry-based pulse sequence to correlate ^{15}N chemical shift frequencies and ^{1}H–^{15}N dipolar coupling frequencies.[56] A pair of π pulses with a phase modulation of 70° (π_{70}–π_{70}) was employed in the $R18^{7}{}_{1}$ scheme.[57]

The chemical shift frequencies were referenced externally to the adamantane methylene ^{13}C resonance at 38.48 ppm and the ammonium sulfate ^{15}N resonance at 26.8 ppm. All of the spectra were processed with the Bruker TopSpin 3.2 software package and analyzed with the program CARA (ETH Zürich).[58]

One-dimensional ^{13}C and ^{15}N MAS solid-state NMR spectra of the polypeptide construct are shown in Figure 18.8. All of the spectra are fully decoupled but were obtained with two different methods that distinguish between "immobile" and "mobile" residues. Figure 18.8A shows a one-dimensional cross-polarization (CP) MAS solid-state NMR spectrum recorded on uniformly ^{13}C- and ^{15}N-labeled 1TM-CXCR1(1–72). The peaks in the spectrum are well dispersed, and a well-resolved carbonyl resonance region results around 172 ppm. We also utilized INEPT-based experiments (insensitive nuclei enhanced by polarization transfer)[59] to identify the mobile regions of 1TM-CXCR1(1–72). Several mobile residues contribute to the ^{1}H/^{13}C and ^{1}H/^{15}N INEPT spectra shown in Figure 18.8B and E. Lipids signals in the ^{13}C-detected INEPT spectrum, which are not of interest here, are marked with an asterisk. A two-dimensional ^{13}C–^{13}C homonuclear correlation spectrum obtained with a 10 ms PDSD mixing time shows well-resolved aliphatic and carbonyl resonances in Figure 18.9A and B. Various amino acid types can be readily identified by their characteristic correlation patterns and unique chemical shifts, including threonine, serine, valine, alanine, and isoleucine. All resonances were further traced using 20, 50, 100, and 500 ms mixing times in the ^{13}C–^{13}C homonuclear correlation spectra. The linewidths measured in the direct dimension, for well-resolved ^{13}C signals, such as those from the Cα, Cβ, and Cγ atoms of the sidechains, are in the range of 0.5–0.8 ppm (90–150 Hz at 750 MHz). In NCA and NCO experiments, polarization is transferred from the backbone amide nitrogen atom to the Cα and carbonyl-carbons, which assists in correlating intraresidue and sequential backbone residues. The NCA transfer is intraresidue, since it connects the amide ^{15}N and ^{13}Cα resonances of the same residue; the NCO transfer is interresidue, since it connects the amide ^{15}N resonance with the CO resonance of the preceding residue, as shown in

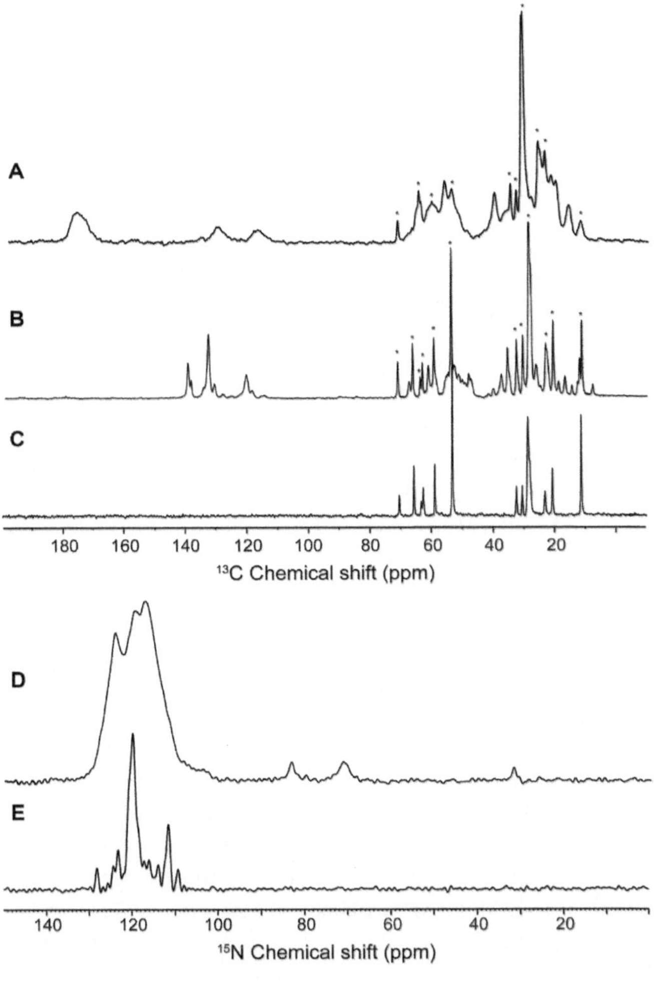

Figure 18.8 Spectra showing signals from mobile and immobile sites of 1TM-CXCR1(1–72). ^{15}N and ^{13}C NMR spectra of uniformly ^{13}C,^{15}N-labeled 1TM-CXCR1(1–72) in DMPC liposomes. The spectra were obtained at a ^{1}H resonance frequency of 750 MHz at 25 °C. (A) and (B) are one-dimensional ^{13}C NMR cross-polarization (solid-state NMR) and refocused INEPT (solution NMR) spectra acquired with 256 scans and 512 scans respectively. (C) ^{13}C NMR spectrum of hydrated DMPC. (D) and (E) are one-dimensional ^{15}N NMR spectra obtained by cross-polarization (solid-state NMR) and refocused INEPT (solution NMR), acquired with 512 and 1000 scans, respectively.

Figure 18.9C and D. A 40 ms PDSD mixing interval, following the SPECIFIC-CP transfer, enabled two-dimensional NCACX and NCOCX spectra to be obtained. This allowed the correlation of signals from sidechain carbons to those of backbone Ca, CO, and amide N atomic sites.

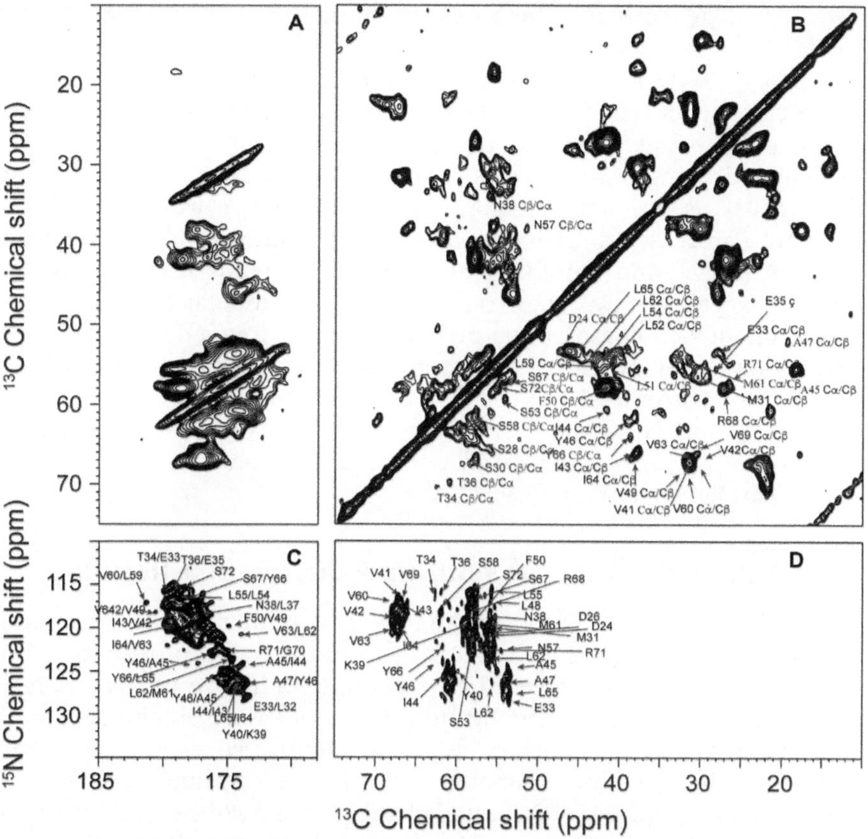

Figure 18.9 Assignments of resonances from 1TM-CXCR1(1–72) in a contour plot of a two-dimensional $^{13}C/^{13}C$ PDSD spectrum. (A) Carbonyl carbon region. (B) Aliphatic carbon region. (C) NCO and (D) NCA spectral regions. (A)–(D) were obtained from a sample of uniformly $^{15}N/^{13}C$ 1TM-CXCR1(1–72) reconstituted in DMPC proteoliposomes. All spectra were recorded at a field of 750 MHz, a spinning frequency of 11.11 kHz, and a sample temperature of 10 °C. For the $^{13}C/^{13}C$ PDSD experiments, magnetization between ^{13}C spins was exchanged using a 10 ms mixing period, resulting in predominantly one-bond correlations and a limited number of two-bond intraresidue correlations. For the two-dimensional NCO and NCA spectra, polarization between ^{1}H and ^{15}N was exchanged using a 1 ms spin-lock cross-polarization (CP) period; selective transfers from ^{15}N to $^{13}C'$ or $^{13}C_{\alpha}$ were achieved with a 4 ms SPECIFIC CP pulse.

With the assistance of spectra from three-dimensional NCACX and NCOCX experiments, and two-dimensional PDSD experiments performed with various contact times, it was possible to assign ~70% of the resonances observed in the 1TM-CXCR1(1–72) spectra, namely the signals from residues 24–72. The remainder of the protein is mobile and thus could not be assigned using CP-based methods. The resonance assignments are shown in

Figure 18.9. The carbon resonances from the mobile regions of 1TM CXCR1(1–72) were assigned using TOBSY (total through-bond correlation spectroscopy). TOBSY utilizes *J*-couplings to correlate ^{13}C atoms, which means it can only be used for residues that are mobile enough to average out the dipolar couplings.[60] Signals from residues 1–23 in 1TM-CXCR1(1–72) were assigned using TOBSY, complemented by spectra from ^{1}H/^{13}C and ^{1}H/^{15}N INEPT experiments (Figure 18.13). Similarly, we repeated the assignment process on samples containing uniformly ^{13}C,^{15}N-labeled 1TM-CXCR1(1–72) and unlabeled IL-8(1–66). Two-dimensional ^{15}N detected SLF experiment was used to measure ^{1}H–^{15}N dipolar couplings from the experimentally measured perpendicular edge frequencies of the respective rotationally averaged powder patterns. Variation in these dipolar couplings results from differences in the alignment of sites in the protein relative to the bilayer normal. The perpendicular edge frequencies were multiplied by four to obtain the full dipolar coupling value, corresponding to twice the parallel edge frequency.

18.3.2 Interactions and Dynamics: Experimental Results

Binding a ligand perturbs the local chemical environment of the interacting amino acids and changes the chemical shifts of the nuclei proximate to the binding site. By detecting spectral changes upon addition of IL-8, atomic-resolution details of the protein–ligand interactions were obtained.

MAS solid-state NMR spectra had sufficient resolution to observe site-specific changes in resonances of the N-terminal domain of 1TM-CXCR1(1–72) upon binding unlabeled IL-8(1–66). Since sections of the protein undergo local motions on different time scales, two different types of experiments were necessary. Cross-polarization-based experiments were used to study the more rigid residues of 1TM-CXCR1(1–72) and INEPT-based experiments captured the signals from the more mobile residues, such as at the N-terminal domain. Figure 18.10 shows two CP-based PDSD ^{13}C/^{13}C correlation spectra, the one on the left was obtained on a sample of uniformly labeled ^{13}C, ^{15}N 1TM-CXCR1(1–72), and the spectrum on the right on uniformly labeled ^{13}C, ^{15}N 1TM-CXCR1(1–72) with unlabeled IL-8(1–66) bound. All signals were assigned to specific residues. Using PDSD only residues 24–72 of 1TM-CXCR1(1–72) were detected, which includes part of the N-terminal domain closest to the transmembrane helix, the transmembrane helix, and the intracellular loop. Differences between the two spectra recorded on 1TM-CXCR1(1–72) with and without IL-8(1–66) bound are highlighted with boxes. Additional signals were detected in the spectrum of 1TM-CXCR1(1–72) with IL-8(1–66) bound, indicating several additional residues were immobilized by the interaction with the chemokine, although complete assignments were not feasible. Comparison between Figure 18.10A and 18.10B also show that upon interaction with IL-8(1–66) several signals in the CP-based PDSD ^{13}C/^{13}C correlation spectrum undergo chemical shift perturbations. These are likely due to changes in the chemical environment

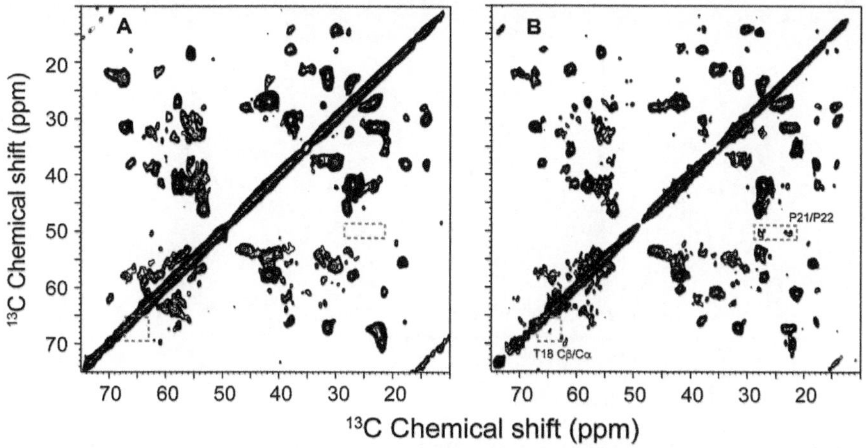

Figure 18.10 Spectral changes upon IL-8(1–66) binding to CXCR1 observed in two-dimensional ^{13}C–^{13}C correlation experiments. (A) Aliphatic region of the two-dimensional ^{13}C, ^{13}C proton-driven spin diffusion (PDSD) spectrum obtained from uniformly ^{13}C/^{15}N-labeled 1TM-CXCR1(1–72). (B) As in (A) in the presence of unlabeled IL-8(1–66). The aliphatic regions of the ^{13}C/^{13}C correlation spectra were obtained with 10 ms mixing intervals. Differences between the two spectra are highlighted within the dotted boxes.

from the protein–ligand interactions. The reproducibility of the solid-state NMR data in terms of chemical shift and line width was evaluated by repeating the measurements multiple times on different samples using the same protocol and experimental conditions. The data sets showed chemical shifts from well-resolved peaks are reproducible within +/− 0.1 ppm.

The first 23 residues of the N-terminal domain of 1TM-CXCR1(1–72) are not observed in CP-based experiments. Therefore, the use of INEPT and TOBSY were essential to characterize changes upon complex formation between 1TM-CXCR1(1–72) and unlabeled IL-8. Significant frequency changes were observed in ^{1}H/^{15}N and ^{1}H/^{13}C INEPT experiments obtained on uniformly ^{15}N,^{13}C-labeled 1TM-CXCR1(1–72) upon interaction with unlabeled IL-8(1–66) (Figure 18.11A and C *vs.* Figure 18.11B and D). A few residues in 1TM-CXCR1(1–72) were immobilized upon binding IL-8(1–66). Immobilization is distinguished by the disappearance of signals due to rotational correlation times incompatible with INEPT-based experiments. For instance, one of the residues easily identified in the spectra due to its distinct chemical shift, the tryptophan 10 side chain, became immobilized upon binding to IL8(1–66) and its corresponding signal could no longer be detected. Intensity observed in the ^{1}H/^{13}C INEPT spectra due to natural abundance ^{13}C in lipids is marked with an asterisk.

Two-dimensional INEPT spectra provided improved resolution (see Figure 18.12). Several well-separated signals in two-dimensional ^{1}H/^{15}N INEPT spectra illustrate the effect of IL-8(1–66) binding to the receptor. For

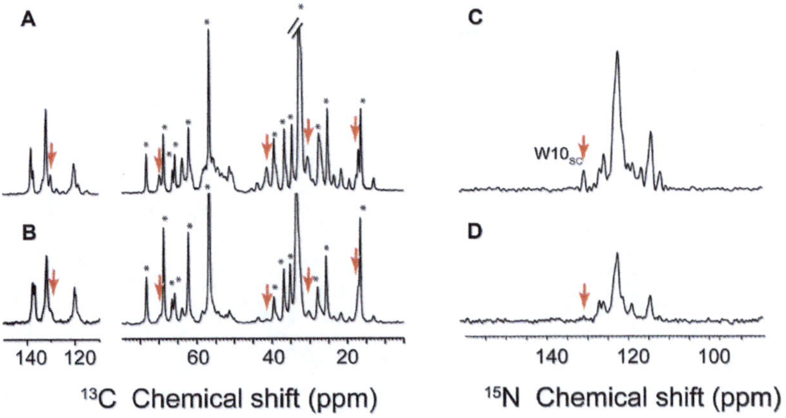

Figure 18.11 INEPT spectra show evidence of reduced dynamics in 1TM-CXCR1 upon binding of IL-8(1–66). (A, C) MAS solid-state NMR INEPT experiments performed on uniformly $^{15}N/^{13}C$-labeled 1TM-CXCR1 reconstituted in DMPC proteoliposomes. (B, D) same as in (A) and (C), but with an equimolar amount of ^{14}N IL-8(1–66) added to the sample. INEPT experiments only show signals associated with mobile residues. The 110–160 ppm region corresponds to signals from aromatic residues. A large loss in signal intensity is observed in this region upon binding to IL-8(1–66). The 10–75 ppm region corresponds to signals from aliphatic protein residues and lipids. Signals from the lipids are indicated with asterisks.

example, the resonances assigned to W10, G19, and T18 are present in the spectra of 1TM-CXCR1(1–72) by itself (Figure 18.12A and C), but missing in the spectra of 1TM-CXCR1(1–72) bound to IL-8(1–66) (Figure 18.12B and D).

TOBSY experiments enabled the analysis of the structural changes in the flexible N-terminal domain of 1TM-CXCR1(1–72) induced by interactions with IL-8(1–66). In TOBSY experiments, intraresidue correlations of aliphatic carbon resonances of mobile residues are correlated using a polarization transfer through ^{13}C–^{13}C resonances *via J*-couplings. The first 23 residues of the flexible N-terminal domain in 1TM-CXCR1(1–72) were assigned using this method (Figure 18.13A). Upon binding IL-8(1–66), only a few signals could be detected in TOBSY spectra of 1TM-CXCR1(1–72), in contrast to those of the apo-receptor. These changes are likely due to the restriction of motions of those residues that are in direct contact with bound IL-8(1–66). The signals that could not be detected in TOBSY spectra upon binding IL-8(1–66) are from residues located in the middle of the N-terminal domain: M9, W10, D11, F12, D13, D14, L15, N16, F17, T18, G19, and M20. Interestingly, residues P21, P22, and A23 could still be observed in the TOBSY spectrum of the complex, indicating they retain significant mobility. Also, the first eight residues of 1TM-CXCR1(1–72) are mobile when bound to IL-8(1–66). Notably, IL-8(1–66) binding to the N-terminal domain of 1TM-CXCR1(1–72) perturbed the environment of these amino acids sufficiently to alter their chemical shifts (Figure 18.14B).

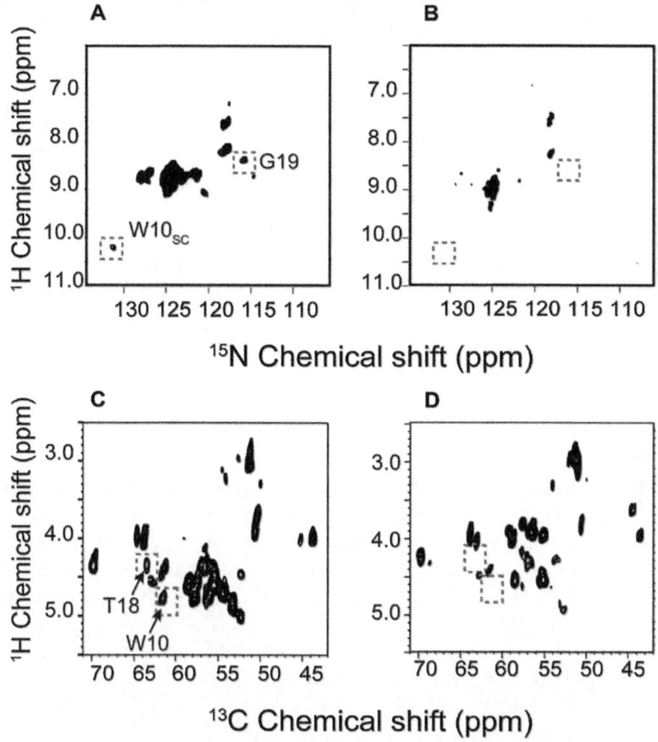

Figure 18.12 Two-dimensional INEPT spectra resolved signals from mobile residues in the 1TM-CXCR1(1–72)–IL-8(1–66) complex. (A, C) 11.11 kHz MAS NMR spectra obtained from uniformly $^{15}N/^{13}C$-labeled 1TM-CXCR1(1–72) reconstituted in DMPC proteoliposomes at 25 °C (B, D) same as in (A) and (C), but with an equimolar amount of ^{14}N IL-8(1–66) added to the sample. (C, D) Spectra are focused on the chemical shift region of aliphatic carbons. The boxed zoomed spectra in the upper-left corner depict the aromatic region of $^{1}H/^{13}C$ INEPT spectra. Lipid signals have been digitally subtracted from the spectra.

18.3.3 Discussion

IL-8 is one of the best-studied chemokines. It recruits neutrophils during the inflammatory immune response by binding and activating the chemokine receptors CXCR1 and CXCR2.[61,62] The protein is generally thought to interact with two main regions of the receptor; one binding site is located within the flexible N-terminal domain and the other is composed of extracellular loops 2 and 3. Previously, the main focus has been on the identification of those residues in IL-8 involved in the interactions with the N-terminal binding site on CXCR1 primarily through mutational analysis and NMR spectroscopy.[17,18,63–66] We used the truncated construct IL-8(1–66) in the experiments. It is now well-established that three regions of the protein interact with the N-terminal domain of CXCR1: the N-loop, the third β-sheet, and the N-terminal part of the α-helix.[18] By contrast, very few studies

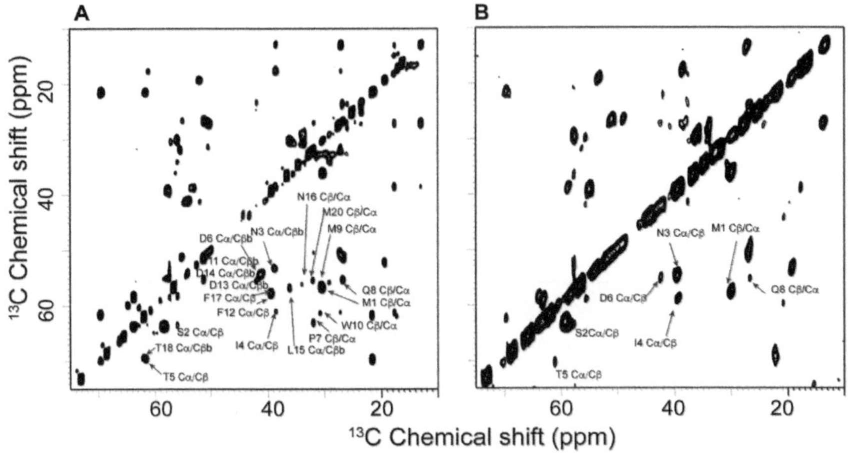

Figure 18.13 Two-dimensional TOBSY $^{13}C/^{13}C$ correlation spectra resolved signals from mobile residues in 1TM-CXCR1(1–72) upon IL-8(1–66) complex formation. TOBSY experiments correlate aliphatic intraresidue carbon signals of dynamic regions by transferring the polarization through ^{13}C–^{13}C bonds *via* J couplings. (A) $^{13}C/^{13}C$ TOBSY spectrum recorded on uniformly $^{15}N/^{13}C$-labeled 1TM-CXCR1(1–72). (B) Same as in (A), but with an equimolar amount of ^{14}N IL-8(1–66) added to the sample. Experiments were carried out at 25 °C.

have been carried out that describe the effects of binding IL-8 on CXCR1. Prior experimental studies that have been carried out have used soluble peptides: one used the first 38 residues of the N-terminal domain of CXCR1 in proteoliposomes,[63] another used a synthetic peptide corresponding to the first 40 residues of the N-terminal domain of CXCR1,[67] and several studies were performed using a construct corresponding to residues 9–29 of CXCR1, where residues 15–19 were replaced with a hexanoic acid moiety.[63,64,68,69] These studies found that residues 9–29 are important for the interaction with IL-8, especially residues D14, G19, P21, P22, D24, E25, D26, and Y27. We have expressed a CXCR1 construct 1TM-CXCR1(1–72), which contains the entire flexible N-terminal domain, the first transmembrane helix and the first intracellular loop of CXCR1. By performing solid-state NMR experiments on 1TM-CXCR1(1–72) in phospholipid bilayers, it was possible to map the interaction of IL-8(1–66) onto the N-terminal binding site of the receptor, identify the changes in dynamics upon binding, and observe tilt angle changes in the helix. This is the first time that changes in dynamics in CXCR1 induced by interactions of IL-8 with the N-terminal binding site have been observed as well as chemical shift perturbations in 1TM-CXCR1(1–72) upon binding to IL-8 in $^{13}C/^{13}C$ correlation spectra. The magnitudes of all chemical shift changes are quantified using the equation $\Delta\delta = 1/2\sqrt{(\Delta C/2)^2 + (\Delta N/5)^2}$, where ΔC and ΔN are the changes in carbon and nitrogen chemical shift frequencies, respectively. The chemical

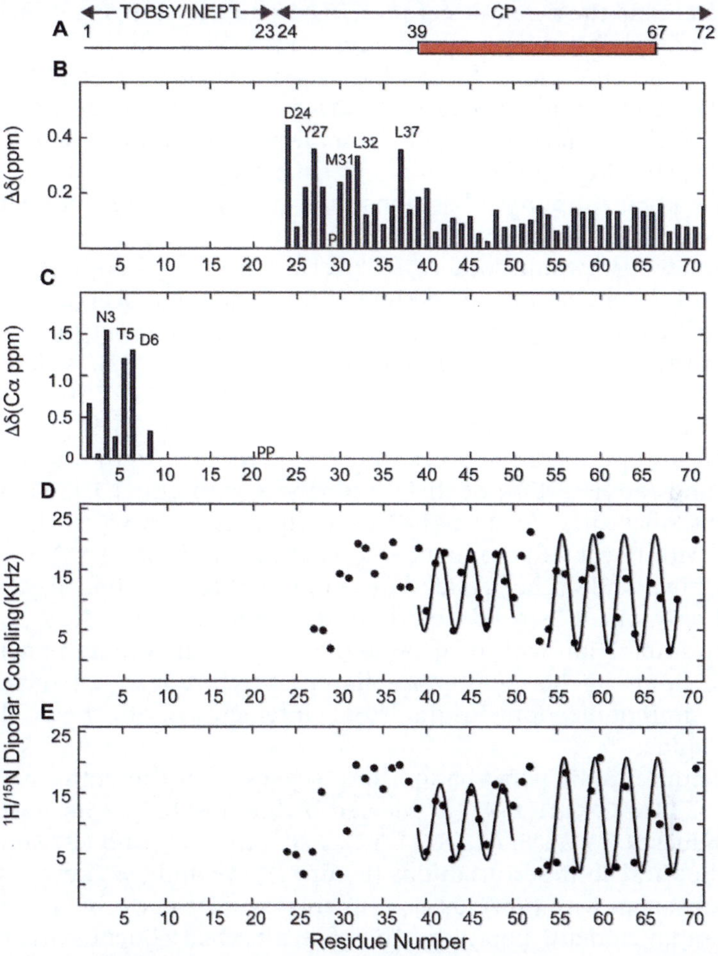

Figure 18.14 (A) Spectral perturbations that occur in spectra of 1TM-CXCR1(1–72) upon interaction with IL-8(1–66) as a function of residue number. (B) Chemical shift perturbations in the immobile region of the protein from NCA data (see Figure 18.9). (C) Chemical shift perturbation in the mobile region of the protein from TOBSY data (see Figure 18.13). Chemical shift perturbations for residues 9–20 are not shown here either because they cannot be detected by these experiments (CP or TOBSY), due to a change in dynamics, or because they cannot be assigned in both samples with and without the ligand. (D, E) Dipolar waves obtained by plotting measured values of the ^{1}H–^{15}N DCs *versus* residue number for (D) 1TM-CXCR1(1–72) and for (E) 1TM-CXCR1(1–72) in the presence of unlabeled IL-8(1–66).

shift perturbations upon interaction with IL-8(1–66) are summarized in Figure 18.14B and C. The largest differences map to the N-terminal domain, in particular residues D24, Y27, M31, L32, and L37, while resonances from amino acids in the transmembrane helix itself remained largely unperturbed.

Characterizing the dynamics of proteins over a wide range of timescales is challenging for all forms of spectroscopy, including solid-state NMR spectroscopy. It is very helpful to be able to generate the observable magnetization through different coupling mechanisms because the methods provide reliable information about molecular dynamics. For example, TOBSY, which relies on *J*-coupling to transfer magnetization in the absence of strong dipolar couplings,[70] demonstrates that the motions of the residues are fast compared to the 10^{-6} s timescale required to average out the dipolar couplings, which would otherwise overwhelm the *J*-couplings, making magnetization transfer by this method impractical. Moreover, this timescale is close to the lower limit of measurements of the relaxation times T_1 and T_2. Based on the resonance assignments of 1TM-CXCR1(1–72), which utilized TOBSY and INEPT spectra, it is clear that residues 1–23 of the N-terminal domain are mobile. The data also show that residues 24–37 undergo motions on an intermediate timescale that interferes with magnetization transfer and the detection of their resonances with either INEPT (solution NMR) or CP-based (solid-state NMR) experiments. Even when the receptor interacts with IL-8(1–66) the first eight residues and residue A23 were found to be mobile, while other resides were immobilized so that they could be detected only with CP-based experiments. Previous studies have suggested that upon interaction with IL-8(1–66), the N-terminal domain is not associated with lipids in the membrane bilayer, therefore it is unlikely that the observed immobilization results from interaction with both IL-8 and phospholipids.[18,63]

The chemical shift perturbation data suggest that the entire N-terminal domain of 1TM-CXCR1(1–72) is affected by binding IL-8(1–66). The signals from residues N3, T5, D6, A23, D24, Y27, S30, M31, L32, and L37 showed the largest chemical shift perturbations (Figure 18.14B and C). These results are largely consistent with previous experimental studies performed on a variety of N-terminal domain peptides. These results also concur with two molecular dynamics simulations that docked IL-8 to an N-terminal domain peptide[71] and with full-length CXCR1 in POPC bilayers.[72] Some residues whose resonances undergo chemical shift perturbations upon interaction with IL-8(1–66) have not been identified as part of the binding site before, *e.g.*, residues 3–6 and leucine 37. It is possible that the chemical shift perturbation of L37 may be due to its proximity to the transmembrane helix, which starts at residue 39, and the presence of a lipid bilayer, rather direct molecular interactions.

No structure of CXCR1 in an active conformation has been described, to date. However, several other class A GPCR structures have been solved in an active conformation and compared to their inactive conformation, which provides insight into the structural rearrangements needed to start the signaling cascade inside the cell. In all of those structures, several transmembrane helices change tilt angles, mainly helix 3, helix 5, and helix 6.[73] Dipolar waves can be obtained by plotting measured values of the $^{1}H–^{15}N$ dipolar couplings as a function of residue number. Dipolar waves have characteristic features that allow helix length, helix orientation in the membrane, and the presence of any kinks to be readily identified.[74] Two separate dipolar waves with distinct amplitudes were found, indicating that the transmembrane helix of 1TM-CXCR1(1–72) has a kink about halfway through the membrane. Upon interaction with IL-8(1–66) no tilt angle changes were observed in 1TM-CXCR1(1–72) (Figure 18.7D and E). Both chemical shift perturbation and dipolar waves analyses indicate that the transmembrane helix is not involved in binding IL-8 and does not undergo conformation changes upon binding. Binding-induced differences were observed in the N-terminal domain residues 9–20, which are immobilized upon binding to IL-8(1–66) and in residues N3, T5, D6, A23, D24, Y27, M31 L32, and L37, which exhibit substantial chemical shift perturbations.

By exploiting the flexibility of solid-state NMR spectroscopy, not only could high-resolution spectra of polypeptides in membrane bilayers be obtained, but also these spectra displayed differences in resonance frequencies resulting from the presence or absence of binding of the chemokine to its receptor. The information about changes in environment resulting from protein–protein interactions that are reflected in chemical shift frequencies is supplemented by measurements that reflect changes in dynamics of both the chemokine and its receptor upon their binding. Notably, all NMR results inherently have atomic resolution. As a result, future studies have the potential to provide complete descriptions of how chemokine binding to its receptor starts the signaling cascade through the receptor to its associated G-protein and subsequently other players in the immune system.

Acknowledgements

This research was supported by grants P41EB002031 and R35GM122501 from the National Institutes of Health and utilized the Biomedical Technology Resource Center for NMR Molecular Imaging of Proteins at the University of California, San Diego.

References

1. S. J. Opella and F. M. Marassi, Applications of NMR to membrane proteins, *Arch. Biochem. Biophys.*, 2017, **628**, 92–101.
2. S. J. Singer and G. L. Nicolson, The fluid mosaic model of the structure of cell membranes, *Science*, 1972, **175**(4023), 720–731.
3. J. S. Waugh, L. M. Huber and U. Haeberlen, Approach to High-Resolution nmr in Solids, *Phys. Rev. Lett.*, 1968, **20**(5), 180–182.
4. A. Pines, M. G. Gibby and J. S. Waugh, Proton-Enhanced Nuclear Induction Spectroscopy. A Method for High Resolution NMR of Dilute Spins in Solids. The, *J. Chem. Phys.*, 1972, **56**(4), 1776–1777.
5. J. Schaefer and E. O. Stejskal, Carbon-13 nuclear magnetic resonance of polymers spinning at the magic angle, *J. Am. Chem. Soc.*, 1976, **98**(4), 1031–1032.

6. D. M. Rosenbaum, S. G. Rasmussen and B. K. Kobilka, The structure and function of G-protein-coupled receptors, *Nature*, 2009, **459**(7245), 356–363.
7. A. S. Hauser, M. M. Attwood, M. Rask-Andersen, H. B. Schioth and D. E. Gloriam, Trends in GPCR drug discovery: new agents, targets and indications, *Nat. Rev. Drug Discovery*, 2017, **16**(12), 829–842.
8. A. J. Kooistra, S. Mordalski, G. Pandy-Szekeres, M. Esguerra, A. Mamyrbekov and C. Munk, *et al.*, GPCRdb in 2021: integrating GPCR sequence, structure and function, *Nucl. Acids Res.*, 2021, **49**(D1), D335–D343.
9. S. J. Opella, Structure determination of membrane proteins in their native phospholipid bilayer environment by rotationally aligned solid-state NMR spectroscopy, *Acc. Chem. Res.*, 2013, **46**(9), 2145–2153.
10. S. H. Park, B. B. Das, F. Casagrande, Y. Tian, H. J. Nothnagel and M. Chu, *et al.*, Structure of the chemokine receptor CXCR1 in phospholipid bilayers, *Nature*, 2012, **491**(7426), 779–783.
11. F. Sallusto and M. Baggiolini, Chemokines and leukocyte traffic, *Nat. Immunol.*, 2008, **9**(9), 949–952.
12. G. M. Clore, E. Appella, M. Yamada, K. Matsushima and A. M. Gronenborn, Three-dimensional structure of interleukin 8 in solution, *Biochemistry*, 1990, **29**(7), 1689–1696.
13. E. T. Baldwin, I. T. Weber, R. Charles, J. C. Xuan, E. Appella and M. Yamada, *et al.*, Crystal structure of interleukin 8: symbiosis of NMR and crystallography, *Proc. Natl. Acad. Sci. U. S. A.*, 1991, **88**(2), 502–506.
14. K. Rajarathnam, I. Clark-Lewis and B. D. Sykes, 1H NMR solution structure of an active monomeric interleukin-8, *Biochemistry*, 1995, **34**(40), 12983–12990.
15. A. Ravindran, P. R. Joseph and K. Rajarathnam, Structural basis for differential binding of the interleukin-8 monomer and dimer to the CXCR1 N-domain: role of coupled interactions and dynamics, *Biochemistry*, 2009, **48**(37), 8795–8805.
16. K. Rajarathnam, G. N. Prado, H. Fernando, I. Clark-Lewis and J. Navarro, Probing receptor binding activity of interleukin-8 dimer using a disulfide trap, *Biochemistry*, 2006, **45**(25), 7882–7888.
17. H. Fernando, C. Chin, J. Rosgen and K. Rajarathnam, Dimer dissociation is essential for interleukin-8 (IL-8) binding to CXCR1 receptor, *J. Biol. Chem.*, 2004, **279**(35), 36175–36178.
18. S. Berkamp, S. H. Park, A. A. De Angelis, F. M. Marassi and S. J. Opella, Structure of monomeric Interleukin-8 and its interactions with the N-terminal Binding Site-I of CXCR1 by solution NMR spectroscopy, *J. Biomol. NMR*, 2017, **69**(3), 111–121.
19. X. Ding, X. Zhao and A. Watts, G-protein-coupled receptor structure, ligand binding and activation as studied by solid-state NMR spectroscopy, *Biochem. J.*, 2013, **450**(3), 443.
20. J. Urbina and J. S. Waugh, Proton-enhanced 13C nuclear magnetic resonance of lipids and biomembranes, *Proc. Natl. Acad. Sci. U. S. A.*, 1974, **71**(12), 5062–5067.

21. M. J. Bogusky, G. C. Leo and S. J. Opella, Comparison of the dynamics of the membrane-bound form of fd coat protein in micelles and in bilayers by solution and solid-state nitrogen-15 nuclear magnetic resonance spectroscopy, *Proteins*, 1988, **4**(2), 123–130.
22. B. A. Lewis, G. S. Harbison, J. Herzfeld and R. G. Griffin, NMR structural analysis of a membrane protein: bacteriorhodopsin peptide backbone orientation and motion, *Biochemistry*, 1985, **24**(17), 4671–4679.
23. A. A. De Angelis, A. A. Nevzorov, S. H. Park, S. C. Howell, A. A. Mrse and S. J. Opella, High-resolution NMR spectroscopy of membrane proteins in aligned bicelles, *J. Am. Chem. Soc.*, 2004, **126**(47), 15340–15341.
24. A. A. De Angelis, S. C. Howell, A. A. Nevzorov and S. J. Opella, Structure determination of a membrane protein with two trans-membrane helices in aligned phospholipid bicelles by solid-state NMR spectroscopy, *J. Am. Chem. Soc.*, 2006, **128**(37), 12256–12267.
25. S. H. Park, S. Prytulla, A. A. De Angelis, J. M. Brown, H. Kiefer and S. J. Opella, High-resolution NMR spectroscopy of a GPCR in aligned bicelles, *J. Am. Chem. Soc.*, 2006, **128**(23), 7402–7403.
26. S. H. Park, S. Berkamp, G. A. Cook, M. K. Chan, H. Viadiu and S. J. Opella, Nanodiscs versus macrodiscs for NMR of membrane proteins, *Biochemistry*, 2011, **50**(42), 8983–8985.
27. S. J. Opella, Solid-state NMR and membrane proteins, *J. Magn. Reson.*, 2015, **253**, 129–137.
28. T. Gopinath, D. Weber, S. Wang, E. Larsen and G. Veglia, Solid-State NMR of Membrane Proteins in Lipid Bilayers: To Spin or Not To Spin?, *Acc. Chem. Res.*, 2021, **54**(6), 1430–1439.
29. M. Hong and T. Doherty, Orientation determination of membrane-disruptive proteins using powder samples and rotational diffusion: A simple solid-state NMR approach, *Chem. Phys. Lett.*, 2006, **432**(1), 296–300.
30. S. D. Cady, C. Goodman, C. D. Tatko, W. F. DeGrado and M. Hong, Determining the orientation of uniaxially rotating membrane proteins using unoriented samples: a 2H, 13C, AND 15N solid-state NMR investigation of the dynamics and orientation of a transmembrane helical bundle, *J. Am. Chem. Soc.*, 2007, **129**(17), 5719–5729.
31. S. H. Park, B. B. Das, A. A. DeAngelis, M. Scrima and S. J. Opella, Mechanically, Magnetically, and "Rotationally Aligned" Membrane Proteins in Phospholipid Bilayers Give Equivalent Angular Constraints for NMR Structure Determination, *J. Phys. Chem. B*, 2010, **114**, 13995–14003.
32. R. Smith and B. A. Cornell, The dynamics of the intrinsic membrane polypeptide gramicidin a in phospholipid bilayers: a solid state carbon-13 NMR study, *Biophys. J.*, 1986, **49**(1), 117–118.
33. B. A. Cornell, F. Separovic, A. J. Baldassi and R. Smith, Conformation and orientation of gramicidin a in oriented phospholipid bilayers measured by solid state carbon-13 NMR, *Biophys. J.*, 1988, **53**(1), 67–76.
34. F. Tian, Z. Song and T. A. Cross, Orientational constraints derived from hydrated powder samples by two-dimensional PISEMA, *J. Magn. Reson.*, 1998, **135**(1), 227–231.

35. S. H. Park, A. A. Mrse, A. A. Nevzorov, A. A. De Angelis and S. J. Opella, Rotational diffusion of membrane proteins in aligned phospholipid bilayers by solid-state NMR spectroscopy, *J. Magn. Reson.*, 2006, **178**(1), 162–165.
36. B. B. Das, H. J. Nothnagel, G. J. Lu, W. S. Son, Y. Tian and F. M. Marassi, *et al.*, Structure determination of a membrane protein in proteoliposomes, *J. Am. Chem. Soc.*, 2012, **134**(4), 2047–2056.
37. B. B. Das, S. H. Park and S. J. Opella, Membrane protein structure from rotational diffusion, *Biochim. Biophys. Acta*, 2015, **1848**(1 Pt B), 229–245.
38. O. Anddronesi, S. Becker, K. Seidel, H. Heise, H. Young and M. Baldus, Determiation of membrane protein structure and dyamics by magi-angle-spinning solid-state NMR spectroscopy, *J. Am. Chem. Soc.*, 2005, **127**, 12965–12974.
39. S. Cady, C. Goodman, C. Tatko, W. DeGrado and M. Hong, Determining the orientation of uniaxially rotating membrane proteins using unoriented samples: a H-2, C-13, and N-15 solid-state NMR investigation of the dynamics and orientation of a transmembrane helical bundle, *J. Am. Chem. Soc.*, 2007, **129**, 5719–5729.
40. G. C. Leo, L. A. Colnago, K. G. Valentine and S. J. Opella, Dynamics of fd coat protein in lipid bilayers, *Biochemistry*, 1987, **26**, 854–862.
41. M. J. Bogusky, R. A. Schiksnis, G. C. Leo and S. J. Opella, Protein backbone dynamics by solid state and solution 15N NMR spectroscopy, *J. Magn. Reson.*, 1987, **72**, 186–190.
42. M. Etzkorn, S. Martell, O. C. Andronesi, K. Seidel, M. Engelhard and M. Baldus, Secondary structure, *dynamics, and topology of a seven-helix receptor in native membranes, studied bby solid-state NMR spectroscopy. Angew Chem Int Ed*, 2007, **46**, 459–462.
43. A. McDermott, Structure and dynamics of membrane proeins by magic angle spinning solid-state NMR, *Annu. Rev. Biophys.*, 2009, **38**, 385–403.
44. B. Wylie, L. Sperling, H. Frericks, G. Shah, W. Franks and C. Rienstra, Chemical-shift anisotropy measurements of amide and carbonyl resonances in a microcrystalline protein with slow magic-angle spinning NMR spectroscopy, *J. Am. Chem. Soc.*, 2007, **129**, 5318–5319.
45. C. R. Sanders, B. J. Hare, K. P. Howard and J. H. Prestegard, Magnetically-oriented phospholipid micelles as a tool for the study of membrane-associated molecules, *Prog. Nucl. Magn. Reson. Spectrosc.*, 1994, **26**(Part 5), 421–444.
46. R. S. Prosser, F. Evanics, J. L. Kitevski and M. S. Al-Abdul-Wahid, Current applications of bicelles in NMR studies of membrane-associated amphiphiles and proteins, *Biochemistry*, 2006, **45**(28), 8453–8465.
47. R. T. Clubb, J. G. Omichinski, G. M. Clore and A. M. Gronenborn, Mapping the binding surface of interleukin-8 complexed with an N-terminal fragment of the type 1 human interleukin-8 receptor, *FEBS Lett.*, 1994, **338**(1), 93–97.

48. P. Loetscher, M. Seitz, I. Clark-Lewis, M. Baggiolini and B. Moser, Both interleukin-8 receptors independently mediate chemotaxis. Jurkat cells transfected with IL-8R1 or IL-8R2 migrate in response to IL-8, GRO alpha and NAP-2, *FEBS Lett.*, 1994, **341**(2–3), 187–192.
49. W. Fu, Y. Zhang, J. Zhang and W. F. Chen, Cloning and characterization of mouse homolog of the CXC chemokine receptor CXCR1, *Cytokine*, 2005, **31**(1), 9–17.
50. L. Skrabanek, F. Campagne and H. Weinstein, Building protein diagrams on the web with the residue-based diagram editor RbDe, *Nucleic Acids Res.*, 2003, **31**(13), 3856–3858.
51. R. S. Thakur, N. D. Kurur and P. K. Madhu, An experimental study of decoupling sequences for multiple-quantum and high-resolution MAS experiments in solid-state NMR, *Magn. Reson. Chem.*, 2008, **46**(2), 166–169.
52. R. S. Thakur, N. D. Kurur and P. K. Madhu, Swept-frequency two-pulse phase modulation for heteronuclear dipolar decoupling in solid-state NMR, *Chem. Phys. Lett.*, 2006, **426**(4), 459–463.
53. N. M. Szeverenyi, M. J. Sullivan and G. E. Maciel, Observation of spin exchange by two-dimensional fourier transform 13C cross polarization-magic-angle spinning, *J. Magn. Reson. (1969)*, 1982, **47**(3), 462–475.
54. M. H. Frey and S. J. Opella, Carbon-13 spin exchange in amino acids and peptides, *J. Am. Chem. Soc.*, 1984, **106**(17), 4942–4945.
55. M. Baldus, A. T. Petkova, J. Herzfeld and R. G. Griffin, Cross polarization in the tilted frame: assignment and spectral simplification in heteronuclear spin systems, *Mol. Phys.*, 1998, **95**(6), 1197–1207.
56. J. S. Waugh, Uncoupling of local field spectra in nuclear magnetic resonance: determination of atomic positions in solids, *Proc. Natl. Acad. Sci. U. S. A.*, 1976, **73**(5), 1394.
57. X. Zhao, M. Edén and M. H. Levitt, Recoupling of heteronuclear dipolar interactions in solid-state NMR using symmetry-based pulse sequences, *Chem. Phys. Lett.*, 2001, **342**(3), 353–361.
58. R. Keller The CARA/Lua Programmers Manual. *DATONALAG*. 2003.
59. G. A. Morris and R. Freeman, Enhancement of Nuclear Magnetic-Resonance Signals by Polarization Transfer, *J. Am. Chem. Soc.*, 1979, **101**(3), 760–762.
60. J. Leppert, O. Ohlenschlager, M. Gorlach and R. Ramachandran, Adiabatic TOBSY in rotating solids, *J. Biomol. NMR*, 2004, **29**(2), 167–173.
61. P. M. Murphy and H. L. Tiffany, Cloning of complementary DNA encoding a functional human interleukin-8 receptor, *Science*, 1991, **253**(5025), 1280–1283.
62. W. E. Holmes, J. Lee, W. J. Kuang, G. C. Rice and W. I. Wood, Structure and functional expression of a human interleukin-8 receptor, *Science*, 1991, **253**(5025), 1278–1280.
63. S. H. Park, F. Casagrande, L. Cho, L. Albrecht and S. J. Opella, Interactions of interleukin-8 with the human chemokine receptor CXCR1 in

phospholipid bilayers by NMR spectroscopy, *J. Mol. Biol.*, 2011, **414**(2), 194–203.

64. A. A. Kendrick, M. J. Holliday, N. G. Isern, F. Zhang, C. Camilloni and C. Huynh, *et al.*, The dynamics of interleukin-8 and its interaction with human CXC receptor I peptide, *Protein Sci.*, 2014, **23**(4), 464–480.
65. P. R. B. Joseph, J. M. Sarmiento, A. K. Mishra, S. T. Das, R. P. Garofalo and J. Navarro, *et al.*, Probing the Role of CXC Motif in Chemokine CXCL8 for High Affinity Binding and Activation of CXCR1 and CXCR2 Receptors, *J. Biol. Chem.*, 2010, **285**(38), 29262–29269.
66. P. R. B. Joseph and R. Krishna, Solution NMR characterization of WT CXCL8 monomer and dimer binding to CXCR1 N-terminal domain, *Protein Sci.*, 2015, **24**(1), 81–92.
67. R. T. Clubb, J. G. Omichinski, G. M. Clore and A. M. Gronenborn, Mapping the binding surface of interleukin-8 complexed with an N-terminal fragment of the Type 1 human interleukin-8 receptor, *FEBS Lett.*, 1994, **338**(1), 93–97.
68. N. J. Skelton, C. Quan, D. Reilly and H. Lowman, Structure of a CXC chemokine-receptor fragment in complex with interleukin-8, *Structure*, 1999, 7(2), 157–168.
69. S. R. Leong, R. C. Kabakoff and C. A. Hébert, Complete mutagenesis of the extracellular domain of interleukin-8 (IL-8) type A receptor identifies charged residues mediating IL-8 binding and signal transduction, *J. Biol. Chem.*, 1994, **269**(30), 19343–19348.
70. J. Leppert, O. Ohlenschläger, M. Görlach and R. Ramachandran, Adiabatic TOBSY in rotating solids, *J. Biomol. NMR*, 2004, **29**(2), 167–173.
71. N. S. Gandhi and R. L. Mancera, Molecular dynamics simulations of CXCL-8 and its interactions with a receptor peptide, heparin fragments, and sulfated linked cyclitols, *J. Chem. Inf. Model.*, 2011, **51**(2), 335–358.
72. J. W. Liou, F. T. Chang, Y. Chung, W. Y. Chen, W. B. Fischer and H. J. Hsu, In silico analysis reveals sequential interactions and protein conformational changes during the binding of chemokine CXCL-8 to its receptor CXCR1, *PLoS One*, 2014, **9**(4), e94178.
73. A. J. Venkatakrishnan, X. Deupi, G. Lebon, F. M. Heydenreich, T. Flock and T. Miljus, *et al.*, Diverse activation pathways in class A GPCRs converge near the G-protein-coupling region, *Nature*, 2016, **536**, 484.
74. M. F. Mesleh and S. J. Opella, Dipolar Waves as NMR maps of helices in proteins, *J. Magn. Reson.*, 2003, **163**(2), 288–299.
75. S. H. Park, F. Casagrande, B. B. Das, L. Albrecht, M. Chu and S. J. Opella, Local and Global Dynamics of the G Protein-Coupled Receptor CXCR1, *Biochemistry*, 2011, **50**, 2371–2380.

CHAPTER 19

High-resolution NMR Studies of Antibiotics in Membranes

MAIK DERKS,[a,b] LEANNA SMID,[a] RAJ KUMAR,[a] LEA MARIE BECKER,[a] RHYTHM SHUKLA,[a,b] EEFJAN BREUKINK[b] AND MARKUS WEINGARTH*[a]

[a] NMR Spectroscopy, Bijvoet Centre for Biomolecular Research, Department of Chemistry, Faculty of Science, Utrecht University, Padualaan 8, 3584 CH Utrecht, The Netherlands; [b] Membrane Biochemistry and Biophysics, Bijvoet Centre for Biomolecular Research, Department of Chemistry, Utrecht University, Padualaan 8, 3584 CH Utrecht, The Netherlands

*Email: m.h.weingarth@uu.nl

19.1 Introduction

The emergence of multidrug-resistant bacteria is a slow, silent but deadly pandemic with major burdens for the public healthcare system.[1] Novel antibiotics can relieve this burden provided the resistance build-up against these new antibiotics can be restricted. This can be achieved by making sure that the novel antibiotic is multitargeting,[2] but alternatively, an essential but non-proteinaceous target can be chosen on which multiple enzymes rely for their proper functioning. Such a target is Lipid II, an essential molecule for bacterial cell wall synthesis that carries the basic building block of the peptidoglycan layer across the plasma membrane and delivers it to the peptidoglycan synthesis machinery located at the extracellular (or periplasmic) side of the membrane (Figure 19.1).[3–5] Several prominent compounds have been

New Developments in NMR No. 26
NMR Spectroscopy for Probing Functional Dynamics at Biological Interfaces
Edited by Anirban Bhunia, Hanudatta S. Atreya and Neeraj Sinha

Published by the Royal Society of Chemistry, www.rsc.org

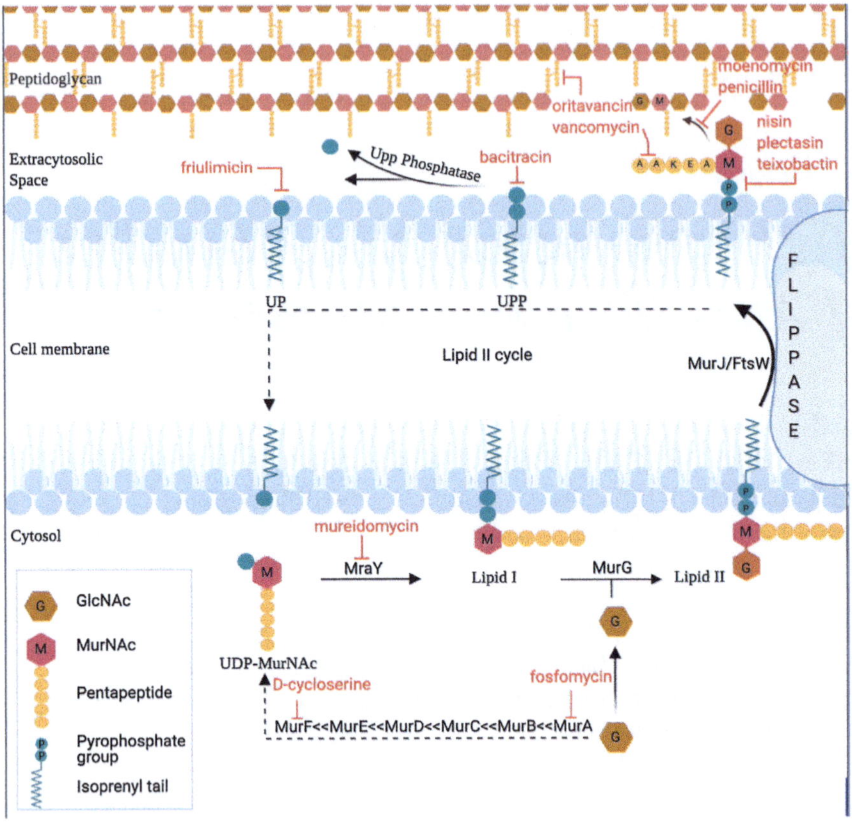

Figure 19.1 Schematic of the bacterial cell wall, the Lipid II cycle, and the construction of the peptidoglycan strands.

discovered that kill an expanded panel of key pathogens by binding to Lipid II. Of these compounds, vancomycin may be the best-known considering its long-term clinical use as an antibiotic.[6] The structural diversity of these compounds likely implicates diverse mechanisms by which they bind to Lipid II. However, despite the unquestionable potency of Lipid II-binding antibiotics and their robustness to resistance development, their binding modes are not well understood, which is the major limitation to the use of these powerful compounds as templates for next-generation antibiotics.

This paucity of structural information is due to the difficulty in studying small (<10 kDa) drug–receptor complexes in membranes at atomic resolution. For instance, X-ray diffraction is infeasible due to a lack of long-range order in membranes,[7] (single particle) cryo-EM typically requires assemblies >100 kDa,[8] and solution NMR is not applicable to immobilized molecules in liposomal preparations because of fast relaxation. In contrast, solid-state NMR (ssNMR) is less restrictive and amendable to insoluble or immobilized samples, emerging as a powerful tool where other methods are not applicable. Nevertheless, some remarkable progress in our understanding of Lipid II-binding drugs has been made using solution NMR in combination with membrane mimetics such as micelles or apolar solvents,[9–12] although it is uncertain if such conditions enable the observations of physiologically relevant drug–receptor states. Recently, our group has reported the first ssNMR studies of antibiotic–Lipid II complexes in model membranes and bacterial cell membranes that mark an important step forward for a detailed understanding of the native binding modes of Lipid II-binding drugs.[13,14] This progress was decisively enabled by the use of modern ssNMR approaches for membrane proteins such as ^{1}H-detection[15–20] and dynamic nuclear polarization.[21–24]

The here discussed peptide antibiotics represent only a small subset of the known Lipid II-binding antibiotics and we refer the reader to excellent reviews for a comprehensive overview of the breadth of Lipid II-binding antibiotics.[3–5] Here, we focus on Lipid II-binding antibiotics for which NMR data, and especially ssNMR data, are available.[13,14] While we, to the best of our knowledge, cover all ssNMR studies, the selection of solution NMR studies is not complete and we do not cover some exciting investigations.[11,25]

19.2 Lipid II Modifications

The structure of Lipid II is not conserved among bacterial species, which can have dramatic consequences for the activity of antibiotics.[26] While the pyrophosphate and the sugars of Lipid II are invariable, a variety of bacterial strains subject the pentapeptide to a broad spectrum of modifications that are linked to an antimicrobial resistance mechanism (Figure 19.2), first

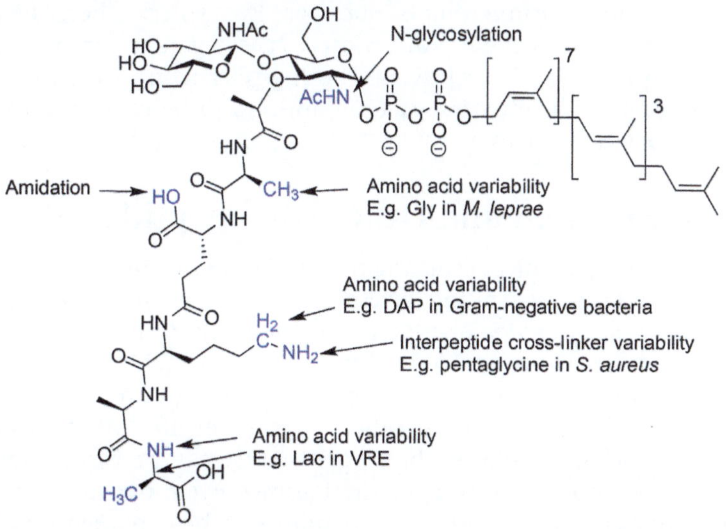

Figure 19.2 Schematic overview of Lipid II modifications encountered in different bacterial strains.

observed in the clinic with vancomycin-resistant *Enterococcus* (VRE).[27] Next to the pentapeptide, in certain bacteria, so-called bridging peptides are present in Lipid II that eventually link neighboring peptidoglycan strands, and these linkers can provide additional binding sites for antibiotics such as oritavancin (see Section 19.3).[26,28] These modifications occur either during Lipid II biosynthesis or on the peptidoglycan level.

The first residue of the pentapeptide moiety is typically an L-alanine that is bound to the carboxyl group of *N*-acetyl-muramic acid (MurNAc) but is replaced with glycine or L-serine in *Mycobacterium leprae*.[29] The amino acid at position two is a conserved D-γ-glutamate residue that can be amidated by the combined action of two enzymes, MurT and GatD.[30] The defensin plectasin is suspected to interact with the anionic carboxyl group of the D-γ-glutamate (see Section 19.5).[10] Additionally, vancomycin resistance has been associated with the degree of peptidoglycan amidation in staphylococci strains.[31] The third amino acid is usually a diamino-acid that varies greatly among bacteria. Typically, Gram-positive bacteria have an L-lysine residue while most Gram-negative bacteria (and some Gram-positive like *Bacillus subtilis*) contain a meso-diaminopimelic acid (DAP) residue. DAP has been shown to be critical for the mode of action of tridecaptin A1,[11] a lipopeptide with the rare quality of activity against clinically relevant Gram-negative strains. MurF-ligases add the final two amino acids, typically a D-Ala-D-Ala dipeptide. Resistance to antibiotics such as vancomycin, which bind to the D-Ala-D-Ala dipeptide, develops relatively quickly as the MurF-ligase accepts a broad spectrum of substrates.

Additionally, there are rare modifications to the sugar backbone that occur before and after incorporation in the mature peptidoglycan. The glycan strands are modified in all bacterial species to avoid degradation by naturally occurring antimicrobials such as lysozymes. There has been speculation over whether the modifications to the sugars occur on Lipid II or after incorporation into the cell wall. The *N*-glycosylation which occurs on the MurNAc sugar in some *Mycobacteria* appears to be present on Lipid II before peptidoglycan polymerization.[28,32]

19.3 Vancomycin and Other Glycopeptides

The heptapeptide vancomycin was the first and remains the most prominent of the class of Lipid II-binding glycopeptides. Vancomycin was isolated from *Amycolatopsis orientalis* in 1956 and came to clinical use in 1958.[6] The vancomycin core structure includes a disaccharide attached to residue four and the non-canonical amino acids hydroxyphenylglycin and dihydroxyphenylglycin, while other, non-neighboring residues are linked together to form a complex ring structure. The ring structure can be further modified yielding various natural and semisynthetic vancomycin derivatives.

Vancomycin binds to the D-Ala-D-Ala terminus of the pentapeptide of Lipid II by forming five hydrogen bonds. This prevents further elongation and building of peptidoglycan and therefore leads to cell death. Vancomycin resistance first occurred in the 1980s. Different mechanisms of resistance are known, some of which are based on the mutation of the pentapeptide as is the case for vancomycin-resistant *S. aureus* (VRSA) or *Enterococcus* (VRE). Replacement of the D-Ala-D-Ala terminus to D-Ala-D-Lac or D-Ala-D-Ser results in reduced binding due to the loss of one hydrogen bond or steric hindrance, respectively. In order to address the increasing problem of vancomycin resistance, a number of semi-synthetic analogs such as oritavancin were developed. These agents contain additional substituents that can compensate for the loss of the hydrogen bond with the pentapeptide.[4]

In a series of pioneering *in situ* ssNMR studies, Jacob Schaefer and coworkers investigated the effects on the peptidoglycan synthesis that different substituents to the core structure of vancomycin have.[33–35] As a key method, they used rotational echo double resonance (REDOR)[36] to measure heteronuclear distances between amino acids in the peptidoglycan network and the glycopeptide derivatives in cell walls and intact *S. aureus* cells. Therefore, glycopeptides were fluorinated at a specific position, while glycine and D-alanine in cell wall components were specifically isotope-enriched by growing bacteria in the presence of ^{15}N- and ^{13}C-labeled amino acids. Of note, thereby, the interpeptide pentaglycine that is used to cross-link the peptidoglycan strands of *S. aureus* also gets isotope-enriched. In a first study, the cell wall synthesis of *S. aureus* was investigated with and without the drug. By measuring the REDOR dephasing curves of D-[1-^{13}C]-alanine and [^{15}N]-glycine as well as [1-^{13}C]-glycine and L-[ε^{15}N]-lysine labeled cells, they showed that vancomycin interferes with transglycosylation rather than transpeptidation in the peptidoglycan.[33]

Furthermore, Schaefer *et al.* characterized the contribution of different substituents to vancomycin's activity against *S. aureus*.[34,35] Therefore, REDOR in combination with ^{19}F-labeled vancomycin analogs was used to probe the position of the substituents with respect to the peptidoglycan network. Most notably, a molecular rationale could be provided for how the vancomycin derivative oritavancin is able to bind to D-Ala-D-Lac in vancomycin-resistant *S. aureus* strains. It could be shown that, in oritavancin, a bulky hydrophobic moiety bound to the disaccharide sterically hinders cell wall synthesis steps, while 4-epi-vancomycine bound to residue six of the peptide core stabilizes the binding of the drug to the pentaglycyl-bridge. These results explain how vancomycin derivatives like oritavancin are still active against vancomycin-resistant *S. aureus* strands by the recruitment of additional binding sites.[35]

Generally, it stands to reason that it would be favorable to target the non-variable part of Lipid II, which would mitigate the resistance problems that vancomycin and its analogs have encountered.

19.4 Nisin and Other Lantibiotics

Lantibiotics are small antimicrobial peptides with characteristic rings formed by the post-translationally modified amino acids lanthionine or methyllanthionine. A number of lantibiotics were shown to target **Lipid II**, with nisin being the best-known example.[5]

Nisin is a 34-residue peptide with five lanthionine rings, dubbed **A–E**, with two mechanisms of action (Figure 19.3A). First, rings A and B sequester the pyrophosphates of Lipid II through intermolecular hydrogen bonds in a cage-like structure (Figure 19.3B), thereby blocking peptidoglycan synthesis.[9] The second proposed mechanism of action is nisin binding to **Lipid II** to form a pore complex in the bacterial membrane (Figure 19.3C). This pore was shown to be composed of eight nisin and four Lipid II molecules that span the plasma membrane[37] and increase the leakage by nisin by orders of magnitude.[38] The structure of the nisin:Lipid II complex was initially elucidated in DMSO using solution NMR and a truncated Lipid II construct.[9]

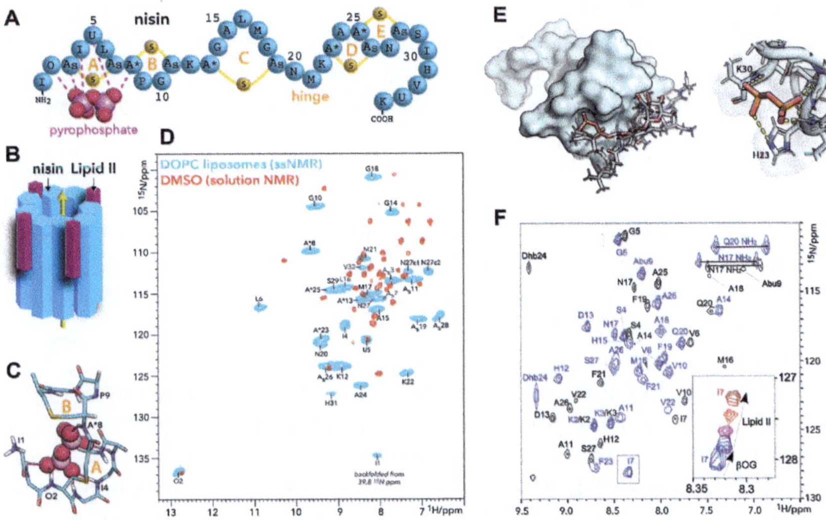

Figure 19.3 (A) Nisin features five thioether rings, of which rings A and B are critical for binding Lipid II. (B) Nisin and Lipid II form a defined pore that spans the bacterial plasma membrane. (C) The pyrophosphate cage structure as determined in DMSO.[9] The thioether rings A and B of nisin are annotated. (D) Overlay of 2D NH spectra of the nisin–Lipid II complex in DOPC liposomes (blue) and in DMSO (red). The ssNMR spectrum was measured at 950 MHz (^{1}H frequency) and 60 kHz magic-angle spinning. The solution NMR spectrum in DMSO was published previously.[9] (A) and (D) reproduced from ref. 14. Copyright: 2018, the authors. (E) Left: Model of the complex formed between LtnA1 and (E,E)-farnesyl Lipid II.[12] LtnA1 is shown as a surface representation, Lipid II is represented as a stick model. Right: Key binding interactions between the C-terminal residues of LtnA1 and pyrophosphate of Lipid II. (F) Two states of nukacin ISK-1 shown by a solution NMR ^{1}H–^{15}N 2D HSQC. The cross-peaks belonging to states A and B are colored blue and black, respectively. Only one of the states is able to bind to Lipid II.[40] (F) Reproduced from ref. 14, https://doi.org/10.1038/s42003-018-0150-3, under the terms of the CC BY 4.0 license, http://creativecommons.org/licenses/by/4.0/.

However, recent ssNMR studies conducted by our group have demonstrated that the nisin–Lipid II complex must be different in lipid membranes, as demonstrated by striking chemical shift changes in 2D ^{15}N–^{1}H spectra (Figure 19.3D).[14] While chemical shift differences were most pronounced in the membrane-spanning rings C–E, it also covered rings A–B, suggesting comprehensive conformational changes throughout the entire peptide.

The interaction between nisin and Lipid II was also studied in cellular membranes of *Micrococcus flavus* using ^{1}H-detected ssNMR and ^{13}C-detected DNP-enhanced ssNMR. The cellular state of Lipid II-bound nisin globally corresponded to one in DOPC, demonstrating that the ssNMR spectra captured a native binding state of nisin. However, some spectral differences were observed in cell membranes for the residues at and around the hinge between rings C and D, suggesting the importance of the plasticity of the hinge for nisin's adaptability to heterogeneous bacterial membranes. This finding provides a molecular rationale for the strain-specific enhanced bioactivity in nisin derivatives with changes to the hinge.[39]

Another lantibiotic that binds to Lipid II is nukacin ISK-1, a 27-residue peptide. Recently, solution NMR studies of nukacin ISK-1 showed that the free peptide exhibits two different conformations.[40] Titrations with Lipid II in β-octylglucoside micelles showed that only one of these conformers was able to bind to Lipid II (Figure 19.3F). Based on chemical shift perturbations on nukacin ISK-1 in the presence of Lipid II, a docking model of the complex was generated. Of note, when dodecylphosphocholine (DPC) micelles were used, nukacin ISK-1 was not able to bind to Lipid II, which is yet another compelling demonstration that non-physiological media such as micelles can heavily perturb the binding mode of Lipid II-binding antibiotics, which is in line with a recent review by Schanda and co-workers on the spurious impact of detergents on membrane protein structure.[41]

Recently, Vederas and co-workers derived a structural modal of lacticin 3147[12] in complex with Lipid II using solution NMR with a truncated Lipid II construct in solvent ($CD_3OH : CD_3CN : H_2O$). Lacticin 3147 is a so-called two-component lantibiotic system of two synergistic peptides, one of which (LtnA1) directly targets Lipid II. Using comprehensive ^{1}H–^{1}H NOE NMR data, a model of the complex formed between LtnA1 and Lipid II could be obtained (Figure 19.3E). Based on chemical shift perturbations, residues of the D-ring (H23,E24, M26, K30) of LtnA1 coordinate to the pyrophosphate with hydrogen bonds, while other residues of LtnA1 interact with the pentapeptide of Lipid II. Comprehensive solution NMR data in DPC micelles are also available on the interaction of the lantibiotic mersacidin with Lipid II in micelles.[42] Mersacidin structurally resembles LtnA1 and, unlike nisin, has increased activity in Ca^{2+}-rich media. However, given the dramatic effect of unnatural media (DMSO or DPC) on the lantibiotics nisin and nukacin ISK-1, the complexes of LtnA1 or mersacidin with Lipid II could be different in membranes.

19.5 Plectasin and Other CS-αβ Defensins

Cysteine-stabilized alpha-beta defensins (CS-αβ defensins) are a subclass of Lipid II-binding peptides that originate from higher organisms such as fungi and vertebrates. They are short (<60 amino acids) and feature a conserved fold with an α-helix and an antiparallel β-sheet, stabilized by three to five disulfide bonds.[43] The structure of several CS-αβ defensins has been solved with solution NMR: lucifensin,[44] copsin,[45] eurocin,[46] MGD-1,[47] and plectasin.[10] All display an identical fold but the loops and termini vary in length and conformation (Figure 19.4C). While the fold is conserved in CS-αβ defensins, the primary sequences of CS-αβ defensins show a low degree of conservation (Figure 19.4A). It is unknown if CS-αβ defensins share a common Lipid II-binding mode, since structural data on their interaction with Lipid II are only available for plectasin, which is discussed below.

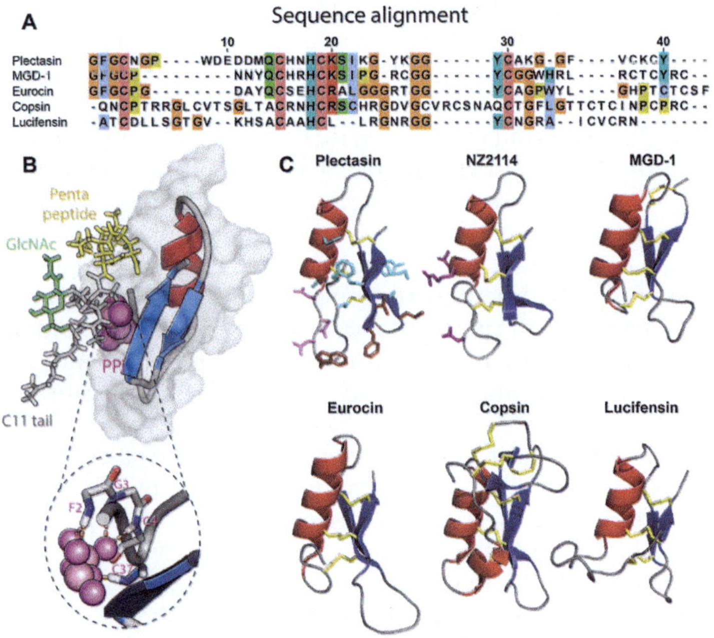

Figure 19.4 (A) Sequence alignment of CS-αβ defensins. Numbering is based on plectasin. (B) NMR CSP-based driven docking model of the plectasin–LII complex in DPC micelles.[10] Interactions are shown between the pyrophosphate and backbone amide protons of N-terminal residues and the β-sheet. (C) Comparison of 3D structures, including disulfide bonds, of several CS-αβ defensins displaying the high fold similarity despite low sequence homology. For plectasin the anionic patch is depicted in violet, the membrane interacting residues in brown, and proposed Lipid II-binding residues in cyan. For NZ2114 the three amino acid substitutions are depicted in magenta. Adapted from ref. 52 with permission from John Wiley & Sons, Copyright © 2019 The Authors.

Plectasin (and its triple mutant NZ2114[48]), discovered by the company Novozymes, is a 40-amino-acid fungal defensin from *Pseudoplectania nigrella* with a broad activity range against Gram-positive bacteria, including common drug-resistant strains such as streptococci, staphylococci, and enterococci.[49–51] Plectasin has been shown to be active in rodent models with low toxicity.[48–50] To date, plectasin is the only CS-αβ defensin for which the binding mode to Lipid II has been studied by NMR.[10] This Lipid II-binding mode was explored by measuring chemical shift perturbations (CSPs) of amide resonances of the peptide in DPC micelles. Upon addition of DPC to free plectasin, strong chemical shift changes were observed on one side of the peptide around W8, K32, and F35 (Figure 19.5C), suggesting that this side is in contact with the membrane. This finding could be corroborated in a similar NMR experiment with eurocin, in which hydrophobic residues on the same side around the short loop between the β-strands also showed CSPs upon addition of DPC micelles.[46]

The plectasin–Lipid II interface was established by comparing solution NMR spectra of plectasin bound to DPC micelles in the absence and presence of Lipid II. Significant CSPs were observed for residues F2, C4, D12, Y29, A31, G33, C37, and K38, which are mostly located in a patch near the proposed membrane-binding site. These CSPs were then used as ambiguous restraints to drive docking studies in HADDOCK.[53] In the resulting docking model, the primary binding site involved the pyrophosphate of Lipid II and amide protons from residues F2, G3, C4, and C37 of plectasin and the side chain of H18 (Figure 19.5B). While not supported by experimental data, docking studies suggested that the N-terminus and the conserved H18 sidechain interact with the carboxyl of E2 in Lipid II (Figure 19.4B). In line with this finding, MRSA strains depleted of enzymes (MurT/GatD) responsible for the amidation of the E2 carboxyl were found to be more susceptible to plectasin.[30] Surprisingly, the three amino acid substitutions (D9N, M13L, Q14R; see Figure 19.4C) of improved plectasin variant NZ2114 are located in a negatively charged patch (D9–D12 in plectasin), opposite to the proposed Lipid II-binding site, with currently unknown functionality, a clear indication that there is more to learn about the mechanism of action of plectasin.

19.6 Teixobactin

In 2015, Kim Lewis and colleagues discovered a soil bacterium, *Eleftheria terrae*, which produced a high-efficacy antibiotic named teixobactin.[54] Teixobactin showed excellent activity against all Gram-positive bacteria, including multidrug-resistant VRE and MRSA superbugs. Teixobactin is considered a breakthrough drug not only because of its exceptional activity but also because of its robustness to resistance development, as no resistant mutants were detected after passage of *S. aureus* in sub-MIC concentrations of teixobactin over 27 days. Teixobactin in itself has a unique structure. It has 11 amino acids and comprises several non-canonical residues including

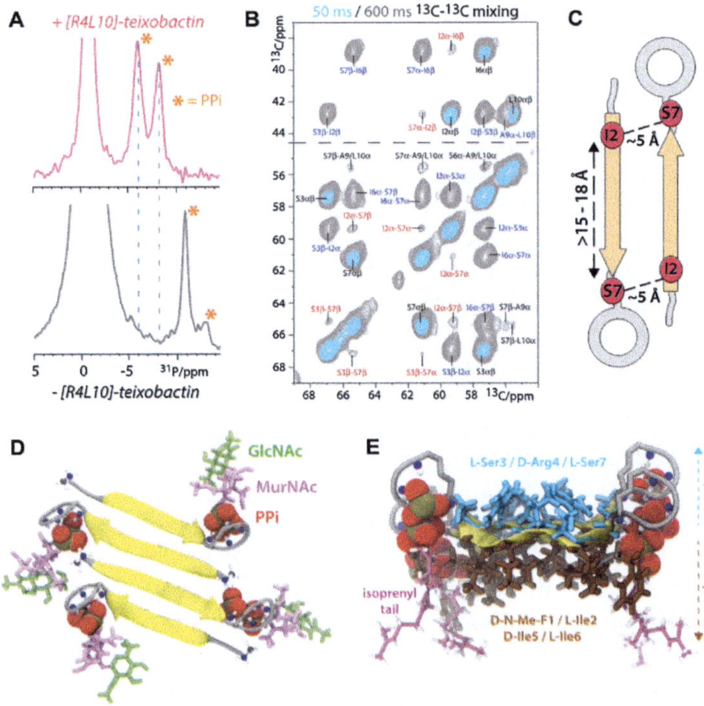

Figure 19.5 (A) D ^{31}P ssNMR spectra of Lipid II in liposomes in the presence (pink) and absence (gray) of [R4L10]-teixobactin. Orange asterisks mark PPi signals; bulk lipids come around 0 ppm. (B) Overlay of 2D ^{13}C^{13}C PARISxy[59] spectra of Lipid II-bound [R4L10]-teixobactin measured with 50 (blue) and 600 ms (gray) magnetization transfer time. Intermolecular peptide–peptide contacts in red, sequential contacts in blue. All NMR spectra were acquired at 950 MHz. (C) Contacts observed in (B) between Ile2–Ser7 relate to intermolecular magnetization transfer between antiparallel β-strands. Distances are Cα–Cα spacings. (D) [R4L10]-teixobactin aligns into antiparallel fiber-like β-sheets (yellow). Lipid II PPi is shown in red, MurNAc in magenta, GlcNAc in green. Backbone nitrogens and amino protons that coordinate PPi are shown explicitly. (E) Hydrophilic residues (Ser3, Arg4, Ser7; in cyan) of [R4L10]-teixobactin are water-exposed; hydrophobic residues (N-Me-F1, Ile2, Ile5, Ile6; in brown) and the Lipid II isoprenyl tails point toward the membrane. Reproduced from ref. 13, https://doi.org/10.1038/s41467-020-16600-2, under the terms of the CC BY 4.0 license, http://creativecommons.org/licenses/by/4.0/.

four D-amino acids and a C-terminal depsi-cycle that contains the rare cationic residue enduracididine. With a unique structure and high potency, teixobactin could provide the template for manufacturing a new class of antibiotics that are likely able to avoid or delay resistance development.

Since its discovery, a number of structural and chemical biology studies have been performed to understand the mode of action of teixobactin.[55–58] These studies have shown that teixobactins target cognate cell wall precursors such as Lipid II and Lipid III and strongly suggested that the depsi-cycle of teixobactin targets the pyrophosphate group. While a solution NMR study in micelles suggested that teixobactin aggregates in the Lipid II-bound state,[55] the molecular details of teixobactin's binding mode were elusive until recently.

To gain insight into the mode of action of teixobactin, we used ssNMR experiments to solve the complex structure formed by the selectively $^{13}C,^{15}N$-labeled teixobactin analog [R4L10]-teixobactin and $^{13}C,^{15}N$-labeled Lipid II in physiologically relevant conditions.[13]

Using an array of biophysical techniques, we elaborated on the mode of action of teixobactin. Unambiguous intermolecular contacts obtained in 2D PARIS ^{13}C–^{13}C spin diffusion spectra showed that teixobactin forms anti-parallel β-sheets in the Lipid II-bound state (Figures 19.5A–D). Fluorescence microscopy experiments established that after treatment with teixobactin fluorescently tagged Lipid II accumulated in μm-sized clusters after a couple of hours.

With $^{13}C,^{15}N$-labeled Lipid II, we could map the complex interface at atomic resolution. Using 2D ^{1}H–^{31}P spectra at fast MAS (60 kHz) and 2D $^{13}C^{13}C$ PARIS spectra, we could show that the backbone amino-protons of the despi-cycle of teixobactin target the pyrophosphate and that the MurNAc sugar is directly present at the complex interface, while the GlcNAc sugar is distal from the interface. Surprisingly, the pentapeptide is not involved in the complex formation and remains highly mobile. The latter could be conclusively demonstrated by a scalar 2D CC TOBSY[60] spectrum, which exclusively showed signals of the mobile pentapeptide. The fact that the pentapeptide is not involved in the interface is one of the major reasons for the robustness of teixobactin to resistance development. Topological insights into the complex were obtained with T_2-edited H(H)C experiments,[61] which unambiguously showed that the long hydrophobic isoleucine side-chains in the elongated N-terminus act as membrane-anchors. All this information provided the basis for structure calculation using HADDOCK,[53] which also yielded insights into the "hardcoded" stereochemistry of the complex, which allows a favorable distribution of hydrophilic and hydrophobic residues (Figure 19.5E).

Interestingly, while the scaffold and the mode of action of teixobactin are unique among Lipid II-targeting antibiotics, it bears resemblance to hypeptin,[62] an octapeptide with good antimicrobial activity against *S. aureus* isolated 30 years ago from a *Pseudomonas* strain. While hypeptin also possesses a four-residue membered depsi-cycle and several D-amino acids, its linear N-terminus is much shorter than in teixobactin.[63]

19.7 Daptomycin

Daptomycin is a last-resort cyclic lipopeptide antibiotic. Being initially considered unsuitable for clinical application due to toxic myopathies, a revised administration scheme was approved by the FDA in 2003 for the treatment of complicated skin infection and bacteremia caused by Gram-positive pathogens such as MRSA. The success of daptomycin, together with the emergence of superbugs, attracted renewed interest in this class of compound. It represents an important addition to the current panel of available antimicrobial agents with its high activity against Gram-positive pathogens, including resistant strains for which limited therapeutic alternatives exist.

Daptomycin is a non-ribosomal lipopeptide produced in *Streptomyces roseosporus*.[66] It consists of 13 amino acids and includes six non-canonical and four cationic residues. The cyclic portion at the C-terminus is composed of 10 residues and the linear portion is linked to a decanoyl fatty acid chain which is indispensable for its antibacterial activity (Figure 19.6A). Daptomycin has a net charge of −3 and its bactericidal activity depends on the presence of Ca^{2+} ions, which facilitate the interaction with membranes. Solution NMR studies have critically advanced our understanding of the binding mode of daptomycin.[64,67–69] Several studies have shown that daptomycin initially only weakly interacted with Ca^{2+} with small conformational changes.[64,67,68] However, as readily visible by spectral line-broadening, Ca^{2+} induces the oligomerization of daptomycin, which facilitates the interaction with membranes. Only minor conformational changes were observed

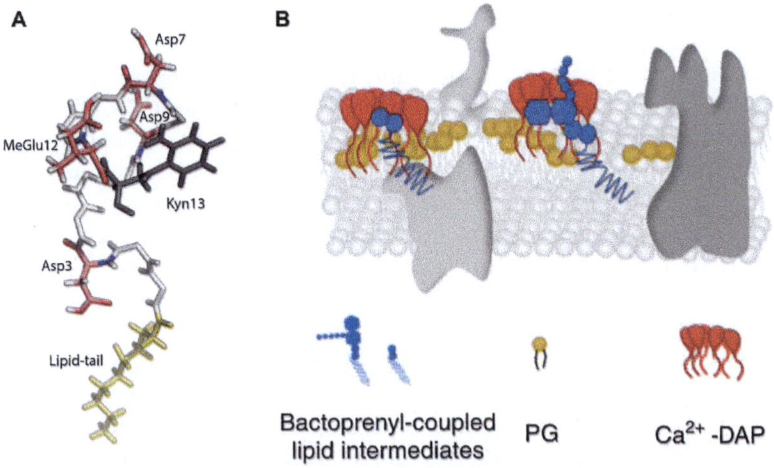

Figure 19.6 (A) Solution NMR structure of daptomycin (pdb 1XT7).[64] The four anionic residues (Asp3, Asp7, Asp9, MeGlu12) are colored in red, Kyn13 at the ring-closure in gray, and the lipid-tail in yellow. (B) The formation of a tripartite complex of Ca^{2+}-daptomycin with PG and bactoprenyl-coupled lipid intermediates blocks cell wall synthesis and triggers the delocalization of the cell wall biosynthetic machinery. Reproduced from ref. 65, https://doi.org/10.1038/s41467-020-15257-1, under the terms of the CC BY 4.0 license, http://creativecommons.org/licenses/by/4.0/.

between daptomycin in the presence or absence of micelles while in the presence of Ca^{2+}.[70]

However, 30 years after its discovery, the mode of action of daptomycin and the importance of Ca^2 are still poorly understood. The prevailing hypothesis is that daptomycin perturbs the membrane integrity, presumably by the formation of oligomeric pores, which requires the presence of anionic PG lipids and Ca^{2+} ions.[71,72] It was also suggested that daptomycin accumulates in PG-rich membrane domains, which would lead to the delocalization of cell division proteins, thereby inhibiting the cell wall synthesis.[73] Recently, Tanja Schneider and co-workers provided evidence that daptomycin specifically targets Lipid II, forming a tripartite complex together with PG in the presence of Ca^{2+} ions (Figure 19.6B).[13] This interaction is followed by delocalization of enzymes of the cell wall synthesis machinery. The specific interaction with Lipid II, for example, was demonstrated by pre-incubation with teixobactin, which resulted in significantly delayed daptomycin-induced killing. However, the molecular details of the interaction of daptomycin with Lipid II are entirely unknown, let alone whether this interaction is direct or conveyed by other binding partners (PG or Ca^{2+}) that daptomycin recruits during its mode of action.

19.8 Conclusions

Solution and solid-state NMR provide powerful avenues to increase our molecular understanding of Lipid II-binding antibiotics and thereby are important tools for drug design. Given the undoubted importance of studying Lipid II-targeting antibiotics in a physiologically relevant environment, we expect that solid-state NMR will play a greater role in the future in improving our knowledge on these potential next-generation drugs. With spectacular advancements in methodologies such as ^1H-detection and DNP as well as in instrumentation such as the advent of magnetic field strength above 1000 MHz, solid-state NMR will enable the study of Lipid II-binding antibiotics and other membrane-active drugs directly in cell membranes[13,14] or even in cells.[74]

Abbreviations

CSP	Chemical shift perturbation
CS-αβ	Cysteine-stabilized alpha-beta defensins
DAP	Meso-diaminopimelic acid
DMSO	Dimethylsulfoxide
DNP	Dynamic nuclear polarization
DOPC	1,2-Dioleoyl-*sn*-glycero-3-phosphocholine
DPC	Dodecylphosphocholine
FDA	Food and Drug Administration
GlcNAc	*N*-acetyl-glucose
HADDOCK	High Ambiguity Driven protein-protein DOCKing

HSQC	Heteronuclear single quantum coherence
MAS	Magic angle spinning
MIC	Minimum inhibitory concentration
MRSA	Meticillin-resistant *Staphylococcus aureus*
MurNAc	*N*-acetyl-muramic acid
NMR	Nuclear magnetic resonance
NOE	Nuclear Overhauser effect
PARIS	Phase-alternated recoupling irradiation scheme
pdb	Protein Data Bank
PG	Phosphatidylglycerol
REDOR	Rotational echo double resonance
ssNMR	Solid-state nuclear magnetic resonance
TOBSY	TOtal through Bond correlation SpectroscopY
VRE	Vancomycin-resistant *Enterococcus*
VRSA	Vancomycin-resistant *Staphylococcus aureus*

Acknowledgements

We are indebted to Prof. John Vederas for providing the coordinates of the NMR structure of the LtnA1–Lipid II complex. This work is part of the research programs ECHO and VIDI with project numbers 723.014.003 and 711.018.001, which are financed by the Dutch Research Council (NWO).

References

1. A. Cassini, L. D. Hogberg, D. Plachouras, A. Quattrocchi, A. Hoxha, G. S. Simonsen, M. Colomb-Cotinat, M. E. Kretzschmar, B. Devleesschauwer, M. Cecchini, D. A. Ouakrim, T. C. Oliveira, M. J. Struelens, C. Suetens, D. L. Monnet and Burden of AMR Collaborative Group, Attributable deaths and disability-adjusted life-years caused by infections with antibiotic-resistant bacteria in the EU and the European Economic Area in 2015: a population-level modelling analysis, *Lancet Infect. Dis.*, 2019, **19**(1), 56–66.
2. L. L. Silver, Multi-targeting by monotherapeutic antibacterials, *Nat. Rev. Drug Discovery*, 2007, **6**(1), 41–55.
3. S. F. Oppedijk, N. I. Martin and E. Breukink, Hit 'em where it hurts: The growing and structurally diverse family of peptides that target lipid-II, *Biochim. Biophys. Acta*, 2016, **1858**(5), 947–957.
4. A. Muller, A. Klockner and T. Schneider, Targeting a cell wall biosynthesis hot spot, *Nat. Prod. Rep.*, 2017, **34**(7), 909–932.
5. E. Breukink and B. de Kruijff, Lipid II as a target for antibiotics, *Nat. Rev. Drug Discovery*, 2006, **5**(4), 321–332.
6. M. H. McCormick, J. M. McGuire, G. E. Pittenger, R. C. Pittenger and W. M. Stark, Vancomycin, a new antibiotic. I. Chemical and biologic properties, *Antibiot. Annu.*, 1955, **3**, 606–611.

7. J. Birch, H. Cheruvara, N. Gamage, P. J. Harrison, R. Lithgo and A. Quigley, Changes in Membrane Protein Structural Biology, *Biology-Basel*, 2020, **9**(11), E401.
8. R. Nygaard, J. Kim and F. Mancia, Cryo-electron microscopy analysis of small membrane proteins, *Curr. Opin. Struct. Biol.*, 2020, **64**, 26–33.
9. S. T. Hsu, E. Breukink, E. Tischenko, M. A. Lutters, B. de Kruijff, R. Kaptein, A. M. Bonvin and N. A. van Nuland, The nisin-lipid II complex reveals a pyrophosphate cage that provides a blueprint for novel antibiotics, *Nat. Struct. Mol. Biol.*, 2004, **11**(10), 963–967.
10. T. Schneider, T. Kruse, R. Wimmer, I. Wiedemann, V. Sass, U. Pag, A. Jansen, A. K. Nielsen, P. H. Mygind, D. S. Raventos, S. Neve, B. Ravn, A. M. Bonvin, L. De Maria, A. S. Andersen, L. K. Gammelgaard, H. G. Sahl and H. H. Kristensen, Plectasin, a fungal defensin, targets the bacterial cell wall precursor Lipid II, *Science*, 2010, **328**(5982), 1168–1172.
11. S. A. Cochrane, B. Findlay, A. Bakhtiary, J. Z. Acedo, E. M. Rodriguez-Lopez, P. Mercier and J. C. Vederas, Antimicrobial lipopeptide tridecaptin A1 selectively binds to Gram-negative lipid II, *Proc. Natl. Acad. Sci. U. S. A.*, 2016, **113**(41), 11561–11566.
12. A. Bakhtiary, S. A. Cochrane, P. Mercier, R. T. McKay, M. Miskolzie, C. S. Sit and J. C. Vederas, Insights into the Mechanism of Action of the Two-Peptide Lantibiotic Lacticin 3147, *J. Am. Chem. Soc.*, 2017, **139**(49), 17803–17810.
13. R. Shukla, J. Medeiros-Silva, A. Parmar, B. J. A. Vermeulen, S. Das, A. L. Paioni, S. Jekhmane, J. Lorent, A. Bonvin, M. Baldus, M. Lelli, E. J. A. Veldhuizen, E. Breukink, I. Singh and M. Weingarth, Mode of action of teixobactins in cellular membranes, *Nat. Commun.*, 2020, **11**(1), 2848.
14. J. Medeiros-Silva, S. Jekhmane, A. L. Paioni, K. Gawarecka, M. Baldus, E. Swiezewska, E. Breukink and M. Weingarth, High-resolution NMR studies of antibiotics in cellular membranes, *Nat. Commun.*, 2018, **9**(1), 3963.
15. J. Medeiros-Silva, D. Mance, M. Daniels, S. Jekhmane, K. Houben, M. Baldus and M. Weingarth, (1) H-Detected Solid-State NMR Studies of Water-Inaccessible Proteins In Vitro and In Situ, *Angew. Chem., Int. Ed. Engl.*, 2016, **55**(43), 13606–13610.
16. S. Jekhmane, J. Medeiros-Silva, J. Li, F. Kummerer, C. Muller-Hermes, M. Baldus, B. Roux and M. Weingarth, Shifts in the selectivity filter dynamics cause modal gating in K(+) channels, *Nat. Commun.*, 2019, **10**(1), 123.
17. D. Lalli, M. N. Idso, L. B. Andreas, S. Hussain, N. Baxter, S. Han, B. F. Chmelka and G. Pintacuda, Proton-Based Structural Analysis of a Heptahelical Transmembrane Protein in Lipid Bilayers, *J. Am. Chem. Soc.*, 2017, **139**(37), 13006–13012.
18. R. Linser, M. Dasari, M. Hiller, V. Higman, U. Fink, J. M. Lopez del Amo, S. Markovic, L. Handel, B. Kessler, P. Schmieder, D. Oesterhelt, H. Oschkinat and B. Reif, Proton-detected solid-state NMR spectroscopy

of fibrillar and membrane proteins, *Angew. Chem., Int. Ed. Engl.*, 2011, **50**(19), 4508–4512.

19. T. Schubeis, T. Le Marchand, C. Daday, W. Kopec, K. Tekwani Movellan, J. Stanek, T. S. Schwarzer, K. Castiglione, B. L. de Groot, G. Pintacuda and L. B. Andreas, A beta-barrel for oil transport through lipid membranes: Dynamic NMR structures of AlkL, *Proc. Natl. Acad. Sci. U. S. A.*, 2020, **117**(35), 21014–21021.
20. C. Pinto, D. Mance, T. Sinnige, M. Daniels, M. Weingarth and M. Baldus, Formation of the beta-barrel assembly machinery complex in lipid bilayers as seen by solid-state NMR, *Nat. Commun.*, 2018, **9**(1), 4135.
21. M. Kaplan, A. Cukkemane, G. C. van Zundert, S. Narasimhan, M. Daniels, D. Mance, G. Waksman, A. M. Bonvin, R. Fronzes, G. E. Folkers and M. Baldus, Probing a cell-embedded megadalton protein complex by DNP-supported solid-state NMR, *Nat. Methods*, 2015, **12**(7), 649–652.
22. M. Kaplan, S. Narasimhan, C. de Heus, D. Mance, S. van Doorn, K. Houben, D. Popov-Celeketic, R. Damman, E. A. Katrukha, P. Jain, W. J. C. Geerts, A. J. R. Heck, G. E. Folkers, L. C. Kapitein, S. Lemeer, P. M. P. van Bergen En Henegouwen and M. Baldus, EGFR Dynamics Change during Activation in Native Membranes as Revealed by NMR, *Cell*, 2016, **167**(5), 1241–1251e11.
23. K. M. Visscher, J. Medeiros-Silva, D. Mance, J. Rodrigues, M. Daniels, A. Bonvin, M. Baldus and M. Weingarth, Supramolecular Organization and Functional Implications of K(+) Channel Clusters in Membranes, *Angew. Chem., Int. Ed. Engl.*, 2017, **56**(43), 13222–13227.
24. L. Joedicke, J. Mao, G. Kuenze, C. Reinhart, T. Kalavacherla, H. R. A. Jonker, C. Richter, H. Schwalbe, J. Meiler, J. Preu, H. Michel and C. Glaubitz, The molecular basis of subtype selectivity of human kinin G-protein-coupled receptors, *Nat. Chem. Biol.*, 2018, **14**(3), 284–290.
25. P. Cudic, J. K. Kranz, D. C. Behenna, R. G. Kruger, H. Tadesse, A. J. Wand, Y. I. Veklich, J. W. Weisel and D. G. McCafferty, Complexation of peptidoglycan intermediates by the lipoglycodepsipeptide antibiotic ramoplanin: minimal structural requirements for intermolecular complexation and fibril formation, *Proc. Natl. Acad. Sci. U. S. A.*, 2002, **99**(11), 7384–7389.
26. D. Munch and H. G. Sahl, Structural variations of the cell wall precursor lipid II in Gram-positive bacteria - Impact on binding and efficacy of antimicrobial peptides, *Biochim. Biophys. Acta*, 2015, **1848**(11 Pt B), 3062–3071.
27. S. Chang, D. M. Sievert, J. C. Hageman, M. L. Boulton, F. C. Tenover, F. P. Downes, S. Shah, J. T. Rudrik, G. R. Pupp, W. J. Brown, D. Cardo, S. K. Fridkin and Vancomycin-Resistant Staphylococcus aureus Investigative Team, Infection with vancomycin-resistant Staphylococcus aureus containing the vanA resistance gene, *N. Engl. J. Med.*, 2003, **348**(14), 1342–1347.

28. W. Vollmer, Structural variation in the glycan strands of bacterial peptidoglycan, *FEMS Microbiol. Rev.*, 2008, **32**(2), 287–306.
29. S. Mahapatra, D. C. Crick, M. R. McNeil and P. J. Brennan, Unique structural features of the peptidoglycan of Mycobacterium leprae, *J. Bacteriol.*, 2008, **190**(2), 655–661.
30. D. Munch, T. Roemer, S. H. Lee, M. Engeser, H. G. Sahl and T. Schneider, Identification and in vitro analysis of the GatD/MurT enzyme-complex catalyzing lipid II amidation in Staphylococcus aureus, *PLoS Pathog.*, 2012, **8**(1), e1002509.
31. H. Hanaki, K. Kuwahara-Arai, S. Boyle-Vavra, R. S. Daum, H. Labischinski and K. Hiramatsu, Activated cell-wall synthesis is associated with vancomycin resistance in methicillin-resistant Staphylococcus aureus clinical strains Mu3 and Mu50, *J. Antimicrob. Chemother.*, 1998, **42**(2), 199–209.
32. S. Mahapatra, T. Yagi, J. T. Belisle, B. J. Espinosa, P. J. Hill, M. R. McNeil, P. J. Brennan and D. C. Crick, Mycobacterial Lipid II Is Composed of a Complex Mixture of Modified Muramyl and Peptide Moieties Linked to Decaprenyl Phosphate, *J. Bacteriol.*, 2005, **187**(8), 2747.
33. L. Cegelski, S. J. Kim, A. W. Hing, D. R. Studelska, R. D. O'Connor, A. K. Mehta and J. Schaefer, Rotational-echo double resonance characterization of the effects of vancomycin on cell wall synthesis in Staphylococcus aureus, *Biochemistry*, 2002, **41**(43), 13053–13058.
34. S. J. Kim, L. Cegelski, M. Preobrazhenskaya and J. Schaefer, Structures of Staphylococcus aureus cell-wall complexes with vancomycin, eremomycin, and chloroeremomycin derivatives by 13C{19F} and 15N{19F} rotational-echo double resonance, *Biochemistry*, 2006, **45**(16), 5235–5250.
35. S. J. Kim, L. Cegelski, D. Stueber, M. Singh, E. Dietrich, K. S. Tanaka, T. R. Parr, A. R. Far and J. Schaefer, Oritavancin exhibits dual mode of action to inhibit cell-wall biosynthesis in Staphylococcus aureus, *J. Mol. Biol.*, 2008, **377**(1), 281–293.
36. T. Gullion and J. Schaefer, Detection of Weak Heteronuclear Dipolar Coupling by Rotational-Echo Double-Resonance Nuclear Magnetic Resonance, in *Advances in Magnetic and Optical Resonance*, ed. W. S. Warren, Academic Press, 1989, vol. 13, pp. 57–83.
37. H. E. Hasper, B. de Kruijff and E. Breukink, Assembly and stability of nisin-lipid II pores, *Biochemistry*, 2004, **43**(36), 11567–11575.
38. E. Breukink, I. Wiedemann, C. van Kraaij, O. P. Kuipers, H. G. Sahl and B. de Kruijff, Use of the cell wall precursor lipid II by a pore-forming peptide antibiotic, *Science*, 1999, **286**(5448), 2361–2364.
39. L. Zhou, A. J. van Heel and O. P. Kuipers, The length of a lantibiotic hinge region has profound influence on antimicrobial activity and host specificity, *Front. Microbiol.*, 2015, **6**, 11.
40. D. Fujinami, A. A. Mahin, K. M. Elsayed, M. R. Islam, J. I. Nagao, U. Roy, S. Momin, T. Zendo, D. Kohda and K. Sonomoto, The lantibiotic nukacin ISK-1 exists in an equilibrium between active and inactive lipid-II binding states, *Commun. Biol.*, 2018, **1**, 150.

41. C. Chipot, F. Dehez, J. R. Schnell, N. Zitzmann, E. Pebay-Peyroula, L. J. Catoire, B. Miroux, E. R. S. Kunji, G. Veglia, T. A. Cross and P. Schanda, Perturbations of Native Membrane Protein Structure in Alkyl Phosphocholine Detergents: A Critical Assessment of NMR and Biophysical Studies, *Chem. Rev.*, 2018, **118**(7), 3559–3607.
42. S. T. Hsu, E. Breukink, G. Bierbaum, H. G. Sahl, B. de Kruijff, R. Kaptein, N. A. van Nuland and A. M. Bonvin, NMR study of mersacidin and lipid II interaction in dodecylphosphocholine micelles. Conformational changes are a key to antimicrobial activity, *J. Biol. Chem.*, 2003, **278**(15), 13110–13117.
43. O. Dias Rde and O. L. Franco, Cysteine-stabilized alphabeta defensins: From a common fold to antibacterial activity, *Peptides*, 2015, **72**, 64–72.
44. M. K. Nygaard, A. S. Andersen, H. H. Kristensen, K. A. Krogfelt, P. Fojan and R. Wimmer, The insect defensin lucifensin from Lucilia sericata, *J. Biomol. NMR*, 2012, **52**(3), 277–282.
45. A. Essig, D. Hofmann, D. Munch, S. Gayathri, M. Kunzler, P. T. Kallio, H. G. Sahl, G. Wider, T. Schneider and M. Aebi, Copsin, a novel peptide-based fungal antibiotic interfering with the peptidoglycan synthesis, *J. Biol. Chem.*, 2014, **289**(50), 34953–34964.
46. J. S. Oeemig, C. Lynggaard, D. H. Knudsen, F. T. Hansen, K. D. Norgaard, T. Schneider, B. S. Vad, D. H. Sandvang, L. A. Nielsen, S. Neve, H. H. Kristensen, H. G. Sahl, D. E. Otzen and R. Wimmer, Eurocin, a new fungal defensin: structure, lipid binding, and its mode of action, *J. Biol. Chem.*, 2012, **287**(50), 42361–42372.
47. Y. S. Yang, G. Mitta, A. Chavanieu, B. Calas, J. F. Sanchez, P. Roch and A. Aumelas, Solution structure and activity of the synthetic four-disulfide bond Mediterranean mussel defensin (MGD-1), *Biochemistry*, 2000, **39**(47), 14436–14447.
48. K. S. Brinch, P. M. Tulkens, F. Van Bambeke, N. Frimodt-Moller, N. Hoiby and H. H. Kristensen, Intracellular activity of the peptide antibiotic NZ2114: studies with Staphylococcus aureus and human THP-1 monocytes, and comparison with daptomycin and vancomycin, *J. Antimicrob. Chemother.*, 2010, **65**(8), 1720–1724.
49. P. H. Mygind, R. L. Fischer, K. M. Schnorr, M. T. Hansen, C. P. Sonksen, S. Ludvigsen, D. Raventos, S. Buskov, B. Christensen, L. De Maria, O. Taboureau, D. Yaver, S. G. Elvig-Jorgensen, M. V. Sorensen, B. E. Christensen, S. Kjaerulff, N. Frimodt-Moller, R. I. Lehrer, M. Zasloff and H. H. Kristensen, Plectasin is a peptide antibiotic with therapeutic potential from a saprophytic fungus, *Nature*, 2005, **437**(7061), 975–980.
50. K. S. Brinch, A. Sandberg, P. Baudoux, F. Van Bambeke, P. M. Tulkens, N. Frimodt-Moller, N. Hoiby and H. H. Kristensen, Plectasin shows intracellular activity against Staphylococcus aureus in human THP-1 monocytes and in a mouse peritonitis model, *Antimicrob. Agents Chemother.*, 2009, **53**(11), 4801–4808.
51. X. Chen, Y. Wen, L. Li, J. Shi, Z. Zhu, Y. Luo, Y. Li and R. Chen, The Stability, and Efficacy Against Penicillin-Resistant Enterococcus

faecium, of the Plectasin Peptide Efficiently Produced by Escherichia coli, *J. Microbiol. Biotechnol.*, 2015, **25**(7), 1007–1014.
52. J. Medeiros-Silva, S. Jekhmane, E. Breukink and M. Weingarth, Towards the Native Binding Modes of Antibiotics that Target Lipid II, *ChemBioChem*, 2019, **20**(14), 1731–1738.
53. C. Dominguez, R. Boelens and A. M. Bonvin, HADDOCK: a protein-protein docking approach based on biochemical or biophysical information, *J. Am. Chem. Soc.*, 2003, **125**(7), 1731–1737.
54. L. L. Ling, T. Schneider, A. J. Peoples, A. L. Spoering, I. Engels, B. P. Conlon, A. Mueller, T. F. Schaberle, D. E. Hughes, S. Epstein, M. Jones, L. Lazarides, V. A. Steadman, D. R. Cohen, C. R. Felix, K. A. Fetterman, W. P. Millett, A. G. Nitti, A. M. Zullo, C. Chen and K. Lewis, A new antibiotic kills pathogens without detectable resistance, *Nature*, 2015, **517**(7535), 455–459.
55. C. Oster, G. P. Walkowiak, D. E. Hughes, A. L. Spoering, A. J. Peoples, A. C. Catherwood, J. A. Tod, A. J. Lloyd, T. Herrmann, K. Lewis, C. G. Dowson and J. R. Lewandowski, Structural studies suggest aggregation as one of the modes of action for teixobactin, *Chem. Sci.*, 2018, **9**(47), 8850–8859.
56. H. Yang, M. Wierzbicki, D. R. Du Bois and J. S. Nowick, X-ray Crystallographic Structure of a Teixobactin Derivative Reveals Amyloid-like Assembly, *J. Am. Chem. Soc.*, 2018, **140**(43), 14028–14032.
57. A. Parmar, A. Iyer, S. H. Prior, D. G. Lloyd, E. T. Leng Goh, C. S. Vincent, T. Palmai-Pallag, C. Z. Bachrati, E. Breukink, A. Madder, R. Lakshminarayanan, E. J. Taylor and I. Singh, Teixobactin analogues reveal enduracididine to be non-essential for highly potent antibacterial activity and lipid II binding, *Chem. Sci.*, 2017, **8**(12), 8183–8192.
58. Y. Zong, F. Fang, K. J. Meyer, L. Wang, Z. Ni, H. Gao, K. Lewis, J. Zhang and Y. Rao, Gram-scale total synthesis of teixobactin promoting binding mode study and discovery of more potent antibiotics, *Nat. Commun.*, 2019, **10**(1), 3268.
59. M. Weingarth, G. Bodenhausen and P. Tekely, Broadband magnetization transfer using moderate radio-frequency fields for NMR with very high static fields and spinning speeds, *Chem. Phys. Lett.*, 2010, **488**(1), 10–16.
60. M. Baldus and B. H. Meier, Total Correlation Spectroscopy in the Solid State. The Use of Scalar Couplings to Determine the Through-Bond Connectivity, *J. Magn. Reson., Ser. A*, 1996, **121**(1), 65–69.
61. T. Doherty and M. Hong, 2D H-1-P-31 solid-state NMR studies of the dependence of inter-bilayer water dynamics on lipid headgroup structure and membrane peptides, *J. Magn. Reson.*, 2009, **196**(1), 39–47.
62. J. Shoji, H. Hinoo, T. Hattori, K. Hirooka, Y. Kimura and T. Yoshida, Isolation and characterization of hypeptin from *Pseudomonas* sp, *J. Antibiot.*, 1989, **42**(10), 1460–1464.
63. F. von Nussbaum and R. D. Sussmuth, Multiple attack on bacteria by the new antibiotic teixobactin, *Angew. Chem., Int. Ed. Engl.*, 2015, **54**(23), 6684–6686.

64. L. J. Ball, C. M. Goult, J. A. Donarski, J. Micklefield and V. Ramesh, NMR structure determination and calcium binding effects of lipopeptide antibiotic daptomycin, *Org. Biomol. Chem.*, 2004, **2**(13), 1872–1878.
65. F. Grein, A. Muller, K. M. Scherer, X. Liu, K. C. Ludwig, A. Klockner, M. Strach, H. G. Sahl, U. Kubitscheck and T. Schneider, Ca(2+)-Daptomycin targets cell wall biosynthesis by forming a tripartite complex with undecaprenyl-coupled intermediates and membrane lipids, *Nat. Commun.*, 2020, **11**(1), 1455.
66. M. Debono, M. Barnhart, C. B. Carrell, J. A. Hoffmann, J. L. Occolowitz, B. J. Abbott, D. S. Fukuda, R. L. Hamill, K. Biemann and W. C. Herlihy, A21978C, a complex of new acidic peptide antibiotics: isolation, chemistry, and mass spectral structure elucidation, *J. Antibiot.*, 1987, **40**(6), 761–777.
67. D. Jung, A. Rozek, M. Okon and R. E. Hancock, Structural transitions as determinants of the action of the calcium-dependent antibiotic daptomycin, *Chem. Biol.*, 2004, **11**(7), 949–957.
68. K. S. Rotondi and L. M. Gierasch, A well-defined amphipathic conformation for the calcium-free cyclic lipopeptide antibiotic, daptomycin, in aqueous solution, *Biopolymers*, 2005, **80**(2–3), 374–385.
69. D. Beriashvili, R. Taylor, B. Kralt, N. Abu Mazen, S. D. Taylor and M. Palmer, Mechanistic studies on the effect of membrane lipid acyl chain composition on daptomycin pore formation, *Chem. Phys. Lipids*, 2018, **216**, 73–79.
70. W. R. Scott, S. B. Baek, D. Jung, R. E. Hancock and S. K. Straus, NMR structural studies of the antibiotic lipopeptide daptomycin in DHPC micelles, *Biochim. Biophys. Acta*, 2007, **1768**(12), 3116–3126.
71. S. D. Taylor and M. Palmer, The action mechanism of daptomycin, *Bioorg. Med. Chem.*, 2016, **24**(24), 6253–6268.
72. S. K. Straus and R. E. Hancock, Mode of action of the new antibiotic for Gram-positive pathogens daptomycin: comparison with cationic antimicrobial peptides and lipopeptides, *Biochim. Biophys. Acta*, 2006, **1758**(9), 1215–1223.
73. A. Muller, M. Wenzel, H. Strahl, F. Grein, T. N. V. Saaki, B. Kohl, T. Siersma, J. E. Bandow, H. G. Sahl, T. Schneider and L. W. Hamoen, Daptomycin inhibits cell envelope synthesis by interfering with fluid membrane microdomains, *Proc. Natl. Acad. Sci. U. S. A.*, 2016, **113**(45), E7077–E7086.
74. S. Narasimhan, S. Scherpe, A. Lucini Paioni, J. van der Zwan, G. E. Folkers, H. Ovaa and M. Baldus, DNP-Supported Solid-State NMR Spectroscopy of Proteins Inside Mammalian Cells, *Angew. Chem., Int. Ed. Engl.*, 2019, **58**(37), 12969–12973.

CHAPTER 20

New Concepts for the Mechanisms of Action of Antimicrobial Peptides from Solid-state NMR Investigations

E. S. SALNIKOV,[a] E. GLATTARD,[a] M. LOINTIER,[a] J. RAYA,[a] D. W. JUHL,[a,†] A. SAAD[a] AND B. BECHINGER*[a,b]

[a] University of Strasbourg/CNRS, UMR7177, Institut de Chimie de Strasbourg, 67070 Strasbourg, France; [b] Institut Universitaire de France (IUF), Paris, France
*Email: bechinge@unistra.fr

20.1 Introduction

20.1.1 Antimicrobial Peptides

Antimicrobial peptides (AMPs) are part of the innate immune system and are released by higher organisms when infections occur as they provide a powerful and fast first line of defence against bacterial, viral and fungal pathogens.[1,2] They have been found in humans and in many species of the plant and animal kingdom,[3] while the earliest discoveries were from microorganisms.[4] Thousands of peptide sequences are listed currently in the corresponding data bases which continuously increase in numbers,[5] while

[†]Present address: Aarhus University, Department of Chemistry, DK-8000 Aarhus C, Denmark.

New Developments in NMR No. 26
NMR Spectroscopy for Probing Functional Dynamics at Biological Interfaces
Edited by Anirban Bhunia, Hanudatta S. Atreya and Neeraj Sinha

Published by the Royal Society of Chemistry, www.rsc.org

more detailed biophysical and cell biological investigations tend to focus on selected peptides with the goal of understanding their mechanisms of action.[6] The antimicrobial activities of magainins from *Xenopus laevis* frogs were described more than 30 years ago,[7] adding an exciting biomedical perspective to much earlier publications where related amphibian peptides have been described.[8]

Magainin and other linear cationic antimicrobial peptides interact with bacterial membranes where they interfere with their barrier function and the transmembrane ionic gradients.[9,10] While the peptide-induced disruption of bacterial and fungal membranes already interferes with microbial growth the peptides can also enter the cell interior where they are able to interact with a variety of anionic biopolymers[11–13] and where they result in a complex pattern of metabolic reactions.[14,15] In many cases a modulation of the immune response of the host organism has been associated with these peptides which are, therefore, also named 'host defence peptides'.[16]

The mechanism of antimicrobial action based on the disruption of the physico-chemical properties of the membrane, rather than specific interactions with chiral proteinaceous receptors, makes the peptides less prone to be neutralized by microbial resistance[2] and has made the study of these peptides particularly interesting in times when the occurrence of multiresistant pathogens rapidly increases.[17] To understand how they develop antimicrobial activities it is essential to know in detail their structure, topology, dynamics and interactions within membranes. Solid-state NMR spectroscopy is probably the only technique that allows high-resolution structural investigations in liquid crystalline lipid bilayers mimicking as closely as possible their natural environment.

Therefore, in this paper, some of the solid-state NMR techniques used during the investigations of peptide–lipid bilayer interactions will be introduced. Using this technique interesting new details about how AMPs interact with lipid bilayers continue to stimulate further research, even after decades of biophysical and structural investigations. Magainins and many other linear peptide sequences are random coil in solution and adopt a folded secondary structure when interacting with membranes. Thereby they cover a large conformational space and their detailed alignment relative to the lipid bilayer surface adds an additional variable to be considered. A detailed description of their membrane interactions often includes exchange processes involving multiple equilibria.[10]

20.1.2 Solid-state NMR Investigations of Polypeptides

Mutidimensional solution NMR is a well-established approach for high-resolution structural investigations of biomolecules when at the same time it allows characterization of their dynamics over a large range of time scales. Early on, peptides were investigated in membrane-mimetic micelles made from deuterated DPC[18,19] and more recently in isotropic bicelles or nanodiscs.[20,21] However, in many cases these membrane-mimicking assemblies are considered too small to fully reproduce the events in natural environments. Therefore, solid-state NMR techniques have been developed to investigate polypeptides associated with extended lipid bilayers.[22–27] The latter approach also works for polypeptides associated with large complexes that rotate too slowly on the NMR time scales to furnish well-resolved 'isotropic' resonances. The chemical shifts, dipolar and quadrupolar interactions all depend on the molecular alignment relative to the magnetic field and this anisotropy results in considerable line broadening when a solid or semi-solid sample is introduced into the magnetic field of the NMR spectrometer. Furthermore, the dipolar interactions between nuclei are pronounced for immobilized molecules and broaden the spectra further.

Fast rotation around the 'magic angle' (54.7°) results in averaging of these anisotropies and dipolar interactions.[28] As a result, magic angle spinning (MAS) solid-state NMR spectra show only the isotropic peak positions and resemble those observed in solution. Although the line shapes are broader than those obtained in the liquid state and thereby the spectra are less well resolved, the strategies for assignment and structural analysis follow similar concepts.[29–31] Magic angle spinning remains a popular solid-state NMR approach and the technique has already been used for the analysis of a considerable number of membrane proteins[22–25,32–34] or of lipids within bilayer assemblies.[35]

An alternative solid-state NMR approach consists in analysing the anisotropies of interactions inherent to NMR spectra of immobilized samples rather than averaging them mechanically by MAS. This technology has found widespread application in the investigation of membranes which can be aligned relative to the magnetic field of the NMR spectrometer (B_o) by mechanical supports but also by magnetic interactions.[36,37] When such lipid bilayer samples are aligned uniaxially with their normal parallel to the magnetic field direction individual molecules exhibit a unique molecular alignment relative to B_o, characterized by orientation-dependent resonance positions where the resulting anisotropic chemical shifts, dipolar and quadrupolar interactions can be exploited to analyse the orientation of bonds, protein domains and polypeptides as a whole. Thereby these data provide information on the structure, dynamics and topology of membrane-associated polypeptides.[29,30,38,39] For example, the ^{15}N chemical shift has been shown to be directly correlated to the approximate tilt angle of helical domains and has thereby provided a most valuable approach in the study of membrane topologies of a large number of peptides[40,41] (Figure 20.1). This approach has been extended to ^{2}H solid-state NMR spectra from methyl deuterated alanines, which give information about the alignment of the alanine C_α–C_β bond relative to the B_o magnetic field of the spectrometer (which coincides with the membrane normal). From the combination of ^{15}N chemical shift and ^{2}H quadrupolar coupling information, both helical tilt and pitch angles can be determined in an accurate manner (*cf.* Section 20.2.2).[42]

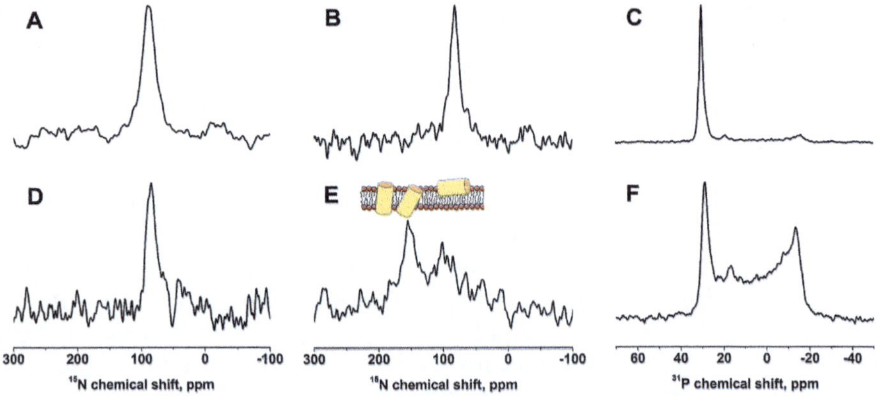

Figure 20.1 Proton-decoupled ^{15}N (A,B,D,E) and ^{31}P (C,F) solid-state NMR spectra of (A, C) 2 mole % [^{15}N-A14]-PGLa in oriented POPC, (B) 2 mole% [^{15}N-A15]-magainin 2 in diC12:0-PC, (D) [^{15}N-A14]-PGLa in the presence of unlabelled magainin 2 in POPE/POPG at the molar ratio 1 : 1 : 75 : 25, and (E,F) 0.5 mole% [^{15}N-A14]-PGLa-GGC-CGG-magainin 2 heterodimer in diC10:0-PC. The POPE/POPG sample (panel D) was measured at 37 °C and all the other spectra were acquired at ambient temperature. An exponential apodization of 200 Hz was applied before Fourier transformation to obtain the spectrum shown in panels (A) and (E), and 100 Hz for all others. The correlation between ^{15}N chemical shift and helix orientation in the membrane is schematically shown in panel (E).

20.2 Solid-state NMR Applications to Antimicrobial Peptides

20.2.1 Magainins and PGLa

The discovery of magainins coincided with a time period where new NMR techniques using oriented samples for the structural analysis of membrane-associated polypeptides were developed. Peptides were prepared that were isotopically labelled with ^{15}N at an amide bond, reconstituted into lipid bilayers which are uniaxially oriented with their normal parallel to the magnetic field direction and the ^{15}N chemical shift spectra recorded.[43] In this configuration transmembrane helices exhibit ^{15}N chemical shifts around 200 ppm, while those that are oriented parallel to the bilayer surface resonate at <100 ppm (Figure 20.1E). Later, the analysis was refined to include the effect of motions typically encountered in liquid crystalline membranes.[26,40,44] Gramicidin A and magainin 2 were among the first peptides where oriented solid-state NMR techniques were developed and applied.[43,45–47]

Figure 20.1 shows unpublished ^{15}N solid-state NMR spectra of magainin antimicrobial peptides (Table 20.1). PGLa in POPC (Figure 20.1A) and magainin 2 in membranes made of phosphatidylcholine carrying two saturated lauric acid chains (di-C12-PC) (Figure 20.1B). The peptides exhibit chemical

Table 20.1 Sequences of selected antimicrobial peptides. The one-letter code is used for peptides made from conventional amino acids only. The alamethicin sequence is given by the three-letter code with the following non-standard residues: Aib: α-aminoisobutyric acid, Phl: L-phenylalaninol, Ac- for acetyl- and -NH_2 for the carboxyamide termini, respectively.

magainin 2	GIGKF LHSAK KFGKA FVGEI MNS
PGLa	GMASK AGAIA GKIAK VALKA L–NH_2
LAH4	KKALL ALALH HLAHL ALHLA LALKK A–NH_2
alamethicin (F50/7)	Ac-Aib-Pro-Aib-Ala-Aib-Aib-Gln-Aib-Val-Aib-Gly-Leu-Aib-Pro-Val-Aib-Aib-Gln-Gln-Phl

shifts of 89 ± 10 and 82 ± 7 ppm, respectively, indicative of orientations parallel to the membrane surface ('in-plane'). This alignment is also observed for PGLa in lipid bilayers made of POPE/POPG 3/1 mole/mole thereby closely mimicking the lipid composition of bacterial membranes[48] and maintained when an equimolar amount of magainin 2 is present (Figure 20.1D).

This is an important observation because PGLa has been shown to adopt more transmembrane alignments in saturated lipid bilayers in particular in the presence of magainin 2[49,50] and/or in thin membranes,[51] observations which have led to the proposition of alternative models for synergistic interactions.[49,52] Indeed, a large proportion of the PGLa ^{15}N chemical shift intensity occurs at 153 ± 12 ppm when the covalently linked heterodimer of PGLa and magainin 2 was investigated in the short chain, fully saturated di-C10:0-PC lipid (Figure 20.1E). For this experiment both peptides were connected through a cystine bond established through carboxy-terminal GGC extensions.[53–55] The remaining intensity at 98 ± 21 ppm probably reflects an equilibrium of the heterodimer between different topologies and/or the presence of PGLa-GGC monomers[54] that may occur through chemical decomposition.

While solid-state NMR experiments indicated for the very first time that PGLa and magainin 2 align parallel to the membrane surface (Figure 20.2A) when reconstituted into supported lipid bilayers made from lipids closely mimicking their natural environment[56] the key findings were later confirmed by alternative approaches such as oriented CD,[57] fluorescence quenching experiments[58] and ATR-FTIR spectroscopy.[59,60] This membrane orientation has been confirmed for magainin 2 in many different lipid compositions.[56] It has also been found for magainin analogues[61,62] and for related linear cationic antimicrobial peptides.[63–67]

In contrast, solid-state NMR investigations on the much more hydrophobic alamethicin antimicrobial peptide (Table 20.1) show the formation of transmembrane helical bundles (Figure 20.2B) in DMPC or POPC membranes,[68,69] a configuration that comes along with very different characteristics in single-channel electrophysiological recordings.[70,71] However, even this hydrophobic sequence can adopt in-plane alignments[72,73] and thus potentially follow a second alternative way to develop antimicrobial activity.

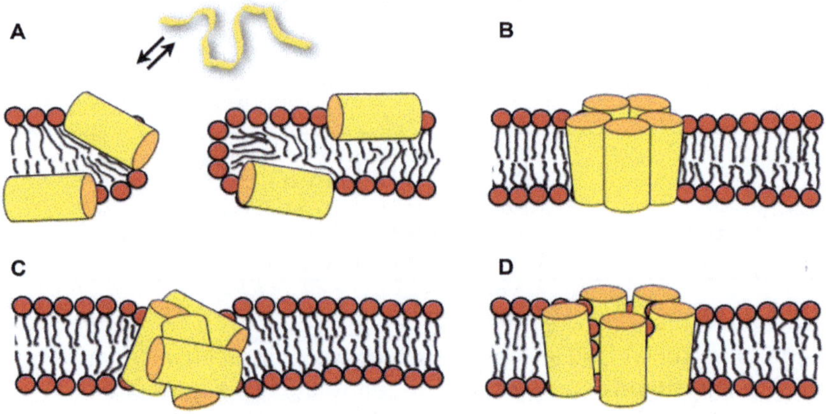

Figure 20.2 Models proposed for the pore-forming activities of antimicrobial peptides. (A) Amphipathic molecules align at the membrane interface where they impose curvature strain that results in openings. At high concentration a peptide carpet results in the disintegration of the membrane. (B) The helical bundle model has been proposed for alamethicin. (C) Micelle-like structures form within the lipid bilayer and (D) toroidal pores which are lined by peptides and lipids. The SMART model includes many of these views where a phase diagram describes the many different supramolecular arrangements a soft lipid membrane can adopt in the presence of peptides.

On the one hand, the alignment of magainin 2 parallel to the membrane surface has been observed in membranes of different compositions.[56] This in-planar topology favours the reversibility of its membrane association (Figure 20.2A) and the possibility to redistribute between different vesicles/bacteria.[56] On the other hand, the membrane-association of more hydrophobic sequences is stabilized by their association with oligomeric structures (Figure 20.2B).[68,70,71]

When compared to magainin 2, the related PGLa peptide (Table 20.1) tends to adopt a wider range of tilt angles provided it is investigated in bilayers made of fully saturated lipids.[51,74] In such environments the PGLa tilt angle depends on the membrane thickness, the peptide-to-lipid ratio, membrane hydration and the presence of magainin 2.[49,51] Interestingly, with the introduction of lipid unsaturations (*e.g.* oleoyl chains) this peptide also remains stably aligned parallel to the bilayer surface (Figure 20.2A).[50,51,75] These experiments emphasize the importance of the correct choice of lipid where head group and fatty acyl chain composition have a major effect on the structural findings.

20.2.2 Static Solid-state NMR Spectroscopy of Lipids

In order to understand the mechanisms of action of cationic amphipathic antimicrobial peptides it has become important to also investigate changes in the lipid packing and morphology.[76,77] A number of solid-state NMR approaches have been introduced to investigate the molecular conformation and supramolecular arrangement of the lipids and thereby provide highly complementary information to the data on the peptide structure, dynamics and topology (see above).[76] Because membrane association changes not only the peptide conformation but also affects the packing and macroscopic phase properties of the lipids, key pieces of information have been obtained by this technology when it comes to understanding the mechanism of action of antimicrobial peptides. First, ^{31}P solid-state NMR spectra provide indications on bilayer, hexagonal, cubic, micellar or other highly curved lipid arrangements.[77,78] In some instances, the ^{31}P solid-state NMR spectra have been characteristic for samples encompassing mixtures of membrane phases.[79] This technique is particularly straightforward because the ^{31}P isotope occurs at 100% natural abundance, is present in all phospholipids and has been well characterized in the past to be used in this context.[76]

Figure 20.1C and F shows ^{31}P solid-state NMR spectra of phosphatidylcholine membranes mechanically supported on glass plates. For pure phosphatidylcholine bilayers at ambient temperature the spectral anisotropy reaches from about 30 to −15 ppm. The spectrum shown in Figure 20.1C is characterized by a predominant intensity at 30 ppm indicative that the majority of phospholipids are aligned with their long axis parallel to the B_0 field of the NMR spectrometer (which is coincident with the sample normal).[76,80] In contrast, the spectrum shown in Figure 20.1F exhibits major contributions reaching up to the −15 ppm region, indicating that the lipid long axis covers all orientations relative to the glass plate surface. This observation indicates that the magainin-PGLa heterodimer, even at the relatively low peptide-to-lipid ratio, induces considerable curvature strain which results in the deformation and misalignment of the di-C10-PC membrane. A detailed analysis of such behaviour has been published in ref. 77. The orientational disorder by the membrane interactions of the antimicrobial aurein peptides has been made apparent by ^{31}P solid-state NMR spectroscopy and found to be strongly dependent on the detailed lipid composition.[81] Further examples of supported lipid bilayers,[77,82] of magnetically oriented samples or of deformed vesicles have been published.[83] In yet another approach, pioneering work by Dufourc and Kothe analysed ^{31}P NMR line shape and relaxation behaviour over a large temperature range to establish in considerable detail the lipid head group dynamics within bilayers.[84]

A further step in the investigation of peptide–lipid interactions arises from ^{2}H solid-state NMR spectra of deuterated fatty acyl chains which provide information on the segmental order parameters. This information has been taken to monitor changes in the packing of the lipid chains and in the hydrophobic thickness of the bilayer[48] also in bacterial preparations.[85] ^{2}H and ^{31}P solid-state NMR spectroscopy have also been calibrated to measure the surface charge at the membrane interface.[86]

In the presence of magainin 2, PGLa or related cationic amphipathic peptides, a pronounced decrease in order parameters of the fatty acyl chains has been monitored by ^{2}H solid-state NMR spectroscopy, in particular for segments that make up the bilayer interior.[48,87,88] This observation agrees with models where an interfacial amphipathic helix inserts between the lipid head groups and glycerol region,[58] thereby creating additional space in the hydrophobic interior. The concomitantly increased disorder of the lipid fatty acyl chains[89,90] is compensated by a reduction of the membrane thickness[77,91] and extends over 10 nm in diameter.[92,93] Thus, the membrane-disruptive properties of in-plane aligned amphipathic helices and the local changes in lipid morphology are part of the active mechanism and have been included in the model illustrated in Figure 20.2A.

For other peptides, preferential interactions with POPG over POPC have been observed by ^{2}H NMR spectroscopy, *e.g.* for maculatin 1.1. The peptide increases the order of the PG fatty acyl chain when this transmembrane peptide is investigated in magnetically oriented PC/PG bicelles.[94] Interestingly, in the absence of peptide the fatty acyl chains of deuterated DMPC decrease in order due to the presence of DMPG.[95] This effect is reverted when maculatin is added to mixed DMPC/DMPG liposomes possibly by partial enrichment of the anionic PG lipid in the proximity of the cationic peptide. While surface-oriented cationic antimicrobial peptides rather show a disordering of the fatty acyl chains their preferential interactions with anionic phosphatidylserine have also been observed.[89,96]

Such NMR studies have been extended to investigation of the interactions of antimicrobial peptides with the membranes of intact bacteria[97,98] or to analyse bacterial biofilms.[99] Furthermore, MAS solid-state NMR experiments have allowed to distinguish the ^{31}P chemical shifts of various phospholipids.[100,101] Changes in shift and line shape have been taken as indicators of electrostatic interactions at the membrane surface.[102]

20.2.3 Solid-state NMR Investigations of the Synergistic Enhancements of AMP Activities

More recently, the solid-state investigations shifted to the synergistic interactions of magainin 2 and PGLa where an equimolar mixture of the peptides exhibits an order of magnitude increase in antimicrobial action and of calcein release from phospholipid liposomes.[6,103] Interestingly, these two peptides are part of the naturally occurring cocktail stored in the skin of *Xenopus laevis* frogs. While early on many studies on biological activity were performed on the mixtures,[104] the biophysical studies that followed focused on investigating one peptide at the time.[6] Because synergism between magainin 2 and PGLa is also observed in pure phospholipid model membranes the interactions between the peptides that are at the foundations of synergistically enhanced pore formation can be studied in model membranes.[55,103,105,106] This is an important simplification over cellular environments where other mechanisms can play a role such as one component facilitating access of the other to its active site.

In early biophysical studies it was proposed that synergism arises from a combination of fast membrane pore formation by PGLa and increased stability due to magainin 2,[105] however, the molecular mechanisms of such synergistic events remains obscure. From first experiments it was suggested that magainin and PGLa assemble as transmembrane helical bundles (Figure 20.2B).[74] However, in the following, using ^{15}N solid-state NMR on oriented membranes, it turned out that both peptides are stably oriented parallel to the membrane surface when investigated in membranes made from phospholipids carrying unsaturated fatty acyl chains including *E. coli* lipid extracts or POPE/POPG bacterial membrane mimetics (Figure 20.1D).[48,50,51,54,107] Therefore, the alignment of these helices remains similar to those when magainin or PGLa are studied alone[51,56] (Figures 20.1A,B and D and 20.2A).

Thus, the careful design of solid-state NMR experiments on supported lipid bilayers has provided important guidance to develop new concepts for how the peptides can interact synergistically within membranes. When viewed in this context additional insights about possible PGLa–magainin 2 interactions come from comparing sequence modifications,[52,54,108,109] FRET investigations,[110] diffraction studies, molecular dynamics simulations[111,112] and most recently fluorescence correlation experiments.[106]

Notably, fluorescence quenching and fluorescence correlation experiments show that in lipid membranes PGLa and magainin 2 arrange in higher order structures, which correlates with a much increased membrane partitioning of both peptides.[6,106] Importantly, this increased membrane affinity is only observed in POPE/POPG but not POPC/POPG membranes and can by itself explain the synergistic increase in antimicrobial action. A related lipid dependence of synergism is also observed in calcein release experiments which is considerably more pronounced in POPE/POPG 3/1 mole/mole and has been related to the high negative intrinsic curvature of this membrane.[55] Such data underline the importance of the choice of lipid composition in such structural and functional studies.

Interestingly, while solid-state NMR studies reveal that PGLa remains stably inserted parallel to the surface of bilayers carrying unsaturations, it shows different membrane topologies in fully saturated membranes where the addition of magainin 2 can even result in a transition to transmembrane alignments.[48,50,51] Although such saturated lipid compositions are not representative of bacterial membranes this transition reveals interesting details on the energetic contributions that determine the membrane alignment of amphipathic peptides.[48] In different contexts such topological rearrangements may be part of regulatory mechanisms of membrane proteins.

20.2.4 NMR Investigations of AMP Dimers

Covalently linked dimers of antimicrobial peptides have been identified in nature or been designed, and some of them have been studied by solid-state NMR spectroscopy and other biophysical methods. The heterodimeric distinction is made of two different polypeptide chains which are covalently linked through a cystine bond. The compact four-helix dimer of heterodimers observed in solution[113] unfolds when interacting with membranes. When contacting the membrane chain 2, which is made up of 25 amino acids forming an amphipathic helix, partitions into the membrane interface with the helix long axis parallel to the surface, in a manner similar to magainin. In contrast, the solid-state NMR data show that chain 1 made of 22 residues associates only loosely with the lipid bilayer.[66,114] The structural findings of two chains that interact rather independently with the bilayer agree with antimicrobial tests where the dimer is only slightly more active when compared to the mixture of monomers.[115]

Homotarsinin is made of two identical 24-residue chains. When investigated by multidimensional solution NMR spectroscopy some NMR crosspeaks are doubled indicative that the dimer is asymmetric.[116] In solution the peptide adopts a tightly packed coiled-coil conformation where the hydrophobic interaction surface encapsulates a hydrophilic cluster with residues from both chains.[116] In antimicrobial assays the dimer is significantly more active than the monomers.[116] This agrees with models predicting an increase in membrane-disruptive activities if two helices partition side-by-side into the membrane interface (illustrated in ref. 109 and 117). This has also been observed for cystine-linked magainin homodimers[118,119] and magainin–PGLa heterodimers.[55,120] A preliminary ^{15}N solid-state NMR spectrum of ^{15}N-PGLa coupled to unlabelled magainin and reconstituted into uniaxially oriented di-C12-PC is shown in Figure 20.1E. In the direct neighbourhood of magainin the PGLa chain shows a tilted orientation in this thin and fully saturated membrane, similar to observations with PGLa alone[51] or the isolated peptide in the presence of magainin 2.[54]

20.2.5 Designed Antimicrobial Model Peptides and Refinement of the Topological Analysis

At the time when magainins were discovered and their antimicrobial activities characterized they were also thought to oligomerize as transmembrane helical bundles (Figure 20.2B).[121] However, solid-state NMR experiments soon indicated that such highly cationic amphipathic peptides reside along the membrane interface (Figure 20.2A) in excellent agreement with their physico-chemical character (Figure 20.3A and B).[71,117] In order to better understand the interactions that determine the alignment of amphipathic helical peptides relative to the membrane, peptides were designed where histidines replace the lysines that are part of the polar helical surface (Figure 20.3C). Thus, the resulting LAH4 peptides were made of a central sequence made of alanines and leucines which is interrupted by four histidines. In addition, two lysines at each terminus enhance their solubility in aqueous buffer, serve as membrane anchors and thus facilitate handling (Table 20.1).

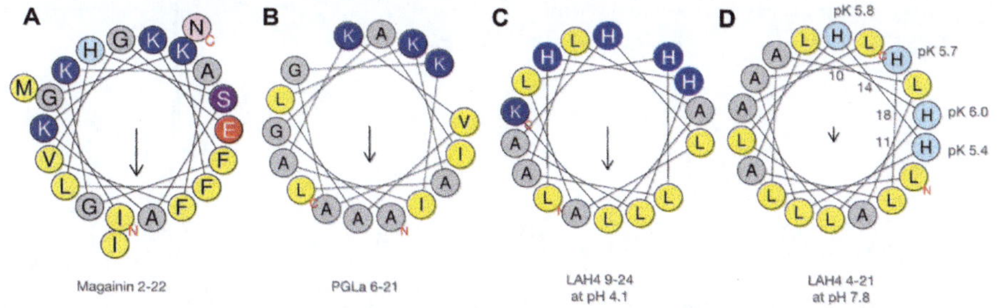

Figure 20.3 Helical wheels of AMPs calculated with the heliquest software.[179] For the calculations only parts of the peptide that show helical conformations in DPC micellar environments are included.[75,180] Two helical structures are shown for LAH4, namely from residues 9–24 at pH 4.1 and from residues 4–21 at pH 7.8.[19] To estimate the hydrophobic moment (arrows) of LAH4 at pH 4.1 the histidines were replaced by lysines. The histidine p*K* values of LAH4 in the presence of DPC micelles are indicated in panel (D). Adapted from ref. 10 with permission from John Wiley & Sons, Copyright © 2015 European Peptide Society and John Wiley & Sons, Ltd.

In this manner the overall change and hydrophobic moment of the peptides can be modified by pH changes in the surrounding medium (Figure 20.3C and D).[122–124] Indeed, solid-state and ATR-FTIR studies indicate that the LAH4 peptides adopt a transmembrane alignment at neutral and basic pH and an alignment parallel to the membrane surface under acidic conditions.[60,122] When LAH4 is reconstituted into POPC bilayers the transition between transmembrane and in-planar alignments has previously been found to occur at $pH_{50} = 6.1$,[122] a value confirmed here by ^{15}N chemical shift measurements of the 16 and 6 positions. Whereas ^{15}N chemical shifts of 78 ± 8 ppm and 73.5 ± 2.5 ppm, respectively, confirm an in-planar alignment in POPC at pH 5.3 (Figure 20.4A and B) the predominant intensity of ^{15}N-Ala16 at pH 7.4 occurs at 224 ± 12 ppm (these new data are not shown but confirm Figure 1C in ref. 123).

The LAH4 peptide was designed based on an amphipathic distribution of the alanines, leucines and the more polar histidine residues, especially when these are charged at low pH (Figure 20.3C and D), thereby being related to the amphipathic amino acid distribution of PGLa or magainins (Figure 20.3A and B). Notably, based on this design[2] the LAH4 and other histidine-rich peptides exhibit pronounced antimicrobial activities, in particular when investigated at pH 5–5.5.[125,126] Therefore, by investigating the LAH4–lipid interactions at slightly acidic pH it is in its biologically most active state. These studies also provide an example of how details on the polypeptide topology can be obtained from a combination of oriented solid-state NMR spectra. Therefore, the peptide was prepared with different ^{15}N labels (Figure 20.4A and B, ref. 123) and with a deuterated methyl group at the alanine 16 position. The corresponding ^{2}H solid-state NMR spectrum at pH 5.3 exhibits a 9 ± 2 kHz quadrupolar splitting after reconstitution into

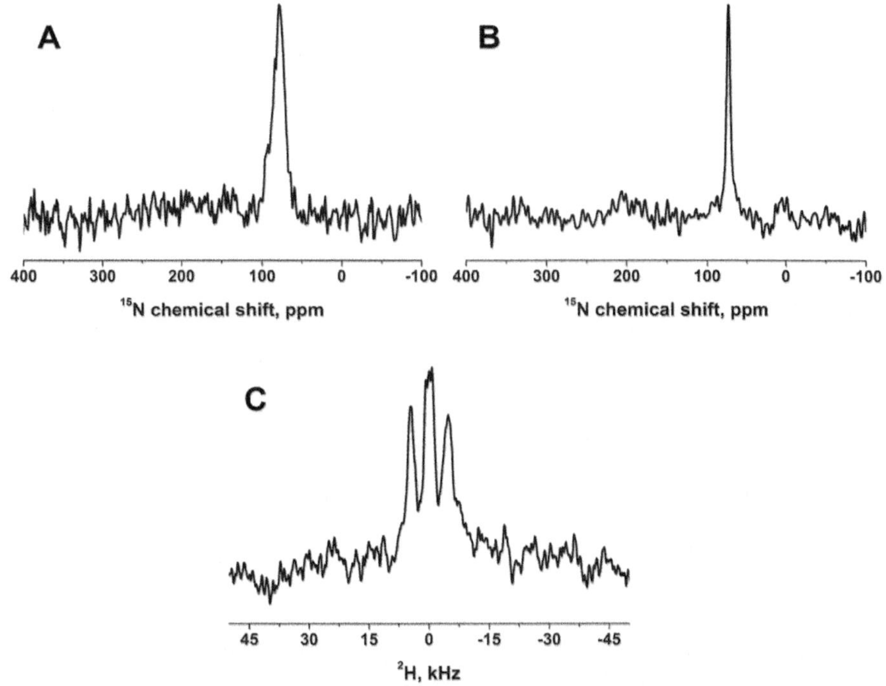

Figure 20.4 Proton-decoupled ^{15}N (A, B) and ^{2}H (C) solid-state NMR spectra of 2 mole% LAH4 in supported POPC at 23 °C. The pH was adjusted to 5.3. The normal of the glass plates was aligned parallel to the external magnetic field of the NMR spectrometer. The spectra are from LAH4 peptides labelled with ^{15}N at (A) alanine 16, (B) alanine 6 or (C) with $^{2}H_3$-alanine at position 16. An exponential apodization function of 100 Hz was applied before Fourier transformation of the free induction decay for the spectrum shown in (A), 50 Hz for (B), and 500 Hz for (C).

POPC (Figure 20.4C). In such deuterium spectra the central intensity with 1 kHz splitting is from residual deuterons of HDO.

In a next step we analysed the topology of the peptide at pH 5.3 (Figure 20.5). While it has been shown that the ^{15}N chemical shift of a ^{15}N-labelled peptide bond reconstituted in uniaxially aligned membranes is a good indicator of the approximate helix topology[40] (Figure 20.1E), a more detailed analysis which includes several solid-state NMR orientational restraints allows one to obtain detailed helical tilt and pitch angles.[44,65,66,127] This approach works by back-calculating the orientation-dependent NMR measurables for all possible peptide orientations.

Starting from a perfect transmembrane alignment (tilt/pitch = 0/0) all combinations of possible tilt/pitch angles are screened at 2° intervals. The coloured traces in the restriction plots indicate the angular pairs which agree with the experimental chemical shifts, dipolar or quadrupolar splittings, respectively, including their error bars. In Figure 20.5A and C the restrictions that arise from the ^{15}N chemical shifts of ^{15}N-Ala16 or ^{15}N-Ala6 are shown in

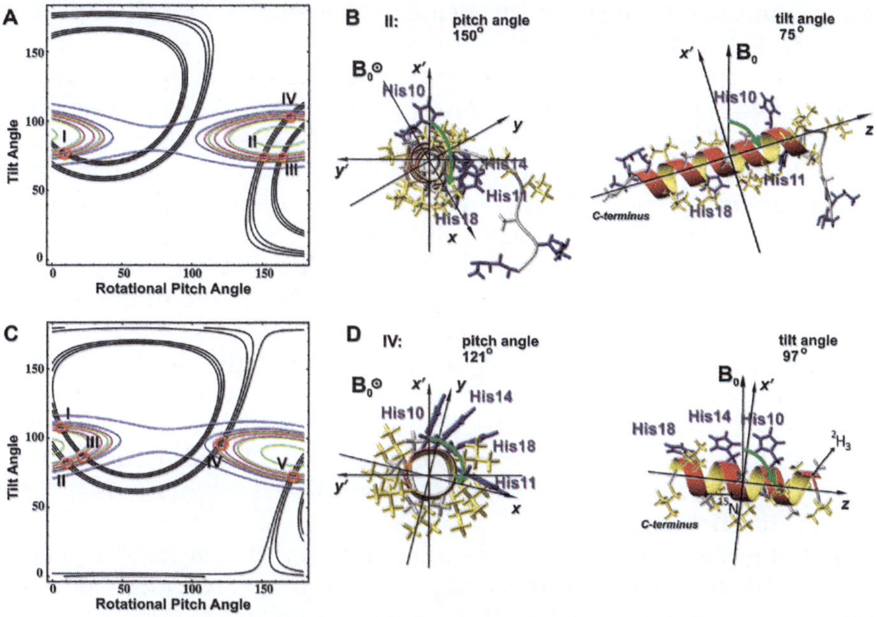

Figure 20.5 Restriction analysis (A,C) and the corresponding most probable peptide alignments (B,D) for data obtained with 2 mole% LAH4 in oriented POPC at pH 5.3. The topological restraints from each NMR measurement are shown for ^{15}N-Ala6 (73.5 ± 2.5 ppm, Figure 20.4B) in purple, ^{15}N-A13 (70 ± 5 ppm, 122) in green, ^{15}N-A16 (78 ± 8 ppm, Figure 20.4A) in blue, and $^{2}H_3$-A16 (9 ± 2 kHz, Figure 20.4C) in black. The restraints were obtained assuming rocking (18° Gaussian distribution) and wobbling (10°) motions of the helix[44]. The LAH4 structure found in the DPC-micelles at pH 4.1 is used for the analysis shown in panels (A) and (B) (pdb:2kjn, model1),[19] and an ideal helix encompassing residues 5–19 ($\varphi = -65°$, $\psi = -45°$) in panels (C) and (D). Leucines are shown in yellow, alanines in grey, and histidines in blue. Four possible intersections are obtained (red circles) when the restriction analysis is performed with the structure obtained in DPC micelles at pH 4.1 (A, red circles). Panel (B) shows the details of the corresponding topology II (pitch 150°/tilt 75°). C. Five possible orientations agree with all restrictions when an ideal helix and of which panel (D) shows topology IV (121°/97°). In (B) and (D) the tilt and pitch angles are illustrated by the green arrows. The relative orientations of the N–H and of the alanyl Cα–Cβ vectors are shown in (D; side view).

blue and red, respectively, and the restraints from the quadrupolar splitting of $^{2}H_3$-Ala16 are in black. In addition, the previously measured ^{15}N chemical shift of ^{15}N-Ala13 of 70 ± 5 ppm is shown in green.[122]

The restriction plots (Figure 20.5A and C) illustrate that on the one hand the ^{15}N chemical shift clearly indicates the in-planar alignment of the helix (*i.e.* tilt around 90° with tilt angle variations of about ±25°). On the other hand, this measurable hardly limits the pitch angle (red and green traces; ≈70° excluded) or doesn't restrict them at all (blue traces). In contrast, the

^2H quadrupolar splitting of methyl deuterated alanine is strongly dependent on tilt and pitch angle (black traces). Because of their very different angular dependence the ^{15}N chemical shift and the alanyl methyl quadrupolar splitting provide highly complementary information. Whereas the first is dominated by the σ_{33} main tensor element which orients close to parallel to the helix main axis, the dominant element of the deuterium quadrupolar splitting orients along the C-CD_3 vector, which covers an angle of about 59° with the helix axis[128] (Figure 20.5D, side view).

There are several LAH4 structures that can be used for the restriction analysis but none of them has been obtained at pH 5.3 and in the presence of lipid bilayers.[19] This raises the question which of the structures previously obtained by multidimensional ^1H solution NMR in the presence of DPC micelles fits best the orientational restraints obtained in oriented membranes. Furthermore, because the p*K* values in DPC micelles have been measured to range between 5.4 and 6.0 some of the histidines may not be fully charged at pH 5.3, in particular residue 11 which exhibits a p*K* of 5.4 in these pH-titration experiments.[19]

When the NMR structure obtained at pH 4.1 in DPC micelles is aligned using the orientational restraints from the solid-state NMR experiments four possible topologies agree with all the data (Figure 20.5A, red circles). The orientation that represents best the amphipathic distribution of amino acids exhibits a tilt angle of 75° and is shown in Figure 20.5B. However, even this energetically most favoured topology positions the C-terminal residues including His 18 ($pK_a = 6$ in DPC micelles, Figure 20.3D) inside the membrane. Notably, the helical conformation of this structure ranges from residues 9–24 with only a few NOEs defining the conformation at the labelled Ala6 position. At pH 6.1 the helix starts at position 4 but there is a kink involving residues 10–13 which positions the labelled sites at the 6 and 16 s position in two different helical domains.[19]

Because the DPC structure may differ in important details from the conformation in lipid bilayers we also calculated the tilt/pitch angular constraints for a continuous helix encompassing residues 5–19 (Figure 20.5C). A continuous helical structure from residues 4–21 is also observed in DPC micelles at pH 7.4.[19] The solid-state NMR data agree with five topologies of this structural model (Figure 20.5C, red circles), of which the most favourable one exhibits a helix tilt angle of 97°, *i.e.* almost flat on the membrane surface with histidine 11 being positioned within the membrane interface and histidines 10, 14 and 18 reaching into the aqueous environment (Figure 20.5D). Thereby, such a topology reflects the pK values measured in the presence of DPC where histidine 11 is the first to adopt a neutral charge more than 0.5 units below the pK typically observed in aqueous environments (Figure 20.3D). Notably, apart from the detailed environment, the mutual interactions between the histidines along the polar face of the helix also modulate the de-protonation equilibrium of these side chains. The experimentally determined pitch angles deviate from a clear positioning of the charged and apolar residues outside and inside the membrane which could indicate that the peptides occur as oligomers or in supramolecular arrangements when interacting with the membrane. A similar observation had been made with the amphipathic membrane-anchor of huntingtin.[44] Indeed, fluorescence quenching experiments show that LAH4 does not distribute randomly at the membrane surface but forms mesophase arrangements under conditions related to those investigated here.[129]

Comparison of Figure 20.5A,B and C,D illustrates the importance of the choice of molecular structure for the topological analysis. In the case reported here none of the known structures corresponds to the bilayer situation and the pH. However, considering the helical outlines obtained at higher pH and the possibility that distortions of the helical structure may arise from the highly curved micellar surface, a continuous helix seems a good choice to fit the angular restraints (Figure 20.5C and D). Ideally, many more angular constraints can be measured, thus the bilayer structure can be calculated independently,[46,47] or additional structural data are obtained in a lipid bilayer environment, *e.g.* by MAS solid-state NMR (*cf.* below).[30,130] Previously, the orientational restraints obtained by solid-state NMR were used for structural refinement of a DPC solution NMR structure in a related manner to the input from residual dipolar couplings in solution NMR spectroscopy.[44]

Other important considerations during the structural analysis are the choice of the tensors (size of main tensor elements and their orientation within the molecular frame), dynamics, chemical shift reference and experimental errors. Because many of these have previously been analysed in considerable detail[44,131–133] they are not discussed here.

Notably, oligourea sequences that mimic the composition of the LAH4 peptides have been prepared and show potent activities when tested in biological assays.[134] Their membrane association has been studied by biophysical methods.[135] The investigation of the membrane topology of these novel compounds was possible due to the detailed analysis of the ^{15}N chemical shift tensor in this different chemical context.[136]

20.2.6 Structural Investigations of Bilayer Lipids and of Membrane-bound Antimicrobial Peptides by MAS Solid-state NMR Spectroscopy

Oriented solid-state NMR spectra have been used to determine the backbone structure of several membrane-associated polypeptides including their topological and conformational changes during biological interactions.[24,44,46] Interestingly, of the solid-state NMR structures of polypeptides in lipid bilayers most have been obtained by the use of oriented solid-state NMR (*cf.* http://www.loquetlab.org/nmr_mpstruc/). The necessary NMR equipment is in the meantime available from commercial sources,[137] but sample preparation can be tricky. Furthermore, spectral assignment of the alignment-dependent resonances requires the preparation of proteins and peptides with a number of labelling schemes.[138] Therefore, only a few laboratories pursue this approach consistently.

Magic angle sample spinning provides alternative methods and is more in-line with the techniques and equipment available in most solid-state NMR laboratories.[23–25,32–34] Whereas sample homogeneity and motional averaging have a pronounced influence on the appearance of solid-state NMR spectra, the inherent dynamics of lipids in liquid crystalline bilayers results in sharp and well-resolved ^{1}H and /or ^{13}C MAS solid-state NMR spectra (Figure 20.6). Indeed, when phospholipid liposomes in their liquid crystalline state are investigated by MAS solid-state NMR spectroscopy the INEPT pulse sequence furnished well-resolved ^{13}C solid-state NMR spectra of the lipids while only a small residual methylene resonance is observed under cross-polarization conditions (Figure 20.6B and C). Whereas the first pulse sequence was initially developed for solution conditions where polarization transfer from the ^{1}H occurs with the help of *J*-coupling interactions, the cross-polarization approach relies on dipolar interactions between ^{1}H and ^{13}C which are characteristic of solid or semi-solid samples. Thereby the data underline the high degree of dynamics in liquid crystalline membranes averaging most of the dipolar interactions and making this signal enhancement approach inefficient (Figure 20.6C). In contrast, under the same conditions the membrane-associated peptide provides reasonable signal intensities (Figure 20.6D). The choice of the NMR pulse sequence can thus be used to distinguish signal intensities from mobile and immobile sites. The lipid spectra have also been analysed to reveal intermolecular NOEs and surprising conformational flexibility of the lipids inside the bilayer[35] or contact sites between peptide and lipids.[139,140]

Furthermore, it is of interest to investigate the membrane-associated peptides using MAS solid-state NMR approaches. Most of these involve the measurement of dipolar couplings to provide information on intra- and intermolecular distances as well as the oligomerization details of peptides within lipid bilayers.[63,68,139,141–143] These can be measured by REDOR, CODEX or two-dimensional correlation experiments. To obtain well-defined data, the conformation and interatomic contacts have to persist over the duration of the experiment. Therefore, many experiments are performed in the gel and/or frozen state.[140,141]

The structure of the stable nisin–lipid II complex has been obtained using MAS solid-state NMR spectroscopy.[142] However, the MAS spectra of other membrane-bound peptides are often not as nicely resolved because the conformational heterogeneity common for many membrane-associated peptides results in broad lines and requires extensive sample optimization. As a consequence, in many cases distance correlation peaks are difficult to resolve within the isotropic chemical shift range, in particular when helical conformations dominate. Therefore, for many membrane polypeptides, the secondary structure dependence of chemical shifts has been a very useful and only tool for conformational analysis.[144]

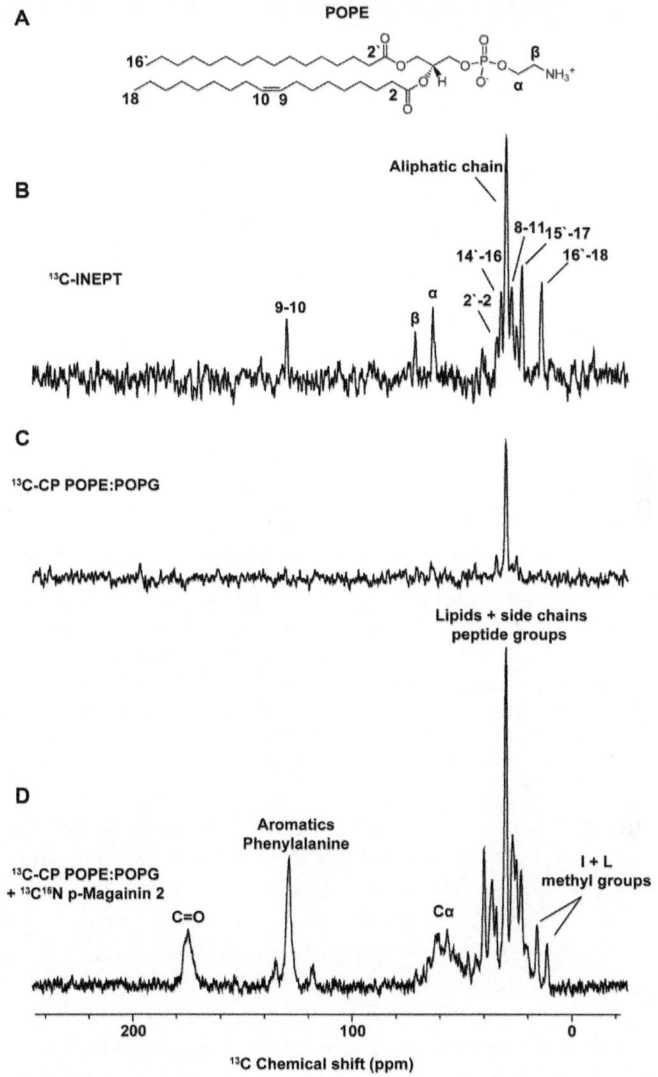

Figure 20.6 Proton-decoupled ^{13}C solid-state NMR spectra of multilamellar vesicles composed of POPE : POPG at 3/1 molar ratio recorded at 320K, 25 kHz magic angle spinning, B_0 11.74 T. (A) POPE (16 : 0, 18 : 1) lipid structure. For acquisition an INEPT (B) or a cross-polarization pulse sequence was used (C). (D) Same as (C) in the presence of [U-^{13}C,^{15}N]-p-magainin 2 at a peptide-to-lipid molar ratio of 1/50.

Because we are interested in the interactions and more detailed conformational properties of backbone and side chains side of magainins in phospholipid bilayers we also started a structural analysis by MAS solid-state NMR spectroscopy complementary to the oriented spectra, as shown in Figures 20.1 and 20.4. In a first step the peptides were overexpressed and

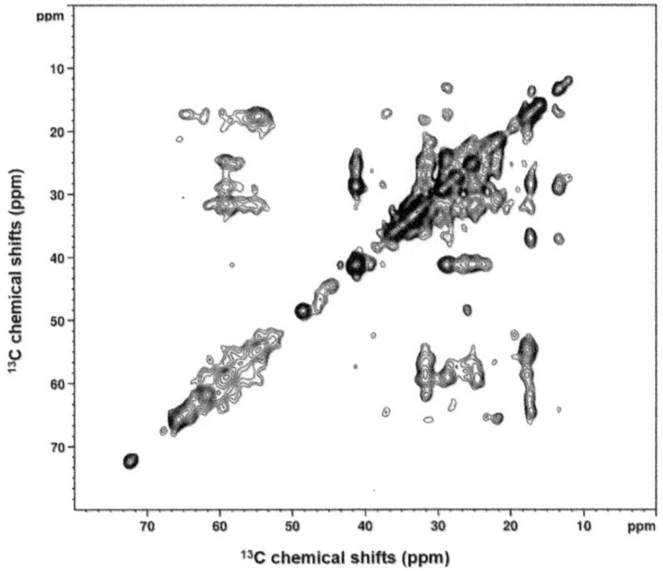

Figure 20.7 Magic angle solid-state NMR spectrum of a 1 : 1 mixture of an equimolar mixture of pPGL and pMagainin 2 in POPC/POPG (3 : 1) at a 1 : 50 peptide-to-lipid ratio. Both peptides are uniformly labelled with ^{13}C and ^{15}N. The aliphatic region of a 2D ^{13}C–^{13}C RFDR spectrum is displayed. The mixing time is 15 ms and the MAS frequency 15 kHz. The temperature of the gas flow was set to 265 K but due to MAS is 10–15 K higher inside the sample.

labelled uniformly with $^{15}N/^{13}C$ and purified by semi-quantitative HPLC as described previously.[145] The production as fusion protein that forms tight inclusion bodies protects the bacteria from the antimicrobial activity of the AMP. Because overexpression produces the free acid at the carboxy terminus and cleavage of the fusion constructs a peptide with an additional proline at the amino terminus (pPGL and pMagainin) the peptide sequences had been tested by us for their antimicrobial activities and synergistic interactions.[146] Although some modifications of the termini reduce the antimicrobial activity the synergism between the peptides remains high. Figure 20.7 exhibits a ^{13}C–^{13}C correlation spectrum of the synergistic mixture reconstituted into POPC/POPG 3/1 mole/mole membranes. Although preliminary, the spectrum shows that after sample optimization the line width is around 1.5–2 ppm and reasonable spectral resolution can be obtained also for such polypeptides in bilayer environments.

20.2.7 Mechanistic Models for AMP Activities

Soon after the discovery of magainins it became clear that they interact with membranes where that adopt amphipathic helical conformations. Indeed, electrophysiological recordings showed membrane openings[147,148] and it was initially proposed that the peptides associate into transmembrane helical bundles similar to alamethicin or the acetylcholine receptor.[121] However, important differences between these peptides exist not only when the electrophysiological characteristics are compared to each other but also in the physico-chemical properties of the sequences (Table 20.1). Whereas alamethicins are made up of mostly hydrophobic amino acids, the assembly of a magainin 2 transmembrane helical oligomer would result in the accumulation of many positive charges within a narrow channel.[117] Therefore, solid-state NMR experiments provided most valuable information in showing that the peptides orient parallel to the membrane surface rather than in a transmembrane fashion.[43,47] As a result new mechanisms had to be taken into consideration including the formation of toroidal pores made of lipids and peptides (Figure 20.2D),[149,150] a dense 'carpet' of peptides covering the membrane surface (Figure 20.2A),[151] or unspecific peptide aggregates within the bilayer (Figure 20.2C).[152] Clearly at high peptide concentrations micellar and bicellar assemblies form.[87,153,154] However, bacterial membranes lose their function probably well before they completely disintegrate.[9,155,156] Therefore, solid-state NMR and other biophysical investigation of the lipid membranes at much lower peptide-to-lipid ratios have provided important insight to complement our view. Indeed ^{2}H and ^{31}P solid-state NMR indicate that at lower peptide concentrations membranes are deformed globally and disordered at the level of the fatty acyl chains.[48,77,109] Indeed an amphipathic molecule that resides in the lipid bilayer interface produces considerable positive curvature strain (Figure 20.2A),[109] which acts in combination with the intrinsic curvatures of the phospholipids.[157] In response, conformational changes of the lipid fatty acyl chains (such as increased *trans-gauche* isomerization) adjust the packing of the hydrophobic region.[153] Furthermore, a preferential association of cationic peptides with anionic lipids has been demonstrated using ^{2}H solid-state NMR of selectively labelled lipids.[95,96]

The conformational and topological freedom of the peptides[81] and the lipids makes predictions of the details of the supramolecular assemblies difficult but clearly in response to peptide disruptions the bilayers can change thickness, morphology and macroscopic phase properties globally or locally. This concept has been summarized in the SMART model for Soft Membranes Adapt and Respond, also Transiently[10] where the lipids and peptides mutually adjust and respond in a flexible manner. At low peptide concentrations the lipid bilayer stays intact and pores only form transiently due to stochastic fluctuations of the peptide density.[147,148] Openings also occur when the peptides equilibrate between the outer and inner leaflet of the bilayer.[158–160] At higher peptide-to-lipid ratios larger structural changes of longer duration occur and ultimately membrane lysis where the transitions depend not only on the peptide-to-lipid ratio but also the membrane composition, temperature, hydration, salt, pH and other environmental factors.[153] Additional considerations are the need to cross the bacterial outer layers in order to reach the plasma membrane,[161] the selective association of peptides with certain lipids,[96,162] or the formation of phase boundaries which are thought to allow the facilitated passage of ions.[163,164]

In a manner related to the earliest models on magainin antimicrobial activity, the first explanations for the synergistic interactions between PGLa and magainin proposed a mixed transmembrane helical bundle.[74] While these first investigations were performed in fully saturated membranes where PGLa indeed adopts a close to transmembrane alignment[51] no data at all were available for magainin in the presence of the other peptide. Solid-state NMR investigations on ^{15}N-labelled peptides reconstituted into uniaxially aligned bilayers made from phospholipids carrying one unsaturation at the second chain showed that magainin 2 and PGLa both remain at the membrane surface regardless of the presence of the other peptide.[51] In the following these observations were confirmed also in other membranes including *E. coli* lipid extracts and POPE/POPG bacterial membrane mimetics indicating that the synergistic enhancement of activities is from interactions of the peptides in such in-planar topologies.[48,50,54,110] Again, using data obtained from oriented solid-state NMR spectroscopy stimulated the development of new concepts for the biological activities of these peptides.

Recently, fluorescence correlation and fluorescence quenching approaches showed that the peptides form higher-order structures in the membrane and when added as a mixture exhibit an order of magnitude increased membrane affinity.[106] Notably, the increase in affinity has been observed in POPE/POPG but not when the zwitterionic PE is replaced by PC. Similarly, synergistic enhancement of calcein release requires that the liposomes are made from lipids of high intrinsic negative curvature (such as PE),[55] while molecular dynamics simulations show peptides associating into extended structures.[112] When mimicking an environment as closely as possible to the natural surrounding solid-state NMR investigations of peptides in lipid bilayers already constitute a major improvement when compared to isotropic solvents, micelles or crystals. In addition, these most recent data indicate that also the detailed lipid composition of the bilayer, including head group and fatty acyl chain, can be of major importance and should be chosen with care. Future solid-state NMR studies aim to decipher the molecular details within these supramolecular assemblies that are the foundations for the much-increased biological action (*cf.* Figure 20.7).

20.3 Conclusions and Outlook

Solid-state NMR spectroscopy of peptides and lipids has made essential contributions to establish models to explain their antimicrobial activity. Experiments are still ongoing when new peptides, the synergistic interactions between AMPs or newly discovered dimeric compounds are investigated. The combination of static oriented and MAS solid-state NMR investigations greatly expands the range of investigations including membrane morphology and dynamics. Thus, over the years a first model of antimicrobial peptides assembling into transmembrane helical bundles

(Figure 20.2B) has evolved into a much more complex view where lipids play an essential role.[43,165] An in-depth understanding of the molecular and supramolecular interaction is required to understand the mechanism of antimicrobial membrane interactions as well as its synergistic enhancement when some selected peptides are added as a mixture.[109] Our corrected understanding of antimicrobial action and the resulting novel concepts have resulted in the design of novel compounds with high antimicrobial efficiency including short peptides and mimetics,[166–168] foldamers,[169] polymers[170] and small organic molecules.[83]

These insights were possible due to new and improved technologies also in the field of solid-state NMR spectroscopy. Therefore, it is of interest that during the last few years dynamic nuclear polarization (DNP) approaches and protocols have emerged also for supported lipid bilayers, where biradicals,[171] NMR probes, supports and protocols were developed.[137] A homogeneous distribution of the biradical source of unpaired electrons but also the efficient cooling and microwave penetration of the sample turned out to be important.[137,172] After improving many technical issues, the DNP signal enhancements could be increased from initially three-fold to two-orders of magnitude for oriented solid-state NMR spectra which reduce the acquisition times of multidimensional spectra from days to under an hour.[107,173,174] Notably, spectra were also obtained for the PGLa antimicrobial peptide which despite the low temperatures and therefore changes in lipid membrane phase remains oriented parallel to the membrane surface.[107] A transmembrane helical domain slightly changes its tilt angle as could be expected from the increase in membrane thickness when passing through the gel-to-liquid crystalline phase transition.[107] Because rotational averaging ceases at the cryogenic conditions, the two-dimensional PISEMA spectrum was much increased in resolution when compared to ambient temperatures.[107] The much-increased sensitivity obtained at low temperature and under DNP conditions promises that these membrane-associated peptides also can be analysed at much increased sensitivity with solid-state NMR approaches.

20.4 Materials and Methods

20.4.1 Materials

DMPC, POPC, POPE, and POPG lipids were from Avanti Polar Lipids Inc. (Alabaster, AL, United States) and used without further purification. Labelled peptides, *i.e.* [^{15}N-A16]-LAH4 and [^{15}N-A6, $^{2}H_3$-A16]-LAH4, were prepared by automated solid-phase synthesis using standard Fmoc chemistry on a Millipore 9050 synthesizer. Identity was verified by MALDI-TOF mass spectrometry and peptides were purified by HPLC.

20.4.2 Sample Preparation for Oriented Solid-state NMR

Peptides of 2.5 mg were dissolved in 0.3 ml milliQ-water and the pH was adjusted by the addition of small drops of NaOH or HCl solutions. The peptide solution was added to 35 mg of lipid, vortexed, bath-sonicated for 10 minutes, and exposed to three freeze/thaw/vortex cycles. The resulting suspension was then deposited on 8×12 mm ultra-thin glass plates (thickness 00, Marienfeld, Lauda-Königshofen, Germany) and left in the hood to dry overnight. Thereafter the sample was equilibrated at 93% relative humidity at 310 K for two days. The glass plates were stacked on top of each other and re-equilibrated, before the sample was wrapped with Teflon tape and sealed in plastic wrapping. The pH value of the samples was again tested from a small portion of the sample that was mixed with 300 μl milliQ water.

20.4.3 Solid-state NMR of Oriented Samples

The solid-state NMR experiments were performed on a Bruker Avance 300 wide-bore spectrometer, equipped with a triple static solid-state NMR probe modified with a flattened coil[175] (Bruker BioSpin, Rheinstetten, Germany). Proton-decoupled ^{15}N spectra were acquired with cross-polarization using the following parameters: contact time of 0.4 ms, B_1 fields of 31 kHz, recycle delay of 3 s, spectral width of 25 kHz, and acquisition time of 10.2 ms. Before Fourier transformation, an exponential apodization function typically corresponding to a line broadening of 100 Hz (if not stated otherwise) was applied. A total of 50 000 transients was typically collected for each spectrum.

A quadrupolar echo pulse sequence was used to record the deuterium solid-state NMR spectra.[176] The ^{2}H 90° pulse was 5.5 μs and the echo delay was 100 μs, with a recycle time of 0.5 s. Around 350 000 transients were added and an exponential apodization function corresponding to a line broadening of 0.5 kHz was applied before Fourier transformation. The spectra were referenced to $^{2}H_2O$ at 0 Hz.

Proton-decoupled ^{31}P spectra were acquired using a Hahn-echo pulse sequence using the following parameters: ^{31}P 90° pulse of 8 μs, echo delay of 100 μs, recycle delay of 3 s, spectral width of 100 kHz, acquisition time of 10.2 ms, and continuous wave proton-decoupling with B_1 field of 31 kHz. Before Fourier transformation, an exponential apodization function corresponding to a line broadening of 50 Hz was applied.

20.4.4 MAS Spectra of Uniformly Labelled Peptides

Peptides were prepared by bacterial over-expression and uniformly labelled with ^{13}C and ^{15}N.[145] Both peptides and lipids were dissolved in trifluoroethanol (TFE) and mixed in a round-bottomed flask. The TFE was evaporated under a stream of nitrogen and a lipid film was formed. The film was dried under vacuum for several hours. The lipid film (15.3 mg of lipid; 20 μmol) was suspended into 1 mL of 50 mM acetate buffer at pH 4.3 and exposed to eight freeze–thaw cycles before being packed into a thin-wall 2.5 mm MAS rotor.

The ^{13}C RFDR solid-state NMR spectrum was recorded on a Bruker Avance spectrometer working at ^{1}H Larmor frequency of 750 MHz equipped with a triple channel 2.5 mm MAS probe (Bruker BioSpin, Rheinstetten, Germany).

One-dimensional ^{13}C CP and INEPT NMR experiments were carried out at 125.7 MHz on a Bruker Avance WB 500 MHz (11.7 T) spectrometer using a triple channel (H/X/Y) 2.5 mm MAS probe (Bruker BioSpin, Rheinstetten, Germany). Samples were spun at 25 kHz at a temperature of 320 K.

^{13}C CP/MAS spectra were acquired using an APHH-CP pulse sequence,[177] with 4k to 10k scans, a 1 s recycling delay and 1.5 ms mixing time. B_1 fields were set to 58 kHz and 80 kHz for ^{13}C and ^{1}H, respectively, with a proton decoupling of 130 kHz during acquisition.

^{13}C INEPT experiments[178] were carried out with 1k scans, a 5 s recycling delay and 1.6 ms mixing time. The 90° high power widths for ^{13}C and ^{1}H were set to 4.31 and 3.125 μs, respectively.

Abbreviations

Aib	α-aminobutyric acid
AMP	antimicrobial peptide
ATR FTIR	attenuated total reflection Fourier transform infrared
CD	circular dichroism
CP	cross-polarization
DMPC	1,2-dimyristoyl-*sn-glycero*-3-phosphocholine
DMPG	1,2-dimyristoyl-*sn-glycero*-3-phospho-(1′-*rac*-glycerol)
DPC	dodecylphosphocholine
HPLC	high-performance liquid chromatography
INEPT	insensitive nuclei enhanced by polarization transfer
MALDI-TOF	matrix-assisted laser-desorption ionization time of flight
MAS	magic angle spinning
NMR	nuclear magnetic resonance
PC	phosphatidylcholine
PE	phosphatidylethanolamine
PG	phosphatidylglycerol
POPC	1-palmitoyl-2-oleoyl-*sn-glycero*-3-phosphocholine
POPE	1-palmitoyl-2-oleoyl-*sn-glycero*-3-phosphoethanolamine
POPG	1-palmitoyl-2-oleoyl -*sn-glycero*-3- phospho-(1′-*rac*-glycerol)
RFDR	radio frequency-driven recoupling
SMART	Soft Membranes Adapt and Respond, also Transiently
TFE	trifluoroethanol

Acknowledgements

We gratefully acknowledge the financial support by the Agence Nationale de la Recherche (projects membrane DNP 12-BSV5-0012, MemPepSyn 14-CE34-0001-01, InMembrane 15-CE11-0017-01, Biosupramol 17-CE18-0033-3 and the LabEx Chemistry of Complex Systems 10-LABX-0026_CSC), the University of Strasbourg, the CNRS, the Région Alsace and the RTRA International Center of Frontier Research in Chemistry. BB thanks the Institut Universitaire de France for providing additional time to be dedicated to research.

References

1. H. G. Boman, *Ann. Rev. Immunol.*, 1995, **13**, 61–92.
2. M. Zasloff, *Nature*, 2002, **415**, 389–395.
3. B. Agerberth, H. Gunne, J. Odeberg, P. Kogner, H. G. Boman and G. H. Gudmundsson, *Proc. Natl. Acad. Sci. U. S. A.*, 1995, **92**, 195–199.
4. R. J. Dubos and R. D. Hotchkiss, *J. Exp. Med.*, 1941, **73**, 629–640.
5. F. Ramos-Martín, T. Annaval, S. Buchoux, C. Sarazin and N. D'Amelio, *Life Sci. Alliance*, 2019, **2**, 1–11.
6. B. Bechinger, D. W. Juhl, E. Glattard and C. Aisenbrey, *Front. Med. Technol.*, 2020, **2**, 615494.
7. M. Zasloff, *Proc. Natl. Acad. Sci. U. S. A.*, 1987, **84**, 5449–5453.
8. G. Kiss and H. Michl, *Toxicon*, 1962, **1**, 33–39.
9. D. Roversi, V. Luca, S. Aureli, Y. Park, M. L. Mangoni and L. Stella, *ACS Chem. Biol.*, 2014, **9**, 2003–2007.
10. B. Bechinger, *J. Peptide Sci.*, 2015, **21**, 346–355.
11. A. Marquette, B. Lorber and B. Bechinger, *Biophys. J.*, 2010, **98**, 2544–2553.
12. N. P. Chongsiriwatana, J. S. Lin, R. Kapoor, M. Wetzler, J. A. C. Rea, M. K. Didwania, C. H. Contag and A. E. Barron, *Sci. Rep.*, 2017, **7**, 16718.
13. S. A. Overall, S. Zhu, E. Hanssen, F. Separovic and M. A. Sani, *Int. J. Mol. Sci.*, 2019, **20**(1), 181.
14. J. Kozlowska, L. S. Vermeer, G. B. Rogers, N. Rehnnuma, S. B. Amos, G. Koller, M. McArthur, K. D. Bruce and A. J. Mason, *PLoS Pathog.*, 2014, **10**, e1004104.
15. M. H. Cardoso, K. C. de Almeida, E. S. Candido, A. M. Murad, S. C. Dias and O. L. Franco, *Sci. Rep.*, 2017, **7**, 4197.
16. H. X. Luong, T. T. Thanh and T. H. Tran, *Life Sci.*, 2020, **260**, 118407.
17. G. Upert, A. Luther, D. Obrecht and P. Ermert, *Med. Drug Discovery*, 2021, **9**, 100078.
18. L. R. Brown, *Biochim. Biophys. Acta*, 1979, **557**, 135–148.
19. J. Georgescu, V. H. O. Munhoz and B. Bechinger, *Biophys. J.*, 2010, **99**, 2507–2515.
20. E. J. Dufourc, *Biochim. Biophys. Acta Biomembr.*, 2021, **1863**, 183478.
21. K. Klöpfer and F. Hagn, *Prog. Nucl. Magn. Reson. Spectrosc.*, 2019, **114–115**, 271–283.
22. R. Shukla, J. Medeiros-Silva, A. Parmar, B. J. A. Vermeulen, S. Das and A. L. Paioni, *et al.*, *Nat. Commun.*, 2020, **11**, 2848.
23. M. T. Eddy, T. Y. Yu, G. Wagner and R. G. Griffin, *J. Biomol. NMR*, 2019, **73**, 451–460.
24. S. Wang, T. Gopinath and G. Veglia, *J. Biomol. NMR*, 2019, **73**, 617–624.

25. J. Paulino, M. Yi, I. Hung, Z. Gan, X. Wang, E. Y. Chekmenev, H. X. Zhou and T. A. Cross, *Proc. Natl. Acad. Sci. U. S. A.*, 2020, **117**, 11908–11915.
26. E. S. Salnikov, C. Aisenbrey, B. Pokrandt, B. Brugger and B. Bechinger, *Front. Mol. Biosci.*, 2019, **6**, 83.
27. J. Lapin, E. O. Awosanya, R. J. A. Esteves and A. A. Nevzorov, *Solid State Nucl. Magn. Reson.*, 2021, **111**, 101701.
28. A. Pines, M. G. Gibby and J. S. Waugh, *J. Chem. Phys.*, 1973, **59**, 569–590.
29. T. Gopinath, K. R. Mote and G. Veglia, *J. Biomol. NMR*, 2015, **62**, 53–61.
30. N. Das, J. Dai, I. Hung, M. R. Rajagopalan, H. X. Zhou and T. A. Cross, *Proc. Natl. Acad. Sci. U. S. A.*, 2015, **112**, E119–126.
31. G. Jaipuria, A. Leonov, K. Giller, S. K. Vasa, L. Jaremko, M. Jaremko, R. Linser, S. Becker and M. Zweckstetter, *Nat. Commun.*, 2017, **8**, 14893.
32. A. Vogel, M. Bosse, M. Gauglitz, S. Wistuba, P. Schmidt, A. Kaiser, V. V. Gurevich, A. G. Beck-Sickinger, P. W. Hildebrand and D. Huster, *Molecules*, 2020, **25**, 5489.
33. V. S. Mandala and M. Hong, *Curr. Opin. Struct. Biol.*, 2019, **58**, 183–190.
34. V. Ladizhansky, *Biochim. Biophys. Acta*, 2017, **1865**, 1577–1586.
35. D. Huster and K. Gawrisch, *J. Am. Chem. Soc.*, 1999, **121**, 1992–1993.
36. C. Aisenbrey, P. Bertani and B. Bechinger, in *Antimicrobial Peptides*, ed. A. Guiliani and A. C. Rinaldi, Humana Press, Springer, New York, 2010, ch. 14, pp. 209–233.
37. B. Bechinger and E. S. Salnikov, in *Solid-state NMR: Applications in Biomembrane Structure*, ed. F. Separovic and M. A. Sani, IOP electronic Book, Biophysical Society, 2020, ch. 7, pp. 322–365.
38. E. S. Salnikov, G. M. Anantharamaiah and B. Bechinger, *submitted*, 2018.
39. A. Itkin, E. S. Salnikov, C. Aisenbrey, J. Raya, V. Raussens, J. M. Ruysschaert and B. Bechinger, *ACS Omega*, 2017, **2**, 6525–6534.
40. B. Bechinger and C. Sizun, *Concepts Magn. Reson.*, 2003, **18A**, 130–145.
41. C. Aisenbrey, E. S. Salnikov and B. Bechinger, *J. Membr. Biol.*, 2019, **252**, 371–384.
42. B. Bechinger, J. M. Resende and C. Aisenbrey, *Biophys. Chem.*, 2011, **153**, 115–125.
43. B. Bechinger, Y. Kim, L. E. Chirlian, J. Gesell, J. M. Neumann, M. Montal, J. Tomich, M. Zasloff and S. J. Opella, *J. Biomol. NMR*, 1991, **1**, 167–173.
44. M. Michalek, E. Salnikov, S. Werten and B. Bechinger, *Biophys. J.*, 2013, **105**, 699–710.
45. B. A. Cornell, F. Separovic, A. J. Baldassi and R. Smith, *Biophys.J.*, 1988, **53**, 67–76.
46. T. A. Cross, *Methods Enzymol.*, 1997, **289**, 672–696.
47. B. Bechinger, M. Zasloff and S. J. Opella, *Protein Sci.*, 1993, **2**, 2077–2084.
48. N. Harmouche and B. Bechinger, *Biophys. J.*, 2018, **115**, 1033–1044.
49. P. Tremouilhac, E. Strandberg, P. Wadhwani and A. S. Ulrich, *Biochim. Biophys. Acta*, 2006, **1758**, 1330–1342.

50. E. Strandberg, J. Zerweck, P. Wadhwani and A. S. Ulrich, *Biophys. J.*, 2013, **104**, L09–L11.
51. E. Salnikov and B. Bechinger, *Biophys. J.*, 2011, **100**, 1473–1480.
52. J. Zerweck, E. Strandberg, O. Kukharenko, J. Reichert, J. Burck, P. Wadhwani and A. S. Ulrich, *Sci. Rep.*, 2017, **7**, 13153.
53. T. Hara, Y. Mitani, K. Tanaka, N. Uematsu, A. Takakura and T. Tachi, *et al.*, *Biochemistry*, 2001, **40**, 12395–12399.
54. E. Glattard, E. S. Salnikov, C. Aisenbrey and B. Bechinger, *Biophys. Chem.*, 2016, **210**, 35–44.
55. R. Leber, M. Pachler, I. Kabelka, I. Svoboda, D. Enkoller, R. Vácha, K. Lohner and G. Pabst, *Biophys. J.*, 2018, **114**, 1945–1954.
56. B. Bechinger, *J. Pept. Sci.*, 2011, **17**, 306–314.
57. S. J. Ludtke, K. He, Y. Wu and H. W. Huang, *Biochim. Biophys. Acta*, 1994, **1190**, 181–184.
58. K. Matsuzaki, O. Murase, H. Tokuda, S. Funakoshi, N. Fujii and K. Miyajima, *Biochemistry*, 1994, **33**, 3342–3349.
59. E. Gazit, I. R. Miller, P. C. Biggin, M. S. P. Sansom and Y. Shai, *J. Mol. Biol.*, 1996, **258**, 860–870.
60. B. Bechinger, J. M. Ruysschaert and E. Goormaghtigh, *Biophys. J.*, 1999, **76**, 552–563.
61. A. Ramamoorthy, S. Thennarasu, D. K. Lee, A. Tan and L. Maloy, *Biophys. J.*, 2006, **91**, 206–216.
62. A. J. Mason, W. Moussaoui, T. Abdelrhaman, A. Boukhari, P. Bertani and A. Marquette, *et al.*, *J. Biol. Chem.*, 2009, **284**, 119–133.
63. M. A. Sani and F. Separovic, *Chemistry*, 2018, **24**, 286–291.
64. R. M. Hayden, G. K. Goldberg, B. M. Ferguson, M. W. Schoeneck, M. D. Libardo and S. E. Mayeux, *et al.*, *J. Phys. Chem. B*, 2015, **119**, 15235–15246.
65. J. M. Resende, R. M. Verly, C. Aisenbrey, C. Amary, P. Bertani, D. Pilo-Veloso and B. Bechinger, *Biophys. J.*, 2014, **107**, 901–911.
66. J. M. Resende, C. M. Moraes, V. H. D. O. Munhoz, C. Aisenbrey, R. M. Verly, P. Bertani, A. Cesar, D. Pilo-Veloso and B. Bechinger, *Proc. Natl. Acad. Sci. U. S. A.*, 2009, **106**, 16639–16644.
67. B. Bechinger and S. U. Gorr, *J. Dental Res.*, 2017, **96**, 254–260.
68. E. S. Salnikov, J. Raya, M. De Zotti, E. Zaitseva, C. Peggion, G. Ballano, C. Toniolo, J. Raap and B. Bechinger, *Biophys. J.*, 2016, **111**, 2450–2459.
69. E. S. Salnikov, M. De Zotti, F. Formaggio, X. Li, C. Toniolo, J. D. O'Neil, J. Raap, S. A. Dzuba and B. Bechinger, *J. Phys. Chem. B*, 2009, **113**, 3034–3042.
70. M. S. Sansom, *Eur. Biophys. J.*, 1993, **22**, 105–124.
71. B. Bechinger, *J. Membr. Biol.*, 1997, **156**, 197–211.
72. K. He, S. J. Ludtke, W. T. Heller and H. W. Huang, *Biophys. J.*, 1996, **71**, 2669–2679.
73. E. Salnikov, C. Aisenbrey, V. Vidovic and B. Bechinger, *Biochim. Biophys. Acta*, 2010, **1798**, 258–265.

74. P. Tremouilhac, E. Strandberg, P. Wadhwani and A. S. Ulrich, *J. Biol. Chem.*, 2006, **281**, 32089–32094.
75. B. Bechinger, M. Zasloff and S. J. Opella, *Biophys. J.*, 1998, **74**, 981–987.
76. B. Bechinger and E. S. Salnikov, *Chem. Phys. Lipids*, 2012, **165**, 282–301.
77. C. Kim, J. Spano, E. K. Park and S. Wi, *Biochim. Biophys. Acta*, 2009, **1788**, 1482–1496.
78. P. R. Cullis and B. De Kruijff, *Biochim. Biophys. Acta*, 1979, **559**, 399–420.
79. B. Bechinger, *Biochim. Biophys. Acta*, 2005, **1712**, 101–108.
80. C. Aisenbrey, M. Michalek, E. S. Salnikov and B. Bechinger, in *Lipid-Protein Interactions: Methods and Protocols*, ed. J. H. Kleinschmidt, Springer, New York, 2013, ch. 16, pp. 357–387.
81. J. T. J. Cheng, J. D. Hale, M. Elliott, R. E. W. Hancock and S. K. Straus, *Biochim. Biophys. Acta*, 2011, **1808**, 622–633.
82. R. M. Verly, C. M. de Moraes, J. M. Resende, C. Aisenbrey, M. P. Bemquemer, D. Pilo-Veloso, A. P. Valente, F. C. Alemida and B. Bechinger, *Biophys. J.*, 2009, **96**, 2194–2203.
83. C. Ghosh, N. Harmouche, B. Bechinger and J. Haldar, *ACS Omega*, 2018, **3**, 9182–9190.
84. E. J. Dufourc, C. Mayer, J. Stohrer, G. Althoff and G. Kothe, *Biophys. J.*, 1992, **61**, 42–57.
85. M. Laadhari, A. A. Arnold, A. E. Gravel, F. Separovic and I. Marcotte, *Biochim. Biophys. Acta*, 2016, **1858**, 2959–2964.
86. P. G. Scherer and J. Seelig, *Biochemistry*, 1989, **28**, 7720–7727.
87. K. J. Hallock, D. K. Lee, J. Omnaas, H. I. Mosberg and A. Ramamoorthy, *Biophys. J.*, 2002, **83**, 1004–1013.
88. S. L. Grage, S. Afonin, S. Kara, G. Buth and A. S. Ulrich, *Front. Cell Dev. Biol.*, 2016, **4**, 65.
89. E. S. Salnikov, A. J. Mason and B. Bechinger, *Biochimie*, 2009, **91**, 734–743.
90. M. Bortolus, A. Dalzini, C. Toniolo, K. S. Hahm and A. L. Maniero, *J. Pept. Sci.*, 2014, **20**, 517–525.
91. S. Ludtke, K. He and H. Huang, *Biochemistry*, 1995, **34**, 16764–16769.
92. F. Y. Chen, M. T. Lee and H. W. Huang, *Biophys. J.*, 2003, **84**, 3751–3758.
93. A. Mecke, D. K. Lee, A. Ramamoorthy, B. G. Orr and M. M. Banaszak Holl, *Biophys. J.*, 2005, **89**, 4043–4050.
94. A. P. Le Brun, S. Zhu, M. A. Sani and F. Separovic, *Front. Chem.*, 2020, **8**, 572.
95. J. D. Gehman, F. Luc, K. Hall, T. H. Lee, M. P. Boland, T. L. Pukala, J. H. Bowie, M. I. Aguilar and F. Separovic, *Biochemistry*, 2008, **47**, 8557–8565.
96. A. J. Mason, A. Martinez, C. Glaubitz, O. Danos, A. Kichler and B. Bechinger, *FASEB J.*, 2006, **20**, 320–322.
97. V. Booth, D. E. Warschawski, N. P. Santisteban, M. Laadhari and I. Marcotte, *Biochim. Biophys. Acta Proteins Proteom.*, 2017, **1865**, 1500–1511.

98. Z. Bouhlel, A. A. Arnold, D. E. Warschawski, K. Lemarchand, R. Tremblay and I. Marcotte, *Biochim. Biophys. Acta Biomembr.*, 2019, **1861**, 871–878.
99. J. Jeffries, W. Thongsomboon, J. A. Visser, K. Enriquez, D. Yager and L. Cegelski, *Biopolymers*, 2021, **112**, e23395.
100. T. J. Pinheiro and A. Watts, *Biochemistry*, 1994, **33**, 2451–2458.
101. D. E. Warschawski, A. A. Arnold and I. Marcotte, *Biophys. J.*, 2018, **114**, 1368–1376.
102. B. Bonev, A. Watts, M. Bokvist and G. Grobner, *Phys. Chem. Chem. Phys.*, 2001, **3**, 2904–2910.
103. A. Marquette and B. Bechinger, *Biomolecules*, 2018, **8**, 18.
104. H. V. Westerhoff, M. Zasloff, J. L. Rosner, R. W. Hendler, A. de Waal, G. Vaz, P. M. Jongsma, A. Riethorst and D. Juretic, *Eur. J. Biochem.*, 1995, **228**, 257–264.
105. K. Matsuzaki, Y. Mitani, K. Akada, O. Murase, S. Yoneyama, M. Zasloff and K. Miyajima, *Biochemistry*, 1998, **37**, 15144–15153.
106. C. Aisenbrey, M. Amaro, P. Pospisil, M. Hof and B. Bechinger, *Sci. Rep.*, 2020, **10**, 11652.
107. E. S. Salnikov, C. Aisenbrey, F. Aussenac, O. Ouari, H. Sarrouj, C. Reiter, P. Tordo, F. Engelke and B. Bechinger, *Sci. Rep.*, 2016, **6**, 20895.
108. T. Hara, H. Kodama, M. Kondo, K. Wakamatsu, A. Takeda, T. Tachi and K. Matsuzaki, *Biopolymers*, 2001, **58**, 437–446.
109. C. Aisenbrey, A. Marquette and B. Bechinger, *Adv. Exp. Med. Biol.*, 2019, **1117**, 33–64.
110. A. Marquette, E. Salnikov, E. Glattard, C. Aisenbrey and B. Bechinger, *Curr. Topics Med. Chem.*, 2015, **16**, 65–75.
111. M. Pachler, I. Kabelka, M. S. Appavou, K. Lohner, R. Vacha and G. Pabst, *Biophys. J.*, 2019, **117**, 1858–1869.
112. I. Kabelka, M. Pachler, S. Prevost, I. Letofsky-Papst, K. Lohner, G. Pabst and R. Vacha, *Biophys. J.*, 2020, **118**, 612–623.
113. D. Raimondo, G. Andreotti, N. Saint, P. Amodeo, G. Renzone, M. Sanseverino, I. Zocchi, G. Molle, A. Motta and A. Scaloni, *Proc. Natl. Acad. Sci. U. S. A.*, 2005, **102**, 6309–6314.
114. R. Verardi, N. J. Traaseth, L. Shi, F. Porcelli, L. Monfregola, S. De Luca, P. Amodeo, G. Veglia and A. Scaloni, *Biochim. Biophys. Acta*, 2011, **1808**, 34–40.
115. M. Dalla Serra, O. Cirioni, R. M. Vitale, G. Renzone, M. Coraiola and A. Giacometti, *et al.*, *Biochemistry*, 2008, **47**, 7888–7899.
116. R. M. Verly, J. M. Resende, E. F. C. Junior, M. T. C. de Magalhães, C. F. C. R. Guimarães and V. H. O. Munhoz, *et al.*, *Sci. Rep.*, 2017, **7**, 40854.
117. B. Bechinger, *Biochim. Biophys. Acta*, 1999, **1462**, 157–183.
118. C. E. Dempsey, S. Ueno and M. B. Avison, *Biochemistry*, 2003, **42**, 402–409.
119. E. N. Lorenzon, N. A. Santos-Filho, M. A. Ramos, T. M. Bauab, I. L. Camargo and E. M. Cilli, *Protein Pept. Lett.*, 2016, **23**, 738–747.
120. M. Nishida, Y. Imura, M. Yamamoto, S. Kobayashi, Y. Yano and K. Matsuzaki, *Biochemistry*, 2007, **46**, 14284–14290.

121. A. Vaz Gomes, A. de Waal, J. A. Berden and H. V. Westerhoff, *Biochemistry*, 1993, **32**, 5365–5372.
122. B. Bechinger, *J. Mol. Biol.*, 1996, **263**, 768–775.
123. B. Perrone, A. J. Miles, E. S. Salnikov, B. Wallace and B. Bechinger, *Eur. Biophys. J.*, 2014, **43**, 499–507.
124. A. J. Mason, B. Bechinger and A. Kichler, *Mini. Rev. Med. Chem.*, 2007, **7**, 491–497.
125. T. C. B. Vogt and B. Bechinger, *J. Biol. Chem.*, 1999, **274**, 29115–29121.
126. A. J. Mason, C. Gasnier, A. Kichler, G. Prevost, D. Aunis, M. H. Metz-Boutigue and B. Bechinger, *Antimicrob. Agents Chemother.*, 2006, **50**, 3305–3311.
127. C. Aisenbrey, C. Sizun, J. Koch, M. Herget, U. Abele, B. Bechinger and R. Tampe, *J. Biol. Chem.*, 2006, **281**, 30365–30372.
128. V. V. Vostrikov, B. A. Hall, D. V. Greathouse, R. E. Koeppe, 2nd and M. S. Sansom, *J Am. Chem. Soc.*, 2010, **132**, 5803–5811.
129. C. Aisenbrey and B. Bechinger, *Langmuir*, 2014, **30**, 10374–10383.
130. M. Gustavsson, R. Verardi, N. J. Traaseth and G. Veglia, *Biophys. J.*, 2012, **102**, 423A.
131. E. Salnikov, P. Bertani, J. Raap and B. Bechinger, *J. Biomol. NMR*, 2009, **45**, 373–387.
132. P. Bertani, J. Raya and B. Bechinger, *Solid-state NMR Spec.*, 2014, **61–62**, 15–18.
133. E. Salnikov, C. Aisenbrey, J. Raya and B. Bechinger, in *New Developments in NMR: Advances in Biological Solid-State NMR, Proteins and Membrane-Active Peptides*, ed. F. Separovic and A. Naito, Royal Society of Chemistry, 2014, ch. 12, pp. 214–234.
134. C. Douat, C. Aisenbrey, S. Antunes, M. Decossas, O. Lambert, B. Bechinger, A. Kichler and G. Guichard, *Angew. Chem., Int. Ed. Engl.*, 2015, **54**, 11133–11137.
135. C. Aisenbrey, C. Douat, A. Kichler, G. Guichard and B. Bechinger, *J. Phys. Chem. B*, 2020, **124**, 4476–4486.
136. C. Aisenbrey, N. Pendem, G. Guichard and B. Bechinger, *Org. Biomol. Chem.*, 2012, **10**, 1440–1447.
137. E. Salnikov, H. Sarrouj, C. Reiter, C. Aisenbrey, A. Purea and F. Aussenac, *et al.*, *J. Phys. Chem. B*, 2015, **119**, 14574–14583.
138. R. Verardi, N. J. Traaseth, L. R. Masterson, V. V. Vostrikov and G. Veglia, *Adv. Exp. Med. Biol.*, 2012, **992**, 35–62.
139. M. Hong and K. Schmidt-Rohr, *Acc. Chem. Res.*, 2013, **46**, 2154–2163.
140. M. R. Elkins, I. V. Sergeyev and M. Hong, *J. Am. Chem. Soc.*, 2018, **140**, 15437–15449.
141. M. Ouellet, J. D. Doucet, N. Voyer and M. Auger, *Biochemistry*, 2007, **46**, 6597–6606.
142. J. Medeiros-Silva, S. Jekhmane, A. L. Paioni, K. Gawarecka, M. Baldus, E. Swiezewska, E. Breukink and M. Weingarth, *Nat. Commun.*, 2018, **9**, 3963.
143. A. A. Shcherbakov and M. Hong, *J. Biomol. NMR*, 2018, **71**, 31–43.

144. G. Cornilescu, F. Delaglio and A. Bax, *J. Biomol. NMR*, 1999, **13**, 289–302.
145. V. Vidovic, L. Prongidi-Fix, B. Bechinger and S. Werten, *J. Pept. Sci.*, 2009, **15**, 278–284.
146. D. W. Juhl, E. Glattard, C. Aisenbrey and B. Bechinger, *Faraday Discuss.*, 2020, **232**, 463.
147. B. Christensen, J. Fink, R. B. Merrifield and D. Mauzerall, *Proc. Natl. Acad. Sci. U. S. A.*, 1988, **85**, 5072–5076.
148. H. Duclohier, G. Molle and G. Spach, *Biophys. J.*, 1989, **56**, 1017–1021.
149. S. J. Ludtke, K. He, W. T. Heller, T. A. Harroun, L. Yang and H. W. Huang, *Biochemistry*, 1996, **35**, 13723–13728.
150. K. Matsuzaki, *Biochim. Biophys. Acta*, 1998, **1376**, 391–400.
151. Y. Shai, *Biochim. Biophys. Acta*, 1999, **1462**, 55–70.
152. H. Jenssen, P. Hamill and R. E. Hancock, *Clin. Microbiol. Rev.*, 2006, **19**, 491–511.
153. B. Bechinger and K. Lohner, *Biochim. Biophys. Acta*, 2006, **1758**, 1529–1539.
154. J. Wolf, C. Aisenbrey, N. Harmouche, J. Raya, P. Bertani, N. Voievoda, R. Süss and B. Bechinger, *Biophys. J.*, 2017, **113**, 1290–1300.
155. H. Choi, N. Rangarajan and J. C. Weisshaar, *Trends Microbiol.*, 2016, **24**, 111–122.
156. F. Savini, M. R. Loffredo, C. Troiano, S. Bobone, N. Malanovic and T. O. Eichmann, *et al.*, *Biochim. Biophys. Acta Biomembr.*, 2020, **1862**, 183291.
157. B. Kollmitzer, P. Heftberger, M. Rappolt and G. Pabst, *Soft Matter*, 2013, **9**, 10877–10884.
158. K. Matsuzaki, O. Murase, N. Fujii and K. Miyajima, *Biochemistry*, 1995, **34**, 6521–6526.
159. Y. Tamba, H. Ariyama, V. Levadny and M. Yamazaki, *J. Phys. Chem. B*, 2010, **114**, 12018–12026.
160. S. A. Wheaten, F. D. Ablan, B. L. Spaller, J. M. Trieu and P. F. Almeida, *J Am Chem Soc*, 2013, **135**, 16517–16525.
161. K. Hall, T. H. Lee, A. I. Mechler, M. J. Swann and M. I. Aguilar, *Sci. Rep.*, 2014, **4**, 5479.
162. N. Voievoda, T. Schulthess, B. Bechinger and J. Seelig, *J. Phys. Chem. B*, 2015, **119**, 9678–9687.
163. L. Cruzeiro-Hansson and O. G. Mouritsen, *Biochim. Biophys. Acta*, 1988, **944**, 63–72.
164. F. Jean-Francois, S. Castano, B. Desbat, B. Odaert, M. Roux, M. H. Metz-Boutigue and E. J. Dufourc, *Biochemistry*, 2008, **47**, 6394–6402.
165. Y. Pouny, D. Rapaport, A. Mor, P. Nicolas and Y. Shai, *Biochemistry*, 1992, **31**, 12416–12423.
166. M. L. Mangoni and Y. Shai, *Cell Mol. Life Sci.*, 2011, **68**, 2267–2280.
167. S. Chou, C. Shao, J. Wang, A. Shan, L. Xu, N. Dong and Z. Li, *Acta Biomater.*, 2016, **30**, 78–93.
168. M. Ahn, P. Gunasekaran, G. Rajasekaran, E. Y. Kim, S. J. Lee and G. Bang, *et al.*, *Eur. J. Med. Chem.*, 2017, **125**, 551–564.

169. H. Yokoo, M. Hirano, T. Misawa and Y. Demizu, *ChemMedChem*, 2021, **16**, 1226–1233.
170. L. A. Rank, N. M. Walsh, R. Liu, F. Y. Lim, J. W. Bok, M. Huang, N. P. Keller, S. H. Gellman and C. M. Hull, *Antimicrob. Agents Chemother.*, 2017, **61**, e00204-17.
171. E. S. Salnikov, S. Abel, G. Karthikeyan, H. Karoui, F. Aussenac, P. Tordo, B. Bechinger and O. Ouari, *Chemphyschem*, 2017, **18**, 2103–2113.
172. E. S. Salnikov, F. Aussenac, S. Abel, A. Purea, P. Tordo, O. Ouari and B. Bechinger, *Solid State Nucl. Magn. Reson.*, 2019, **100**, 70–76.
173. E. Salnikov, M. Rosay, S. Pawsey, O. Ouari, P. Tordo and B. Bechinger, *J. Am. Chem. Soc.*, 2010, **132**, 5940–5941.
174. B. Bechinger, *eMagRes*, 2018, **7**, emrstm1558.
175. B. Bechinger and S. J. Opella, *J. Magn. Reson.*, 1991, **95**, 585–588.
176. J. H. Davis, K. R. Jeffrey, M. Bloom, M. I. Valic and T. P. Higgs, *Chem. Phys. Lett.*, 1976, **42**, 390–394.
177. S. Hediger, B. H. Meier, N. D. Kurur, G. Bodenhausen and R. R. Ernst, *Chem. Phys. Lett.*, 1994, **223**, 283–288.
178. B. Elena, A. Lesage, S. Steuernagel, A. Bockmann and L. Emsley, *J. Am. Chem. Soc.*, 2005, **127**, 17296–17302.
179. R. Gautier, D. Douguet, B. Antonny and G. Drin, *Bioinformatics*, 2008, **24**, 2101–2102.
180. J. Gesell, M. Zasloff and S. J. Opella, *J. Biomol. NMR*, 1997, **9**, 127–135.

CHAPTER 21

Structure and Dynamics of Native Biological Materials by Solid-state NMR Spectroscopy

RICHA DUBEY AND NEERAJ SINHA*

Centre of Biomedical Magnetic Resonance, SGPGIMS Campus, Raibarelly Road, Lucknow 226014, India
*Email: neeraj.sinha@cbmr.res.in

21.1 Introduction

Biological systems are notably complex, and they belong to a group of undertakings that work together to perform a particular function. Biological materials are mainly produced by biological systems.[1] Native biological materials or natural materials, such as blood, bone, skin, hair, muscle, nails, and tooth enamel, various body fluids, and other organic materials and different natural substances are produced with the aid of living organisms. The structures and properties of biological material are majorly dependent upon the physical and chemical properties of the components present in the system.[2] They consist of many groups including protein-based biological materials (collagen, keratin, elastin, gelatin, silk), and polysaccharide-based biological materials (glycogen, cellulose, chitin/chitosan, chondroitin). In recent years, these biological materials have become a great choice of interest for biomaterials designing for both indispensable research and clinical applications. In order to probe the structural details of these important classes of materials, there are several advanced spectroscopic techniques, for example Raman spectroscopy and Fourier transform infrared

New Developments in NMR No. 26
NMR Spectroscopy for Probing Functional Dynamics at Biological Interfaces
Edited by Anirban Bhunia, Hanudatta S. Atreya and Neeraj Sinha

Published by the Royal Society of Chemistry, www.rsc.org

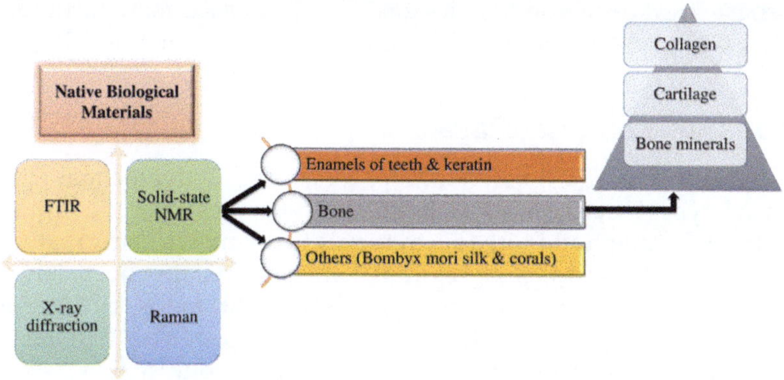

Figure 21.1 Details of spectroscopic techniques for structural studies of various native biological materials.

(FTIR) methods, for these biological materials (Figure 21.1). However, for the molecular- and atomic-level structural determination in its native state, solid-state nuclear magnetic resonance (ssNMR) spectroscopy is a vital technique, that is properly applicable for the investigation of the structure and dynamics of various native biological materials derived from vertebrates. The ssNMR techniques furnished the most insightful view and detailed information on the individual component or the interaction among the same/distinctive additives inside a biological system. Among the different biological materials (for example, tooth enamel and keratin) studied using ssNMR by many groups over the past few years, human mineralized bone tissues and enamel have long been of great interest. Therefore, it is crucial to understand the molecular and structural atomic details of other biological materials such as keratin, as a way to assist in the layout of biomaterials for complex biological requirements for therapeutic applications. Additionally, little literature is available for the structural studies of biological materials obtained from other species (such as *Bombyx mori* silk, and tropical- and deep-water corals) using ssNMR spectroscopy. Therefore, this chapter provides a thorough overview of the up-to-date information about advanced ssNMR techniques used for the structural evaluation and dynamics properties of biological materials in their native state.

21.2 ssNMR Studies of Native Biological Materials

NMR is a spectroscopic method to measure the compositional aspects of biological materials. The atomic scale structure and dynamic information of various biological materials are hallmarks of NMR methodologies. This spectroscopic technique is generally used to study the structural characterization of a diverse array of other biological assemblies that include bacteriophages, viruses, amyloid fibrils, biopolymers, biomaterials, and mammalian cytoskeleton viral and membrane-associated proteins.[3] Along with these, ssNMR methodologies also are being used for structural studies in other biological materials, which are described in the following sections.

21.2.1 Bone and Cartilage

Bone is a living composite biomaterial, consisting of organic mass (proteins, primarily type I collagen, which offers it tensile strength, lipids, polysaccharides), water and inorganic minerals (primarily hydroxyapatite, which provide its stiffness to compression).[4] It supports the body structurally, provides locomotion, protects the soft tissues and helps in the storage of calcium and phosphorus.[5] This mineralized connective tissue consists of four types of cells: osteoblasts (synthesize new bone matrix), osteocytes, bone-lining cells (maintaining and remodeling) and osteoclasts (resorb the calcified matrix), that maintain the mechanical properties in the bone matrix.[6] Bone formation and remodeling is a notably complicated process, wherein new bone is formed by replacing old bone, and this cycle is constituted of three phases: (1) initiation of bone resorption, (2) the transformation from resorption to new bone formation, and (3) finally, bone formation.[7] The entire process takes place due to coordinated actions of bone cells such as osteoclasts, osteoblasts, osteocytes, and bone-lining cells, as depicted in Figure 21.2. It has also been postulated that age-associated bone fragility fractures are a major problem for public health. Bone remodeling is an essential phenomenon for fracture healing and mechanical use.[8]

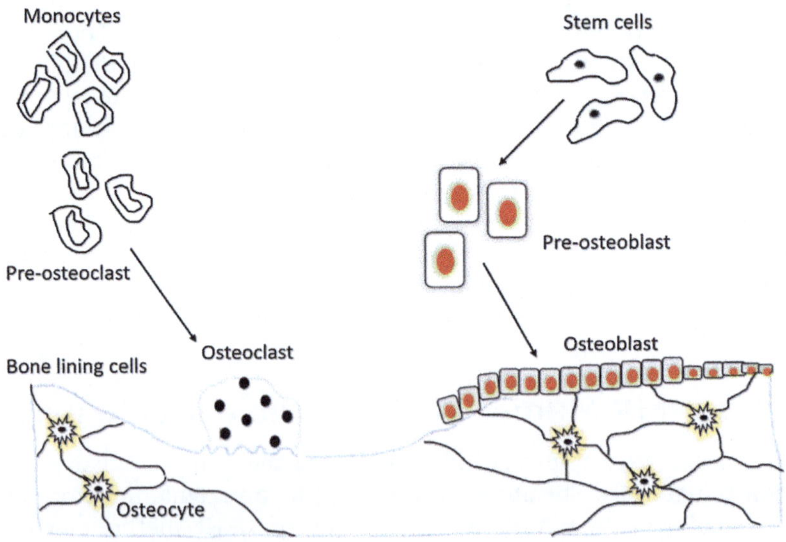

Figure 21.2 An illustration of the physiology of bone formation and remodeling. Osteoblasts develop from undifferentiated cells, while osteocytes form from osteoblasts. Osteoclasts have a separate stem cell line: blood-borne monocytes.

Bone has a complex structure that spans over a couple of periods extending from the macro- to nanostructure level, wherein bone includes cells, matrix and minerals, to the nanometer level, wherein the crystals and collagen fibrils can be seen (see Figure 21.3). Inside the bone matrix, along with inorganic minerals and type 1 collagen, quite a few non-collagenous proteins or NCPs such as phosphoproteins, pyrophosphate, osteocalcin, osteopontin, osteonectin, proteolipids and alkaline phosphatase (implicated to play a role in bone quality and fragility) are also found.[9,10] These components play an essential role in the processes of normal and pathological bone resorption and formation, collagen fibrillogenesis and mineralization. In addition, the matrix quality and mechanical properties during the bone matrix mineralization process are influenced by NCPs.[11] Any changes in NCPs can have an effect on bone material properties including hardness through their impact on mineralization.[10,12] As well as all the mentioned components, citrate ions are also quite abundant in bone and as a result present more COO^- groups for binding to calcium of apatite.[13,14] By means of advanced ssNMR and distance measurement, the group of Schmidt-Rohr has shown that citrate is strongly bound to the apatite nanocrystals in bone.[15] It was also evidenced that citrate anions bridge the mineral platelets in bone.[16] Due to mineralization, the collagen fibril architecture is dramatically affected in various orthopedic diseases such as osteoporosis, ontogenesis imperfecta, hypophosphatasia and also bone cancer. The process of bone mineralization has been intensively studied over the last few decades, but the exact molecular mechanism of mineralization of bone collagen is not yet entirely elucidated. Similarly, the biochemical mechanisms mediated by NCPs in biomineralization remain elusive. In bone, their formation and mineralization are highly dependent on intermolecular interactions. Therefore, in the last few years various research groups have tried to understand the exact mechanism of collagen mineralization through various interaction studies including organic–inorganic interface interactions, the role of water in the mineralization process and the water–lipid interactions, *etc.*

A variety of spectroscopic techniques such as X-ray diffraction (XRD), infrared (IR), Raman spectroscopy, transmission electron microscopy (TEM)

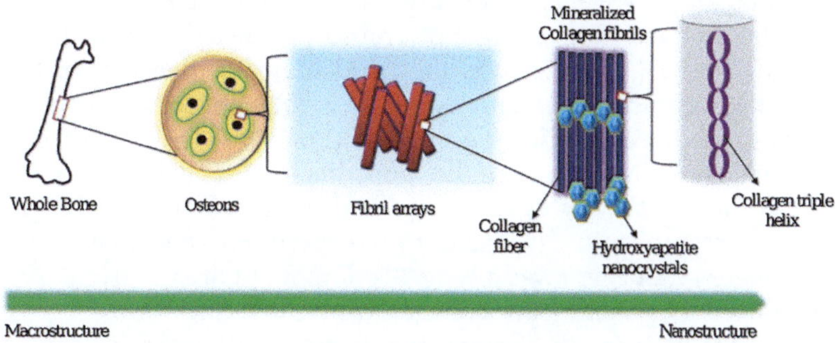

Figure 21.3 A schematic representation of the hierarchical structure of bone.

and scanning electron microscopy (SEM) have been used widely to elucidate the structure and motional properties of different components of bone matrix. In recent years, ssNMR has also had a major impact on elucidating the native structural details and dynamic information on bone mineralization. Nowadays, this approach presents a useful and flexible technique for the nanoscopic structural analysis and dynamics study. In this continuation, the contribution of ssNMR for the understanding of bone mineralization has been broadly reviewed by Melinda J. Duer in 2015.[17] Traditionally, proton (^{1}H) and phosphorus (^{31}P) NMR have been widely used to characterize the structure and dynamics inside bone matrix. ^{1}H–^{31}P cross-polarization (CP) and 2D ^{1}H–^{31}P heteronuclear correlation (HETCOR) NMR have also been used as a basic set of ssNMR experiments to study the structural analysis of collagen,[18] nanocrystalline hydroxyapatite[19] or other bone components. In 2017, Mroue *et al.* discussed an in-depth view of all the ssNMR methodologies studied for the bone mineral, organic matrix and organic–inorganic interface.[20]

In brief, the identification and quantification of various mineral ions (*e.g.*, PO_4^{3-}, OH^-, CO_3^{2-}), metal cations (*e.g.*, Ca^{2+}, Na^+, Mg^{2+} and K^+) in bone have been carried out with ssNMR spectroscopy. Therefore, to detect the NMR active nuclei, ^{43}Ca magic angle spinning (MAS) NMR,[21] ^{23}Na MAS NMR of equine bone and bovine enamel specimens were studied.[22] ^{31}P MAS NMR has been employed on hydroxyapatite to reveal the chemical structure and composition of bone minerals. To obtain more detailed information about the distribution of phosphate units in the crystalline apatite core and surface layer in bone mineral crystals has been studied with the aid of 2D ^{1}H–^{31}P HETCOR ssNMR experiments.[23–25] The prominent cross-peak at 0 ppm (OH^-) and broader cross-peaks at 5.4 ppm corresponding to the phosphate groups with hydrogen-bonded water molecules on the mineral surface have been identified. Along with these two peaks, weaker and less intense cross-peaks at 8–13 ppm ascribed to hydrogen phosphate anion (HPO_4^{2-}) ions were also observed.

Further detailed information about the bone organic matrix has been obtained using ssNMR methods. As described earlier in the literature, the bone organic matrix is composed of type I collagen (approximately 85%–90%), along with NCPs, lipids, polysaccharides, water and citrates.[26,27] Type 1 collagen is a well-studied organic component in cartilage, which provides the characteristic signals of amino acid associated with collagen. Additionally, water (~5%–25%) is the third major component of the bone matrix. Dehydration-induced and D_2O/H_2O-exchanged structural changes in bone also have been studied using ^{13}C MAS NMR.[28,29] Water dynamics generally measures the bone quality and mechanical behavior of bone. In this way, to obtain the water dynamics and its distribution, a 2D ^{2}H spin–lattice (T_2) or spin–spin (T_1) relaxation experiment has been employed, which indicated that there was a long T_2 component arising from pore water and short-lived T_2 arising from collagen/mineral-bound water.[30]

Various joint diseases like osteoarthritis, joint inflammation and trauma are related to the degeneration of cartilage. Therefore, to apprehend the atomic-scale structural modifications and molecular mobility of collagen in the process of dehydration and rehydration of cartilage, a ramp CP and refocused insensitive nuclei enhanced by polarization transfer (INEPT) have been used, which distinguished the cartilage components at the different time scale motions.[31] In CPMAS, magnetization is transferred from high-abundance nuclei ^{1}H to low-abundance nuclei ^{13}C through strong dipolar couplings, while in refocused INEPT, the dipolar couplings are averaged out by ^{1}H–^{13}C scalar *J*-coupling during the RF-free delays. In addition, the recent advances in NMR methodology and instrumentation have boosted the sensitivity and signal enhancement in bone, including through paramagnetic doping,[32,33] ultrafast MASNMR[34] and dynamic nuclear polarization (DNP).[35–37] In this respect, the recent ssNMR developments are discussed in the subsequent sections.

21.2.1.1 ssNMR Studies of Bone Triglyceride

Along with water molecules, lipids (mainly triglycerides) are also present in bone mineralized tissues, and these play an important role in the deposition of calcium phosphate onto collagen fibers during the bone mineralization process.[38] They constitute about 2% of the dry weight of the bone matrix that influences the metabolic functions of bone osteoblasts and osteocytes.[39] They regulate the exchange and transport of nutrients between cells and their environment and hence help in the critical signaling pathway.[40] To understand the implications of lipids in bone mineralization, an understanding of the structural–functional mechanism of lipids in bone mineralization is needed.

21.2.1.1.1 Detection and Identification of Lipid (Triglyceride) in Bone by High-resolution ssNMR. The lipid content is found to vary with age, health, diet and tissue location. It has been suggested that a layer of lipid particles overlaid the collagen fibrils of cortical bone of bovine tibia,[39] however, the precise additives of lipids has not been clearly identified. Later, a ^{1}H-based 2D NMR experiment test was used to selectively identify and represent triglycerides found in cortical bone.[41] To identify the exact type of lipids found in the bone matrix, a correlation spectroscopy (COSY) experiment using constant-time uniform-sign cross-peak (CTUC) COSY sequence was performed on bovine cortical bone, as shown in Figure 21.4. Generally, the CTUC COSY sequence is a changed model of ^{13}C MAS NMR that provides much a higher resolution structure in comparison with a normal COSY experiment with short T_2 (transverse relaxation times). To gain unique information about proton–proton proximities (up to a few Å) inside the molecular framework, the assignment of ^{1}H resonances was performed using this COSY-type sequence together with a ^{1}H–^{1}H double quantum–single quantum (DQ–SQ) being carried out. In this study, sharp ^{1}H resonances arising from triglyceride lipid cortical bone were obtained.

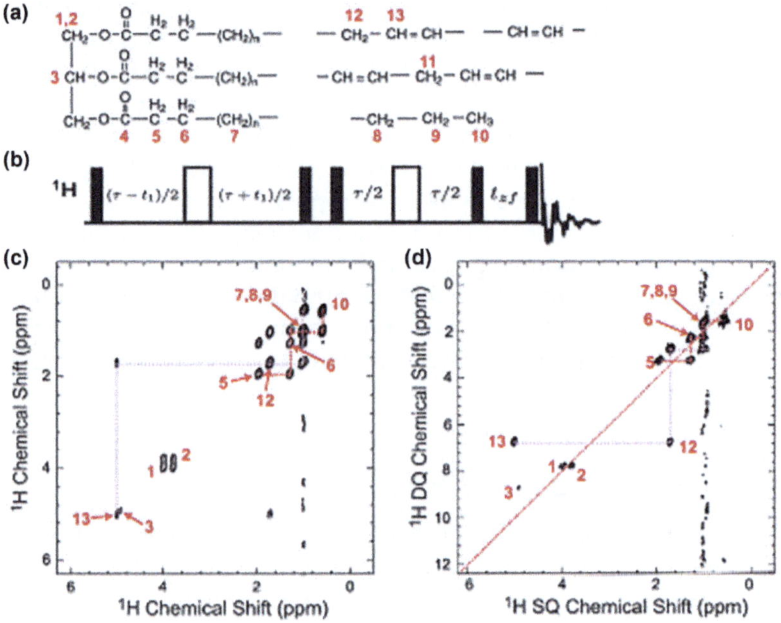

Figure 21.4 (a) Molecular fragments of triglyceride and fatty acids, together with the carbon labels. (b) ^1H-detected CTUC 2D COSY-type NMR pulse sequence. The solid and blank rectangles denote π/2 and π pulses, respectively. (c) 2D $^1H/^1H$ NMR chemical shift correlation spectrum of bovine cortical bone recorded at 10 kHz MAS rate using the pulse sequence shown in (b). (d) 2D $^1H/^1H$ NMR DQ–SQ correlation spectrum of bovine cortical bone at 10 kHz MAS rate. Reproduced from ref. 41 with permission from the Royal Society of Chemistry.

21.2.1.1.2 Water–Lipid Interaction in Native Bone. The stiffness, strength and toughness of bone are majorly associated with the water content (mobile and bound) inside the bone matrix and this varies significantly with the individual's age and disease condition.[42] Previously, it has been demonstrated that the ^1H NMR relaxation parameter, *i.e.* spin–lattice (T_2), can be used for obtaining proton signals from collagen, collagen-bound water and pore water.[43] Different components of bone such as organic matrix, minerals and water compartments have an impact on the mechanical behavior of bone, thereby it is essential to identify the characteristics that are important to fracture risk.[44] It has been observed that magnetic resonance imaging (MRI) has great potential to better diagnose bone fracture risk than x-ray absorptiometry (DXA).[45] Lipid is also an important component of bone tissues that helps in the process of bone mineralization. Therefore, observing these interactions is of great interest for understanding of the bone mineralization process in healthy as well as in disease conditions. It has been earlier illustrated that water affected the collagen dynamics in cartilage and bone, as observed in ^{13}C, ^2H and ^{15}N NMR studies.[46,47] Various additives of the organic mass of bone matrix have distinct dynamic/motional behaviors. Therefore, any structural changes to the lipid chain inside bone matrix reflect a change in the relaxation parameters. The transverse relaxation parameters (T_2) of lipid molecules were found to be much larger than collagen because it possesses greater motional rigidity. In this way, to check the water accessibility at different triglyceride sites, measurement of the relaxation parameters has been achieved for fresh native, H/D-exchanged and dehydrated native bone samples in our group,[48] as shown in Figure 21.5. The perturbations in these interactions were observed due to the hydration network dependencies of T_2 values and as a result slow- (see Figure 21.5B) and fast-decaying (see Figure 21.5C) components can be seen (see Figure 21.5), suggesting that each group of triglycerides interacts in a different way with water molecules and the perturbation in the local environment; the hydration level (hydrogen-bonding network) across the triglycerides has been determined upon dehydration and H/D exchange.

21.2.1.2 Detection of Nucleic Acid in Native Bone by Isotope Enrichment and DNP-enhanced NMR

The extracellular bone matrix (ECM) is complex, and ssNMR contributes an excellent picture for understanding of the ECM in isotope-enriched native biomaterials blended with unexpectedly evolving DNP techniques. In general, the DNP approach is used to improve the sensitivity of NMR signals by transferring considerably larger electronic spin polarization to the NMR active nuclei at lower temperatures using a microwave laser. Typically, the ssNMR spectrum of large molecules consists of an overlapping of many broad anisotropic line-shapes that cannot be resolved in one dimension. Therefore, selective isotopic labeling has become a better choice of study.

Goldberga *et al.* reported a DNP-enhanced NMR spectrum of ^{13}C–^{15}N amino-acid-enriched mouse bone and assigned each signal (see Figure 21.6A; Table 21.1) with a number of minor resonances (see Figure 21.6A) from low-abundance species that have been intently associated with the common (deoxy) nucleotide base ring nitrogen.[49]

Additionally, signals from the nitrogen atoms of protonated histidine (His) imidazolium (180 ppm) and unprotonated imidazole (250 ppm) ring forms were also seen (see Figure 21.6A).[49] The His signal assignment was compared with the selectively amino-acid-rich osteoblast model (see Figure 21.6B). Further, to see the ^{15}N–^{13}C correlation spectrum, a 2D dipolar-assisted rotational resonances (DARR) experiment has been carried out which showed a cross-peak (52 ppm ^{15}N and 65 ppm ^{13}C), which ascribes to the $(CH_3)_3N^+$–CH_2–CH_2–O–) correlation in the head groups of phosphatidylcholine (PC) lipids, because phospholipids are abundant in the mineralized tissues (see Figure 21.6C).[49] In a DARR experiment, magnetization is transferred from ^1H to ^{13}C nuclei, then different ^{13}C nuclei which might be close. In Figure 21.6C is also a distinctive fingerprint correlation; the close

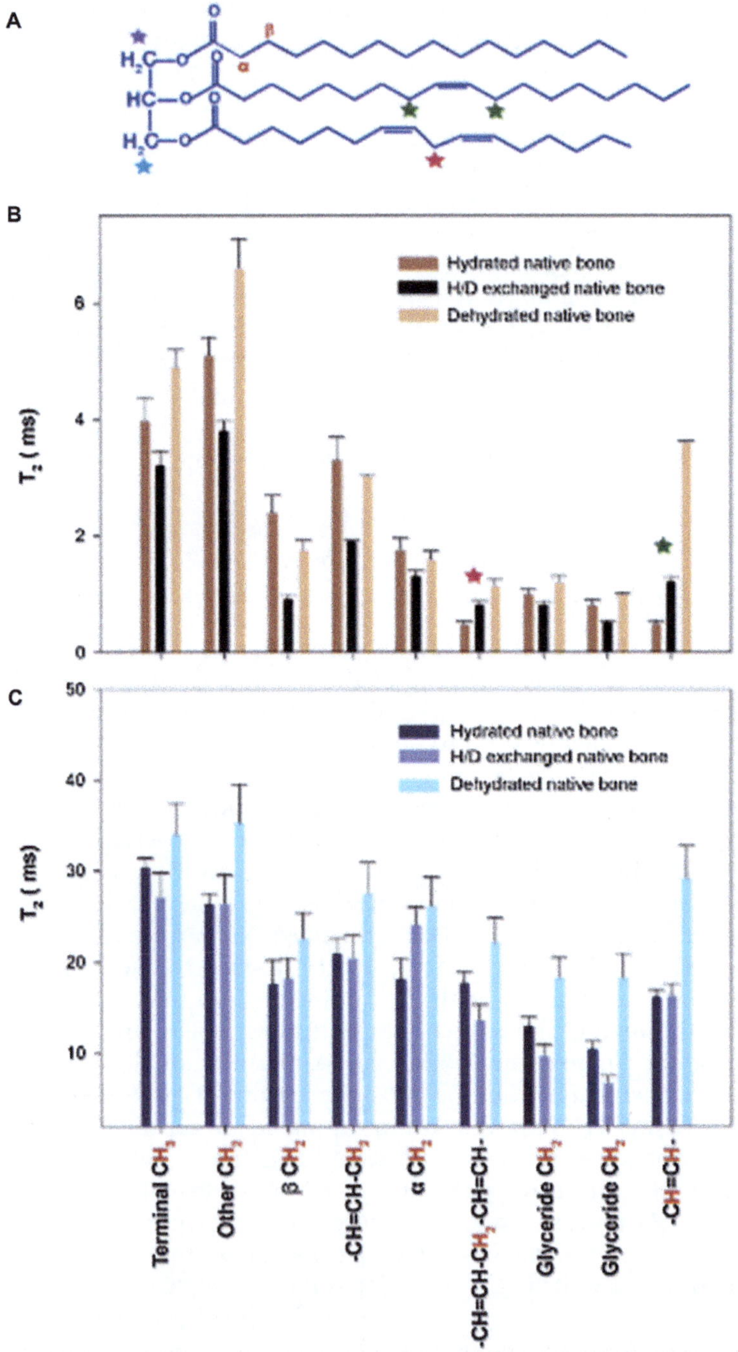

Figure 21.5 (A) Chemical structure of triglyceride and variations of [1]H T_2 values. (B) Short and (C) long components of different resonances of triglycerides of hydrated native bone, dehydrated native bone and H/D exchanged native bone at 100 µs interpulse delay. Reproduced from ref. 48 with permission from Elsevier, Copyright 2020.

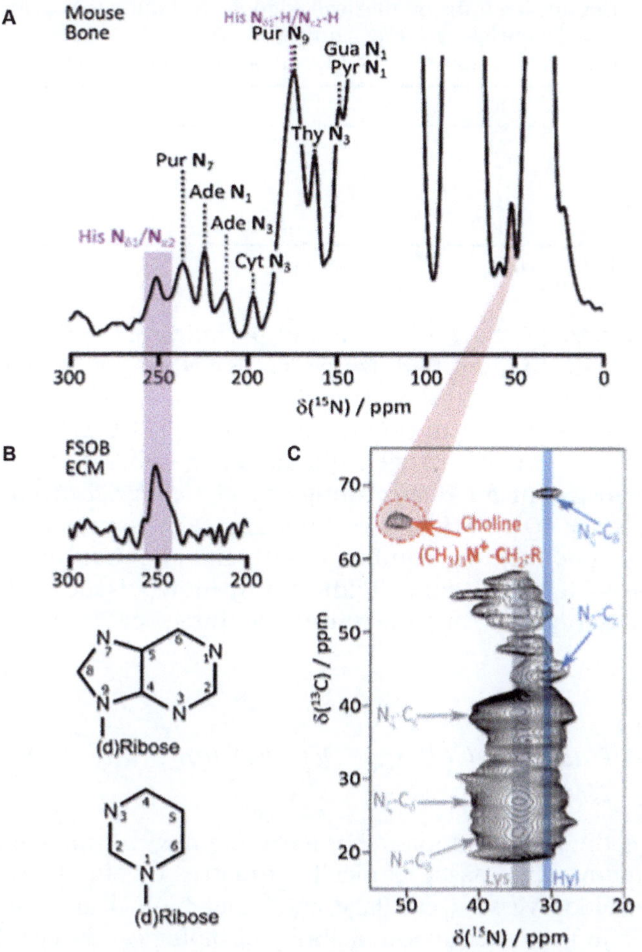

Figure 21.6 (A) High-frequency region of the ^{15}N DNP-NMR spectrum showing assignments of resolved base signals of nucleic acids, and ring His signals; (B) ^{15}N spectrum of His* fetal sheep osteoblast (FSOb) ECM, highlighting correspondence with the signal from the imidazole form of the His ring; (C) DARR-assisted ^{15}N–^{13}C correlation spectrum of labeled bone, transmitting magnetization along the hydroxylysine (Hyl) and lysine (Lys) sidechain to emphasize correlations between Hyl (as well as the intense Lys), Nz and Cd. Also highlighted is the signal from the $(CH_3)_3N^+–CH_2–CH_2–O–$ correlation from phosphatidylcholine headgroups. Atom numbering of the nucleic acid purine (top) and pyrimidine skeletons is also shown. Reproduced from ref. 49 with permission from the Royal Society of Chemistry.

correlations with lysine-enriched ECM from vascular smooth muscle cells were also observed. 2D DARR also confirmed the presence of pyrimidine signals in cytosine and thymine, and uracil from RNA.

Table 21.1 Chemical shifts of received signals from nucleic acid detected in a $^{13}C,^{15}N$-enriched bone sample by DNP NMR. Adapted from ref. 49 with permission from the Royal Society of Chemistry.

	N_1	N_3	N_7	N_9
Ade	224	212	225[a]	174[a]
Gua	148		225[a]	174[a]
Cyt	148	197		197
Thy/Ura	162	148		

[a]Single unresolved signal.

Furthermore, the purine C4–C5 and pyrimidine C4–C5 cross peak resonances have been identified using ^{13}C DNP NMR, *i.e.*, Incredible Natural Abundance Double Quantum Transfer Experiment (INADEQUATE) NMR and ^{13}C–^{15}N Transfer Echo Double Resonance (TEDOR) have been used to map the short carbon–nitrogen distances. In general, 2D INADEQUATE NMR is also useful for determining the signals arise from neighboring carbons and TEDOR NMR is used for distance measurement. Overall, this targeted isotope labeling combined with signal enhancement DNP NMR comes across low-abundance additives in intact tissue and provides information about the organic–inorganic interface in native biological materials.

21.2.1.3 Cartilage Collagen Degradation Study by Solid-state NMR

Collagen is the most abundant fibrillar protein with triple-helical structure found in mammalian tissue extracellular matrix. It is found in skins, bones, ligaments, blood vessels, cartilage and other biological tissues.[50] It has a crucial role in the function and stability of biological tissues. Earlier it was observed that this protein exhibits dynamic fluctuations and undergoes structural changes upon binding with proteins. Collagen forms a rigid network in cartilage and provides the mechanical stability of biological tissues.[51] Degradation of the collagen moiety of cartilage causes various rheumatic diseases that lead to pathological thinning of the cartilage layer of the bones. Zernia *et al.* elaborated a static and MAS ssNMR methods that have been applied to investigate the dynamics of cartilage collagen.[52] At the molecular level, this triple helical protein has repetitive glycine followed by hydroxyproline and proline residues. These amino acids provide stability to native collagens. ssNMR spectroscopy has confirmed that proline and hydroxyproline provide a dynamic equilibrium among endo and exo states of prolyl rings in collagen.[53,54]

Alkaptonuria (AKU) is a rare inherited disease, which occurs because of loss of an enzyme known as homogentisic dioxygenase (HGD) or excessive tiers of poisonous homogentisic acid (HGA). People with alkaptonuria typically develop osteoarthritis, particularly in the spine and joints, and it also induces joint cartilage degradation. HGA has a tendency to oxidize to benzoquinone acetate (BQA), which finally leads to a darkish brown pigmentation in joint cartilage known as ochronosis. Earlier ssNMR studies on AKU cartilage have shown broadening in the NMR signals.[55] Recently, Chow *et al.* carried out ^{1}H–^{13}C HETCOR DNP ssNMR cartilage samples and saw the structural change in collagen proteins upon pigmentation in AKU.[56] To identify the pigment NMR signals, a comparison study of non-pigmented and pigmented cartilage from patients with AKU has been carried out, as shown in the Figure 21.7 below. As expected, the spectra mainly featured collagen signals, although there was a difference in the ^{13}C aromatic area (110–140 ppm) in the spectrum. Further HGA-derived pigment was investigated through EPR spectroscopy to identify the redox activity hydroquinone observed at 120 ppm in the ssNMR spectrum. Two distinct radical species (air- and water-stable radicals; semiquinone) were found within the HGA-derived pigment in the EPR spectrum. These radical species in cartilage tissues can cause modifications to cartilage material properties in AKU patients. Furthermore, pigmentation was observed to impact the collagen structure, with the generation of transient radicals disrupting intra-strand hydrogen bonding in collagen, indicating local unravelling or damage to the collagen triple helix inside the coloured AKU cartilage sample.

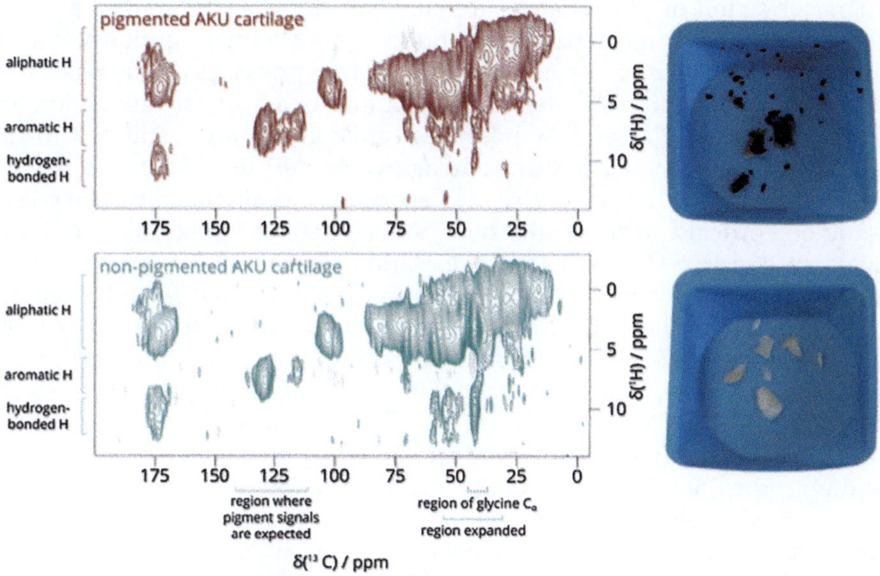

Figure 21.7 2D ^{1}H–^{13}C HETCOR DNP-enhanced ssNMR spectrum of pigmented (red) and non-pigmented (green) human AKU cartilage at 50 ms contact time. Reproduced from ref. 56, https://doi.org/10.1002/anie.202000618, under the terms of the CC BY 4.0 license, https://creativecommons.org/licenses/by/4.0/.

21.2.1.4 *Probing Dynamics in Bone Mineral Using ssNMR Spectroscopy*

The adult skeleton contains 80% cortical (compact) and 20% trabecular (spongy) bone. To probe the mineralization process inside the bone matrix diverse researches have been carried out, as described in the literature. The bone mineral components like hydroxyapatite are dominated by calcium phosphate deposited on an organic matrix, mainly on collagen. Two fundamental ssNMR techniques, MAS and cross-polarization (CP), have been widely studied to obtain a clear picture of the mineralization process. MAS NMR is used to reduce line broadening because of chemical shift anisotropy (CSA) and dipolar coupling.[57] CP is typically used to enhance low-intensity ssNMR signals from dilute spins with the aid of polarization transfer from highly abundant spins.[58] In addition, ^{31}P ssNMR provides us with completely unique information about the mineral additives inside the bone matrix, without requiring any structural or chemical preprocessing. It gives single, usually featherless ^{31}P peaks (appearing at 3.1 ppm) in the spectrum, which is difficult for the spectroscopist to analyze. Therefore, various ssNMR methodologies have been developed to study its behavior. First, Kaflak *et al.* reported the usefulness of the ^{1}H–^{31}P CP kinetics to the MAS NMR characterization of a human bone sample (mainly trabecular bone)[59] and established an experimental procedure for the analysis of the phosphate component of bone mineral.[60] Furthermore, its efficiency is suitable for the characterization of phosphate materials.[61] In the NMR relaxation process, return of nuclear magnetization is done to its thermal equilibrium with various molecular motions and all through this process, various relaxation mechanisms like dipolar interaction, CSA and quadrupolar interaction are involved. In bone, T_2 and T_1 relaxation phenomena are attributed to the molecular mobility of the extracellular matrix and used to probe the dynamics. Next, the ^{31}P spin–lattice (T_2) relaxation was discussed for the role of water on cortical and trabecular bone tissue in static conditions in order to avoid an effect of MAS on CP and relaxation. Further, $T_{1\rho}$ relaxation measurements can also be used for evaluation of bone quality and strength. Previously, $T_{1\rho}$ relaxation time was measured through ^{1}H–^{31}P CP by increasing the ^{31}P spin lock and it has been observed that the dehydrated sample had longer $T_{1\rho}$ ($\sim$250 ms), while those containing structural or adsorbed water had shorter $T_{1\rho}$ ($\sim$120 ms).[60] Overall, it has been observed that water has a significantly stronger effect on $T_{1\rho}$ than it does on T_1.[62] In addition, the ^{31}P relaxation rate (T_1 and T_2) was found to be dependent on the magnetic field strength[63] and a significant decrease in the relaxation rates with increasing deuterium concentrations were also observed upon chemical exchange.[64] It was shown that bone mineral particles found in collagenous fibers contain a crystalline apatite core covered by an amorphous hydrated surface layer. Furthermore, Kaflak *et al.* studied for the first time approximately discrete morphological regions of nanoparticles of human bone mineral and capabilities related to bone diseases, including in postmenopausal women suffering from osteoarthritis and also with osteoporosis or osteopenia.[65] It has been observed that the amorphous hydrated surface layer is more hydrated and disordered than the crystalline apatite core and it comprises as much as 24% of the phosphorus of bone mineral nanoparticles. Along with these, ^{1}H MAS NMR corroborated the OH^- peak at 0 ppm in hydroxyapatite, however the underlying problem with detecting the OH^- peak was that it interfered with resonances from the bone chemical elements. To overcome this problem numerous solid-state MAS NMR techniques and strategies have been proposed and examined on bone. In this context, reverse ^{31}P/^{1}H CP, 2D HETCOR based on the ^{1}H/^{31}P CP (conventional CP HETCOR), inverse CP, and double CP techniques have been proposed to remove unwanted protons from organic substances. While this process is rapid due to short recycle delays, it is difficult to quantify. As a result, the previously employed ssNMR techniques have various limitations, therefore, the development of a reliable, convenient NMR methodology to continue research on the OH^- content of biological apatite is needed. Recently, the hydroxyl (OH^-) content of bone apatite mineral has been measured using proton ssNMR with a multiple-pulse dipolar filter under slow MAS.[66] Figure 21.8A shows the pulse-sequence used to pick out the proton magnetization from relatively mobile groups with corresponding long transverse relaxation times. By making use of this, the apatite proton OH^- peaks have become higher visible and organized for quantitative analysis. As shown in the case of the hydroxyapatite/chitosan composite, the regular spectrum (Bloch decay; denoted as BD signal) (see Figure 21.8B) reveals a tiny peak at 0.0 ppm from apatite structural OH^- present on the lower-frequency shoulder of a huge water signal (5.2 ppm).

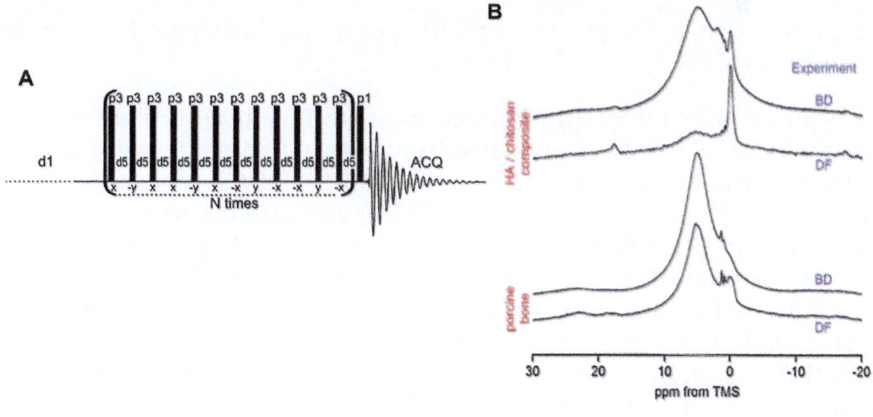

Figure 21.8 (A) The dipolar filter pulse sequence used to acquire proton ssNMR spectra of apatite biomaterials. (B) Comparison of the dipolar filter (DF) and conventional pi/2 pulse-acquired (BD) proton spectra of synthetic and natural materials containing apatite. The spectra of each sample have been scaled to the same intensity at 0 ppm, that is, at the position of the hydroxyl peak. Reproduced from ref. 66 with permission from the Royal Society of Chemistry.

The water signal is overpowering and the apatite OH^- peak is infrequently observable in the spectra of bone (porcine cranial bone). With the double filter (DF) approach the OH^- peak will become significantly enlarged and appropriate for quantitative assessment in all cases. Overall, ^1H MAS NMR spectroscopy used with the dipolar filter may be appropriate for tracking the OH^- level inside the apatite minerals of calcified tissues.

Furthermore, it has been suggested that doping with paramagnetic reagents such as Cu^{2+} effectively increases the longitudinal ^1H relaxation rate and speeds up data acquisition. To this end, Mroue *et al.* illustrated the utility of paramagnetic ion doping (Cu-EDTA) at high magnetic fields (900 MHz ^1H Larmor frequency) in bovine cortical bone and suggested the relative proximity of three amino acids in the protein backbone (glycine, alanine, and proline) to the bone mineral surface.[33] In addition, gadolinium(III) DTPA is also used as a contrast enhancer to improve the signal-to-noise ratio by increasing the longitudinal relaxation of protons and to shorten the experimental time.[32] Additionally, the recent development of MAS NMR technology is used to rotate solid samples at frequencies up to 110 kHz to overcome poor spectral resolution and sensitivity while preserving the details of the high-resolution structure of the bone.[67] Mroue *et al.* first reported the ^1H-detected NMR measurements on bone under ultrafast MAS conditions and showed that water has a bipolar ^1H–^1H coupling network that differs from the collagen backbone and side-chain regions.[67] These results showed the promising potential of proton-detected ultrafast MAS-NMR for monitoring structural and dynamic changes inside the bone matrix in various bone diseases.

21.2.1.5 Non-collagenous Proteins (NCPs) – Minerals Interaction Study Using ssNMR Spectroscopy

The regulation of mineralization became additionally controlled through the activity of NCPs. Many NCPs had been proven to directly be involved in controlling the hydroxyapatite crystallization;[68] and others can inhibit or promote crystal nucleation or growth.[69] It also contributes to cell signaling, has activity in bone and enamel-mimetic mineralization, and maintains ion homeostasis. NCPs, such as osteocalcin and osteopontin, function as regulators in the bone mineralization process. Osteocalcin (OC; 49 amino-acid) is one of the most abundant NCPs (~20%) found in bone matrix (mainly in mineralized cortical bone),[70,71] however osteopontin (OPN) is a glycoprotein found in mineralized extracellular matrices of bones and teeth. Recently, the regulatory impact of unmodified OC in the *in vitro* crystallization of bone-mimetic apatite has been studied using an ssNMR technique.[72] For this study, apatite was co-precipitated with the unmodified OC protein molecules; consequently a biomaterial complex and similarly structural feature within the mineral phase were deduced. 2D ^{13}C DARR measurements and 3D NCACX and NCOCX ssNMR experiments were used to deduce the structural properties of uniformly $^{13}C,^{15}N$ classified unmodified OC (95–98%) inside the complex. The carbon atoms that had been close in the space were visible as narrower cross-peaks inside the 2D DARR (mixing time 200 ms) spectrum, as proven in Figure 21.9, showing the proximity of carbons and constraints at the molecular structure.

Several assigned peaks are shown in Figure 21.10. The carboxylate peaks of glutamic acid and aspartic acid (marked with red color) side chains were proven to have interactions with the mineral ions. All the constraints obtained after assignment of a 2D DARR spectrum suggested that compact packing of the two helices is preserved. The aromatic amino-acid cross-peaks indicate that the pitch of the α3 helix is preserved and the center of the α1 helix is close to the α3 helix. This morphological study confirmed the higher degree of crystallinity in the hydroxyapatite-unmodified OC complex. Furthermore, the partially ordered/disordered mineral interphase formed with unmodified OC has much less water (*i.e.* in unmodified OC, the hydrated surface layer may be very thin) in comparison to the same phase in hydroxyapatite, as suggested by the 2D $^{1}H–^{31}P$ HETCOR experiments.[72] Overall, the work describes its influence on the morphology and properties of basic inorganic components in bone. Similarly, the same research group defined the regulatory impact of recombinant osteopontin (a type of NCP) on biomimetic apatite crystallite which was coated by disordered layers.[73] Earlier NMR studies suggested that OPN showed a flexible structure in solution[74] and behaved like an intrinsically disordered protein (IDP). The effects of recombinant OPN including residues 45–264 from the quail Coturnix

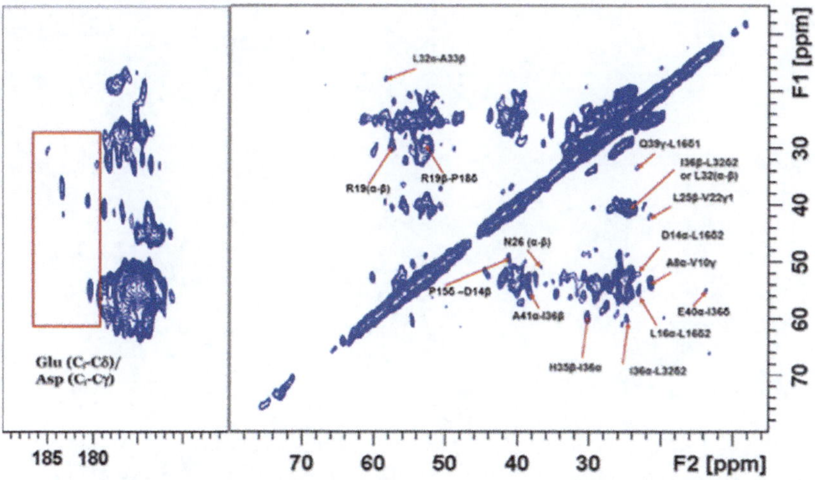

Figure 21.9 2D ^{13}C DARR spectra of [U-$^{13}C,^{15}N$] HAP· unmodified OC complex measured at a mixing time of 200 ms. Acidic sidechain cross-peaks are shown within the red box. Also, peaks marked with red arrows indicate identified carbon–carbon correlations in unmodified OC. Reproduced from ref. 72 with permission from Elsevier, Copyright 2019.

japonica on bone-like mineral production were also investigated in this study. To obtain the information about mineral content, 2D ^1H–^{31}P HETCOR experiments, utilizing ^1H–^1H decoupling during ^1H evolution in t_1 and using

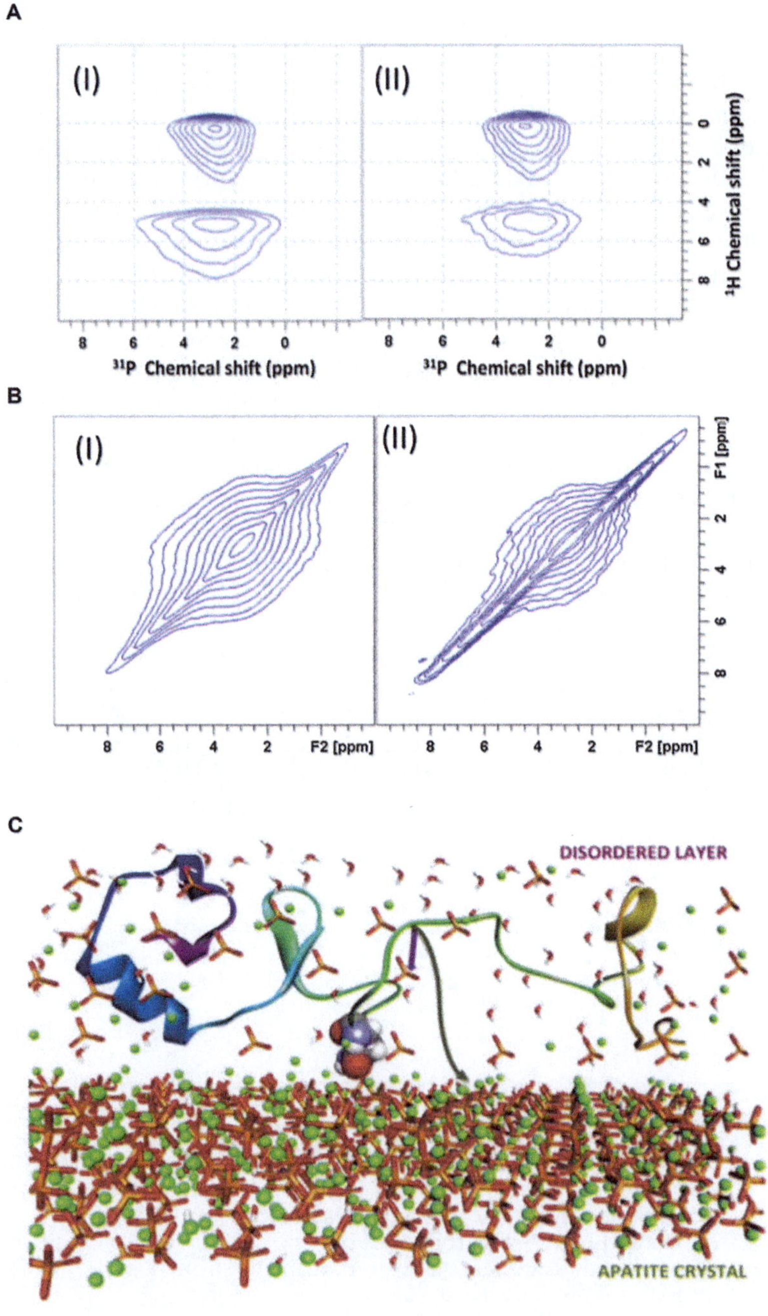

phase-modulated Lee–Goldburg (PMLG) coupling, have been carried out. Sample spinning at 10 kHz is enough to average out ^1H–^1H dipolar couplings between proton-containing mineral ions (OH^-, HPO_4^{2-}) and mobile water, but not between protons on quail OPN and very stationary water, therefore signals from protein protons adjacent to phosphate groups were not visible in the spectra. Thus, two cross-peaks (OH^- at 0 ppm of apatite crystalline state and H_2O at 5.2 ppm of amorphous surface layer) of minerals in 2D HETCOR were observed (see Figure 21.10). The disordered phosphate peak in the hydroxyapatite quail OPN complex is narrower than in hydroxyapatite, indicating that the protein acts to reduce the degree of disorder in the surface layer, similar to where unmodified OC is situated. Further evidence for the protein effect on cross-phase magnetization transfer among phosphates inside the mineral is inferred from a 2D ^{31}P DARR experiment (see Figure 21.10B). The phosphate diagonal peaks were found to be narrower in hydroxyapatite·quail OPN complex as compared with the hydroxyapatite, suggesting the involvement mainly from the disordered mineral at 3.2 ppm and up to a chemical shift value ~3.5 ppm. Finally, the CHHP, 2D ^{31}P–^{13}C double CP, and 2D ^{13}C–^{31}P z-filtered TEDOR studies revealed that the protein is primarily found in disordered coating/surface layers, with one or a few side chains reaching the faces of the apatite crystalline core region. (see Figure 21.10C).

21.2.2 Dentin

Teeth are hard and calcified tissues that mainly consist of enamel, dentine complex and cementum. Dentine complex occupies the largest portion of teeth. It contains 40–45% mineralized tissues by volume, with the organic content accounting for 30% of the extracellular matrix, and the remaining 20–25% water molecules.[75] The natural composite material inside the extracellular organic matrix constitutes both organic (~90% type 1 collagen and a small amount of types III and V collagen) and inorganic components (calcium phosphate or hydroxyapatite; HA) in addition to non-collagenous proteins (~50% dentine phosphoprotein). It is believed that the inorganic

Figure 21.10 (A) 2D ^1H–^{31}P HETCOR spectra of hydroxyapatite (I) and hydroxyapatite·quail OPN complex (II) recorded using the PMLG pulse scheme in t_1 for ^1H homonuclear decoupling with a CP contact time of 0.8 ms. (B) ^{31}P 2D DARR spectra of hydroxyapatite (I) and hydroxyapatite·quail OPN complex (II) recorded at a contact time of 2 ms, a proton mixing time of 500 ms, 32 scans, and a recycle delay of 2 s. (C) NMR-based model showing quail OPN (structure predicted using the ROSETTA algorithm) located in a disordered layer covering the apatite crystal and having an Asn sidechain able to reach in and interact with the surface of the crystal. Purple and green arrows represent magnetization transfers first to disordered PO_4^{3-} ions and later to apatite PO_4^{3-} ions. Reproduced from ref. 73, https://doi.org/10.1038/s41598-020-72786-x, under the terms of the CC BY 4.0 license, https://creativecommons.org/licenses/by/4.0/.

components, *i.e.* HA enclosed within and surrounded by organic matrix, present inside teeth and type 1 collagen protein provide the protective sheet underlying the inorganic HA crystallite lattice. Type 1 collagen protein adopts a triple helical structure and spontaneously self-assembles into the fibrillar structure. It has been suggested that these collagen fibrils accommodate ~56% of minerals inside the pores of the fibril architecture in dentin. In addition to these components, the enamel matrix proteins, *i.e.* amelogenin at the dentin surface, help in the enamel mineralization process. The progressive self-aggregation behavior of amelogenin is an important system in tooth formation. *In vitro* studies have suggested that amelogenin aggregates in a stepwise way through numerous intermediates, *i.e.* dimer, oligomers, nanospheres and nanosphere chains.[76–79] At the atomic level, the mineral phase in enamel is discovered to be complicated as it includes exclusive ions, for example Na^+, Mg^{2+}, F^- and $CO_3{}^{2-}$. Recent advancements in ssNMR, along with other biophysical studies such as CD, FTIR and AFM imaging, are now emerging for use in detailed structural studies. Previously, a limited study has been carried out to obtain the insight details of teeth mineralization using ssNMR[80,81] but with improved ssNMR methodologies in recent times, it has emerged as an effective method for the study of biomineralization in teeth. To this end, for the understanding of the exact mechanism of teeth mineralization, studying the local structure along with intermolecular interactions among the various components of the teeth has become a great necessity.

21.2.2.1 *The Picture of Biomineralization in Human Dentin by ssNMR Spectroscopy*

Biomineralization is a complex and lifelong process, in which the inorganic minerals occupy the organic matrix and form hybrid biological tissues, for example enamel, dentin and cementum,[82] and therefore understanding this process is important for curing mineralization-related diseases in vertebrates. In the preceding work, a sequence of research has been achieved for the structure calculation of different components of enamel at the atomic level. Some of the problems related to biomineralization have formerly been studied by means of ^{17}O, ^{31}P and ^{1}H CP ssNMR,[83] which provide a significant understanding of the complexity of the mineral structure. Furthermore, to understand the nature of phosphate units and the existence of $HPO_4{}^{2-}$ and OH^- in the mineral surfaces of rat dentin, ^{31}P–^{1}H heteronuclear correlation, *i.e.* HETCOR, has been studied. A two-dimensional HETCOR is appropriate to probe phosphorus species in close proximity to hydrogen [NMR investigations of the other major components of the enamel have been additionally analyzed by means of ^{23}Na MAS NMR, high-resolution 3Q-MAS NMR, in addition to $^{23}Na\{^{31}P\}$ rotational echo double resonance (REDOR) and $^{1}H\{^{23}Na\}$ R-HMQC correlation experiments] to assess the local environment around various metal cations inside the mineral structure of teeth.[22] Similarly, other ssNMR techniques, *i.e.* $^{31}P\{^{1}H\}$ double-quantum (DQ) filtered Lee–Goldburg (LG) CP HECTOR, have been reported to measure the teeth mineralization on Wistar rats of different age groups.[84] In general, the LG-CP method may be used to gain polarization transfer with efficient suppression of ^{1}H–^{1}H spin diffusion.[85] The resulting studies have proven that the water content and quantity of phosphorus considerably vary. The phosphorus-bearing species in human dentin majorly consists of orthophosphate and hydrogen phosphate. It is thought that orthophosphate is located in a disordered segment in which the phosphate ions are both hydrogen-bonded to structural water, and present in the apatite structure. Tsai *et al.*[86] reported a series of ssNMR experiments using the LG-CP HETCOR technique, as well as powdered XRD of third molar teeth of human dentin, as a result of which three mineral phases were identified and quantified, and the amount of phosphorus species in human dentin is no longer thought to vary with age. Unlike bone-like matrix, it has been extensively confirmed that the mineral component of human dentin carries amorphous calcium phosphate that coats the crystalline hydroxyapatite core region. The amorphous surface layer in human dentin represents HPO_4^{2-} and structural water molecules. Earlier studies have shown that ssNMR characterization of water molecules along with their dynamic study has contributed a huge amount of expertise on the physical properties of these composite biological materials. Most ssNMR studies have focused on the structural characterization of the organic interface including citrate and CO_3^{2-} using ^{13}C MAS ssNMR.[15,16,87] The ^{1}H ssNMR spectrum was found to be very complicated due to the presence of many hydrogenated species in these types of biological materials. In particular, the ^{1}H ssNMR spectrum of water in different environments in human dentin remains not well understood. However, the role of water in bone matrix has been extensively studied in the literature. More information about the ^{1}H signal can be obtained from various 2D ssNMR techniques, *i.e.* ^{1}H–^{13}C, ^{1}H–^{31}P heteronuclear correlations, ^{1}H–^{1}H dipolar correlations along with 1D version of these correlations with DCP experiments. Recently, Coppel *et al.* proposed an ssNMR methodology to study the importance of hydrogenated species, mainly water molecules, in human dentin of whole teeth.[88] The authors suggested a series of ssNMR experiments for ^{1}H characterization of diverse human dentin additives, demonstrating the importance of water molecules in dentinogenesis. As illustrated in Figure 21.11, multiple ssNMR pulse sequences were utilized for this. The experiment (^{1}H DF-SE MAS) highlighted the structural role of water molecule at short and long filtering times in human dentin, using demineralized human dentin with and without vacuum drying, that evidenced a dynamics (slower dipolar relaxation) of amide hydrogen followed by a reduction in the H-bonded water molecules in the collagen triple helical structure in dentin. Furthermore, to obtain more information about ^{1}H signals from the organic phase, the 1D $^{1}H/^{31}P$ DCP NMR experiment ($^{1}H \rightarrow {}^{31}P \rightarrow {}^{1}H$) provides ^{31}P-filtered ^{1}H NMR spectra of

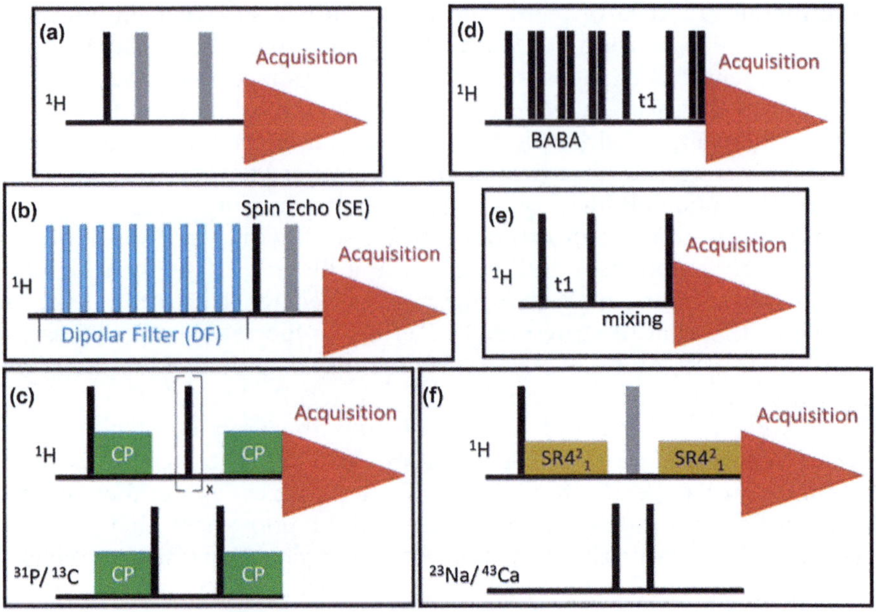

Figure 21.11 NMR pulse sequences used to study human dentin using ^{1}H signals: (a) DEPTH pulse sequence (recycle delay 10 s), (b) ^{1}H MAS rotor synchronous dipolar filtering and spin echo excitation; DF-SE (filtering time between 0.76 and 6 ms and a recycle delay 10 s), (c) $^{1}H/^{31}P$ double cross-polarization; DCP (first CP transfer between 3 and 6 ms, a second CP transfer between 0.2 and 2 ms and a recycle delay of 3 s), (d) 2D ^{1}H–^{1}H double-quantum–single quantum; DQ–SQ MAS [2 periods recoupling using the back-to-back (BABA) sequence], (e) 2D ^{1}H–^{1}H single quantum-single quantum; SQ–SQ, NOESY (mixing time 18 ms), (f) 1D $^{1}H/^{23}Na$ and $^{1}H/^{43}Ca$ D-HMQC. Reproduced from ref. 88 with permission from Elsevier, Copyright © 2020 Acta Materialia Inc.

human dentin. The next ^{1}H characterization step includes 2D DQ–SQ MAS NMR experiments, which identified a hydrogenated species in close proximity to the dipolar coupled network or rigid species. Additionally, in a similar study by Ramamoorthy's group on bone,[67] the SQ–SQ (NOESY) NMR spectra of human dentin also confirmed a strong correlation between water hydrogen molecules to NH and collagen. Finally, 1D ^{1}H–^{43}Ca and ^{1}H–^{23}Na D-HMQC MAS were performed to study the inorganic minerals in human dentin.

Therefore, based upon the above study in combination with the previous literature, a schematic view of the ^{1}H chemical shift range of main hydrogenated species is shown in Figure 21.12. ssNMR studies were carried out on dentin and enamel separately, in a manner similar to the structural statistical analysis of the mineral tissue of human enamel and the origin of the ^{1}H–^{31}P nuclear Overhauser enhancement (NOE) effect, and it was concluded that the NOE build-up rate in dentin is faster than in teeth.

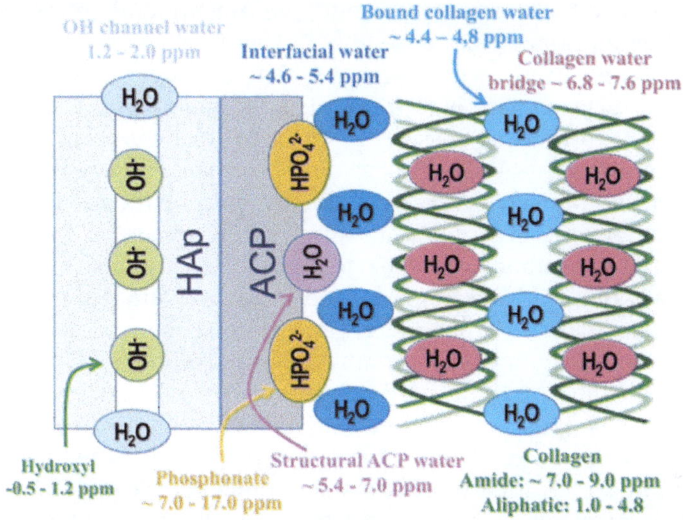

Figure 21.12 Schematic representation of the ^{1}H chemical shift range of human dentin. Reproduced from ref. 88 with permission from Elsevier, Copyright 2021.

21.2.2.2 Probing the Structure of Mammalian Tooth Enamel Protein: Amelogenin

Human full-length amelogenin includes 175 amino acid residues. The predominant tooth protein amelogenin is an intrinsically disordered protein (IDP) and aggregates in a stepwise manner *via* various intermediates such as dimers, oligomers, nanospheres, *etc.* during enamel formation.[89,90] It has been suggested that amelogenin regulates the growth of hydroxyapatite during enamel mineralization. It was also believed that amelogenin interacted with other proteins present inside the enamel matrix and helps in the growth, mineralization and organization of enamel. A primary sequence analysis indicated that the N- and hydrophilic C-terminal of amelogenin is highly conserved, however variations are found in the central region (HQP-rich region) and mutations in the central region of the sequence lead to deformation in the enamel architecture.[91,92] Therefore, to understand the exact function of amelogenin, the structural studies of this protein are needed. Previously, the structure of amelogenin of various species, *i.e.* porcine and mice, has been studied using various biophysical and NMR techniques.[75,93,94] The secondary structure of amelogenin has been studied using a variety of biophysical techniques, including CD, FTIR, and Raman spectroscopy.[95,96] The N-terminal region of monomeric amelogenin forms a prolonged β-sheet/strand structure, whilst the central region adopts a spiral/polyproline II (PPII) structure.[97–99] Buchko *et al.* suggested a possible stepwise interaction mechanism, beginning with the N-terminus and observed through the C-terminus, leading to amelogenin nanosphere assembly.[76] The conserved amelogenin N- and C-termini are responsible for protein–protein and protein–mineral interactions. Furthermore, the supramolecular assembly of the amelogenin nanosphere was studied by Zhang *et al.*, who suggested that amelogenin molecules form complex 3D-structures with N-terminal helix segments and C-terminal PPII-helices, which further self-assemble through ipsilateral interactions on the N-terminus of the molecule.[100] The dimer form of amelogenin was studied by cryo-electron microscopy.[77] Some experimental evidence has suggested that the enamel matrix contains an amyloid-like structure[101,102] and these highly ordered assemblies of amelogenin can be studied through ssNMR spectroscopy. Congo red staining was used to validate amyloid assemblies in a mouse enamel matrix, while atomic force microscopy and a dot-blot assay were used to visualize and analyze amyloids in growing porcine enamel, respectively (see Figure 21.13A).[101] Initially, the structural feathers of the chosen residues in full-length amelogenin mineralized with hydroxyapatite were studied.[103,104] Arachchige *et al.* for the first time provided the molecular-level information and identified the tertiary and quaternary intermolecular interactions among full-length amelogenin bound to hydroxyapatite.[94] The presence of intermolecular salt bridges in the dimer subunits of amelogenin has been studied in full-length murine amelogenin by a residue-specific labeling scheme with threonine and arginine (T&R) residues. Further, the dipolar interactions between T&R residues have been assigned and explained by 2D CP ^{13}C–^{13}C DAAR on three mutated samples (M179T, M179R and both) (see Figure 21.13B) and it was demonstrated that the IDP amelogenin becomes established whilst interacting with hydroxyapatite under physiological conditions that mimic the early stages of tooth development. The secondary structure calculation studied with ssNMR confirmed that T and R residues predominantly adopt β-strand structure in the amelogenin–hydroxyapatite complex and antiparallel β-sheet alignment was observed in the C-termini of the two amelogenin (see Figure 21.13C). Furthermore, the high-resolution structure of Leucine-Rich Amelogenin Protein (LRAP; 59 amino-acid), a natural variant of amelogenin, has also been studied by several groups.[105–107] Recent studies have shown that LRAP assemblies consist of a β-sheet secondary structure (residues 12–27) which was further assembled into amyloid-like structures.[106] The correlation spectrum obtained from various ssNMR spectra (2D ^{13}C–^{13}C DARR, 2D and 3D ^{13}C–^{15}N NCACX, NCOCX) for the sequential assignment was used to define the structure of LRAP assemblies.[106] The structure of LRAP self-assemblies facilitates understanding of the full-length amelogenin self-assemblies involved in biomineralization.

21.2.3 Keratin

Keratin or scleroprotein is a fibrous structural protein that is found in scales, hair, nails, horns, feathers, fur and the outer layer of the skin in vertebrates. These keratinous tissues exert an important role in structural and protection phenomena. Depending upon the physicochemical properties and nature of cells/tissues, keratins are distinguished and typically classified into two different categories: α-keratin, with an intermediate filament diameter ~7 nm, for example, wool, fingernails, horns, stratum corneum, quills, hooves, and β-keratin, with an intermediate filament diameter ~3 nm, for example, feathers, avian beaks, reptilian claws and scales.[108] Both types of keratins share the same amino acid sequence and fundamental structure in diverse vertebrates such as teleost fishes, amphibians, reptiles, birds, and mammals. [109] The amino acid sequence of a specific keratin provides information about secondary, tertiary and quaternary structures along with the nature of the bonds (*i.e.* ionic or covalent) to other components of the cytoskeleton.[110,111] The secondary structure of keratins includes an N-terminal head domain, a central α-helical rod domain and C-terminal tail domain and they have a tendency to self-assemble into a filament-like structure. Molecular structural studies of α-keratins were very limited, probably due to their complexity. The predicted secondary structure content of keratin showed heterogeneity with α-helical rod domains co-existing with β-sheets, random coils and turns in the matrix proteins. Initially, the structural organization and molecular dynamics of isotopically labeled mouse keratin were studied by ^{13}C ssNMR experiments. Further, to understand the motion of peptide backbone, measurement of the carbon relaxation time using NOE has been carried out, suggesting the average correlation times were distributed over the range of 0.2–20 ns.[112]

The α-keratinous protein present in the hair is comprised of two components: cortex (85–90% of hair) and cuticle. The spindle-shaped macrofibril of cortex further contains two structural components: microfibril and matrix or keratin intermediate filament-associated proteins (KIFap). The matrix or KIFap contains significant amounts of disulfide bonds with cross-linked structure through an inter- or intramolecular covalent bond.[110] Kricheldorf and Mullar characterized the structure of intact keratin substances in horse hair and horn along with human hair and parrot feather using ssNMR spectroscopy.[113] Nishikawa *et al.* further studied the structural details of the matrix of human hair with ^{13}C ssNMR, especially ^{13}C-enriched methyl carbon labeling in the thiol group formed by a reduction of the disulfide bond in the matrix.[114] The finding in the above-mentioned study suggested that there are at least two kinds of domains (two Cys S–^{13}CH$_3$ peaks) in the matrix region in ^{13}C CP/MAS spectrum. These differences may be due to the molecular motion or chemical shift distribution in the local electronic state. To overcome this possibility, ^{13}C spin–lattice relaxation times were measured using Torchia's pulse sequence which suggested the ^{13}C T_1 values (3.6 and 3.7s) of the two domains. The non-significance difference in ^{13}C T_1 values confirmed the distinction in the chemical shift distribution in place of the distinction in molecular motion.[114]

Duer *et al.* studied the material properties of equine hoof α-keratin with ssNMR experiments, *i.e.* CP, MAS in conjunction with two-dimensional ^{13}C–^{1}H wideline separation spectroscopy experiment (WISE) and ^{2}H NMR.[115] A ^{13}C–^{1}H WISE experiment provides the insight details of molecular motions (correlation times less than about 10^4–10^5 s) present in the

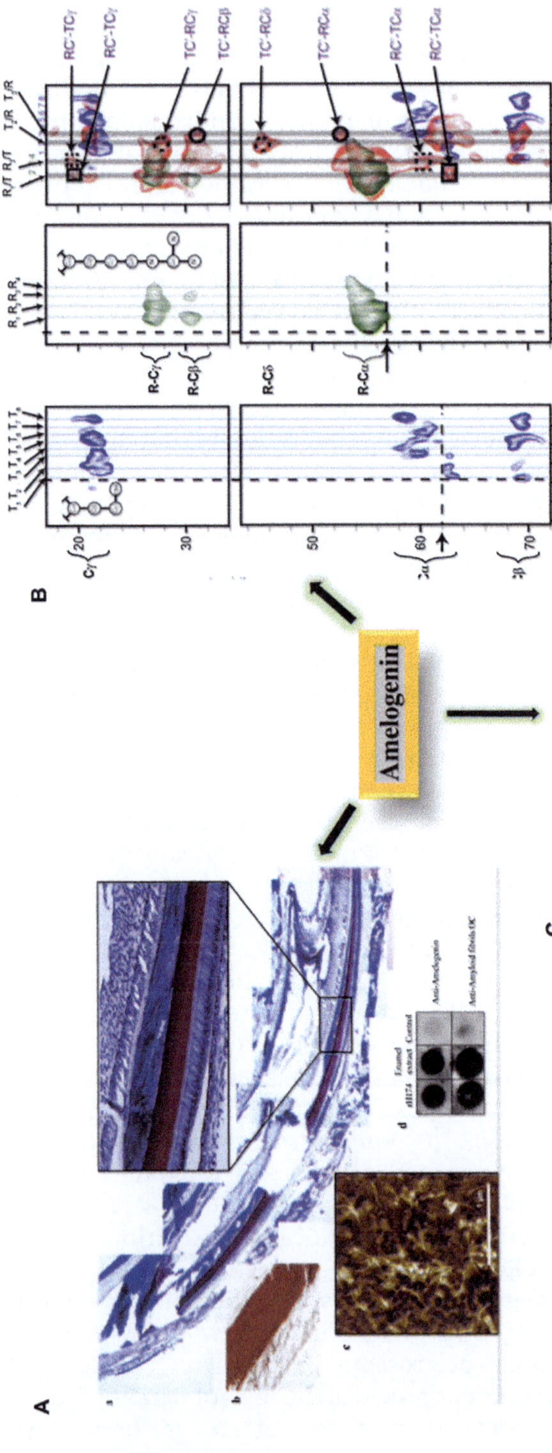
A
B
C
Amelogenin
R176
T174
T171
T174'
R176'
T171'
Intermolecular interaction
R-Cγ
R-Cβ
R-Cδ
R-Cα
^{13}C (ppm)
Anti-Amelogenin
Anti-Amyloid fibrils OC

particular solid sample. This technique provides information about carbon nuclei closest to the respective directly bonded ^{1}H spins. Using this technique, Duer *et al.* revealed a strong dependence of molecular conformation and molecular dynamics on the degree of hydration of the material.[115] The presence of a high degree of ordering upon dehydration showed breaking of the disulfide linkages among cysteine residues and reduced mobility of the sidechains inside the proteins. Additionally, the loss of sidechain mobility in the proteins was apparent and this was presumably due to the change in intermolecular interactions (especially H-bonding). However, in hydrated hoof wall keratins, there was no proof for motion of the backbone of the keratin molecules with correlation times much less than around 10^{3} s. Furthermore, the impact of hydration on hard α-keratin was also investigated by ^{1}H ssNMR.[116] Later, the investigation of the morphology variations and changes in molecular dynamics prompted by means of uniaxial compressive force on hard α-keratin from human nail was studied by Demco *et al.*[117] In this study, ^{13}C CPMAS spectroscopy and longitudinal relaxation in

Figure 21.13 (A) In vivo studies indicate the presence of an amyloid-like assembly in developing enamel. AFM analysis of enamel matrix extracted from unerupted porcine teeth shows fibrillar structures similar to recombinant amelogenin ribbons *in vitro.* Dot blot analysis of rH174, porcine enamel extract and bovine serum albumin (BSA) with anti-amelogenin and anti-amyloid OC antibodies. rH174 and porcine enamel extract samples stain positive with both antibodies, suggesting that amelogenin and amyloids are present. Reproduced from ref. 101 with permission from Springer Nature, Copyright 2021. (B) CP-^{13}C–^{13}C-DARR coherence spectra for left: M179-T (blue); and middle: M179-R (green). M179-T and M179-R were collected at a 500 MHz ^{1}H resonance frequency at 25 °C with a 100 ms mixing time. The blue or green lines coincide with C^{0} resonances and highlight the cross-peaks observed with these resonances. The bold vertical and horizontal dashed lines indicate the random coil values of the associated C^{0} and Ca resonances, respectively. The chemical shifts upfield of the random coil values indicate that the majority of the T and R residues are in a β-sheet environment. Right: M179-T/R (red) overlaid with M179-T and M179-R is shown. M179-T/R was collected at an 850 MHz ^{1}H resonance frequency at 25 °C with a 100-ms mixing time. Cross-peaks are observed between two back-bone T-C0 and arginine side-chain carbons R-Ca, R-Cb, R-Cd and R-Cg, and the complementary R-C^{0} backbone with threonine side-chain carbons T-Ca and T-Cg, establishing that there is an intermolecular interaction between the C-termini of amelogenin molecules in the protein–HAP complex. (C) Representative MD structure of two amelogenin C-termini in a fully aligned, antiparallel β-sheet alignment that is consistent with the current and previous structural characterizations. The arginine and threonine residues are shown as element-colored sticks (carbon = cyan, nitrogen = blue, oxygen = red). The backbone residues are shown as blue (T) and green (R) ribbon. One set of the intermolecular interactions observed by ssNMR is highlighted with a pink-shaded ellipse. Reproduced from ref. 94 with permission from Elsevier, Copyright 2021.

rotating frame ($T_{1\rho}$) were employed to study the transitions, sidechain and backbone dynamics. The changes inside the sidechain dynamics of various keratin residues with increasing pressure revealed the transition of α-helix to β-sheet structure.[117]

Previously, ssNMR-based approaches were used to investigate the molecular-level arrangement of lipids and keratin in adhesive setae of gecko toe and skin shedding. The keratin-dominated spectra in CP/MAS confirmed the presence of peaks associated with lipids in the aliphatic region for toe and skin sheds. Further, the mobility assessment of lipids present in the toe sheds has been studied by a DP/MAS approach and the acquired spectrum suggested the presence of mobile molecular segments in the samples which were determined to be more mobile than keratin, as suggested earlier in the study of stratum corneum.[118] This is possibly due to structural differences in the specialized adhesive setae compared with the rest of the epidermis. Overall, this study can help in understanding the lipid–protein interactions in various other biological systems and will further be used in the design of synthetic adhesives.

In addition to the above-mentioned techniques, ^{13}C double quantum–single quantum (DQ/SQ) correlation *via* Permutationally Offset Stabilized C7 (POST-C7) sequence and proton-driven spin diffusion (PDSD) techniques were used to study ^{13}C,^{15}N-labeled murine fur.[119] In POST-C7, correlation of two nuclear sites which can be at once dipolar coupled to each other has been identified that allows to map the networks of ^{13}C–^{13}C bonds. PDSD experiments used to correlate ^{13}C spins through their mutual dipolar coupling to intervening ^{1}H nuclei and the resulting 2D ^{13}C–^{13}C correlation spectra thus show correlations up to 10 Å apart in distance. The DQ–SQ experiments confirmed many correlations (one-bond correlations related to amide, aromatic and α-carbon) in the low-frequency vicinity of the spectrum that can be assigned for specific amino acid residue types which had been directly bonded. Further, to recognize the chemical surroundings close by carbons, which might not be immediately bonded, PDSD has been employed wherein ^{13}C–^{13}C spin diffusion utilized both ^{13}C–^{13}C and ^{13}C–^{1}H dipolar couplings to transfer magnetization among close by carbons. In general, a shorter mixing time maps single bond proximity, and at the same time a longer mixing time reveals longer ^{13}C–^{13}C proximities which include $C_δ$-Cg. Similarly, in this study, at variable mixing times, proximities between N–H and α-carbons, along with more distant carbon atoms and between side-chain carbons have been assigned for complete structural calculation.[119]

Recently, a comparative study of structure and dynamics of natural and regenerated β-keratin isolated from chicken feather has been done using CP/MAS, ^{13}C and ^{1}H spin–lattice relaxation time measurements, and ^{13}C 2D phase-adjusted spinning sidebands (2D PASS) MAS ssNMR measurements.[120] 2D PASS MAS ssNMR[121] provides information about CSA parameters (correlation anisotropic spectrum in the indirect dimension with a purely isotropic spectrum in the indirect dimension at low MAS frequency). The ^{13}C MAS study confirmed a decrement of the α-helix content in pretreated keratin in comparison with natural keratin. This is possibly due to the breakage of disulfide cross-linkage within the polysaccharide chains on chemical treatment in synthetic keratin. The relaxation measurement with the aid of ^{13}C-spin–lattice, anisotropy parameters and molecular correlation time confirmed excessive levels of breakage in disulfide cross-linkage and a greater degree of disorderliness in regenerated keratins.[120]

21.2.4 Corals and Silk

Corals are a marine invertebrate consisting of calcium carbonate-rich cytoskeletons, which are generally known as aragonite. Coral's aragonite cytoskeletons play an important role in the storage of atmospheric CO_2. Corals are one of many marine organisms that have a tendency to biomineralize, and the process of biomineralization is important in understanding how corals respond to the paleoenvironment.[122] The skeleton growth in corals involves conversion of the precipitated amorphous calcium carbonate into the aragonite fibrils.[123] The microscopic aragonite fibrils turn into a cytoskeletal framework containing macromolecules, polysaccharides, proteins and their glycosylated derivatives, usually named skeletal organic matrix (SOM).[124,125] Previously, the process of biomineralization and morphological studies in corals have been extensively studied.[123,124,126,127] However, the underlying mechanism of the aragonite mineral formation in corals is not clearly defined. The inorganic minerals present in the aragonite crystals (such as Ba, Sr, Cd and P) are generally probed to study the structural details of coral skeletons. Several groups use ^{31}P ssNMR spectroscopy currently to provide information on the distribution of P in coral aragonite. In this chapter, the recent ssNMR methodologies used for the structural determination of coral skeleton materials have been reviewed. Mason *et al.* used single- and double-resonance ssNMR experiments to determine the nature of P in coral skeleton.[128] It has been suggested that the study of P/Ca ratio in tropical- and deep-water corals is needed to gain knowledge of the processes by which phosphorus is taken into the coral skeletal aragonite. With the use of various ssNMR such as ^{31}P single-pulse magic angle spinning (SP/MAS), ^{31}P{^{1}H} CP MAS NMR, ^{1}H{^{31}P} REDOR, ^{31}P{^{1}H} HETCOR, Carr–Purcell–Meiboom–Gill (CPMG) experiments, Mason *et al.* reported that the majority of phosphorus atoms present as phosphate defects in the aragonite structure, but in some cases crystalline hydroxyapatite inclusions also dominated.[128] Further, Von Euw *et al.* studied the calcareous skeleton of stony corals and demonstrated that mineral deposition is biologically driven.[129] To identify the bioorganic molecules associated with the minerals and obtain knowledge about the nucleation phenomena of mineral phase, the group used ^{1}H–^{13}C CP and single-pulse ^{13}C NMR techniques in combination with ultrahigh-resolution 3D imaging.[129] In this study, CP magnetization transfer from ^{1}H nuclei to nearby ^{13}C nuclei, that detects carbon species localized in a hydrogen-rich chemical environment (primarily the exposed surface of the mineral/organic interface), and single-pulse ^{13}C

NMR depicts all ^{13}C spins present in the skeleton, *i.e.*, mostly carbonate ions. Therefore, this study confirmed that the noticeably disordered environments in coral skeleton originate from inorganic carbonates. Moreover, the presence of HCO_3^- and OH^- ions in the mineral phase related to the interfacial areas of coral skeleton has additionally been confirmed. In addition, ^{31}P–^{1}H HETCOR experiments also suggested that the skeleton proteins strongly bound to highly disordered calcium carbonate environments of aragonite crystals. A recent study by Akiva *et al.* showed that the minerals are first precipitated into calcium-carbonate-rich species and small aragonite crystallites, which are further converted into mature aragonite fibers found in stony coral skeleton.[130] Further, for the examination of mineral phases and the protein–mineral interfaces, corals polyps were grown in ^{13}C-glucose-enriched seawater and ssNMR experiments were recorded and as an end result the group confirmed the involvement of various proteins and ions in the two distinct developmental levels of the planula, exposing information of the intimate interplay of mineralization proteins (CARP1, CARP3 and CARP4) with amorphous calcium carbonate (ACC).

In addition, ssNMR studies were also used for the structure determination of silkworm and spider silks. It has been suggested that silk peptides consist of a mixture of secondary structures including helix, random coil, and β-sheet and turn. This kind of structure is difficult to analyze by XRD and TEM but can be conveniently examined by ssNMR. Several newly advanced ssNMR strategies and procedures are being used to accurately analyze silk structure. The structural characterization of various types of silk, recombinant silk proteins and silk peptide mimics with ssNMR spectroscopy was extensively reviewed by Asakura *et al.* in 2013.[131] In this chapter, we report the recent ssNMR methodologies used for the molecular-level structural elucidation of silk products. An earlier report suggested that the dragline silk of various spiders has distinct mechanical properties and the differences and similarities in dragline fibers from three arachnid families (four different geniuses) have been studied with cross-polarization and direct ^{13}C detection NMR spectroscopy.[132] Conclusively, these types of ssNMR techniques can be used to distinguish different types of spider silk and help in the characterization of their structure and hydration dynamics. More recently, silk fibroin (aa sequence 5263 residues; mol wt 391.637 kDa) from *Bombyx mori* has received a lot of interest and is also a good example to study with ssNMR spectroscopy because the fibers (fiber diameter 1 micron) of silk fibroin contain a highly ordered β-sheet structure. Silk fibroin is composed of a mixture of crystalline, semi-crystalline and amorphous regions which are responsible for the water solubility and elasticity behavior. Various research groups have explored its mechanical properties for the development of biomaterials as a new medical scaffold for soft tissues in the last few years. Characterizing the morphology and miscibility of mixed substances is likewise one of the beauties of ssNMR spectroscopy. Previous studies have proven that silk fibroin from *Bombyx mori* adopts two dimorphic structures: silk I (much less stable) and silk II. The ssNMR study carried out for the development of non-woven biomaterials based on silk fibroin and polyurethane showed an ordered structure with inter- and intramolecular H-bonds in the ordered structure of pure silk fibroin.[133] Furthermore, the structure of pure silk fibroin, pellethane and a silk fibroin/pellethane-blended non-woven sheet were investigated using ssNMR and no spectrum changes were found between the pure and blended samples. The miscibility of blended materials was further studied utilizing proton spin-lattice relaxation periods in the laboratory frame (T_1^H), which demonstrated that the molecular chains of silk fibroin and pellethane are within tens of nanometers of each other.[133] For biomedical applications, it is crucial to understand the structure and dynamics of silk fibroin in the hydrated as well as in the dry state, due to the fact that biomaterials are usually used in a hydrated state. The study of silk fibroin in a hydrated state is needed for understanding its function and in designing novel biomaterials. Previously, various biophysical studies such as Raman, IR, XRD, DSC and NMR have been used to describe the conformational changes in silk fibroin and dynamics of water interaction with it.[134–137] In this context, to obtain information about the effect of hydration, Asakura *et al.* studied residue-specific conformational changes in "crystalline and noncrystalline domains" in silk fibroin with [3-^{13}C] Ser, [3-^{13}C] Tyr and [3-^{13}C] Ala labeling.[138] In this work, a ^{13}C DD (dipolar decoupling)/MAS NMR technique is used over ^{13}C CP/MAS NMR for structural analysis of hydrated silk fibroin, because the dipole–dipole interactions needed for CP are averaged out due to molecular motion, and as result the signal intensity is found to be decreased. As a conclusion, the study concluded that the fraction of β-sheet and two random coil distributions with rapid and gradual chain dynamics have been determined for various ^{13}C labeling sites distributed in the crystalline and noncrystalline domains of the silk fibroin.[138] Moreover, ^{13}C chemical shift and selective labeling provide extremely important information about several local conformations in a residue-specific manner. The dynamics of water molecules interacting with silk fibroin have been investigated using ^{2}H solution NMR relaxation, particularly ^{2}H T_1–T_2 and T_2–T_2 NMR and exchange measurements[139], and the four compartments of water molecules have been established; A: bulk water outside, B: water molecules trapped weakly on the surface of silk fibroin, C: bound water molecules on the inner surface, and D: bound water molecules on the inside of silk fibroin.[139] Next, the local conformation and dynamics of the hydrated *Bombyx mori* silk fibers, which include silk cocoon, silk sericin and silk fibroin, have been additionally analyzed through the use of ^{13}C refocused INEPT NMR to probe the fast motion component (for mobile domain), ^{13}C CP/MAS NMR (for rigid domain) and ^{13}C DD/MAS NMR (for both domain).[140] After studying the ssNMR spectrum, it has been concluded that, upon hydration, the appearance of the random coil components increased with fast motion that corresponds to a decrease in the breaking strength and an increase in the elongation at break.[140] Therefore, the above-mentioned techniques give a different prospective on the dynamic properties of silk fibroin. To increase the hydrophobicity for improved dimensional stability in silk fibroin, a chemical modification like acetylation has been carried out.[141] For the first time, this study clarified the structure and dynamics of the acetylated silk fibroin from the crystalline and non-crystalline domains independently in the presence of water. The performed ssNMR experiments provide amino-acid-specific dynamics in the ^{13}C-labeled silk fibroin. The mobilities of tyrosine and alanine in the amorphous region of acetylated silk fibroin were found to be increased by hydration, however, no impact of water at the β-sheet area of those residues have been observed.

21.3 Conclusion and Future Directions

ssNMR spectroscopy has been proven to be a first-rate method in understanding of the structural organization and dynamics of amyloid proteins, membrane proteins and nanocrystalline peptides and proteins, *etc.* at the atomic level. In the past few years, this technique has characterized the structure of mineralized tissues, organic–inorganic interface study, structure of ordered fibrils of collagen triple-helical molecules, inorganic components of bone and other related biological materials. Besides the extensive study of bone matrix, nowadays this technique is also used to study residue-specific structural and dynamical conformational aspects of other biomaterials at the atomic level. In this chapter, we have highlighted the recent advancements in ssNMR studies applied to biological materials and how this technique can provide insight into the structural organization and dynamical conformations for various biological materials. One of the most recent studies has shown by isotope enrichment and DNP-enhanced (signal enhancement factors of 50–100) that NMR is used for the detection and identification of nucleic acid. Along with this, detection and identification using high-resolution ssNMR have also been shown to be of great importance in future studies. Further advancements include the development of ultra-fast MAS NMR and a technique involving a multiple-pulse dipolar filter under slow MAS, which have been used and may be suitable for monitoring the OH^- level in the apatite mineral of calcified tissues. In addition, ^{1}H MAS NMR spectroscopy with the dipolar filter can be used for proton signal enlargement. In addition, detection of diseases caused by collagen degradation of cartilage can also be studied by ^{1}H–^{13}C-HETCOR-DNP ssNMR. Moreover, this spectroscopic technique was also found to be useful for an NCPs–mineral interaction study. Apart from bone, dentin mineralization, structural characterization of enamel proteins like amelogenin has been characterized by advanced ssNMR spectroscopy. Furthermore, the structural evaluation of keratinous proteins using PDSD, WISE, DQ–SQ and PASS NMR has been characterized in order to offer deeper understanding of the structural foundation of industrial and cosmetic processes and to assist in the design of new synthetic materials. An NMR relaxation measurement study detected the nature of disulfide cross-linkage and the degree of disorderliness in regenerated keratins. This technique also gives the different prospectives about the structure and dynamics in biological materials obtained from other species like corals and silk. In summary, the techniques presented in this chapter and accumulated insights will undoubtedly be of use in future studies for elucidating the complex structure and dynamic properties of other biological materials, and help in any progress in ssNMR methodologies in understanding the physiological and pathological functions in various bone diseases which include osteoporosis and arthritis and type 2 diabetes.

Abbreviations

ssNMR	solid-state nuclear magnetic resonance
CP	cross-polarization
DCP	double cross-polarization
MAS	magic angle spinning
HETCOR	heteronuclear correlation
DNP	dynamic nuclear polarization
MRI	magnetic resonance imaging
DARR	dipolar assisted rotational resonance
INEPT	insensitive nuclei enhanced by polarization transfer
INADEQUATE	incredible natural abundance double quantum transfer experiment
TEDOR	transfer echo double resonance
REDOR	rotational echo double resonance
DQ–SQ	double quantum–single quantum
CTUC	constant-time uniform-sign cross-peak
PDSD	proton-driven spin diffusion
WISE	wideline separation spectroscopy experiment
PASS	phase-adjusted spinning sidebands
POST-C7	permutationally offset stabilized C7
HMQC	heteronuclear multiple quantum coherence
NCSs	non-collagenous proteins
AKU	alkaptonuria
IDP	intrinsically disordered proteins
OC	osteocalcin
OPN	osteopontin
CD	circular dichroism
FTIR	Fourier transform infrared
LRAP	leucin-rich amelogenin protein
KIFap	keratin intermediate filament-associated proteins
SOM	skeleton organic matrix

Acknowledgements

The authors acknowledge financial support from the Centre of Biomedical Research, SGPGIMS campus, Lucknow, Uttar Pradesh.

References

1. H. Ehrlich Biomaterials and Biological Materials, Common Definitions, History, and Classification. *Biol. Mater. Marine Origin*. 2010, 3–22.
2. J. B. Park and R. S. Lakes, Structure–property relationships of biological materials, *Biomaterials.*, 2007, 225–263.
3. C. M. Quinn and T. Polenova, Structural biology of supramolecular assemblies by magic-angle spinning NMR spectroscopy, *Q. Rev. Biophys.*, 2017, **50**, 13.
4. J. Buckwalter and R. Cooper, Bone structure and function, *Instruct. Course Lect.*, 1987, **36**, 27–48.
5. H. Datta, W. Ng, J. Walker, S. Tuck and S. Varanasi, The cell biology of bone metabolism, *J. Clin. Pathol.*, 2008, **61**(5), 577–587.
6. P. A. Downey and M. I. Siegel, Bone biology and the clinical implications for osteoporosis, *Phys. Therapy*, 2006, **86**(1), 77–91.
7. K. Matsuo and N. Otaki, Bone cell interactions through Eph/ephrin: bone modeling, remodeling and associated diseases, *Cell Adhesion Migration*, 2012, **6**(2), 148–156.
8. R. Florencio-Silva, G. R. d. S. Sasso, E. Sasso-Cerri, M. J. Simões and P. S. Cerri, Biology of bone tissue: structure, function, and factors that influence bone cells, *BioMed. Res. Int.*, 2015, **2015**, 421746.
9. F. Nudelman, A. J. Lausch, N. A. Sommerdijk and E. D. Sone, In vitro models of collagen biomineralization, *J. Struct. Biol.*, 2013, **183**(2), 258–269.
10. S. Morgan, A. A. Poundarik and D. Vashishth, Do non-collagenous proteins affect skeletal mechanical properties?, *Calcified Tissue Int.*, 2015, **97**(3), 281–291.
11. A. L. Boskey, Noncollagenous matrix proteins and their role in mineralization, *Bone Mineral*, 1989, **6**(2), 111–123.
12. A. Boskey, L. Spevak, E. Paschalis, S. Doty and M. McKee, Osteopontin deficiency increases mineral content and mineral crystallinity in mouse bone, *Calcified Tissue Int.*, 2002, **71**(2), 145–154.
13. F. Dickens, The citric acid content of animal tissues, with reference to its occurrence in bone and tumour, *Biochem. J.*, 1941, **35**(8–9), 1011–1023.
14. R. Hartles, Citrate in mineralized tissues, *Adv. Oral Biol.*, 1964, **1**, 225–253.
15. Y.-Y. Hu, A. Rawal and K. Schmidt-Rohr, Strongly bound citrate stabilizes the apatite nanocrystals in bone, *Proc. Natl. Acad. Sci. U. S. A.*, 2010, **107**(52), 22425–22429.
16. E. Davies, K. H. Müller and W. C. Wong, *et al.*, Citrate bridges between mineral platelets in bone, *Proc. Natl. Acad. Sci.*, 2014, **111**(14), E1354–E1363.
17. M. J. Duer, The contribution of solid-state NMR spectroscopy to understanding biomineralization: atomic and molecular structure of bone, *J. Magn. Reson.*, 2015, **253**, 98–110.

18. D. Huster, Solid-state NMR studies of collagen structure and dynamics in isolated fibrils and in biological tissues, *Ann. Rep. NMR Spectrosc.*, 2008, **64**, 127–159.
19. C. Jäger, T. Welzel, W. Meyer-Zaika and M. Epple, A solid-state NMR investigation of the structure of nanocrystalline hydroxyapatite, *Magn. Reson. Chem.*, 2006, **44**(6), 573–580.
20. K. H. Mroue, A. Viswan, N. Sinha and A. Ramamoorthy, Solid-state NMR spectroscopy: the magic wand to view bone at nanoscopic resolution, *Ann. Rep. NMR Spectrosc.*, 2017, **92**, 365–413.
21. J. Xu, P. Zhu and Z. Gan, *et al.*, Natural-abundance ^{43}Ca solid-state NMR spectroscopy of bone, *J. Am. Chem. Soc.*, 2010, **132**(33), 11504–11509.
22. D. Laurencin, A. Wong and W. Chrzanowski, *et al.*, Probing the calcium and sodium local environment in bones and teeth using multinuclear solid state NMR and X-ray absorption spectroscopy, *Phys. Chem. Chem. Phys.*, 2010, **12**(5), 1081–1091.
23. Y. Wang, S. Von Euw and F. M. Fernandes, *et al.*, Water-mediated structuring of bone apatite, *Nat. Mater.*, 2013, **12**(12), 1144–1153.
24. S. Maltsev, M. J. Duer, R. C. Murray and C. Jaeger, A solid-state NMR comparison of the mineral structure in bone from diseased joints in the horse, *J. Mater. Sci.*, 2007, **42**(21), 8804–8810.
25. M. J. Duer, T. Friščić, R. C. Murray, D. G. Reid and E. R. Wise, The mineral phase of calcified cartilage: its molecular structure and interface with the organic matrix, *Biophys. J.*, 2009, **96**(8), 3372–3378.
26. A. L. Boskey and P. G. Robey, The composition of bone, *Primer Metabolic Bone Diseases and Disorders of Mineral Metabolism*, 2013, 49–58.
27. M. E. Launey, M. J. Buehler and R. O. Ritchie, On the mechanistic origins of toughness in bone, *Ann. Rev. Mater. Res.*, 2010, **40**, 25–53.
28. R. K. Rai and N. Sinha, Dehydration-induced structural changes in the collagen–hydroxyapatite interface in bone by high-resolution solid--state NMR spectroscopy, *J. Phys. Chem. C*, 2011, **115**(29), 14219–14227.
29. P. Zhu, J. Xu, N. Sahar, M. D. Morris, D. H. Kohn and A. Ramamoorthy, Time-resolved dehydration-induced structural changes in an intact bovine cortical bone revealed by solid-state NMR spectroscopy, *J. Am. Chem. Soc.*, 2009, **131**(47), 17064–17065.
30. F. Gul-E-Noor, C. Singh, A. Papaioannou, N. Sinha and G. S. Boutis, Behavior of water in collagen and hydroxyapatite sites of cortical bone: fracture, mechanical wear, and load bearing studies, *J. Phys. Chem. C*, 2015, **119**(37), 21528–21537.
31. J. Xu, P. Zhu, M. D. Morris and A. Ramamoorthy, Solid-state NMR spectroscopy provides atomic-level insights into the dehydration of cartilage, *J. Phys. Chem. B*, 2011, **115**(33), 9948–9954.
32. K. H. Mroue, R. Zhang and P. Zhu, *et al.*, Acceleration of natural-abundance solid-state MAS NMR measurements on bone by paramagnetic relaxation from gadolinium-DTPA, *J. Magn. Reson.*, 2014, **244**, 90–97.

33. K. H. Mroue, N. MacKinnon and J. Xu, *et al.*, High-resolution structural insights into bone: a solid-state NMR relaxation study utilizing paramagnetic doping, *J. Phys. Chem. B*, 2012, **116**(38), 11656–11661.
34. R. Zhang, K. H. Mroue and A. Ramamoorthy, Proton-based ultrafast magic angle spinning solid-state NMR spectroscopy, *Acc. Chem. Res.*, 2017, **50**(4), 1105–1113.
35. T. Maly, G. T. Debelouchina and V. S. Bajaj, *et al.*, Dynamic nuclear polarization at high magnetic fields, *J. Chem. Phys.*, 2008, **128**(5), 02B611.
36. A. J. Rossini, A. Zagdoun, M. Lelli, A. Lesage, C. Coperet and L. Emsley, Dynamic nuclear polarization surface enhanced NMR spectroscopy, *Acc. Chem. Res.*, 2013, **46**(9), 1942–1951.
37. C. Singh, R. K. Rai, F. Aussenac and N. Sinha, Direct evidence of imino acid–aromatic interactions in native collagen protein by DNP-enhanced solid-state NMR spectroscopy, *J. Phys. Chem. Lett.*, 2014, **5**(22), 4044–4048.
38. J. Ennever, L. Riggan and J. Vogel, Proteolipid and collagen calcification, in vitro, *Cytobios*, 1984, **39**(155–156), 151–157.
39. S. Xu and J. Y. Jianqing, Beneath the minerals, a layer of round lipid particles was identified to mediate collagen calcification in compact bone formation, *Biophys. J.*, 2006, **91**(11), 4221–4229.
40. A. During, G. Penel and P. Hardouin, Understanding the local actions of lipids in bone physiology, *Progress Lipid Res.*, 2015, **59**, 126–146.
41. K. H. Mroue, J. Xu, P. Zhu, M. D. Morris and A. Ramamoorthy, Selective detection and complete identification of triglycerides in cortical bone by high-resolution ^{1}H MAS NMR spectroscopy, *Phys. Chem. Chem. Phys.*, 2016, **18**(28), 18687–18691.
42. J. S. Nyman, Q. Ni, D. P. Nicolella and X. Wang, Measurements of mobile and bound water by nuclear magnetic resonance correlate with mechanical properties of bone, *Bone*, 2008, **42**(1), 193–199.
43. R. A. Horch, J. S. Nyman, D. F. Gochberg, R. D. Dortch and M. D. Does, Characterization of 1H NMR signal in human cortical bone for magnetic resonance imaging, *Magn. Reson. Med.*, 2010, **64**(3), 680–687.
44. M. Unal, A. Creecy and J. S. Nyman, The role of matrix composition in the mechanical behavior of bone, *Curr. Osteoporosis Rep.*, 2018, **16**(3), 205–215.
45. R. A. Horch, D. F. Gochberg, J. S. Nyman and M. D. Does, Non-invasive predictors of human cortical bone mechanical properties: T_2-discriminated ^{1}H NMR compared with high resolution X-ray, *PLoS One*, 2011, **6**(1), e16359.
46. C. Singh and N. Sinha, Mechanistic insights into the role of water in backbone dynamics of native collagen protein by natural abundance ^{15}N NMR spectroscopy, *J. Phys. Chem. C*, 2016, **120**(17), 9393–9398.
47. M. A. Fernández-Seara, S. L. Wehrli, M. Takahashi and F. W. Wehrli, Water content measured by proton-deuteron exchange NMR predicts bone mineral density and mechanical properties, *J. Bone Mineral Res.*, 2004, **19**(2), 289–296.

48. N. Tiwari, R. Rai and N. Sinha, Water-lipid interactions in native bone by high-resolution solid-state NMR spectroscopy, *Solid State Nucl. Magn. Reson.*, 2020, **107**, 101666.
49. W. YingáChow, Detection of nucleic acids and other low abundance components in native bone and osteosarcoma extracellular matrix by isotope enrichment and DNP-enhanced NMR, *RSC advances*, 2019, **9**(46), 26686–26690.
50. P. Halling *Proteins: Structures and molecular properties. by Thomas E. Creighton, WH Freeman, New York, 1992, xiii+ 512 pp, price£ 22.95. ISBN 0-7167-7030-X.* Wiley Online Library; 1995.
51. D. Huster, L. Naji, J. Schiller and K. Arnold, Dynamics of the biopolymers in articular cartilage studied by magic angle spinning NMR, *Appl. Magn. Reson.*, 2004, **27**(3), 471–487.
52. G. Zernia and D. Huster Investigation of collagen dynamics by solid-state NMR spectroscopy. *Modern Magnetic Resonance.* Springer, 2008, pp. 87–92.
53. W. Y. Chow, C. J. Forman and D. Bihan, *et al.*, Proline provides site-specific flexibility for in vivo collagen, *Sci. Rep.*, 2018, **8**(1), 1–13.
54. W. Y. Chow, D. Bihan and C. J. Forman, *et al.*, Hydroxyproline ring pucker causes frustration of helix parameters in the collagen triple helix, *Sci. Rep.*, 2015, **5**(1), 1–11.
55. W. Y. Chow, A. M. Taylor, D. G. Reid, J. A. Gallagher and M. J. Duer, Collagen atomic scale molecular disorder in ochronotic cartilage from an alkaptonuria patient, observed by solid state NMR, *J. Inherited Metabolic Dis.*, 2011, **34**(6), 1137–1140.
56. W. Y. Chow, B. P. Norman and N. B. Roberts, *et al.*, Pigmentation chemistry and radical-based collagen degradation in alkaptonuria and osteoarthritic cartilage, *Angew. Chem., Int. Ed.*, 2020, **59**(29), 11937–11942.
57. E. R. Andrew, The narrowing of NMR spectra of solids by high-speed specimen rotation and the resolution of chemical shift and spin multiplet structures for solids, *Prog. Nucl. Magn. Reson. Spectrosc.*, 1971, **8**(1), 1–39.
58. A. Pines, M. Gibby and J. Waugh, Proton-enhanced nuclear induction spectroscopy. A method for high resolution NMR of dilute spins in solids, *J. Chem. Phys.*, 1972, **56**(4), 1776–1777.
59. A. Kaflak, D. Chmielewski, A. Górecki and W. Kolodziejski, Kinetics of $^{1}H \rightarrow {}^{31}P$ cross-polarization in human trabecular bone, *Solid State Nucl. Magn. Reson.*, 1998, **10**(4), 191–195.
60. A. Kaflak-Hachulska, A. Samoson and W. Kolodziejski, ^{1}H MAS and $^{1}H \rightarrow {}^{31}P$ CP/MAS NMR Study of Human Bone Mineral, *Calcified Tissue Int.*, 2003, **73**(5), 476–486.
61. A. Kaflak, D. Chmielewski, A. Górecki, A. Slosarczyk and W. Kolodziejski, Efficiency of $^{1}H \rightarrow {}^{31}P$ NMR cross-polarization in bone apatite and its mineral standards, *Solid State Nucl. Magn. Reson.*, 2006, **29**(4), 345–348.

62. A. Kaflak and W. Kolodziejski, Phosphorus-31 spin–lattice NMR relaxation in bone apatite and its mineral standards, *Solid State Nucl. Magn. Reson.*, 2007, **31**(4), 174–183.
63. A. C. Seifert, A. C. Wright, S. L. Wehrli, H. H. Ong, C. Li and F. W. Wehrli, ^{31}P NMR relaxation of cortical bone mineral at multiple magnetic field strengths and levels of demineralization, *NMR Biomed.*, 2013, **26**(9), 1158–1166.
64. S. Tadimalla and K. I. Momot, Effect of Partial H_2O D_2O Replacement on the Anisotropy of Transverse Proton Spin Relaxation in Bovine Articular Cartilage, *PLoS One*, 2014, **9**(12), e115288.
65. A. Kaflak, D. Chmielewski and W. Kolodziejski, Solid-state NMR study of discrete environments of bone mineral nanoparticles using phosphorus-31 relaxation, *J. Applied Biomed.*, 2016, **14**(4), 321–330.
66. A. Kaflak, S. Moskalewski and W. Kolodziejski, The solid-state proton NMR study of bone using a dipolar filter: apatite hydroxyl content versus animal age, *RSC Adv.*, 2019, **9**(29), 16909–16918.
67. K. H. Mroue, Y. Nishiyama and M. K. Pandey, *et al.*, Proton-detected solid-state NMR spectroscopy of bone with ultrafast magic angle spinning, *Scientific reports*, 2015, **5**(1), 1–10.
68. G. K. Hunter, P. V. Hauschka, R. A. Poole, L. C. Rosenberg and H. A. Goldberg, Nucleation and inhibition of hydroxyapatite formation by mineralized tissue proteins, *Biochem. J.*, 1996, **317**(1), 59–64.
69. J. Sodek, B. Ganss and M. McKee, Osteopontin, *Crit. Rev. Oral Biol. Med.*, 2000, **11**(3), 279–303.
70. C. Chenu, S. Colucci and M. Grano, *et al.*, Osteocalcin induces chemotaxis, secretion of matrix proteins, and calcium-mediated intracellular signaling in human osteoclast-like cells, *J. Cell Biol.*, 1994, **127**(4), 1149–1158.
71. P. A. Price, J. W. Poser and N. Raman, Primary structure of the gamma-carboxyglutamic acid-containing protein from bovine bone, *Proc. Natl. Acad. Sci.*, 1976, **73**(10), 3374–3375.
72. T. Iline-Vul, A. Kulpanovich, M. Nadav-Tsubery, A. Semionov, K. Keinan-Adamsky and G. Goobes, How does osteocalcin lacking γ-glutamic groups affect biomimetic apatite formation and what can we say about its structure in mineral-bound form?, *J. Struct. Biol.*, 2019, **207**(2), 104–114.
73. T. Iline-Vul, R. Nanda and B. Mateos, *et al.*, osteopontin regulates biomimetic calcium phosphate crystallization from disordered mineral layers covering apatite crystallites, *Sci. Rep.*, 2020, **10**(1), 1–16.
74. L. Fisher, D. Torchia, B. Fohr, M. Young and N. Fedarko, Flexible structures of SIBLING proteins, bone sialoprotein, and osteopontin, *Biochem. Biophys. Res. Commun.*, 2001, **280**(2), 460–465.
75. M. Goldberg, A. B. Kulkarni, M. Young and A. Boskey, Dentin: Structure, Composition and Mineralization: The role of dentin ECM in dentin formation and mineralization, *Front. Biosci. (Elite Edn.)*, 2011, **3**, 711.

76. G. W. Buchko, B. J. Tarasevich, J. Bekhazi, M. L. Snead and W. J. Shaw, A solution NMR investigation into the early events of amelogenin nanosphere self-assembly initiated with sodium chloride or calcium chloride, *Biochemistry*, 2008, **47**(50), 13215–13222.
77. P.-A. Fang, J. F. Conway, H. C. Margolis, J. P. Simmer and E. Beniash, Hierarchical self-assembly of amelogenin and the regulation of biomineralization at the nanoscale, *Proc. Natl. Acad. Sci. U. S. A.*, 2011, **108**(34), 14097–14102.
78. C. Du, G. Falini, S. Fermani, C. Abbott and J. Moradian-Oldak, Supramolecular assembly of amelogenin nanospheres into birefringent microribbons, *Science.*, 2005, **307**(5714), 1450–1454.
79. K. M. Bromley, A. S. Kiss and S. B. Lokappa, *et al.*, Dissecting amelogenin protein nanospheres: characterization of metastable oligomers, *J. Biol. Chem.*, 2011, **286**(40), 34643–34653.
80. A. Roufosse, W. Aue, J. Roberts, M. Glimcher and R. Griffin, Investigation of mineral phases of bone by solid-state phosphorus-31 magic-angle sample-spinning nuclear magnetic resonance, *Biochemistry*, 1984, **23**(25), 6115–6120.
81. L. C. Bonar, M. Shimizu, J. E. Roberts, R. G. Griffin and M. J. Glimcher, Structural and composition studies on the mineral of newly formed dental enamel: A chemical, X-ray diffraction, and 31p and proton nuclear magnetic resonance study, *J. Bone Mineral Res.*, 1991, **6**(11), 1167–1176.
82. E. A. Abou Neel, A. Aljabo and A. Strange, *et al.*, Demineralization–remineralization dynamics in teeth and bone, *Int. J. Nanomed.*, 2016, **11**, 4743.
83. F. Pourpoint, C. Coelho Diogo and C. Gervais, *et al.*, High-resolution solid state NMR experiments for the characterization of calcium phosphate biomaterials and biominerals, *J. Mater. Res.*, 2011, **26**(18), 2355–2368.
84. Y.-H. Tseng, Y.-L. Tsai and T. W. Tsai, *et al.*, Characterization of the phosphate units in rat dentin by solid-state NMR spectroscopy, *Chem. Mater.*, 2007, **19**(25), 6088–6094.
85. V. Ladizhansky and S. Vega, Polarization transfer dynamics in Lee–Goldburg cross polarization nuclear magnetic resonance experiments on rotating solids, *J. Chem. Phys.*, 2000, **112**(16), 7158–7168.
86. Y.-L. Tsai, M.-W. Kao, S.-J. Huang, Y.-L. Lee, C.-P. Lin and J. C. C. Chan, Characterization of Phosphorus Species in Human Dentin by Solid-State NMR, *Molecules*, 2020, **25**(1), 196.
87. K. Beshah, C. Rey, M. Glimcher, M. Schimizu and R. Griffin, Solid state carbon-13 and proton NMR studies of carbonate-containing calcium phosphates and enamel, *J. Solid State Chem.*, 1990, **84**(1), 71–81.
88. Y. Coppel, Y. Prigent and G. Grégoire, Characterization of hydrogenated dentin components by advanced ^{1}H solid-state NMR experiments, *Acta Biomater.*, 2021, **120**, 156–166.
89. A. Fincham, J. Moradian-Oldak and T. Diekwisch, *et al.*, Evidence for amelogenin" nanospheres" as functional components of secretory-stage enamel matrix, *J. Struct. Biol.*, 1995, **115**(1), 50–59.

90. J. Moradian-Oldak, C. Du and G. Falini, On the formation of amelogenin microribbons, *Eur. J. Oral Sci.*, 2006, **114**, 289–296.
91. S. Delgado, M. Girondot and J.-Y. Sire, Molecular evolution of amelogenin in mammals, *J. Mol. Evol.*, 2005, **60**(1), 12–30.
92. M. L. Paine, Y.-P. Lei, K. Dickerson and M. L. Snead, Altered amelogenin self-assembly based on mutations observed in human X-linked amelogenesis imperfecta (AIH1), *J. Biol. Chem.*, 2002, **277**(19), 17112–17116.
93. K. Delak, C. Harcup and R. Lakshminarayanan, *et al.*, The tooth enamel protein, porcine amelogenin, is an intrinsically disordered protein with an extended molecular configuration in the monomeric form, *Biochemistry*, 2009, **48**(10), 2272–2281.
94. R. J. Arachchige, S. D. Burton and J.-X. Lu, *et al.*, Solid-state NMR identification of intermolecular interactions in amelogenin bound to hydroxyapatite, *Biophys. J.*, 2018, **115**(9), 1666–1672.
95. V. Renugopalakrishnan, E. Strawich, P. Horowitz and M. J. Glimcher, Studies of the secondary structures of amelogenin from bovine tooth enamel, *Biochemistry*, 1986, **25**(17), 4879–4887.
96. S. Zheng, A. T. Tu, V. Renugopalakrishnan, E. Strawich and M. J. Glimcher, A mixed β-turn and β-sheet structure for bovine tooth enamel amelogenin: Raman spectroscopic evidence, *Biopolym.: Original Res. Biomol.*, 1987, **26**(10), 1809–1813.
97. R. Lakshminarayanan, D. Fan, C. Du and J. Moradian-Oldak, The role of secondary structure in the entropically driven amelogenin self-assembly, *Biophys. J.*, 2007, **93**(10), 3664–3674.
98. H. Margolis, E. Beniash and C. Fowler, Role of macromolecular assembly of enamel matrix proteins in enamel formation, *J. Dental Res.*, 2006, **85**(9), 775–793.
99. N. Matsushima, Y. Izumi and T. Aoba, Small-angle X-ray scattering and computer-aided molecular modeling studies of 20 kDa fragment of porcine amelogenin: does amelogenin adopt an elongated bundle structure?, *J. Biochem.*, 1998, **123**(1), 150–156.
100. X. Zhang, B. E. Ramirez, X. Liao and T. G. Diekwisch, Amelogenin supramolecular assembly in nanospheres defined by a complex helix-coil-PPII helix 3D-structure, *PLoS One*, 2011, **6**(10), e24952.
101. K. Carneiro, H. Zhai and L. Zhu, *et al.*, Amyloid-like ribbons of amelogenins in enamel mineralization, *Sci. Rep.*, 2016, **6**, 23105.
102. O. Martinez-Avila, S. Wu and S. J. Kim, *et al.*, Self-assembly of filamentous amelogenin requires calcium and phosphate: from dimers via nanoribbons to fibrils, *Biomacromolecules*, 2012, **13**(11), 3494–3502.
103. G. Chung, S. Jung and S. Oh, Cellular and molecular mechanisms of dental nociception, *J. Dental Res.*, 2013, **92**(11), 948–955.
104. W. J. Shaw, Solid-state NMR studies of proteins immobilized on inorganic surfaces, *Solid State Nucl. Magn. Reson.*, 2015, **70**, 1–14.

105. W. J. Shaw and K. Ferris, Structure, orientation, and dynamics of the C-terminal hexapeptide of LRAP determined using solid-state NMR, *J. Phys. Chem. B*, 2008, **112**(51), 16975–16981.
106. C.-W. Ma, J. Zhang, X.-Q. Dong and J.-X. Lu, Amyloid structure of high-order assembly of Leucine-rich amelogenin revealed by solid-state NMR, *J. Struct. Biol.*, 2019, **206**(1), 29–35.
107. S. M. Althaus, K. Mao, J. A. Stringer, T. Kobayashi and M. Pruski, Indirectly detected heteronuclear correlation solid-state NMR spectroscopy of naturally abundant 15N nuclei, *Solid State Nucl. Magn. Reson.*, 2014, **57**, 17–21.
108. B. Wang, W. Yang, J. McKittrick and M. A. Meyers, Keratin: Structure, mechanical properties, occurrence in biological organisms, and efforts at bioinspiration, *Progress Mater. Sci.*, 2016, **76**, 229–318.
109. H. H. Bragulla and D. G. Homberger, Structure and functions of keratin proteins in simple, stratified, keratinized and cornified epithelia, *J. Anatomy*, 2009, **214**(4), 516–559.
110. P. M. Steinert, A. C. Steven and D. R. Roop, The molecular biology of intermediate filaments, *Cell.*, 1985, **42**(2), 411–419.
111. P. A. Coulombe and M. B. Omary, 'Hard'and 'soft'principles defining the structure, function and regulation of keratin intermediate filaments, *Curr. Opin. Cell Biol.*, 2002, **14**(1), 110–122.
112. J. W. Mack, D. A. Torchia and P. M. Steinert, Solid-state NMR studies of the dynamics and structure of mouse keratin intermediate filaments, *Biochemistry*, 1988, **27**(15), 5418–5426.
113. H. Kricheldorf and D. Müller, Secondary structure of peptides 16 th. Characterization of proteins by means of ^{13}C NMR CP/MAS spectroscopy, *Colloid Polym. Sci.*, 1984, **262**(11), 856–861.
114. N. Nishikawa, Y. Horiguchi, T. Asakura and I. Ando, Carbon-13 solid-state nmr study of ^{13}C-enriched human hair keratin, *Polymer*, 1999, **40**(8), 2139–2144.
115. M. J. Duer, N. McDougal and R. C. Murray, A solid-state NMR study of the structure and molecular mobility of α-keratin, *Phys. Chem. Chem. Phys.*, 2003, **5**(13), 2894–2899.
116. C. Melian, D. E. Demco and M. Istrate, *et al.*, Morphology and side-chain dynamics in hydrated hard α-keratin fibres by ^{1}H solid-state NMR, *Chem. Phys. Lett.*, 2009, **480**(4–6), 300–304.
117. D. E. Demco, L. Utiu, W. Tillmann, B. Blümich and C. Popescu, Morphology and molecular dynamics of hard α-keratin under pressure by ^{1}H and ^{13}C solid-state NMR, *Chem. Phys. Lett.*, 2011, **509**(1–3), 62–66.
118. S. Björklund, A. Nowacka, J. A. Bouwstra, E. Sparr and D. Topgaard, Characterization of stratum corneum molecular dynamics by natural-abundance ^{13}C solid-state NMR, *PLoS One*, 2013, **8**(4), e61889.
119. W. C. V. Wong, A. Narkevicius and W. Y. Chow, *et al.*, Solid state NMR of isotope labelled murine fur: a powerful tool to study atomic level keratin structure and treatment effects, *J. Biomol. NMR*, 2016, **66**(2), 93–98.

120. M. Ghosh, B. P. Prajapati, N. Kango and K. K. Dey, A comprehensive and comparative study of the internal structure and dynamics of natural β – keratin and regenerated β – keratin by solid state NMR spectroscopy, *Solid State Nucl. Magn. Reson.*, 2019, **101**, 1–11.
121. W. T. Dixon, Spinning-sideband-free and spinning-sideband-only NMR spectra in spinning samples, *J. Chem. Phys.*, 1982, **77**(4), 1800–1809.
122. J. W. Beck, R. L. Edwards and E. Ito, *et al.*, Sea-surface temperature from coral skeletal strontium/calcium ratios, *Science*, 1992, **257**(5070), 644–647.
123. T. Mass, A. J. Giuffre and C.-Y. Sun, *et al.*, Amorphous calcium carbonate particles form coral skeletons, *Proc. Natl. Acad. Sci. U. S. A.*, 2017, **114**(37), E7670–E7678.
124. T. Mass, J. L. Drake, E. C. Peters, W. Jiang and P. G. Falkowski, Immunolocalization of skeletal matrix proteins in tissue and mineral of the coral Stylophora pistillata, *Proc. Natl. Acad. Sci. U. S. A.*, 2014, **111**(35), 12728–12733.
125. L. Addadi and S. Weiner, Interactions between acidic proteins and crystals: stereochemical requirements in biomineralization, *Proc. Natl. Acad. Sci. U. S. A.*, 1985, **82**(12), 4110–4114.
126. V. B. Pearse, Incorporation of metabolic CO_2 into coral skeleton, *Nature.*, 1970, **228**(5269), 383.
127. A. L. Cohen and T. A. McConnaughey, Geochemical perspectives on coral mineralization, *Rev. Mineral. Geochem.*, 2003, **54**(1), 151–187.
128. H. E. Mason, P. Montagna, L. Kubista, M. Taviani, M. McCulloch and B. L. Phillips, Phosphate defects and apatite inclusions in coral skeletal aragonite revealed by solid-state NMR spectroscopy, *Geochim. Cosmochim. Acta*, 2011, **75**(23), 7446–7457.
129. S. Von Euw, Q. Zhang and V. Manichev, *et al.*, Biological control of aragonite formation in stony corals, *Science*, 2017, **356**(6341), 933–938.
130. A. Akiva, M. Neder and K. Kahil, *et al.*, Minerals in the pre-settled coral Stylophora pistillata crystallize via protein and ion changes, *Nat. Commun.*, 2018, **9**(1), 1–9.
131. T. Asakura, Y. Suzuki, Y. Nakazawa, G. P. Holland and J. L. Yarger, Elucidating silk structure using solid-state NMR, *Soft Matter.*, 2013, **9**(48), 11440–11450.
132. M. S. Creager, J. E. Jenkins and L. A. Thagard-Yeaman, *et al.*, Solid-state NMR comparison of various spiders' dragline silk fiber, *Biomacromolecules*, 2010, **11**(8), 2039–2043.
133. C. T. Nakazawa, A. Higuchi, A. Asano, T. Kameda, D. Aytemiz and Y. Nakazawa, Solid-state NMR studies for the development of non-woven biomaterials based on silk fibroin and polyurethane, *Polym. J.*, 2017, **49**(7), 583–586.
134. X. Hu, D. Kaplan and P. Cebe, Effect of water on the thermal properties of silk fibroin, *Thermochim. Acta*, 2007, **461**(1–2), 137–144.
135. K. Yazawa, K. Ishida, H. Masunaga, T. Hikima and K. Numata, Influence of water content on the β-sheet formation, thermal stability, water

removal, and mechanical properties of silk materials, *Biomacromolecules*, 2016, **17**(3), 1057–1066.
136. T. Asakura, M. Demura, Y. Watanabe and K. Sato, ^{1}H pulsed NMR study of Bombyx mori silk fibroin: dynamics of fibroin and of absorbed water, *J. Polym. Sci. Part B: Polym. Phys.*, 1992, **30**(7), 693–699.
137. Z. Zhu, D. Gong, L. Liu and Y. Wang, Microstructure elucidation of historic silk (Bombyx mori) by nuclear magnetic resonance, *Analytic. Bioanalytic. Chem.*, 2014, **406**(11), 2709–2718.
138. T. Asakura, K. Isobe, A. Aoki and S. Kametani, Conformation of crystalline and noncrystalline domains of [3-13C] Ala-,[3-^{13}C] Ser-, and [3-^{13}C] Tyr-Bombyx mori silk fibroin in a hydrated state studied with ^{13}C DD/MAS NMR, *Macromolecules.*, 2015, **48**(22), 8062–8069.
139. T. Asakura, K. Isobe, S. Kametani, O. T. Ukpebor, M. C. Silverstein and G. S. Boutis, Characterization of water in hydrated Bombyx mori silk fibroin fiber and films by ^{2}H NMR relaxation and ^{13}C solid state NMR, *Acta Biomater.*, 2017, **50**, 322–333.
140. T. Asakura, M. Endo, Y. Tasei, T. Ohkubo and T. Hiraoki, Hydration of Bombyx mori silk cocoon, silk sericin and silk fibroin and their interactions with water as studied by ^{13}C NMR and ^{2}H NMR relaxation, *J. Mater. Chem. B*, 2017, **5**(8), 1624–1632.
141. T. Asakura, H. Matsuda and A. Naito, Acetylation of Bombyx mori silk fibroin and their characterization in the dry and hydrated states using ^{13}C solid-state NMR, *Int. J. Biol. Macromol.*, 2020, **155**, 1410–1419.

CHAPTER 22

The Lipid Phase of the Stratum Corneum Studied by Solid-state NMR: A Not So Rigid Barrier

OSKAR ENGBERG AND DANIEL HUSTER*

Institute of Medical Physics and Biophysics, University of Leipzig, Härtelstr. 16-18, D-04107 Leipzig, Germany
*Email: daniel.huster@medizin.uni-leipzig.de

22.1 Introduction

Skin is the largest organ in the human body. Its main function is to prevent excessive water loss and maintain a barrier function against all undesired environmental influences. The outermost skin layer, the epidermis, consists of several distinctive layers, of which the *stratum corneum* (SC) is the outermost one to which the skin's barrier function is ascribed.[1] The SC comprises a thickness of 10–20 μm in humans.[2] It consists of two major components, traditionally referred to as "brick and mortar":[3] the "bricks" are formed by flattened and keratin-filled dead corneocytes and are embedded in the "mortar," a unique intercellular layer of lipids aligned approximately parallel to the surface of the cells. In humans, the lipid phase of the SC is formed by three lipid classes mixed in an approximately equal molar ratio: ceramides (Cer), free fatty acids (FFA), and cholesterol (Chol).

New Developments in NMR No. 26
NMR Spectroscopy for Probing Functional Dynamics at Biological Interfaces
Edited by Anirban Bhunia, Hanudatta S. Atreya and Neeraj Sinha

Published by the Royal Society of Chemistry, www.rsc.org

Ceramides are a family of lipid molecules which can be found in a large chemical variety in nature. The backbone of these lipids is constituted by a sphingoid base, which can be a dihydrosphingosine [dS], a sphingosine [S], a phytosphingosine [P], or a 6-hydroxy sphingosine moiety [H]. A fatty acid is attached *via* the free amide, which can be a non-hydroxylated [N], an α-hydroxylated [A], or an esterified ω-hydroxyl [EO] fatty acid. The carbon chain lengths of both moieties vary between 16 and 20 for the sphingoid base and 14 to 32 for the acyl chain. The acyl chains are typically saturated, the most common representative being lignoceric acid (LA, C24:0). As a notable exception, the ω-hydroxyl fatty acid Cer subclasses are typically ultra-long, comprising a fully saturated C30:0 upper part and a polyunsaturated C18:2 lower part. In contrast to phospholipids, the main building blocks of cellular membranes, ceramides have a much smaller and less hydrophilic head group, constraining the water uptake of the SC lipid layers to only about 1–2 water molecules per Cer molecule.[4] The combination of the letters for the amidated fatty acid [N], [A], or [EO] with those of the sphingoid moiety [dS], [S], [P], or [H] defines the nomenclature of the 12 ceramide subclasses.[5] Examples of the chemical structures for Cer[NS] and Cer[EOS] are given in Scheme 22.1.

The FFAs of the lipid phase of the SC are mostly saturated, the most prominent representatives in human skin being C18:0, C24:0, and C26:0.[6] The chemical structure of LA is shown in Scheme 22.1. Given the low pH of skin (~5.5), it is assumed that the FFAs are protonated.

Finally, the sterol cholesterol (Chol) accounts for the last third of the lipid species in the SC. Chol is the only lipid of the SC that is also ubiquitously found in mammalian cellular membranes. There, it is known as an essential component that favorably interacts with saturated lipid chains[6,7] and has a profound impact on the plasma membrane's domain structure.[8,9] A small fraction of the Chol in the SC is sulfated.[10] The structure of Chol is also displayed in Scheme 22.1.

The characteristic composition of lipid molecules of specific chemical structure plays the decisive role in the physicochemical properties of the SC and its barrier function is a consequence of the composition and organization of the SC lipids. To understand the permeability of SC membranes,

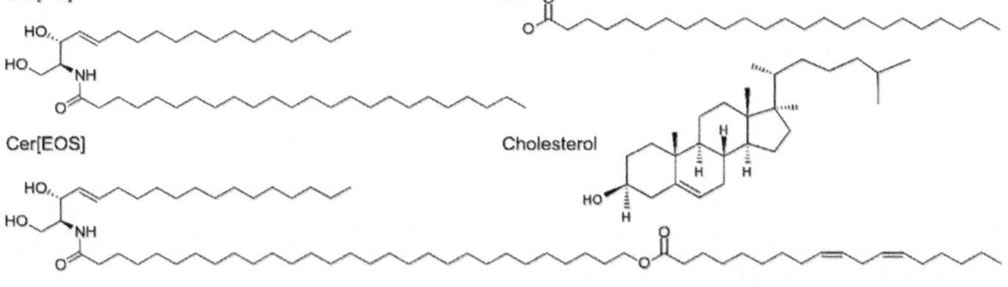

Scheme 22.1 Chemical structures of some essential molecules of the lipid phase of the human SC.

the lipid packing density is of particular importance.[11] An exact high-resolution structure of this complex interface has not been determined yet, however, relatively well-defined models have been developed. The multilamellar structure of the SC was first imaged by freeze fracture electron microscopy.[12] With improved fixation methods, lamellar lipid sheets stacked on top of each other could be found, exhibiting a width of 12 nm.[13] X-ray diffraction studies have further analyzed the structure of the lipid layers and identified the coexistence of a long periodicity phase (LPP) with a Bragg spacing of approximately 13 nm and a short periodicity phase (SPP) with a spacing of approximately 6 nm.[14] These data could not be explained by a simple lipid bilayer structure as is known for cellular membranes. Furthermore, the heterogeneous distribution of Cer and FFA chain lengths renders a straightforward bilayer structure unlikely. In addition, Cer molecules may assume an extended and/or hairpin structure in the lipid arrangement,[15] further complicating the structural analysis. Taken together, a unique lipid structure of the SC emerges, which has to this day not been fully understood. In addition to the highly heterogeneous structure of the lipids in the SC, the very low hydration level in the SC deserves specific attention. Over the last three decades, extensive biophysical studies have been conducted involving X-ray and neutron diffraction,[16,17] Fourier transform infrared,[18] and solid-state NMR spectroscopy[19–22] in combination with electron microscopy[23] to characterize in atomistic detail the exact arrangement of the lipid phase in the SC. So far, these methods have provided data that describe the lipid phase of the SC as an arrangement of Cer and FFA in a crystalline orthorhombic phase at skin temperature or 32 °C. The crystalline lipid phase formed by completely rigid lipid molecules was associated with the low permeability of the SC under these conditions. With elevated temperature, or in the absence of FFAs in artificial mixtures of the SC, a hexagonal phase with somewhat more mobile lipids and even a liquid phase with very mobile lipids have been described.[24] In these phases, intermediate and high permeability of water have been found, respectively.

Over the past two decades, this model of the rigid SC's lipid phase has served well as the basis for understanding the barrier function of skin as well as its modification by permeation enhancers or altered lipid distribution as encountered in skin diseases. However, very recently, new experimental data have been reported, where highly mobile lipids were detected within the SC, challenging this model of a very rigid lipid phase in the SC.[25–27] Molecular dynamics (MD) simulations of the SC's lipid phase have also described a fluid slab phase in the interior of the lipid structure.[28] All these results have modified our understanding of the SC's lipid phase, including the suggestion that it may exhibit a domain structure as is known from the plasma membrane of cells.[25] These recent observations have actively started the debate on the correct model of the SC lipid phase, requiring a review of previous data and their interpretation.

In the current chapter, we will briefly review the toolbox of biophysical methods for studying the structure and dynamics of the SC lipid phase, with special emphasis on solid-state NMR methods for characterizing the molecules that form this very important interface. Then, we will review the major findings which have contributed to the traditional model of the rigid SC lipid phase, before we discuss the recent findings that have helped identify highly mobile lipids within the SC lipid phase. Finally, an outlook is given on which questions are still open and how they can be addressed in further research work.

22.2 Biophysical Methods to Study the Structure and Dynamics *of Stratum Corneum* Lipid Assemblies

22.2.1 Solid-state NMR Spectroscopy

In this section, we focus on solid-state NMR spectroscopy as one of the most central methods in lipid research. Stationary ^{2}H NMR was developed in the 1970s by the groups of J. Seelig[29] and J. H. Davis[30] and remains highly important in membrane research and particularly the investigation of the lipid moiety of the SC. The requirement of deuterated lipid molecules due to the low natural abundance of ^{2}H represents both a blessing and a curse: it allows for specific investigation of individual compounds in a complex lipid mix, but restricts applications of ^{2}H NMR to model mixtures, and is limited to available deuterated lipids. As the gyromagnetic ratio of ^{2}H is only 15.4% of that of ^{1}H and due to the low natural abundance of ^{2}H of 0.0115%, isotopic ^{2}H labeling is imperative. Native SC samples can only be studied using ^{13}C magic-angle spinning (MAS) NMR at natural abundance. Ultrahigh MAS frequencies that allow ^{1}H detection of rigid solids have not been applied in SC research thus far.

22.2.1.1 Stationary ^{2}H NMR Spectroscopy

^{2}H is a quadrupolar nucleus with a spin of 1 ($I = 1$), which has three energy levels in the magnetic field allowing for two transitions. The transition frequencies differ from each other due to the interaction of the quadrupole moment of the nucleus with the electric field gradient of the ^{2}H–C bond, which is referred to as quadrupolar interaction. Consequently, stationary ^{2}H NMR spectra feature a characteristic doublet structure. The moderate size quadrupolar coupling constant of $e^2qQ/h = 167$ kHz for aliphatic C–^{2}H bonds[29,30] renders ^{2}H a very suitable nucleus for investigating lipid molecules in restricted geometries and fast dynamics.

Solid-state ^{2}H NMR spectra are powder patterns with complex line shapes, as each ^{2}H–C bond assumes a specific orientation with respect to the external magnetic field described by the powder angle θ. In a standard powder sample, all molecules are randomly oriented in the magnetic field, which gives rise to a spherical-weighted sum over all orientations (θ) of the C–^{2}H bonds in space, resulting in the so-called Pake spectrum.[29,30] The peaks of the powder line shape correspond to the 90° angle of the C–^{2}H bond relative to the external B_0 field, and the shoulders correspond to the less abundant 0° angle of the C–^{2}H bond to B_0. The splitting observed for the ^{2}H NMR signal of an individual site ($\Delta\upsilon_Q$) is proportional to the second Legandré polynomial according to:

$$\Delta\upsilon_Q = \frac{3}{2}\left(\frac{e^2qQ}{h}\right)\left(\frac{3\cos^2\theta - 1}{2}\right) \tag{22.1}$$

The quadrupolar splitting determined from the ^{2}H NMR spectra is defined as the distance between the two "horns" of the Pake spectra representing the doublet for $\theta = 90°$. In a rigid ^{2}H–C bond, this amounts to $\Delta\upsilon_Q = 125.25$ kHz.[29,30]

Fast molecular motions scale down the NMR interaction strength and lead to ^{2}H NMR motionally averaged powder patterns characterized by narrower quadrupolar splittings. For instance, the rotation of the terminal methyl group in a free fatty acid reduces the observed quadrupolar splitting to 1/3, yielding a ^{2}H NMR powder pattern with a quadrupolar splitting of $\Delta\upsilon_Q = 41.75$ kHz.[31] Thus, the ^{2}H NMR spectrum of a rigid perdeuterated FFA is described by the superposition of the Pake spectra of n methylene segments with a quadrupolar splitting of 125.25 kHz and a narrower Pake spectrum of the three ^{2}H nuclei with a reduced quadrupolar splitting of 41.75 kHz, as shown in Figure 22.1A.

^{2}H NMR powder spectra are very useful for characterizing molecular motions and anisotropy for lipids.[32] For lipids undergoing fast axially symmetric rotational diffusion in the membrane ($\tau < 6$ μs) the effect of motional averaging is conveniently described by an order parameter (S_{CD}) that scales the rigid limit quadrupolar splitting according to:[29,30]

$$\Delta\upsilon_Q = \frac{3}{2}\left(\frac{e^2qQ}{h}\right)\left(\frac{3\cos^2\theta - 1}{2}\right)S_{CD} \tag{22.2}$$

$$\text{with } S_{CD} = \frac{1}{2}\langle 3\cos^2\beta - 1\rangle \tag{22.3}$$

The angular brackets indicate a time average, β indicates the unique time-dependent angle between the membrane normal and the C–^{2}H bond in a methylene or methyl group.[29] The order parameters for lipids in a fluid phase undergoing axially symmetric rotational diffusion amount to ≤ 0.5. An order parameter of 0 indicates isotropic motion. Geometry factors can also narrow the ^{2}H NMR line shape by conformational constrains, as observed for instance for the C–^{2}H bond of the C2 methylene group in phospholipid bilayers.[33] Fluid lipid membranes feature an order parameter profile with the highest values for the upper-chain methylene segments, which decrease toward the chain ends.[34]

The ^{2}H NMR line shapes observed for different SC lipid phases are summarized in Figure 22.1. At a skin temperature of 32 °C and below, the

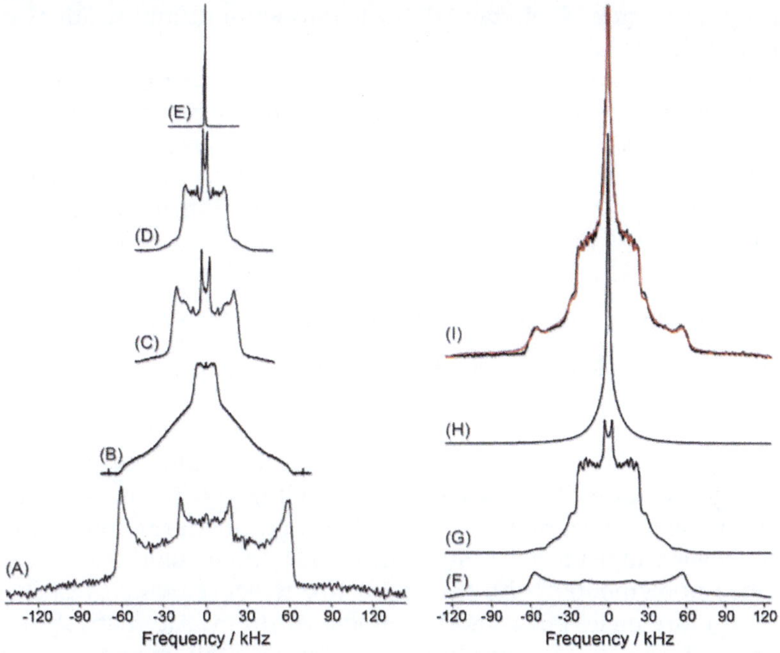

Figure 22.1 (Left) ^{2}H NMR spectra of pure lipid phases from various SC and phospholipid mixtures: (A) orthorhombic phase, (B) gel phase, (C) liquid ordered phase, (D) liquid disordered phase, (E) isotropic phase. (Right) Deconvolution of a ^{2}H NMR spectrum of a SC lipid mix consisting of an equimolar mixture of Cer[NS] with a C24:0 acyl chain, LA-d_{47}, and cholesterol at a temperature of 65 °C. The experimental line shape [black line in panel (I)] can be well fitted (red line) by simulating a superposition of 22% orthorhombic phase (F), 56% fluid phase (G), and 22% isotropic (H) ^{2}H NMR spectral line shapes. The figure was assembled from previously published data.[20,39,84]

lipids typically form a rigid orthorhombic phase (A). Lipids are organized in a hexagonal lattice and the molecules are completely rigid, with the exception of the terminal methyl group, which undergoes the aforementioned rotation leading to a broad Pake pattern (A). At elevated temperature, lipid motion will increase, leading to partial averaging of the quadrupole interaction and narrower ^{2}H NMR spectra. This is the result of transition to the gel phase, where lipids undergo motions with correlation times in the μs range leading to ^{2}H NMR spectra that are more difficult to describe[30,35] (B). Alternatively, rather fluid phases occur, which are referred to as liquid ordered (lo, C) and liquid disordered phases (ld, D).[8,9,36] In the lo phase, lipids undergo axially symmetric rotations, but most methylene chains are found in the *trans* conformation.[37,38] In the ld phase, the fraction of methylene segments that are found in the *gauche* conformation is much higher, leading to lower order parameters. Finally, lipid molecules in highly curved geometries such as cubic phases or completely isotropically mobile lipids feature a narrow ^{2}H NMR line without any residual quadrupolar splitting (E).

A frequently observed phenomenon is that SC lipid mixtures are not phase pure, as indicated by a superposition of ^{2}H NMR spectra characteristic for more than one phase.[20,22,26,39] Using numerical line shape simulations,[20,40] these ^{2}H NMR powder spectra can be deconvoluted to quantitatively determine the relative proportion of each phase, yielding a quantitative description of the structure of the lipid mix as illustrated in Figure 22.1F–I.

22.2.1.2 Magic-angle Spinning ^{13}C NMR Spectroscopy

While static ^{2}H NMR powder patterns provide a wealth of structural and dynamical information on SC lipid mixes, the analysis suffers from the fact that the broad powder patterns occupy a significant portion of the NMR spectral window and severe signal superposition results. In the previous section, we have described how isotopic labeling can help overcome this problem at the expense that only simplified lipid mixtures containing isotopically labeled molecules are studied. Usually, intact skin tissue samples cannot be supplemented with the respective isotopes, which restricts the method to more or less complicated model mixtures. Alternatively, advantage can be taken of high-resolution solid-state NMR methods involving magic-angle spinning (MAS).[41] Here, the sample is placed in a cylindrical rotor, oriented at the magic angle (54.7°) with respect to the external magnetic field and spun at frequencies between ~5 to 100 kHz. This leads to an efficient averaging of all NMR interactions described by second-rank tensors, which are responsible for the broad line shapes in stationary NMR. For heteronucleus detection, the cross-polarization (CP) magic-angle spinning (CPMAS) technique[42] combines MAS, CP, and high-power decoupling, yielding relatively narrow signals that are observed at their respective chemical shifts. For lipid membranes, ^{13}C MAS NMR spectra were first reported by the E. Oldfield group, who showed well-resolved NMR spectra of multilamellar dispersions of phospholipids and cholesterol.[43] As the lipids in liquid crystalline membranes are highly mobile, even ^{1}H MAS NMR spectra showed astonishingly well-resolved peaks with line widths below 20 Hz,[44–46] which allowed comprehensive studies of the structure and lateral organization of lipids in fluid membranes by the NOESY technique.[44,47] This astonishing effect that ^{1}H MAS spectra of lipid dispersions could be detected even at low MAS frequencies of 5 kHz is explained by the fast axially symmetric reorientation of the molecules in the bilayer, leading to significant averaging of the homogeneous ^{1}H–^{1}H dipolar tensor.[48]

Due to the dense packing and high rigidity of the molecules in the lipid phase of the SC, ^{13}C-detected MAS NMR techniques are the method of choice. Depending on the residual mobility of the lipids, different excitation schemes can be employed. Excitation of ^{13}C NMR coherences by CP has the advantage that ^{13}C is detected with higher sensitivity and the accumulation of the NMR signal is determined by the T_1 relaxation time of ^{1}H nuclei, which is shorter than for ^{13}C. However, CP introduces a dynamic bias toward rigid sites, as it relies on ^{1}H–^{13}C dipolar couplings which are subject to motional scaling that can be different for individual molecules and segments of a molecule.[19,49] In contrast, the INEPT method uses scalar couplings for the polarization transfer, which restricts the method to only highly mobile molecules/segments. Only directly excited ^{13}C MAS NMR spectroscopy detects both rigid and mobile sites equally.

A big advantage of the ^{13}C MAS NMR method in SC research is its applicability to intact skin from any source, as it relies on the natural abundance of ^{13}C. However, due to the high structural similarity between the individual Cer species and the highly repetitive nature of the Cer acyl chains and FFA, a significant signal superposition in the aliphatic region of the NMR spectra is observed. Due to the large number of chain methylene segments with identical or very similar chemical shift, dynamic range problems of the detector can render detection of weak signals from the chemically unique head groups of the Cer molecules difficult. Furthermore, correlation spectroscopy to increase the dimensionality of the NMR spectra is hampered by the low natural abundance of ^{13}C (1.1%). The introduction of ^{13}C-enriched molecules into SC model mixtures can be advantageous in this case.

In SC research, ^{13}C MAS NMR methods have been pioneered by D. Topgaard, E. Sparr, and co-workers to investigate lipid dynamics[50] and their modification by moisturizers,[51] penetration enhancers,[52] and skin hydration,[53] and to track solvents[17] in intact skin or isolated SC from natural sources. Especially at skin temperature, the achieved resolution of the ^{13}C MAS NMR spectra is limited, mainly due to the interference of the relatively slow molecular dynamics with the frequencies of MAS and high-power proton decoupling,[54] resulting in relatively broad lines and limited resolution. In an attempt to quantify the mobile and rigid segments of the SC, the groups of D. Topgaard and E. Sparr have developed a method based on the INEPT pulse sequence to determine the proportion of mobile sites in intact SC samples.[55] Figure 22.2 shows a typical ^{13}C MAS NMR spectrum of a hydrated SC lipid mix consisting of Cer[NS]/Cer[EOS], a FFA mix, and Chol detected at 32 °C.

22.2.2 Other Biophysical Techniques

22.2.2.1 X-ray and Neutron Diffraction Techniques

X-ray diffraction techniques have contributed most data for the development of the structural model of the SC lipid phase.[14,16,24] As the lipid phase of the SC is organized in many layers of highly similar structure, it represents a one-dimensional crystal for scattering work. The lamellar lipid organization can be deduced from small-angle X-ray diffraction providing the repeat distance as a parameter for the periodicity of the one-dimensional lattice. In cases where two lamellar phases are present, for instance in samples with

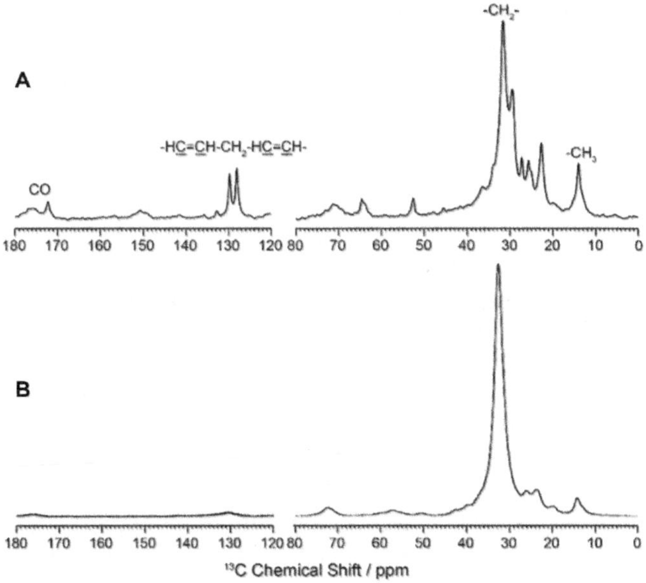

Figure 22.2 Proton-decoupled natural abundance ^{13}C MAS NMR spectra of a SC mimicking mix of Cer[NS] (0.7 mol), Cer[EOS] (0.3 mol), a FFA mix (1 mol), and Chol (1 mol) excited by (A) direct excitation and (B) CP. NMR spectra were acquired at a MAS frequency of 7 kHz and a temperature of 32 °C.

coexistence of the SPP and the LPP, the X-ray diffraction patterns show two sets of Bragg diffraction peaks.

The lateral packing of the SC lipids can be studied by wide-angle X-ray diffraction, where the short-range structure and order are investigated on a length scale of ~3–15 Å. Wide-angle X-ray diffraction allows discrimination between the liquid phase (with a broad diffraction peak at ~4.6 Å), the hexagonal phase (with a strong peak at 4.1 Å), and the orthorhombic phase (with two diffraction peaks at 3.7 and 4.1 Å).

With the introduction of neutron diffraction, the field of SC lipid phase research has seen an important improvement in the quality of the structural models. Neutron diffraction was first applied to SC lipid mixtures by the group of R. Neubert.[56] Carried out on oriented stacks of multilamellar SC lipid layers, it provides two important advantages over X-ray diffraction experiments.[57] First, due to the contrast variation between H_2O and D_2O, the sign of the structure factors can be determined, which results in a quantitative evaluation of the scattering length density profile. Second, by using specifically deuterated lipid molecules, their localization and distribution parallel to the membrane normal can be determined, taking advantage of the large difference in the coherent scattering amplitudes of ^1H *vs.* ^2H leading to well-defined distribution functions of specifically deuterated segments of ceramides, FFA, and cholesterol.

22.2.2.2 *Fourier Transform Infrared Spectroscopy*

Fourier transform infrared spectroscopy (FTIR) is very useful because the deuteration of specific lipids or their segments allows their exclusive detection in complicated SC lipid mixtures.[18,58] Differences in wavelengths for the $C{-}^1H_2$ and $C{-}^2H_2$ stretching band modes can be used to distinguish the lipid order caused by the *trans-gauche* isomerization in the hydrocarbon chains and can therefore also be used to determine phase transition temperatures.[59] Individual lipid phases also exhibit different spectra, *e.g.* in a mixture of fatty acids and ceramides. For studying intermolecular lipid packing in SC lipid mixtures, the CH_2 scissoring vibrations are very useful, which exhibit a doublet when the sample is in an orthorhombic phase at lower temperatures.[60] At higher temperatures, the splitting of the FTIR band disappears, indicating the phase transition from the orthorhombic to a hexagonal phase. Information on intramolecular chain conformation is obtained from the symmetric and asymmetric CH_2 stretching modes, which shift to higher wave numbers in the conformationally disordered state.

22.2.2.3 *Molecular Dynamics Simulations*

Molecular dynamic (MD) simulations provide a wealth of structural information with atomistic detail and have very successfully been carried out for various lipid systems,[61] including SC mixtures.[28,62] Usually, MD simulations provide higher spatial and temporal resolution than experimental studies.[63] Provided the MD force fields are correct (for instance CHARMM36[28]), numerous experimental parameters can be obtained from simulations, in particular NMR order parameters and cross-relaxation rates, probability distributions, area per lipid, membrane thickness, rotational diffusion rates, *etc.*[64–66] MD simulations have been used to simulate models of both the SPP[67] and the LPP.[28] However, the rigid lipid components of the SC require much longer simulation times than are currently possible. In order to overcome the limitations in simulation time, course-grained models have been applied to SC lipid mixtures,[68] too. In addition, the MD approach is critically dependent on building up a well-defined structure as a starting configuration. While this has been well established for symmetrical phospholipid bilayers,[63] the current uncertainties of the SC unit cell and the specific spatial organization of the lipids in the SC render it difficult to find realistic initial conditions for successful MD simulations of SC models.

22.3 Results

22.3.1 The LPP Model of the *Stratum Corneum*: Rigid Lipid Layer with a Highly Mobile Core

The current model of the SC lipid matrix was developed on the basis of scattering data. Early X-ray experiments identified the coexistence of the LPP and the SPP in the SC.[14] Subsequent very comprehensive work by the group of J. Bouwstra proposed the most accepted sandwich model of the LPP, consisting of three lipid layers (Figure 22.3).[24] While the outer layers consist of a crystalline phase of FFA and Cer in hairpin conformation, the central layer is formed by a liquid phase, which mostly contains the linoleic chain of the ultralong ω-acylceramide Cer[EOS] along with some cholesterol.[24]

This model could be confirmed and further refined using neutron diffraction experiments capable of localizing specific deuterated segments within the diffraction unit cell.[56] Due to the large difference between the neutron scattering lengths of ^{1}H and ^{2}H, such experiments are particularly well suited to understand the localization and distribution of individually deuterated segments of SC lipids within the lipid layers. In addition, structural features of individual Cer molecules (*i.e.* hairpin *vs.* extended structure) can be addressed by this method, as deuterated segments can be precisely localized in the unit cell. The current model of the LPP unit cell comprises all ceramides in fully extended conformation and a three-layered lipid structure.[4,69] The lineolate of Cer[EOS] was found to localize on either end of the central third of the LPP unit cell.[69] Interestingly, the cholesterol head group also was found to co-localize with the lineolate moieties (Figure 22.3).

Scattering methods are extremely useful to determine the structural features of highly ordered structures with long periodicity, but are less suited for the detection of mobile phases of more dynamic molecules, as long-range order is lost in such phases. The mobility of the lipids can be very well

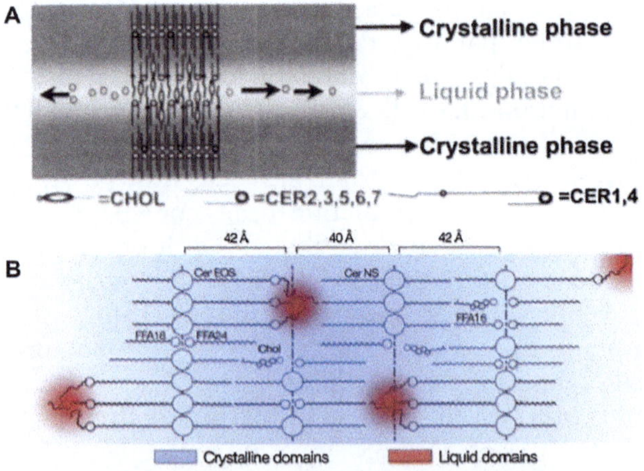

Figure 22.3 Comparison of the first[24] (A) and most recent[26] (B) illustrations of the sandwich models of the unit cell of the LPP of complex SC lipid mixtures based on X-ray and neutron diffraction work as well as ^{2}H solid-state NMR studies. (A) Reproduced from ref. 24, https://doi.org/10.1016/S0022-2275(20)31502-9, (B) Reproduced from ref. 26, https://doi.org/10.1194/jlr.M080432,

addressed by spectroscopic tools, in particular by NMR methods. Support for the sandwich model with a highly mobile lineolate also came from a recent ^{2}H NMR and FTIR study[26] as well as from all-atom MD simulations.[28] In the group of M. Lafleur, a Cer[EOS] molecule with a deuterated oleic acid ω-hydroxyl [EO] fatty acid was studied in an LPP SC lipid model.[26] In the Cer[EOS]/Cer[NS/FFA mix/Chol mixture, the deuterated free lignoceric acid (LA) and the deuterated LA of Cer[NS] showed orthorhombic ^{2}H NMR spectra at 25 °C, which became gel-like at 37 °C and more liquid-ordered at 45 °C (Figure 22.4). In stark contrast, the deuterated oleic acid of Cer[EOS] showed a very narrow and completely isotropic ^{2}H NMR spectrum at all temperatures, which is indicative of a completely disordered state undergoing very large amplitude motions on a fast timescale with correlation times much faster than ~6 μs.

The ^{2}H NMR powder spectra showed similar line shapes as found in earlier studies of SC lipid mixtures for Cer[NS]-d_{47} and LA-d_{47}, which featured the orthorhombic phase state at lower temperatures. Formation of gel and fluid phases with lipids undergoing fast rotational diffusion about their long axis but in mostly all-*trans* conformation was observed upon temperature increase (see Figure 22.4). Also in agreement with previous results, the Cer[NS] and LA formed an isotropic phase above 60 °C. The deuterated oleate moiety of Cer[EOS], however, showed an isotropic ^{2}H NMR spectrum at all temperatures (even at −20 °C), indicating the highly fluid nature of the ω-acyl chain. To rule out that the isotropic NMR signal of the ω-acyl chain was caused by the formation of Cer[EOS] micelles during sample preparation, the authors performed centrifugation and discarded the supernatant. The isotropic nature of the Cer[EOS] oleate was confirmed by identical ^{2}H NMR spectra obtained after the centrifugation procedure. These measurements directly confirmed the proposed nanodroplets in the LPP of SC lipid mixtures, as illustrated in Figure 22.3B. The high mobility and disorder of the ω-hydroxyl fatty acid of Cer[EOS] are clearly related to its unsaturation and not to the length of the ultralong chain of Cer[EOS]. A ^{2}H NMR study of a SC lipid mixture containing Cer[EOS] with a saturated palmitic acid exhibited a much higher order.[70]

FTIR spectra were also recorded to probe the conformational order of the lipid chains in the aforementioned Cer[EOS]/Cer [NS]/FFA mix/Chol mixture.[26] Both the free LA and the LA of Cer[NS] showed a symmetric stretching band ($\nu_S(CH_2)$) of ~2088 cm^{-1} for the deuterated methylene groups at temperatures up to 60 °C and at higher temperatures abruptly shifted to higher wave numbers of 2096 cm^{-1} indicating the transition to a liquid disordered state. In contrast, for the oleic acid of Cer[EOS] the same band was detected at 2096 cm^{-1} at all temperatures, not undergoing any phase transition.

While mobile lipid components of the SC can be detected by including a more complete set of deuterated lipid segments, ^{13}C MAS NMR spectroscopy represents an alternative (see Section 22.2.1.2). ^{13}C CPMAS NMR experiments have been performed on an SC model mix of Cer[EOS]/Cer[NS]/FFA mix/Chol.[27] It was found that in the LPP model mixture, the INEPT MAS

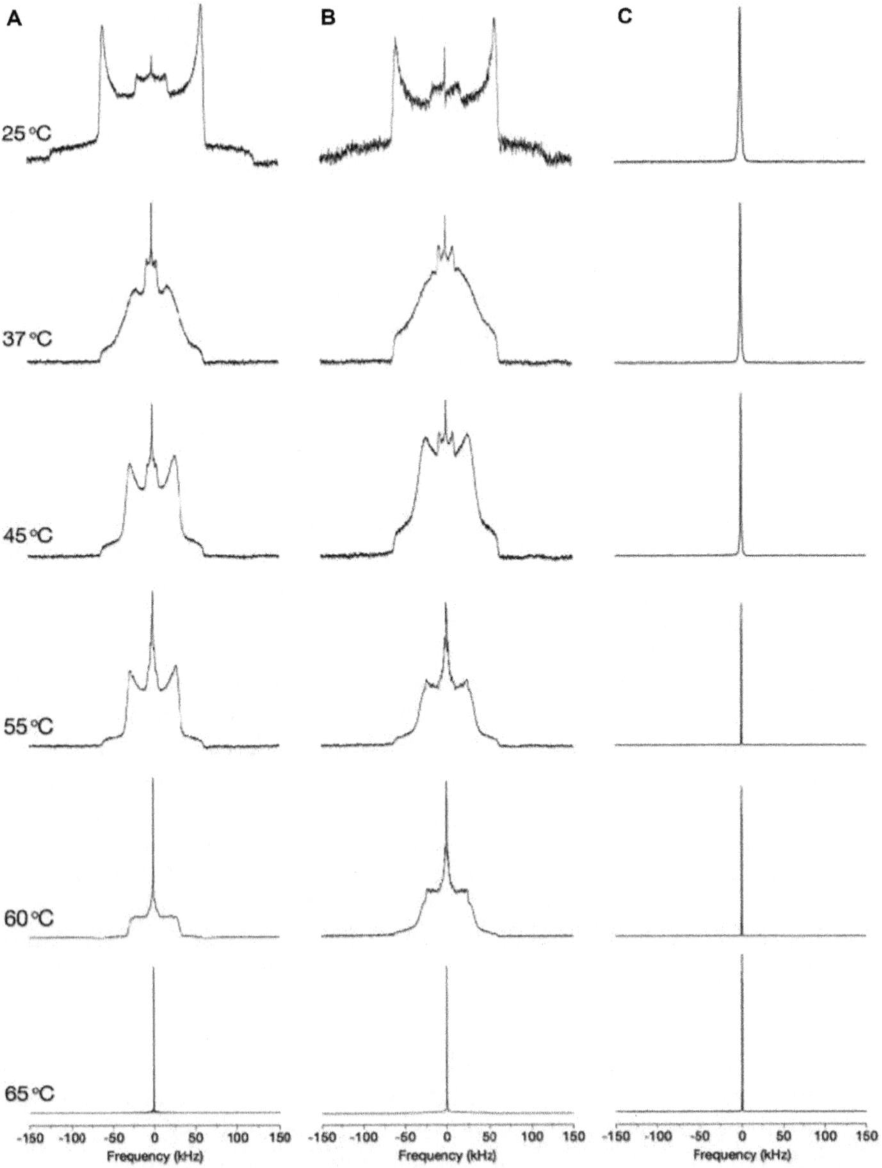

Figure 22.4 Temperature dependence of the ^{2}H NMR spectra of LA-d-$_{47}$ (A), Cer[NS]-d_{47} (B), and the perdeuterated oleic acid chain of Cer[EOS]-d_{33} (C) in a Cer[EOS]/Cer [NS]/FFA mix/Chol mixture. The ^{2}H NMR spectra are normalized to the same intensity. Reproduced from ref. 26, https://doi.org/10.1194/jlr.M080432, under the terms of the CC BY 4.0 license, https://creativecommons.org/licenses/by/4.0/.

NMR spectrum showed peaks for the chain segments of the linoleate, indicating fast ($\tau_c < 10$ ns) C–H bond reorientations, while low-order parameters ($S_{CD} \leq 0.08$) were estimated for these lipid segments.[27] These results also support the presence of highly fluid segments inside the LPP of the SC ascribed to the linoleoyl moiety, which is pronounced to be conformationally disordered and may even fold back into the hydrophilic head group region.[27]

Interestingly, the exclusive formation of the SPP, even in the presence of the ultralong Cer[EOS], has also been reported.[57] Cer[NS] and FFA are predominantly present in the center of the unit cell, featuring interdigitated chains. Cer[EOS] extends from the unit cell border into the center, where Chol and the highly disordered linoleate of Cer[EOS] are localized.[69]

22.3.2 The SPP Model of the *Stratum Corneum*: An Interface Containing Highly Mobile Lipids

A lot of biophysical work has been carried out to study the structural and dynamical features of the SPP.[20–22,39,40,70–77] For some specific SC lipid mixtures, the SPP is also formed in the presence of the ultralong Cer[EOS],[57] but most mixtures investigated do not contain this very special ω-acylceramide.

Our group has studied the structure and dynamics of several SC lipid mixtures representing the SPP using ^{2}H NMR spectroscopy. As an example, we will briefly discuss the equimolar Cer[NS]/LA/Chol mixture using the canonical Cer[NS] with a C24:0 acyl chain.[20] The ^{2}H NMR spectra of the mixture with either deuterated Cer[NS]-d_{47}, LA-d_{47}, or Chol-d_6 are shown in Figure 22.5. By switching the ^{2}H label between these individual components of the mixture, the thermotropic phase behavior of each molecule in the mixture can be studied comprehensively. At 25 °C and at physiological skin temperature (32 °C), the orthorhombic phase dominates the ^{2}H NMR spectra. However, smaller proportions of the Cer[NS] (15%) and the LA (21%) show fluid phases. Significant restructuring of the mixture occurs upon temperature increase. At 50 °C, both Cer[NS] and LA are almost entirely fluid/liquid ordered, while 62% of the Chol remains crystalline. Further temperature increase yields ^{2}H NMR spectra that almost entirely indicate the fluid liquid ordered phase. For Cer[NS] and LA, order parameter profiles could be determined under these conditions, allowing to calculate structural features of the LA chain.[20] The ^{2}H NMR spectra of Chol show the characteristic features of an upright orientation of the molecule in the lipid arrangement undergoing axially symmetric reorientations.[78] Raising the temperature to 80 °C converts the mixture into a completely isotropic phase.

This example further illustrates the power of the ^{2}H NMR analysis in characterizing SC lipid mixtures by examining the structure and dynamics of each deuterated lipid component of the SC lipid mix. In particular, the NMR results indicate that SC lipid mixtures are often not phase pure and phases of lipids with different structures and dynamics coexist. This suggests

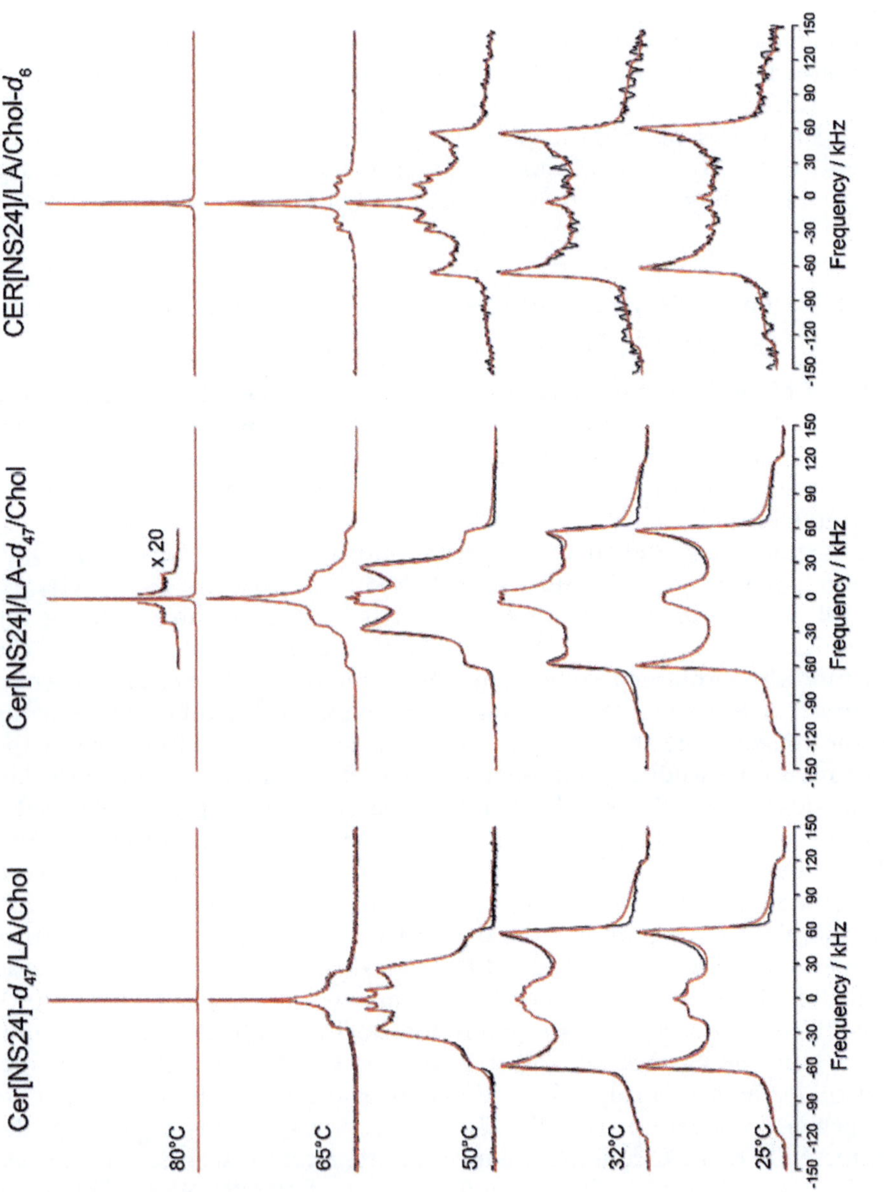
Cer[NS24]-d47/LA/Chol
Cer[NS24]/LA-d47/Chol
CER[NS24]/LA/Chol-d6
80°C
65°C
50°C
32°C
25°C
x 20
-150 -120 -90 -60 -30 0 30 60 90 120 150
Frequency / kHz

segregation and macroscopic demixing of these individual phases, as also proposed in diffraction[71] and FTIR work.[79,80] Like FTIR, ^{2}H NMR is very powerful in identifying mobile phases that are particularly underrepresented in diffraction studies due to the lack of long-range order and provides higher site resolution. While the thermotropic phase behavior is very rich for ceramide-containing lipid mixtures, most studies indicate predominantly orthorhombic phases at physiological skin temperature. However, significant increase in lipid mobility is observed in the presence of shorter chain ceramides,[20,22] unsaturated or short-chain FFAs,[74,75,77] or other penetration enhancers.[70]

An important component of the SC lipids that had not been deuterated and investigated by ^{2}H NMR before is the sphingosine backbone of the ceramides. This is of utmost importance, because it has been speculated that the ceramides could either form an extended or a hairpin conformation.[15] However, without direct study of the sphingosine backbone, it has been difficult to establish a consensus. Recent success in synthesis methods allowed synthesizing a Cer[NS] molecule with a perdeuterated sphingosine-d_{28} chain (see Scheme 22.1). With this molecule, we prepared NMR samples with Cer[NS]/LA/Chol at varying Chol concentrations to avoid phase-separated crystalline cholesterol.[76,81] One sample contained Cer[NS] with a perdeuterated acyl chain (Cer[NS]-acyl-d_{47}) and another one contained a perdeuterated sphingosine chain (Cer[NS]-sph-d_{28}). The ^{2}H NMR spectra for these mixtures are shown in Figure 22.6.

NMR spectra were measured between 25 °C and 80 °C to study the thermotropic phase behavior of the mixture. Usually, ^{2}H NMR spectra with multiple components were detected. To determine the exact proportions of each lipid phase and quantify the structural and dynamic parameters of the individual chains, we used numerical line shape simulations to reproduce the ^{2}H NMR spectra. At low temperature ($\leq$32 °C), the Cer[NS]-acyl-d_{47} NMR spectrum indicated an orthorhombic phase as indicated by the very large quadrupolar splitting of ~125 kHz, confirming that the acyl chains were in an extended all-*trans* conformation. The inner narrower Pake spectrum corresponds to the rotating terminal methyl groups.[31] In notable contrast, the Cer[NS]-sph-d_{28} chain shows a drastically different ^{2}H NMR spectrum under these conditions. The ^{2}H NMR line shape was simulated by combining a broad component characteristic for a rigid lipid phase and a narrow component featuring a superposition of Pake doublets scaled by relatively low-order parameters as known from fluid lipid phases. The broad component resembles a gel phase observed for phosphatidylcholines below their

Figure 22.5 Temperature dependence of the ^{2}H NMR spectra of Cer[NS]-d_{47} (left column), LA-d-$_{47}$ (middle column), and Chol-d_6 (right column) of an equimolar Cer [NS]/LA /Chol mixture. The NMR spectra are normalized to the same intensity. Experimental spectra are shown in black, best-fit numerical line shape simulations in red. Reproduced from ref. 20 with permission from American Chemical Society, Copyright 2015.

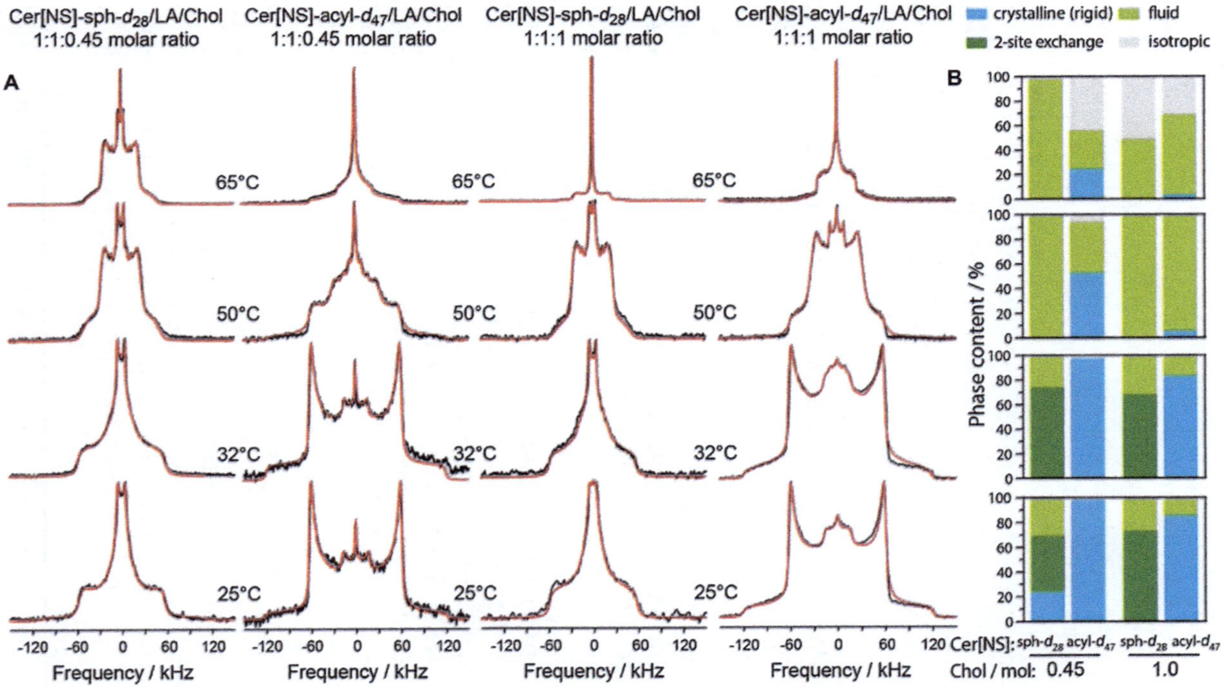

Figure 22.6 (A) Stationary ^{2}H NMR spectra (black) of the SC lipid model mixture composed of Cer[NS]/LA/Chol at molar ratios of 1 : 1 : 0.45 (left two columns) and 1 : 1 : 1 (right two columns) at varying temperatures. The first and third columns display the ^{2}H NMR spectra of the Cer[NS]-sph-d_{28} (deuterated sphingosine-d_{28}). Columns 2 and 4 show the NMR spectra of the Cer[NS]-acyl-d_{47} (perdeuterated lignoceroyl-d_{47}). Simulated spectra are shown in red. (B) Relative proportions of the individual phases observed in the mixtures. Reproduced from ref. 25, https://doi.org/10.1002/anie.202003375, under the terms of the CC BY 4.0 license, https://creativecommons.org/licenses/by/4.0/.

transition temperatures[30] and also in SC mixtures containing short acyl chain Cer[NS] with 16 carbons in a ternary SC model membrane.[20] This spectral component was simulated using a two-site exchange model in which the deuterated all-*trans* sphingosine chain alternates between two conformational states, assuming a jump angle of 180° about the long axis of the lipid.[20] This yielded a reasonable description of the ^{2}H NMR spectra acquired at low temperature. Quantification of the individual phases could be done from the line shape simulations (Figure 22.6B). The majority (84–98%) of the lignoceryl-d_{47} chain of Cer[NS] was found in the orthorhombic phase at 25 °C or 32 °C in membranes with both low and high cholesterol contents. In stark contrast, the sphingosine-d_{28} chains of the same molecule were found to constitute a large fraction (45–72%) that undergoes two-site exchange and a smaller part (26–32%) that forms a highly fluid fraction. In the fluid state, the sphingosine-d_{28} chains would undergo rapid *trans-gauche* isomerizations.

To gain insight into the various chain dynamics, one can calculate the estimated chain length from the experimental-order parameter profiles, using the mean-torque model.[82] This was done at physiological skin temperature (32 °C) for both chains (see Figure 22.7). The lignoceroyl chains of Cer[NS] are in an all-*trans* state, yielding an average length of $l = 29.2$ Å. The fraction of the sphingosine chain that is in the all-*trans* conformation undergoing two-site exchange is naturally shorter ($l = 17.8$ Å) than the acyl chain. The same sphingosine chain of Cer[NS] found in the fluid lipid phase has a projected chain length of 12.7–13.0 Å due to several *gauche* defects introduced in the fluid chains. From the calculation of the chain lengths, we can estimate the sphingosine chain to contain 4.3 or 4.6 *gauche* defects at skin temperature, which results in a significantly shorter chain, as one *gauche* defect shortens the lipid chain length by 1.1 Å.[83]

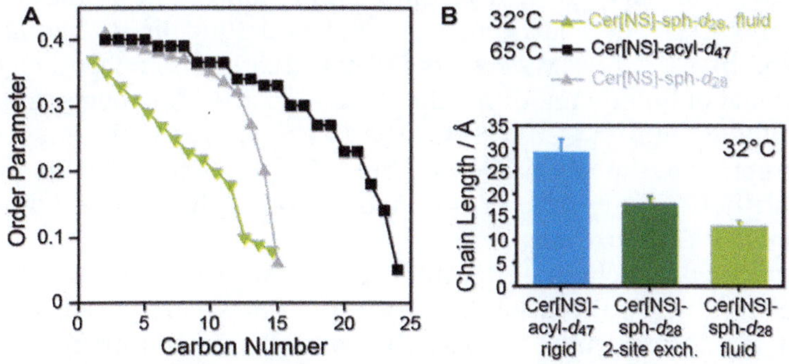

Figure 22.7 Order parameter profiles (A), projected chain lengths (B) of both lipid chains of Cer[NS] in mixture with LA and Chol from ^{2}H NMR measurements. Reproduced from ref. 25, https://doi.org/10.1002/anie.202003375, under the terms of the CC BY 4.0 license, https://creativecommons.org/licenses/by/4.0/.

At 50 °C, both chains undergo a dramatic restructuring. Both chains show much narrower ^{2}H NMR spectra, which means that they are in fluid phases. Only the lignoceroyl chain of Cer[NS] is found to have a substantial amount in the orthorhombic phase in lipid mixtures with low and high cholesterol content. At 65 °C, the Cer[NS] lignoceroyl chains are mostly in the fluid phase which, however, is characterized by a high-order parameter profile reminiscent of liquid ordered phases in phospholipid membranes.[84,85] The sphingosine chain is almost entirely fluid. The order parameter profile of the sphingosine chain at 65 °C is similar to that of the lignoceroyl chain in the plateau region close to the top of the chain (see Figure 22.7). This is in remarkable contrast to 32 °C, where the fluid sphingosine chain has a much lower order parameter and a steeper decline in order along the chain than at 65 °C, suggesting that some major restructuring of the SC lipid mix occurs at higher temperature. At 80 °C, both deuterated chains of Cer[NS] form a completely isotropic phase, indicating that the multilamellar structure of the SC lipid phase is completely abolished. Isotropic phase formation is completely reversible by cooling the samples.

These results strongly advocate a revision of the rigid model of the *stratum corneum* SPP lipid phase at physiological temperature. At 32 °C, the lignoceroyl chain of Cer[NS] displays a very characteristic rigid all-*trans* chain indicating an orthorhombic phase, similar to earlier studies, where ceramides found in the SC[17,20–22,39,40] were used. This is in remarkable contrast to the behavior of the sphingosine chain of the same Cer[NS] molecule, which undergoes rapid *trans-gauche* isomerizations, resulting in a 36% larger cross-sectional area than an all-*trans* chain. These order differences, which also result in significant alterations of the lipid chain lengths, will cause packing defects in the SC lipid layers and hydrophobic mismatching. This strongly suggests that the lipids need to adopt some kind of domain structure, where very mobile and immobile lipid segments separate laterally, as frequently found in membrane research involving several lipid components[8,9,36] and also in neutron diffraction[71] and FTIR studies[79,80] of SC model mixtures. The coexistence of solid and fluid lipids has also been confirmed by ^{13}C MAS NMR results.[50] This situation is conceptually similar to the model of liquid nanodrops that J. Boustra and M. Lafleur proposed for the mobile ω-chain of Cer[EOS] for the LPP.[26] However, the suggested domain structure of the SPP proposed here refers to a lateral fluctuation of the lipid distribution in each layer, as opposed to the fluid phase found in the center of LPP Cer mixtures.

As a sphingosine chain either undergoes a two-site exchange or is in the fluid state subject to rotational diffusion, an axis is required for the motions of the Cer[NS] molecule. At the same time, we find the lignoceryl chain of the same molecule in the orthorhombic phase. Therefore, the hairpin conformation of Cer[NS] seems highly unlikely. Consequently, the extended confirmation is the only possible structure for Cer[NS] to accommodate the independent movements of the chains observed in the ^{2}H NMR results.

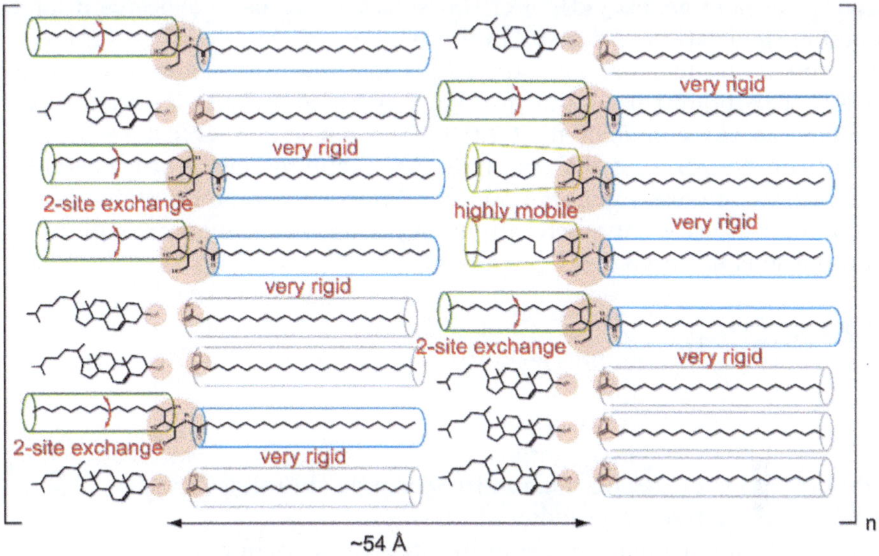

Figure 22.8 Model of the lipid arrangement in the SPP of a SC mixture composed of an equimolar ratio of Cer[NS]/LA/Chol at physiological skin temperature. Reproduced from ref. 25, https://doi.org/10.1002/anie.202003375, under the terms of the CC BY 4.0 license, https://creativecommons.org/licenses/by/4.0/.

In summary, the ^{2}H NMR results for the two lipid chains of Cer[NS] lead us to propose a modified model for the SPP of the *stratum corneum* lipid phase at skin temperature (see Figure 22.8). According to this model, the Cer[NS] molecule is exclusively found in the extended conformation with immobile and mobile chain moieties clustering together, giving rise to fluctuations and heterogeneities in the lateral lipid distribution of the SC. While some molecular rigidity is required to prevent dehydration and excessive permeability of the skin, mobile moieties of SC lipids like Cer[EOS] and the sphingosine chain of Cer[NS] could induce the translocation of endogenous molecules like water, antimicrobial peptides, or cytokines. It would also provide some membrane elasticity and agree with the sensory and defensive function of the SC.

22.4 Conclusions

While the lipid phase of the *stratum corneum* has traditionally been described as a very rigid interface of highly saturated long-chain lipids, the liquid core of the LPP was appreciated very early[24] and later explicitly confirmed by X-ray diffraction,[56] ^{2}H NMR,[26] and MD studies.[28] Furthermore, also the model of the SPP, which has traditionally been considered a fully rigid lipid interface, has to be reconsidered to also incorporate quite mobile sphingosine chains. Not only does the discovery of the sphingosine mobility strongly suggest an extended Cer[NS] structure, it also advocates a lateral domain organization of the SPP in the SC. Clearly, more work is required in order to fully understand the barrier function of the SC in both the LPP and the SPP. Open questions concern the dynamics of the sphingosine chain in ceramide species characteristic for the LPP, such as Cer[EOS] or Cer[EOP], as well as the dynamics of the sphingoid backbone moiety of other ceramide species. To completely understand the barrier function of the lipid phase of the human SC, a comprehensive investigation of the dynamics of the SC lipid phase is required.

Abbreviations

Cer	ceramide
Chol	cholesterol
CP	cross-polarization
CPMAS	cross-polarization magic-angle spinning
FFA	free fatty acid
FTIR	Fourier transform infrared spectroscopy
INEPT	insensitive nuclei enhancement by polarization transfer
LA	lignoceric acid
LPP	long periodicity phase
MAS	magic-angle spinning
SPP	short periodicity phase
SC	*stratum corneum*

Acknowledgements

This work was supported by the German Research Foundation (HU 720/18-1). O E. acknowledges the Sigrid Juselius Foundation, Ruth and Nils-Erik Stenbäck's Foundation. The authors thank Ferdinand Fandrei for the ^{13}C MAS NMR spectra shown in Figure 22.2.

References

1. R. J. Scheuplein and I. H. Blank, *Physiol. Rev.*, 1971, **51**, 702.
2. J. van Smeden, M. Janssens, G. S. Gooris and J. A. Bouwstra, *Biochim. Biophys. Acta*, 2014, **1841**, 295.
3. P. M. Elias, *J. Invest. Dermatol.*, 1983, **80**, 44s.
4. E. H. Mojumdar, G. S. Gooris, D. J. Barlow, M. J. Lawrence, B. Deme and J. A. Bouwstra, *Biophys. J.*, 2015, **108**, 2670.
5. S. Motta, M. Monti, S. Sesana, R. Caputo, S. Carelli and R. Ghidoni, *Biochim. Biophys. Acta*, 1993, **1182**, 147.
6. D. Huster, K. Arnold and K. Gawrisch, *Biochemistry*, 1998, **37**, 17299.
7. H. A. Scheidt, T. Meyer, J. Nikolaus, D. J. Baek, I. Haralampiev, L. Thomas, R. Bittman, A. Herrmann, P. Müller and D. Huster, *Angew. Chem., Int. Ed.*, 2013, **52**, 12848.

8. T. Bartels, R. S. Lankalapalli, R. Bittman, K. Beyer and M. F. Brown, *J. Am. Chem. Soc.*, 2008, **130**, 14521.
9. A. Bunge, P. Müller, M. Stöckl, A. Herrmann and D. Huster, *Biophys. J.*, 2008, **94**, 2680.
10. M. A. Lampe, A. L. Burlingame, J. Whitney, M. L. Williams, B. E. Brown, E. Roitman and P. M. Elias, *J. Lipid Res.*, 1983, **24**, 120.
11. L. E. Uche, G. S. Gooris, J. A. Bouwstra and C. M. Beddoes, *Biochim. Biophys. Acta*, 2021, **1863**, 183487.
12. A. S. Breathnach, T. Goodman, C. Stolinski and M. Gross, *J. Anat.*, 1973, **114**, 65.
13. K. C. Madison, D. C. Swartzendruber, P. W. Wertz and D. T. Downing, *J. Invest. Dermatol.*, 1987, **88**, 714.
14. S. H. White, D. Mirejovsky and G. I. King, *Biochemistry*, 1988, **27**, 3725.
15. K. Vavrova, A. Kovacik and L. Opalka, *Eur. Pharm. J.*, 2017, **64**, 28.
16. J. A. Bouwstra and M. Ponec, *Biochim. Biophys. Acta*, 2006, **1758**, 2080.
17. D. Kessner, A. Ruettinger, M. A. Kiselev, S. Wartewig and R. H. Neubert, *Skin Pharmacol. Physiol.*, 2008, **21**, 58.
18. R. Mendelsohn and D. J. Moore, *Methods Enzymol.*, 2000, **312**, 228.
19. Q. D. Pham, D. Topgaard and E. Sparr, *Proc. Natl. Acad. Sci. U. S. A.*, 2017, **114**, E112.
20. S. Stahlberg, B. Skolova, P. K. Madhu, A. Vogel, K. Vavrova and D. Huster, *Langmuir*, 2015, **31**, 4906.
21. N. Kitson, J. Thewalt, M. Lafleur and M. Bloom, *Biochemistry*, 1994, **33**, 6707.
22. E. Brief, S. Kwak, J. T. Cheng, N. Kitson, J. Thewalt and M. Lafleur, *Langmuir*, 2009, **25**, 7523.
23. I. Iwai, H. Han, H. L. den, S. Svensson, L. G. Ofverstedt, J. Anwar, J. Brewer, M. Bloksgaard, A. Laloeuf, D. Nosek, S. Masich, L. A. Bagatolli, U. Skoglund and L. Norlen, *J. Invest. Dermatol.*, 2012, **132**, 2215.
24. J. A. Bouwstra, G. S. Gooris, F. E. Dubbelaar and M. Ponec, *J. Lipid Res.*, 2001, **42**, 1759.
25. O. Engberg, A. Kovacik, P. Pullmannová, M. Juhašcik, L. Opálka, D. Huster and K. Vávrová, *Angew. Chem., Int. Ed.*, 2020, **59**, 17383.
26. A. Paz Ramos, G. Gooris, J. Bouwstra and M. Lafleur, *J. Lipid Res.*, 2018, **59**, 137.
27. Q. D. Pham, E. H. Mojumdar, G. S. Gooris, J. A. Bouwstra, E. Sparr and D. Topgaard, *Q. Rev. Biophys.*, 2018, **51**, e7.
28. E. Wang and J. B. Klauda, *J. Am. Chem. Soc.*, 2019, **141**, 16930.
29. J. Seelig, *Q. Rev. Biophys.*, 1977, **10**, 353.
30. J. H. Davis, *Biochim. Biophys. Acta*, 1983, **737**, 117.
31. K. Beshah, E. T. Olejniczak and R. G. Griffin, *J. Chem. Phys.*, 1987, **86**, 4730.
32. T. R. Molugu, S. Lee and M. F. Brown, *Chem. Rev.*, 2017, **117**, 12087.
33. A. K. Engel and D. Cowburn, *FEBS Lett.*, 1981, **126**, 169.
34. M. Lafleur, B. Fine, E. Sternin, P. R. Cullis and M. Bloom, *Biophys. J.*, 1989, **56**, 1037.

35. R. J. Wittebort, E. T. Olejniczak and R. G. Griffin, *J. Phys. Chem.*, 1987, **86**, 5411.
36. S. L. Veatch, O. Soubias, S. L. Keller and K. Gawrisch, *Proc. Natl. Acad. Sci. U. S. A.*, 2007, **104**, 17650.
37. I. V. Polozov and K. Gawrisch, *Biophys. J.*, 2006, **90**, 2051.
38. J. M. Hutchison, K. C. Shih, H. A. Scheidt, S. M. Fantin, K. F. Parson, G. A. Pantelopulos, H. R. Harrington, K. F. Mittendorf, S. Qian, R. A. Stein, S. E. Collier, M. G. Chambers, J. Katsaras, M. W. Voehler, B. T. Ruotolo, D. Huster, R. L. McFeeters, J. E. Straub, M. P. Nieh and C. R. Sanders, *J. Am. Chem. Soc.*, 2020, **142**, 12715.
39. S. Stahlberg, S. Lange, B. Dobner and D. Huster, *Langmuir*, 2016, **32**, 2023.
40. A. Kovacik, A. Vogel, J. Adler, P. Pullmannova, K. Vavrova and D. Huster, *Biochim. Biophys. Acta*, 2018, **1860**, 1162.
41. E. R. Andrew, A. Bradbury and R. G. Eades, *Nature*, 1958, **182**, 1659.
42. E. O. Stejskal and J. Schaefer, *J. Am. Chem. Soc.*, 1976, **98**, 1031.
43. J. Forbes, C. Husted and E. Oldfield, *J. Am. Chem. Soc.*, 1988, **110**, 1059.
44. D. Huster, K. Arnold and K. Gawrisch, *J. Phys. Chem. B*, 1999, **103**, 243.
45. F. Volke and A. Pampel, *Biophys. J.*, 1995, **68**, 1960.
46. Z. Zhou, B. G. Sayer, R. E. Stark and R. M. Epand, *Chem. Phys. Lipids*, 1997, **90**, 45.
47. D. Huster and K. Gawrisch, *J. Am. Chem. Soc.*, 1999, **121**, 1992.
48. J. H. Davis, M. Auger and R. S. Hodges, *Biophys. J.*, 1995, **69**, 1917.
49. P. Schmidt, L. Thomas, P. Müller, H. A. Scheidt and D. Huster, *Chemistry*, 2014, **20**, 4986.
50. S. Bjorklund, A. Nowacka, J. A. Bouwstra, E. Sparr and D. Topgaard, *PLoS One*, 2013, **8**, e61889.
51. S. Björklund, J. M. Andersson, Q. D. Pham, A. Nowacka, D. Topgaard and E. Sparr, *Soft Matter*, 2014, **10**, 4535.
52. Q. D. Pham, S. Björklund, J. Engblom, D. Topgaard and E. Sparr, *J. Controlled Release*, 2016, **232**, 175.
53. E. H. Mojumdar, Q. D. Pham, D. Topgaard and E. Sparr, *Sci. Rep.*, 2017, **7**, 15712.
54. W. P. Rothwell and J. S. Waugh, *J. Chem. Phys.*, 1981, **74**, 2721.
55. Q. D. Pham, G. Carlstrom, O. Lafon, E. Sparr and D. Topgaard, *Phys. Chem. Chem. Phys.*, 2020, **22**, 6572.
56. M. A. Kiselev, N. Y. Ryabova, A. M. Balagurov, S. Dante, T. Hauss, J. Zbytovska, S. Wartewig and R. H. Neubert, *Eur. Biophys. J.*, 2005, **34**, 1030.
57. A. Schroter, D. Kessner, M. A. Kiselev, T. Hauss, S. Dante and R. H. Neubert, *Biophys. J.*, 2009, **97**, 1104.
58. D. J. Moore and M. E. Rerek, *Acta Derm Venereol. Suppl.*, 2000, **208**, 16.
59. R. Mendelsohn and D. J. Moore, *Chem. Phys. Lipids*, 1998, **96**, 141.
60. D. J. Moore, M. E. Rerek and R. Mendelsohn, *J. Phys. Chem. B*, 1997, **101**, 8933.

61. R. M. Venable, Y. Zhang, B. J. Hardy and R. W. Pastor, *Science*, 1993, **262**, 223.
62. E. Wang and J. B. Klauda, *J. Phys. Chem. B*, 2019, **123**, 2525.
63. S. Moradi, A. Nowroozi and M. Shahlaei, *RSC Adv.*, 2019, **9**, 4644.
64. S. E. Feller, D. Huster and K. Gawrisch, *J. Am. Chem. Soc.*, 1999, **121**, 8963.
65. A. Vogel, G. Reuther, M. B. Roark, K. T. Tan, H. Waldmann, S. E. Feller and D. Huster, *Biochim. Biophys. Acta*, 2010, **1798**, 275.
66. A. Vogel, H. A. Scheidt, D. J. Baek, R. Bittman and D. Huster, *Phys. Chem. Chem. Phys.*, 2016, **18**, 3730.
67. E. Wang and J. B. Klauda, *J. Phys. Chem. B*, 2018, **122**, 2757.
68. M. Podewitz, Y. Wang, P. Gkeka, S. von Grafenstein, K. R. Liedl and Z. Cournia, *J. Phys. Chem. B*, 2018, **122**, 10505.
69. E. H. Mojumdar, G. S. Gooris, D. Groen, D. J. Barlow, M. J. Lawrence, B. Deme and J. A. Bouwstra, *Biochim. Biophys. Acta*, 2016, **1858**, 1926.
70. A. Eichner, S. Stahlberg, S. Sonnenberger, S. Lange, B. Dobner, A. Ostermann, T. E. Schrader, T. Hauss, A. Schroeter, D. Huster and R. H. Neubert, *Biochim. Biophys. Acta*, 2017, **1859**, 745.
71. A. Schroeter, S. Stahlberg, B. Skolova, S. Sonnenberger, A. Eichner, D. Huster, K. Vavrova, T. Hauss, B. Dobner, R. H. Neubert and A. Vogel, *Soft Matter*, 2017, **13**, 2107.
72. S. Stahlberg, A. Eichner, S. Sonnenberger, A. Kovácik, S. Lange, T. Schmitt, B. Deme, T. Hauß, B. Dobner, R. H. H. Neubert and D. Huster, *Langmuir*, 2017, **33**, 9211.
73. T. Schmitt, S. Lange, B. Dobner, S. Sonnenberger, T. Hauss and R. H. H. Neubert, *Langmuir*, 2018, **34**, 1742.
74. X. Chen, S. Kwak, M. Lafleur, M. Bloom, N. Kitson and J. Thewalt, *Langmuir*, 2007, **23**, 5548.
75. A. C. Rowat, N. Kitson and J. L. Thewalt, *Int. J. Pharm.*, 2006, **307**, 225.
76. B. Skolova, B. Janusova, J. Zbytovska, G. Gooris, J. Bouwstra, P. Slepicka, P. Berka, J. Roh, K. Palat, A. Hrabalek and K. Vavrova, *Langmuir*, 2013, **29**, 15624.
77. T. N. Engelbrecht, A. Schröter, T. Hauss, B. Deme, H. A. Scheidt, D. Huster and R. H. H. Neubert, *Soft Matter*, 2012, **8**, 6599.
78. E. J. Dufourc, E. J. Parish, S. Chitrakorn and C. P. Smith, *Biochemistry*, 1984, **23**, 6062.
79. R. Mendelsohn, I. Selevany, D. J. Moore, M. C. Mack Correa, G. Mao, R. M. Walters and C. R. Flach, *J. Phys. Chem. B*, 2014, **118**, 4378.
80. R. Mendelsohn, E. Rabie, R. M. Walters and C. R. Flach, *J. Phys. Chem. B*, 2015, **119**, 9740.
81. J. A. Bouwstra, G. S. Gooris, M. A. Salomons-d Vries, J. A. van der Spek and W. Bras, *Int. J. Pharm.*, 1992, **84**, 205.
82. H. I. Petrache, S. W. Dodd and M. F. Brown, *Biophys. J.*, 2000, **79**, 3172.

83. S. E. Feller, K. Gawrisch and A. D. MacKerell, Jr, *J. Am. Chem. Soc.*, 2002, **124**, 318.
84. M. Bosse, J. Sibold, H. A. Scheidt, L. J. Patalag, K. Kettelhoit, A. Ries, D. B. Werz, C. Steinem and D. Huster, *Phys. Chem. Chem. Phys.*, 2019, **21**, 15630.
85. O. Engberg, A. Bochicchio, A. F. Brandner, A. Gupta, S. Dey, R. A. Bockmann, S. Maiti and D. Huster, *Front. Physiol.*, 2020, **11**, 578868.

Subject Index